Discovering the Universe:
From the Stars to the Planets

DISCOVERING THE UNIVERSE

FROM THE STARS TO THE PLANETS

Neil F. Comins   *University of Maine*

W. H. Freeman and Company
New York

To my colleagues in the Department of Physics and Astronomy at the University of Maine, T. Alan Clark and William J. F. Wilson, University of Calgary, and Marcel Bergman.

Publisher: Clancy Marshall
Acquisitions Editor: Anthony Palmiotto
Senior Developmental Editor: Kharissia Pettus
Marketing Director: John Britch
Project Editor: Kerry O'Shaughnessy
Media and Supplements Editor: Amy Thorne
Design Manager: Blake Logan
Production Coordinator: Paul W. Rohloff
Senior Illustration Coordinator: Bill Page
Illustrations: Imagineering Media Services
Photo Editor: Bianca Moscatelli
Photo Researchers: Christina Micek, Christine Buese
Assistant Editor: Dave Quinn
Composition: Black Dot Group
Printing and Binding: RR Donnelley

ISBN-13: 978-1-4292-3042-1/ISBN-10: 1-4292-3042-8
©2009 by W. H. Freeman and Company

Printed in the United States of America

First printing

W. H. Freeman and Company
41 Madison Avenue
New York, NY 10010
Houndmills, Basingstoke
RG21 6XS, England
www.whfreeman.com

CONTENTS OVERVIEW

CONTENTS

Astronomy and astronomy education are both changing rapidly. In the realm of astronomical discovery, astronomers have answered the age-old question of the fate of the universe; Pluto has been reclassified from being a planet to being a dwarf planet; many planets have been discovered orbiting other stars, including planets with properties similar to the Earth; observations have determined that the Milky Way is a barred spiral galaxy, rather than a normal spiral, as previously believed; and, while extraterrestrial life has *not* been discovered, even that holy grail is becoming for many a mainstream part of astronomical research.

In the realm of astronomy education, educators are continuing to develop methods to help students understand how to think like scientists and grasp the core concepts, even when the concepts are at odds with scientists' prior beliefs. In-class interactivities, where students respond to questions with "clickers," enrich the classroom experience. Online materials provide tutorials and practice questions that turn students from passive into active learners.

Discovering the Universe: From the Stars to the Planets presents the big picture of how the stars, the galaxies, and the entire universe formed before discussing the planets and other components of our solar system. This approach explores and explains the core areas of understanding in a scientific progression; the order follows the temporal development of the objects that make up our universe. By understanding the formation and evolution of stars, galaxies, and the universe, students are better prepared to understand the details of how our solar system (the Sun and everything that orbits it) developed. The pedagogy includes

- presenting the observations and underlying physical *concepts* needed to connect astronomical observations to theories that explain them coherently and meaningfully.

- presenting information about the planets both as individual objects and how they compare to each other.

- using both textual and graphical information to present concepts for students who learn in different ways.

- helping students compare their beliefs with the findings of modern science and understand why the scientific view is correct.

- using analogies from everyday life to make cosmic phenomena more concrete.

- providing visually rich timelines that connect astronomical discoveries with other events throughout history.

- expanding student perspectives by exploring plausible alternative situations (asking "What if . . . ?" questions).

- linking material presented in the book with enhanced material offered electronically.

BRINGING THE UNIVERSE INTO CLEARER FOCUS

Margin questions about important concepts are presented in most sections of the book. These questions encourage students to frequently test themselves and correct their beliefs before errors accumulate. For example, after learning about inertia in Section 2-7, students are asked how, while driving in a car, they can show that their

Sitting in a moving car, how can you experimentally verify that your body has inertia?

bodies have inertia. (They could put on the brakes and feel themselves being restrained by the seatbelts). Answers to approximately one-third of the margin questions appear at the end of this text.

Margin photos provide a connection between the concepts being presented and their applications in everyday life. For example, during the discussion of solar eclipses, a photo shows a student using a telescope to view an eclipse.

Margin charts, showing the location in the sky of important astronomical objects cited in the text, are provided. Sufficient detail in the margin charts allows students to locate the objects with either the unaided eye or a small telescope, as appropriate. In this example, a star chart of the constellation Monoceros is shown with the photo of the Rosette Nebula.

New coverage of the planets Astronomers have created a new classification scheme for the objects in the solar system. These *planets, dwarf planets,* and *small solar system objects* are explained and reconciled with the existing classes of objects, including planets, moons, asteroids, meteoroids, and comets. Also explained is how Pluto fits more comfortably with the dwarf planets than with the eight planets.

Astrobiology coverage The final chapter in the book contains expanded coverage on astrobiology, so that students can explore this exciting area of discovery.

New dynamic art Summary figures appear throughout the book to show either the interactions between important concepts or the evolution of important objects introduced. For example, the location of the Sun in the sky, which varies over the seasons, as does the corresponding intensity of the light and the appropriate ground cover, is shown in a sequence of drawings combined into one figure.

PROVEN FEATURES SUPPORT LEARNING

What Do You Think? and **What Did You Think?** questions in each chapter ask students to consider their present beliefs and actively compare them with the correct science presented in the book. Numbered icons mark the places in the text where each concept is discussed. Encouraging students to think about what they believe

is true and then work through the correct science step-by-step has proved to be an effective teaching technique, especially when time constraints prevent instructors from working with students individually or in small groups as they try to reconcile incorrect beliefs with proper science.

Chapter opening narratives introduce and launch the chapter's topics and provide students with a context for understanding the material.

Learning objectives underscore the key chapter concepts.

Section headings are brief sentences that summarize section content and serve as a quick study guide to the chapter when reread.

Icons link the text to web material

- *Starry Night Enthusiast™* icons link the text to a specific interactivity in the *Starry Night Enthusiast™* observing programs, which is an optional CD with the textbook.

- Video icons link the text to relevant video clips available on the textbook's Web site.

- Animation icons link the text to animated figures available on the textbook's Web site.

- Active Integrated Media Modules (AIMM) and Interactive Exercise icons link the text to interactive exercises available on the textbook's Web site.

- Web link icons direct the reader to further information on a particular topic.

Guided Discovery boxes offer hands-on experience with astronomy. Many use *Starry Night Enthusiast™* software.

An Astronomer's Toolbox introduces equations used in astronomy. Most of the material in the book is descriptive, so setting off essential equations in numbered boxes to maintain the flow of the material. The toolboxes also contain worked examples, additional explanations, and practice doing calculations; answers are given at the end of the book. All the equations are summarized in Appendix C.

Insight into Science boxes are brief asides that relate topics to the nature of scientific inquiry and encourage critical thinking.

Wavelength tabs with photographic images show whether an image was made with radio waves (R), infrared radiation (I), visible light (V), ultraviolet light (U), X rays (X), or gamma rays (G).

Review and practice material

- **Summary of Key Ideas** is a bulleted list of key concepts.

- **What Did You Think?** questions at the end of each chapter answers the What Do You Think? questions posed at the beginning of the chapter.

- **Key Words, Review Questions,** and **Advanced Questions** help the students understand the chapter material.

- **Discussion Questions** offer interesting topics to spark lively, insightful debate.

- **Web Questions** take students to the Web for further study.

- **AstroPortal** is a flexible and comprehensive online resource containing interactive, multi-step problem-solving tutorials, animations, the complete eBook, as well as powerful assignment and grading capabilities.

- **Observing Projects,** featuring *Starry Night Enthusiast*™ activities, ask students to be astronomers themselves.

What If . . .

Selected chapters conclude with a "What If . . ." essay that encourages critical thinking by speculating on how changes in the universe could have profound effects on Earth.

An Astronomer's Almanac, a dynamic timeline that relates discoveries in astronomy to other historical events, opens each Part. The almanac provides strong context for the information presented.

MEDIA AND SUPPLEMENTS PACKAGE

FOR STUDENTS

The eBook: Affordable, Innovative, Customizable
The eBook for *Discovering the Universe: From the Stars to the Planets* is a complete online version of the textbook. It provides a rich learning experience by integrating a wealth of multimedia resources and adding unique features.

- Animations, videos, activities, and other visualization aids are delivered within the relevant sections of the electronic textbook, encouraging students to access the resources as the concepts are explained.

- Access from any Internet-connected computer is easy via a standard Web browser.

- Navigation to any section or subsection or to any printed book page number is quick and intuitive.

- In-text self-quiz questions and links to all glossary entries are provided.

- Students can highlight, take notes, and bookmark pages.

- Full-text search is built in, with an option to also search the complete glossary.

The eBook is available with printed copies of the text, or it can be purchased online at http://ebooks. bfwpub.com.

AstroPortal

AstroPortal is the most flexible and student-focused resource available for the Introductory Astronomy course, providing multimedia learning and assessment tools that build on the teaching strategies presented the textbook. For consistency and convenience, AstroPortal offers a **wide array of teaching assets in one location, with one login.**

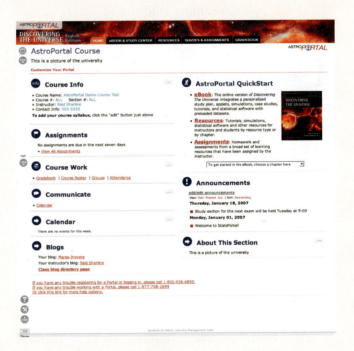

Features of the AstroPortal

- **Personalized study plan** offers self-quizzes to gauge student understanding. Quiz results generate a student-specific set of additional exercises and eBook sections keyed to the questions students missed. Instructors receive a class-wide view of the results, so they can adjust their lectures to help students succeed.

- **Tutorials** engage students in the scientific process of discovery and interpretation. Dozens of these concept-driven, experiential walkthroughs allow students to make observations, draw conclusions, and apply their knowledge. Interwoven with multimedia, activities, and questions, students receive a deep, self-guided exploration of the concepts.

• **The multimedia-enhanced eBook** contains the complete text with a wealth of helpful functions. All student multimedia is linked directly from the eBook pages, enabling students to access animations and tutorials within the context of the topic at hand. Powerful customization features are available, including instructor and student notes, highlighting, and editing capabilities.

• The *Scientific American* Newsfeed provides continual chemistry-related articles, links, and podcasts selected by *Scientific American* editors.

• **Animations** of key concepts and images from the text.

• **NASA videos, interactive drag-and-drop exercises**

• **Automatically graded homework/quiz problems** allow instructors to assess student comprehension and corrects student misconceptions immediately. Instructors can build assignments using their own problems or provided end-of-chapter, tutorial-based, and test bank problems.

• **Course management tools** such as instructor and class blogs, discussion forums, email, announcements, calendar, and powerful gradebook.

If you would like more information or have any questions about AstroPortal please contact your local representative or visit www.whfreeman.com/astronomy.

Starry Night Enthusiast

Starry Night Enthusiast™ 6.2 is a brilliantly realistic planetarium software package. It is designed for easy use by anyone with an interest in the night sky. See the sky from anywhere on Earth or lift off and visit any solar system body or any location up to 20,000 light-years away. View 2,500,000 stars along with more than 170 deep-space objects such as galaxies,

star clusters, and nebulae. Travel 15,000 years in time, check out the view from the International Space Station, and see planets up close from any one of their moons. Included are stunning OpenGL graphics. Handy star charts can be printed to explore outdoors. The *Starry Night Enthusiast* CD is available at no extra charge with the text upon request. Use ISBN 1-4292-3644-2.

Observing Projects Using *Starry Night Enthusiast*™

by T. Alan Clark and William J. F. Wilson, University of Calgary, and Marcel Bergman
ISBN 1-4292-1866-5
Available for packaging with the text and compatible with both PC and Mac, this workbook contains a variety of comprehensive lab activities for use with *Starry Night Enthusiast*™ 6.2. The Observing Projects workbook can also be packaged with the *Starry Night* software. Use ISBN 1-4292-2011-2.

Free Student Companion Web Site

The Companion Web site at www.whfreeman.com/dtustars features a variety of study and review resources designed to help students understand the concepts. The open-access Web site includes the following:

• **Online quizzing** offers questions and answers with instant feedback to help students study, review, and prepare for exams. Instructors can access results through an online database or they can have results e-mailed directly to their accounts.

• **Animations and videos** both original and NASA-created, are keyed to specific chapters.

• **Web links** provide a wealth of online resources to the student.

• **Active integrated media modules** take students deeper into key topics from the text.

- **Interactive drag-and-drop exercises** based on text illustrations help students grasp the vocabulary in context

- **Flashcards** offer help with vocabulary and definitions

FOR INSTRUCTORS

Instructor's eBook

The eBook for the first edition of *Discovering the Universe: From the Stars to the Planets* offers instructors flexibility and customization options not previously possible with any printed textbook. Instructors have access to:

- **eBook Customization** Instructors can choose the chapters that correspond with their syllabus, and students will get a custom version of the eBook with the selected chapters only.

- **Instructor Notes** Instructors can create an annotated version of the eBook by adding notes to any page. When students registered to this instructor's course log in, they see the notes.

- **Custom Content** Instructor content can include text, Web links, and images, allowing instructors to place any content they choose exactly where they want it. This feature can be used to augment, edit, or enrich the textbook. Instructor media resources are easily integrated, and eBook pages from other sites can be directly linked.

Online Course Materials (WebCT, Blackboard)

As a service for adopters, we will provide content files in the appropriate online course format, including the instructor and student resources for the textbook. The files can be used as is or can be customized to fit specific needs. Prebuilt quizzes, links, and activities, among other materials, are included.

Classroom Presentation and Interactivity

A set of online lecture presentations created in PowerPoint allows instructors to tailor their lectures to suit their own needs using images and notes from the textbook. These presentations are available on the instructor portion of the companion Web site.

Clicker Questions

Written by Neil Comins, these questions can be used as lecture launchers with or without a classroom response system such as iClicker. Each chapter includes questions relating to figures from the text and common misconceptions, as well as writing questions for instructors who would like to add a writing or class discussion element to their lectures.

ACKNOWLEDGMENTS

I am deeply grateful to the astronomers and teachers who reviewed the manuscript for this edition. This is a stronger and better book because of their conscientious efforts:

Kurt S. J. Anderson, *New Mexico State University*
Nadine G. Barlow, *Northern Arizona University*
Ralph L. Benbow, *Northern Illinois University*
John C. Brandt, *University of New Mexico*
James S. Brooks, *Florida State University*
Eric D. Carlson, *Wake Forest University*
Jennifer L. Cash, *South Carolina State University*
Erik N. Christensen, *South Florida Community College*
Stephen S. Eikenberry, *University of Florida*
Richard E. Griffiths, *Carnegie Mellon University*
Thomasanna C. Hail, *University of Mississippi*
Javier Hasbun, *University of West Georgia*
Scott S. Hildreth, *Chabot College*
Thomas Hockey, *University of Northern Iowa*
Mark Hollabaugh, *Normandale Community College*
J. Christopher Hunt, *Prince George's Community College*
Rob Klinger, *Parkland College*
H. S. La, *Clayton State University*
Claud Lacy, *University of Arkansas*
Steve May, *Walla Walla Community College*
Ken Menningen, *University of Wisconsin–Stevens Point*
Scott Miller, *Pennsylvania State University*
Jerome A. Orosz, *San Diego State University*
Lawrence Pinsky, *University of Houston*
Eric Preston, *Indiana State University*
Itai Seggev, *University of Mississippi*
Jack Sulentic, *University of Alabama*
Richard Williamon, *Emory University*
J. Wayne Wooten, *Pensacola Junior College*

I would also like to thank the many people whose advice on previous editions has had an ongoing influence:
Gordan Baird, *University of Arizona*
Henry E. Bass, *University of Mississippi*
J. David Batchelor, *Community College of Southern Nevada*
Jill Bechtold, *University of Arizona*
Peter A. Becker, *George Mason University*
John Bieging, *University of Arizona*
Greg Black, *University of Virginia*
John W. Burns, *Mt. San Antonio College*
Gene Byrd, *University of Alabama*
Eugene R. Capriotti, *Michigan State University*
Michael W. Castelaz, *Pisgah Astronomical Research Institute*
Gerald Cecil, *University of North Carolina*
David S. Chandler, *Porterville College*
David Chernoff, *Cornell University*
Tom Christensen, *University of Colorado, Colorado Springs*
Chris Clemens, *University of North Carolina*
Christine Clement, *University of Toronto*
Halden Cohn, *Indiana University*
John Cowan, *University of Oklahoma*
Antoinette Cowie, *University of Hawaii*

Charles Curry, *University of Waterloo*
James J. D'Amario, *Harford Community College*
Purnas Das, *Purdue University*
Peter Dawson, *Trent University*
John M. Dickey, *University of Minnesota, Twin Cities*
John D. Eggert, *Daytona Beach Community College*
Bernd Enders, *College of Marin*
Robert Frostick, *West Virginia State College*
Martin Gaskell, *University of Nebraska*
Bruce Gronich, *University of Texas–El Paso*
Siegbert Hagmann, *Kansas State University*
David Hedin, *Northern Illinois University*
Chuck Higgins, *Penn State University*
James L. Hunt, *University of Guelph*
Nathan Israeloff, *Northeastern College*
Kenneth Janes, *Boston University*
William C. Keel, *University of Alabama*
William Keller, *St. Petersburg Junior College*
Marvin D. Kemple, *Indiana University–Purdue University Indianapolis (IUPUI)*
Pushpa Khare, *University of Illinois at Chicago*
F. W. Kleinhaus, *Indiana University–Purdue University Indianapolis (IUPUI)*
Rob Klinger, *Parkland College*
Patick M. Len, *Cuesta College*
John Patrick Lestrade, *Mississippi State University*
C. L. Littler, *University of North Texas*
M. A. K. Lohdi, *Texas Tech University*
Michael C. LoPresto, *Henry Ford Community College*
Phyllis Lugger, *Indiana University*
R. M. MacQueen, *Rhodes College*
Robert Manning, *Davidson College*
Paul Mason, *University of Texas-El Paso*
P. L. Matheson, *Salt Lake Community College*
Rahul Mehta, *University of Central Arkansas*
J. Scott Miller, *University of Louisville*
L. D. Mitchell, *Cambria County Area Community College*
J. Ward Moody, *Brigham Young University*
Siobahn M. Morgan, *University of Northern Iowa*
Steven Mutz, *Scottsdale Community College*
Gerald H. Newsom, *Ohio State University*
Bob O'Connell, *College of the Redwoods*
William C. Oelfke, *Valencia Community College*
Richard P. Olenick, *University of Dallas*
John P. Oliver, *University of Florida*

Melvyn Jay Oremland, *Pace University*
David Patton, *Trent University*
Jon Pedicino, *College of the Redwoods*
Sidney Perkowitz, *Emory University*
David D. Reid, *Wayne State University*
Adam W. Rengstorf, *Indiana University*
James A. Roberts, *University of North Texas*
Henry Robinson, *Montgomery College*
Dwight P. Russell, *University of Texas in El Paso*
Barbara Ryden, *Ohio State University*
Larry Sessions, *Metropolitan State University*
C. Ian Short, Florida *Atlantic University*
John D. Silva, *University of Massachusetts at Dartmouth*
Michael L. Sitko, *University of Cincinnati*
Earl F. Skelton, George F. Smoot, *University of California at Berkeley*
Alex G. Smith, *University of Florida*
David Sturm, *University of Maine, Orono*
Paula Szkody, *University of Washington*
Michael T. Vaughan, *Northeastern University*
Andreas Veh, *Kenai Peninsula College*
John Wallin, *George Mason University*
William F. Welsh, *San Diego State University*
R. M. Williamon, *Emory University*
Edward L. (Ned) Wright, *University of California at Los Angeles*
Jeff S. Wright, *Elon College*
Nicolle E. B. Zellner, *Rensselaer Polytechnic Institute*

I would like to add my special thanks to the wonderfully supportive staff at W. H. Freeman and Company who make the process so enjoyable. Among others, these people include Anthony Palmiotto, Kharissia Pettus, Kerry O'Shaughnessy, Blake Logan, Bianca Moscatelli, Bill Page, Paul Rohloff, and Dave Quinn. Thanks also go to the copyeditors, proofreaders, and indexers at Black Dot Group. Warm thanks also to David Sturm, University of Maine, Orono, for his help in collecting current data on objects in the solar system; to my UM astronomy colleague David Batuski; and to my wife, Sue, and my sons, James and Josh, for their patience and support while preparing this book.

Neil F. Comins
neil.comins@umit.maine.edu

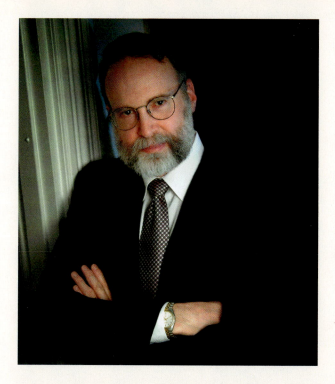

Professor **Neil F. Comins** is on the faculty of the University of Maine. Born in 1951 in New York City, he grew up in New York and New England. He earned a bachelor's degree in engineering physics at Cornell University, a master's degree in physics at the University of Maryland, and a Ph.D. in astrophysics from University College, Cardiff, Wales, under the guidance of Bernard F. Schutz. Dr. Comins's work for his doctorate, on general relativity, was cited in Subramanyan Chandrasekhar's Nobel laureate speech. He has done theoretical and experimental research in general relativity, observational astronomy, computer simulations of galaxy evolution, and science education. The fourth edition of *Discovering the Universe* was the first textbook in this series that Dr. Comins wrote for W. H. Freeman and Company, having taken over following the death of Bill Kaufmann in 1994. This text has been followed by others. He is also the author of three trade books, *What If the Moon Didn't Exist?*, *Heavenly Errors,* and *The Hazards of Space Travel: A Tourist's Guide*. *What If the Moon Didn't Exist?* has been made into planetarium shows, been excerpted for television and radio, and translated into several languages and is the theme for the Mitsubishi Pavilion at the World Expo 2005 in Aichi, Japan. *Heavenly Errors* explores misconceptions people have about astronomy, why such misconceptions are so common, and how to correct them. Dr. Comins has appeared on numerous television and radio shows and gives many public talks. Although he has jumped out of airplanes while in the military, today his activities are a little more sedate: He is a licensed pilot and avid sailor, having once competed against Prince Philip, Duke of Edinburgh.

PART I

Understanding Astronomy

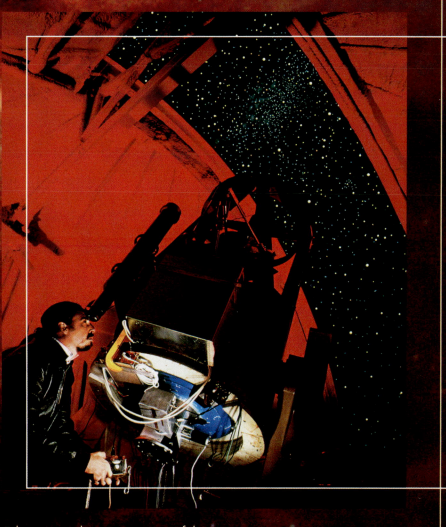

Telescopes enhance our views of the cosmos.
(David Parker/Photo Researchers, Inc.)

AN ASTRONOMER'S ALMANAC

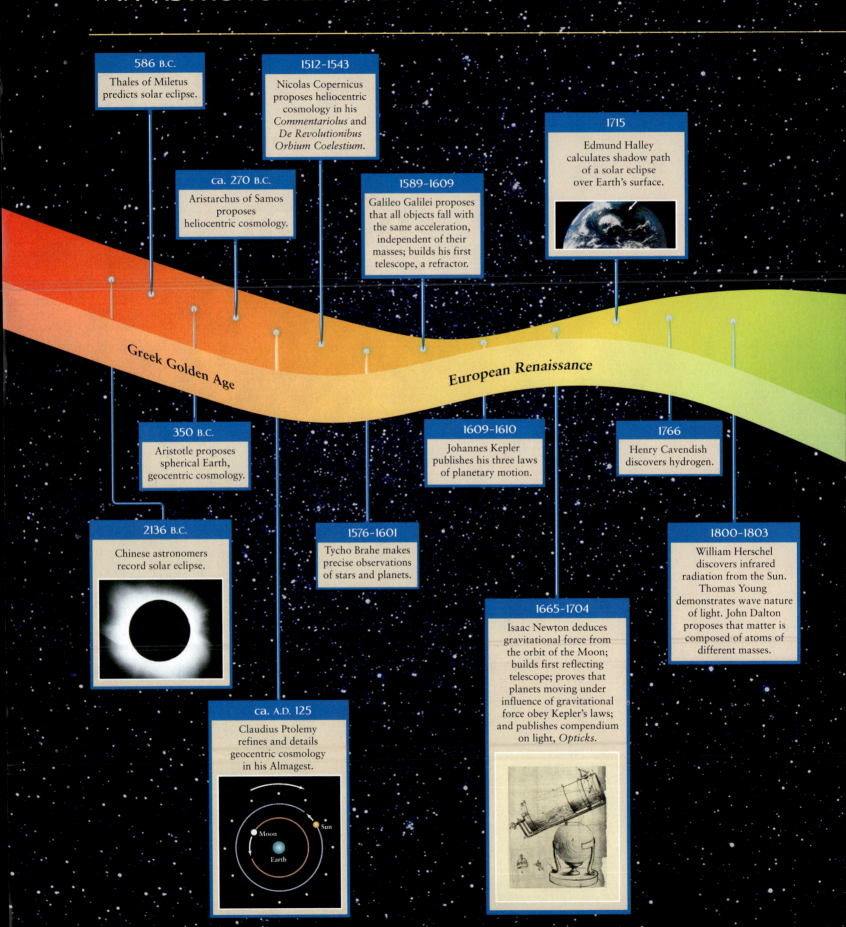

586 B.C.
Thales of Miletus predicts solar eclipse.

1512–1543
Nicolas Copernicus proposes heliocentric cosmology in his *Commentariolus* and *De Revolutionibus Orbium Coelestium*.

ca. 270 B.C.
Aristarchus of Samos proposes heliocentric cosmology.

1589–1609
Galileo Galilei proposes that all objects fall with the same acceleration, independent of their masses; builds his first telescope, a refractor.

1715
Edmund Halley calculates shadow path of a solar eclipse over Earth's surface.

Greek Golden Age

European Renaissance

350 B.C.
Aristotle proposes spherical Earth, geocentric cosmology.

1609–1610
Johannes Kepler publishes his three laws of planetary motion.

1766
Henry Cavendish discovers hydrogen.

2136 B.C.
Chinese astronomers record solar eclipse.

1576–1601
Tycho Brahe makes precise observations of stars and planets.

1800–1803
William Herschel discovers infrared radiation from the Sun. Thomas Young demonstrates wave nature of light. John Dalton proposes that matter is composed of atoms of different masses.

1665–1704
Isaac Newton deduces gravitational force from the orbit of the Moon; builds first reflecting telescope; proves that planets moving under influence of gravitational force obey Kepler's laws; and publishes compendium on light, *Opticks*.

ca. A.D. 125
Claudius Ptolemy refines and details geocentric cosmology in his Almagest.

Moon

Sun

Earth

1871-1873

Dimitri Mendeleyev develops periodic table of the elements. Henry Draper develops spectroscopy. James Clerk Maxwell asserts that light is an electromagnetic phenomenon.

1885-1888

Johann Balmer expresses spectral lines of hydrogen mathematically. Heinrich Hertz detects radio waves.

1900

Max Planck explains blackbody radiation. Paul Villard discovers gamma rays.

1942-1949

J. S. Hey detects radio waves from the Sun. First astronomical telescope launched into space. Herbert Friedman detects X-rays from the Sun. 200-in optical reflecting telescope begins operation on Mt. Palomar, California.

1990-1996

Hubble Space Telescope launched. Keck 10-m optical/infrared telescopes begin operation at Mauna Kea, Hawaii. SOHO solar observatory launched.

1930-1934

Karl Jansky builds first radio telescope. James Chadwick discovers the neutron. Bernhard Schmidt builds his Schmidt optical reflecting telescope.

1963-1967

Largest single-dish radio telescope, 300 m across, begins operation at Arecibo, Puerto Rico. First Very Long Baseline Interferometer (VLBI) images.

2004-present

Two rovers travel on Mars, detectors search for gravitational radiation.

Industrial Revolution

Information Age

1840-1849

J. W. Draper invents astronomical photography; takes first photographs of the Moon. Christian Doppler proposes that wavelength is affected by motion. Lord Rosse completes 60-in reflecting telescope at Birr Castle in Ireland. Armand Fizeau and Jean-Bernard Foucault measure speed of light accurately.

1895-1897

Wilhelm Roentgen discovers X-rays. Joseph Thomson detects the electron. Yerkes 40-in optical refracting telescope completed.

1913

Niels Bohr proposes quantum theory of the atom.

1975

First Charge Coupled Device (CCD) astronomical observations.

1980

Very Large Array (VLA) radio observatory completed, Socorro, New Mexico.

1999

Chandra X-ray Telescope launched.

CHAPTER 1

Discovering the Night Sky

WEB LINK 1.1 Circumpolar Star Trails behind the Anglo-Australian Telescope, Siding Springs Mountain, New South Wales, Australia (Anglo-Australian Observatory/David Malin Images)

R I V U X G

WHAT DO YOU THINK?

1 Is the North Star—Polaris—the brightest star in the night sky?

2 Do astronomers regard the familiar patterns of stars in the sky as constellations?

3 What causes the seasons?

4 When is Earth closest to the Sun?

5 How many zodiac constellations are there?

6 Does the Moon have a dark side that we never see from Earth?

7 Is the Moon ever visible during the daytime?

8 What causes lunar and solar eclipses?

Answers to these questions appear in the text beside the corresponding numbers in the margins and at the end of the chapter.

You have picked an exciting time to study astronomy. Our knowledge about the cosmos (or the universe) is growing as never before. Current telescope technology makes it possible for astronomers to observe objects that were invisible to them just a few years ago. These new observations have deepened our understanding of virtually every aspect of the universe. We can now watch it expand and see stars explode in distant galaxies; we have discovered planets orbiting nearby stars; we have seen newly born stars enshrouded in clouds of gas and dust; and we have identified black holes and other remnants of stellar evolution. Many of these objects are so far away that the light we see from them began its journey to Earth millions and even billions of years ago. Thus, as we look farther and farther out into the universe—defined as everything that we can see or that can be seen—we also see farther and farther back into time.

Telescopes are not the only means by which we are deepening our understanding of the skies. We have also begun the process of physically exploring our neighborhood in space. In just the past half century, humans have walked on the Moon, and space probes have rolled over and dug into Martian soil. Other space missions have landed on an asteroid, brought back debris from a comet, discovered active volcanoes and barren ice fields on the moons of Jupiter, visited murky Titan and the shimmering rings of Saturn, and traveled beyond the realm of the planets in our solar system, to mention a few accomplishments. We are also witnessing the dawn of space tourism.

In the best locations, the night sky is truly breathtaking (Figure 1-1a). Even if you can't see the thousands of stars visible in clear locations (Figure 1-1b), software such as *Starry Night Enthusiast*™ can show them to you. The night sky can draw you out of yourself, inviting you to understand what is happening beyond Earth and inspiring you to think about our place in the universe.

Hundreds of years ago, the explanations of what people saw in the sky were based on beliefs that had to be accepted on faith—there was no way to test ideas of what the stars are, or whether the Moon really has liquid water oceans, or how the planets move, or why the Sun shines. Times have changed. We are fortunate to be living in an era when science has answers to many of the questions that the sky inspires.

Beautiful, intriguing, and practical, astronomy has something for everyone. This course and this book will help you better understand the universe by sharing what we have learned about some of these questions. As you proceed through this book, I hope that you will gain a new appreciation of the awesome power of the human mind to reach out, to observe, to explore, and to comprehend. One of the great lessons of modern astronomy is that by gaining, sharing, and passing on knowledge, we transcend the limitations of our bodies and the brevity of human life.

In this chapter you will discover

- how astronomers organize the night sky to help them locate objects in it
- that Earth's spin on its axis causes day and night
- how the tilt of Earth's axis of rotation and Earth's motion around the Sun combine to create the seasons
- that the Moon's orbit around Earth creates the phases of the Moon and lunar and solar eclipses
- how the year is defined and how the calendar was developed

SCALES OF THE UNIVERSE

In learning a new field it is often useful to see the "big picture" before exploring the details. For this reason, we begin by surveying the major types of objects in the universe, along with their ranges of size and the scale of distances between them.

1-1 Astronomical distances are, well, astronomical

One of the thrills and challenges of studying astronomy is becoming familiar with and comfortable with the ranges of sizes that are used. In our everyday lives we typically deal with distances ranging from millimeters to thousands of kilometers. (The metric system of units is standard in science and will be used throughout this book; however,

a

b

R I V U X G

FIGURE 1-1 The Night Sky Without and With Light Pollution (a) Sunlight is a curtain that hides virtually everything behind it. As the Sun sets, places with little smog or light pollution treat viewers to beautiful panoramas of stars that can inspire the artist or scientist in many of us. This photograph shows the night sky during a power outage seen from Goodwood, Ontario, Canada. (b) This photograph shows the same sky with normal city lighting. (© Todd Carlson/SkyNews Magazine)

we will often provide the equivalent in British units. Table E-10 lists conversion factors between the two sets of units.) A hundredth of a meter or a thousand kilometers are numbers that are easy to visualize and write. In astronomy, we deal with particles as small as a millionth of a billionth of a meter and systems of stars as large as a thousand billion billion kilometers across. Similarly, the speeds of some things, like light, are so large as to be cumbersome if you have to write them out in words each time. To deal with numbers much larger and much smaller than 1, we use a short-hand throughout this book called **scientific notation** or "powers of ten." (Read Appendix A if you are unfamiliar with scientific notation.)

The size of the universe that we can observe and the range of sizes of the objects in it are truly staggering. Figure 1-2 summarizes the range of sizes from atomic particles up to the diameter of the entire universe visible to us. Unlike linear intervals measured on a ruler, moving up 0.5×10^{-2} m (0.5 cm) along the arc of this figure brings you to objects 10 times larger. Because of this, going from the size of a proton (roughly 10^{-15} m) up to the size of an atom (roughly 10^{-10} m) takes about the same space along the arc as going from the distance between Earth and the Sun to the distance between Earth and the nearby stars.

This wide range of sizes underscores the fact that astronomy *synthesizes* or brings together information from many other fields of science. We will need to understand what atoms are composed of and how they behave; the nature and properties of light; the response of matter and energy to the force of gravity; the generation of energy by fusing particles together in stars; the

ability of carbon—and carbon alone—to serve as the foundation of life; as well as other information. These concepts will all be introduced as they are needed.

What, then, have astronomers seen of the universe? Figure 1-3 presents examples of the types of objects we will explore in this text. An increasing number of planets like Jupiter, rich in hydrogen and helium (Figure 1-3a), as well as rocky planets not much larger than the Earth, have been discovered orbiting other stars. Much smaller pieces of space debris—some of rock and metal called **asteroids** or **meteoroids** (Figure 1-3b), and others of rock and ice called **comets** (Figure 1-3c)—orbit the Sun (Figure 1-3d) and other stars. Vast stores of interstellar gas and dust are found in many galaxies; these are often the incubators of new generations of stars (Figure 1-3e). Stars by the millions, billions, or even trillions are held together in galaxies by the force of gravity (Figure 1-3f). Galaxies like our own Milky Way often contain large amounts of that interstellar gas and dust, as well as regions of space where matter is so dense that it cannot radiate light; these regions are called **black holes** (Figure 1-3g). Groups of galaxies are held together by gravity in clusters (Figure 1-3h), and clusters of galaxies are held together by gravity in superclusters. Huge quantities of intergalactic gas are often found between galaxies (Figure 1-3i).

Every object in astronomy is constantly changing—each has an origin, an active period you might consider as its "life," and each will have an end. We will study these processes along with the important physical concepts upon which they are based. You will also discover that all the matter astronomers see in stars and galaxies is but the

FIGURE 1-2 The Scales of the Universe This curve gives the sizes of objects in meters, ranging from subatomic particles at the bottom to the entire observable universe at the top. Every 0.5 cm up along the arc represents a factor of 10

larger. (Top to bottom: R. Williams and the Hubble Deep Field Team [STScI] and NASA; AAT; L. Golub, Naval Observatory, IBM Research, NASA; Richard Bickel/Corbis; Scientific American Books; Jose Luis Pelaez/Getty Images; Rothamsted Research Centre for Bioimaging; Courtesy of Florian Banhart/University of Mainz)

tip of the cosmic iceberg—there is much more in the universe, but astronomers do not yet know its nature.

PATTERNS OF STARS

When you gaze at the sky on a clear, dark night where the air is free of pollution and there is not too much light, there seem to be millions of stars twinkling overhead. In reality, the unaided human eye can detect only about 6000 stars over the entire sky. At any one time, you can see roughly 3000 stars in dark skies, because only half of the stars are above the *horizon*—the boundary between Earth and the sky. In very smoggy or light-polluted cities, you may see only a tenth of that number, or less (Figure 1-1).

In any event, you probably have noticed patterns, technically called *asterisms*, formed by bright stars, and you are familiar with some common names for these patterns, such as the ladle-shaped Big Dipper and broad-shouldered Orion. These recognizable patterns of stars are informally called *constellations* in everyday conversation and they have names derived from ancient legends (Figure 1-4a).

1-2 Constellations make locating stars easy

You can orient yourself on Earth with the help of easily recognized constellations. For instance, if you live in the northern hemisphere, you can use the Big Dipper to find the direction north. To do this, locate the Big Dipper and imagine that

FIGURE 1-3 Inventory of the Universe Pictured here are examples of the major categories of objects that have been found throughout the universe. You will discover more about each type in the chapters that follow. (a: NASA/Hubblesite; b: NASA; c: Peter Stättmayer/European Southern Observatory; d: Big Bear Observatory; e: NASA/Jeff Hester & Paul Scowen; f: Anglo-Australian Observatory; g: NOAO; h: NASA; i: N. F. Comins & F. N. Owen/NRAO)

its bowl is resting on a table (Figure 1-5). If you see the dipper upside down in the sky, as you frequently will, imagine the dipper resting on an upside-down table above it. Locate the two stars of the bowl farthest from the Big Dipper's handle. These are called the *pointer stars.* Draw a mental line through these stars in the direction away from the table, as shown in Figure 1-5. The first *moderately bright* star you then encounter is Polaris, also called the North Star because it is located almost directly over Earth's North Pole. So, while Polaris is not even among the 20 brightest stars (see Appendix E-6), it is easy to locate. Whenever you face Polaris, you are facing north. East is then on your right, south is behind you, and west is on your left. There is no equivalent star over the South Pole.

The Big Dipper example also illustrates the fact that being familiar with constellations makes it easy to locate other stars. The most effective way to do this is to use vivid visual connections, especially those of your own devising. For example, imagine gripping the handle of the Big Dipper and slamming its bowl straight down onto the head of Leo (the Lion). Leo comprises the first group of bright stars your dipper encounters. As shown in Figure 1-5, the brightest star in this group is Regulus, the dot of the backward question mark that traces the lion's mane. As another example, follow the arc of the Big Dipper's handle away from its bowl. The first bright star you encounter along that arc beyond the handle is Arcturus in Boötes (the Herdsman). Follow the same arc farther to the prominent bluish star Spica in Virgo (the Virgin). Spotting these stars and remembering their names is easy if you remember the saying "Arc to Arcturus and speed on to Spica."

During the winter months in the northern hemisphere, you can see some of the brightest stars in the sky. Many of them are in the vicinity of the "winter triangle," which connects bright stars in the constellations of Orion

a

b

R I V U X G

FIGURE 1-4 The Constellation Orion (a) The pattern of stars (asterism) called Orion is prominent in the winter sky. From the northern hemisphere, it is easily seen high above the southern horizon from December through March. You can see in this photograph that the various stars have different colors, something to watch for when you observe the night sky.

(b) Technically, constellations are entire regions of the sky. The constellation called Orion and parts of other nearby constellations are depicted in this photograph. All the stars inside the boundary of Orion are members of that constellation. The celestial sphere is covered by 88 constellations of differing sizes and shapes. (© 2004 Jerry Lodriguss/www.astropix.com)

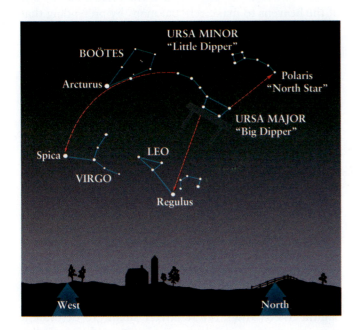

FIGURE 1-5 The Big Dipper as a Guide In the northern hemisphere, the Big Dipper is an easily recognized pattern of seven bright stars. This star chart shows how the Big Dipper can be used to locate the North Star as well as the brightest stars in three other constellations. While the Big Dipper appears right side up in this drawing of the sky shortly before sunrise, at other times of the night it appears upside down.

(the Hunter), Canis Major (the Larger Dog), and Canis Minor (the Smaller Dog), as shown in Figure 1-6. The winter triangle passes high in the sky at night during the middle of winter. It is easy to find Sirius, the brightest star in the night sky, by locating the belt of Orion and following a straight mental line from it to the left (as you face Orion). The first bright star that you encounter is Sirius.

INSIGHT INTO SCIENCE

Flexible Thinking Part of learning science is learning to look at things from different perspectives. For example, when learning to identify the prominent constellations, be sure to view them from different orientations (that is, with the star chart rotated at different angles), so that you can find them at different times of the night and the year.

The "summer triangle," which graces the summer sky as shown in Figure 1-7, connects the bright stars Vega in Lyra (the Lyre), Deneb in Cygnus (the Swan), and Altair in Aquila (the Eagle). A conspicuous portion of the Milky Way forms a

Create a story with which you can remember the connection between the constellations Orion and Taurus, which is up and to the right of Orion.

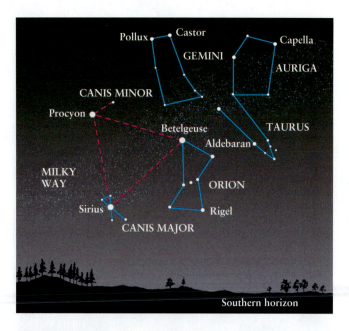

FIGURE 1-6 The Winter Triangle This star chart shows the southern sky as it appears during the evening in December. Three of the brightest stars in the sky make up the winter triangle. In addition to the constellations involved in the triangle, Gemini (the Twins), Auriga (the Charioteer), and Taurus (The Bull) are also shown.

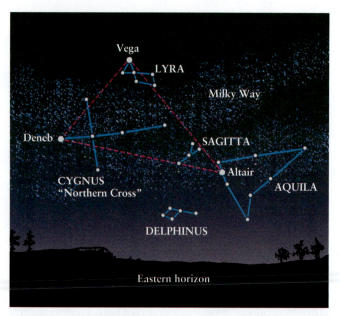

FIGURE 1-7 The Summer Triangle This star chart shows the northeastern sky as it appears in the evening in June. In addition to the three constellations involved in the summer triangle, the faint stars in the constellations Sagitta (the Arrow) and Delphinus (the Dolphin) are also shown.

*Students using a celestial sphere.
(JUPITERIMAGES/Thinkstock/Alamy)*

beautiful background for these constellations, which are nearly overhead during the middle of summer at midnight. For more on the constellations, see *Guided Discovery: The Stars and Constellations.*

Astronomers require more accuracy in locating dim objects than is possible simply by moving from constellation to constellation. They have therefore created a sky map, called the **celestial sphere,** and applied a coordinate system to it, analogous to the coordinate system of north-south latitude and east-west longitude used to navigate on Earth. If you know a star's celestial coordinates, you can locate it quickly. For such a sky map to be useful in finding stars, the stars must be fixed on it, just as cities are fixed on maps of Earth.

1-3 The celestial sphere aids in navigating the sky

If you look at the night sky year after year, you will see that the stars do indeed appear fixed relative to one another. Furthermore, through-

out each night the entire pattern of stars appears to rigidly orbit Earth. We employ this artificial, Earth-based view of the heavens to make celestial maps by pretending that the stars are attached to the inside of an enormous hollow shell, the celestial sphere, with Earth at its center (Figure 1-8). Visualized another way, you can imagine the half of the celestial sphere that is visible at night as a giant bowl covering Earth.

Whereas asterisms such as the Big Dipper are often called "constellations" in normal conversation, astronomers use the word **constellation** to describe an entire area of the sky and all the objects in it (see Figure 1-4b). The celestial sphere is divided into 88 constellations of differing sizes and shapes. (Keep in mind that most constellations and their asterisms—the recognizable star patterns—have the same name; for example, Orion the asterism and Orion the constellation, which sometimes complicates conversation.) The boundaries of the constellations are straight lines that meet at right angles (see Figure 1-4b). Some constellations, like Ursa Major (the Large Bear), are very large, while others, like Sagitta (the Arrow), are relatively small. To describe a star's location, we might say "Albireo in the constellation Cygnus (the Swan)," much as we would refer to "Chicago in the state of Illinois," "Melbourne in the state of New South Wales," or "Ottawa in the province of Ontario."

2

GUIDED DISCOVERY

The Stars and Constellations

Many students take an introductory astronomy course expecting to be taught the familiar stars and constellations. Often, instructors just do not have time to cover this material. You can, however, learn them on your own in several ways.

First, use the two steps for memorizing the night sky discussed in the text:

1. Observe easily identified constellations.

2. Associate these constellations with nearby, lesser-known stars and constellations.

For example, you may not remember where the star Aldebaran in Taurus is in the sky, but if you remember that Orion is fighting Taurus, you can locate Aldebaran by following a line defined by the belt of Orion to the right (away from Sirius). The first bright star you encounter is Aldebaran.

Make your own connections between the constellations. Have fun while you're doing it. Chances are that you will remember the phrase "slam the Big Dipper's bowl downward to hit Leo the Lion on the head" rather than "the first bright group of stars directly below the Big Dipper is Leo."

Second, study print or software sky materials. A very efficient way to learn the constellations is to use the board and card game *Stellar 28*. In addition, astronomy computer programs such as *Starry Night Enthusiast*™ are available. *Starry Night Enthusiast*™ shows many things in addition to what constellations are up at night, including the motion, location, and phases of the planets; the location of deep sky objects, such as nebulae; and the sky as seen from any location on Earth on any date.

A third way to familiarize yourself with the night sky is to go out on clear nights armed with star charts to see which constellations are up. You will find a set of such charts from the *Griffith Observer* magazine at the back of this book. To use the charts, first select the one that best corresponds to the date and time of your observation. Hold the chart vertically and turn it so that the direction you are facing outdoors shows at the bottom. The patterns and connections between stars and constellations will soon become evident. Using a flashlight with a red plastic coating over the light will make it easier to read the chart without ruining your night vision. You will find stargazing a truly enjoyable experience.

The stars seem fixed on the celestial sphere only because of their remoteness. In reality, they are at widely varying distances from Earth, and they do move relative to one another. But we neither see their motion nor perceive their relative distances because the stars are so far from here. You can understand this by imagining a jet plane just one kilometer overhead traveling at 1000 km (620 mi) per hour across the sky. Its motion is unmistakable. However, a plane moving at the same speed and altitude, but appearing along the horizon, seems to be moving about a hundred times more slowly. And an object that is at the distance of the Sun, traveling at the same speed and moving across the sky, would appear to be going nearly a hundred million times more slowly than the plane overhead.

The stars (other than the Sun) are all more than 40 trillion km (25 trillion mi) from us. Therefore, although the patterns of stars in the sky do change, their great distances prevent us from seeing those changes over the course of a human lifetime. Thus, as unrealistic as it is, the celestial sphere is so useful for navigating the heavens that it is used by astronomers even at the most sophisticated observatories around the world.

As shown in Figure 1-8, we can project key geographic features from Earth out into space to establish directions and bearings. If we expand Earth's equator onto the celestial sphere, we obtain the **celestial equator**, which divides the sky into northern and southern hemispheres, just as Earth's equator divides Earth into two hemispheres. We can also imagine projecting Earth's North Pole and South Pole out into space along Earth's axis of rotation. Doing so gives us the **north celestial pole** and the **south celestial pole**, also shown in Figure 1-8.

Using the celestial equator and poles as reference features, astronomers divide up the surface of the celestial sphere in precisely the same way that the latitude and longitude grid divides Earth. The equivalent to latitude on Earth is called **declination (dec)** on the celestial sphere. It is measured from 0° to 90° north or south of the celestial equator. The equivalent of longitude on Earth is called **right ascension (r. a.)** on the celestial sphere, measured from 0 h to 24 h around the celestial equator (see Figure 1-8). The boundaries of the constellations, introduced above, run along lines of constant right ascension and constant declination.

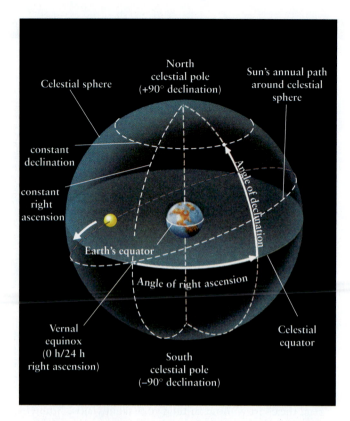

FIGURE 1-8 The Celestial Sphere The celestial sphere is the apparent "bowl" or hollow sphere of the sky. The celestial equator and poles are projections of Earth's equator and axis of rotation onto the celestial sphere. The north celestial pole is therefore located directly over Earth's North Pole, while the south celestial pole is directly above Earth's South Pole. Analogous to longitude and latitude, the coordinates in space are right ascension (r.a.) and declination (dec), respectively. The star in this figure has the indicated r.a. and dec.

Just as the location of Greenwich, England, defines the *prime meridian,* or zero of longitude on Earth, we need to establish a zero of right ascension. It is defined as one of the places where the Sun's annual path across the celestial sphere intersects the celestial equator. (We will explore later in this chapter why the Sun appears to move in a circle around the celestial sphere during the course of a year.) The celestial equator and the Sun's path intersect at two points. The equivalent on the celestial sphere of Earth's prime meridian is where the Sun crosses the celestial equator moving northward. Angles of right ascension are measured from this point, called the **vernal equinox** (Figure 1-8).

In navigating on the celestial sphere, astronomers measure the distance between objects in terms of angles. If you are not familiar with angular separation, read *An Astronomer's Toolbox 1-1.*

> The planets move through constellations, while the stars do not. What two possible properties of the planets could allow them to do that?

EARTHLY CYCLES

The Sun systematically rises and sets at different times and in different places throughout the year. Similarly, the Moon rises and sets at different times each day and repeats its cycle roughly once every 29½ days. Furthermore, as noted in Section 1-2, we do not see the same constellations up in the sky every night of the year, but the cycle of constellations we can see at night repeats each year. The daily and annual rhythms of the sky, Earth, and all life on it arise from three celestial motions: Earth's spinning, which causes day and night and also causes the apparent daily motion of the celestial sphere and all the objects on it; Earth's orbit around the Sun, which creates the seasons, the year, and the change in times at which constellations are up at night; and the Moon's orbit around Earth, which creates the lunar phases, the cycle of tides, and the spectacular phenomena we call eclipses.

1-4 Earth's rotation creates the day-night cycle and its revolution defines a year

Earth spins on its axis. Such motion is called **rotation.** We do not feel Earth's rotation because our planet is so physically large compared to us that its gravitational attraction holds us firmly on Earth's surface. Earth's rotation causes the stars—as well as the Sun, Moon, and planets—to appear to rise on the eastern horizon, move across the sky, and set on the western horizon. Earth's daily rotation, causing the Sun to rise and set, thereby creates day and night. The **diurnal motion,** or daily motion, of the celestial bodies is apparent in time-exposure photographs, such as that shown in Figure 1-9.

Take a friend outside on a clear, warm night to observe the diurnal motion of the stars for yourselves. Soon after dark, find a spot away from bright lights, and note the constellations in the sky relative to some prominent landmarks near you on Earth. A few hours later, check again from the same place. You will find that the entire pattern of stars (as well as the Moon, if it is visible) has shifted. New constellations will have risen above the eastern horizon, while other constellations will have disappeared below the western horizon. If you check again just before dawn, you will find the stars that were just rising in the east when the night began are now low in the western sky.

Different constellations are visible at night during different times of the year. This occurs because Earth orbits, or *revolves,* around the Sun. **Revolution** is the motion of any astronomical object around another astronomical object. Earth takes one year, or about 365¼ days, to go once around the Sun. A year on Earth

Observational Measurements Using Angles

Ancient mathematicians invented a system of angles and angular measure that is still used today to denote the relative positions and apparent sizes of objects in the sky. To locate stars on the celestial sphere, for example, we do not need to know their distances from Earth. All we need to know is the angle from one star to another in the sky, a property that remains fixed over our lifetimes because the stars are all so far away.

An **arc angle**, often just called an **angle**, is the opening between two lines that meet at a point. Angular measure is a method of describing the size of an angle. The basic unit of angular measure is the **degree**, designated by the symbol °. A full circle is divided into 360°. A right angle measures 90°. As shown in the figure below, the angle between the two "pointer stars" in the Big Dipper is about 5°.

Astronomers also use angular measure to describe the apparent sizes of celestial objects. For example, imagine the full Moon. As seen from Earth, the angle across the Moon's diameter is nearly ½°. We therefore say that the **angular diameter**, or **angular size**, of the Moon is ½°. Alternatively, astronomers say that the Moon "subtends" an angle of ½°. In this context, subtend means "to extend across."

An adult human hand held up at arm's length provides a means of estimating angles. For example, five fingers extended at arm's length covers an angle of about 8°, whereas the tip of your finger is about 1½° wide. Various segments of your index finger extended to arm's length can be similarly used to estimate angles a few degrees across, as shown in the figure above on the right.

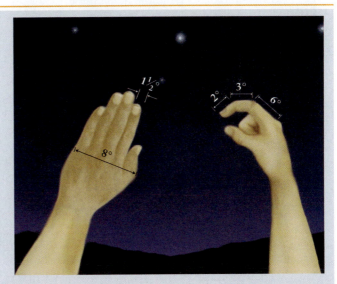

Estimating Angles with the Human Hand Various parts of the adult human hand extended to arm's length can be used to estimate angular distances and sizes in the sky.

To talk about smaller angles, we subdivide the degree into 60 arcminutes (abbreviated 60 arcmin or 60′). An arcminute is further subdivided into 60 arcseconds (abbreviated 60 arcsec or 60″). A dime viewed face-on from a distance of 1.6 km (1 mile) has an angular diameter of about 2 arcsec.

From everyday experience, we know that an object looks bigger when it is nearby than when it is far away. The angular size of an object therefore does not necessarily tell you anything about its actual physical size. For example, the fact that the Moon's angular diameter is ½° does not tell you how big the Moon really is. But if you also happen to know the distance to the Moon, then you can calculate the Moon's physical diameter. In general, the physical diameter of an object can be calculated from the equation:

$$\text{physical diameter} = \text{distance} \times \tan (\text{angular diameter})$$

where tan (angular diameter) means the tangent of the angle denoted "angular diameter." In the Moon's case, using a measured distance (see Appendix E-3) of 384,400 km and an angular diameter of ½°, we find the diameter to be roughly 3350 km. The difference between this value and the exact diameter of 3476 km is due primarily to the approximate value of ½° that we have used.

The Big Dipper The angular distance between the two "pointer stars" at the front of the Big Dipper is about 5°. For comparison, the angular diameter of the Moon is about ½°.

Try these questions: The Sun is 1.5×10^8 km away and has a diameter of 1.4×10^6 km. How large an angle does it make in our sky? How does that compare to the arc angle of the Moon in our sky? What arc angle would the Moon make in our sky if it were twice as far away? Half as far?

(Answers appear at the end of the book.)

R I V U X G

FIGURE 1-9 Circumpolar Star Trails This long exposure, taken from Australia's Siding Spring mountain and aimed at the south celestial pole, shows the rotation of the sky. The stars that pass between the pole and the ground are all circumpolar stars. (Anglo-Australian Observatory/David Malin Images)

is measured by the motion of our planet relative to the stars. For example, draw a straight line from the Sun through Earth to some star on the opposite side of Earth from the Sun. As Earth revolves, that line inscribes a straight path on the celestial sphere and returns to the original star 365¼ days later. The length of any cycle of motion, such as Earth's orbit around the Sun, that is measured with respect to the stars is called a **sidereal period**.

If Earth were spinning over a fixed place on the Sun, rather than revolving around it, then every star would rise and set at the same times throughout the year. As a result of Earth's motion around the Sun, however, the stars rise approximately 4 min earlier each day than they did the day (or night) before. This effect accumulates, bringing different constellations up at night throughout the year. Figure 1-10 summarizes this motion. When the Sun is within the boundaries of (colloquially, "in") Virgo (September 18–November 1), for example, the hemisphere containing the Sun and the constellations around Virgo are in daylight (see Figure 1-10a). When the Sun is up, so are Virgo and the surrounding constellations, so we cannot see them. During that time of year, the constellations on the other side of the celestial sphere, centered on the constellation Pisces, are in darkness. Thus, when the Sun is "in" Virgo,

a

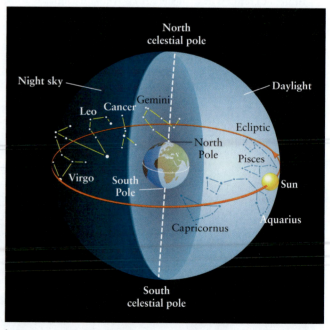

b

FIGURE 1-10 Why Different Constellations Are Visible at Different Times of the Year (a) On the autumnal equinox each year, the Sun is in the constellation Virgo. As seen from Earth, that part of the sky is then daylight and we see stars only on the other half of the sky, centered around the constellation Pisces. (b) Six months later, the Sun is in Pisces. This side of the sky is then bright, while the side centered on Virgo is in darkness.

FIGURE 1-11 *Motion of Stars at the Poles* Because the Earth rotates around its poles, stars seen from these locations appear to move in huge, horizontal circles. This is the same effect you would get by standing up in a room and spinning around; everything would appear to move in circles around you. At the North Pole stars move left to right, while at the South Pole they move right to left.

FIGURE 1-12 *Rising and Setting of Stars at the Equator* Standing on the equator, you are perpendicular to the axis around which the Earth rotates. As seen from there, the stars rise straight up on the eastern horizon and set straight down on the western horizon. This is the same effect you get when driving straight over the crest of a hill; the objects on the other side of the hill appear to move straight upward as you descend.

R I V U X G

FIGURE 1-13 *Rising and Setting of Stars at Middle North Latitudes* Unlike the motion of the stars at the poles (see Figure 1-11), the stars at all other latitudes do change angle above the ground throughout the night. This time-lapse photograph shows stars setting. The latitude determines the angle at which the stars rise and set. (David Miller/DMI)

Pisces and the constellations around it are high in our sky at night.

Six months later, when the Sun is "in" Pisces, that half of the sky is filled with daylight, while Virgo and the constellations around it are high in the night sky (see Figure 1-10b). These arguments apply everywhere on Earth at the same time because the Sun moves along its path very slowly as seen from Earth, taking a year to make one complete circuit.

We spoke earlier of stars rising on the eastern horizon and setting on the western horizon. Depending on your latitude, some of the stars and constellations never disappear below the horizon. Instead, they trace complete circles in the sky over the course of each night (see Figure 1-9). To understand why this happens, imagine that you are standing on Earth's North Pole at night. Looking straight up, you see Polaris. Because Earth is spinning around its axis directly under your feet, all the stars appear to move from left to right (counterclockwise) in horizontal rings around you. The exception is Polaris, which always remains at the North Pole's **zenith.** (Every place has a different zenith, which is the point directly overhead anywhere on Earth.) As seen from the North Pole, no stars rise or set (Figure 1-11). They just seem to revolve around Polaris in horizontal circles. Stars and constellations that never go below the horizon are called **circumpolar.** While there is no bright south pole star equivalent to Polaris, all stars seen from the South Pole are also circumpolar and move from right to left (clockwise).

> Why are asterisms like the Big Dipper sometimes seen upright and sometimes upside down?

If you live in the northern hemisphere, Polaris is always located above your northern horizon at an angle equal to your latitude. Only the stars and constellations that pass between Polaris and the land directly below it are circumpolar. As you go farther south in the northern hemisphere, the number of stars and constellations that are circumpolar decreases. Likewise, as you go farther north in the southern hemisphere, the number of stars and constellations that are circumpolar decreases.

Now visualize yourself at the equator. All the stars appear to rise straight up in the eastern sky and set straight down in the western sky (Figure 1-12). Polaris is barely visible on the northern horizon. While Polaris never sets, all the other stars do, and therefore none of the stars are circumpolar as seen from the equator.

As you can see from these last two mental exercises, the angle at which the stars rise and set depends on your viewing latitude. Figure 1-13 shows stars setting at 35° north latitude. Polaris is fixed at 35° above the horizon to the right of this figure, not at the zenith, as it is at the North Pole, nor on the horizon, as seen from the equator. As another example, except for those stars in the corners, all the stars whose paths are shown in Figure 1-9 are circumpolar because they are visible all night, every night.

1-5 The seasons result from the tilt of Earth's rotation axis combined with Earth's revolution around the Sun

Imagine that you could see the stars even during the day, so that you could follow the Sun's apparent motion against the background constellations throughout the year. (The Sun appears to move among the stars, of course, because Earth orbits around it.) From day to day, the Sun traces a straight path on the celestial sphere. This path is called the **ecliptic**. As you can see in Figure 1-14a, the ecliptic makes a closed circle bisecting the celestial sphere. The ecliptic is precisely the loop labeled "Sun's annual path around celestial sphere" in Figure 1-8.

The term "ecliptic" has a second use in astronomy. Earth orbits the Sun in a plane also called the ecliptic. The two ecliptics exactly coincide: Imagine yourself on the Sun watching Earth move day by day. The path of Earth on the celestial sphere as seen from the Sun is precisely the same as the path of the Sun as seen from Earth (Figure 1-14b). Recall also the discussion of the line running from the Sun through Earth to the celestial sphere in Section 1-4.

INSIGHT INTO SCIENCE

Define Your Terms As with interpersonal communication, using the correct words is very important in science. Usually, scientific words each have a specific meaning, such as "rotation" denoting spin and "revolution" meaning one object orbiting another. Be especially careful to understand the context in which words with more than one meaning, such as "ecliptic" and "constellation," are used.

Equinoxes and Solstices   The ecliptic and the celestial equator are different circles tilted 23½° with respect to each other on the celestial sphere. This occurs because Earth's rotation axis is tilted 23½° away from a line perpendicular to the ecliptic (Figures 1-14 and 1-15). These two circles intersect at only two points, which are exactly opposite each other on the celestial sphere (Figure 1-16). Each of these two points is called an **equinox** (from the Latin words meaning "equal night"), because when the Sun appears at either point, it is directly over Earth's equator, resulting in 12 h of daytime and 12 h of nighttime everywhere on Earth on that day.

Except for tiny changes each year, Earth maintains this tilted orientation as it orbits the Sun. Therefore, Polaris is above the North Pole throughout the year. For half the year, the northern hemisphere is tilted toward the Sun, and, as a result, the Sun rises higher in the northern hemisphere's sky than it does during the other half of the year (Figure 1-17). Equivalently, when the southern hemisphere is tilted toward the Sun, the Sun rises higher in the southern hemisphere's sky.

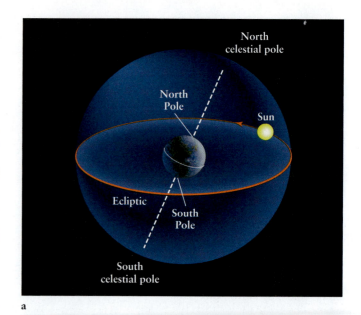

a

b

FIGURE 1-14 The Ecliptic (a) The ecliptic is the apparent annual path of the Sun on the celestial sphere. (b) The ecliptic is also the plane described by Earth's path around the Sun. The planes created by the two ecliptics exactly coincide. As in (a), the rotation axis of Earth is shown here tilted 23½° from being perpendicular to the ecliptic.

Consider the location of the Sun as seen from the northern hemisphere throughout the year. The day that the Sun rises farthest south of east is around December 22 each year (Figure 1-17) and is called the **winter solstice**. The winter solstice is the point on the ecliptic farthest south of the celestial equator (see Figure 1-16). It is also the day when the Sun rises to the lowest height at noon (Figure 1-17a) and it signals the day of the year in the northern hemisphere with the fewest number of daylight hours.

As the Sun moves along the ecliptic after the winter solstice, it rises earlier, is more northerly on the eastern horizon, and passes higher in the sky at mid-day than it did on preceeding days. Three months later, around March 21, the Sun crosses the celestial equator heading northward. This is called the vernal equinox and is one of the two days on which the Sun rises due east and sets due

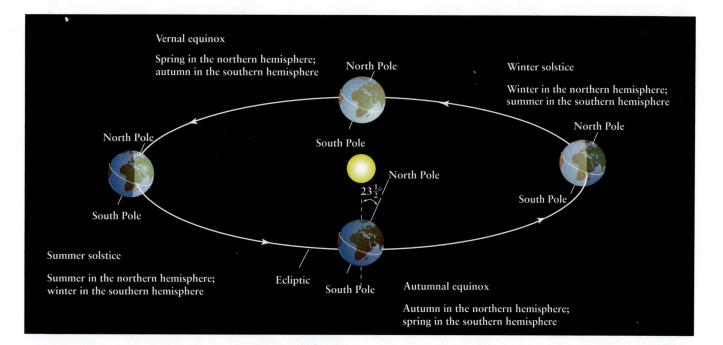

FIGURE 1-15 The Tilt of Earth's Axis Earth's axis of rotation is tilted 23½° from being perpendicular to the plane of Earth's orbit. Earth maintains this orientation (with its North Pole aimed at the north celestial pole near the star Polaris) throughout the year as it orbits the Sun. Consequently, the amount of solar illumination and the number of daylight hours at any location on Earth vary in a regular fashion with the seasons.

west (Figure 1-17b). The vernal equinox is the "prime meridian" of the celestial sphere, as discussed earlier. Three months after the vernal equinox, around June 21, the Sun rises farthest north of east and passes highest in the sky (Figure 1-17c). This is the **summer solstice** (see Figure 1-16), the day of the year in the northern hemisphere with the most daylight.

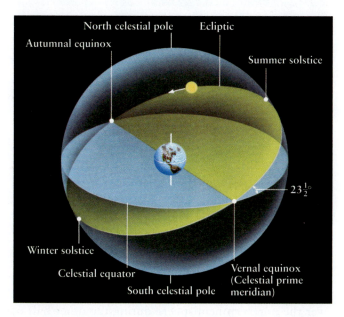

FIGURE 1-16 The Seasons Are Linked to Equinoxes and Solstices The ecliptic is inclined to the celestial equator by 23½° because of the tilt of Earth's axis of rotation. The ecliptic and the celestial equator intersect at two points called the equinoxes. The northernmost point on the ecliptic is called the summer solstice; the southernmost point is called the winter solstice.

From June 21 through December 21, the Sun rises farther south than it did the preceding day. Its highest point in the sky is lower each succeeding day—the cycle of the previous 6 months reverses. The **autumnal equinox** occurs around September 22 (Figure 1-17d), with the Sun heading southward across the celestial equator, as seen from Earth.

The higher the Sun rises during the day, the more daylight hours there are. During the days with longer periods of daylight, more light and heat from the Sun strike that hemisphere. Furthermore, when the Sun is higher in the sky, its energy is more concentrated on Earth's surface (see the ovals of sunlight in each of the illustrations in Figure 1-17). Thus, during these days more energy is deposited on each square meter of the surface, thereby warming them more, than when the Sun is lower in the sky. The temperature and, hence, the seasons are determined by the duration of daylight at any place and the height of the Sun in the sky there. (Bear in mind that winds and clouds greatly affect the weather throughout the year—we ignore these effects here.)

To summarize, the Sun is lowest in the northern sky on the winter solstice. This marks the beginning of winter in the northern hemisphere. As the Sun moves northward, the amount of daylight increases daily. The vernal equinox marks a midpoint in the amount of light and heat from the Sun onto the northern hemisphere and is the beginning of spring. When the Sun reaches the summer solstice, it is highest in the northern sky and is above the horizon for the most hours of any day of the year. This is the beginning of summer. Returning southward, the Sun crosses the celestial equator once again on the autumnal equinox, the beginning of fall.

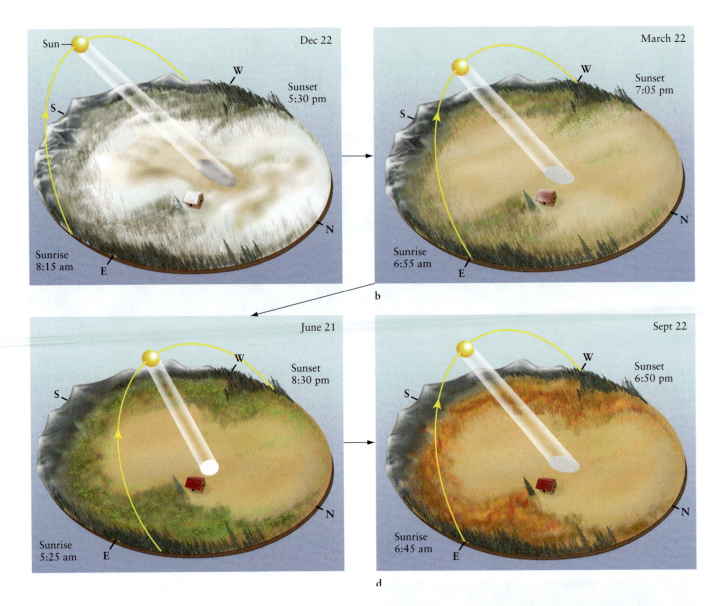

Dec 22

Sun—

W

Sunset
5:30 pm

S—

N

Sunrise
8:15 am   E

March 22

W

Sunset
7:05 pm

S—

N

Sunrise
6:55 am   E

b

June 21

W

Sunset
8:30 pm

S—

N

Sunrise
5:25 am   E

Sept 22

W

Sunset
6:50 pm

S—

N

Sunrise
6:45 am   E

d

FIGURE 1-17 The Sun's Daily Path and the Energy It Deposits Here (a) On the winter solstice, the first day of winter, the Sun rises farthest south of east, is lowest in the noontime sky, stays up the shortest time, and its light and heat are least intense (most spread out) of any day of the year in the northern hemisphere. (b) On the vernal equinox, the first day of spring, the Sun rises precisely in the east and sets precisely in the west. Its light and heat have been growing more intense, as shown by the brighter oval of light than in (a). (c) On the summer solstice, first day of summer, the Sun rises farthest north of east of any day in the year, is highest in the noontime sky, stays up the longest time, and its light and heat are most intense of any day in the northern hemisphere. (d) On the autumnal equinox, the same astronomical conditions exist as on the vernal equinox.

Earth's orbit around the Sun is elliptical (we will discuss this oval shape in detail in Chapter 2). While the distance between Earth and Sun changes by 5 million km (3 million mi) throughout the year, this variation has only a minor effect on the seasons. If the seasons *were* caused by the changing distance from Earth to the Sun, all parts of Earth should have the same seasons at the same time. In fact, the northern and southern hemispheres have exactly opposite seasons. Furthermore, Earth is closest to the Sun on or around January 3 of each year—the dead of winter in the northern hemisphere!

The variation in Earth's distance from the Sun over the year would have a greater effect if it were not for the fact that the southern hemisphere has more area covered by oceans than does the northern hemisphere. As a result, when Earth is closer to the Sun (and the Sun is high in the southern hemisphere's sky), the southern oceans scatter more light and heat directly back into space than occurs when the Sun is higher over the northern hemisphere during the other half of the year. Had the extra energy sent back into space when we are closer to the Sun been absorbed by our planet, Earth would, indeed, heat more during this time than when the Sun is over the northern hemisphere.

Explain why Figure 1-13 must have been taken facing west. *Hint:* Examine Figure 1-17.

The Sun's Path Across the Sky   During the northern hemisphere's summer months, when the northern hemisphere is tilted toward the Sun, the Sun rises in the northeast and sets in the northwest. The Sun provides more than 12 h of daylight in the northern hemisphere and passes high in the sky. At the summer solstice, the Sun is as far north as it gets, giving the greatest number of daylight hours to the northern hemisphere.

During the northern hemisphere's winter months, when the northern hemisphere is tilted away from the Sun, the Sun rises in the southeast. Daylight lasts for fewer than 12 h, as the Sun skims low over the southern horizon and sets in the southwest. Night is longest in the northern hemisphere when the Sun is at the winter solstice.

The Sun's maximum angle above the southern horizon is different at different latitudes. The farther north you are, the lower the Sun is in the sky at any time of day than it is on that day at more equatorial locations. At latitudes above 66½° north latitude or below 66½° south latitude, the Sun does not rise at all during parts of their fall and winter months. During their spring and summer months, those same regions of Earth have continuous sunlight for weeks or months (Figure 1-18), hence the name "Land of the Midnight Sun."

The Sun takes one year to complete a trip around the ecliptic. Since there are about 365¼ days in a year and 360° in a circle, the Sun appears to move along the ecliptic at a rate of slightly less than 1° per day. The constellations through which the Sun moves throughout the year as it travels along the ecliptic are called the **zodiac** constellations. We cannot see the stars of these constellations when the Sun is among them, of course, but we can plot the Sun's path on the celestial sphere to determine through which constellations it moves. Traditionally, there were 12 zodiac constellations whose borders were set in antiquity. In 1930, the boundaries were redefined by astronomers, and the Sun now moves through 13 constellations throughout the year. (The thirteenth zodiac constellation is Ophiuchus, the Serpent Holder. The Sun passes through Ophiuchus from December 1 to December 19 each year.) Table 1-1 lists all the zodiac constellations and the dates the Sun passes through them. You may not have the "sign" that you think you have.

1-6 Clock times based on the Sun's location created scheduling nightmares

The Sun's daily motion through the sky provided our distant ancestors' earliest reference for time because the Sun's location determines whether it is day or night and roughly whether it is before or after midday. The Sun's motion through the sky came to determine the length of the **solar day,** upon which our 24-h day is based. Ideally, this is the interval of time between when the Sun is highest in the sky on one day until the time it is highest in the sky on the next day. However, the length of the solar day actually varies throughout the year. This occurs because Earth's orbit around the Sun is not perfectly circular—our planet speeds up as it approaches the Sun and slows as it moves away—and because Earth's rotation axis is tilted 23½° from being perpendicular to the ecliptic. These two effects change the apparent speed of the Sun across the sky throughout the year. The *average* time interval between consecutive noontimes throughout the year is 24 h, which determines the time we use on our clocks. This is called the *mean (or average)* solar day.

> What region of Earth has the smallest range of seasonal temperature changes and why?

Traditionally, noontime was taken to be the instant when the Sun is highest in the sky. But as we have just seen, the interval from one noontime to the next is not exactly 24 h, so the Sun is not always highest at noon. The difference between clock noontime and astronomical noontime (when the Sun is highest) is as much as 16 min.

R I V U X G

FIGURE 1-18 **The Midnight Sun** This time-lapse photograph was taken on July 19, 1985, at 69° north latitude in northeastern Alaska. At that latitude, the Sun is above the horizon continuously from mid-May until the end of July. (Doug Plummer/Science Photo Library)

Because Earth is rotating eastward, the Sun is highest at different longitudes (measured east and west along lines running between Earth's poles) at different times. For example, astronomical noon in New York City occurs earlier than it does a little farther west, in Philadelphia. Before the advent of time zones, local time was based on astronomical noon. To travel west from New York to Philadelphia by train, for example, you had to know the departure time at New York, using New York time, as well as the arrival time in Philadelphia (say, if someone was going to meet you) in Philadelphia time. Such time considerations became very confusing and burdensome as society became more complex. **Time zones** were established in the late nineteenth century to alleviate this problem. In a time zone, everyone agrees to set their clocks alike. Time zones are based on the time at 0° longitude in Greenwich, England, a location called the prime meridian, as mentioned earlier. With some variations due to geopolitical boundaries, every 15° of longitude around the globe begins a new time zone. The resulting 24 time zones are shown in Figure 1-19. Going from one time zone to the next usually requires you to change the time on your wristwatch by exactly 1 h.

There is also a *sidereal day,* the length of time from when a star is in one place in the sky until it is next in the same place. The solar and sidereal days differ from each other in length because while Earth rotates, it revolves around the Sun. This motion of Earth in its orbit day by day changes the location of the stars, bringing them back to their original positions 4 min earlier each day. Therefore, the sidereal day is 23 h, 56 min long, while the solar day is 24 h long.

> Why would a calendar based on sidereal days not be satisfactory?

1-7 Calendars based on equal-length years also created scheduling problems

Just as the day is caused by Earth's rotation, the year is the unit of time based on Earth's revolution about the Sun. Earth does not take exactly 365 days to orbit the Sun, so the year is not exactly 365 days long. Basing the year on a 365-day cycle led to important events occurring on the wrong day. To resolve this problem, Roman statesman Julius Caesar implemented a new calendar in the year 46 B.C. Because measurement revealed to ancient astronomers that the length of a year is approximately 365¼ days, this "Julian" calendar established a system of leap years to accommodate the extra quarter of a day. By adding an extra day to the calendar every 4 years, Caesar hoped to ensure that seasonal astronomical events, such as the beginning of spring, would occur on the same date year after year.

TABLE 1-1 The 13 Constellations of the Zodiac

Constellation	Dates of the Sun's passage through
Pisces	March 13–April 20
Aries	April 20–May 13
Taurus	May 13–June 21
Gemini	June 21–July 20
Cancer	July 20–August 11
Leo	August 11–September 18
Virgo	September 18–November 1
Libra	November 1–November 22
Scorpius	November 22–December 1
Ophiuchus	December 1–December 19
Sagittarius	December 19–January 19
Capricorn	January 19–February 18
Aquarius	February 18–March 13

The Julian calendar would have worked just fine if a year were exactly 365¼ days long and if Earth's rotation axis (now pointing toward Polaris, as discussed earlier) never changed direction. Neither assumption is correct. Thus, over time, a discrepancy accumulated between the calendar and the actual time—astronomical and cultural events began to fall on different dates each year. To straighten things out, a committee established by Pope Gregory XIII recommended a refinement, thus creating the Gregorian calendar in 1582. Pope Gregory began by dropping 10 days (October 5, 1582, was proclaimed to be October 15, 1582), which brought the first day of spring back to March 21. Next, he modified Caesar's system of leap years. Caesar had added February 29 to every calendar year that is evenly divisible by four. For example, 1996, 2000, 2004 and 2008 were all leap years with 366 days. But this system produces an error of about 3 days every 4 centuries. To solve the problem, Pope Gregory decreed that century years would be leap years only if evenly divisible by 400. For example, the years 1700, 1800, and 1900 were not leap years under the improved Gregorian system. But the year 2000—which can be divided evenly by 400—was a leap year.

We use the Gregorian system today. It assumes that the year is 365.2425 mean solar days long, which is very close to the length of the *tropical year,* defined as the time interval from one vernal equinox to the next. In fact, the error is only 1 day in every 3300 years. That won't cause any problems for a long time.

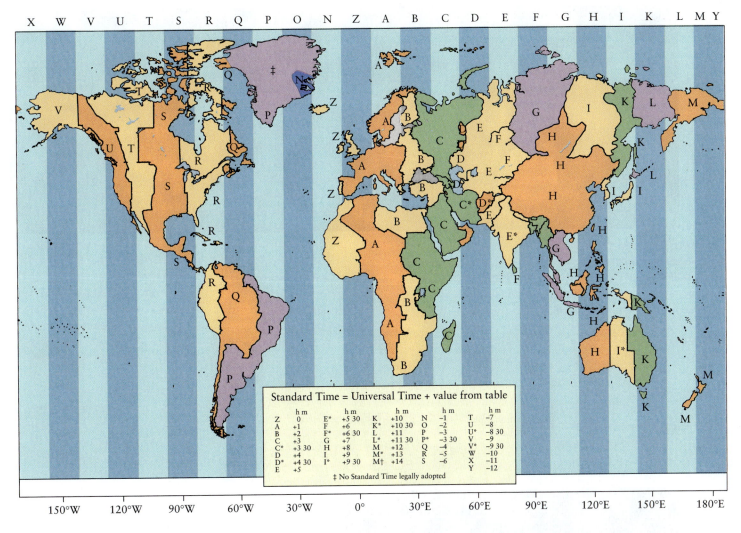

X W V U T S R Q P O N Z A B C D E F G H I K L M Y

Standard Time = Universal Time + value from table

	h m		h m		h m		h m		h m
Z	0	E*	+5 30	K	+10	N	−1	T	−7
A	+1	F	+6	K*	+10 30	O	−2	U	−8
B	+2	F*	+6 30	L	+11	P	−3	U*	−8 30
C	+3	G	+7	L*	+11 30	P*	−3 30	V	−9
C*	+3 30	H	+8	M	+12	Q	−4	V*	−9 30
D	+4	I	+9	M*	+13	R	−5	W	−10
D*	+4 30	I*	+9 30	M†	+14	S	−6	X	−11
E	+5							Y	−12

‡ No Standard Time legally adopted

150°W   120°W   90°W   60°W   30°W   0°   30°E   60°E   90°E   120°E   150°E   180°E

FIGURE 1-19 Time Zones of the World For convenience, Earth's 360° circumference is divided into 24 time zones. Ideally, each time zone would run due north-south. However, political considerations make many zones irregular. Indeed, there are even a few zones only a half hour wide.

1-8 Precession is a slow, circular motion of Earth's axis of rotation

As noted just above, the orientation of Earth's axis of rotation changes slightly with respect to the celestial sphere (that is, it "points" in a slightly different direction). While this change is small over our lifetimes, it eventually causes the north celestial pole to drift in a circle taking it far from Polaris. This major change in orientation is caused by gravitational forces from the Moon and Sun pulling on the slight equatorial bulge at Earth's equator created by Earth's rotation—our planet's diameter is about 43 km (27 mi) greater at the equator than from pole to pole. **Gravitation (gravity)** is the universal force of attraction between all matter. The strength of the gravitational force between two objects depends on the amounts of mass the objects have and the distance between them, as we explain in more detail in Chapter 2.

Because of Earth's tilted axis of rotation, the Sun and Moon are usually not located directly over Earth's equator. As a result, their gravitational attractions on Earth try to force the equatorial bulge to be as close to them as possible. However, Earth does not respond to these forces from the Sun and Moon by straightening up perpendicular to them on its axis. Instead, it changes the direction in which its axis of rotation points on the celestial sphere— a motion called **precession**. This is exactly the same behavior exhibited by a spinning top (Figure 1-20a). If the top were not spinning, gravity would pull it over on its side. But when it is spinning, the combined actions of gravity and rotation cause the top's axis of rotation to precess or wobble in a circular path. As with the toy top, the combined actions of gravity from the Sun and Moon plus rotation cause Earth's axis to trace a circle in the sky while remaining tilted about 23½° away from the perpendicular. In the mid-1990s, astronomers simulated the

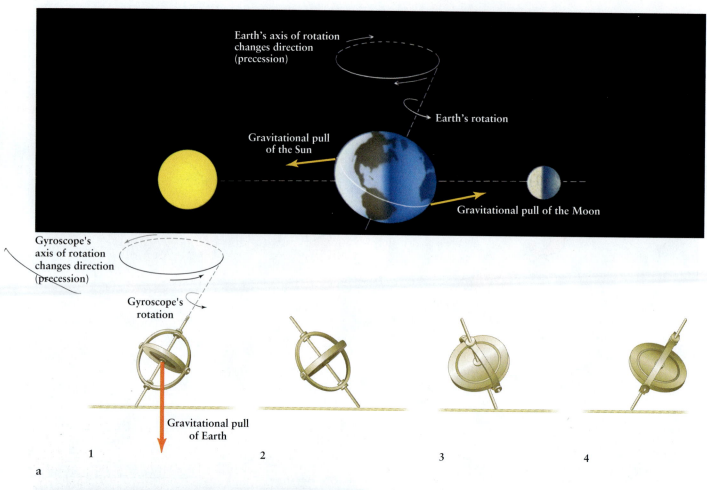

Earth's axis of rotation
changes direction
(precession)

Earth's rotation

Gravitational pull
of the Sun

Gravitational pull of the Moon

Gyroscope's
axis of rotation
changes direction
(precession)

Gyroscope's
rotation

Gravitational pull
of Earth

1   2   3   4

a

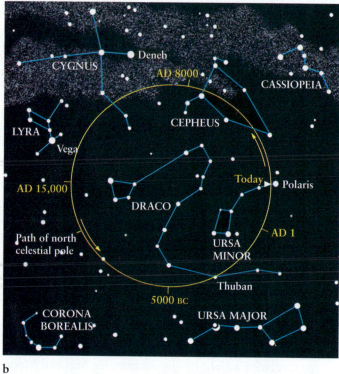

CYGNUS   Deneb

AD 8000

CASSIOPEIA

LYRA

Vega   CEPHEUS

AD 15,000

Today ▷ ● Polaris

DRACO

AD 1

Path of north
celestial pole   URSA
MINOR

Thuban

5000 BC

CORONA   URSA MAJOR
BOREALIS

b

FIGURE 1-20 Precession and the Path of the North Celestial Pole (a) The gravitational pulls of the Moon and the Sun on Earth's equatorial bulge cause Earth to precess. The situation is analogous to that of a spinning top. The top of the toy top shows the motion of Earth's North Pole or South Pole, while the point on which the top spins represents the center of Earth. As the top spins, Earth's gravitational pull causes the top's axis of rotation to move in a circle—to precess. (b) As Earth precesses, the north celestial pole slowly traces out a circle among the northern constellations. At the present time, the north celestial pole is near the moderately bright star Polaris, which serves as the pole star. The total precession period is about 26,000 years.

behavior of Earth and discovered that without a large Moon, Earth would not keep to a 23½° tilt, but rather would change angle dramatically.

Earth's rate of precession is slow compared to human time scales. It takes about 26,000 years for the north celestial pole to trace out a complete circle around the sky, as shown in Figure 1-20b. (The south celestial pole executes a similar circle in the southern sky.) At the present time, Earth's north axis of rotation points within 1° of the star Polaris. In 3000 B.C., it was pointing near the star Thuban in the constellation Draco (the Dragon). In A.D. 14,000, the pole star will be near Vega in Lyra.

As Earth's axis of rotation precesses, its equatorial plane also moves. Because Earth's equatorial plane defines the location of the celestial equator in the sky, the celestial equator also changes over time. Recall that the intersections of the celestial equator and the ecliptic define the equinoxes, so these key locations in the sky shift slowly from year to year. This entire phenomenon is often called the **precession of the equinoxes.** This change was discovered by the great Greek astronomer Hipparchus in the second century B.C. Today, the vernal equinox is located in the constellation Pisces (the Fishes). Two thousand years ago, it was located in Aries (the Ram). Around the year A.D. 2600, the vernal equinox will move into Aquarius (the Water Bearer).

1-9 The phases of the Moon originally inspired the concept of the month

While the Moon's precessional effect on Earth unfolds over millennia, many lunar effects are noticeable every day. As the Moon orbits Earth, it moves from west to east, changing position among the background stars. Its position relative to the Sun also changes, and, as a result, we see different **lunar phases.**

The Sun illuminates half of the Moon at all times. The Moon's phase that we see depends on how much of its sunlit hemisphere is facing Earth. When the Moon is closest to the Sun in the sky, its dark hemisphere faces us. This phase, during which the Moon is at most a tiny crescent, is called the *new* Moon (Figure 1-21).

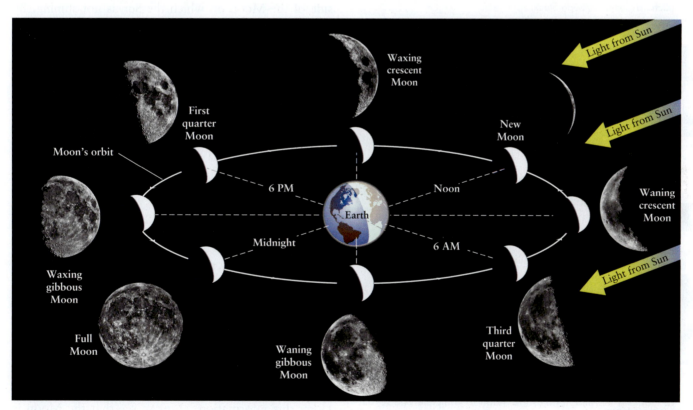

RIVUXG

FIGURE 1-21 The Phases of the Moon The diagram shows the Moon at eight locations on its orbit as viewed from far above Earth's North Pole. The corresponding photographs show the resulting lunar phases *as seen from Earth.* The waning and third quarter photographs look backward, but they are correctly oriented as seen from Earth. Light from the Sun illuminates one-half of the Moon at all times, while the other half is dark. It takes about 29½ days for the Moon to go through all its phases. (Yerkes Observatory and Lick Observatory)

R I V U X G

FIGURE 1-22 The Moon during the Day The Moon is visible at some time during daylight hours virtually every day. The time of day or night it is up in our sky depends on its phase. (Richard Cummins/SuperStock)

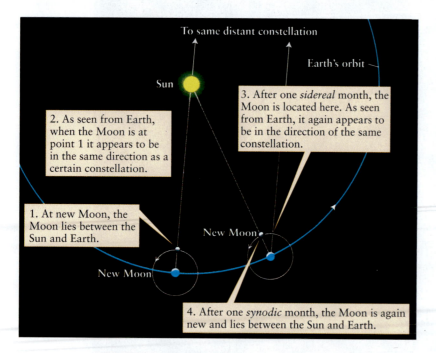

To same distant constellation

Earth's orbit

Sun

3. After one *sidereal* month, the Moon is located here. As seen from Earth, it again appears to be in the direction of the same constellation.

2. As seen from Earth, when the Moon is at point 1 it appears to be in the same direction as a certain constellation.

1. At new Moon, the Moon lies between the Sun and Earth.

New Moon

New Moon

4. After one *synodic* month, the Moon is again new and lies between the Sun and Earth.

FIGURE 1-23 The Sidereal and Synodic Months The sidereal month is the time it takes the Moon to complete one revolution with respect to the background stars. However, because Earth is constantly moving in its orbit about the Sun, the Moon must travel through more than 360° to get from one new Moon to the next. The synodic month is the time between consecutive new Moons or consecutive full Moons. Thus, the synodic month is slightly longer than the sidereal month.

During the 7 days following the new phase, more of the Moon's illuminated hemisphere becomes exposed to our view, resulting in a phase called the *waxing crescent* Moon. At the *first quarter* Moon, we see half of the illuminated hemisphere and half of the dark hemisphere. "Quarter Moon" refers to how far in its cycle the Moon has gone, rather than what fraction of the Moon appears lit by sunlight.

During the next week still more of the illuminated hemisphere can be seen from Earth, giving us the phase called the *waxing gibbous* Moon. "Gibbous" means "rounded on both sides." When the Moon arrives on the opposite side of Earth from the Sun, we see virtually all of the fully illuminated hemisphere. This phase is the full Moon. Over the following 2 weeks, we see less and less of the illuminated hemisphere as the Moon continues along its orbit. This movement produces the phases called the *waning gibbous* Moon, *third quarter* Moon, and the *waning crescent* Moon. The Moon completes a full cycle of phases in 29½ days.

Confusion often occurs over the terms "far side" and "dark side" of the Moon. The far side is the side of the Moon facing away from Earth. The dark side is the side of the Moon on which the Sun is not shining. By examining the photographs in Figure 1-21, you can see that the same side of the Moon faces Earth all the time. The half of the Moon that never faces Earth is the far side. However, the far side is not always the dark side, since we see part of the dark side whenever we see less than a full Moon.

Figure 1-21 shows the Moon at various positions in its orbit. Remember that the bright side of the Moon is on the right (west) side of the waxing Moon, while the bright side is on the left (east) side of the waning Moon. This information can tell you at a glance whether the Moon is waxing or waning. When looking at the Moon through a telescope, the best place to see details is where the shadows are longest. This occurs at the boundary between the bright and dark regions, called the **terminator.**

Figure 1-21 also shows local time around the globe, from noon, when the Sun is highest in the sky, to midnight, when it is on the opposite side of Earth. These time markings roughly indicate when the Moon is highest in the sky. For example, at first quarter, the Moon is 90° east of the Sun in the sky; hence, the Moon is highest at sunset. At full Moon, the Moon is opposite the Sun in the sky; thus, the Moon is highest at midnight. Using this information, you can see that the Moon is visible during the daytime (Figure 1-22) for a part of most days of the year.

Since the dawn of civilization, people have sought accurate timekeeping systems. Ancient Egyptians wanted to know when the Nile would flood, and farmers everywhere needed to know when to plant crops. Migratory tribes wanted to know when the weather would change. Religious leaders scheduled observances in accordance with celestial events. Thus, astronomers have traditionally been responsible for telling time. Indeed, of the four ways in which time cycles are set, three are astronomical in origin: Time is determined by the positions of the Moon, Sun, or stars or, in our own age, by technological means, such as atomic clocks.

The approximately 4 weeks that the Moon takes to complete one cycle of its phases inspired our ancestors to invent the concept of a month. Astronomers find it useful to define two types of months, depending on whether the Moon's motion is measured relative to the stars or to the Sun. Neither type corresponds exactly to the months of our usual calendar, which have different lengths.

The **sidereal month** is the time it takes the Moon to complete one full orbit of 360° around Earth (Figure 1-23). As with the sidereal day, the length of the sidereal month is determined by the location of the Moon in its orbit around Earth as measured with respect to the stars. Equivalently, this is the time it takes the Moon to

start at one place on the celestial sphere and return to exactly the same place again. The sidereal orbital period of the Moon takes approximately 27.3 days. The **synodic month**, or **lunar month**, is the time it takes the Moon to complete one 29½-day cycle of phases (that is, from new Moon to new Moon or from full Moon to full Moon) and thus is measured with respect to the Sun rather than the stars.

The synodic month is longer than the sidereal month because Earth is orbiting the Sun while the Moon goes through its phases. As shown in Figure 1-23, the Moon must travel *more* than 360° along its orbit to complete a cycle of phases (for example, from one new Moon to the next), which takes about 2.2 days longer than the sidereal month.

Both the sidereal month and synodic month vary somewhat, because the gravitational pull of the Sun on the Moon affects the Moon's speed as it orbits Earth. The sidereal month can vary by as much as 7 h, while the synodic month can change by as much as 12 h.

Is the Moon waning or waxing in figure 1-22?

FIGURE 1-24 Conditions for Eclipses The Moon must be very nearly on the ecliptic at new Moon for a solar eclipse to occur. A lunar eclipse occurs only if the Moon is very nearly on the ecliptic at full Moon. When new Moon or full Moon phases occur away from the ecliptic, no eclipse is seen, because the Moon and Earth do not pass through each other's shadow.

The terms *synodic* and *sidereal* are also used in discussing the motion of the other bodies in the solar system. The synodic period of a planet is the time between consecutive straight alignments of the Sun, Earth, and that planet (during which time interval the planet also goes through a cycle of phases, as seen from Earth). Recall that any orbit measured with respect to the stars is called "sidereal," including orbits of the planets around the Sun, as well as orbits of moons around their planets. Earth's sidereal year is 365.2564 days. Our sidereal year differs from the time between consecutive vernal equinoxes (called a *tropical year*) primarily due to Earth's precession (see Section 1-8).

ECLIPSES

8 Eclipses are among the most spectacular natural phenomena. During a **lunar eclipse,** the brilliant full Moon often darkens to a deep red. A lunar eclipse occurs when the Moon passes through Earth's shadow. This can happen only when the Sun, Earth, and Moon are in a straight line at full Moon. During a **solar eclipse,** broad daylight is transformed into an eerie twilight, as the Sun seems to be blotted from the sky. A solar eclipse occurs when the Moon's shadow moves across Earth's surface. As seen from Earth, the Moon moves in front of the Sun.

1-10 Eclipses occur only when the Moon crosses the ecliptic during the new or full phase

At first glance, it would seem that eclipses should happen at every new and full Moon, but, in fact, they occur much less often because the Moon's orbit is tilted 5° from the plane of the ecliptic (Figure 1-24). Consequently, the new Moon and full Moon usually occur when the Moon is either above or below the plane of Earth's orbit. In such positions, a true alignment between the Sun, Moon, and Earth is not possible and so an eclipse cannot occur.

Indeed, because its orbit is tilted 5° from the ecliptic the Moon is usually above or below the plane of our orbit around the Sun. The Moon crosses the ecliptic at what is called the **line of nodes** (see Figure 1-24). When the Moon crosses the plane of the ecliptic during its new or full phase, an eclipse takes place. By calculating the number of times a new Moon takes place on the line of nodes, we find that at least two and no more than five solar eclipses occur each year. Lunar eclipses occur just about as frequently as solar eclipses, with the maximum number of eclipses (solar plus lunar) possible in a year being seven.

1-11 Three types of lunar eclipses occur

Earth's shadow has two distinct parts, as shown in Figure 1-25a. The **umbra** is the part of the shadow where all direct sunlight is blocked by Earth. If you were in Earth's umbra looking at Earth, you would not see the Sun behind it at all. The **penumbra** of the shadow is where Earth blocks only some of the sunlight. If you were in Earth's penumbra looking at Earth, you would see a crescent Sun behind it. The Moon has an analogous umbra and penumbra.

Depending on how the Moon travels through Earth's shadow, three kinds of lunar eclipses may occur. A **penumbral eclipse,** when the Moon passes

TABLE 1-2 Lunar Eclipses, 2008–2011

Date	Visible from	Type	Duration of totality (h:min)
2009 February 9	Eastern Europe, Asia, Australia, Pacific, western North America	Penumbral	
2009 July 7	Australia, Pacific, Americas	Penumbral	
2009 August 6	Americas, Europe, Africa, western Asia	Penumbral	
2009 December 31	Europe, Africa, Asia, Australia	Partial	
2010 June 26	Eastern Asia, Australia, Pacific, western Americas	Partial	
2010 December 21	Eastern Asia, Australia, Pacific, Americas, Europe	Total	1:13
2011 June 15	South America, Europe, Africa, Asia, Australia	Total	1:41
2011 December 10	Europe, eastern Africa, Asia, Australia, North America	Total	:52
2012 June 4	Asia, Australia, Pacific, Americas	Partial	
2012 November 28	Europe, eastern Africa, Asia, Australia, North America	Penumbral	

a

b

through only Earth's penumbra, is easy to miss. The Moon still looks full, just a little dimmer than usual and sometimes slightly reddish in color (path 1 in Figure 1-25a).

When just part of the lunar surface passes through the umbra, a bite seems to be taken out of the Moon, and we see a **partial eclipse** (path 2 in Figure 1-25a). When the Moon travels completely into the umbra, we see a **total eclipse** of the Moon (path 3 in Figure 1-25a). Total lunar eclipses with the maximum duration, lasting for up to 1 h and 47 min, occur when the Moon is closest to Earth and it travels directly through the center of the umbra. Table 1-2 lists all the total and partial lunar eclipses from 2009 through 2012.

Even during a total eclipse, the Moon does not completely disappear. A small amount of sunlight passing through Earth's atmosphere is bent into Earth's umbra. The light deflected into the umbra is primarily red and orange, and thus the darkened Moon glows faintly in rust-colored hues (Figure 1-25b). At sunrise and sunset the sky appears red or orange for the same reason, because at those times more of Earth's atmosphere deflects red and orange light from the Sun toward you.

Everyone on the side of Earth over which a lunar eclipse occurs can see it, provided that clouds don't obscure the event. Lunar eclipses are perfectly safe to watch with the naked eye.

Why does the new moon sometimes appear as a crescent?

FIGURE 1-25 Three Types of Lunar Eclipses (a) People on the nighttime side of Earth see a lunar eclipse when the Moon moves through Earth's shadow. The umbra is the darkest part of the shadow. In the penumbra, only part of the Sun is covered by Earth. The inset shows the various lunar eclipses that occur, depending on the Moon's path through Earth's shadow. (b) This sequence of nine photographs was taken over a 3-h period during the total lunar eclipse of January 20, 2000. During the total phase, the Moon has a distinctly reddish color. (Fred Espenak, NASA/Goddard Space Flight Center; © Fred Espenak, MrEclipse.com)

1-12 Three types of solar eclipses also occur

Because of their different distances from Earth, the Sun and the Moon have nearly the same angular diameter as seen from Earth—about ½°. When the Moon completely covers the Sun, the result is a total solar eclipse. You must be at a location within the Moon's umbra to see a total solar eclipse. During those few precious moments, hot gases (the **solar corona**) surrounding the Sun can be observed and photographed (Figure 1-26). In this way, astronomers have been able to learn more about the Sun's temperature, chemistry, and atmospheric activity.

You can see in Figure 1-27 that only the tip of the Moon's umbra ever reaches Earth's surface. As Earth turns and the Moon orbits, the tip traces an **eclipse path** across our planet. Figure 1-28 shows the paths over Earth of all the total eclipses through 2020. Only people within these areas are treated to the spectacle of a total solar eclipse. *Viewing the Sun directly for more than a moment at any time without an approved filter causes permanent eye damage.* It is only safe to look at a total solar eclipse without a filter during the brief time when the Moon completely blocks the Sun. At all other times during the eclipse, you must view it either through an approved filter or in the image made on a flat surface by a telescope or a pinhole camera (see margin photographs.)

Earth's rotation and the orbital motion of the Moon cause the umbra to race along the eclipse path at speeds in excess of 1700 km/h (1050 mph). For this reason, a total eclipse never lasts for more than 7½ min at any one location on the eclipse path, and it usually lasts for only a few moments.

The Moon's umbra is also surrounded by a penumbra (see Figure 1-27). The photograph in Figure 1-27 shows the dark spot produced by the Moon's umbra and penumbra on Earth's surface during a total solar eclipse. During a solar eclipse, the Moon's penumbra extends over a large portion of Earth's surface. When only the penumbra sweeps across Earth's surface, as happens in high latitude regions, the Sun is only partly covered by the Moon. This circum-

Looking at a solar eclipse through a filter (left) and a filtered telescope.
(AP Photo/Mahmoud Tawil)

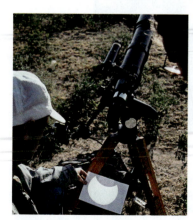

Looking at a solar eclipse projected by a telescope
(Guillermo Gonzalez/Visuals Unlimited)

R I V U X G

 FIGURE 1-26 A Total Eclipse of the Sun During a total solar eclipse, the Moon completely covers the Sun's disk, and the solar corona can be photographed. This halo of hot gases extends for millions of kilometers into space. This gorgeous image is a composite of several taken in Chisamba, Zambia, during the June 21, 2001, solar eclipse. (F. Espenak)

stance results in a *partial eclipse of the Sun*. Similarly, people in the penumbra of a total eclipse see a partial eclipse. In either case, the Sun looks crescent as seen from Earth.

The Moon's orbit around Earth is not quite a perfect circle. The distance between Earth and the Moon, which averages 384,400 km (238,900 mi), varies by a few percent as the Moon goes around Earth. The width of the eclipse path depends primarily on the Earth-Moon distance during an eclipse. The eclipse path is widest—up to 270 km (170 mi)—when the new Moon happens to be at the point in its orbit nearest Earth. Usually, however, the path is much narrower.

If a solar eclipse occurs when the Moon is farthest from Earth, then the Moon's umbra falls short of Earth and no one sees a total eclipse. From Earth's surface, the Moon then appears too small to cover the Sun completely, and a thin ring or "annulus" of light is seen around the edge of the Moon at mid-eclipse. This type of eclipse is called an **annular eclipse** (Figure 1-29). The length of the Moon's umbra is nearly 5000 km (3100 mi) shorter than the average distance between the Moon and Earth's surface. Thus, the Moon's shadow often fails to reach Earth, making annular eclipses slightly

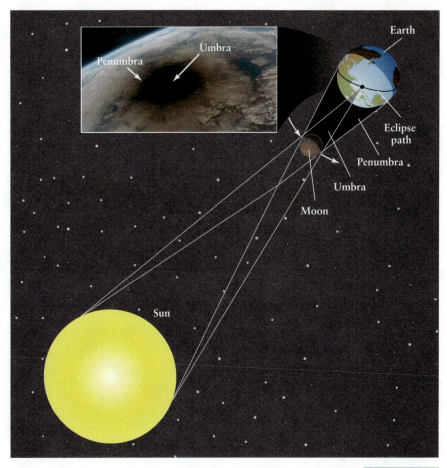

FIGURE 1-27 The Geometry of a Total Solar Eclipse
During a total solar eclipse, the tip of the Moon's umbra traces an eclipse path across Earth's surface. People inside the eclipse path see a total solar eclipse, whereas people inside the penumbra see only a partial eclipse. The photograph in this figure shows the Moon's shadow on Earth. It was taken from the *Mir* space station during the August 11, 1999, total solar eclipse. The Moon's umbra appears as the very dark spot on the eastern coast of the United States. It is surrounded by the penumbra. (Jean-Pierre Haigneré, Centre National d'Etudes Spatiales, France/GSFS)

R I V U X G

more common than total eclipses. Table 1-3 lists all the total, partial, and annular solar eclipses from 2009 through 2012. It is *never* safe to look directly at a partial or annular eclipse.

A total solar eclipse is a dramatic event. The sky begins to darken, the air temperature falls, and the winds increase as the Moon's umbra races toward you. All nature responds: Birds go to roost, flowers close their petals, and crickets begin to chirp as if evening had arrived. As the moment when the Sun becomes totally eclipsed approaches, the landscape is bathed in shimmering bands of light and dark as the last few rays of sunlight peek out from behind the edge of the Moon. Finally, the corona blazes forth in a star-studded daytime sky. It is an awesome sight, but always take care in viewing it.

1-13 Frontiers yet to be discovered

In this chapter we have examined several factors that explain the changing seasons on Earth. Geologists have discovered a variety of cycles of global temperature change that have occurred over the history of Earth. Indeed, they have found that much of Earth once suffered a global freezing. Conversely, we are now undergoing a global warming. Historically, these changes occur over tens of thousands of years, hundreds of thousands of years, and possibly longer cycles. Most scientists are now convinced that the present climate change is a result of a combination of human and other causes, some of which have yet to be discovered.

> Why aren't there any annular lunar eclipses?

> Why are some of the paths in Figure 1-28 wider than others?

WEB LINK 1.13

TABLE 1-3 Solar Eclipses, 2008–2011

Date	Type	Visible from	Total/annular eclipse time (min:s)
2009 January 26	Annular	Africa, southeast Asia, Australia	7:54
2009 July 22	Total	Eastern Asia, Pacific, Hawaii	6:39
2010 January 15	Annular	Africa, Asia	11:08
2010 July 11	Total	Southern South America	5:20
2011 January 11	Partial	Europe, Africa, central Asia	
2011 June 1	Partial	Eastern Asia, northern North America, Iceland	
2011 July 1	Partial	Southern Indian Ocean	
2011 November 25	Partial	Southern Africa, Tasmania, New Zealand	
2012 May 20	Annular	Asia, Pacific, North America	5:46
2012 November 13	Total	Australia, New Zealand, southern South America	4:02

Total and Annular Solar Eclipse Paths: 2001–2020

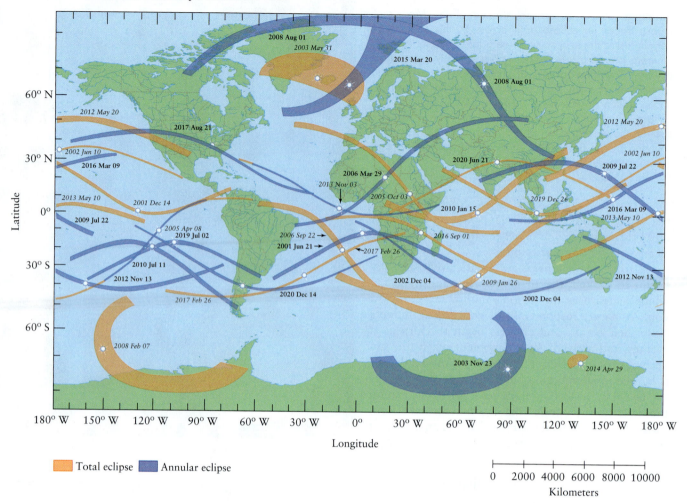

Total eclipse Annular eclipse

0 2000 4000 6000 8000 10000
Kilometers

FIGURE 1-28 Eclipse Paths for Total and Annular Eclipses 2001–2020 This map shows the eclipse paths for the 14 total solar and 13 annular eclipses that occur between 2001 and 2020. In each eclipse, the Moon's shadow travels along the eclipse path in a generally eastward direction across the Earth's surface. (Courtesy of Fred Espenak, NASA/Goddard Space Flight Center)

R I V U X G

SUMMARY OF KEY IDEAS

Sizes in Astronomy
• Astronomy examines objects that range in size from the parts of an atom ($\sim 10^{-15}$ m) to the size of the observable universe ($\sim 10^{26}$ m).

• Scientific notation is a convenient shorthand for writing very large and very small numbers.

FIGURE 1-29 An Annular Eclipse of the Sun This composite of five exposures taken at sunrise in Costa Rica shows the progress of an annular eclipse of the Sun that occurred on December 24, 1974. Note that at mid-eclipse the edge of the Sun is visible around the Moon. (Dennis Di Cicco)

Patterns of Stars

• The surface of the celestial sphere is divided into 88 unequal areas called constellations.

• The boundaries of the constellations run along lines of constant right ascension or declination.

Earthly Cycles

• The celestial sphere appears to revolve around Earth once in each day-night cycle. In fact, it is Earth's rotation that causes this apparent motion.

• The poles and equator of the celestial sphere are determined by extending the axis of rotation and the equatorial plane of Earth out onto the celestial sphere.

• Earth's axis of rotation is tilted at an angle of 23½° from a line perpendicular to the plane of Earth's orbit (the plane of the ecliptic). This tilt causes the seasons.

• Equinoxes and solstices are significant points along Earth's orbit that are determined by the relationship between the Sun's path on the celestial sphere (the ecliptic) and the celestial equator.

• Earth's axis of rotation slowly changes direction relative to the stars over thousands of years, a phenomenon called precession. Precession is caused by the gravitational pull of the Sun and Moon on Earth's equatorial bulge.

• The length of the day is based upon Earth's rotation rate and the average motion of Earth around the Sun. These effects combine to produce the 24-h day upon which our clocks are based.

• The phases of the Moon are caused by the relative positions of Earth, Moon, and Sun. The Moon completes one cycle of phases in a synodic month, which averages 29½ days.

• The Moon completes one orbit around Earth with respect to the stars in a sidereal month, which averages 27.3 days.

Eclipses

• The shadow of an object has two parts: the umbra, where direct light from the source is completely blocked; and the penumbra, where the light source is only partially obscured.

• A lunar eclipse occurs when the Moon moves through Earth's shadow. During a lunar eclipse, the Sun, Earth, and Moon are in alignment with Earth between the Sun and the Moon, and the Moon is in the plane of the ecliptic.

• A solar eclipse occurs when a strip of Earth passes through the Moon's shadow. During a solar eclipse, the Sun, Earth, and Moon are in alignment with the Moon between Earth and the Sun, and the Moon is in the plane of the ecliptic.

• Depending on the relative positions of the Sun, Moon, and Earth, lunar eclipses may be penumbral, partial, or total, and solar eclipses may be annular, partial, or total.

WHAT DID YOU THINK?

1 *Is the North Star—Polaris—the brightest star in the night sky?* No. Polaris is a star of medium brightness compared with other stars visible to the naked eye.

2 *Do astronomers regard the familiar patterns of stars in the sky as constellations?* Astronomers sometimes use the common definition of a constellation as a pattern of stars. Formally, however, a constellation is an entire area of the celestial sphere and all the stars and other objects in it. Viewed from Earth, the entire sky is covered by 88 different-sized constellations. If there is any room for confusion, astronomers refer to the patterns as asterisms.

3 *What causes the seasons?* The tilt of Earth's rotation axis with respect to the ecliptic causes the seasons. They are not caused by the changing distance from Earth to the Sun that results from the shape of Earth's orbit.

4 *When is Earth closest to the Sun?* On or around January 3 of each year.

5 *How many zodiac constellations are there?* There are 13 zodiac constellations, the lesser-known one being Ophiuchus.

6 *Does the Moon have a dark side that we never see from Earth?* Half of the Moon is always dark. Whenever we see less than a full Moon, we are seeing part of the Moon's dark side. So, the dark side of the Moon is not the same as the far side of the Moon, which we never see from Earth.

7 *Is the Moon ever visible during the daytime?* The Moon is visible at some time during daylight hours almost every day of the year. Different phases are visible during different times of the day.

8 *What causes lunar and solar eclipses?* When the Moon is crossing the ecliptic in the full or new phase, the shadows of Earth or Moon, respectively, then fall on the Moon or Earth. These shadows on the respective surfaces are eclipses.

Key Terms for Review

angle, 13

angular diameter (angular size), 13

annular eclipse, 28

arc angle, 13

asteroid, 6

autumnal equinox, 17

black hole, 6

celestial equator, 11

celestial pole, 11

celestial sphere, 10

circumpolar star, 15

comet, 6

constellation, 10

declination, 11

degree (°), 13

diurnal motion, 12

eclipse path, 28

ecliptic, 16

equinox, 16

gravitation, 21

line of nodes, 26

lunar eclipse, 26

lunar phase, 23

meteoroid, 6

north celestial pole, 11

partial eclipse, 27

penumbra, 26

penumbral eclipse, 26

precession, 21

precession of the equinoxes, 23

revolution, 12

right ascension, 11

rotation, 12

scientific notation, 6

sidereal month, 25

sidereal period, 14

solar corona, 28

solar day, 19

solar eclipse, 26

south celestial pole, 11

summer solstice, 17

synodic month (lunar month), 25

terminator, 24

time zone, 20

total eclipse, 27

umbra, 26

vernal equinox, 12

winter solstice, 16

zenith, 15

zodiac, 19

Review Questions

1. Where is the horizon? a. directly overhead, b. along the celestial equator, c. the boundary between land and sky, d. along the path that the Sun follows throughout the day, e. the line running from due north, directly overhead, ending due south.

2. How many constellations are there? a. 2, b. 12, c. 13 d. 56, e. 88.

3. Which of the following lies on the celestial sphere directly over Earth's equator? a. ecliptic, b. celestial equator, c. north celestial pole, d. south celestial pole, e. horizon.

4. The length of time it takes Earth to orbit the Sun is a. an hour, b. a day, c. a month, d. a year, e. a century.

5. In Figure 1-8, what is another name for the "Sun's annual path"?

6. How are constellations useful to astronomers?

7. What is the celestial sphere, and why is this ancient concept still useful today?

8. What is the celestial equator, and how is it related to Earth's equator? How are the north and south celestial poles related to Earth's axis of rotation?

9. What is the ecliptic, and why is it tilted with respect to the celestial equator?

10. By about how many degrees does the Sun move along the ecliptic each day?

11. Through how many constellations does the Sun move every day?

12. Through how many constellations does the Sun move every year?

13. Why does the tilt of Earth's axis relative to its orbit cause the seasons as Earth revolves around the Sun? Draw a diagram to illustrate your answer.

14. What are the vernal and autumnal equinoxes? What are the summer and winter solstices? How are these four points related to the ecliptic and the celestial equator?

15. What is precession, and how does it affect our view of the heavens?

16. How does the daily path of the Sun across the sky change with the seasons?

17. Why is it warmer in the summer than in the winter?

18. Why is it convenient to divide Earth into time zones?

19. Why does the Moon exhibit phases?

20. What is the difference between a sidereal month and a synodic month? Which is longer? Why?

21. What is the line of nodes, and how is it related to solar and lunar eclipses?

22. What is the difference between the umbra and the penumbra of a shadow?

23. What is a penumbral eclipse of the Moon? Why is it easy to overlook such an eclipse?

24. Which type of eclipse—lunar or solar—have most people seen? Why?

25. How is an annular eclipse of the Sun different from a total eclipse of the Sun? What causes this difference?

26. When is the next leap year?

27. At which phase(s) of the Moon does a solar eclipse occur? A lunar eclipse?

28. Is it safe to watch a solar eclipse without eye protection? A lunar eclipse?

29. During what phase is the Moon "up" least in the daytime?

Advanced Questions

The answers to all computational problems, which are preceded by an asterisk (*), appear at the end of the book.

30. During what phase(s) does the Moon rise after sunrise and before sunset? After sunset and before sunrise? At sunset? At sunrise?

31. Why can't a person in Australia use the Big Dipper to find north?

32. Are there any stars in the sky that are not members of a constellation?

33. At what places on Earth is Polaris seen on the horizon?

34. Where do you have to be on Earth to see the Sun at your zenith? If you stay at one such location for a full year, on how many days will the Sun pass through the zenith?

35. Where do you have to be on Earth to see the south celestial pole at your zenith? What is the maximum possible elevation (angle) of the Sun above the horizon at that location? On what date is this maximum elevation achieved?

36. Where on the horizon does the Sun rise at the time of the vernal equinox?

37. Consult a star map of the southern hemisphere and determine which, if any, of the bright southern stars could someday become south celestial pole stars.

38. Are there stars in the sky that never set where you live? Are there stars that never rise where you live? Does your answer depend on your location on Earth? Why or why not?

39. Using a diagram, demonstrate that your latitude on Earth is equal to the altitude of the north celestial pole above your northern horizon.

40. Using a star map, determine which bright stars, if any, could someday mark the location of the vernal equinox. Give the approximate years when this should happen.

41. What is the phase of the Moon if it **a.** rises at 3 A.M.? **b.** sets at 9 P.M.? At what time does **c.** the full Moon set? **d.** the first quarter Moon rise?

42. What is the phase of the Moon if, on the first day of spring, the Moon is located at the position of **a.** the vernal equinox, **b.** the summer solstice, **c.** the autumnal equinox, **d.** the winter solstice?

*43. How many more sidereal months than synodic months are there in a year? Why?

44. How do we know that the phases of the Moon are not due to the Moon moving in Earth's shadow?

45. Do the paths of total solar eclipses fall more frequently on oceans or on land? Explain.

46. Can one ever observe an annular eclipse of the Moon? Why or why not?

47. Which of the five images of the Sun in Figure 1-29 is the first and which is the last in the sequence shown? Justify your answer.

48. During a lunar eclipse, does the Moon enter Earth's shadow from the east or the west? Explain your answer.

49. Do we see all of the Moon's surface from Earth? *Hint:* Carefully examine the photographs in Figure 1-21.

50. Explain why the waning gibbous Moon, third quarter Moon, and waning crescent photographs in Figure 1-21 are correctly oriented as seen from Earth.

51. Why is a small crescent of light often observed on the Moon when it is exactly in the new phase?

52. Make a drawing of an annular solar eclipse as seen from space similar to Figure 1-27. Be sure to make clear how it differs from a total solar eclipse.

*53. Assuming that the Sun makes an angle of ½° in our sky and is at a distance of 1.496×10^{11} m, what is the Sun's diameter? Divide this value by 2 to find the Sun's radius and explain why this result is slightly different from the value given in Appendix E-8 at the back of the book.

54. How long was the exposure for the photograph of circumpolar star trails in Figure 1-9?

55. Determine which stars whose paths are shown in Figure 1-9 are *not* circumpolar. An easy way to write the answer is in terms of distance on the photograph from the location of the south celestial pole.

56. How does Figure 1-9 show that there is no bright "south pole" star?

Discussion Questions

57. Examine the list of the 88 constellations in Appendix E-7. Are there any constellations whose names obviously date from modern times? Where are these constellations located? Why do they not have ancient names?

58. In his novel *King Solomon's Mines,* H. Rider Haggard described a total solar eclipse that was seen in both South Africa and the British Isles. Is such an eclipse possible? Why or why not?

59. Describe how a lunar eclipse would look if Earth had no atmosphere.

60. Examine a listing of total solar eclipses over the next several decades. What are the chances that you might be able to travel to one of the eclipse paths? Do you think you might go through your entire life without ever seeing a total eclipse of the Sun?

What if...

61. The Moon moved about Earth in an orbit perpendicular to the plane of Earth's orbit? What would the cycle of lunar phases be? Would solar and lunar eclipses be possible under these circumstances?

62. Earth's axis of rotation were tilted at a different angle? What would the seasons be like where you are now if the axis of rotation is **a.** 0° and **b.** 45° to its

orbital plane? What would be different about the seasons and the day-night cycle if you lived at one of the Earth's poles in these two situations?

63. You watched Earth from the Moon? What would you see for Earth's **a.** daily motion, **b.** motion along the celestial sphere, and **c.** cycle of phases?

64. The Moon didn't rotate? Describe how its surface *features* would appear from Earth—that is, would we see all sides of it over time? Why or why not? (Ignore the change in phases when discussing its appearance.) *Carefully* study the photographs in Figure 1-21 and state whether the same features are visible at all times or whether we see different features over time. (Again, ignore the phases.) What can you conclude about whether the Moon actually rotates?

Web Questions

***65.** Work through the AIMM (Active Integrated Media Module) called *A Small Angle Toolbox* in Chapter 1 of the *Discovering the Universe: From the Stars to the Planets* Web site. Use it to determine the diameters in kilometers of the Sun, Saturn, and Pluto given the following distances and angular sizes:

Object	Distance (km)	Angular size (″)
Sun	1.5×10^8	1800
Saturn	1.5×10^9	16.5
Pluto	6.3×10^9	0.06

66. Search the Web and identify at least four cultures whose stories are used to name modern constellations. Briefly relate the stories of a constellation from each culture.

67. Search the Web for information about the Great Nebula of Orion (see label on Figure 1-4; it is also called the Orion Nebula). Can the Great Nebula be seen with the naked eye? Does it exist alone in space or is it part of a larger system of interstellar material? What has been learned by examining the Great Nebula with telescopes sensitive to infrared light?

68. Search the Web for information about the national flags of Australia, New Zealand, and Brazil, and the state flag of Alaska. What stars are depicted on these flags? Explain any similarities or differences among these flags.

69. Search the Web for the English meaning of the Japanese word *Subaru*. Make a drawing of the Subaru car's emblem and explain it.

70. Use the U.S. Naval Observatory's Web site to determine the times of sunset and sunrise on **a.** your birthday and **b.** the date this assignment is due. Are the times the same? Explain why or why not.

71. Use data in the book or search the Web for information about the next total solar eclipse. Through which major cities, if any, does the path of totality pass? What is the maximum duration of totality? Find a location where this maximum duration is observed. Will the eclipse be visible (even as a partial eclipse) from your present location?

72. Search the Web for information about the next total lunar eclipse. Will the total phase of the eclipse be visible from your present location? If not, will the penumbral phase be visible? Draw a picture of the Sun, Earth, and Moon at totality and indicate your location on the drawing of Earth.

73. Access the animation "The Moon's Phases" in Chapter 1 of the *Discovering the Universe: From the Stars to the Planets* Web site. This shows the Earth-Moon system as seen from a vantage point above Earth's North Pole. **a.** Describe where you would be on the diagram if you are on the equator and the time were 6:00 P.M. **b.** If it were 6:00 P.M. and you were standing on Earth's equator, would a third-quarter Moon be visible? Why or why not? If it would be visible, describe its appearance.

Observing Projects

Many of the following projects are based on the planetarium program, *Starry Night Enthusiast*™. If assigned them, it is likely that your teacher had the software included with this text. Install this program on a suitable computer and run it. You can learn to use the program with the help of the User's Guide, which can be accessed under **Help > User's Guide** on the main menu. A brief guide is also presented in Appendix G. You can also experiment on your own or with the help of a friend, or use the on-line help. **Start each Project by clicking the Home button in the toolbar to reset the program to the current time and to your home location.**

74. On a clear, cloud-free night, use the star charts at the end of this book to see how many constellations of the zodiac you can identify. Which ones are easy to find? Which are difficult?

75. Use *Starry Night Enthusiast*™ to see if you can identify the zodiac constellations that are visible in the sky this evening. Stop the advance of time by clicking the **Stop** button (■) in the toolbar and then set the time to some time after sunset. Now use the Hand Tool to "grab" and move the sky until you recognize the zodiac constellations. To see if you are correct, right click over the region if you are using a PC (**Ctrl-click** on a Mac) of the constellation and its name or the name of a star in it (often with the root name of the constellation on the end of it) will appear on the top line of the box. Identifying constellations is often a challenge on the computer because the sky is so compressed. You might find it helpful to grab and move the sky around until you find some of the more familiar constellations, such as Orion or Ursa Major and use these for reference. Also, once you find a zodiac constellation, others will be adjacent to it. If you still have difficulty finding zodiac constellations, click on the **Options** tab to open the **Options** pane, expand the *Guides* layer and click on *The Ecliptic* to highlight the part of the sky along which the zodiac is located. If you still have trouble, open the *Constellations* layer and click on *Stick Figures (Astronomical)*. If you need even more help, click the **Auto Identify** option. Then, as you center a constellation on the screen, its name and classical image will appear. Make a list of the zodiac constellations that are visible to you today. Click the **Home** button in the toolbar to return to your home location and to the present time. Click the **Stop** button once again. To determine which zodiac constellation the Sun is passing through today, change the time in the toolbar to *12:00:00 PM* and use the Hand Tool to find the Sun in the sky. Then select **View > Hide** *Daylight* from the menu. Find the Sun and determine in which constellation it is located. The steps described above may help.

76. Use *Starry Night Enthusiast™* to observe the diurnal motion of the sky. **a.** Click the *Home* button in the toolbar and use the Hand Tool to center your field of view on the northern horizon (if you live in the northern hemisphere) or southern horizon (if you live in the southern hemisphere). Adjust the view so that the horizon appears approximately flat. Select **View > Hide** *Daylight* from the menu. Set the Time Flow Rate to *3 minutes* and click on the **Run time forward** button in the toolbar to start time flow. (Adjust the rate of the sky motion if it is too slow or fast to follow the general motion of the stars.) Do the stars appear to revolve clockwise or counterclockwise? Explain this in terms of Earth's rotation. Are any of the stars circumpolar? Explain why or why not. **b.** Click the **Stop** button in the toolbar to stop time flow and re-center your field of view on the southern horizon (if you live in the northern hemisphere) or the northern horizon (if you live in the southern hemisphere). Set the sky in motion again. Explain what you see. Are any of these stars circumpolar? Explain.

77. Use the *Starry Night Enthusiast™* program to observe the Sun's motion on the celestial sphere. Click the **Home** button in the toolbar. Open the *Options* pane and expand the **Local View** layer. Turn off the *Daylight* and *Local Horizon* options. Under the Guides layer, turn on *The Ecliptic* and the *Celestial Grid* options. Note that the Celestial Equator is between the −15° and +15° lines of declination. Open the **Find** pane and double-click the entry for the Sun to center the Sun in the view. Make sure you can see at least 90° of the celestial sphere. **a.** Under *Constellations* layer of the **Options** pane, select **Auto Identify**. In which constellation is the Sun located today? Is this the same constellation as the astrological sign for today's date (see Table 1-1)? **b.** Be sure the view is

locked on the Sun by opening the contextual menu (*right-click* over the Sun on a PC or *Ctrl-click* on a Mac) and selecting *Center*. Set the **Time Flow Rate** to *1 day* and start time flow (adjust flow rate, as necessary). Observe the Sun for a full year of simulated time (or more). Note the constellations through which it passes. How many are there? In which constellations does it cross the celestial equator? What path does it follow?

78. Examine the star charts that are published monthly in such popular astronomy magazines as *Sky & Telescope* and *Astronomy*. How do they differ from the star charts within the covers of this book? On a clear, cloud-free night, use one of these star charts to locate the celestial equator and the ecliptic on the night sky. Note the inclination of the Milky Way to the ecliptic and celestial equator. What do your observations tell you about the orientation of Earth and its orbit around the Sun relative to the rest of the Galaxy?

79. Observe the Moon on each clear night over the course of a month. Note the Moon's location among the constellations and record that location on a star chart that also shows the ecliptic. After a few weeks, your observations will begin to trace the Moon's orbit. Identify the orientation of the line of nodes by marking the points where the Moon's orbit and the ecliptic intersect. On what dates is the Sun near to the nodes marked on your star chart? Compare these dates with the dates of the next solar and lunar eclipses.

80. Use *Starry Night Enthusiast™* to study the Moon's path in the sky and eclipses. Open the **Options** pane and turn off the *Local Horizon* and *Daylight* options in the **Local View** layer. Under the *Guides* layer, turn on *The Ecliptic* and *Celestial Grid*. Open the *Find* pane and double-click the entry for *The Moon*. Change the field of view of the observed sky to about 15° using the Zoom buttons on the upper right of the toolbar. Note whether the Moon lies on the ecliptic. Set the **Time Flow Rate** to *6 minutes* (adjust this if sky motion is too fast or too slow). Keeping the Moon locked in the center of your screen, set time in motion. **a.** In which direction does the Moon move against the background stars? Keep in mind that the celestial equator runs east-west (east is to the left) and the ecliptic intersects it at 23½°. Ignore the Moon's rocking motion, which is due to its rising and setting. Does it ever change direction relative to the stars? Why or why not? Describe its path relative to the ecliptic. **b.** Change the **Time Flow Rate** to *1 hour* (adjust as necessary). Determine how many days elapse between successive times that the Moon is on the ecliptic. **c.** Set the time and date to 9 AM on July 21, 2009. Running time forward at a rate of 1 min, stop time when the Moon is closest to the Sun. Sketch the Sun and Moon on a piece of paper. Even though the Moon may not be directly over the Sun on the screen, an eclipse is occurring. What type of eclipse is it? Why might the Moon *not* be directly over the Sun at that time? **d.** Click on the *Events* tab, expand the *Event Filters* layer and deselect all but the *Lunar and Solar Eclipse Events*. Expand the *Events Browser* layer, set the *Start* and *End* dates to cover the period from 1/21/2009 to 1/21/2010 and click *Find Events*. Right click on each solar eclipse in turn, choose *View Event,* zoom back out to about 15°, and *Run time forward* and *backward* at a rate of 1 min until the Moon is again closest to the Sun. Draw the Sun and Moon. Are they separated by the same amount as in part c? If not, why not? If there is a difference, what effect does it have on the eclipse?

81. It is quite possible that a lunar eclipse will occur while you are taking this course. If so, look up on the Internet the precise time that it will happen; in the current issue of a reference from the U.S. Naval Observatory, such as the *Astronomical Almanac or Astronomical Phenomena;* or in such magazines as *Sky & Telescope* and *Astronomy,* which generally run articles about eclipses a month or two before they happen. Make arrangements to observe this eclipse. Note the times at which the Moon enters and exits Earth's umbra.

82. Sunrise, sunset, moonrise, moonset. Use *Starry Night Enthusiast*™ to determine when the Sun rises and sets today. Similarly, find out when the Moon rises and sets today. Because of the scenery, you may have to approximate these times. Starting with the time set to midnight this morning, determine over what period the Moon is visible during daylight hours today and over what period it is visible at night.

WHAT IF... Earth's Axis Lay on the Ecliptic?

Imagine Earth is tilted so that its axis of rotation lies in the plane of its orbit about the Sun (see the accompanying figure). We'll call it NeoEarth. NeoEarth still rotates once every 24 h. We arbitrarily fix its north pole to point toward the star Tau Tauri, a star nearly on the ecliptic, just north of the bright star Aldebaran in the constellation Taurus (see Figure 1-4b or 1-6).

Neo's Seasons Let's see how the seasons unfold on NeoEarth. It is March 21, the date of the spring equinox. The Sun is directly over the equator, as shown in the figure below. For the next three months, the Sun rises higher in the northern hemisphere. Unlike on our Earth, the Sun does not stop moving northward when it is over 23½° north latitude. Rather, the Sun rises farther and farther north day by day until it appears over the north pole of NeoEarth at the summer solstice, around June 22. Three months later, at the autumnal equinox, the Sun again rises over the equator, and day and night have the same length everywhere.

The Sun appears over the south pole of NeoEarth at the winter solstice, around December 22. The Sun's apparent motion through NeoEarth's sky completes the seasonal cycle by moving north, appearing over the equator once again around March 21.

Neo's Climate Day and night take on new meanings for inhabitants of NeoEarth. On our Earth, the regions above the Arctic Circle and below the Antarctic Circle have days or weeks of continual light in summer and days or weeks of constant darkness in winter. But on NeoEarth, *every place* has extended winter periods of constant darkness followed by extended summer periods of constant daylight. Spring and fall on NeoEarth have daily cycles of daylight and darkness, which separate the periods of continual daylight and darkness.

A Neo Day At the latitude of Atlanta, Georgia (33°46′), on NeoEarth, the day-night cycle occurs for only seven and a half months out of the year. During the other four and a half months, there is continuous day or continuous night, coupled with harsh summers and winters. With variations, this sequence of events occurs everywhere on NeoEarth.

The seasonal cycle on NeoEarth prevents the formation of permanent polar ice caps. In summer, polar regions experience the same tropical heating and high temperatures as the equatorial regions of our Earth. The polar regions of NeoEarth in winter are exceptionally cold, so seasonal polar ice caps may form. Because the southern polar cap resting on Antarctica is not permanent, the oceans, and the shorelines on the continents, are higher than those on our Earth.

If seasonal polar ice caps form, the dominant force controlling weather may shift from the jet streams that circle our Earth along lines of latitude to a pole-to-pole flow. Thermal flows created by intense heating at one location and cooling at others may replace our Earth's trade winds and other east-west winds. How else would NeoEarth differ from our world?

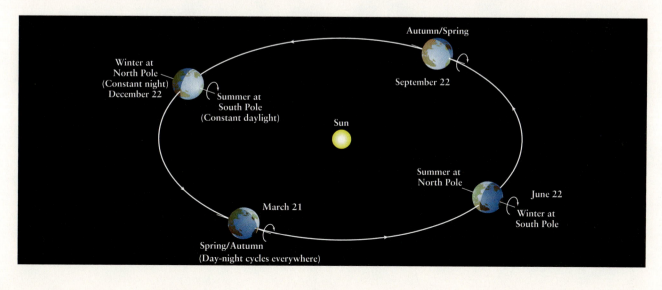

CHAPTER 2

Gravitation and the Motion of the Planets

Gravitation, the only universal force of attraction in nature, holds objects onto surfaces and in orbit. Our Moon, smaller and less massive than Earth, has only 1/6 the gravitational force that we feel here. Therefore, astronauts as shown in this *Apollo 15* image, can easily manage the spacesuits they wore. On Earth the suits each weighed 180 lb, while on the Moon they were only 30 lb. (NASA)

R I V U X G

WHAT DO YOU THINK?

1. What makes a theory scientific?

2. What is the shape of Earth's orbit around the Sun?

3. Do the planets orbit the Sun at constant speeds?

4. Do all of the planets orbit the Sun at the same speed?

5. How much force does it take to keep an object moving in a straight line at a constant speed?

6. How does an object's mass differ when measured on Earth and on the Moon?

7. Do astronauts orbiting Earth feel the force of gravity from our planet?

Answers to these questions appear in the text beside the corresponding numbers in the margins and at the end of the chapter.

Science enables us to understand and manipulate an awesome range of nature's properties. Scientists are a lot like detectives, and the process they follow when trying to explain scientific phenomena has a good deal in common with the activities of sleuths as they try to solve mysteries. For example, an investigator might suddenly realize that his prime suspect, the person around whom his whole case has revolved, could not have committed the crime because he was actually baking a soufflé at the time. Similarly, the astronomers and mathematicians who studied the motions of heavenly bodies made their biggest leap forward when the evidence forced them to look beyond the suspect they believed to be at the center of everything—Earth.

This chapter traces how we moved from an Earth-centered view of the universe to a Sun-centered one and how we came to understand the physical principles by which the basic motions of the universe operate. The process of this discovery initially involved the efforts of a few determined scientist-sleuths. To unravel the mysteries that puzzled them—such as why planets appear at times to change direction on the celestial sphere or why the assumption that Earth is the center of the universe kept failing to predict the locations of certain bodies—they used careful observations and a willingness to question their own and others' assumptions.

The groundwork for modern science was set down by Greek mathematicians and philosophers beginning around 2500 years ago, when Pythagoras and his followers began using mathematics to describe natural phenomena. About 200 years later, Aristotle asserted that the universe is comprehensible: It is governed by regular laws. The Greeks typically did not, however, perform experiments to test their ideas, an essential part of the scientific method used today. Nevertheless, they were among the first to leave a written record of their ideas, allowing succeeding generations to develop, criticize, and test their conclusions.

These Greek ideas, rediscovered in the seventeenth century, led to the development of the scientific method of examining, understanding, and predicting how things work. Science provides explanations for activities and events, and it makes predictions about things that have not yet happened or that have not yet been observed. These are incredibly powerful tools that enable us to understand what we see without having to accept events on faith or to fear that things like the force of gravity will change on a whim. Science simplifies life and takes some of the uncertainty out of the world.

Consider the topic of this chapter, our understanding of gravity. Until Isaac Newton made the conceptual leap that the force holding Earth in orbit around the Sun is the same force that holds us onto the Earth, these two effects were considered to be separate and unrelated. Once they were connected, and an equation was written to describe the behavior of Earth and of falling objects, people had, for the first time, the ability to predict the future motion of projectiles and other falling objects reliably. Albert Einstein (Chapter 9) plumbed the depths of gravitational behavior even further, developing equations that make even more accurate predictions. The fundamentals of gravity we study here lead to our understanding of its effects on the universe, including how it causes stars and planets to form (Chapter 15), enables stars to shine (Chapter 5), and holds galaxies together (Chapter 10), among many other things.

In this chapter you will discover

- what makes a theory scientific
- the scientific revolution that dethroned Earth from its location at the center of the universe
- Copernicus's argument that the planets orbit the Sun
- why the direction of motion of the planets on the celestial sphere sometimes appears to change
- that Kepler's determination of the shapes of planetary orbits depended on the careful observations of his mentor Tycho Brahe
- how Isaac Newton formulated an equation to describe the force of gravity and how he thereby explained why the planets and moons remain in orbit

SCIENCE: KEY TO COMPREHENDING THE COSMOS

Understanding how nature works enables us to manipulate the matter and energy that comprise our environment and thereby to create new things to make our lives better. Improvements in technology lead, in turn, to better research equipment, enabling us to make even deeper discoveries about space, time, matter, energy, and the relationships among them. This spiral of understanding and application began centuries ago. In this chapter we will explore the nature of science and use it to see how gravity keeps planets and other objects orbiting the Sun as well as the moons orbiting their respective planets.

2-1 Science is both a body of knowledge *and* a process of learning about nature

Science is actually two things. First, it is a body of knowledge that we acquire by observations and experiments. The details of the motions of the Moon, planets, and Sun on the celestial sphere, described in Chapter 1, are examples of that knowledge. While nature can be discussed descriptively, as it is for the most part in this book, science also provides mathematical equations that quantify the effects being studied.

Second, science is a process for gaining more knowledge in a way that ensures that the information can be tested and thereby accepted by everyone. Science as a process is also called the **scientific method,** and it describes how scientists ideally go about observing, explaining, and predicting physical reality. The scientific method (Figure 2-1) can begin in a variety of places, but most often it starts by people making observations or doing experiments. For example, the observation that some objects (such as, planets) move along the celestial sphere, while others (stars) remained fixed on it, demanded explanation. The results of observations or experiments are compared with the predictions of any preexisting theories that are supposed to explain them. If the new data and old ideas are not consistent, then a *hypothesis* that modifies or replaces the existing explanation is proposed. Hypotheses on related topics that make accurate predictions are incorporated together as a **scientific theory** (often just called a **theory**).

In everyday conversation, a theory is an idea based on common sense, intuition, or deep-seated personal beliefs. Such theories neither originate in equations nor lead to rigorous predictions. The word *theory* in science has a very different connotation. It is an explanation of observations or experimental results that can be described quantitatively and tested formally. The mathematical description of a scientific theory is considered a **model** of the real system. For example, Newton's *theory* (or, in earlier usage, *law*) of gravitation is written as an equation that predicts how bodies attract each other. The word *gravity* is often used as shorthand for *gravitation*, and both are used in this book.

As just noted, to be considered scientific, a theory must make *testable* predictions that can be verified using new observations and experiments. Testing is a crucial aspect of the scientific method, which requires that the theory accurately forecast the results of new observations in its realm of validity. Newton's law of gravitation predicts that the Sun's gravitational force makes the planets move in elliptical orbits, and it predicts how long it should take each planet to orbit the Sun. As we will see shortly, observations have confirmed most of these predictions.

Scientists who develop new or more accurate models are going where no person has gone before. Many of them find this process of discovery as satisfying as an artist creating a masterpiece, an athlete breaking a world record, or an astronaut going into space. Scientists

FIGURE 2-1 The Scientific Method This flow chart shows the basic processes by which scientists study nature and develop new scientific theories. Different scientists start at different places on this chart, including making observations or doing experiments; creating or modifying scientific theories; or making predictions from theories. Anyone interested in some aspect of science and willing to learn the tools of that science can participate in the adventure. (© Neil F. Comins)

who make observations or do experiments that reveal previously unknown facts about nature often have similar reactions to their discoveries.

INSIGHT INTO SCIENCE

Science Is Inclusive Science is intended as an inclusive endeavor. In principle, a scientific theory can be created, modified, or tested by anyone inclined to do so. In practice, however, being involved in the scientific enterprise requires that you understand the mathematical tools of science. Assuring that theories are written in terms of equations so that they can be carefully analyzed and tested by others is part of the process intended to prevent the scientific method from being derailed.

For a theory to be considered scientific, it must also be potentially possible to disprove it. For example, Newton's law of gravitation can be tested and potentially disproved by observations, and thus qualifies as a scientific theory. The idea that Earth was created in 6 days cannot be tested, much less disproved. It is not a scientific theory, but rather a matter of faith.

If the predictions of a theory are inconsistent with observations, the theory is modified, applied in more limited circumstances, or discarded in favor of a more accurate explanation. For example, Newton's law of gravitation is entirely adequate for describing the motion of an apple falling to Earth, the flight of a soccer ball, or the path of Earth orbiting the Sun; however, it is inaccurate in the vicinity of a black hole, where matter is especially dense. In this case, Newton's law of gravitation is replaced by Einstein's theory of general relativity, which describes gravitational behavior more accurately and over a much wider range of conditions than Newton's law, but at the cost of much greater mathematical complexity.

> Give an example of one scientific hypothesis and one nonscientific hypothesis.

INSIGHT INTO SCIENCE

Theories and Beliefs New theories are personal creations, but science is not a personal belief system. As stated in the previous Insight into Science, scientific theories make predictions that can be tested independently. If everyone who performs tests of the theory's predictions gets results consistent with the theory, the theory is considered valid in that realm. In comparison, belief systems—such as which sports team or political system is best—are personal matters. People will always hold differing opinions about such issues.

Science also strives to explain as many things as possible with as few theories as possible. We see billions upon billions of objects in the universe. It would be virtually impossible to study all of them separately so that we could create detailed descriptions of each one. Fortunately, individual theories explaining each object are not necessary. Scientists overcome this problem by noting that many of the bodies in space appear similar to each other. By categorizing them suitably and then applying the scientific method to these groups of objects, we form a few theories that describe many objects and how they have evolved. These few theories can then be tested and refined as necessary. Such groupings of objects have proven invaluable, and they give us insights into the structure and organization of billions of stars and galaxies that are, indeed, very similar to one another.

INSIGHT INTO SCIENCE

WEB LINK 2.1

Keep It Simple When several competing theories describe the same concepts with the same accuracy, scientists choose the simplest theory—namely, the one that contains the fewest unproven assumptions. That basic tenet, formally expressed by William of Occam in the fourteenth century, is known as **Occam's razor.** Indeed, the original form of the Sun-centered cosmology, which we are about to explore, was appealing because it made the same predictions within a simpler model than did the Earth-centered cosmology. *Remember Occam's razor.*

While the vast majority of scientists carefully and scrupulously follow the rules of scientific research, we acknowledge that some experiments are run poorly. Some scientists have ignored experimental data or observations that don't mesh with cherished beliefs or have even fudged data or stolen data from others. Virtually all of these oversights and misdeeds are eventually discovered because most theories and their predictions are tested by several independent researchers.

The scientific method can be summarized in six words: *observe, hypothesize, predict, test, modify, economize.* I urge you to watch for applications of the scientific method throughout this book. Our first encounter with it is the discovery that Earth orbits the Sun.

CHANGING OUR EARTH-CENTERED VIEW OF THE UNIVERSE

Early Greek astronomers tried to explain the motion of the five then-known planets: Mercury, Venus, Mars, Jupiter, and Saturn. Most people at that time held a *geocentric* view of the universe: Based on the observed

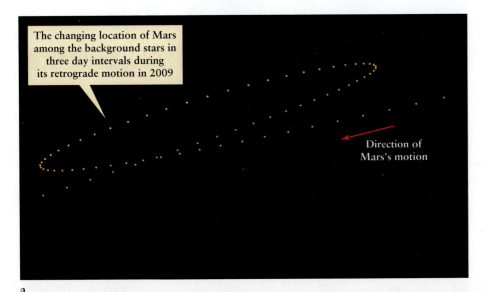

The changing location of Mars among the background stars in three day intervals during its retrograde motion in 2009

Direction of Mars's motion

a

ANIMATION 2.2 **FIGURE 2-2** Paths of Mars (a) The retrograde motion of Mars as it would be seen in a series of images taken on the same photographic plate. (b) To help visualize this motion on the celestial sphere, the same time interval from September 2009 through June 2010, is plotted in cartoon form. From December 23, 2009, through March 12, 2010, Mars is in retrograde motion. The retrograde loop is sometimes north of the normal path and sometimes south of it (see Figure 2-3).

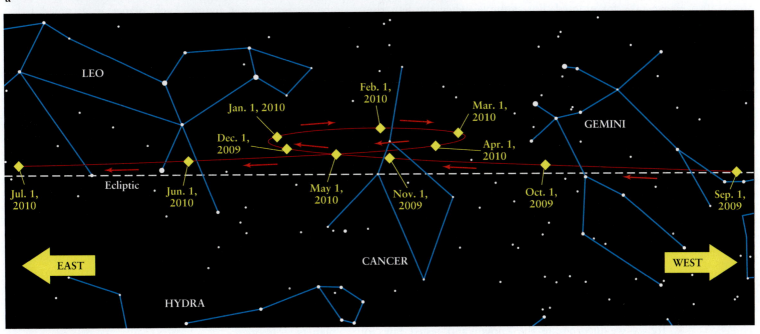

b

motion of the celestial sphere, they believed that the Sun, the Moon, the stars, and the planets revolve around Earth. A theory of the overall structure and evolution of the universe is called a **cosmology,** so the prevailing Earth-centered cosmology was called *geocentric.* Geocentric cosmology, consistent as it is with casual observation of the sky, held sway for more than 2000 years.

2-2 The belief in a Sun-centered cosmology formed slowly

Explaining the motions of the five planets in a geocentric universe was one of the main challenges facing the astronomers of antiquity. The Greeks knew that the positions of the planets slowly shift relative to the "fixed" stars in the constellations. In fact, the word *planet* comes from a Greek term meaning "wanderer."

They also observed that planets do not move at uniform rates through the constellations. From night to night, as viewed in the northern hemisphere, the planets usually move slowly to the left (eastward) relative to the background stars. This movement is called **direct motion.** Occasionally, however, a planet seems to stop and then back up for several weeks or months. This reverse movement (to the west relative to the background stars) is called **retrograde motion.** Both direct and retrograde motions are best observed by plotting or photographing the nightly position of a planet against the background stars over a long period (Figure 2-2a and b).

All planetary motions on the celestial sphere are much slower than the apparent daily movement of the entire sky caused by Earth's rotation, and so they are superimposed on it. Therefore, the planets always rise in the eastern half of the sky and set in the western half, as the stars do.

Earth–centered Universe

As we move through the twenty-first century, most of us find it hard to understand why anyone would believe that the Sun, planets, and stars orbit Earth. After all, we *know* that Earth spins on its axis. We *know* that the gravitational force from the Sun holds the planets in orbit, just as Earth's gravitational force holds the Moon in orbit. These facts have become part of our understanding of the motions of the heavenly bodies and we are taught these things from the time we are children.

Psychologists call this background information that we use to help explain things *a conceptual framework.* Any conceptual framework contains all of the information we take for granted. For example, when the Sun rises, moves across the sky, and sets, we take for granted that it is Earth's rotation that causes the Sun's apparent motion.

Our ancestors possessed a different conceptual framework for understanding the cosmos. They did not know that Earth rotates. They did not know that the then-mysterious force that held them to the ground is the same force that attracts Earth to the Sun and the Moon to Earth. They did not know that the Sun is a star, just like the fixed points of light in the sky, and they did not know any of the other laws of motion that we take for granted.

Because they did not feel Earth move under their feet, or see any other indication that Earth is in motion, our forebears sensed nothing to support the belief that we are in motion. The obvious conclusion for one who has a prescientific conceptual framework, even today, is that Earth stays put while objects in the heavens move around it.

This prescientific conceptual framework for understanding the motions of the heavenly bodies was based on the senses and on common sense. That is, people observed motions and drew "obvious," commonsense conclusions. Today, we incorporate the known and tested laws of physics in our understanding of the natural world. Many of these realities are utterly counterintuitive, and, there-fore, the conceptual frameworks that we possess are less consistent with common sense than those held in the past. Studying science helps us develop intuition that is consistent with the actual workings of nature.

Geocentric Explanation of the Planets' Retrograde Motion   The early Greeks developed many theories to account for the occasional retrograde motion of the planets and the resulting loops that the planets trace out against the background stars. One of the most successful ideas was expounded by the last of the great ancient Greek astronomers, Ptolemy, who lived in Alexandria, Egypt, 1900 years ago. His basic concepts are sketched in the accompanying figure. Each planet is assumed to move in a small circle called an *epicycle,* the center of which moves in a larger circle called a *deferent,* whose center is offset from Earth. As viewed from Earth, the epicycle moves eastward along the deferent, and both it and the planet on it revolve in the same direction (counterclockwise).

Most of the time, the motion of the planet on its epicycle adds to the eastward motion of the epicycle on the deferent. Thus, the planet is seen from Earth to be in direct motion (to the left or eastward) against the background stars throughout most of the year (Figure GD2.1a). However, when the planet is on the part of its epicycle nearest Earth, its motion along the epicycle subtracts from the motion of the epicycle along the deferent. The planet thus appears to slow and then halt its usual movement to the left (eastward motion) among the constellations, and then seems to move to the right (westward) among the stars for a few weeks or months (Figure GD2.1b). This concept of epicycles and deferents enabled Greek astronomers to explain the retrograde loops of the planets.

Using the wealth of astronomical data in the library at Alexandria, including records of planetary positions

---

The effort to understand planetary motion—and especially to explain retrograde motion—using geocentric cosmology resulted in an increasingly contrived and complex model (see *Guided Discovery: Earth-Centered Universe*). The ancient Greek astronomer Aristarchus proposed a more straightforward explanation of planetary motion, namely, that all of the planets, including Earth, revolve around the Sun. The retrograde motion of Mars in this **heliocentric (Sun-centered) cosmology** occurs just because the faster-moving Earth overtakes and passes the red planet (Figure 2-3). The occasional retrograde movement of a planet is merely the result of our changing viewpoint as we orbit the Sun—an idea that is beautifully simple compared to the geocentric system with all of its complex planetary motions. (The word *heliocentric* is misleading. Although the local planets, moons, and small pieces of space debris do orbit the Sun, the stars and innumerable other objects in space don't. In fact, the Sun and the bodies that orbit the Sun all orbit the center of our Milky Way Galaxy.)

Because simplicity and accuracy are hallmarks of science, the complex geocentric model eventually gave way to the simpler, more elegant heliocentric cosmology, but dethroning the geocentric model did not occur immediately. In the first place, Earth just does not seem

> Why do you think Mars is seen sometimes above the ecliptic and sometimes below it?

The Earth–centered Universe *(continued)*

covering hundreds of years, Ptolemy deduced the sizes of the epicycles and deferents and the rates of revolution needed to produce the recorded paths of the planets. After years of arduous work, Ptolemy assembled his calculations in the *Almagest*, in which the positions and paths of the Sun, Moon, and planets were described with unprecedented accuracy. In fact, the *Almagest* was so successful that it became the astronomer's bible. For more than 1000 years, Ptolemy's cosmology endured as a useful description of the workings of the heavens.

Eventually, however, the commonsense explanation of the Earth-centered cosmology began to go awry.

Errors and inaccuracies that were unnoticeable in Ptolemy's day compounded and multiplied over the years, especially errors due to precession, the slow change in the direction of Earth's axis of rotation. Fifteenth-century astronomers made some cosmetic adjustments to the Ptolemaic system. However, the system became less and less satisfactory as more fanciful and arbitrary details were added to keep it consistent with the observed motions of the planets. After Newton's time, scientists knew that orbital motion required a force to be acting on the body. However, nothing in Ptolemy's epicycle theory produced such a force.

a

b

FIGURE GD2-1 A Geocentric Explanation of Planetary Motion Each planet revolves around an epicycle, which, in turn, revolves around a deferent centered approximately on Earth. As seen from Earth, the speed of the planet on the epicycle alternately (a) adds to or (b) subtracts from the speed of the epicycle on the deferent, thus producing alternating periods of direct and retrograde motions.

to move! This observation, along with strict geocentric religious teachings and the human desire to be at the center of everything, outweighed the simpler plan proposed by Aristarchus. Not until 1300 years later did anyone seriously reconsider the advantages of a heliocentric cosmology.

2-3 Copernicus devised the first comprehensive heliocentric cosmology

Over the centuries, increasingly accurate observations of the planets' locations revealed errors in the predic-

tions of geocentric cosmology. To reconcile that cosmology with the data, more and more complex motions were attributed to the planets. By the mid-1500s, the geocentric cosmology had become truly unwieldy in its efforts to predict the motions of the planets accurately. It was then that the Polish mathematician, lawyer, physician, economist, cleric, and artist Nicolaus Copernicus resurrected Aristarchus's theory. Copernicus (see *Guided Discovery: Astronomy's Foundation Builders*) was motivated by an effort to simplify the celestial scheme.

After assuming that the planets orbit the Sun rather than Earth, Copernicus, through observations, determined which

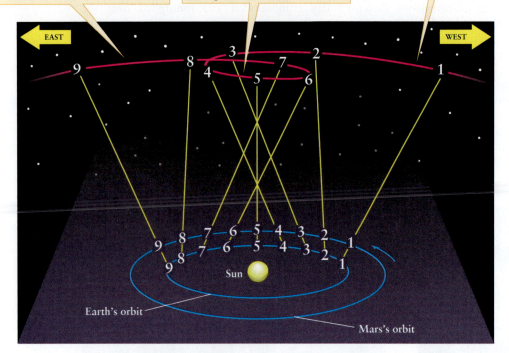

3. From point 6 to point 9, Mars again appears to move eastward against the background stars as seen from Earth (direct motion).

2. As Earth passes Mars in its orbit from point 4 to point 6, Mars appears to move westward against the background stars (retrograde motion).

1. From point 1 to point 4, Mars appears to move eastward against the background stars as seen from Earth (direct motion).

FIGURE 2-3 A Heliocentric Explanation of Planetary Motion Earth travels around the Sun more rapidly than does Mars. Consequently, as Earth overtakes and passes this slower-moving planet, Mars appears (from points 4 through 6) to move backward among the background stars for a few months.

elongation varies from 0° to a maximum value, depending upon where we see it in its orbit around the Sun. At *greatest eastern* or *greatest western elongation,* Mercury and Venus are as far from the Sun in angle as they can be. This is about 28° for Mercury and about 47° for Venus. When either Mercury or Venus rises before the Sun, it is visible in the eastern sky as a bright "star" and is often called the "morning star." Similarly, when either of these two planets sets after the Sun, it is visible in the western sky and is then called the "evening star." Because these two planets are not always at their greatest elongations, they are often very close in angle to the Sun. This is especially true of Mercury, often making it hard to see from Earth. Venus is often nearly halfway up the sky at sunrise or sunset and therefore quite noticeable during much of its orbit. Because they are so bright and sometimes appear to change color due to the motion of Earth's atmosphere, Venus and Mercury are often mistaken for UFOs. (The same motion of the air causes the road in front of your car to shimmer on a hot day.)

Planets farther from the Sun than Earth have different configurations. When one of them is located behind the Sun, as seen from Earth, it is said to be in **conjunction.** When it is opposite the Sun in the sky, the planet is at **opposition.** It is not difficult to determine when a planet happens to be located at one of the key positions in Figure 2-4. For example, when Mars is at opposition, it appears high and bright in the sky at midnight.

It is relatively easy to follow a planet as it moves from one configuration to another. However, these observations alone do not tell us the planet's actual orbit, because Earth, from which we make the observations, is also moving. Copernicus was therefore careful to distinguish between two characteristic time intervals, or *periods,* of each planet.

Recall from your study of the Moon in Chapter 1 that the sidereal period of orbit is the true orbital period of any astronomical body. A planet's **sidereal period** is the time it takes that body to make one complete orbit of the Sun. The sidereal period is the length of a year for each planet.

planets are closer to the Sun than Earth and which are farther away. Because Mercury and Venus are always observed fairly near the Sun, he correctly concluded that their orbits must lie inside Earth's. The other planets visible to Copernicus—Mars, Jupiter, and Saturn— can sometimes be seen high in the sky in the middle of the night, when the Sun is far below the horizon. This can occur only if Earth comes between the Sun and a planet. Copernicus therefore concluded (correctly) that the orbits of Mars, Jupiter, and Saturn lie outside Earth's orbit.

The geometric arrangements among Earth, another planet, and the Sun are called **configurations.** For example, when Mercury or Venus is directly between Earth and the Sun (Figure 2-4), we say the planet is in a configuration called an **inferior conjunction;** when either of these planets is on the opposite side of the Sun from Earth, its configuration is called **a superior conjunction.**

The angle between the Sun and a planet as viewed from Earth is called the planet's **elongation.** A planet's

GUIDED DISCOVERY

Astronomy's Foundation Builders

In the two centuries between 1500 and 1700, human understanding of the motion of celestial bodies and the nature of the gravitational force that keeps them in orbit surged forward as never before. Theories related to this subject were developed by brilliant thinkers, whose work established and verified the heliocentric model of the solar system and the role of gravity.

(E. Lessing/Art Resource)

Nicolaus Copernicus (1473–1543) Copernicus, the youngest of four children, was born in Torun, Poland. He pursued his higher education in Italy, where he received a doctorate in canon law and studied medicine. Copernicus developed a heliocentric theory of the known universe and just before his death in 1543 published this work under the title *De Revolutionibus Orbium Coelestium*. His revolutionary theory was flawed in that he assumed that the planets had circular orbits around the Sun. This was corrected by Johannes Kepler.

(Painting by Jean-Leon Huens, courtesy of National Geographic Society)

Tycho Brahe (1546–1601) and ***Johannes Kepler (1571–1630)*** Tycho (depicted within the portrait of Kepler) was born to nobility in the Danish city of Knudstrup, which is now part of Sweden. At age 20 he lost part of his nose in a duel and wore a metal replacement thereafter. In 1576 the Danish king Frederick II built Tycho an astronomical observatory that Tycho named Uraniborg (after Urania, Greek muse of astronomy). Tycho rejected both Copernicus's heliocentric theory and the Ptolemaic geocentric system. He devised a halfway theory called the *Tychonic system*. According to Tycho's theory, Earth is stationary, with the Sun and Moon revolving around it, while all the other planets revolve around the Sun. Tycho died in 1601.

Kepler was educated in Germany, where he spent 3 years studying mathematics, philosophy, and theology. In 1596, Kepler published a booklet in which he attempted to mathematically predict the planetary orbits. Although his theory was altogether wrong, its boldness and originality attracted the attention of Tycho Brahe, whose staff Kepler joined in 1600. Kepler deduced his three laws from Tycho's observations.

(Art Resource)

Galileo Galilei (1564–1642) Born in Pisa, Italy, Galileo studied medicine and philosophy at the University of Pisa. He abandoned medicine in favor of mathematics. He held the chair of mathematics at the University of Padua, and eventually returned to the University of Pisa as a professor of mathematics. There Galileo formulated his famous law of falling bodies: All objects fall with the same acceleration regardless of their weight. In 1609 he constructed a telescope and made a host of discoveries that contradicted the teachings of Aristotle and the Roman Catholic Church. He summed up his life's work on motion, acceleration, and gravity in the book *Dialogues Concerning the Two Chief World Systems*, published in 1632.

(National Portrait Gallery, London)

Isaac Newton (1642–1727) Newton delighted in constructing mechanical devices, such as sundials, model windmills, a water clock, and a mechanical carriage. He received a bachelor's degree in 1665 from the University of Cambridge. While there, he began developing the mathematics that later became calculus (developed independently by the German Gottfried Leibniz). While pursuing experiments in optics, Newton constructed a reflecting telescope and also discovered that white light is actually a mixture of all colors. His major work on forces and gravitation was the tome *Philosophiae Naturalis Principia Mathematica*, which appeared in 1687. In 1704, Newton published his second great treatise, *Opticks*, in which he described his experiments and theories about light and color. Upon his death in 1727, Newton was buried in Westminster Abbey, the first scientist to be so honored.

The other useful time interval that Copernicus used is the **synodic period**. The synodic period is the time that elapses between two successive identical configurations as seen from Earth. It can be from one opposition to the next, for example, or from one conjunction to the next (Figure 2-5). It tells us, among other things, when to expect a planet to be closest to Earth and, therefore, most easily studied.

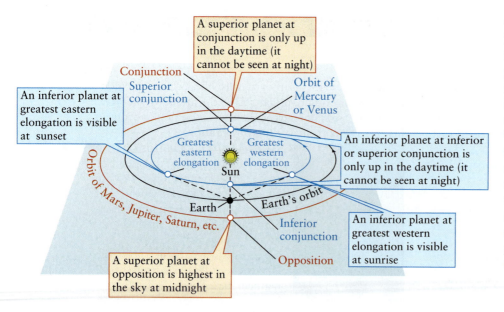

An inferior planet at greatest eastern elongation is visible at sunset

A superior planet at conjunction is only up in the daytime (it cannot be seen at night)

Conjunction
Superior conjunction

Orbit of Mercury or Venus

Greatest eastern elongation

Sun

Greatest western elongation

An inferior planet at inferior or superior conjunction is only up in the daytime (it cannot be seen at night)

Orbit of Mars, Jupiter, Saturn, etc.

Earth

Earth's orbit

Inferior conjunction

An inferior planet at greatest western elongation is visible at sunrise

A superior planet at opposition is highest in the sky at midnight

Opposition

FIGURE 2-4 Planetary Configurations Key points along a planet's orbit have names, as shown. These points identify specific geometric arrangements between Earth, another planet, and the Sun. Knowing where a planet is with respect to the Sun helps astronomers know when and where to look for the planet.

INSIGHT INTO SCIENCE

Take a Fresh Look When a scientific concept is hard to visualize, try another perspective. For example, a planet's sidereal period of orbit is easy to understand when viewed from the Sun but more complicated as seen from Earth. The synodic period of each planet, on the other hand, is easily determined from Earth. As we will see, especially when we study Einstein's theories of relativity, each of these perspectives is called a *frame of reference.*

Thus, nearly 500 years ago, Copernicus was able to obtain the first six entries shown in Table 2-1 (the others are contemporary results included for completeness). Copernicus was then able to devise a straightforward geometric method for determining the distances of the planets from the Sun. His answers turned out to be remarkably close to the modern values, as shown in Table 2-2. From these two tables it is apparent that the farther a planet is from the Sun, the longer it takes to complete its orbit.

Copernicus presented his heliocentric cosmology, including supporting observations and calculations, in a book entitled *De Revolutionibus Orbium Coelestium* (*On the Revolutions of the Celestial Spheres*), which

was published in 1543, the year of his death. His great insight was the conceptual simplicity of a heliocentric cosmology compared to geocentric views, especially in explaining retrograde motion. However, Copernicus incorrectly assumed that the planets travel along circular paths around the Sun. Without using epicycles similar to those used in geocentric theory (Figure GD2.1), many of his predictions were no more accurate than those of the earlier theory! As we will see shortly, by changing the shape of the orbits, Kepler was able to do away with epicycles and make even more accurate predictions than either the geocentric or original Copernican theory.

2-4 Tycho Brahe made astronomical observations that disproved ancient ideas about the heavens

In November 1572, a bright star suddenly appeared in the constellation Cassiopeia. At first, it was even brighter than Venus, but then it began to grow dim. After 18 months it faded from view.

Modern astronomers recognize this event as a supernova explosion, the violent death of a certain type of star (see Chapter 8). In the sixteenth century, however, the prevailing opinion was quite different. Teachings dating back to Aristotle and Plato argued that the heavens were

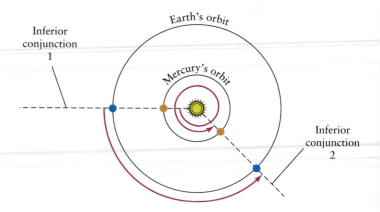

FIGURE 2-5 Synodic Period The time between consecutive conjunctions of Earth and Mercury is 116 days. Typical of synodic periods for all planets, the location of Earth is different at the beginning and end of the period. You can visualize the synodic periods of the exterior planets by putting Earth in Mercury's place in this figure and putting one of the outer planets in Earth's place.

TABLE 2-1 Synodic and Sidereal Periods of the Planets (in Earth Years)

	Synodic (year)	Sidereal (year)
Mercury	0.318	0.241
Venus	1.599	0.616
Earth	—	1.0
Mars	2.136	1.9
Jupiter	1.092	11.9
Saturn	1.035	29.5
Uranus	1.013	84.0
Neptune	1.008	164.8

permanent and unalterable. From that perspective, the "new star" of 1572 could not really be a star at all, because the heavens do not change. Many astronomers and theologians of the day argued that the sighting must be some sort of bright object quite near Earth, perhaps not much farther away than the clouds overhead. A 25-year-old Danish astronomer named Tycho Brahe (see *Guided Discovery: Astronomy's Foundation Builders*) realized that straightforward observations might reveal the distance to this object.

Consider what happens when two people look at a nearby object from different places—they see it in different positions relative to the things behind it. Furthermore, their heads face at different angles when looking at it. This variation in angle that occurs when viewing a nearby object from different locations is called **parallax** (Figure 2-6).

Tycho reasoned as follows: If the new star is nearby, its position should shift against the background

TABLE 2-2 Average Distances of the Planets from the Sun

	Measurement (AU)	
	By Copernicus	Modern
Mercury	0.38	0.39
Venus	0.72	0.72
Earth	1.00	1.00
Mars	1.52	1.52
Jupiter	5.22	5.20
Saturn	9.07	9.54
Uranus	Unknown	19.19
Neptune	Unknown	30.06

stars over the course of a night (Figure 2-7a). His careful observations, done in the spirit of the scientific method, failed to disclose any parallax, and so the new star had to be far away, farther from Earth than anyone had imagined (Figure 2-7b). Tycho summarized his findings in a small book, *De Stella Nova* (*On the New Star*), published in 1573.

Tycho's astronomical records were soon to play an important role in the development of a heliocentric cosmology. From 1576 to 1597, Tycho made comprehensive observations, measuring planetary positions with an accuracy of 1 arcmin, about as precise as is possible with the naked eye. [Arcminute (arcmin) is defined in *An*

FIGURE 2-6 Parallax Nearby objects are viewed at different angles from different places. These objects also appear to be in different places with respect to more distant objects when viewed at the same time by observers located at different positions. Both effects are called parallax, and they are used by astronomers, surveyors, and sailors to determine distances. (Tobi Zausner)

a   b

FIGURE 2-7 The Parallax of a Nearby Object in Space
Tycho thought that Earth does not rotate and that the stars revolve around it. From our modern perspective, the changing position of the supernova would be due to Earth's rotation as shown. (a) Tycho argued that if an object is near Earth, its position relative to the background stars should change over the course of a night. (b) Tycho failed to measure such changes for the supernova in 1572. This is illustrated in *(b)* by the two telescopes being parallel to each other. He, therefore, concluded that the object was far from Earth.

To Kepler's delight, the ellipse turned out to be the curve for which he had been searching. Predictions of the locations of planets based on elliptical paths were in very close agreement with where the planets actually were. He published this discovery in 1609 in a book known today as *New Astronomy*. This important discovery is now considered the first of **Kepler's laws:**

Kepler's first law: *The orbit of a planet around the Sun is an ellipse with the Sun at one focus.*

The shapes of ellipses have two extremes. The roundest ellipse, occurring when the two foci merge, is a circle. The most elongated ellipse approaches being a straight line. The shape of a planet's orbit around the Sun is described by its *orbital eccentricity*, designated by the letter *e*, which ranges from 0 (circular orbit) to just under 1.0 (nearly a straight line). Figure 2-8b shows a sequence of ellipses and their associated eccentricities. Observations have revealed that there is no object at the second focus of each elliptical planetary orbit.

Tycho's observations also showed Kepler that planets do not move at uniform speeds along their orbits. Rather, a planet moves most rapidly when it is nearest the Sun, a point on its orbit called **perihelion.** Conversely, a planet moves most slowly when it is farthest from the Sun, called its **aphelion.**

After much trial and error, Kepler discovered a way to describe how fast a planet moves anywhere along its orbit. This discovery, also published in *New Astronomy*, is illustrated in Figure 2-9. Suppose that it takes 30 days for a planet to go from point A to point B.

> Describe a simple experiment to demonstrate that your eyes (and, implicitly, your brain) use parallax to determine distances.

Astronomer's Toolbox 1-1.] Upon Tycho's death in 1601, most of these invaluable records were given to his gifted assistant Johannes Kepler (see *Guided Discovery: Astronomy's Foundation Builders*).

KEPLER'S AND NEWTON'S LAWS

WEB LINK 2.4 Until Kepler's time, astronomers had assumed that heavenly objects move in circles. For philosophical and aesthetic reasons, circles were considered the most perfect and most harmonious of all geometric shapes. However, using circular orbits failed to yield accurate predictions for the positions of the planets. For years, Kepler tried to find a shape for orbits that would fit Tycho's observations of the planets' positions against the background of distant stars. Finally, he began working with a geometric form called an **ellipse.**

2-5 Kepler's laws describe orbital shapes, changing speeds, and the lengths of planetary years

You can draw an ellipse as shown in Figure 2-8a. Each thumbtack is at a **focus** (plural: **foci**). The longest diameter (major axis) across an ellipse passes through both foci. Half of that distance is called the **semimajor axis.** In astronomy, the length of the semimajor axis is also the average distance between a planet and the Sun.

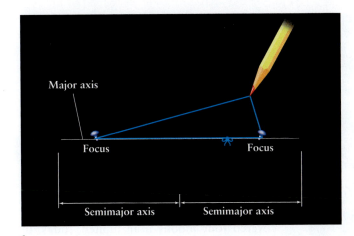

FIGURE 2-8 Ellipses (a) The construction of an ellipse: At all places along an ellipse, the sum of the distances to the two foci is a constant. An ellipse can be drawn with a pencil, a loop of string, and two thumbtacks, as shown. If the string is kept taut, the pencil traces out an ellipse. The two thumbtacks are located at the two foci of the ellipse. (b) A series of ellipses with different eccentricities, *e*. Eccentricities range between 0 (circle) to just under 1.0 (almost a straight line). Note that all eight planets have eccentricities less than 0.21.

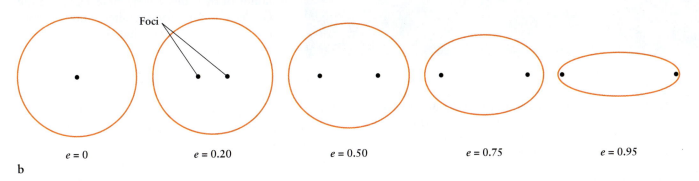

During that time, the line joining the Sun and the planet sweeps out a nearly triangular area (shaded in Figure 2-9). Kepler discovered that the line joining the Sun and the planet sweeps out the same area during any other 30-day interval. In other words, if the planet also takes 30 days to go from point C to point D, then the two shaded segments in Figure 2-9 are equal in area. Kepler's second law, also called the **law of equal areas**, can be stated thus:

Kepler's second law: *A line joining a planet and the Sun sweeps out equal areas in equal intervals of time.*

A consequence of Kepler's second law is that each planet's speed decreases as it moves from perihelion to aphelion. The speed then increases as the planet moves from aphelion toward perihelion.

Kepler was also able to relate a planet's year to its distance from the Sun. This discovery, published in 1619, is Kepler's third law. It predicts the planet's sidereal period if we know the length of the semimajor axis of the planet's orbit:

Kepler's third law: *The square of a planet's sidereal period around the Sun is directly proportional to the cube of the length of its orbit's semimajor axis.*

The relationship is easiest to use if we let P represent the sidereal period in Earth years and a represent the length of the semimajor axis measured in **astronomical units** (AU). One astronomical unit is the average distance from Earth to the Sun. It is commonly used in discussing distances to various objects in the solar system, because no powers of ten need be used with it, as they

would if distances between planets were referred to in kilometers or miles. (*See An Astronomer's Toolbox 2-1* for more details of astronomical distance units.) Now we can write Kepler's third law as

$$P^2 = a^3$$

This equation says that a planet closer to the Sun has a shorter year than does a planet farther from the Sun. Using this equation with Kepler's second law reveals that planets closer to the Sun move more rapidly than those farther away. Using data from Tables 2-1 and 2-2, we can demonstrate Kepler's third law as shown in Table 2-3.

When Newton derived Kepler's third law using the law of gravitation, discussed later in this chapter, he discovered that the mass of the planet affects the period of its orbit around the Sun. However, this effect of the planet's mass is vanishingly small for all the planets in the solar system, which is why the equation for Kepler's third law, as shown in Table 2-3, gives such good results for the planets' orbits even though it does not take the masses into account. When calculating the motion of pairs of stars orbiting each other, the effects of the masses must be taken into account, as described in *An Astronomer's Toolbox 6-4*.

Kepler's three laws apply not only to the planets orbiting the Sun, but also whenever any object orbits another under the influence of their mutual gravitational attraction. Thus, Kepler's laws apply to moons orbiting planets, artificial satellites orbiting Earth, and (with the above caveat) even two stars revolving around each other.

We saw in Chapter 1 that the Moon's orbit around Earth is not circular. Where in its orbit is the Moon moving fastest, and where is it moving slowest?

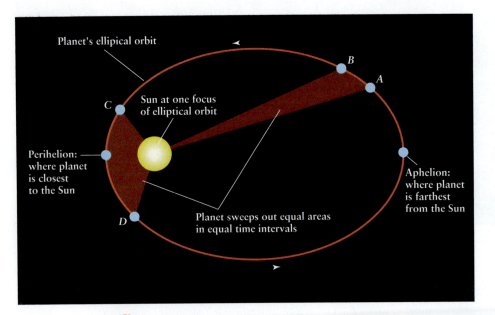

Planet's ellipical orbit

B

A

C

Sun at one focus
of elliptical orbit

Perihelion:
where planet
is closest
to the Sun

Aphelion:
where planet
is farthest
from the Sun

D

Planet sweeps out equal areas
in equal time intervals

FIGURE 2-9 Kepler's First and Second Laws
According to Kepler's first law, every planet travels
around the Sun along an elliptical orbit with the Sun
at one focus. According to his second law, the line joining the
planet and the Sun sweeps out equal areas (the burgandy-
colored regions) in equal intervals of time (time from A to B
equals time from C to D). Note: This drawing shows a highly
elliptical orbit, with $e = 0.74$. Even though this is a much
greater eccentricity than that of any planet in the solar system,
the concept still applies to all planets and other orbiting bodies.

2-6 Galileo's discoveries strongly supported a heliocentric cosmology

While Kepler was in central Europe working on the laws of planetary orbits, an Italian physicist was making dramatic observations in southern Europe. Galileo Galilei did not invent the telescope, but he was one of the first people to point the new device toward the sky and publish his observations. He saw things that no one had ever imagined—mountains on the Moon and spots on the Sun. He also discovered that the apparent size of Venus as seen through his telescope was related to the planet's phase (Figure 2-10). Venus appears smallest at gibbous phase and largest at crescent phase. These observations were a big chink in the geocentric cosmology's armor, as it could not explain why Venus has phases or changes size, but a heliocentric cosmology explains both. Galileo's observations, therefore, supported the conclusion that Venus orbits the Sun, not Earth.

In 1610, Galileo (see *Guided Discovery: Astronomy's Foundation Builders*) also discovered four moons near Jupiter. Today, in honor of their discoverer, these are called the **Galilean moons** (or **satellites**, another term for moon). Galileo concluded that the moons orbit Jupiter because they move across from one side of the planet to the other. Confirming observations were made in 1620 (Figure 2-11). These observations all provided further proof that Earth is not at the center of the universe. Like Earth in orbit around the Sun, Jupiter's four moons obey Kepler's third law: The square of a moon's orbital period around Jupiter is directly proportional to the cube of its average distance from the planet.

Galileo's telescopic observations constituted the first fundamentally new astronomical data since humans began recording what they saw in the sky. In contradiction to then-prevailing opinions, these discoveries strongly supported a heliocentric view of the universe. Because Galileo's ideas could not be reconciled with certain passages in the Bible or with the writings of Aristotle and Plato, the Roman Catholic Church condemned him, and he was forced to spend his later years under house arrest "for vehement suspicion of heresy." In 1992, Pope John Paul II stated that the church erred in condemning him.

A major stumbling block prevented seventeenth-century thinkers from accepting Kepler's laws and Galileo's conclusions about the heliocentric cosmology.

TABLE 2-3 A Demonstration of Kepler's Third Law

	Sidereal period P (year)	Semimajor axis a (AU)	P^2	$= a^3$
Mercury	0.24	0.39	0.06	0.06
Venus	0.61	0.72	0.37	0.37
Earth	1.00	1.00	1.00	1.00
Mars	1.88	1.52	3.53	3.51
Jupiter	11.86	5.20	140.7	140.6
Saturn	29.46	9.54	867.9	868.3
Uranus	84.01	19.19	7,058	7,067
Neptune	164.79	30.06	27,160	27,160

AN ASTRONOMER'S TOOLBOX 2-1

Units of Astronomical Distance

Throughout this book we will find that some of our traditional units of measure become cumbersome. It is fine to use kilometers to measure the diameters of craters on the Moon or the heights of volcanoes on Mars. However, it is as awkward to use kilometers to express the large distances to planets, stars, or galaxies as it is to talk about the distance from New York City to San Francisco or Paris to London in millimeters. Astronomers have therefore devised new units of measure.

When discussing distances across the solar system, astronomers use a unit of length called the astronomical unit (AU), which is the average distance between Earth and the Sun:

$$1 \text{ AU} \approx 1.5 \times 10^8 \text{ km} \approx 9.3 \times 10^7 \text{ mi}$$

Jupiter, for example, is an average of 5.2 times farther from the Sun than is Earth. Thus, the average distance between the Sun and Jupiter can be conveniently stated as 5.2 AU. This can be converted into kilometers or miles using the previous relationship.

When talking about distances to the stars, astronomers choose between two different units of length. One is the **light-year (ly)**. A light-year is the *distance* that light travels in a year through a vacuum (that is, in the absence of air). Do keep in mind that the word *year* in this unit helps describe a separation between two objects rather than representing a unit of time.

$$1 \text{ ly} \approx 9.46 \times 10^{12} \text{ km} \approx 63{,}000 \text{ AU}$$

The space between the planets, stars, and galaxies are nearly ideal vacuums. One light-year is roughly equal to six trillion miles. Proxima Centauri, the closest star to Earth, other than the Sun, is just over 4.2 light-years from us.

The second commonly used unit of length is the **parsec (pc)**, the distance at which two objects separated by 1 AU make an angle of 1 arcsec. Imagine taking a journey far into space, beyond the orbits of the outer planets. Watching the solar system as you move away, the angle between the Sun and Earth becomes smaller and smaller. When they are side by side from your perspective, and you measure the angle between them as $1/3600°$ (called 1 arcsec), you have reached a distance that astronomers call 1 parsec, as shown in the figure below. The parsec turns out to be longer than the light-year, specifically,

$$1 \text{ pc} \approx 3.09 \times 10^{13} \text{ km} \approx 3.26 \text{ ly}$$

Thus, the distance to the nearest star can be stated as 1.3 pc as well as 4.2 ly. Whether one uses light-years or parsecs is a matter of personal taste.

For larger distances, *kilolight* years (kly), *megalight* years (Mly), *kiloparsecs* (kpc), and *megaparsecs* (Mpc) are used. The prefixes "kilo" and "mega" simply mean "thousand" and "million," respectively:

$$1 \text{ kly} = 10^3 \text{ ly}$$
$$1 \text{ Mly} = 10^6 \text{ ly}$$
$$1 \text{ kpc} = 10^3 \text{ pc}$$
$$1 \text{ Mpc} = 10^6 \text{ pc}$$

For example, the distance from Earth to the center of our Milky Way Galaxy is about 8.6 kpc, and the rich cluster of galaxies in the direction of the constellation Virgo is 20 Mpc away.

Try these questions: The nearest star (other than the Sun) is 4.22 ly away. How many miles away is it? How many kilometers? How many parsecs?

(Answers appear at the end of the book.)

A Parsec The parsec, a unit of length commonly used by astronomers, is equal to 3.26 ly. The parsec is defined as the distance at which 1 AU perpendicular to the observer's line of sight makes an angle of 1 arcsec.

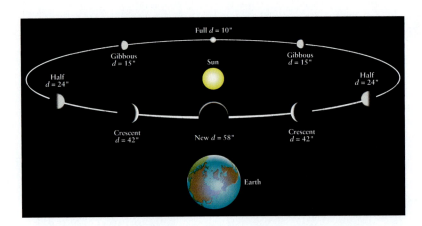

FIGURE 2-10 The Changing Appearance of Venus This figure shows how the appearance (phase) of Venus changes as it moves along its orbit. The number below each view is the angular diameter (*d*) of the planet as seen from Earth, in arcseconds. Note that the phases correlate with the planet's angular size and its angular distance from the Sun, both as seen from Earth. These observations clearly support the idea that Venus orbits the Sun.

Once anything on Earth is put in motion, it quickly comes to rest. Why don't the planets orbiting the Sun stop, too?

What is the shape of the orbit around Earth of the International Space Station?

The scientific method clarified most of the issues surrounding planetary orbits, leading to the equations and laws developed by the brilliant and eccentric (for example, he believed in alchemy) scientist Isaac Newton, who was born on Christmas Day in 1642, less than a year after Galileo died. In the decades that followed, Newton revolutionized science more profoundly than any person before him, and in doing so, he found physical and mathematical proofs of the heliocentric cosmology.

INSIGHT INTO SCIENCE

Theories and Explanations Scientific theories (or laws) based on observations can be useful for making predictions even if the reasons that these theories work are unknown. The explanation for Kepler's laws came decades after Kepler deduced them, when Newton derived them in 1665 them with his mathematical expression for gravitation, the force that holds the planets in their orbits.

2-7   Newton formulated three laws that describe fundamental properties of physical reality

Until the mid-seventeenth century, virtually all mathematical astronomy was done empirically. That is, astronomers from Ptolemy to Kepler created equations directly from data and observations.

Isaac Newton (see *Guided Discovery: Astronomy's Foundation Builders*) introduced a new approach. He began with three physical assumptions, now called **Newton's laws of**

motion, which led to equations that have since been tested and shown to be correct in many everyday situations. He also found a formula for the force of **gravity** (or **gravitation**), the attraction between all objects due to their masses. Putting the assumptions into mathematical form and combining them with the equation for gravity, Newton was able to derive Kepler's three laws and use them to predict the orbits of bodies such as comets and other objects in the solar system. Newton also was able to use these same equations to predict the motions of bodies on and near Earth, such as the path of a projectile or the speed of a falling object.

Newton's first law—the law of inertia: *Inertia is the property of matter that keeps an object at rest or moving in a straight line at a constant speed unless acted upon by a net external force.*

If all of the external forces acting on an object don't cancel each other out, then there is a *net* external force acting on the object. This is the same as saying that there is an unbalanced external force. Here is an example: If you put a soccer ball between your hands and press on it so that it doesn't move, your hands represent a balanced pair of forces acting on the ball. In that case, you are exerting no net external force on it. Conversely, when your foot hits a soccer ball and the ball sails away, your foot *has* exerted a net external force on the ball.

At first, this law might seem to conflict with your everyday experience. For example, if you shove a chair, it does not move at a constant speed forever but comes to rest after sliding only a short distance. From Newton's viewpoint, however, a net external force does indeed act on the moving chair—namely, friction between the chair's legs and the floor. Without friction, the chair would continue in a straight path at a constant speed. A net external force changes the motion of an object.

Newton's first law tells us that there must be an outside force acting on the planets to continually change their directions and keep them in orbit. If there were no

a

b

R I V U X G

FIGURE 2-11 *Jupiter and Its Largest Moons* In 1610, Galileo discovered four "stars" that move back and forth across Jupiter. He concluded that they are four moons that orbit Jupiter just as our Moon orbits Earth. (a) Observations made by Jesuits in 1620 of Jupiter and its four visible moons. (b) Photograph, taken by amateur astronomer C. Holmes, shows the four Galilean satellites alongside an overexposed image of Jupiter. Each satellite would be bright enough to be seen with the unaided eye were it not overwhelmed by the glare of Jupiter. (Courtesy of C. Holmes)

force acting on them, they would move away from the Sun along straight-line paths at constant speeds. Because this does not happen, Newton concluded that some force confines the planets to their elliptical orbits. As we shall see, that force is gravity.

Newton's second law describes quantitatively how a force changes the motion of an object. To better appreciate the concepts of force and motion, we must first understand two related quantities: velocity and acceleration.

Imagine an object motionless in space. Push on it and it begins to move. At any moment, you can describe the object's motion by specifying both its speed and direction. Speed and direction of motion together constitute an object's **velocity.** If you continue to push on the object, its speed will increase—it will accelerate.

Acceleration is the rate at which velocity changes with time. Because velocity involves both speed and direction, a slowing down, a speeding up, or a change in direction are all forms of acceleration.

Suppose an object revolved around the Sun in a perfectly circular orbit. As this object moved along its orbit, its speed would remain constant, but its direction of motion would be continuously changing. This body would have acceleration that involved only a change of direction.

Newton's second law—the force law: *The acceleration of an object is directly proportional to the net*

force acting on it and is inversely proportional to its mass.

In other words, the harder you push on something that can move, the faster it will accelerate. Also, an object of greater mass accelerates more slowly when acted on by a force than does an object of lesser mass acted on by the same force. That is why you can accelerate a child's wagon faster than you can accelerate a car by pushing on them equally hard.

Newton's second law can be succinctly stated as an equation. If a **force** acts on an object, the object will experience an acceleration such that

$$\text{Force} = \text{mass} \times \text{acceleration}$$

The **mass** of an object is a measure of the total number of particles that it contains and is expressed in units of kilograms. For example, the mass of the Sun is 2×10^{30} kg, the mass of a hydrogen atom is 1.7×10^{-27} kg, and the mass of the author of this book is 83 kg. At rest, the Sun, a hydrogen atom, and I have these same masses regardless of where we happen to be in the universe. It is important not to confuse the concept of mass with the concept of weight. Your **weight** is the force with which you push down on a scale due to the gravitational attraction of the world on which you stand.

6

Sitting in a moving car, how can you experimentally verify that your body has inertia?

Energy and Momentum

Scientists identify two types of energy that are available to any object. The first, called **kinetic energy**, is associated with the object's motion. For speeds much less than the speed of light, we can write the amount of kinetic energy, KE, of an object as

$$KE = \tfrac{1}{2}mv^2$$

where m is the object's mass and v is its speed. Kinetic energy is a measure of how much work the object can do on the outside world or, equivalently, how much work the outside world has done to give the object this speed.

Work is also a rigorously defined concept that often is at odds with our intuition. It is defined as the product of the force, F, acting on an object and the distance, d, over which the object moves in the direction of the force:

$$W = Fd$$

For example, if I exert a horizontal force of 50 N (N is the unit newtons and is the metric unit of force) and thereby move an object 10 m in that direction, then I have done 50 N × 10 m = 500 J of work. (I have used the relationship that 1 newton × 1 meter = 1 joule.)

The second type of energy is called **potential energy**. It represents how much energy is available to an object as a result of its location in space. For example, if you hold a pencil above the ground, the pencil has potential energy that can be converted into kinetic energy by Earth's gravitational force. How does that conversion get underway? Just let go of the pencil.

There are various kinds of potential energy, such as the potential energy stored in a battery and the potential energy stored in objects under the influence of gravity. We will focus on *gravitational potential energy*. Far from extremely massive objects, like stars, or extremely dense objects, like black holes, gravitational potential energy, PE, can be written as

$$PE = \frac{GmM}{r}$$

where the constant $G = 6.6683 \times 10^{-11}$ N m²/kg², m is the mass of the object whose gravitational potential energy you are measuring, M is the mass of the object generating the gravitational attraction, and r is the distance between the centers of mass of these two objects.

Near the surface of Earth, this equation simplifies to

$$PE = mgh$$

where $g = 9.8$ m/s² (32 ft/s²) is the gravitational acceleration at Earth's surface, and h is the height of the object above Earth's surface.

Potential energy can be converted into kinetic energy and vice versa. By dropping the pencil, its gravitational potential energy begins to decrease while its kinetic energy begins to increase at the *same rate*. The pencil's total energy is conserved. Conversely, if you throw a pencil up in the air, the kinetic energy you give it will immediately begin to decrease, while its potential energy increases at the same rate.

Related to the motion of an object, and hence to its kinetic energy, are the concepts of *linear momentum*, usually just called **momentum**, and **angular momentum**. Momentum, p, is described by the equation,

$$p = mv$$

where v is the velocity of the object. Both p and v are in boldface to indicate that they both represent motion in some direction or another, as well as a numeric value. Simple algebra reveals that kinetic energy and momentum are related by

$$KE = \frac{p^2}{2m}$$

Linear momentum, then, indicates how much energy is available to an object because of its motion in a straight line (linear motion).

Angular momentum, L, can be expressed mathematically as

$$L = I\omega$$

where I is the **moment of inertia** of an object, and ω (lowercase Greek omega) is the angular speed and direction of the rotating object. Just as an object's mass indicates how hard it is to change an object's straight-line motion, the moment of inertia indicates how hard it is to change the rate at which an object rotates or revolves. The moment of inertia depends on an object's mass and shape. Kinetic energy due to angular motion can be written as

$$KE = \frac{L^2}{2I}$$

Newton's first law can also be expressed in terms of **conservation of linear momentum:**

A body maintains its linear momentum unless acted upon by a net external force.

Equivalently, for angular motion we can write the **conservation of angular momentum:**

A body maintains its angular momentum unless acted upon by a net external torque.

Torques are created when a force acts on an object in some direction other than toward the center of the object's angular motion, as shown in the accompanying figure.

continued on next page

Energy and Momentum (continued)

Earth has angular momentum and keeps spinning on its rotational axis and orbiting the Sun. Likewise, the Moon has angular momentum and keeps spinning on its rotation axis and orbiting Earth. Virtually all objects in astronomy have angular momentum, and it is probably fair to say that conservation of angular momentum is among the most important laws in the cosmos. After all, conservation of angular momentum is what keeps the planets in orbit around the Sun, the moons in orbit around the planets, and astronomical bodies rotating at relatively constant rates, as well as causing many other rotation-related effects that we will encounter throughout this book.

Try these questions: How does tripling the linear momentum of an object change its kinetic energy? How does halving the angular momentum of an object change its kinetic energy? How much work would you do if you pushed on a desk with a force of 100 N and moved it 20 m? How much work would you do if you pushed on a desk with a force of 500 N and moved it 0 m? What two things can you vary to change the angular momentum of an object?

(Answers appear at the end of the book.)

Angular Momentum and Torque (a) When a force acts through an object's rotation axis or toward its center of mass, the force does not exert a torque on the object. (b) When a force acts in some other direction, then it exerts a torque, causing the body's angular momentum to change. If the object can spin around a fixed axis, like a globe, then the rotation axis is the rod running through it. If the object is not held in place, then the rotation axis is in a line through a point called object's *center of mass*. The center of mass of any object is the point that follows a smooth, elliptical path as the object moves in response to a gravitational field. All other points in the spinning object wobble as it moves.

Force is usually expressed in pounds or newtons. For example, the force with which I am pressing down on the ground is 183 lb. But I weigh 183 lb only on Earth. I would weigh 30.5 lb on the Moon, which has less mass and so pulls me down with less gravitational force. Orbiting in the Space Shuttle, my apparent weight (measured by standing on a scale in the shuttle), would be zero, but my mass would be the same as when I am on Earth. Because I still have inertia in the shuttle, an astronaut there would have to push me with a force to get me to float across the cabin. Whenever we describe the properties of planets, stars, or galaxies, we speak of their masses, never of their weights.

Newton's final assumption, called Newton's third law, is the law of action and reaction.

Newton's third law—the law of action and reaction: *Whenever one object exerts a force on a second object, the second object exerts an equal and opposite force on the first object.*

For example, I weigh 183 lb on Earth, and so I press down on the floor with a force of 183 lb. Newton's third law says that the floor is also pushing up against me with an equal force of 183 lb. (If it were less, I would fall through the floor, and if it were more, I would be lifted upward.) In the same way, Newton realized that

because the Sun is exerting a force on each planet to keep it in orbit, each planet must also be exerting an equal and opposite force on the Sun. As each planet accelerates toward the Sun, the Sun in turn accelerates toward each planet.

Because the Sun is pulling on the planets, why don't they fall onto it? Conservation of angular momentum provides the answer. Angular momentum is a measure of how much energy is stored in an object due to its rotation and revolution. The details are presented in *An Astronomer's Toolbox 2-2*. As the orbiting planets fall toward the Sun, their angular momentum provides them with motion perpendicular to that infall, meaning that the planets continually fall toward the Sun, but they continually miss it. Because their angular momentum is conserved, planets neither spiral into the Sun nor fly away from it. Angular momentum remains constant unless acted on by an external torque (also defined in *An Astronomer's Toolbox 2-2*).

It seems plausible that astronauts in the International Space Station don't feel any force of gravity from Earth, but they do. Orbiting 330 km (approximately 200 mi) above Earth's surface, they feel 90% as much gravitational force from the planet as we do standing on it. They are weightless, however, because as they fall

7

If you are on a freely spinning merry-go-round, what will happen to it as you move toward the center?

a   b   R I V U X G

FIGURE 2-12 Conservation of Angular Momentum As this skater brings her arms and outstretched leg in, she must spin faster to conserve her angular momentum. (Getty Images)

earthward, their angular momentum carries them around the planet at just the right rate to continually miss it.

Angular momentum depends on three things: how fast an object rotates or revolves, how much mass it has, and how spread out that mass is. The greater an object's angular motion or mass, or the more the mass is spread out, the greater its angular momentum. Consider, for example, a twirling ice skater. She rotates with a constant mass, practically free of outside forces. When she wishes to rotate more rapidly, she decreases the spread of her mass distribution by pulling her arms and outstretched leg in closer to her body (Figure 2-12). According to conservation of angular momentum, as the spread of mass decreases, the rotation rate must increase. In astronomy, we encounter many instances of the same law, as giant objects, such as stars, contract.

We have now reconstructed the central relationships between matter and motion. Scientific explanation of the heliocentric cosmology still requires a force to hold the planets in orbit around the Sun and the moons in orbit around the planets. Newton identified that, too.

2-8 Newton's description of gravity accounts for Kepler's laws

Isaac Newton did not invent the idea of gravity. An observant seventeenth-century person would understand that some force pulls things down to the ground.

It was Newton, however, who gave us a quantitative description of the action of gravity, or *gravitation*, as it is more properly called. Using his first two laws, Newton proved mathematically that the force acting on each of the planets is directed toward the Sun. He expanded this result to the idea that the nature of the force pulling a falling apple straight down to the ground is the same as the nature of the force on the planets from the Sun. More generally, the gravitational force from every object acts to pull every other object directly toward it.

Newton succeeded in formulating a mathematical model that describes the behavior of the gravitational force that keeps the planets in their orbits (presented in *An Astronomer's Toolbox 2-3*).

Newton's law of universal gravitation: *Two objects attract each other with a force that is directly proportional to the product of their masses and inversely proportional to the square of the distance between them.*

In other words, gravitational force decreases with distance: Move twice as far away from an object and you feel only one-quarter of the force from it that you felt before. Despite its weakening, the force of gravity from each object extends throughout the universe. Also, an object with twice the mass of another object exerts twice the gravitational force as the less massive object.

Using his law of gravity along with his three laws stated earlier, Newton found that he could mathematically explain Kepler's three laws. For example, whereas Kepler discovered by trial and error that the period of orbit, P, and average distance between the Sun and planet, a, are related by $P^2 = a^3$, Newton mathematically derived this equation (corrected by including a tiny contribution due to the mass of the planet, as mentioned earlier). Bodies in elliptical orbits are bound by the force of gravity to remain in orbit.

Newton also discovered that objects can have nonelliptical, unbound orbits. His equations led him to conclude that orbits can also be **parabolas** or **hyperbolas** (Figure 2-13). In both cases, such bodies would make only one pass close to the Sun and then travel out of the solar system, never to return. To date, all of the objects observed in the solar system have begun their existence with elliptical orbit, but some comets (small bodies of rock and ice) have received enough energy from the pull of planets or from jets of gas shooting out of them to develop parabolic or hyperbolic orbits.

Using the equations Newton derived, the orbits of the planets and their satellites could be calculated with unprecedented precision. Using his laws, mathematicians proved that Earth's axis of rotation must precess because of the gravitational pull of the Moon and the

Gravitational Force

From Newton's law of gravitation, if two objects that have masses, m_1 and m_2, are separated by a distance, r, then the gravitational force, F, between them is

$$F = \frac{Gm_1m_2}{r^2}$$

In this formula, G is the **universal constant of gravitation**, whose value has been determined from laboratory experiments:

$$G = 6.668 \times 10^{-11} \text{ N m}^2 \text{ kg}^{-2}$$

where N is the unit of force, a newton.

The equation $F = G(m_1m_2/r^2)$ gives, for example, the force from the Sun on Earth and, equivalently, from Earth on the Sun. If m_1 is the mass of Earth (6.0×10^{24} kg), m_2 is the mass of the Sun (2.0×10^{30} kg), and r is the distance from the center of Earth to the center of the Sun (1.5×10^{11} m):

$$F = 3.6 \times 10^{22} \text{ N}$$

This number can then be used in Newton's second law, $F = ma$, to find the acceleration of Earth due to the Sun. This yields

$$a_{\text{Earth}} = F/m_1 = 6.0 \times 10^{-3} \text{ m/s}^2$$

Newton's third law says that Earth exerts the same force on the Sun, so the Sun's acceleration due to Earth's gravitational force is

$$a_{\text{Sun}} = F/m_2 = 1.8 \times 10^{-8} \text{ m/s}^2$$

In other words, Earth pulls on the Sun, causing the Sun to move toward it. Because of the Sun's greater mass, however, the amount that the Sun accelerates Earth is more than 300,000 times greater than the amount that Earth accelerates the Sun.

Try these questions: Earth's radius is 6.4×10^6 m and 1 kg of mass is equivalent to 2.2 lb of weight on Earth. What is the force that Earth exerts on you? What is the force that you exert on Earth? What is Earth's acceleration on you? What would the Sun's force be on Earth if our planet were twice as far from the Sun as it is? How does that force compare to the force from the Sun at our present location?

(Answers appear at the end of the book.)

Sun on Earth's equatorial bulge (recall Figure 1-20). In the spirit of the scientific method, Newton's laws and mathematical techniques were used to predict new phenomena. For example, Edmond Halley was intrigued by historical records of a comet that was sighted about every 76 years. Using his friend Newton's methods, Halley worked out the details of the comet's orbit and predicted its return in 1758. It was first sighted on Christmas night of that year, and to this day the comet bears Halley's name (Figure 2-14).

Perhaps the most dramatic early use of the scientific method with Newton's ideas was their role in the discovery of the eighth planet in our solar system. The seventh planet, Uranus, had been discovered by William Herschel in 1781 during a systematic telescopic survey of the sky. Fifty years later, however, it was clear that Uranus was not following the orbit predicted by Newton's laws. Two mathematicians, John Couch Adams in England and Urbain-Jean-Joseph Leverrier in France, independently calculated that the deviations of Uranus from its predicted orbit could be explained by the gravitational pull of a then-unknown, more distant planet. Each man predicted that the planet would be found at a certain location in the constellation of Aquarius in

FIGURE 2-13 Conic Sections A conic section is any one of a family of curves obtained by slicing a cone with a plane, as shown. The orbit of one body around another can be an ellipse, a parabola, or a hyperbola. Circular orbits are possible because a circle is just an ellipse for which both foci are at the same point.

R I V U X G

FIGURE 2-14 Halley's Comet Halley's Comet orbits the Sun with an average period of about 76 years. During the twentieth century, the comet passed near the Sun twice—once in 1910 and again, as shown here, in 1986. The comet will pass close to the Sun again in 2061. Although dim in 1986, it nevertheless spread more than 5° across the sky, or 10 times the diameter of the Moon. (Harvard College Observatory/Photo Researchers, Inc.)

September 1846. A telescopic search on September 23, 1846, revealed Neptune less than 1° from its calculated position. Although sighted with a telescope, Neptune was really discovered with pencil and paper.

INSIGHT INTO SCIENCE

Quantify Predictions Mathematics provides a language that enables science to make quantitative predictions that can be checked by anyone. For example, in this chapter, we have seen how Kepler's third law and Newton's universal law of gravitation correctly predict the motion of objects under the influence of the Sun's gravitational attraction.

It is a testament to Newton's genius that his three laws were precisely the basic ideas needed to understand so much about the natural world. Newton's process of deriving Kepler's laws and the universal law of gravitation helped secure the scientific method as an invaluable tool in our process of understanding the universe. Figure 2-15 shows some of the effects of gravity at the scales of planets, stars, and galaxies.

2-9   Frontiers yet to be discovered

The science related to forces and orbits described in this chapter was well established by the beginning of the nineteenth century. However, questions remain. Why are the inertial mass (defined in Newton's second law) and the mass used in his law of gravitation identical? Experiments are underway at high-energy particle accelerators such as CERN, near Geneva, Switzerland, and the Fermi National Accelerator near Chicago, Illinois, to explain this. In another vein, careful observations have revealed that Newton's law of gravitation gives very slightly inaccurate predictions for the orbital path of Mercury. We will explore this further in Chapter 9.

SUMMARY OF KEY IDEAS

Science: Key to Comprehending the Cosmos

• The ancient Greeks laid the groundwork for progress in science by stating that the universe is comprehensible.

• The scientific method is a procedure for formulating theories that correctly predict how the universe behaves.

• A scientific theory must be testable, that is, capable of being disproved.

• Theories are tested and verified by observation or experimentation and result in a process that often leads to their refinement or replacement and to the progress of science.

• Observations of the cosmos have led astronomers to discover some fundamental physical laws of the universe.

Origins of a Sun-centered Universe

• Early Greek astronomers devised a geocentric cosmology, which placed Earth at the center of the universe.

• Copernicus's heliocentric (Sun-centered) theory simplified the general explanation of planetary motions compared to the geocentric theory.

• The heliocentric cosmology refers to motion of planets and smaller debris orbiting the Sun. Other stars do not orbit the Sun.

• The sidereal orbital period of a planet is measured with respect to the stars. It determines the length of the planet's year. Its synodic period is measured with respect to the Sun as seen from the moving Earth (for example, from one opposition to the next).

Kepler's and Newton's Laws

• Ellipses describe the paths of the planets around the Sun much more accurately than do the circles used in previous theories. Kepler's three laws give important details about elliptical orbits.

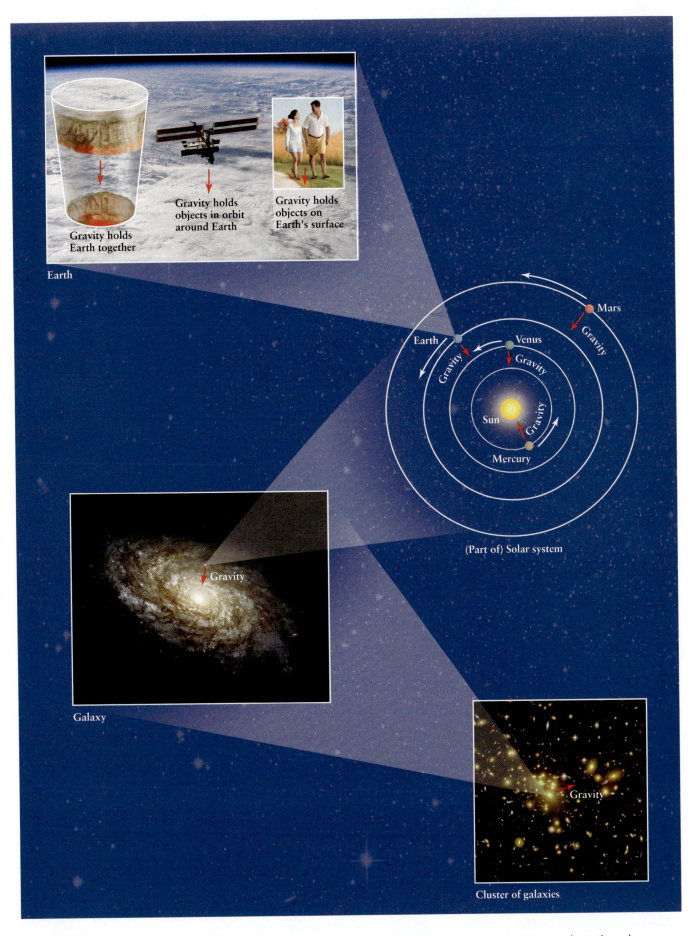

Gravity holds Earth together

Gravity holds objects in orbit around Earth

Gravity holds objects on Earth's surface

Earth

Mars

Gravity

Earth

Gravity

Venus

Gravity

Sun

Gravity

Mercury

(Part of) Solar system

Gravity

Galaxy

Gravity

Cluster of galaxies

FIGURE 2-15 Gravity Works at All Scales This figure shows a few of the effects of gravity here on Earth, in the solar system, in our Milky Way Galaxy, and beyond. Top: Space station (NASA); Couple holding hands (Paul Burns/Digital Vision/Getty Images); Center: Black hole (NASA); Bottom: Galaxy cluster (ESA, NASA, J.-P. Kneib [Caltech/Observatoire Midi-Pyrénées] and R. Ellis [Caltech])

• The invention of the telescope led Galileo to new discoveries, such as the phases of Venus and the moons of Jupiter, that supported a heliocentric view of the universe.

• Newton based his explanation of the universe on three assumptions, now called Newton's laws of motion. These laws and his law of universal gravitation can be used to deduce Kepler's laws and to describe most planetary motions with extreme accuracy.

• The mass of an object is a measure of the amount of matter in it; weight is a measure of the force with which the gravity of a world pulls on an object's mass when the two objects are at rest with respect to each other (or, equivalently, how much the object pushes down on a scale).

• The path of one astronomical object around another, such as that of a comet around the Sun, is an ellipse, a parabola, or a hyperbola. Ellipses are bound orbits, while objects with parabolic and hyperbolic orbits fly away, never to return.

WHAT DID YOU THINK?

1 *What makes a theory scientific?* A theory is an idea or set of ideas proposed to explain something about the natural world. A theory is scientific if it makes predictions that can be objectively tested and potentially disproved.

2 *What is the shape of Earth's orbit around the Sun?* All planets have elliptical orbits around the Sun.

3 *Do the planets orbit the Sun at constant speeds?* No. The closer a planet is to the Sun in its elliptical orbit, the faster it is moving. The planet moves fastest at perihelion and slowest at aphelion.

4 *Do all of the planets orbit the Sun at the same speed?* No. A planet's speed depends on its average distance from the Sun. The closest planet moves fastest, the most distant planet moves slowest.

5 *How much force does it take to keep an object moving in a straight line at a constant speed?* Unless an object is subject to an outside force, like friction, it takes no force at all to keep it moving in a straight line at a constant speed.

6 *How does an object's mass differ when measured on Earth and on the Moon?* Assuming the object doesn't shed or collect pieces, its mass remains constant whether on Earth or on the Moon. Its weight, however, is less on the Moon.

7 *Do astronauts orbiting the Earth feel the force of gravity from our planet?* Yes. They are continually pulled earthward by gravity, but they continually miss it because of their motion around it. Because they are continually in free-fall, they feel weightless.

Key Terms for Review

acceleration, 55	light-year, 53
angular momentum, 56	mass, 55
aphelion, 50	model, 41
astronomical unit, 51	moment of inertia, 56
configuration (of a	momentum, 56
planet), 46	Newton's laws of
conjunction, 46	motion, 54
conservation of angular	Occam's razor, 42
momentum, 56	opposition, 46
conservation of linear	parabola, 58
momentum, 56	parallax, 49
cosmology, 43	parsec, 53
direct motion, 43	perihelion, 50
ellipse, 50	potential energy, 56
elongation, 46	retrograde motion, 43
focus (of an ellipse), 50	scientific method, 41
force, 55	scientific theory, 41
Galilean moons	semimajor axis (of an
(satellites), 52	ellipse), 50
gravity, 54	sidereal period, 46
heliocentric cosmology, 44	superior conjunction, 46
hyperbola, 58	synodic period, 47
inferior conjunction, 46	theory, 41
Kepler's laws, 50	universal constant of
kinetic energy, 56	gravitation, 59
law of equal areas, 51	velocity, 55
law of inertia, 54	weight, 55
law of universal	work, 56
gravitation, 58	

Review Questions

1. Who wrote down the equation for the law of gravitation? **a.** Copernicus **b.** Brahe **c.** Newton **d.** Galileo **e.** Kepler

2. Which of the following most accurately describes the shape of Earth's orbit around the Sun? **a.** circle **b.** ellipse **c.** parabola **d.** hyperbola **e.** square

3. Of the following planets, which takes the longest time to orbit the Sun? **a.** Earth **b.** Uranus **c.** Mercury **d.** Jupiter **e.** Venus

4. What is a Sun-centered model of the solar system called?

5. How long does it take Earth to complete a sidereal orbit of the Sun?

6. How did Copernicus explain the retrograde motions of the planets?

7. Which planets can never be seen at opposition? Which planets never pass through inferior conjunction?

8. At what configuration (superior conjunction, greatest eastern elongation, etc.) would it be best to observe Mercury or Venus with an Earth-based telescope? At what configuration would it be best to observe Mars, Jupiter, or Saturn? Explain your answers.

9. What are the synodic and sidereal periods of a planet?

10. What are Kepler's three laws? Why are they important?

11. In what ways did the astronomical observations of Galileo support a heliocentric cosmology?

12. How did Newton's approach to understanding planetary motions differ from that of his predecessors?

13. What is the difference between mass and weight?

14. Why was the discovery of Neptune a major confirmation of Newton's universal law of gravitation?

15. Why does an astronaut have to exert a force on a weightless object to move it?

Advanced Questions

The answers to all computational problems, which are preceded by an asterisk (*), appear at the end of the book.

16. From the definition $KE = \frac{1}{2} mv^2$, derive the equation $KE = p^2/2m$, as discussed in *An Astronomer's Toolbox 2-2*.

*17. Convert: **a.** 8.3 pc (parsec) into light-years **b.** 6.52 ly into parsecs **c.** 8450 AU into kilometers **d.** 2.7×10^3 Mpc into kiloparsecs.

18. Is it possible for an object in the solar system to have a synodic period of exactly 1 year? Explain your answer.

19. Describe why there is a systematic decrease in the synodic periods of the planets from Mars outward, as shown in Table 2-1.

*20. A line joining the Sun and an asteroid was found to sweep out 5.2 square astronomical units of space in all of 2006. How much area was swept out in 2007? in the 5 years from 2002 to 2007?

*21. A comet moves in a highly elongated orbit around the Sun, with a period of 1,000 years. What is the length of the semimajor axis of the comet's orbit? What is the farthest that the comet can get from the Sun?

*22. The orbit of a spacecraft around the Sun has a perihelion distance of 0.5 AU and an aphelion distance of 3.5 AU. What is the spacecraft's orbital period?

23. Make diagrams of Jupiter's phases as seen from Earth and as seen from Saturn.

24. In what direction (left or right, eastward or westward) across the celestial sphere do the planets normally appear to move as seen from Australia? In what direction is retrograde motion as seen from there?

25. The dictionary defines *astrology* as "the study that assumes and attempts to interpret the influence of the heavenly bodies on human affairs." Based on what you know about scientific theory, is astrology a science? Why or why not? Feel free to further explore astrology, if you wish, before answering this question.

Discussion Questions

26. Which planet would you expect to exhibit the greatest variation in apparent brightness as seen from Earth? Explain your answer.

27. Use two thumbtacks (or pieces of tape), a loop of string, and a pencil to draw several ellipses. Describe how the shapes of the ellipses vary as you change the distance between the thumbtacks.

What if . . .

*28. Earth were 2 AU from the Sun? What would be the length of the year? Assuming that such physical properties as rotation rate were as they are today, what else would be different here?

*29. Earth were ½ AU from the Sun? What would be the length of the year? Assuming that such physical properties as rotation rate were as they are today, what else would be different here?

*30. Earth were 10 AU from the Sun? How much stronger or weaker would the Sun's gravitational pull be than it is on Earth today?

*31. Earth had twice its present mass? Assume that all other properties of Earth and its orbit remain the same. What would be the acceleration of the more massive Earth due to the Sun compared to the present acceleration of Earth from the Sun? *Hint:* Try combining $F = m_1a$ and the force equation in *An Astronomer's Toolbox 2-3*, where m_1 is the mass of Earth in both equations. Given that acceleration determines the period of the planet's orbit, how would the year on the more massive Earth compare to a year today?

32. The Sun suddenly disappeared? What would Earth's path in space be in response to such an event? Describe how Earth would change, as a result, and how humans might survive on a sunless planet.

33. The skies of Earth were perpetually cloudy? How might that have changed the history of our

understanding of the cosmos, and how might humans under such conditions eventually learn what is really "out there"?

34. Scientists remained believers in the first theory of the cosmos that they decided was correct? How might that change the dynamics by which science evolves in the face of new data that conflict with earlier theories?

Web Questions

35. Search the Web for information about Galileo. What were his contributions to physics? Which of Galileo's new ideas were later used by Newton to construct his laws of motion? What incorrect beliefs about astronomy did Galileo hold?

36. Search the Web for information about Kepler. Before he realized that the planets move on elliptical paths, what other models of planetary motion did he consider? What was Kepler's idea of the "music of the spheres"?

37. Search the Web for information about Newton. What were some of the contributions that he made to physics other than developing his laws of motion? What contributions did he make to mathematics?

Observing Projects

 38. It is quite probable that, within a few weeks of your reading of this chapter, one of the planets visible to the unaided eye (all the planets out as far as Saturn) will be in opposition or at greatest eastern elongation, making it visible in the evening sky from your location. Using the *Starry Night Enthusiast*™ computer program, the Internet, or a reference book, such as the current issue of the *Observer's Handbook* of the Royal Astronomical Society of Canada, the *Astronomical Almanac*, or the pamphlet *Astronomical Phenomena* (both of the latter published by the U.S. Naval Observatory), select a planet that is at or near such a configuration. To use *Starry Night Enthusiast*™, launch the program and click the *Sunset* button in the toolbar. Open the *Options* side pane and expand the *Solar System* layer. Click on both boxes in the *Planets-Moons* row to show and label the planets in the view and click on left-hand boxes for all other objects to remove them from view. Use the *Hand Tool* to search the sky for planets as far away as Saturn. If none are in your sky, change the date by a few days and search again. Repeat until you have found one or two visible planets up at night. Plan to make your observations at that time. Make a sketch to show the positions of planets with respect to nearby stars, noting the time and date of this observation. If possible, confirm that your observation is a planet by observing a few days later and showing that the object has moved. (This step is vital in confirming discoveries of new bodies in the solar system.) If a small

telescope or spotting scope is available, make a drawing of what you see through it, being careful to include planet phases, moons, surface features, and nearby stars.

39. If Jupiter's moons are visible in the evening sky, observe them with a small telescope on five consecutive clear nights. Record their positions by making nightly drawings, just as the Jesuit priests did in 1620 (see Figure 2-11a). From your drawings, can you tell which moon orbits closest to Jupiter and which orbits farthest? Are there nights when you saw fewer than four of the Galilean moons? What happened to the other moons on those nights?

 40. Use the *Starry Night Enthusiast*™ software to observe the moons of Jupiter. Select *Favourites > Guides > Atlas* from the menu. Open the **Find** pane and double click the entry for Jupiter to center the planet in the view. Close the **Find** pane and use the **Zoom** buttons in the toolbar to set the field of view to about 15 arcmin (15′) wide. **a.** Note the positions of the moons. Using the single step button ▶, step time forward in increments of six hours and draw the positions of the moons for ten consecutive timesteps. **b.** From your drawings, can you tell which moon orbits closest to Jupiter and which orbits farthest away? Explain your reasoning. **c.** Determine the periods of orbits of these moons (change the length of the timestep, if necessary). **d.** Are there times when you see fewer than four Galilean moons? What has happened to the other moons at those times?

41. If Venus is visible in the evening sky, observe the planet with a telescope once a week for a month. On each night, make a drawing of the phase that you see. Can you determine from your drawings if the planet is nearer or farther from Earth than the Sun is? Do your drawings show any changes in the phase from 1 week to the next? If so, can you deduce whether Venus is coming toward us or moving away from us? Explain.

 42. Use your *Starry Night Enthusiast*™ software to observe the phases of Venus. Select *Favourites > Guides > Atlas* from the menu. Open the **Find** pane and double-click on the entry for *Venus*. Close the **Find** pane and use the **Zoom** buttons in the toolbar to adjust the field of view to about 7 arcmin (7′) wide. **a.** Draw the current shape (phase) of Venus. Set the **Time Flow Rate** to *30 days* and single step forward in time ▶, drawing Venus to scale at each step. Try to be accurate in both size and shape. Make a total of 20 timesteps and drawings. **b.** From your drawings, determine when the planet is nearer or farther from the Earth than is the Sun. **c.** Deduce from your drawings when Venus is coming toward us or is moving away from us. **d.** Explain why Venus goes through this particular cycle of phases.

43. Perform observing project 43 using Mars instead of Venus with a field of view about 1 arcmin (1′) wide and at least 30 timesteps. If you have done project 43, compare your results for the two planets. Why are the cycles of phases as seen from Earth different for the two planets?

 44. Orbital Eccentricities of the Planets. In this exercise, you will observe which planets have the greatest orbital eccentricities. Using *Starry Night Enthusiast™*, select *Favourites > Solar System > Inner Planets > Inner Solar System* from the menu. Stop the advance of time. You will see the orbits of the inner four planets. Use the Location Scroller (hold down the **Shift** key while holding the mouse button [left button on PC] and moving the mouse) to adjust the view so that the orbits are seen face on, in the plane of the screen. Zoom in to about 50°. Using the fact that Venus's orbital eccentricity is 0.007 (very nearly circular), note which two of the inner four planets have the least circular orbits. Now select *Favourites > Solar System > Outer Planets > Outer Solar System* from the menu and stop time advance to view the orbits of the outer planets. Again, adjust the orbits to be face-on in the plane of the screen and zoom to about 70°. How do the orbits of Jupiter, Saturn, Uranus, and Neptune compare in ellipticity to those of the four inner planets? Are the orbits of the outer planets more like the two nearly-circular inner planets or the two less-circular ones? Run time forward and list the order of orbital periods of the outer planets.

CHAPTER 3

Light and Telescopes

Three of the four buildings containing the Very Large Telescope optical telescopes at the Paranal Observatory, Atacama, Chile. (ESO)

R I V U X G

WHAT DO YOU THINK?

1 What is light?

2 Which type of electromagnetic radiation is most dangerous to life?

3 What is the main purpose of a telescope?

4 Why do all research telescopes use mirrors, rather than lenses, to collect light?

5 Why do stars twinkle?

Answers to these questions appear in the text beside the corresponding question numbers in the margins and at the end of the chapter.

With our eyes alone we can see visible light from several thousand stars. Until the seventeenth century, few people even dreamed that there were more of them. It was then that telescopes revolutionized human understanding of the universe, showing for the first time how little of the cosmos we normally see. The process of discovery continues, as telescopes reveal new things in space nearly every day. We can think of the radiation emitted by objects out there as the medium of natural cosmic communication and telescopes as the means by which we gather and read those cosmic messages.

We have also discovered that visible light is only a tiny fraction of the energy emitted by objects in space. Indeed, such things as interstellar clouds of gas and dust; the bodies lying behind these clouds; newly forming stars; intergalactic gas clouds; and a variety of exotic objects, such as black holes and neutron stars are nearly invisible to even our best optical telescopes. However, these things strongly emit a variety of nonvisible radiations (namely radio waves, infrared and ultraviolet radiations, X rays, and gamma rays) that we now have the technology to detect. Today, we use telescopes on the ground, floating in the air, orbiting Earth, or traveling elsewhere in the solar system to see these myriad "stealth" objects in space.

In this chapter you will discover

- the connection between visible light, radio waves, and other types of electromagnetic radiation
- the debate over what light is and how Einstein resolved it
- how telescopes collect and focus light
- why different types of telescopes are used for different types of research
- the limitations of telescopes, especially those that use lenses to collect light
- what the new generations of land-based and space-based high-technology telescopes being developed can do
- how astronomers use the entire spectrum of electromagnetic radiation to observe the stars and other astronomical objects and events

THE NATURE OF LIGHT

So far in this text we have used the word *light* in its everyday sense—the stuff to which our eyes are sensitive. This is more properly called *visible light,* and it is a form of **electromagnetic radiation**, which has properties of both particles, called **photons**, and waves. Detecting electromagnetic radiation using telescopes is the essence of observational astronomy. Although human perception of objects here on Earth and in space comes primarily from the visible light that our eyes detect, this is only a tiny fraction of all the electromagnetic radiation emitted by objects in the universe. The rest of this radiation is invisible to our eyes, so it has to be observed through specially designed telescopes. In this chapter, we examine telescopes for all types of electromagnetic radiation. To understand how telescopes work, we begin by exploring the properties of the electromagnetic radiation that they collect.

3-1 Newton discovered that white light is not a fundamental color and proposed that light is composed of particles

From the time of Aristotle until the late seventeenth century, most people believed that white is the fundamental color of light. The colors of the rainbow (or, equivalently, the colors created by light passing through a prism) were believed to be added or created somehow as white light went from one medium through another. Isaac Newton performed experiments during the late 1600s that disproved these beliefs. He started by passing a beam of sunlight through a glass prism, which spread the light out into the colors of the rainbow (Figure 3-1a). The change in direction as light travels from one medium into another is called **refraction,** and the resulting spread of colors (complete or with colors missing) is called a **spectrum** (plural, **spectra**).

Then Newton selected a single color and sent it through a second prism (Figure 3-1b). The light that emerged from the first prism was refracted by the second prism, but *it remained the same color.* The fact that individual colors of refracted light were unchanged by the second prism led Newton to conclude that the colors of a full spectrum (shades of red, orange, yellow, green, blue, and violet) were in fact properties of the light itself, and that white light is a mixture of colors. To prove this last point, he recombined all the spectrum colors, thereby re-creating white light. Thus, different colors of the spectrum are different entities. But, Newton wondered, what was the nature of light that it could break apart into distinct colors and then be completely reconstituted?

Back in the mid-1600s, the Dutch scientist Christiaan Huygens proposed that light travels in the form of waves. Newton, on the other hand, performed

White light

Refraction

A prism breaks light into its
component colors, or spectrum

Wavelength of
different colors
of light

700 nm

600 nm

500 nm

400 nm

The spectrum falls on a screen

a

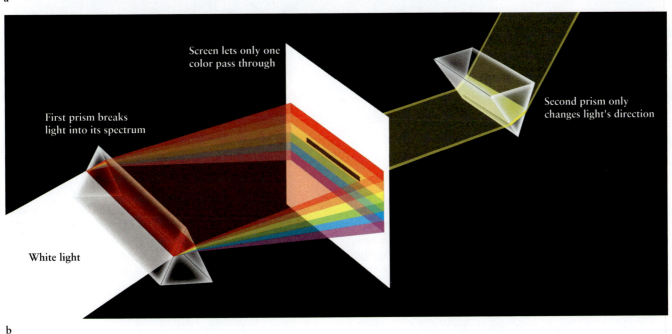

Screen lets only one
color pass through

First prism breaks
light into its spectrum

Second prism only
changes light's direction

White light

b

FIGURE 3-1 Prisms and a Spectrum (a) When a beam of white light passes through a glass prism, the light is separated or refracted into a rainbow-colored band called a spectrum. The numbers on the right side of the spectrum indicate wavelengths in nanometers (1 nm $= 10^{-9}$ m). (b) Drawing of Newton's experiment showing that glass does not add to the color of light, only changes its direction. Because color is not added, this experiment shows that color is an intrinsic property of light.

many experiments in optics that convinced him that light is composed of tiny particles of energy. It turns out that both ideas were right.

In 1801, the English physicist Thomas Young demonstrated that light is indeed composed of waves. Young sent light of a single color through two parallel slits (Figure 3-2a). He reasoned that if the light were waves, then these waves would behave the way waves on the surface of water behave, flowing through similar gaps (Figure 3-2b). In particular, the light waves from each slit would interact with light waves from the other. For example, when two light waves meet, where one wave is going up and the other is going down, they would interfere with each other and partially or totally cancel each other out, leaving dark regions on the screen. When two waves that are both moving up or both moving down meet, they would reinforce each other and create bright regions. As you can see in Figure 3-2, this is precisely what happened. (If light were just random particles going through the slits, they would not interfere with each other in this way, and the pattern on the screen would just yield two bright

This region has waves from 2 slits interfering with each other.

These 5 bumps show liquid making an interference pattern on the water.

This region has waves running parallel to each other.

Direction of water flow

Two parallel slits

Screen

From light source

Interference pattern

a

b

FIGURE 3-2 Wave Travel (a) Electromagnetic radiation travels as waves. Thomas Young's interference experiment shows that light of a single color passing through a barrier with two slits behaves as waves that create alternating light and dark patterns on a screen. (b) Water waves passing through two slits in a ripple tank create interference patterns. As with light, the water waves interfere with each other, creating constructive interference (crests) and destructive interference (troughs) throughout the right side of the tank and on the far right wall. (University of Colorado, Center for Integrated Plasma Studies, Boulder, CO)

regions, one behind each slit.) The analogy with water waves ends here. For example, light waves do not have to travel through a medium, unlike water waves, which travel on water.

Further insight into the wave character of light came from calculations by the Scottish physicist James Clerk Maxwell in the 1860s. Maxwell unified the descriptions of the basic properties of electricity and magnetism into four equations. By combining these equations, he demonstrated that electric and magnetic effects should travel through space together in the form of coupled waves (Figure 3-3) that have equal amplitudes. Maxwell's suggestion that some of these waves, now called electromagnetic radiation, are observed as light was soon confirmed by a variety of experiments. Despite the name electromagnetic radiation, visible light is not electrically charged.

Newton showed that sunlight is composed of all the colors of the rainbow. Young, Maxwell, and others showed that light travels as waves. What makes the colors of the rainbow distinct from each other? The answer is surprisingly simple: Different colors are waves with different wavelengths. A **wavelength,** usually designated by λ, the lowercase Greek letter lambda, is the distance between two successive wave crests.

The wavelengths of all colors are extremely small, less than a thousandth of a millimeter. To express these tiny distances conveniently, scientists use a unit of length called the nanometer (nm), where 1 nm = 10^{-9} m. Another unit you might encounter when talking to astronomers is the angstrom, Å, where 1 Å = 0.1 nm = 10^{-10} m. Experiments demonstrate that visible light has wavelengths ranging from about 400 nm for the shortest wavelength of violet light to about 700 nm for the

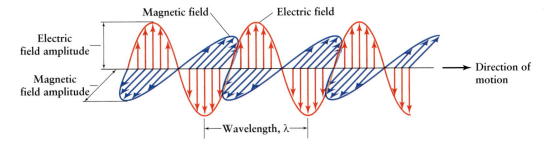

FIGURE 3-3 Electromagnetic Radiation All forms of electromagnetic radiation (radio waves, infrared radiation, visible light, ultraviolet radiation, X rays, and gamma rays) consist of electric and magnetic fields oscillating perpendicular to each other and to the direction they move. In empty space this radiation travels at a speed of 3×10^5 km/s. These fields are the mathematical description of the electric and magnetic effects. The distance between two successive crests, denoted by λ, is called the wavelength of the light.

longest wavelength of red light. Intermediate colors of the rainbow fall between these wavelengths (Figure 3-1a). The complete spectrum of colors from the longest wavelength to the shortest is red, orange, yellow, green, blue, and violet. Referring back to Figure 3-1, you can also see that the amount of refraction that different colors undergo depends on their wavelengths: The shorter the wavelength, the more the light is refracted.

3-2 Light travels at a finite, but incredibly fast, speed

The fact that we see lightning before we hear the accompanying thunderclap tells us that light travels faster than sound. But does that mean that light travels instantaneously from one place to another, or does it move with a measurable speed?

The first evidence for the finite speed of light came in 1675, when Ole Rømer, a Danish astronomer, carefully timed eclipses of Jupiter's moons (Figure 3-4). Rømer discovered that the moment at which a moon enters Jupiter's shadow depends on the distance between Earth and Jupiter. When Jupiter is in opposition—that is, when Jupiter and Earth are on the same side of the Sun—the Earth-Jupiter distance is relatively short compared to when Jupiter is near conjunction. At opposition, Rømer found that eclipses occur slightly earlier than predicted by Kepler's laws, and they occur slightly later than predicted when Jupiter is near conjunction.

Rømer correctly concluded that light travels at a finite speed, and so it takes more time to travel longer distances across space. The greater the distance to Jupiter, the longer the image of an eclipse takes to reach our eyes. From his timing measurements, Rømer concluded that it takes 162 min for visible light to traverse the diameter of Earth's orbit (2 AU). Incidentally, Rømer's interpretation of the data requires a heliocentric cosmology—that both Earth and Jupiter orbit the Sun.

Rømer's subsequent calculation of the speed of light was off by 25% because the value for the astronomical unit (the average distance from Earth to the Sun) that existed at that time was highly inaccurate. Nevertheless, he proved his main point—light travels at a finite speed. The first accurate laboratory measurements of the speed of visible light were performed in the mid-1800s.

> What color is refracted least?

Maxwell's equations also reveal that light of all wavelengths travels at the same speed in a vacuum (a region that contains no matter), and, despite a few atoms per cubic meter, the space between planets and stars is a very good vacuum. The constant speed of light in a vacuum, usually designated by the letter c, has been measured to be 299,792.458 km/s, which we generally round to

$$c = 3.0 \times 10^5 \text{ km/s} = 1.86 \times 10^5 \text{ mi/s}$$

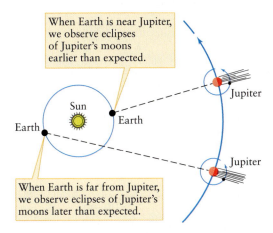

FIGURE 3-4 Evidence that Light Travels at a Finite Speed The times of the eclipses of Jupiter's moons as seen from Earth depend on the relative positions of Jupiter, Earth, and the Sun. Rømer correctly attributed the variations in these times to the variations in the time that it takes light from these events to reach Earth.

(Standard abbreviations for units of speed, such as *km/s* for kilometers per second and *mi/s* for miles per second, will be used throughout the rest of this book.) Light that travels through air, water, glass, or any other substance always moves more slowly than it does in a vacuum.

The value *c* is a fundamental property of the universe. The speed of light appears in equations that describe, among other things, atoms, gravity, electricity, magnetism, distance, and time. It has extraordinary universal properties. For example, if you were traveling in space at 99% the speed of light, 0.99*c*, you would still measure the speed of any light beam moving toward you as *c*, which is also the speed you would measure for any light beam moving away from you!

> The speed of sound is about 0.34 km/s (0.21 mi/s). How can this value and the information in this section be used to determine a person's distance from a lightning strike?

3-3 Einstein showed that light sometimes behaves as particles that carry energy

By 1905, scientists were comfortable with the wave nature of light. However, in that year, Albert Einstein proposed that light sometimes acts as particles, creating what is now called the *wave-particle duality*. He used this idea to explain the *photoelectric effect*. Physicists knew that electrons are bound onto a metal's surface by electric forces and that it takes energy to overcome those forces. Shorter wavelengths of light can knock some electrons off the surfaces of metals, while longer wavelengths of light cannot, no matter how intense the beam of long-wavelength light. Because some colors (or, equivalently, wavelengths) can remove the electrons and others cannot, the electrons must receive different amounts of energy from different

---

AN ASTRONOMER'S TOOLBOX 3-1

Photon Energies, Wavelengths, and Frequencies

All photons with the same energy are identical. This energy depends solely on the photon's wavelength or, equivalently, its frequency. The wavelength, λ, and frequency, f, of a photon are related by the simple equation

$$c = f\lambda \text{ or, equivalently, } f = \frac{c}{\lambda}$$

where c is the speed of light. Knowing either the wavelength or frequency, you can calculate the other with these equations. Also, if you know either of them, you also know the photon's energy, as introduced in the equation from Section 3-3,

$$E = \frac{hc}{\lambda} \text{ or } E = hf$$

Planck's constant, h, is 6.67×10^{-34} J s and the speed of light, c, is 300,000 km/s.

Example: A photon of red light has a wavelength 700 nm. What is its energy? *Note:* All distances in the following equation must be converted to the same units, such as meters.

$$E_{red} = \frac{(6.67 \times 10^{-34} \text{ J s})(300,000 \text{ km/s})}{700 \text{ nm}}$$

Simplifying, we find

$E_{red} = 2.86 \times 10^{-19}$ J, meaning that each photon of red light with wavelength 700 nm has an energy of 2.86×10^{-19} J.

An aside on units: Just as the numbers in the numerator of an equation multiply together, the numbers in the denominator multiply together, and the resulting numerator is divided by the denominator, and units are combined in the same way. The same units found in both the numerator and denominator cancel. For example, a result such as 7 km s/s would have the seconds cancel top and bottom, leaving 7 km as the answer.

Furthermore, in any equation, the units of the same type (such as length) must all be converted to the same standard, so that they can be combined or canceled. In our case, we would convert both nanometers and kilometers to meters, combine the resulting powers of 10, and cancel the units on the top and bottom. For example, 1 km/1 nm = 10^3 m/10^{-9} m = 10^{+12} (with no units). The cancellation of units in the equation left just *Joules*, as it should, since we were calculating an energy.

Compare: In 1 s, a 25-watt lightbulb emits 25 J.

Try these questions: A photon has energy 4.90×10^{-19} J. Calculate its wavelength in nanometers. Referring to Figure 3-1: What is this photon's color? What is the wavelength of a photon with twice this energy? What is the energy of a green photon?

(Answers appear at the end of the book.)

colors of light. But how? Einstein proposed that light travels as waves enclosed in discrete packets, now called **photons**, and that photons with different wavelengths have different amounts of energy. Specifically, *the shorter the wavelength, the higher a photon's energy.*

$$\text{Photon energy} = \frac{\text{Planck's constant} \times \text{the speed of light}}{\text{wavelength}}$$

where Planck's constant (named for the German physicist Max Planck) has the value 6.67×10^{-34} J s, where J is the unit of energy called a joule (see *An Astronomer's Toolbox 3-1*), and wavelength, the distance between wave crests or troughs, is shown in Figure 3-3. Einstein's concept of light, confirmed in numerous experiments, means that light can act both as waves (as when going through slits) and as particles (as when striking matter).

R I V U X G

FIGURE 3-5 Experimental Evidence for Infrared Radiation This photograph shows the visible colors separated by a prism. The two thermometers in the region illuminated by visible light have temperatures less than the thermometer to the right of red. Therefore, there must be more radiation energizing (i.e., heating) the warmest thermometer. This energy is what we call infrared radiation—invisible to the human eye but detectable as heat. (NASA/JPL-Caltech)

The waves shown in Figure 3-3 are moving to the right. If you count the number of wave crests that pass a given point per second, you have found the **frequency** of the photon. The unit of frequency is *hertz,* named in honor of German physicist Heinrich Hertz. One hertz means that one cycle per second—or that one wave crest per second—passes any point. A thousand hertz means a thousand cycles per second, and so on. The frequency is used, among many other things, to identify radio stations. For example, WCPE radio in Wake Forest, North Carolina, has a frequency of 89.7 megahertz (a megahertz is a million hertz or a million cycles per second). Frequency is discussed further in *An Astronomer's Toolbox 3-1.*

All photons with the same wavelength are identical to each other, and, therefore, every photon of a given wavelength carries the same amount of energy as every other photon with that wavelength. The energy delivered by a photon is either enough to eject an electron from the surface of the metal or it is not; there is no middle ground. Extensive testing in the twentieth century confirmed both the wave and particle properties of light.

While the energy of a single photon is fixed by its wavelength, the total number of photons passing per second from that source with a given energy determines the *intensity* of the electromagnetic radiation at that wavelength (that is, how bright the object appears to be). The more photons detected, the higher the intensity, and vice versa. However, the intensity of light does not change the energy *per* photon. If one photon is unable to eject an electron from a metal, then billions of photons with that energy will still be unable to remove that electron.

> How are the frequencies of ocean waves determined?

3-4 Visible light is only one type of electromagnetic radiation

We have said that visible light has a narrow range of wavelengths, from about 400 to 700 nm. However, Maxwell's equations place no length restrictions on the wavelengths of electromagnetic radiation. What lies on either side of this interval? Around 1800, the British astronomer William Herschel discovered **infrared radiation** in an experiment with a prism. When he held a thermometer just beyond the red end of the visible spectrum, the thermometer registered a temperature increase, indicating that it was being heated by an invisible form of energy (Figure 3-5). Infrared radiation, discovered before Maxwell's equations were formulated, was later identified as electromagnetic radiation with wavelengths slightly longer than red light. Our bodies detect infrared radiation as heat.

Shortest
wavelength
(highest energy)

10⁻⁶ nm
10⁻⁵ nm
10⁻⁴ nm
10⁻³ nm
10⁻² nm
10⁻¹ nm
1 nm
10 nm
100 nm
10³ nm = 1 μm
10 μm
100 μm
1000 μm = 1 mm
10mm = 1 cm
10 cm
100 cm = 1 m
10 m
100 m
1000 m = 1 km
10 km
100 km

Longest
wavelength
(lowest energy)

Gamma
rays

X rays

Visible light

Infrared
radiation

Microwaves

Radio waves

Cancer
radiotherapy:
gamma rays

Medical imaging:
X rays

Identifying minerals
using ultraviolet radiaton

Ultraviolet radiation

Violet — 400 nm
Blue
Green — 500 nm
Yellow
Orange — 600 nm
Red — 700 nm

TV remote:
infrared light

Microwave oven:
microwaves

Mobile phone:
radio waves

FIGURE 3-6 The Electromagnetic Spectrum The full array of all types of electromagnetic radiation is called the electromagnetic spectrum. It extends from the longest-wavelength radio waves to the shortest-wavelength gamma rays. Visible light forms only a tiny portion of the full electromagnetic spectrum. Note that 1 μm (micrometer) is 10⁻⁶ m, and 1 nm (nanometer) is 10⁻⁹ m. The insets show how we are now using all parts of the electromagnetic spectrum here on Earth. (Will and Deni McIntyre/Science Photo Library; Edward Kinsman/Photo Researchers, Inc.; Chris Martin-Bahr/Science Photo Library; Bill Lush/Taxi/Getty; Michael Porsche/Corbis; Ian Britton, Royalty-Free/Corbis)

WEB LINK 3.1

In experiments with electric sparks in 1888, Hertz succeeded in producing electromagnetic radiation a few centimeters in wavelength, now known as **radio waves**.

At wavelengths shorter than those of visible light, **ultraviolet (UV) radiation** extends from about 400 nm to 10 nm. In 1895, Wilhelm Roentgen invented a machine that produces electromagnetic radiation with wavelengths shorter than 10 nm, now called **X rays**. Modern versions of Roentgen's machine are found in medical and dental offices and airport security checkpoints. X rays have wavelengths between about 10 and 0.01 nm. Shorter still are **gamma rays**. These boundaries are all arbitrary and are primarily used as convenient divisions in the electromagnetic spectrum, which is actually continuous.

We now know that visible light occupies only a tiny fraction of the full range of possible wavelengths, collectively called the **electromagnetic spectrum**. As shown in Figure 3-6, the electromagnetic spectrum stretches from the longest-wavelength radio waves, through infrared radiation, visible light, ultraviolet radiation, and X rays, to the shortest-wavelength photons, gamma rays. On the long-wavelength side of the visible spectrum, infrared radiation covers the range from about 700 nm to 1 mm. Astronomers interested in infrared radiation often express wavelength in *micrometers* or *microns* (abbreviated μm), where 1 μm = 1000 nm = 10⁻⁶ m. From roughly 1 mm to 10 cm is the range of microwaves. These are sometimes considered as a separate class of photons and sometimes categorized as infrared radiation or radio waves. Formally, radio waves are all electromagnetic waves longer than 10 cm.

The various types of electromagnetic radiation share many basic properties. For example, they are all photons, they all travel at the same speed, and they all sometimes behave as particles and sometimes as waves. But, because of their different wavelengths (and therefore different energies), they interact very differently with matter. For example, X rays penetrate deeply into your body tissues, while visible light is mostly stopped and scattered by the surface layer of skin; your eyes respond to visible light but not to infrared radiation; and your radio detects radio waves but not ultraviolet radiation.

Earth's atmosphere is relatively transparent to visible light, radio waves, short-wavelength infrared, and long-wavelength ultraviolet. As a result, these radiations pass through the atmosphere without much loss and can be detected by ground-based telescopes sensitive to them. Astronomers say that the atmosphere has *windows* for these parts of the electromagnetic spectrum (Figure 3-7).

The longest-wavelength ultraviolet radiation, called UVA, causes tanning and sunburns. Ozone (O_3) in Earth's atmosphere normally screens out intermediate-wavelength ultraviolet radiation, or UVB. Until recently, the ozone in the *ozone layer* high in the atmosphere was being depleted by human-made chemicals, such as chlorofluorocarbons (CFCs) and bromine-rich gases. As a result, more UVB is reaching Earth's surface, and these highly energetic photons severely damage living tissue, causing skin cancer and glaucoma, among other diseases.

Earth's atmosphere is completely opaque to the other types of electromagnetic radiation, meaning that they do not reach Earth's surface. (This is a good thing, because short-wavelength ultraviolet radiation [UVC], X rays, and gamma rays are devastating to living tissue. Gamma rays, packing the highest energies, are the deadliest.) Direct observations of these wavelengths must be performed high in the atmosphere or, ideally, from space.

As noted earlier, photons with different energies interact with matter in different ways. Higher-energy photons will pass through or rip apart material from which lower-energy photons will bounce off. Telescope designs for collecting and focusing radiation, therefore, differ depending on the energy or, equivalently, the wavelength of interest. Knowing the energies of photons and their effects on matter enables astronomers to design both the telescopes to collect them and the devices used to record their presence. In the next section we will consider the lengths to which astronomers have gone to capture visible and invisible (or nonoptical) electromagnetic radiation.

FIGURE 3-7 "Windows" Through the Atmosphere Earth's atmosphere allows different types of electromagnetic radiation to penetrate into it in varying amounts. Visible light, radio waves, short-wavelength infrared, and long-wavelength ultraviolet reach all the way to Earth's surface. The other types of radiation are absorbed or scattered by the gases in the air at different characteristic altitudes (indicated by heights of windows). Although the atmosphere does not have actual "windows," astronomers use the term to characterize the passage of radiation through it.

OPTICS AND TELESCOPES

Since the time of Galileo, astronomers have been designing instruments to collect more light than the human eye can gather on its own. Collecting more light enables us to see things more brightly, in more detail, and at a greater distance. There are two basic types of telescopes—those that collect light through lenses, or **refracting telescopes**, and those that collect it from mirrors, or **reflecting telescopes**. The earliest telescopes, such as Galileo's, used lenses, which have a variety of shortcomings as light-gathering devices. Consequently, all modern research telescopes use mirrors to collect light. Lenses are still used in the eyepieces of home telescopes to straighten the gathered light, so when we look through these telescopes directly, our brains can accurately interpret what we see. Lenses are also used to collect light in binoculars and cameras. Reflecting and refracting telescopes have the same main purpose—to collect as many photons as possible to better observe the sky (see Section 3-6). We begin exploring telescopes by discussing how reflecting telescopes work. Then we consider how lenses collect light and, finally, how astronomers have developed telescopes to see nonvisible electromagnetic radiation.

3-5 Reflecting telescopes use mirrors to concentrate incoming starlight

The first reflecting telescope was built in the seventeenth century by Isaac Newton (Figure 3-8). To understand how these telescopes work, consider a flow of photons, more commonly called a *light ray*, moving toward a flat mirror. In Figure 3-9a, the light ray strikes the mirror, and we imagine a perpendicular line coming out of the mirror at that point. According to the principle of **reflection**, the angle between the incoming light ray and the perpendicular (dashed line) is always equal to the angle between the outgoing, reflected light ray and the perpendicular. This principle is often stated as "The angle of incidence equals the angle of reflection." This rule also applies if the mirror is curved (Figure 3-9b).

> Which has more energy, an infrared photon or an ultraviolet photon?

Using this principle, Newton determined that a concave (hollowed-out) mirror, ground in the shape of a parabola, causes all parallel incoming light rays that strike the mirror to converge to a **focal point** (Figure 3-9b). The distance between this **primary mirror** and the focal point, where the image of the distant object is formed, is called the **focal length** of the mirror. Focal

FIGURE 3-8 Replica of Newton's Reflecting Telescope Built in 1672, this reflecting telescope has a spherical primary mirror 3 cm (1.3 in) in diameter. Its magnification was 40×. (Royal Greenwich Observatory/Science Photo Library)

points exist for light from sources that are extremely far away, like the stars. (Figure 3-10 shows why stars can be considered "far away.") If the object is larger than a point, such as the Moon or a planet, then the light will converge to a plane, called the **focal plane**, located at the distance of the focal length.

To view the image, Newton placed a small, flat mirror at a 45° angle between the primary mirror and the focal point, as sketched in Figure 3-9c and Figure 3-11a. This **secondary mirror** reflects the light rays to one side of the telescope, and the astronomer views the image through an **eyepiece lens**. We will discuss how this lens works in Section 3-8. A telescope with this optical design is still called a **Newtonian reflector**.

Newtonian telescopes are popular with amateur astronomers because they are convenient to use while the observer is standing up. However, they are not used

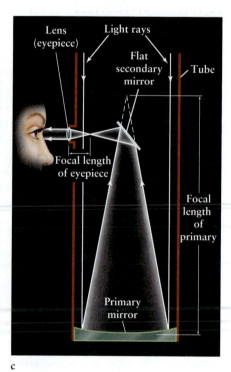

a   b   c

FIGURE 3-9 Reflection (a) The angle at which a beam of light strikes a mirror (the angle of incidence, i) is always equal to the angle at which the beam is reflected from the mirror (the angle of reflection, r). (b) A concave, parabolic mirror causes parallel light rays to converge and meet at the focal point. The distance between the mirror and focal point is the focal length. (c) A Newtonian telescope uses a flat mirror, called the secondary mirror, to send light toward the side of the telescope. The light rays are made parallel again by passing through a lens, called the eyepiece. The dashed line shows where the focal point of this primary mirror would be if the secondary mirror were not in the way.

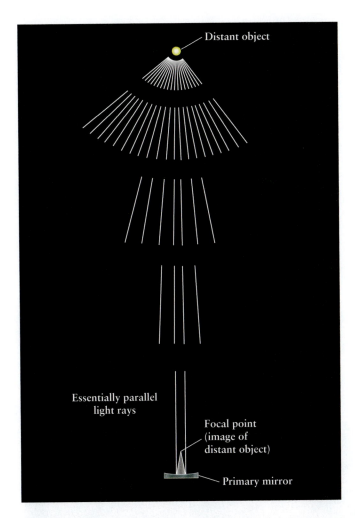

for equipment that has an extremely long focal length or is too heavy to mount on the telescope. The detection equipment is mounted at the *Nasmyth focus* (after Scottish engineer James Nasmyth, who developed it) or the coudé focus (after the French word meaning "bent like an elbow"). In both designs (Figure 3-11c), a curved tertiary (third) mirror channels the light rays away from the telescope to a remote focal point. Equipment that profits from the use of the Nasmyth focus and coudé focus includes a variety of *spectrographs*, instruments that separate light from objects into its individual colors to determine the objects' chemistry, surface temperature, and motion toward or away from us. (We discuss spectrographs in more detail in Chapter 4.)

In the third design, an observing device is located at the undeflected focal point, directly in front of the primary mirror. This arrangement is called the **prime focus** (Figure 3-11d). It has the advantages of making the brightest image (for a given exposure) and having the fewest reflections that can cause light loss and distortion (i.e., no secondary mirror).

> Where in Figure 3-9c does the light from the primary mirror actually come to a focal point?

FIGURE 3-10 Parallel Light Rays from Distant Objects As light travels away from any object, the light rays, all moving in straight lines, separate. By the time light has traveled trillions of kilometers or miles, only the light rays moving in virtually parallel tracks are still near each other.

in research observatories because they are lopsided. If astronomers attach their often heavy and bulky research equipment onto the side of a Newtonian telescope, the telescope twists and distorts the image in unpredictable ways.

Three basic designs exist for the reflecting telescopes used in research. In the first, a hole is drilled directly through the center of the primary mirror. A convex (outwardly curved), secondary mirror placed between the primary mirror and its focal point reflects the light rays back through the hole (Figure 3-11b). This design is called the **Cassegrain focus**. This secondary mirror extends the telescope's focal length. Compact, relatively low-weight equipment is bolted to the bottom of the telescope, and the light is brought into focus down there. This design has an advantage over Newtonian telescopes in that the attached equipment is balanced and does not distort the telescope frame and, hence, the image.

The second design that astronomers use has two variations, depending on the telescope design. It is used

a Newtonian focus
b Cassegrain focus
c Nasmyth focus or coudé focus
d Prime focus

FIGURE 3-11 Reflecting Telescopes Four of the most common optical designs for reflecting telescopes: (a) Newtonian focus (popular among amateur astronomers) and the three major designs used by researchers—(b) Cassegrain focus, (c) Nasmyth focus or coudé focus, and (d) prime focus.

3-6 Telescopes brighten, resolve, and magnify

As mentioned earlier, a telescope's most important function is to provide the astronomer with as bright an image as possible. The brighter an object is, the more information about it the astronomer can extract. The observed brightness of any object depends on the total number of photons collected from it, which, in turn, depends on the area of the telescope's primary mirror. A larger primary mirror intercepts and collects more light than does a smaller one. For exposures of equal times, a telescope with a large primary mirror produces brighter images and detects fainter objects than a telescope with a smaller primary mirror (Figure 3-12). Because the light striking the focal plane or leaving the eyepiece is now more concentrated, objects are brighter as seen through a telescope. No wonder telescopes are sometimes called "photon buckets."

The **light-gathering power** of a telescope is directly related to the area of the telescope's primary mirror.

INSIGHT INTO SCIENCE

Costs and Benefits Science now relies heavily on technology to conduct experiments or to make observations. The cost of cutting-edge astronomical observations may run to hundreds of millions of dollars or more. The return on such investments is a better understanding of how the universe works, how we can harness its capabilities, and our place in it.

Recall that the area and diameter of a circle are related by the formula

$$\text{Area} = \frac{\pi d^2}{4}$$

where d is the diameter of the mirror and π (pi) is about 3.14. Consequently, *a mirror with twice the diameter of another mirror has 4 times the area of the smaller mirror and, therefore, collects 4 times as much light as the smaller one in the same amount of time*. For

R I V U X G

FIGURE 3-12 Light-Gathering Power Because a large primary mirror collects more starlight than does a smaller one, a large telescope produces a brighter image than a smaller one, all other things being equal. The same principle applies to telescopes that collect light using an objective lens rather than a primary mirror. The two photographs of the Andromeda Galaxy were taken through telescopes with different diameters and were exposed for equal lengths of time at equal magnification. (AURA) example, a 36-cm-diameter mirror has 4 times the area of an 18-cm-diameter mirror. Therefore, the 36-cm telescope has 4 times the light-gathering power of a telescope half its size.

The second most important function of any telescope is to reveal greater detail of objects that are more than just points of light, often called *extended objects*. A large telescope increases the sharpness of the image and the degree of detail that can be seen. **Angular resolution** (often called just *resolution*) measures the clarity of images (Figure 3-13). The angular resolution of a telescope is measured as the arc angle between two adjacent stars whose images can just barely be distinguished by the telescope. The smaller the angle, the sharper the image. Large, modern telescopes, such as the Keck telescopes in Hawaii, have angular resolutions better than 0.1 arcsec. *As a general rule, a telescope with a primary mirror twice the diameter of another telescope's primary will be able to see twice as much detail as the smaller telescope.*

The final function of a telescope is to make objects appear larger. This property is called **magnification**. Magnification is associated with resolution, because the larger the image, the more detail of the image you can potentially see (Figure 3-14a & b). The magnification of a reflecting telescope is equal to the focal length of the primary mirror divided by the focal length of the eyepiece lens

$$\text{Magnification} = \frac{\text{Focal length of primary}}{\text{Focal length of eyepiece}}$$

For example, if the primary mirror of a telescope has a focal length of 100 cm and the eyepiece has a focal length of 0.5 cm, then the magnifying power of the telescope is

$$\text{Magnification} = \frac{100 \text{ cm}}{0.5 \text{ cm}} = 200$$

This property is usually expressed as 200×.

There is a limit to the magnification of any telescope. Try to magnify beyond that limit and the image becomes distorted. As a rule, *a telescope with a primary mirror twice the diameter of another telescope's primary will have twice the maximum magnification of the smaller telescope.*

Normal eye glasses or contact lenses are designed to improve which aspect of vision presented in this section?

3-7 Storing and analyzing light from space is key to understanding the cosmos

The invention of photography during the nineteenth century was a boon for astronomy. By taking a long exposure with a camera mounted at the focal plane of a telescope, an astronomer could record extremely faint features that could not be seen by just looking through the telescope. The reason photography is better is that our eyes clear the images in them several times a second, whereas film adds up the intensity of all the photons that affect its emulsion. By keeping telescopes aimed precisely, so that images do not blur, exposures of an hour or more became quite routine. Astrophotographs and their modern electronic equivalents, discussed in the next few paragraphs, make all objects appear brighter and reveal greater detail in extended objects—such as galaxies, star clusters, and planets—than we could otherwise see.

a

b   R I V U X G

FIGURE 3-13 Resolution The larger the diameter of a telescope's primary mirror, the greater the detail the telescope can resolve. These two images of the Andromeda Galaxy, taken through telescopes with different diameters, show this effect. Increasing the exposure time of the smaller diameter telescope (a), will only brighten the image, not improve the resolution. (AURA) a   b   R I V U X G

FIGURE 3-14 *Magnification* The same telescope can magnify by different amounts, depending on the focal length of the eyepiece. (a) An image of the Moon taken with magnification 4 times greater than image (b). Note in this case that the increased magnification leads to increased resolution (i.e., more detail can be seen in the larger image). (a & b: NASA)

Astronomers have long known, however, that a photographic plate is an inefficient detector of light because it depends on a chemical reaction to produce an image. Typically, only 2% of the light striking film triggers a reaction in the photosensitive material. Thus, roughly 98% of the light falling onto a photographic plate is wasted.

Rapidly evolving technology has changed all that. We have replaced photographic film with highly efficient electronic light detectors called **charge-coupled devices (CCDs)**. CCDs typically respond to 70% of the light falling on them; their resolution is better than that of film, and they respond more uniformly to light of different colors. Divided into an array of small, light-sensitive squares called picture elements or, more commonly, **pixels**, each CCD is a square or rectangle a few centimeters on a side. Several of them are often used together to create a larger image (Figure 3-15). This is the same technology used in digital cameras and cellphones. The largest grouping of CCDs used on a telescope has 378 megapixels (A megapixel is 1 million pixels.), compared to most digital cameras, which typically have between 6 and 12 megapixels.

When an image from a telescope is focused on the CCD, an electric charge builds up in each pixel in direct proportion to the intensity of the light (number of photons) falling on that pixel. When the exposure is finished, the charge on each pixel is read into a computer. Figure 3-16 shows one photograph and two CCD images of the same region of the sky, all taken with the same telescope. You can see how many details

Why does the human eye/brain clear many times a second the images that it receives?

visible in the CCD images are absent in the ordinary photograph.

3-8 Eyepieces, refracting telescopes, binoculars, and eyeglasses use lenses to focus incoming light

Although light travels at about 300,000 km/s in a vacuum, it moves more slowly through any medium, such as glass. As light enters the glass, it slows abruptly, much like a person walking from a boardwalk onto a sandy beach. The person's pace suddenly decreases as he or she steps from the smooth, hard pavement onto the sand; and just as he or she steps back onto the boardwalk and resumes the original pace, light that exits a piece of glass resumes its original speed.

As a result of changing speed, light also changes direction as it passes from one transparent medium into another. As noted earlier, this is called refraction. You see refraction every day when looking through windows. Imagine a stream of photons from a star that enters a window, as shown in Figure 3-17a. As a light ray goes from the air into the glass, the light ray's direction changes so that it is more perpendicular to the surface of the glass than it was before entering. Once inside the glass, the light ray travels in a straight line. Upon emerging from the other side, the light ray bends once again, resuming its original direction and speed. The net effect is only a slight, uniform displacement of the objects beyond the glass.

Unlike a window, lenses have surfaces of varying thickness (Figure 3-17b). These curved surfaces force

a   R I V U X G   b   R I V U X G

FIGURE 3-15 Mosaic of Charge-Coupled Devices (CCDs)
(a) These 40 CCDs combine to provide up to 378 million light-sensitive pixels that store images collected by the CFHT telescope on the dormant volcano Mauna Kea, Hawaii. Electronic circuits transfer the data to a waiting computer.

(b) This image of the Rosette Nebula, a region of star formation 5000 ly away in the constellation Monoceros (the Unicorn), was taken with this CCD. It shows the incredible detail that can be recorded by large telescopes and high resolution CCDs. (J. C. Cuillandre/Canada-France-Hawaii Telescope)

a   b   c   R I V U X G

FIGURE 3-16 Photography Versus CCD Images These three views of the same part of the sky, each taken with the same 4-m telescope, compare CCDs to photographic plates. (a) A negative print (black stars and white sky) of a photographic image. (b) A negative CCD image. Notice that many faint stars and galaxies that are invisible in the ordinary photograph can be seen clearly in this CCD image. (c) This (positive) color view was produced by combining a series of CCD images taken through colored filters. (Patrick Seitzer, NOAO) a   b   c   R I V U X G

FIGURE 3-17 Refraction through Uniform and Variable Thickness Glasses (a) Refraction is the change in direction of a light ray when it passes into or out of a transparent medium such as glass. A light ray that enters a denser medium, like going from air into water or glass, is bent or refracted to an angle more perpendicular to the surface than the angle at which it was originally traveling. If the glass is flat, then the light leaving it is refracted back to the direction it had before entering the glass. There is no overall change in the direction in which the light travels. (b) If the glass is in the shape of a suitable convex lens, parallel light rays converge to a focus at the focal point. As with parabolic mirrors, the distance from the lens to the focal point is called the focal length of the lens. (c) The straw as seen through the side of the liquid is magnified and offset from the straw above the liquid because the liquid is given a curved shape by the side of the glass. The straw, as seen through the top of the liquid, is refracted but does not appear magnified because the surface of the water is flat and the beaker has uniform thickness. (Ray Moller/Dorling Kindersley)

the light rays to emerge from the lens in different directions than they had before entering the lens. Lenses that are thicker at their centers than at their edges are called *convex* lenses and force the light that enters them to converge as it passes through. Conversely, lenses thinner at their centers than around their edges are called *concave* lenses. They cause light rays to diverge. Large convex lenses called **objective lenses** were used in telescopes (and still are in many built for home use) instead of primary mirrors, and small lenses are still used as eyepieces to bring the light rays back to being parallel, so that when we view directly through the telescope, our eyes can make sense of what we see.

Parallel light rays that enter an objective lens converge and meet at the focal point of the lens. The distance between the focal point and the lens is the lens' focal length (Figure 3-17b). As with reflecting telescopes, if the object is close enough or large enough to be more than just a dot as seen through the telescope, all the light from it does not converge at the focal point, but rather focuses along the focal plane (Figure 3-18).

A refracting telescope or **refractor** (Figure 3-19), is an arrangement of two lenses used to gather light. The objective lens at the top of the telescope has a large diameter and long focal length. Like a primary mirror, its purpose is to collect as much light as possible. The eyepiece lens, at the bottom of the telescope, is smaller and has a short focal length. The mathematics of magnification for a refracting telescope are the same as for a reflecting telescope; however, the focal length of the objective lens substitutes for the focal length of the primary mirror. Likewise, all of the rules for the limits on telescopes are the same for refractors as for reflectors.

The largest refracting telescope in the world, completed in 1897 and located at the Yerkes Observatory in Williams Bay, Wisconsin, near Chicago (Figure 3-20), has an objective lens that is 102 cm (40 in) in diameter and has a focal length of 19⅓ m (63.5 ft). The second largest refracting telescope, located at the Lick Observatory near San Jose, California, has an objective lens of 91 cm (36 in) in diameter. No major refracting telescopes were constructed in the twentieth century or are planned for this century.

Refracting telescopes suffer from a variety of problems that have limited their use as research instruments. These problems include the following:

• Different colors of light are refracted by different amounts, so different colors have different focal lengths. As a result, objects look blurred. This is called *chromatic aberration*, which is not a problem for reflectors because the reflecting surface is on the top of the mirror so that light never enters the glass.

• It is hard to grind a lens to the very complicated shape necessary to have all parallel light rays passing through it come to the same focal point. This is less of a problem with reflectors because the surface is a simple parabola.

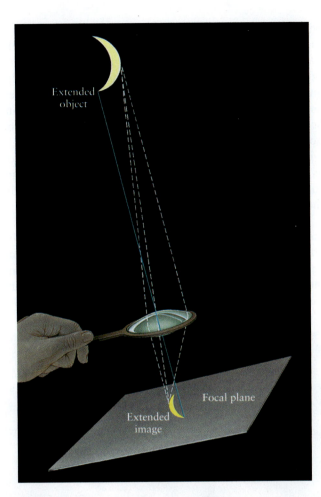

FIGURE 3-18 Extended Objects Create a Focal Plane Light from objects larger than points in the sky does not all converge to the focal point of a lens. Rather, the object creates an image at the focal length in what is called the focal plane.

• A lens must be supported only around its edges to avoid blocking the light. The weight of a large lens can cause it to sag and, thereby, distort the image. This is not a problem with reflectors because the entire underside of the mirror can be supported, as necessary.

• Air bubbles in the glass cause unwanted refractions and, hence, distorted images.

• Glass is opaque to certain ranges of wavelength, meaning that the light does not get through the glass.

These last two points are not problems with reflectors because the light never enters the glass. The combination of these problems makes images from refractors less accurate than those obtained from reflecting telescopes, which explains why modern research telescopes are all reflectors. The mirrored surfaces of research telescopes are polished so smoothly that the highest bumps are less than 0.002 times the thickness of a human hair.

Although they are better light collectors than refracting telescopes, reflecting telescopes are not perfect. Two of the most important prices to pay are blocked light and spherical aberration. We consider these problems in the next two sections.

> How do human eyes focus light?

3-9 Secondary mirrors dim objects but do not create holes in them

You have probably noticed (for example, in looking at Figure 3-11) that the secondary mirror of a reflector blocks some incoming light—one price that astronomers must pay in most reflectors. A secondary mirror typically prevents about 10% of the incoming light from reaching the primary mirror. This problem is addressed by constructing primary mirrors with sufficiently large surface areas to compensate for the loss of light. You might also think that, because light is missing from the center of the telescope due to blockage by the secondary mirror, a corresponding central "hole" appears in the images. However, this problem does not occur, because light from all parts of each object being observed enters all parts of the telescope (Figure 3-21). Indeed, covering any part of the primary mirror or objective lens darkens the image but does not limit which parts of the object you can see through the telescope.

FIGURE 3-19 Essentials of a Refracting Telescope A refracting telescope consists of a large, long-focal-length objective lens that collects and focuses light rays and a small, short-focal-length eyepiece lens that re-straightens the light rays. The lenses work together to brighten, resolve, and magnify the image formed at the focal plane of the objective lens.

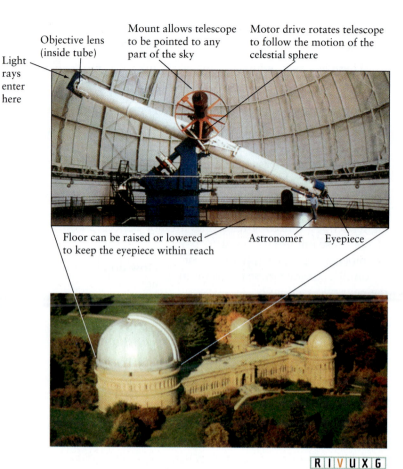

Light rays enter here

Objective lens (inside tube)

Mount allows telescope to be pointed to any part of the sky

Motor drive rotates telescope to follow the motion of the celestial sphere

Floor can be raised or lowered to keep the eyepiece within reach

Astronomer   Eyepiece

R I V U X G

FIGURE 3-20 The Largest Refracting Telescope This giant refracting telescope, built in the late 1800s, is housed at Yerkes Observatory near Chicago. The objective lens is 102 cm (40 in) in diameter, and the telescope tube is 19⅓ m (63½ ft) long. (Yerkes Observatory)

3-10 Shaping telescope mirrors and lenses is an evolving science

To make a reflector, an optician traditionally grinds and polishes a large slab of glass into a concave, spherical surface. Before computer control, grinding a spherical surface was much easier than grinding the ideal parabolic surface. However, light that enters a spherical telescope mirror at different distances from the mirror's center comes into focus at different focal lengths. Images taken directly from such telescopes (Figure 3-22a) appear at least slightly blurry, an effect called **spherical aberration**.

We can avoid spherical aberration in reflecting telescopes by making the mirror parabolic (Figure 3-22b) or by using a thin correcting lens, called a **Schmidt corrector plate**, with a spherical mirror. The corrector plate is located at the top of the telescope (Figure 3-22c). The light coming into the telescope is refracted by the plate just enough to compensate for spherical aberration and to bring all of the light into focus at the same focal length. These correctors have the added benefit of focusing light from a larger angle in the sky than would be in focus

without the plate. A Schmidt corrector plate enables astronomers to map large areas of the sky with relatively few photographs at moderately high magnification. In other words, the Schmidt corrector plate acts like a wide-angle lens on a camera. Schmidt corrector plates are often used in conjunction with Cassegrain mirrors to create *Schmidt-Cassegrain* telescopes, which are also popular for home use because they are compact and give relatively wide-angled images. However, the plate does not allow for as much magnification as a telescope with a parabolic mirror. (If you are interested in buying a telescope, you might find helpful the *Guided Discovery: Buying a Telescope*.)

Although some parabolic primary mirrors have been meticulously hand-ground over the past century, the advent of computer-controlled grinding and rotating furnaces (Figure 3-23), in which the liquid glass actually spins into a parabolic shape, has now made it economical to cast parabolic mirrors with diameters of several meters.

Secondary mirror

Eyepiece

Complete image

Primary mirror

FIGURE 3-21 The Secondary Mirror Does Not Create a Hole in the Image Because the light rays from distant objects are parallel, light from the entire object reflects off all parts of the mirror. Therefore, every part of the object sends photons to the eyepiece. This figure shows the reconstruction of the entire Moon from light passing through just part of this telescope. The same drawing applies everywhere on the primary mirror that is not blocked by the secondary mirror.

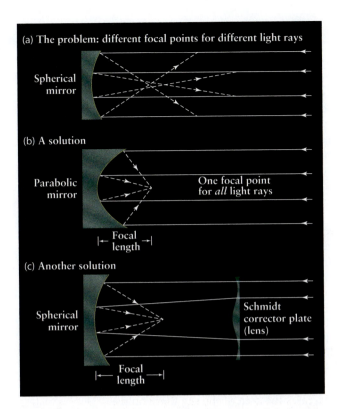

FIGURE 3-22 Spherical Aberration (a) Different parts of a spherically concave mirror reflect light to slightly different focal points. This effect, spherical aberration, causes image blurring. This problem can be overcome by (b) using a parabolic mirror or (c) using a Schmidt corrector plate (a specially curved lens) in front of the telescope.

3-11 Earth's atmosphere hinders astronomical research

Earth's atmosphere affects the light from objects in space before it reaches ground-based telescopes. The air is turbulent and filled with varying amounts of impurities and moisture. You have probably seen turbulence while driving in a car on a hot day, when the road ahead appears to shimmer. Blobs of air, heated by Earth, move upward to create this effect. Light passing through such a blob is refracted, because each hot air mass has a different density than the cooler air around it. Because each blob behaves like a lens, images of objects beyond them appear distorted.

The atmosphere over our heads is similarly moving and changing density, and the starlight passing through it is similarly refracted. Because the air density changes rapidly, the resulting changes in refraction make the stars appear to change brightness and position rapidly, an effect we see as **twinkling**. When photographed from Earth for more than a few seconds, twinkling smears out a star's image, causing it to look like a disk rather than a pinpoint of light (Figure 3-24a). Astronomers use the expression "seeing" to describe how steady the

atmosphere is; when the seeing is bad, much twinkling is occurring and, therefore, telescopic images are spread out and blurry.

The angular diameter of a star's smeared-out image, called the **seeing disk**, is a realistic measure of the best possible resolution. The size of the seeing disk varies from one observatory site to another. At Palomar Observatory in California, the seeing disk is roughly 1 arcsec (1"). The best conditions on Earth have been recorded at the observatory on the 14,000-ft summit of Mauna Kea, the tallest volcano on the island of Hawaii. The seeing disk there has been as small as 0.2".

Without the effects of Earth's atmosphere, stars do not twinkle. As a result, photographs taken from telescopes in space reveal stars as much finer points (Figure 3-24b), and extended objects, such as planets and galaxies, appear in greater detail.

a

b   R I V U X G

FIGURE 3-23 Rotating Furnace for Making Parabolic Telescope Mirrors (a) To make each 8.4-m primary mirror for the Large Binocular Telescope II on Mount Graham in Arizona, 40,000 lb of glass are loaded into a rotating furnace and heated to 1450 K (2150°F). This image shows glass fragments loaded into the cylindrical furnace. (b) After melting, spinning, and cooling, the mirror's parabolic surface is ready for final smoothing and coating with a highly reflective material. (a:Roger Ressmeyer/Corbis; b:The University of Arizona, Steward Observatory)

Buying a Telescope

If you feel elated by discovering treasures, you may want to consider viewing the night sky through powerful binoculars or a telescope. Both types of instrument enable you to find many beautiful, exciting objects in space. Binoculars are nice because you can quickly scan the heavens to see craters on the Moon, star clusters, the Great Nebula of Orion, and even the Andromeda Galaxy. Telescopes open new vistas and even enable amateur astronomers to discover new comets, among other things, adding to our scientific database and immortalizing the discoverers' names.

There are four basic types of telescope mounts that steer the telescopes around the sky: The Fork Equatorial Mount, the German Equatorial Mount, the Altitude-Azimuth (Alt-Azimuth) Mount, and the Dobsonian Mount (see figures). The table on the next page presents the advantages and disadvantages of each mount.

Reflecting and refracting telescopes both have pros and cons, as discussed in the text. Refractors and Newtonian reflectors tend to be longer and bulkier than any other types. Newtonians are the least expensive and simplest telescopes to build. As a result, they are often used with the inexpensive Dobsonian mounts. If you plan to take photographs, a Cassegrain telescope is the best instrument because it remains balanced as the telescope tracks objects across the sky.

If you want to see especially large areas of the sky, you may want to purchase a telescope with a *Schmidt corrector plate*. The most common wide-field telescopes are of the Schmidt-Cassegrain design. These telescopes have a Schmidt corrector plate on top and a Cassegrain mirror on the bottom, with their eyepieces located underneath.

If you build or buy a telescope with the eyepiece positioned so that you have to crawl under it or climb a ladder to see through it, you will find the experience less satisfying than if you can observe in a comfortable position. If the eyepiece is not easily accessible, you can buy a *diagonal mirror* (a right-angle mirror that goes between the telescope body and the eyepiece) to help correct this problem.

You will also want to get a few different eyepieces with different focal lengths (three is a good number to start with), so that you can look at large areas of the sky under low magnification and details of small areas under high magnification. As discussed in the text, the magnification of a telescope depends on the focal lengths of the objective lens (or primary mirror) and the eyepiece. The objective lens or primary mirror is fixed in the telescope. However, all telescopes come with removable eyepieces. Change the eyepiece to another of different focal length, and you can change the telescope's magnification. Keep in mind, however, that increasing the magnification decreases the area of the sky that you see.

Unless your telescope is computerized, it is also essential to have a *finder scope*, which is a small, low-magnification, very-wide-field telescope, with cross hairs, attached directly onto your main telescope. If the finder and your main scope are well aligned, you can quickly zero in on an object.

If you plan to take photographs with a CCD camera or other instrument on your telescope or to show the cosmos to large groups of people, you will need a *tracking motor* to enable the telescope to follow the stars automatically. Tracking motors require their own

a Fork Equatorial Mount
(Andy Crawford/Dorling Kindersley)

b German Equatorial Mount
(Andy Crawford/Dorling Kindersley)

c Alt-Azimuth Mount
(Celestron Images)

d Dobsonian Mount
(Andy Crawford/Dorling Kindersley)

Buying a Telescope *(continued)*

power from batteries, a 12-volt outlet in a car, or a 110-V outlet. The motor should also be able to run at high speed so you can slew (turn the telescope rapidly) from one object to the next. For photography, you will also need a *camera adapter*.

Viewing the Sun is very exciting, especially when you can observe sunspots. If you want to observe the Sun, it is essential that you buy a good *Sun filter*. **Never,** **ever look at the Sun directly, either through a telescope or with your naked eye. Doing so for even a second can lead to partial blindness!**

Finally, you will need a flashlight with a red plastic film over the lens or one with a red LED light. Because red light does not cause the pupils of your eyes to contract, they will remain dilated (wide open) while you use the red flashlight to inspect your equipment and your star charts.

Telescope Mounts

Type of Telescope	Pros	Cons
Fork Equatorial	Can track objects in the sky with a clock drive Useful for taking CCD or film photographs Not too heavy or cumbersome	Eyepiece is in an uncomfortable position near celestial poles
German Equatorial	Can track objects in the sky with a clock drive Useful for taking CCD or film photographs Easy to change direction telescope points	Hard at first to learn to set up Takes a long time to set up Very heavy compared to other types of mounts
Alt-Azimuth	Compact, easy to set up, store, and carry Eyepiece is convenient for viewing	Must be computerized to track objects in the sky
Dobsonian	Least expensive for a given diameter primary mirror Easy to set up Easy to use	Eyepiece is often in inconvenient position Cannot track objects in the sky or take photographs without specialized equipment

a

b

R I V U X G

FIGURE 3-24 Effects of Twinkling The same star field photographed with (a) a ground-based telescope, which is subject to poor seeing conditions that result in stars twinkling, and (b) the Hubble Space Telescope, which is free from the effects of twinkling. (NASA/ESA)

INSIGHT INTO SCIENCE

Research Requires Patience Seeing conditions—indeed, most observing situations in science—are rarely ideal. Besides such natural phenomena, which are beyond their control, scientists must also contend with equipment failures, late deliveries of parts, and design flaws. Patience is not a virtue in cutting-edge science; it is a necessity. Next time you visit an observatory, ask astronomers how the seeing has been lately, but be prepared for a series of expletives!

Light pollution from cities poses another problem for Earth-based telescopes (Figure 3-25). Keep in mind that the larger the primary mirror, the more light it gathers and, therefore, the more information astronomers can obtain from its images or spectra. The 5-m (200-in) telescope at the Palomar Observatory between San Diego and Los Angeles is the first truly great large telescope, providing astronomers with invaluable insights into the universe for decades. However, light pollution

> Where in a typical house can the seeing problems described in this section be observed?

R I V U X G

FIGURE 3-25 Light Pollution These two images of Tucson, Arizona, were taken from the Kitt Peak National Observatory which is 38 linear miles away. They show the dramatic growth in ground light output between 1959 *(top)* and 1989 *(bottom)*. Since 1972, light pollution, a problem for many observatories around the world, has been at least partially controlled by a series of local ordinances. (NOAO/AURA/NSF/Galaxy)

from the two cities now fills the night sky, seriously reducing the ability of that telescope to collect light from objects in space. Not surprisingly, the best observing sites in the world are high on mountaintops—above smog, water vapor, and clouds—and far from city lights. An even better location, astronomers have discovered, is to observe space from space, eliminating the interference of both the lights of civilization and Earth's atmosphere. This is why Hubble and other orbiting telescopes achieve such magnificent resolution in the images they take.

3-12 The Hubble Space Telescope provides stunning details about the universe

For decades, astronomers dreamed of observatories in space. Such facilities would eliminate the image distortion created by twinkling and by poor atmospheric transparency due to pollution, volcanic debris, and water vapor. They could operate 24 h per day and over a wide range of wavelengths—from the infrared through the visible range and far out into the gamma ray part of the spectrum. Over the past 30 years, NASA and other space agencies have launched a variety of space telescopes, including four of what NASA calls its Great Observatories. The first Great Observatory to go up was the Hubble Space Telescope (HST).

Soon after HST was placed in orbit from the space shuttle in 1990, astronomers discovered that the telescope's 2.4-m primary mirror lacked the proper curvature. Therefore, it suffered from spherical aberration, which caused its images to be surrounded by a hazy glow. During a repair mission in December 1993, astronauts installed corrective optics that eliminated the problem (Figure 3-26). The telescope was further upgraded in 2002. Now HST has a resolution of better than 0.1", which is better than can be obtained by telescopes on Earth's surface without the use of advanced technology (see Section 3-13).

The observations taken by HST continue to stagger the imagination. It has made discoveries related to the planets in our solar system, planetary systems around other stars, other galaxies and the distances to them, black holes, quasars, the formation of the earliest galaxies, the age of the universe, and many other things. The Hubble Space Telescope has developed technical difficulties since 2002. A final repair mission to it occurred in October, 2008, bringing it back to pristine condition. Hubble's success, coupled with rapidly improving technology, has spurred scientists and engineers to develop the 6.5-m-diameter James Webb Space Telescope to replace the Hubble. This new telescope is scheduled for launch in 2013.

RIVUXG

FIGURE 3-26 The Hubble Space Telescope (HST) This photograph of HST hovering above the Space Shuttle's cargo bay was taken in 1993, at the completion of the first servicing mission. HST has studied the heavens at infrared, visible light, and ultraviolet wavelengths. (NASA)

3-13 Advanced technology is spawning a new generation of superb ground-based telescopes

The clarity of images taken by the HST may suggest that ground-based observational astronomy is a dying practice. However, two exciting techniques, called *active optics* and *adaptive optics*, enable telescopes on

the ground to match the quality of the Hubble—or better it!

Active optics finds the best orientation for the primary mirror in response to changes in temperature and the shape of the telescope mount. It adjusts the mirror every few seconds to help keep the telescope aimed at its target. With active optics, the New Technology Telescope in Chile and the Keck telescopes in Hawaii routinely achieve resolutions as fine as 0.3", whereas the resolution is much worse for telescopes without active optics at the same sites.

Adaptive optics uses sensors to determine the amount of twinkling created by atmospheric turbulence. The stellar motion is neutralized by computer-activated, motorized supports that actually reshape either the primary mirror or a smaller mirror installed farther down the optical path of the telescope. Adaptive optics effectively eliminate atmospheric distortion and produce remarkably sharp images (Figure 3-27). Many large ground-based telescopes now use adaptive optics on at least some observations, resulting in images comparable to those from HST (Figure 3-27c).

Until the 1980s, telescopes with primary mirrors of between 2 and 6 m were the largest and most powerful in the world. Now, new technologies in mirror building and computer control allow us to construct much larger telescopes. At least 67 reflectors around the world today have primary mirrors measuring 2 m or more in diameter. This number is up by 12 in only 3 years. Among these 67 are 12 that have mirrors between 8 and 10 m in diameter, with more than a dozen other very large telescopes under construction.

Because the cost of building very large mirrors is enormous, astronomers have devised less expensive

a

b

c

RIVUXG

FIGURE 3-27 Images from Earth and Space (a) Image of Neptune from an Earth-based telescope without adaptive optics. (b) Image of Neptune from the same Earth-based telescope with adaptive optics. (c) Image of Neptune from the Hubble Space Telescope, which does not incorporate adaptive optics technology. (a: Courtesy of Center for Adaptive Optics, University of California; b: Courtesy of Center for Adaptive Optics, University of California); c: NASA, L. Sromovksy, and P. Fry [University of Wisconsin-Madison])

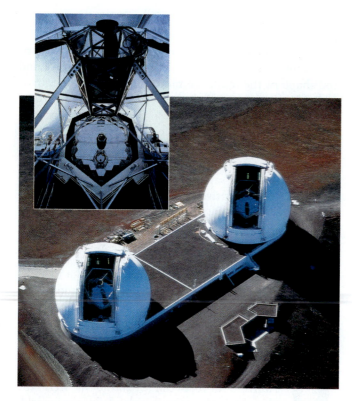

WEB LINK 3.8 **FIGURE 3-28** The 10-m Keck Telescopes Located on the dormant (and, hopefully, extinct) Mauna Kea volcano in Hawaii, these huge twin telescopes each consist of 36 hexagonal mirrors measuring 1.8 m (5.9 ft) across. Each Keck telescope has the light-gathering, resolving, and magnifying ability of a single mirror 10 m in diameter. *Inset:* View down the Keck I telescope. The hexagonal apparatus near the top of the photograph shows the housing for the 1.4-m secondary mirror. (W. M. Keck Observatory, Courtesy of Richard J. Wainscoat)

ways to collect the same amount of light. One approach is to make a large mirror out of smaller pieces, fitted together like floor tiles. The largest examples of this segmented-mirror technique are the 10-m (400-in) Keck I and Keck II telescopes on the summit of Mauna Kea. Thirty-six hexagonal mirrors are mounted side by side in each telescope to collect the same amount of light as a single primary mirror 10 m in diameter (Figure 3-28). Another method, called **interferometry**, combines images from different telescopes. For example, used together to observe the same object, the two Keck telescopes have the resolving power of a single 85-m telescope. The four 8.2-m reflectors of the Very Large Telescope at the Paranal Observatory (see the image at the beginning of this chapter) combine to create the resolution that would come from a single telescope 200 m in diameter. It has an ideal resolution of about 0.002 arcsec.

Why are military and spy agencies around the world so interested in cutting-edge telescope technologies?

NONOPTICAL ASTRONOMY

Looking back at Figure 3-6, you can see that visible light represents a very tiny fraction of the electromagnetic spectrum. As late as the 1940s, astronomers had no idea how much nonvisible radiation is emitted by objects in space. However, today we know that many objects in space that emit undetectable amounts of visible light release large amounts of nonvisible radiation. Specially designed telescopes gather electromagnetic energy in all of the nonvisible parts of the spectrum. From the radio waves and infrared radiation being emitted by vast interstellar gas clouds to the ultraviolet radiation and X rays from the remnants of stars to bursts of gamma rays of extraordinary power from merging black holes and other sources, our growing ability to see the entire electromagnetic spectrum is revealing intriguing phenomena. To get a sense of how much we do not see in the visible part of the spectrum, look at Figure 3-29, which shows ultraviolet, infrared, and visible images of the familiar constellation Orion.

3-14 A radio telescope uses a large concave dish to collect radio waves

The first evidence of nonvisible radiation from outer space came from the work of a young radio engineer, Karl Jansky, at Bell Laboratories. Using radio antennas, Jansky was investigating the sources of static that affect short-wavelength radiotelephone communication. In 1932, he realized that a certain kind of radio noise is strongest when the constellation Sagittarius is high in the sky. Because the center of our Galaxy is located in the direction of Sagittarius, Jansky correctly concluded that he was detecting radio waves from elsewhere in the Galaxy. Jansky's accidental discovery of radio signals from space led to a worldwide effort to "see" what nonvisible wavelengths could teach us about the cosmos.

INSIGHT INTO SCIENCE

Think "Outside the Box" Observations and experiments often require making connections between seemingly unrelated concepts. Jansky's proposal that some radio waves originate in space is an example of connecting apparently disparate fields—astronomy and radio engineering.

Radio telescopes record radio signals from the sky. Just as a mirror reflects visible light, the metal surfaces of radio telescopes reflect radio waves. Each radio telescope has a large concave dish (Figure 3-30) that focuses radio photons in the same way that an optical telescope mirror focuses visible photons. A small antenna tuned to the

a   R I V U X G

Betelgeuse

Alnitak

Sword

b   R I V U X G

c   R I V U X G

FIGURE 3-29 Orion as Seen in Ultraviolet, Infrared, and Visible Wavelengths (a) An ultraviolet view of the constellation Orion obtained during a brief rocket flight on December 5, 1975. The 100-s exposure captured wavelengths ranging from 125 to 200 nm. (b) A false-color view from the Infrared Astronomical Satellite uses color to display the intensity of infrared wavelengths. (c) An ordinary optical photograph. (a: G. R. Carruthers, NRL; b: NASA/JPL-Caltech; c: R. C. Mitchell, Central Washington University)

desired wavelength is located at the telescope's prime focus, or Cassegrain focus. The incoming signal is relayed to amplifiers and recording instruments.

Radio waves have the longest wavelengths of all electromagnetic radiation. Because the angular resolution of any telescope decreases as the wavelengths it collects increases, a radio telescope gives a fuzzier picture than any other type of telescope with the same diameter. Indeed, the first, small radio telescopes produced blurry, indistinct images.

VIDEO 3.1   Very large radio telescopes create sharper radio images because, as with optical telescopes, the bigger the dish, the better the angular resolution. For this reason, most modern radio telescopes have reflectors more than 25 m in diameter. Nevertheless, even the largest radio dish in existence (305-m-diameter in Arecibo, Puerto Rico) cannot come close to the resolution of the best optical telescopes. For example, a 6-m optical telescope has 2000 times better resolution than a 6-m radio telescope that detects radio waves of 1-mm wavelength.

To overcome the limitation on resolution set by telescope diameter, radio astronomers often use interferometry to produce high-resolution radio images. Recall from Section 3-13 that interferometry combines the data received simultaneously by two or more telescopes. The telescopes can be kilometers, continents, or even worlds apart. The radio signals received by all of the dishes are made to "interfere," or blend together, and with suitable computer-aided processing, the combined image of the source is sharp and clear. The results are impressive: The resolution of such a system is equivalent to that of one gigantic dish with a diameter equal to the distance between the farthest telescopes in the array.

Interferometry, exploited for the first time in the late 1940s, gave astronomers their first detailed views of "radio objects" in the sky. One of the most complex systems of radio telescopes used in this way began operating in 1980 on the plains of San Agustin near Socorro, New Mexico. Called the Very Large Array (VLA), it consists of 27 concave dishes, each 26 m (85 ft) in diameter (Figure 3-31). The 27 telescopes are positioned

R I V U X G

FIGURE 3-30 A Radio Telescope Recall that the secondary mirror or prime focus on most telescopes blocks incoming light or other radiation. This radio telescope at the National Radio Astronomy Observatory in Green Bank, West Virginia, has its prime focus hardware located off-center from the telescope's 100-m by 110-m oval reflector. By using this new design, there is no such loss of signal. Such configurations are also common on microwave dishes used to receive satellite transmissions for home televisions. (NRAO/AUI/NSF)

along the three arms of a gigantic Y that can span a distance of 36 km (22 mi). Working together, they can create radio images with 0.1" resolution. This system produces radio views of the sky with resolutions comparable to those of the very best optical telescopes.

More recently, radio telescopes separated by thousands of kilometers have been linked together in **very-long-baseline interferometry (VLBI)**. The Very Long Baseline Array (VLBA) consists of ten 25-m radio telescopes located across the United States, from Hawaii to New Hampshire. With a maximum baseline of 8000 km, VLBA has a resolving power of 0.001".

The best angular resolution on Earth is obtained by combining radio data from telescopes on opposite sides of our planet. In that case, features as small as 0.00001" can be distinguished at radio wavelengths—10,000 times sharper than the best views obtainable from single optical telescopes. Radio telescopes are also being put into space and used in even longer-baseline interferometers.

To make radio images more comprehensible, radio astronomers often use false colors or gray scales to display

their radio views of astronomical objects. An example of the use of false colors is shown in Figure 3-32. The most intense radio emission is shown in red, the least intense in blue. Intermediate colors of the rainbow represent intermediate levels of radio intensity. Black indicates that there is no detectable radio radiation. Astronomers who work at other nonvisible-wavelength ranges also frequently use false-color techniques to display views obtained from their instruments.

Consider an example of how nonoptical astronomy, such as is done with these radio telescopes, can overcome limitations of optical telescopes: Late in the eighteenth century, the astronomer Sir William Herschel observed regions of our Milky Way from which no visible light was emitted. "Surely, there is a hole in the heavens," he reported on seeing the first such region. (He was not referring to a modern "black hole," but rather just to an area that was especially dark to his eyes.) In the twentieth century, astronomers discovered that these regions are actually clouds of interstellar gas and dust that prevent visible light from stars beyond them from reaching us, just as thick clouds of water vapor in our atmosphere can obscure the Sun and darken the sky. However, radio waves and some other nonvisible wavelengths pass through interstellar clouds. With the advent of radio telescopes, astronomers got their first glimpses of the variety of objects that lie beyond these clouds.

R I V U X G

WEB LINK 3.9

FIGURE 3-31 The Very Large Array (VLA) The 27 radio telescopes of the VLA system are arranged along the arms of a Y in central New Mexico. Besides being able to change the angles at which they observe the sky, these telescopes can be moved by train cars so that the array can detect either wide areas of the sky (when they are close together, as in this photograph) or small areas with higher resolution (when they are farther apart). The inset shows the traditional secondary mirror assembly in the center of each of these antennas. (Jim Sugar/Corbis; inset: David Nunuk/Science Photo Library/Photo Researchers)

a   R I V U X G

b   R I V U X G

FIGURE 3-32 Visible and Radio Views of Saturn (a) This picture was taken by a camera on board a spacecraft as it approached Saturn. The view was produced by sunlight scattered from the planet's cloud tops and rings. (b) This false-color picture, taken by the VLA, shows radio emission from Saturn at a wavelength of 2 cm. (a: NASA; b: Image courtesy of NRAO/AUI)

On the boundary between radio and infrared radiation, several microwave telescopes have been orbited around Earth. Notable among these are the Cosmic Background Explorer (COBE), launched in 1989, and the *Wilkinson Microwave Anisotropy Probe* (*WMAP*), launched in 2001. They have been used to take the temperature of the universe, results which we will discuss in Chapter 13.

3-15 Infrared and ultraviolet telescopes also use reflectors to collect their electromagnetic radiation

As with radio and microwave telescopes, infrared and ultraviolet telescopes are all reflectors. Indeed, using suitable CCDs, optical reflecting telescopes can also detect infrared and ultraviolet photons with wavelengths near visible light. The most sensitive infrared-detecting CCDs must be cooled to prevent the heat (and hence infrared photons) of the telescope from overwhelming the infrared radiation received from objects in space.

Because water vapor is the main absorber of infrared radiation from space, locating infrared observatories at sites of low humidity can overcome much of the atmosphere's hindrance. For example, the summit of Mauna Kea is exceptionally dry (most of the moisture in the air is below the height of this volcano). Infrared observations are made by the primarily optical-wavelength Subaru and two Keck telescopes, as well as by NASA's 3-m Infrared Telescope Facility (IRTF), all located there.

The best way to avoid the obscuring effects of water vapor is to place a telescope in orbit around Earth. The 1983 Infrared Astronomical Satellite (IRAS), Hubble Space Telescope's Near Infrared Camera and Multi-Object Spectrometer (NICMOS), the Infrared Space Observatory (ISO) (launched in 1995), and the Spitzer Space Telescope (launched in 2003; see Figure 3-33) have done much to reveal the full richness and variety of the infrared sky. Spitzer is the infrared equivalent to HST, one of NASA's Great Observatories.

Infrared telescopes detect radiation from small bodies in the solar system, the bands of dust in our Galaxy, newly formed stars, the dust disks around stars at various stages of their evolution, and the most distant galaxies, whose radiation arrives here primarily at infrared wavelengths, among other infrared-emitting objects. Astronomers have located more than a half million infrared sources in the sky, most of which are invisible at optical wavelengths. As with radio waves, some infrared radiation passes through interstellar clouds of dust and gas, allowing astronomers to see, for example, individual

R I V U X G

(NASA/JPL-Caltech)   R I V U X G

a

b

FIGURE 3-33 Spitzer Space Telescope (a) The mirror assembly for the Spitzer Space Telescope showing the 85-cm objective mirror. (b) Launched in 2003, this Great Observatory is taking images and spectra of planets, comets, gas and dust around other stars and in interstellar space, galaxies, and the large-scale distribution of matter in the universe. *Inset:* An infrared image of a region of star formation invisible to optical telescopes. (a: Balz/SIRTF Science Center; b: NASA/JPL-Caltech; inset: NASA/JPL)

stars at the center of our Galaxy some 245 trillion km (153 trillion mi) away (Figure 3-34).

During the early 1970s, ultraviolet astronomy got off the ground—literally and figuratively. Both Apollo and Skylab astronauts used small telescopes above Earth's atmosphere to give us some of our first views of the ultraviolet sources in space. Small rockets have also been used to place ultraviolet telescopes briefly above Earth's atmosphere. A typical ultraviolet view is shown in Figure 3-29a.

Some of the finest early ultraviolet astronomy was accomplished by the International Ultraviolet Explorer (IUE), which was launched in 1978 and functioned until 1996. The Space Shuttle was transformed into an orbiting observatory twice in the 1990s, carrying aloft and then returning to Earth three ultraviolet telescopes. In 1992 the Extreme Ultraviolet Explorer (EUVE) was launched. As with infrared observations, ultraviolet images reveal sights previously invisible and often

Where in a typical house would infrared detectors indicate most activity?

unexpected. Many objects in space emit ultraviolet radiation that astronomers can use to study the chemistries of these cosmic bodies. Therefore, in 1999, astronomers launched the Far Ultraviolet Spectroscopic Explorer (FUSE). It is providing us with information about such things as how much deuterium (hydrogen nuclei with one neutron) was created when the universe formed, the location of particularly hot interstellar gas and dust, and the chemical evolution of galaxies.

Telescopes dedicated to studying the Sun have observed it from the ground since 1941, and solar telescopes have studied it continuously from space since 1984. Of the latter, SOHO (the Solar and Heliospheric Observatory) and Hinode have ultraviolet detectors and are still in operation. They observe the Sun's outer layers rising and sinking, along with a wide range of energetic activity emanating from the Sun, including particles and radiation racing outward from it, among other things. In Chapter 5 we will discuss these observations and their physical origins.

a R I V U X G

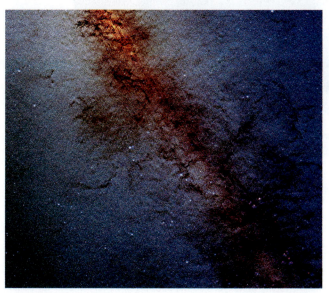

b R I V U X G

FIGURE 3-34 Views of the Milky Way's Central Regions (a) An optical image in the direction of Sagittarius, toward the Milky Way's center. The dark regions are interstellar gas and dust clouds that prevent visible light from beyond them from reaching us. (b) An infrared image of the same area of the sky, showing many more distant stars whose infrared radiation passes through the clouds and is collected by our telescopes. (a & b: Howard McCallon and Gene Kopan/2 MASS Project)

3-16 X-ray and gamma-ray telescopes cannot use normal reflectors to gather information

X rays and gamma rays from space interact strongly with the particles in Earth's atmosphere, preventing these dangerous radiations from reaching our planet's surface. Therefore, direct observations of astronomical sources that emit these extremely short wavelengths must be made from space. Astronomers got their first look at X-ray sources during brief rocket flights in the late 1940s. Several small satellites, launched during the early 1970s, viewed the sources of both X rays and gamma rays in space, revealing hundreds of previously unknown objects, including several black holes (see Chapter 9). X-ray telescopes have also been carried on Space Shuttle missions. The insets in Figure 3-35 show how different our Sun appears when seen through X-ray and visible-light "eyes."

X-ray photons are tricky to collect. Because of their high energies, X rays penetrate nearly head on even highly polished surfaces that they meet. Therefore, normal reflecting mirrors cannot be used to focus them. Instead, X-ray telescopes are designed to deflect photons at a fairly shallow angle because X rays that are barely skimming or grazing a surface can be reflected and thereby focused. This is analogous to skipping a flat rock on water; throw it at a steep angle and it immediately sinks; throw it at a shallow angle and it will skip off the surface. Figure 3-36 shows the design of such "grazing incidence" X-ray telescopes.

After being focused, X rays are detected in several ways. As at infrared, visible, and UV wavelengths, CCDs can detect these photons. Other devices, called *scintillators,* detect the visible light created as X rays pass through them. Still other instruments, called *calorimeters,* detect the heat generated by X rays passing through the detector.

At least seven X-ray telescopes are orbiting Earth today. Among them, the Chandra X-ray Observatory, a NASA Great Observatory named after the Nobel laureate astronomer Subrahmanyan Chandrasekhar, provides images with better than 1" resolution, and the European Space Agency's XMM-Newton has 6" resolution. Other X-ray telescopes are carried high into the atmosphere by balloons. They float at altitudes up to 40 km (25 mi) above Earth's surface and often stay up for several weeks. Whereas low energy X rays do not penetrate that far into the atmosphere, higher energy X rays do, giving these balloon-borne telescopes a wide range of objects to study.

Over 10,000 X-ray sources have been discovered all across the sky. Among these are planetary atmospheres, stars (see the Sun in Figure 3-35), stellar remnants, vast clouds of intergalactic gas, jets of gas emitted by galaxies, black holes, quasars, clusters of galaxies, and a diffuse X-ray glow that fills the universe.

The electromagnetic radiation with the shortest wavelengths and the most energy is the gamma ray. In 1991, the Compton Gamma Ray Observatory, also a NASA Great Observatory, was carried aloft by the Space Shuttle. Named in honor of Arthur Holly Compton, an American physicist who made important discoveries about gamma rays, this orbiting observatory carried four instruments that performed a variety of observations, giving us tantalizing views of the gamma-ray sky until the telescope failed in 2000. At least 10 gamma-ray telescopes are presently in orbit, and several have been flown on balloon missions.

R I V U X G

R I V U X G

a   R I V U X G   b   R I V U X G

FIGURE 3-35 Nonvisible and Visible Radiation (a) This X-ray telescope was carried aloft in 1994 by the Space Shuttle. The inset shows an X-ray image of the Sun. (b) The McMath-Pierce Solar Telescope at Kitt Peak Observatory near Tucson, Arizona (the inverted V-shaped structure), takes visible-light photographs of the Sun, such as the one shown in the inset. Comparing the images in the two insets reveals how important observing nonvisible radiation from astro-nomical phenomena is to furthering our understanding of how the universe operates. (*Top left and right,* L. Golub, Naval Observatory, IBM Research, NASA; *bottom left,* The solar X-ray image is from the Yohkoh mission of ISAS, Japan. The X-ray telescope was prepared by the Lockheed-Martin Solar and Astrophysics Laboratory, the National Astronomical Observatory of Japan, and the University of Tokyo with the support of NASA and ISAS; *bottom right,* NOAO)

Several thousand gamma-ray sources have been discovered. However, gamma rays are too powerful even for grazing incidence telescopes. Therefore, astronomers have devised other methods of detecting them and determining their origin. These include absorbing them in crystals; allowing them to pass through tiny holes called collimators whose directions are well determined; and using chambers in which the gamma rays transform into electrons and positrons (positively charged electrons), leaving a track whose direction can be determined. These techniques are not nearly as precise as those used in other parts of the spectrum, and the best resolution gamma-ray instruments are only accurate to about 5 arcmin.

Astronomers are also taking advantage of the interaction between very high-energy gamma rays and Earth's atmosphere to build telescopes on the ground that indirectly detect the gamma rays. When these photons strike the atmosphere, they often cause particles in the air to move faster than the speed of light in air (but still slower than the speed of light in empty space, the ultimate speed limit). These particles quickly slow down, emitting short bursts of blue light that travel in the same direction as the incoming gamma rays; by detecting this light, it is possible to determine the direction from which the gamma rays came.

ANIMATION 3.3 We now have telescopes with which we can see the energy in the universe from virtually all parts of the electromagnetic spectrum (Figure 3-37). Telescopes provide us with more than just stunning images of objects in space. They also provide information about the chemistry of stars and interstellar gas and dust; the motion of objects toward or away from us; whether stars are rotating; and whether they are alone in space or orbiting a companion, among other things. We will explore how we get this information when we study the various objects in space.

Until recently, photons were the only sources of detailed astronomical information that we had. However, in the past 4 decades, physicists have built detectors for waves and particles whose nature is not electromagnetic. Chapters 5 and 9 discuss two of these devices: neutrino detectors and gravity wave detectors.

3-17 Frontiers yet to be discovered

Before the twentieth century, astronomers were like the blind men in the fable who are trying to describe an elephant. Their perceptions were piecemeal; we could describe parts of the universe but not the whole. Our ancestors did not have the technology that could enable them to see the big picture. However, we are beginning to see it, and, as you will learn in the chapters that follow, our understanding of the cosmos is therefore increasing dramatically. A vast amount of observational information remains to be gathered. Indeed, literally every planet, moon, piece of interplanetary debris, star, stellar remnant, gas cloud, galaxy, quasar, cluster of galaxies, and supercluster of galaxies have stories to tell. Observational astronomy is so new an activity that we are still making new and often unexpected discoveries almost daily. There is still a lot to discover about the elephant.

SUMMARY OF KEY IDEAS

The Nature Of Light

• Photons, units of vibrating electric and magnetic fields, all carry energy through space at the same speed, the speed of light (300,000 km/s in a vacuum, slower in any medium).

• Radio waves, microwaves, infrared radiation, visible light, ultraviolet radiation, X rays, and gamma rays are the forms of electromagnetic radiation. They travel as photons, sometimes behaving as particles, sometimes as waves.

• Visible light occupies only a small portion of the electromagnetic spectrum.

• The wavelength of a visible light photon is associated with its color. Wavelengths of visible light range from about 400 nm for violet light to 700 nm for red light.

• Infrared radiation and radio waves have wavelengths longer than those of visible light. Ultraviolet radiation, X rays, and gamma rays have wavelengths that are shorter.

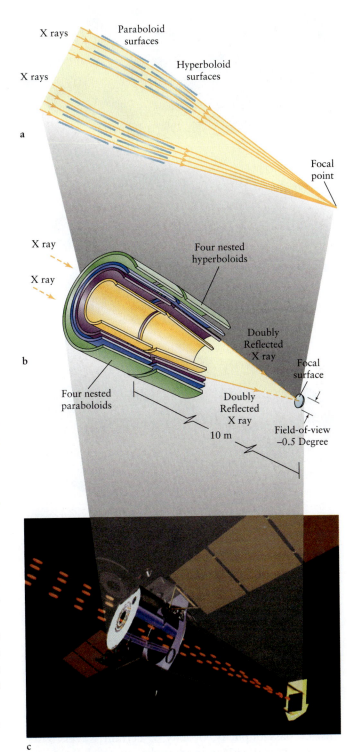

FIGURE 3-36 *Grazing Incidence X-ray Telescopes* X rays penetrate objects they strike head on. To focus them, X rays have to be gently nudged by skimming off cylindrical "mirrors." The shapes of the mirrors optimize the focus. The bottom diagram shows how X rays are focused in the Chandra X-ray Telescope. (a & b: NASA/JPL-Caltech; c: NASA/Chandra X-ray Observatory Center/Smithsonian Astrophysical Observatory)

a   R I V U X G

b   R I V U X G

d   R I V U X G

c   R I V U X G

e   R I V U X G

Optics and Telescopes

• A telescope's most important function is to gather as much light as possible. When possible, it resolves (reveals details) and magnifies an object.

• Reflecting telescopes, or reflectors, produce images by reflecting light rays from concave mirrors to a focal point or focal plane.

• Refracting telescopes, or refractors, produce images by bending light rays as they pass through glass lenses. Glass impurity, opacity to certain wavelengths, and structural difficulties make it inadvisable to build extremely large refractors. Reflectors are not subject to the problems that limit the usefulness of refractors.

• Earth-based telescopes are being built with active and adaptive optics. These advanced technologies yield resolving power comparable to the Hubble Space Telescope.

Nonoptical Astronomy

• Radio telescopes have large, reflecting antennas (dishes) that are used to focus radio waves.

• Very sharp radio images are produced with arrays of radio telescopes linked together in a technique called interferometry.

• Earth's atmosphere is fairly transparent to most visible light and radio waves, along with some infrared and ultraviolet radiation arriving from space, but it absorbs much of the electromagnetic radiation at other wavelengths.

• For observations at other wavelengths, astronomers mostly depend upon telescopes carried above the atmosphere by rockets. Satellite-based observatories are giving us a wealth of new information about the universe and permitting coordinated observation of the sky at all wavelengths.

• Charge-coupled devices (CCDs) record images on many telescopes used between infrared and X-ray wavelengths.

WHAT DID YOU THINK?

1   *What is light?* Light—more properly "visible light," is one form of electromagnetic radiation. All electromagnetic radiation (radio waves, microwaves, infrared radiation, visible light, ultraviolet radiation, X rays, and gamma rays) has both wave and particle properties.

2   *Which type of electromagnetic radiation is most dangerous to life?* Gamma rays have the highest energies of all photons, so they are the most dangerous to life. However, ultraviolet radiation from the Sun is the most common everyday form of dangerous electromagnetic radiation that we encounter.

3   *What is the main purpose of a telescope?* A telescope is designed primarily to collect as much light as possible.

4   *Why do all research telescopes use mirrors, rather than lenses, to collect light?* Telescopes that use lenses have more problems, such as chromatic aberration, internal defects, complex shapes, and distortion from sagging, than do telescopes that use mirrors.

5   *Why do stars twinkle?* Rapid changes in the density of Earth's atmosphere cause passing starlight to change direction, making stars appear to twinkle.

Key Terms for Review

active optics, 89
adaptive optics, 89
angular resolution
   (resolution), 79
Cassegrain focus, 77
charge-coupled device
   (CCD), 80
coudé focus, 77
electromagnetic
   radiation, 68
electromagnetic
   spectrum, 74
eyepiece lens, 76
focal length, 75
focal plane, 76
focal point, 75
frequency, 73
gamma ray, 74
infrared radiation, 73
interferometry, 90
light-gathering power, 78
magnification, 79
Newtonian reflector, 76
objective lens, 82
photon, 68

pixel, 80
primary mirror, 75
prime focus, 77
radio telescope, 90
radio wave, 74
reflecting telescope
   (reflector), 75
reflection, 75
refracting telescope, 75
refraction, 68
refractor, 82
Schmidt corrector
   plate, 84
secondary mirror, 76
seeing disk, 85
spectrum (*plural*
   spectra), 68
spherical aberration, 84
twinkling, 85
ultraviolet (UV)
   radiation, 74
very-long-baseline inter-
   ferometry (VLBI), 92
wavelength (λ), 70
X ray, 74

Review Questions

The answers to all computational problems, which are preceded by an asterisk (*), appear at the end of the book.

1. Describe reflection and refraction. How do these processes enable astronomers to build telescopes?

2. Give everyday examples of refraction and reflection.

3. Describe a reflecting telescope by doing Interactive Exercise 3-1, and transcribe the drawing and correct labels to paper, if requested.

*4. How much more light does a 3-m-diameter telescope collect than a 1-m-diameter telescope?

5. Explain some of the advantages of reflecting telescopes over refracting telescopes.

6. What are the three major functions of a telescope?

7. What is meant by the angular resolution of a telescope?

8. What limits the ability of the 5-m telescope at Palomar Observatory to collect starlight? There are several correct answers to this question.

9. Why will many of the very large telescopes of the future make use of multiple mirrors?

10. What is meant by adaptive optics? What problem does adaptive optics overcome?

11. Compare an optical reflecting telescope to a radio telescope. What do they have in common? How are they different?

12. Why can radio astronomers observe at any time of the day or night whereas optical astronomers are mostly limited to observing at night?

13. Why must astronomers use satellites and Earth-orbiting observatories to study the heavens at X-ray wavelengths?

14. What are NASA's four Great Observatories, and in what parts of the electromagnetic spectrum do (or did) they observe?

15. Why did Rømer's observations of the eclipses of Jupiter's moons support the heliocentric, but not the geocentric, model of the solar system?

Advanced Questions

16. Advertisements for home telescopes frequently give a magnification for the instrument. Is this a good criterion for evaluating such telescopes? Explain your answer.

*17. The observing cage in which an astronomer sits at the prime focus of the 5-m telescope at Palomar Observatory is about 1 m in diameter. What fraction of the incoming starlight is blocked by the cage? *Hint:* The area of a circle of diameter d is $\pi d^2/4$, where $\pi \sim 3.14$.

*18. Compare the light-gathering power of the Palomar Observatory's 5-m telescope to that of the fully dark-adapted human eye, which has a pupil diameter of about 5 mm.

19. Show by means of a diagram why the image formed by a simple refracting telescope is "upside down."

*20. Suppose your Newtonian reflector has a mirror with a diameter of 20 cm and a focal length of 2 m. What magnification do you get with eyepieces whose focal lengths are a. 9 mm, b. 20 mm, and c. 55 mm?

21. Why does no major observatory have a Newtonian reflector as its primary instrument, whereas Newtonian reflectors are popular among amateur astronomers?

22. From the ground, how can astronomers detect gamma-ray sources in space?

23. Why will many of the very large telescopes of the future have ultrathin primary mirrors?

Discussion Questions

24. Discuss the advantages and disadvantages of using a relatively small visible-light telescope in Earth's orbit (for example, the 2.4-m Hubble Space Telescope) versus a large visible-light telescope on a mountaintop (for example, the 8.3-m Subaru telescope on Mauna Kea, Hawaii).

25. If you were in charge of selecting a site for a new observatory, what factors would you consider?

26. Consider two identical Cassegrain telescope mirrors. One is set up as a prime focus telescope, whereas the other is used in a Cassegrain telescope. a. Sketch both telescopes. b. What are the differences between the two that make each useful in different observing situations?

What If . . .

27. Telescopes were first invented today? What objects or areas of the sky would you recommend that astronomers explore first? Why?

28. We had eyes sensitive to radio waves? How would we be different, and how would our visual perceptions of the world be different?

29. Humans were unable to detect any electromagnetic radiation? How would that change our lives, and what alternatives might evolve (some species indeed have) to provide information about distant objects?

Web Questions

30. Several telescope manufacturers build Schmidt-Cassegrain telescopes. These use a correcting lens in an arrangement like that shown in Figure 3-22c. Consult advertisements on the Web and list the dimensions, weights, and costs of some of these telescopes. Why are they popular among amateur astronomers?

31. Access the Active Integrated Media Module "Telescope Magnification" in Chapter 3 of the *Discovering the Universe: From the Stars to the Planets* Web site. A common telescope found in department stores is a 3-in (76-mm) diameter refractor with an $f_{obj} = 750$ mm that boasts a magnification of 300 times. Use the magnification calculator to determine the magnifications that are achieved by using each of the following commonly found eyepieces on that telescope: Eyepiece A with focal length 40 mm; Eyepiece B with focal length 25 mm; Eyepiece C with focal length 12 mm; and Eyepiece D with focal length 2.5 mm. Which eyepiece was used in the advertisement and why was that one chosen?

32. Discuss the advantages and disadvantages of setting up an observatory on the Moon. *Hint:* To get a broad perspective on this question, you might find it useful to explore the Web for the challenges of living on the Moon.

33. The Large Zenith Telescope (LZT) in British Columbia, Canada, uses a 6-m *liquid* mirror made of mercury. Use the Web to investigate this technology. How can a liquid metal be formed into the necessary shape of a telescope mirror? What are the advantages and disadvantages of a liquid mirror?

Observing Projects

34. During the daytime, obtain a telescope and several eyepieces of differing focal lengths. If you can determine the focal length of the telescope's objective lens or mirror (often printed on it), calculate and record the magnifying power for each eyepiece. Focus the telescope on some familiar object, such as a distant lamppost or tree. **DO NOT FOCUS ON THE SUN! Looking directly at the Sun through a telescope will cause blindness.** Describe the image you see through the telescope. Is it upside down? How does the image move as you slowly and gently shift the telescope left and right or up and down? Examine the distant objects under different magnifications. How does the field of view change as you go from low magnification to high magnification?

35. **Determine what fraction of stars visible to the naked eye you can see from your location.** This exercise is best done outdoors with a laptop computer or through a window at night in a very dark room. On a dark, clear night, run *Starry Night Enthusiast*™ . On the Menu, select **View > Hide Daylight**. Open the **Options** side pane, then **Local View**, and click on **Local Light Pollution**. Place the cursor over

the words **Local Light Pollution** and click on the **Local Light Pollution Options** button that appears. Slide the pollution level bar until the view matches your night sky. The goal is to estimate what fraction of the visible stars you are seeing compared to the total number of stars you could be seeing under ideal conditions. Carefully do the following on the screen: Using a roughly 8 cm (3 in) square section of the sky on the screen (a square about half the length of a typical pen on each side), count and record the number of stars with your present setting. Now set the **Local Light Pollution** to **less** (which essentially gives you ideal conditions) and repeat the count of visible stars. Divide the first number you count by the second. What fraction of the stars were you seeing? To see how much light pollution occurs in large cities, adjust the slide bar in the **Local Light Pollution Options** to the far right, for maximum light pollution and repeat the star-counting within the same limited sky region. What fraction of the stars are large city viewers seeing?

36. Observe the stars when the Moon is either full, new, or in a quarter phase. Record the phase of the Moon and note, qualitatively, whether you see many more stars than those just visible in the well-known asterisms such as the Big Dipper or Orion. You are making these estimates so you can compare them with observations on a later date. Repeat these observations on a clear night about a week later, again noting the phase of the Moon and the numbers of visible stars. Compare the numbers of stars on the two nights. On which night did you see more stars? Why? During which three phases of the Moon do you expect that astronomers most like to make their observations?

37. On a clear night, view the Moon, a planet, and a star through a telescope using eyepieces of various focal lengths. (Use your *Starry Night Enthusiast™* program or consult a source on the Internet or such magazines as *Sky & Telescope* or *Astronomy* to determine the phase of the Moon and the locations of the planets.) How do the images change as you view with increasing magnification? Do they become distorted at any level of magnification?

38. In this exercise, you will explore the appearance of a nearby galaxy, M31, the Andromeda Galaxy, under conditions that simulate the use of binoculars or a telescope to enhance your unaided eye view. You can also evaluate the effects of greater light-gathering power and improved resolution that comes from the use of these optical aids. Click **Home** on the toolbar to return to your local view and stop the flow of time. In **Local View** unclick **Local Horizon**. Open the *Find* pane and start typing *"Andromeda Galaxy"* in the *Query* box. Double click on the name when it appears below the box. The sky will center on the galaxy,

which will also be labeled. Check that the field of view is 100°. This is the approximate view that one has of this galaxy with the unaided eye. **a.** Noting that you may have to look at your screen carefully to find it, sketch this galaxy along with a background of several bright neighboring stars, adding notes about its appearance. **b.** Now use the **Zoom** facility on the toolbar to adjust the field of view to about 10° to match that of a pair of powerful binoculars. Again, sketch and describe the galaxy's appearance. **c.** Finally, adjust the field of view to about 2°, to match that of a small telescope. Here, the galaxy will extend across the full field. Sketch and describe this galaxy again, noting particularly any details now apparent at this higher magnification. Locate and identify two galaxies M32 and M110 in this image. **d.** Is the *resolution* of this telescope image better, the same, or worse than that seen through binoculars or with the unaided eye?

39. Extra solar Planets In this exercise we will use *Starry Night Enthusiast™* to observe the distribution of known stars with planets orbiting them.

Display the stars with extra solar planets by clicking on **Favourites > Stars > Extra Solar Planets**. This will take you to a distance of about 90 ly from the solar system, so that you can see the stars with extra solar planets (marked so by circles) in our vicinity. Stop the flow of time. **a.** Open the **Info** pane and expand the **Other Data** panel. Select and double-click on a star with an extra-solar planet to update the *Other Data* information. (This may take some practice, as you must hit the dot in the middle of that star's circle.) Note that the *Other Data* panel contains a list of data relevant to the planet. Form a table to list the star's *Name*, along with the *Extrasolar mass*, the *Extrasolar semi-major axis* and the *Extrasolar period* of the associated planet from the *Other Data* table for comparison with the Sun's planets. Repeat this process for about 10 other stars with extra-solar planets. What general conclusions can you draw about these planets from your list? **b.** Open the **Find** pane and type "Sun" or double-click on the **Sun** to center the view upon our Sun. Increase the distance from Earth to about 300 ly, using the **Up** button below the *Viewing Location* on the toolbar. Hold down the **Shift** key while clicking, holding and moving the mouse to rotate the view in 3-dimensional space. As you pivot the star field around the Sun, examine the distribution of stars with extrasolar planets. Describe this distribution. Are these stars clustered on one side of it? Are they located in a few clumps? Are they located in a thin plane? Explain why they are in the distribution you see, keeping in mind that the Milky Way is a flattened disk of stars, gas, and dust and allowing for the difficulty of detecting these very distant planets.

WHAT IF... Humans Had Infrared-Sensitive Eyes?

Our eyes are sensitive to less than a trillionth of 1% of the electromagnetic spectrum—what we call *visible light*. But this minuscule resource provides an awe-inspiring amount of information about the universe. We interpret visible-light photons as the six colors of the rainbow—red, orange, yellow, green, blue, and violet. These colors combine to form all of the others that make our visual world so rich. But the Sun actually emits photons of all wavelengths. So, what would happen if our eyes had evolved to sense another part of the spectrum?

A Darker Vision Gamma rays, X rays, and most ultraviolet radiation do not pass through Earth's atmosphere. Because the world is illuminated by sunlight, Earth and the sky would look dark, indeed, if our eyes were sensitive only to these wavelengths. Radio waves, in contrast, easily pass through our atmosphere. But to see the same detail from radio waves that we now see from visible-light photons, our eyes would require diameters 10,000 times larger. Each would be the size of a baseball infield!

What about infrared radiation? Although not all incoming infrared photons get through our atmosphere, short-wavelength ("near") infrared radiation passes easily through air. Depending on wavelength, the Sun emits between one-half and a ten-billionth as many infrared photons as red-light photons. Fortunately, most of these are in the near infrared.

Heat-Sensitive Vision To see infrared photons, human eyes would need to be only 5 to 10 times larger. Some snakes have evolved infrared vision. Portable infrared "night vision" cameras and goggles are available to us humans. Because everything that emits heat emits infrared photons, infrared sight would be very useful. Also, not everything we see with infrared sight would be due just to reflected sunlight—hotter objects would be intrinsically brighter than cool ones. For example, seeing infrared would allow us to observe changes in a person's emotional state. Someone who is excited or angry often has more blood near the skin and, thus, releases more infrared radiation (heat) than normal. Conversely, someone who is scared has less blood near the skin and, thus, emits less heat.

Night Vision The night sky would be a spectacular sight through infrared-sensitive eyes. Gas and dust clouds in the Milky Way absorb visible light, thus preventing the light of distant stars from getting to Earth. However, because most infrared radiation passes through these clouds, unaided, we would be able to see distant stars that we cannot see today. On the other hand, the white glow of the Milky Way, which is caused by the scattering of starlight by interstellar clouds, would be dimmer, because the gas and dust clouds do not scatter infrared light as much as they do visible light. (The haze created by the Milky Way would not vanish, however, because when gas and dust clouds are heated by starlight, they emit their own infrared radiation.)

Our concept of stars would be different, too. Many stars, especially young, hot ones, are surrounded by cocoons of gas and dust that emit infrared radiation. This dust is heated by the nearby stars. Instead of appearing as pinpoints, many stars would appear to be surrounded by wild strokes of color, and we would have an Impressionist sky.

The infrared (heat) from this kissing couple has been converted into visible light colors so that we can interpret the invisible radiation. The hottest regions are white, with successively cooler areas shown in yellow, orange, red, green, sky blue, dark blue, and violet. (D. Montrose/Custom Medical Stock Photo)

CHAPTER 4

Atomic Physics and Spectra

Visible spectrum of the Sun (N.A. Sharp, NOAO/ AURA/NSE)

R I V U X G

WHAT DO YOU THINK?

1 Which is hotter, a "red-hot" or a "blue-hot" object?

2 What color does the Sun emit most brightly?

3 How can we determine the age of space debris found on Earth?

Answers to these questions appear in the text beside the corresponding numbers in the margins and at the end of the chapter.

"A star, galaxy, or interstellar gas cloud was discovered yesterday," is a statement that is true virtually every day of the year. The first thing that astronomers want to know about that newly discovered object is—everything. Much of this information is gained by focusing electromagnetic radiation from the object on a diffraction grating that separates its wavelengths. For astronomers, visible light is only the proverbial tip of the electromagnetic radiation iceberg. Probing other wavelengths, scientists in the past two centuries have discovered that most objects in space emit varying amounts of all types of electromagnetic radiation. This spe ctral analysis, along with information gleaned from the shapes and locations of objects, provides us with virtually all of the information we have about the cosmos. Furthermore, these observations provide the data needed to confirm the models of astronomical objects that scientists have developed.

In particular, you will see in this chapter that the resulting *spectrum* enables us to learn a variety of the physical properties of the object, including its surface temperature, chemical composition, mass, size, rotation rate, motion through space, and sometimes its distance from Earth. We begin this ongoing adventure of discovery by discussing the nature of matter and how it interacts with electromagnetic radiation. In subsequent chapters, we consider related issues, such as why stars shine (that is, emit electromagnetic radiation), where different elements found on Earth and throughout the universe come from, and why objects have the motions they do, among many other things.

In this chapter you will discover

- the origins of electromagnetic radiation
- the structure of atoms
- that stars with different surface temperatures emit different intensities of electromagnetic radiation

- that astronomers can determine the chemical compositions of stars and interstellar clouds by studying the wavelengths of electromagnetic radiation that they absorb or emit
- how to tell whether an object in space is moving toward or away from Earth

BLACKBODY RADIATION

We begin our study of why we see what we see with a familiar property of matter—temperature. We will discover that for many astronomical objects, their temperatures determine the relative numbers of photons that they emit at all different wavelengths.

4-1 An object's peak color shifts to shorter wavelengths as it is heated

Imagine that you have mounted an iron rod in a vise in a completely darkened room. You cannot see the rod because it emits too few visible photons, and you cannot feel its warmth without touching it because it is at the same temperature as the air in the room. Now imagine that there is also a propane torch in the room. You light it and heat the iron rod for several seconds, until it begins to glow. At the same time, you can feel the rod warm up, meaning that it is also emitting more infrared radiation.

The rod's first visible color is red—it glows "red hot" (Figure 4-1a). Heated a little more, the rod appears orange and brighter, and feels hotter (Figure 4-1b). Heated more, it appears yellow and brighter still, and feels even hotter (Figure 4-1c). After even more heating, the rod appears white-hot and brighter yet. If it does not melt, further heating will make the rod appear blue and even brighter, and feel even hotter. As confirmed by the photographs in Figure 4-1, this experiment shows how the amount of electromagnetic radiation that any object emits changes with its temperature. A "red-hot" object is the coolest of all glowing bodies.

If you were to put the light emitted by a heated object, such as the iron rod, through a prism or diffraction grating, you would discover that all wavelengths are present and that one is brightest. This peak wavelength is the color that the object appears to have. The experiment with the rod shows two things about the electromagnetic radiation emitted by objects as their temperatures change:

1. As an object heats up, it gets brighter, emitting more electromagnetic radiation at all wavelengths.

2. The brightest color (most intense wavelength) of the emitted radiation changes with temperature.

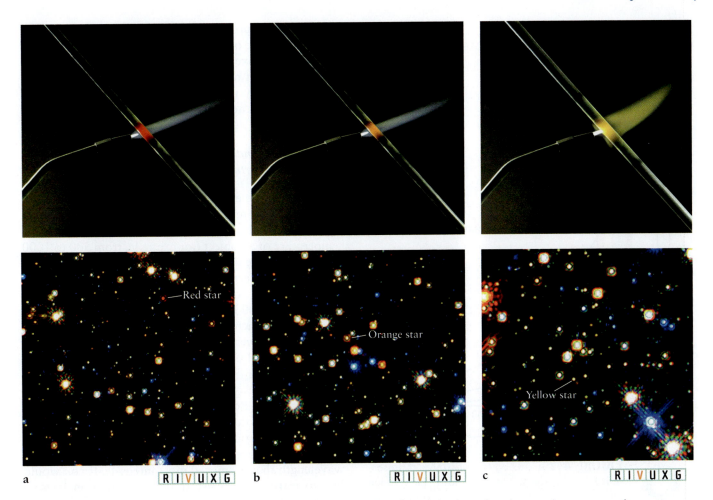

a R I V U X G   b R I V U X G   c R I V U X G

FIGURE 4-1 Objects at different temperatures have different colors and brightnesses This sequence of photographs shows the changing appearance of a piece of iron as it is heated. As the temperature increases, the amount of energy radiated by the bar increases, and so it appears brighter. The apparent color of the bar also changes because, as the temperature increases, the peak wavelength of light emitted by the bar decreases. The stars shown have roughly the same temperatures as the bars above them. (top photos: © 1984 Richard Megna Fundamental Photographs; bottom star photos: NASA)

Most objects big enough to be seen with the naked eye continuously emit electromagnetic radiation as just described. The peak wavelength of any object's radiation, that is, where it emits most intensely, is denoted λ_{max}. When the object is relatively cool, such as a rock or an animal, λ_{max} is a radio, microwave, or infrared wavelength. When it is hot enough, such as a fire or the Sun, λ_{max} is in the range of visible light, giving a hot object its characteristic color. Exceptionally hot objects, such as some stars, emit λ_{max} in the ultraviolet part of the electromagnetic spectrum.

 Stars, molten rock, and iron bars are approximations of an important class of objects that astrophysicists call **blackbodies**. An ideal blackbody absorbs all of the electromagnetic radiation that strikes it. The incoming radiation heats up the blackbody, which then reemits the energy it has absorbed, but with different intensities at each wavelength than it received. Furthermore, the pattern of radiation emitted by blackbodies is independent of their chemical compositions. By measuring the intensity of radiation emitted by a blackbody at several wavelengths, it is possible to plot a curve of its emissions over all wavelengths (Figure 4-2). Ideal blackbodies have smooth **blackbody curves**, whereas objects that approximate blackbodies, such as the Sun, have more jagged curves whose variations from the ideal blackbody are caused by other physics.

The total amount of radiation emitted by a blackbody at each wavelength depends only on the object's temperature and how much surface area it has. The bigger it is, the brighter it is at all wavelengths. However, the *relative* amounts of different wavelengths (for example, the intensity of light at 750 nm compared to the intensity at 425 nm) depend on just the body's temperature. So, by examining the relative intensities of an object's blackbody curve, we are able to determine its temperature, regardless of how big or how far away it is. This is analogous to how a thermometer tells your temperature no matter how big you are. (Temperatures throughout this book are expressed in Kelvins. If you are not familiar with the Kelvin temperature scale, review Appendix D: Temperature Scales.)

The higher the temperature of a blackbody, the shorter the wavelength of maximum emission (the wavelength at which the curve peaks).

Visible light

The higher the temperature of a blackbody, the more light it emits at *all* wavelengths.

12,000 K

6000 K

3000 K

Intensity

Wavelength (nm)

FIGURE 4-2 Blackbody Curves Three representative blackbody curves are shown here. Each curve shows the intensity of radiation over a wide range of wavelengths emitted by a blackbody at a particular temperature.

Rather than absorb and re-radiate light from an outside source like our iron rod, stars radiate electromagnetic radiation that is generated inside them. Even so, stars behave nearly like blackbodies, and the self-generated radiation they emit follows the idealized blackbody curves fairly well.

If a blackbody's peak wavelength is in the infrared, it is often visible to our eyes. What color will it have?

4-2 The intensities of different emitted colors reveal a star's temperature

In 1893, the German physicist Wilhelm Wien showed that

The peak wavelength of radiation emitted by a blackbody is inversely proportional to its temperature.

In other words, the hotter any object becomes, the shorter its λ_{max}, and vice versa. **Wien's law** is the mathematical relationship between the location of the peak for each curve in Figure 4-2 and that blackbody's temperature.

Wien's law proves very useful in computing the surface temperature of a star. It says that all we need to find is the peak wavelength of its electromagnetic radiation—we do not need to know the star's size, distance, or any other physical property.

In 1879, the Austrian physicist Josef Stefan observed that

An object emits energy per unit area at a rate proportional to the fourth power of its temperature in Kelvins.

Think of the electric heating element on a stove. As this device becomes hotter, it gets brighter and emits more energy. Stefan's result says that if you double the temperature of an object (for example, from 500 to 1000 K), the energy emitted each second from each square meter of the object's surface increases by a factor of 2^4, or 16 times. If you triple the temperature (for example, from 500 to 1500 K), the rate at which energy is emitted increases by a factor of 3^4, or 81 times. Stefan's experimental results were supported by Ludwig Boltzmann's theory in 1885. In their honor, the intensity-temperature relationship for blackbodies is named the **Stefan-Boltzmann law.**

The Stefan-Boltzmann law and Wien's law are powerful tools that describe two basic properties of blackbody radiation, namely, the relative rate at which photons of different wavelengths are emitted and the wavelength that is emitted most intensely. If you would like to learn how to use these two radiation laws to make predictions, see *An Astronomer's Toolbox 4-1*. Blackbody curves, such as those shown in Figure 4-2, provide the same information as Wien's law and the Stefan-Boltzmann law, as well as show the intensity of each blackbody at all wavelengths. Note that the blackbody curves for objects with different temperatures never touch or cross. Therefore, when astronomers measure the intensity of radiation from a star at a few wavelengths, they can determine the object's temperature by fitting these measurements to the appropriate, unique blackbody curve.

Although intensity curves of stars "closely follow" blackbody curves, stars are not ideal blackbodies. The *differences* between an ideal blackbody curve and the curves seen from actual stars reveal stellar chemistries, the presence of companion stars and planets too dim to see, and the motions of stars toward or away from us, among other things.

Figure 4-3 shows how the intensity of sunlight varies with wavelength. The blackbody curve for a body with a temperature of 5800 K is also plotted in this figure. Note how the observed intensity curve for the Sun nearly reproduces the ideal blackbody curve at most wavelengths. Because the observed intensity curves for most stars and the idealized blackbody curves are so closely correlated, the laws of blackbody radiation can be applied to starlight. The peak of the intensity curve

AN ASTRONOMER'S TOOLBOX 4-1

The Radiation Laws

Wien's law can be stated as a simple equation. If λ_{max}, the wavelength of maximum intensity, is measured in meters and T is the temperature of the blackbody measured in Kelvins, then

$$\lambda_{max} = \frac{2.93 \times 10^{-3}}{T}$$

Example: The Sun's maximum intensity is at a wavelength of about 500 nm, or 5×10^{-7} m. From Wien's law, we can calculate the Sun's surface temperature as

$$T = \frac{2.9 \times 10^{-3}}{5 \times 10^{-7}} = 5800K$$

The total energy emitted from each square meter of an object's surface each second is called the **energy flux**, F. In this context, flux means "rate of flow." Using this concept, T as the temperature of the object in Kelvins, and σ (lowercase Greek sigma) as a number called the Stefan-Boltzmann constant, we can write the Stefan-Boltzmann law as

$$F = \sigma T^4$$

where

$$\sigma = 5.67 \times 10^{-8} \text{ J/(m}^2 \cdot \text{K}^4 \cdot \text{s)}$$

with J being the energy unit joules, m the unit meters, K the unit degrees Kelvin, and s the unit seconds.

Because F is the energy emitted per second from each square meter of an object, multiplying F by the surface area, $4\pi r^2$, where r is the object's radius, yields the total energy emitted by a spherical object each second. This quantity, denoted L, is called **luminosity**:

$$L = F \cdot 4\pi r^2 = \sigma T^4 4\pi r^2$$

Luminosity is the total energy emitted per second by an entire object.

Compare! We consider one application of the Stefan-Boltzmann law. Suppose you observe two stars of equal size—one with a surface temperature of 10,000 K and the other with the same surface temperature as the Sun (5800 K). You can use the Stefan-Boltzmann law to determine how much brighter the hotter star is

$$\frac{F_{hotter}}{F_{colder}} = \frac{\sigma \times 10,000^4}{\sigma \times 5800^4} \approx 8.8$$

So, the hotter star emits about 8.8 times as much energy from each square meter of its surface than does the cooler star. Because the stars are the same size, the hotter one is therefore 8.8 times brighter than the cooler one.

Try these questions: The color yellow is centered around 550 nm. If the Sun were actually yellow, what would be its surface temperature? (To find out why it is not yellow, see *Guided Discovery: The Color of the Sun* on page 108.) What is the peak wavelength in nanometers given off by a blackbody at room temperature of 300 K? Referring to Figure 3-6, what part of the electromagnetic spectrum is that in?

(Answers appear at the end of the book.)

---

for the Sun is at a wavelength of about 500 nm, which is in the blue-green part of the visible spectrum.

The shapes of ideal blackbody curves were first derived mathematically in 1900 by the German physicist Max Planck. To do this, he had to assume that objects receive electromagnetic radiation as individual packets of energy. This surprising assumption was reinforced in 1905 by Albert Einstein, who suggested that light—a wave—behaves as a beam of particles in some situations, as discussed in Section 3-3.

We can now explain our experiment with the iron bar at the beginning of this chapter in terms of the energies and intensities of photons. Recall from Section 3-3 that the energy carried by a photon is inversely

FIGURE 4-3 **The Sun as a Blackbody** This graph shows the similarity between the intensity of sunlight and the intensity of radiation from a 5800 K blackbody over a wide range of wavelengths. Earth's atmosphere scatters shorter-wavelength photons more than it scatters longer-wavelength ones. As seen from Earth, this scattering lowers the intensity of sunlight and shifts the blackbody peak to the right from where it actually occurs. To avoid these effects, the measurements of the Sun's intensity in this figure were made in space.

GUIDED DISCOVERY

The Color of the Sun

Different people perceive the Sun to have different colors. To some it appears white, to others yellow. Still others, who notice it at sunset, believe it to be orange or even red. But we have seen that the Sun actually emits all colors. Moreover, the peak in the Sun's spectrum falls between blue and green. Why doesn't the Sun appear turquoise? Several factors affect our perception of its color.

Before reaching our eyes, visible sunlight passes through Earth's atmosphere. All wavelengths are absorbed and reemitted by the molecules in the air, a process called *scattering*. Violet light is scattered most strongly, followed in decreasing order by blue, green, yellow, orange, and red. That means that more violet, blue, and green photons are scattered by Earth's atmosphere than are yellow photons. The intense scattering of violet, blue, and green has the effect of shifting the peak of the Sun's intensity entering our eyes from blue-green toward yellow. (The sky is blue because of the strong scattering of blue light—it isn't violet because the Sun emits many fewer violet photons than blue ones.)

The perception of a yellow Sun is further enhanced by our eyes themselves. Our eyes do not see all colors equally well. Rather, the light-sensitive cones in our eyes each respond to one of three ranges of colors, which are centered on red, yellow, and blue wavelengths. None of the cones is especially sensitive to blue-green photons. By adding together the color intensities detected by the three types of cones, our brains *re-create* color. After combining all of the light that it can, the eye is most sensitive to the yellow-green part of the spectrum. We see blue and orange less well and violet and red most poorly. Although the eye sees yellow and green light about equally well, the dominant color from the Sun is yellow because the air scatters green light, and our eyes are relatively insensitive to blue-green. Therefore, a casual glance at the Sun often leaves the impression of a yellow object.

Above Earth's atmosphere, the Sun's light is so intense that it saturates the color-sensitive cones in our eyes. The Sun appears "blindingly white," in space, an astronaut I know told me.

At sunrise and sunset we see an orange or red Sun. This occurs because close to the horizon the Sun's violet, blue, green, and even yellow photons are all strongly scattered by the thick layer of atmosphere through which they travel, leaving the Sun looking redder and redder as it sets.

(Tim Henry and the OSPAN team, U.S. Air Force Research Laboratory   R I V U X G

ImageGap/Alamy   R I V U X G

Roger Ressmeyer/CORBIS)   R I V U X G

TABLE 4-1 Some Properties of Electromagnetic Radiation

	Wavelength (nm)	Photon energy (eV)*	Blackbody temperature (K)
Radio	$>10^7$	$<10^{-4}$	<0.03
Microwave**	10^7 to 4×10^5	10^{-4} to 3×10^{-3}	0.03 to 30
Infrared	4×10^5 to 7×10^2	3×10^{-3} to 2	30 to 4,100
Visible	7×10^2 to 4×10^2	2 to 3	4,100 to 7,300
Ultraviolet	4×10^2 to 10^1	3 to 10^3	7,300 to 3×10^6
X ray	10^1 to 10^{-2}	10^3 to 10^5	3×10^6 to 3×10^8
Gamma ray	$<10^{-2}$	$>10^5$	$>3 \times 10^8$

Note: > means greater than; < means less than.

*1 eV = 1.6×10^{-19} J.

**Microwaves, listed here separately, are often classified as radio waves or infrared radiation.

proportional to its wavelength. In other words, long-wavelength photons, such as radio waves, carry little energy, while short-wavelength photons, such as X rays and gamma rays, each carry much more energy. When the rod was cool, it mostly emitted lower-energy infrared and radio photons; when it was hot enough to glow, it emitted mostly visible photons and more of all types of photons, which is why it looked brighter and felt hotter. The relationship between the energy of a photon and its wavelength is called **Planck's law,** as explored in *An Astronomer's Toolbox 3-1.*

Together, Planck's law and Wien's law relate the temperature of an object to the energy of the photons that it emits (Table 4-1). A cool object emits primarily long-wavelength photons that carry little energy, while a hot object emits mostly short-wavelength photons that carry much more energy. In later chapters, we will find these ideas invaluable for understanding how stars of various temperatures interact with gas and dust in space.

IDENTIFYING THE ELEMENTS BY ANALYZING THEIR UNIQUE SPECTRA

In 1814, the German optician Joseph von Fraunhofer repeated Newton's classic experiment of shining a beam of sunlight through a prism (recall Figure 3-1). However, Fraunhofer magnified the resulting rainbow-colored spectrum. He discovered that the solar spectrum contains hundreds of fine dark lines, which became known as **absorption lines** because the light of these colors has been absorbed by gases between the Sun and the viewer on Earth. Fraunhofer counted more than 600 such lines, and today physicists have detected more than 30,000 of them. Thousands of spectral lines are visible in the photograph of the Sun's spectrum shown in Figure 4-4.

By the mid-1800s, chemists discovered that they could produce spectral lines in the laboratory. Around 1857, the German chemist Robert Bunsen invented a special gas burner that produces a clean, constant flame. Certain chemicals are easy to identify by the distinctive colors emitted when bits of the chemical are sprinkled into the flame of a Bunsen burner.

R I V U X G

FIGURE 4-4 Solar Spectra Starlight passing through a prism or diffraction grating spreads into its component colors. From such spectra we can learn an incredible amount about stars, including their masses, surface temperatures, diameters, chemical compositions, rotation rates, and motions toward or away from us. This image, called a *spectrogram*, shows the spectrum of the Sun sliced and stacked to fit on this page. In it you can see thousands of absorption lines. As you study spectra, also note the distinct differences between the intensities of the various colors emitted by each star and by different stars. (N. A. Sharp, NOAO/AURA/NSF)

Bunsen's colleague Gustav Kirchhoff suggested that light from the colored flames could best be studied by passing it through a prism. By separating the colors, the chemists could see exactly which ones were present. The process of analyzing the spectrum of an object is called *spectroscopy*. Bunsen and Kirchhoff collaborated in designing and constructing the first **spectroscope**. This device consists of a narrow slit, a prism, and several lenses that straighten the light rays and magnify the spectrum so that it can be closely examined (Figure 4-5).

The chemists discovered that the spectrum from a flame consists of a pattern of thin, bright spectral lines, called **emission lines**, against a dark background (not a blackbody spectrum!). They next found that *the number of lines produced and their colors are unique to the element or compound that produces them.* Thus, in 1859, the technique of **spectral analysis**, the identification of chemical substances by their spectral lines, was born.

> Where are radio waves relative to the horizontal axis of Figure 4-2?

4-3 Each chemical element produces its own unique set of spectral lines

A chemical **element** is composed of atoms that cannot be broken down into more basic units and still retain the element's properties. By the mid-1800s, chemists had already identified such familiar elements as hydrogen, oxygen, carbon, iron, gold, and silver. Spectral analysis promptly led to the discovery of additional elements, many of which are quite rare.

After Bunsen and Kirchhoff recorded the prominent spectral lines of all of the known elements, they began to discover other spectral lines

Gustav Kirchhoff (left) and Robert Bunsen (Reproduced courtesy of the Library and Information Centre, The Royal Society of Chemistry.)

1. Add a chemical substance to a flame.

2. Send light from the flame through a narrow slit, then through lenses and a prism.

3. Bright lines in the spectrum show that the substance emits light at specific wavelengths only.

FIGURE 4-5 Early Spectroscope In the mid-1850s, Kirchhoff and Bunsen discovered that when a chemical substance is heated and vaporized, the resulting spectrum exhibits a series of bright spectral lines when passed through a slit, lenses, and a prism. This device is called a spectroscope. In addition, they found that each chemical element produces its own characteristic pattern of spectral lines. The lenses focus and magnify the spectrum.

in mineral samples. In 1860, for example, they found a new line in the blue portion of the spectrum of mineral water. After chemically isolating the previously unknown element responsible for the line, they named it *cesium* (from the Latin *caesius*, meaning "gray-blue"). The next year, a new spectral line in the red portion of the spectrum of a mineral sample led to the discovery of the element rubidium (from *rubidus*, for "red").

During a solar eclipse in 1868, astronomers found a new spectral line in the light coming from the upper atmosphere of the Sun when the main body of the Sun was hidden by the Moon. This line was attributed to a new element, which was named *helium* (from the Greek *helios*, meaning "Sun"). Helium was not actually discovered on Earth until 1895, when it was identified in gases obtained from a uranium compound.

Absorption spectrum of the Sun ⟶

Emission spectrum of iron (in a laboratory on Earth) ⟶

For each emission line of iron, there is a corresponding absorption line in the solar spectrum; hence, there must be iron in the Sun's atmosphere.

R I V U X G

FIGURE 4-6 Iron in the Sun's Atmosphere The upper (absorption) spectrum is a portion of the Sun's spectrum from 425 to 430 nm. Numerous dark spectral lines are visible. The lower (emission) spectrum is a corresponding portion of the spectrum of vaporized iron. Several emission lines can be seen against the black background. The fact that the iron lines coincide with some of the solar absorption lines proves that there is some iron (albeit a very tiny amount) in the Sun's atmosphere. (Carnegie Observatories)

The chemical elements are most conveniently displayed in the form of a **periodic table** (Appendix H). In addition to the 92 naturally occurring elements, there are artificially-produced ones. All of the human-made elements are heavier than uranium (U), and all are highly radioactive, meaning that they spontaneously decay into lighter elements shortly after being created in the laboratory.

INSIGHT INTO SCIENCE

Seek Relationships Finding how properties of different things relate to one another often provides invaluable insights into how they work. The periodic table, for example, enables scientists to determine the properties of similar chemical elements. Keep an eye out for other such relationships throughout this book.

During some experiments, Bunsen and Kirchhoff examined the spectra of hot sources whose light they passed through cooler gas. Instead of seeing a continuum of wavelengths, as they would have seen had the cool gas not been there, they saw dark *absorption lines* among the colors of the rainbow created by the hot source. (This is what Fraunhofer had seen earlier from the Sun—the cooler gases in his results were the atmospheres of Earth and the Sun.) In other experiments, when looking just at the gas without a hot source behind it, they saw bright *emission lines* against an otherwise dark background. Combining the results from both sets of experiments, they discovered that *the emission lines of a particular gas occur at exactly the same wavelengths as the absorption lines of that gas.*

Because each chemical element produces its own unique pattern of spectral lines, scientists can determine the chemical composition of a remote astronomical object by identifying the lines in its spectrum. For example, Figure 4-6 shows a portion of the Sun's absorption spectrum along with the emission spectrum of an iron sample taken here on Earth. The spectral lines of iron also appear in the Sun's spectrum, so we can reasonably conclude that the Sun's atmosphere contains some vaporized iron. This gas absorbs certain wavelengths from the continuum emitted below it.

After photography was invented, scientists preferred to do spectroscopy by making a permanent photographic record of spectra. A device for photographing a spectrum is called a **spectrograph,** and this instrument,

a

b   R I V U X G

c   R I V U X G

FIGURE 4-7 A Grating Spectrograph (a) The diffraction grating in a grating spectrograph has many parallel lines on its surface that reflect light of different colors in different directions. This separation of colors (wavelengths) allows the object's spectrum to be analyzed. (b) This peacock feather contains numerous natural diffraction gratings. The role of the parallel lines etched in a human-made diffraction grating is played by parallel rods of the protein melanin in the feathers. (c) CDs and DVDs store information on closely spaced bumps located on a set of nearly parallel tracks. Light striking these tracks systematically reflects different colors in different directions—a CD or a DVD behaves like a diffraction grating. (b: Paul Silverman/Fundamental Photos; c: W. Cody/CORBIS) in its numerous variations, is the astronomer's most important tool.

In its earliest form, a spectrograph was attached directly to a telescope. It consisted of a slit, two lenses, and a prism arranged to focus the spectrum of an object on film, similar to the device sketched in Figure 4-5. Although conceptually straightforward, this early type of spectrograph had severe drawbacks. A prism does not spread colors evenly: The blue and violet portions of the spectrum are spread out more than the red portion. In addition, because the blue and violet wavelengths must pass through more glass than the red wavelengths (see Figure 3-1), light is absorbed unevenly across the spectrum. Indeed, a glass prism is opaque to ultraviolet wavelengths.

Practical spectrographs used in research today (shown conceptually in Figure 4-7a) separate light from objects in space into the colors of the rainbow using a **diffraction grating**, which is a piece of glass on which thousands of closely spaced parallel grooves are cut. Some of the finest diffraction gratings have more than 10,000 grooves per centimeter. The spacing of the grooves must be very regular. Light waves are diffracted by the grooves in the diffraction grating, just as light or water waves are diffracted when passing through slits (see Figure 3-2). Different wavelengths are diffracted by different amounts. Combining this spreading of the colors in the diffraction grating with the interference between the waves (see Figure 3-2b) yields a spectrum.

Figure 4-7a shows the design of a modern diffraction grating spectrograph with a CCD (charge-coupled device; see Figure 3-15) that has replaced film to record the spectra. This optical device typically mounts at the Nasmyth, coudé, or Cassegrain focal point of a telescope. The image of the object to be examined is focused on the slit. After the spectrum of a star, galaxy, or other object has been recorded, the CCD collects an emission spectrum from a known source, such as a gas composed of helium, neon, and argon, that is also focused on the slit. This "comparison spectrum" is placed next to the spectrum of the object, as in Figure 4-6. Because the wavelengths of the spectral lines in the comparison spectrum are already known from laboratory experiments, these lines can be used to identify and measure the wavelengths of the lines in the spectrum of the star or galaxy under study. Figures 4-7b and 4-7c show diffraction spectra occurring on Earth.

The spectral data are converted by computer to a graph that plots light intensity against wavelength. Dark lines in the rainbow-colored spectrum appear as depressions or valleys on the graph, while bright lines in the spectrum appear as peaks. For example, Figure 4-8 shows both a picture and a plot of a spectrum for hydrogen, in which five dark spectral lines appear.

From what we have just discussed, what do you think causes fireworks to have their distinctive colors?

FIGURE 4-8 Spectrum of Hydrogen Gas (a) When a CCD is placed at the focus of a spectrograph, the spectrum is recorded. In this case, the spectrum is from a hot object whose light shines through hydrogen gas. (b) This spectrum is converted by computer into a graph of intensity versus wavelength. Note that the absorption lines appear as dips in the intensity-versus-wavelength curve. Conversely, when a gas emits only a few wavelengths, (c) its emission spectrum appears as a series of bright lines, which are converted into peaks (d) on a graph of intensity versus wavelength.

4-4 The various brightness levels of spectral lines depend on conditions in the spectrum's source

By the early 1860s, Kirchhoff had discovered the conditions under which different types of spectra are observed. His description is summarized today as **Kirchhoff's laws:**

Law 1 A solid, liquid, or dense gas produces a **continuous spectrum** (*also called a* **continuum**)—*a complete rainbow of colors without any spectral lines. This is a blackbody spectrum.*

The light emitted by a hot iron rod and by an incandescent lightbulb are examples of continuous spectra that are bright enough for us to see.

Law 2 A rarefied (opposite of dense) gas produces an **emission line spectrum**—*a series of bright spectral lines against a dark background.*

The light emitted by neon lights and by low-pressure sodium vapor lights are examples of emission line spectra. The neon has bright-red emission lines, while the low-pressure sodium has bright-yellow lines.

Law 3 The light from an object with a continuous spectrum that passes through a cool gas produces an **absorption line spectrum**—*a series of dark spectral lines among the colors of the rainbow.*

Sunlight and the light from other stars pass through several cooler gases on their way to us. Hence, light from stars produces absorption line spectra.

In summary (Figure 4-9), a continuum is seen if there is no gas between a blackbody and the observer; absorption lines are seen if the background object is hotter than the gas between the observer and it; and emission lines are seen if the background is cooler than such a gas.

Consider, for example, the spectrum of the Sun. We know that the Sun's surface emits a continuous blackbody spectrum, but here on Earth many absorption lines are seen in it (see Figure 4-4). Kirchhoff's third law explains why. There must be a cooler gas between the surface of the Sun and Earth. In fact,

there are two: the Sun's lower atmosphere and Earth's entire atmosphere. Using the catalogs of spectra of different elements, the absorption lines in the Sun's spectrum tell us the chemical composition of the Sun's lower atmosphere and that of Earth's atmosphere.

If you have younger siblings or are around children for a good deal of time, you are likely to hear them go through a stage of development in which they frequently ask, "Why?" Scientists are among those who retain that curiosity, and the appearance of spectra is one of a myriad of things about which they ask that question. Fortunately, we can answer it, and the insights that it provides will be useful in comprehending concepts in virtually every chapter of this book. Our first step in understanding how spectra arise is to explore the building blocks of matter: atoms.

> Diffraction gratings can either reflect or transmit light to create the spectra. Are the peacock feathers reflection or transmission diffraction gratings?

ATOMS AND SPECTRA

An **atom** is the smallest particle of a chemical element that still has the properties of that element. At the time of Kirchhoff's discoveries, scientists knew that all matter is composed of atoms, but they did not know how atoms were structured. Furthermore, scientists saw that atoms of a gas somehow extract light of specific wavelengths from white light that passes through the gas, leaving dark absorption lines, and they perceived that the atoms then radiate light of precisely the same wavelengths—the bright emission lines (see Figure 4-6). But traditional theories of electromagnetism could not explain these phenomena. The answer came early in the twentieth century, with the development of nuclear physics and quantum mechanics.

FIGURE 4-9 Continuous, Absorption Line, and Emission Line Spectra This schematic diagram summarizes how different types of spectra are produced. The prisms are added for conceptual clarity, but in real telescopes diffraction gratings are used to separate the colors. A hot, glowing object emits a continuous spectrum. If this source of light is viewed through a cool gas, dark absorption lines appear in the resulting spectrum. When the same gas is viewed against a cold, dark background, its spectrum consists of just bright emission lines. The animation link shows the details of the activity drawn here.

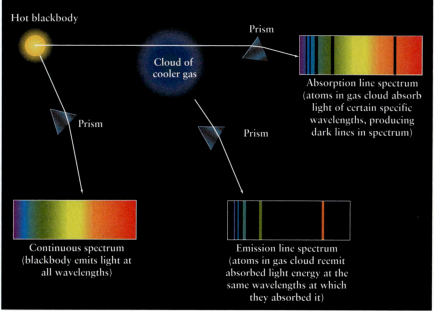

4-5 An atom consists of a small, dense nucleus surrounded by electrons

The internal structure of atoms first came into focus in 1908 when Nobel laureate Ernest Rutherford and his colleagues at the University of Manchester in England were investigating the recently discovered phenomenon of radioactivity. Over time, a **radioactive** element naturally and spontaneously transforms into another element by emitting particles. Certain radioactive elements, such as uranium and radium, were known to emit such particles with considerable speed. It seemed plausible that a beam of these high-speed particles would penetrate a thin sheet of gold. Rutherford and his associates found that almost all of the particles did pass through the gold sheet with little or no deflection. To their surprise, however, an occasional particle bounced right back (Figure 4-10). It must have struck something very dense, indeed.

Within a few decades, the structure of atoms became evident. That dense "something" is now called the **nucleus** of the atom, and it consists of particles called **protons** and **neutrons**. Surrounding the nucleus, one or more **electrons** normally. Newton's second law (see Section 2-7) explains that electrons move much more than nuclei because the electrons have only 1/2000 the mass of a proton or neutron.

Unlike the planets, whose gravitational interaction keeps them orbiting the Sun, electrons orbit nuclei because electrons and protons have a property called *electric charge*. All protons have the exact same positive charge, while all electrons have a negative charge equal in strength to the proton's charge. The terms *positive* and *negative* are arbitrary and just indicate that they are opposite to each other. Particles with opposite charges attract each other. Therefore, protons attract electrons, and it is this attraction that keeps electrons in orbit around nuclei. The interaction between charged particles is said to be a result of the **electromagnetic force**, the second of four fundamental forces in nature. (Gravitation is the other fundamental force we have discussed.)

Particles with the same type of charge, such as a pair of protons or a pair of electrons, repel each other. For atoms that have more than one proton in their

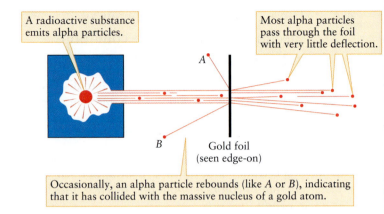

A radioactive substance emits alpha particles.

Most alpha particles pass through the foil with very little deflection.

Gold foil (seen edge-on)

Occasionally, an alpha particle rebounds (like *A* or *B*), indicating that it has collided with the massive nucleus of a gold atom.

FIGURE 4-10 Rutherford Scattering Experiment Most helium nuclei (originally called alpha particles) entering a thin foil scatter slightly as they pass through the medium, but some scatter backward, indicating that they have encountered very dense, compact objects. Such experiments were the first evidence that most matter is concentrated in what are now called atomic nuclei.

nuclei (and many do), the protons are pushing away from each other. For nuclei to have more than one proton, there must be an attractive force stronger than the repulsion of protons to keep them glued together. The electrically neutral neutrons in the nucleus help provide that attractive force, called the **strong nuclear force**. This is the third of the four fundamental forces.

Many scientists find it fascinating that all of the interactions between matter and energy in nature occur as a result of just four forces, these three plus the **weak nuclear force**. The weak nuclear force is involved in some radioactive decays, such as when a neutron transforms into a proton. In Chapter 13, we will further explore the weak nuclear force in our study of the evolution of the whole universe. The properties of all four fundamental forces are summarized in Table 4-2.

The number of protons in an atom's nucleus, called the **atomic number**, determines the element of that atom. All hydrogen nuclei have one proton; all helium nuclei have two, and so forth. There are 92 different types of elements that form naturally. Uranium is the most massive, with 92 protons (see periodic table, Appendix H).

TABLE 4-2 The Four Fundamental Forces of Nature

Name	Strength (compared to the strong force)	Range of effect (from each object)
Strong force	1	Inside atomic nuclei
Electromagnetic force	1/137	Throughout the universe
Weak force	10^{-5}	Inside atomic nuclei
Gravitational force	6×10^{-39}	Throughout the universe

In contrast, the nuclei of most elements can have different numbers of neutrons. For example, hydrogen always has 1 proton, but it can have 0, 1, or 2 neutrons; oxygen, with an atomic number of 8, always has 8 protons, but it can have 8, 9, or 10 neutrons. Each different combination of protons and neutrons is called an **isotope**. There are three isotopes of hydrogen and three isotopes of oxygen. Hydrogen with no neutrons is the most common hydrogen isotope, while oxygen with 8 neutrons is by far the most abundant isotope of oxygen.

Some isotopes are stable, meaning that the numbers of protons and neutrons in their nuclei do not change. However, many elements have isotopes that are unstable and come apart spontaneously. These isotopes are called radioactive. For example, carbon with 6 neutrons, ^{12}C ("carbon twelve"), is stable, while carbon with 8 neutrons, ^{14}C ("carbon fourteen"), is unstable. ^{14}C decays into nitrogen with 7 neutrons, ^{14}N. To learn how radioactive decay is used to determine the ages of different objects, see *An Astronomer's Toolbox 4-2*. Using radioactive age-dating techniques on the oldest rocks brought back from the Moon, astronomers are able to determine roughly when it formed (4.5 billion years ago). Applying the same technique to space debris found on Earth (meteorites), we can determine that the solar system, consisting of the Sun and everything orbiting it, formed some 4.62 billion years ago.

Normally, the number of electrons that orbit an atom is equal to the number of protons in its nucleus, thus making the atom electrically neutral. Astronomers denote neutral atoms by writing the atomic symbol

AN ASTRONOMER'S TOOLBOX 4-2

Radioactivity and the Ages of Objects

The isotopes of many elements are radioactive, meaning that the elements spontaneously transform into other elements. Each radioactive isotope has a distinctive *half-life*, which is the time that it takes half of the initial concentration of the isotope to transform into another element. After two half-lives, a radioactive isotope is reduced to $1/2 \times 1/2$ or 1/4 of its initial concentration (see the accompanying figure). Among the most important radioactive elements for determining the age of objects in astronomy is the isotope of uranium with 146 neutrons, ^{238}U ("U two-thirty-eight"). The half-life of ^{238}U decaying into lead is 4.5 billion years.

The Transformation of Uranium into Lead This figure shows the rate that 1 kg of uranium decays into lead, as described in the text. It contains ⅛ kg of uranium after 13.5 billion years.

To determine the time since an object, such as a piece of space debris discovered on Earth, solidified, scientists estimate how much lead it had when it formed. Then they measure the amounts of uranium and lead that it contains now. Subtracting the amount of original lead, they use the amount of uranium and lead, together with the graph of radioactive decay, to determine the object's age.

Example: Suppose a piece of space debris discovered on Earth was determined to have equal amounts of lead and uranium. How long ago did this debris form? Assuming that it originally had no lead, we see from the chart that a 1-to-1 mix of lead to uranium occurs 4.5 billion years after the object formed. This is one half-life of uranium.

Compare! This process works with any radioactive isotope. However, some isotopes have such short half-lives that they are not useful in astronomy. For example, the well-known carbon dating used to determine the ages of ancient artifacts on Earth is of little use in astronomy because ^{14}C has a half-life of only 5730 years. Because so much of the carbon in a sample has decayed away by then, carbon dating is useful only for time intervals shorter than 100,000 years, usually a period over which little of astronomical importance occurs.

Try these questions: What fraction of a kilogram of radioactive material remains after three half-lives have passed? Using the accompanying figure, estimate how much uranium will remain after 6 billion years. Approximately how many half-lives of ^{14}C pass in 100,000 years?
(Answers appear at the end of the book.)

followed by the Roman numeral I. For example, neutral hydrogen is written as H I and neutral iron is Fe I.

When an atom contains a different number of electrons than protons, the atom is called an **ion**. The process of creating an ion is called **ionization**. Ions are denoted by the atomic symbol followed by a Roman numeral that is one greater than the number of missing electrons. Positively ionized hydrogen (missing its one electron) is denoted H II, while positively ionized iron with seven electrons missing is denoted Fe VIII. It is worth noting that negative ions also exist, in which nuclei have more electrons orbiting than they have protons.

Atoms can share electrons and, by doing so, become bound together. Such groups of atoms are called **molecules**. They are the essential building blocks of all complex structures, including life.

4-6 Spectra occur because electrons absorb and emit photons with only certain wavelengths

Because protons and neutrons have masses about 2000 times greater than the mass of an electron, over 99.95% of the mass of any atom is concentrated in its nucleus. The electron orbits are far from the nucleus, typically 10,000 times farther away than the radius of the nucleus. This is why you may have heard the statement that matter is mostly empty space.

Be careful not to imagine electrons as miniature planets orbiting the nucleus as a "miniature Sun." Protons, neutrons, and electrons are not tiny solid bodies. Rather, like photons, they all have both wave and particle properties. The science that accurately describes their complex behavior is called **quantum mechanics.**

Quantum mechanics explains electrons in atoms can exist in only certain *allowed orbits* around their nuclei, except when they are making a **transition** from one allowed orbit to another. These orbital conditions are completely unlike planets, which can exist at any distance (in any orbit) around the Sun. Each allowed electron orbit has a well-defined energy associated with it, and every different type of atom and molecule has a unique set of allowed orbits. These orbits and the transitions between them are the key to understanding the spectra that we have been discussing.

Consider an atom of the simplest hydrogen isotope, which contains just a single proton in its nucleus orbited by one electron. (This discussion generalizes directly to all of the other atoms and isotopes, made more complex only because they have more than one electron in orbit.) Figure 4-11 shows this hydrogen atom's lowest energy levels.

Normally, the electron is in the lowest-energy allowed orbit or energy level, commonly called the **ground state**; this is labeled $n = 1$ in Figure 4-11. Each

FIGURE 4-11 Energy Level Diagram of Hydrogen The activity of a hydrogen atom's electron is conveniently displayed in a diagram showing some of the energy levels at which it can exist. A variety of electron jumps, or transitions, are also shown, including those that produce the most prominent lines in the hydrogen spectrum. For another perspective on the same information, you can try the linked Interactive Exercise.

allowed orbit with successively higher energy is labeled $n = 2, 3, 4$, and so on. When an electron is in an orbit with more energy than the lowest energy state available to it, it is said to be in an **excited state.**

Electrons change orbits by absorbing or emitting photons. However, electrons cannot absorb just any photon that they encounter. Electrons can absorb only those photons with energies exactly enough to boost them up to a higher-energy allowed orbit. That is, the photons that are absorbed are those with energies equal to the difference between the energies of two allowed orbits. All photons that do not satisfy this condition pass straight through the atom. If the electron starts in its ground state, then it must get exactly the energy necessary to move it to an excited state. If it is already in an excited state, then by absorbing a photon, it must transition to a higher-energy excited state.

For example, referring to Figures 4-11 and 4-12, transitions between the $n = 2$ and the $n = 3$ energy levels require the electron to absorb a photon with energy equal to 12.1 eV –10.2 eV = 1.9 eV (electron volt, a measure of energy). Recall from *An Astronomer's Toolbox 3-1* that a photon's energy corresponds to a certain wavelength. In this case, the photon absorbed by the electron has a wavelength of 656.3 nm (Figure 4-12a); it is a red photon (recall Figure 3-6).

Although some absorption occurs at optical wavelengths, it also happens in many other parts of the electromagnetic spectrum. Indeed, most of the transitions in hydrogen, shown in Figure 4-11 as the Lyman series (up from and down to $n = 1$) and the Paschen series (up from and down to $n = 3$), are nonvisible photons. Only a few of the Balmer series (up from and down to $n = 2$) transitions are visible, with the rest being ultraviolet. We will refer to Balmer transitions when we study stars. They are named after Johann Balmer, who first calculated their wavelengths. The longest-wavelength Balmer line is called H_α *(H alpha)*, the second H_β *(H beta)*, the third H_γ *(H gamma)*, and so forth, ending with the shortest-wavelength Balmer line, H_∞ *(H infinity)*. (The first dozen lines of the series have Greek-letter subscripts; the remainder are identified by numerical subscripts.) Part of the spectrum of a star having Balmer absorption lines H_α through H_θ *(H theta)* is shown in Figure 4-13.

Absorption lines are therefore caused by photons being taken out of the stream of light by electrons, which thereby move into higher-energy allowed orbits. An observer on the right in Figure 4-9 looking at the hot blackbody through the cloud of cooler gas sees dark absorption lines where photons have been absorbed by the cooler gas.

Electrons in excited states are unstable. They lose energy by bumping into other particles and suddenly having either too much or too little energy than is necessary to be in that state. When this happens, the electron is forced to descend to a lower-energy allowed state, either the ground state or some allowed state between where it starts and the ground state. In either case, to make this transition down, the electron must *emit* a photon with energy once again equal to the difference between the energies of the starting and final allowed orbits (see Figure 4-12b).

The emitted photons have the same set of wavelengths as the absorbed photons. Furthermore, the emitted photons are sent out in all directions, which create the glowing, emission-line spectra of hot objects. The color of the gas depends on the atoms and molecules that it contains. Hydrogen-rich gas clouds glow red because of the H_α emission (Figures 4-13 and 4-14a). Oxygen emits many green photons, so oxygen-rich gas clouds glow green (Figure 4-14b).

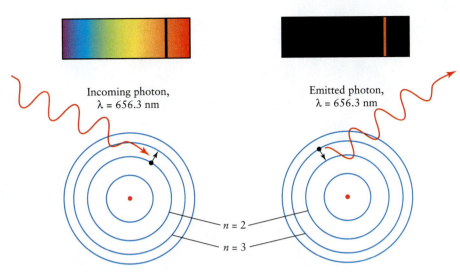

a Atom absorbs a 656.3-nm photon; absorbed energy causes electron to jump from the $n = 2$ orbit up to the $n = 3$ orbit

b Electron falls from the $n = 3$ orbit to the $n = 2$ orbit; energy lost by atom goes into emitting a 656.3-nm photon

FIGURE 4-12 The Absorption and Emission of an H_α Photon This schematic diagram of hydrogen's four lowest allowed orbits shows what happens when a hydrogen atom absorbs or emits an H_α photon, which is red and has a wavelength of 656.28 nm. The spectral lines by such events are also shown. (a) A photon is absorbed by the electron, causing the electron to transition from orbit $n = 2$ up to orbit $n = 3$. (b) A photon is emitted as the electron makes a transition from orbit $n = 3$ down to orbit $n = 2$.

If an electron orbiting in a hydrogen atom encounters a photon with more than 13.6 eV, that photon is absorbed and knocks the electron completely out of orbit and away from the atom. This process is called *photoionization*. Each type of atom and molecule has a different photoionization energy, above which all electrons are kicked out of the atom or molecule. Photoionization occurs in stars and interstellar nebulae (Figure 4-14b). Indeed, many spectral lines from stars and nebulae are those of ionized atoms that retain at least one electron.

Which Balmer line in Figure 4-11 is H_α?

Shorter wavelength

H_θ H_η H_ζ H_ϵ H_δ H_γ H_β H_α

R I V U X G

FIGURE 4-13 Balmer Lines in the Spectrum of a Star This portion of the spectrum of the star Vega shows eight Balmer lines, from H_α at 656.3 nm through H_θ at 388.9 nm. (NOAO)

FIGURE 4-14 Emission Spectra from Interstellar Gas Clouds (a) Stars in this interstellar gas cloud (NGC 2363 in the constellation Camelopardus, the Giraffe) emit blackbody spectra. Electrons in the cloud's hydrogen gas absorb and reemit the red light from these stars. NGC 2363 is located some 10 million light-years away. (b) Part of the Rosette Nebula (NGC 2237), an interstellar gas cloud in the constellation Monoceros (the Unicorn). The green glow is generated by doubly ionized oxygen atoms (O III; oxygen atoms missing two electrons) in the cloud that emit 501-nm photons. The Rosette is 3000 light-years away. (a: L. Drissen, J.-R. Roy, and C. Robert/Département de Physique and Observatoire du Mont Mégantic, Université Laval; and NASA; b: T. A. Rector, B. Wolpa, M. Hanna, AURA/NOAO/NSF)

4-7 Spectral lines shift due to the relative motion between the source and the observer

 In 1842, Christian Doppler, a professor of mathematics in Prague, deduced that wavelength, hence color, is affected by motion. As objects move toward or away from you, they change color, in analogy to how the pitch of a siren changes as it passes you. As shown for the observer on the left in Figure 4-15, the wavelengths of electromagnetic radiation from an approaching source are compressed. The circles represent consecutive wave peaks (S1 through S4) emitted in all directions as the source moves along. Because each successive wave is sent out from a position slightly closer to her, she sees a shorter wavelength than she would if the source were stationary. All of the colors in the spectrum of an approaching source, such as a star, are therefore shifted toward the short-wavelength (blue) end of the spectrum, regardless of its distance. This phenomenon is called a **blueshift.**

Conversely, electromagnetic waves from a receding source are stretched out. The observer on the right in Figure 4-15 sees a longer wavelength than he would if the source were stationary. All of the colors in the spectrum of a receding source, regardless of its distance, are shifted toward the longer-wavelength (red) end of the spectrum, producing a **redshift.** A blueshift or a redshift is also called a **Doppler shift.** The amount of Doppler shift varies directly with approaching or receding speed: When the speeds are small compared to the speed of light, an object that approaches twice as fast as another has all of its colors (or wavelengths) blueshifted twice as much as does the slower-moving object. An object moving away twice as fast as another object has its colors redshifted twice as much as the slower-moving object. Doppler shift applies to spectral lines as well. See *An Astronomers Toolbox 4-3* for mathematical details of the Doppler shift.

The speed of an object toward or away from us is called its **radial velocity** (Figure 4-16) because the motion is along our line of sight or, put another way, along the "radius" drawn from Earth to the object as seen on the celestial sphere. Of course, the star or other object may also have a velocity perpendicular or transverse to our line of sight, such as a car passing a pedestrian. Both the radial velocity and the **transverse velocity** are measured in kilometers per second.

Astronomers cannot measure the transverse velocity directly. Instead they measure the angle that the object moves among the stars on the celestial sphere as seen from Earth. This is called its **proper motion** and is often measured in arcseconds per year or arcseconds per century. The star with the greatest proper motion as seen from Earth is Barnard's star (Figure 4-17). Proper motion does not affect the perceived wavelength and cannot be determined by Doppler shift. It must be measured by

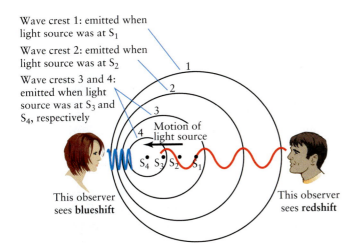

Wave crest 1: emitted when light source was at S_1

Wave crest 2: emitted when light source was at S_2

Wave crests 3 and 4: emitted when light source was at S_3 and S_4, respectively

Motion of light source

This observer sees **blueshift**

This observer sees **redshift**

FIGURE 4-15 The Doppler Shift Wavelength is affected by motion between the light source and the observer. A source of light is moving toward the left. The four circles (numbered 1 through 4) indicate the location of light waves that were emitted by the moving source when it was at points S_1 through S_4, respectively. Note that the waves are compressed in front of the source but stretched out behind it. Consequently, wavelengths appear shortened (blueshifted) if the source is moving toward the observer and lengthened (redshifted) if the source is moving away from the observer. Motion perpendicular to the observer's line of sight does not affect wavelength.

comparing the locations of objects with higher proper motion to the positions of objects with lower proper motions. Planets moving among the relatively fixed stars on the celestial sphere are examples of bodies that have large proper motions.

Stellar proper motions are so small that they can be measured for only relatively nearby stars in our Galaxy. However, the radial velocity of virtually every object in space can be determined. Indeed, even the slow motions of rising and sinking gases on the Sun's surface can be measured by their spectral Doppler shifts. Doppler shift measurements of stars in double-star systems give crucial data about the speeds of the stars orbiting each other. Doppler measurements of planets, the Sun, and other stars reveal that many of these bodies are rotating. This is determined when the Doppler shift shows that half of a surface is heading toward us, while the other half is simultaneously moving away from us. Furthermore, Doppler measurements reveal that all of the very distant galaxies are moving away from us, meaning that the universe is expanding. The spectra of distant galaxies enable us to determine the rate of that expansion. In later chapters, we will refer to the Doppler shift whenever we need to convert an observed wavelength shift into a speed toward or away from us.

> In three dimensions, what shape will the circular wave crests (as shown in Figure 4-15) have?

> Does a police siren approaching you sound higher or lower in pitch than the siren at rest relative to you?

INSIGHT INTO SCIENCE

Remote Science Astronomical objects are so remote and their activities are often so complex that astrophysicists must use a tremendous amount of physics to interpret observations. For example, a single spectrum can contain information about stars, extrasolar planets, interstellar gas, Earth's motion and atmosphere, and the performance of the observing telescope and the equipment attached to it. All of these factors must be understood theoretically and accounted for.

Actual velocity

Transverse velocity

Radial velocity

Line of sight

Earth

FIGURE 4-16 Radial and Transverse Velocities of a Star The actual velocity of a star (or other object) can be separated into radial and transverse motions. The speed of the star toward or away from Earth in kilometers per hour is its radial velocity. This motion creates a Doppler shift in the star's spectrum. The speed in kilometers per hour perpendicular to the radial velocity is called its transverse velocity, which does not affect the star's spectrum. The *angular* motion of the star across the celestial sphere (that is, in the direction of its transverse velocity) is called proper motion.

With the knowledge of spectroscopy, atomic and nuclear physics, and the Doppler shift, we can now list many of the vital properties of matter in space that are available to us. These include the chemical compositions of stars, interstellar gas clouds, and other objects; the temperatures of these objects; the rotation of objects in space; the motion of objects toward or away from Earth; the presence of planets and dim companion stars; and the rate at which the universe is expanding. Knowing these properties, astrophysicists have developed models that provide numerous insights into the ages, masses, distances, internal activity, rotation rates, and companion objects of stars, among other things.

The Doppler Shift

Suppose that λ_0 (lambda naught) is the wavelength of a spectral line from a stationary source. This is the wavelength that a reference book would list or a laboratory experiment would yield. If the source is moving toward or away from you, this particular spectral line is shifted to a different wavelength, λ. The size of the wavelength shift is usually written as $\Delta\lambda$ (delta lambda), where $\Delta\lambda = \lambda - \lambda_0$. This $\Delta\lambda$ is the difference between the wavelength that you actually observe in the spectrum of a star or galaxy and the wavelength listed in astronomical reference books.

Doppler proved that the wavelength shift is governed by the simple equation

$$\frac{\Delta\lambda}{\lambda_0} = \frac{v}{c}$$

where v is the speed of the source toward or away from the observer. As usual, c is the speed of light (3×10^5 km/s).

Example: The spectral lines of hydrogen appear in the spectrum of the bright star Vega, as shown in Figure 4-13. The hydrogen line H_α has a normal wavelength of 656.285 nm, but in Vega's spectrum the line is located at 656.255 nm. The wavelength shift is -0.030 nm. The minus sign indicates that the star is moving toward us, so the star is approaching us with a speed of -14 km/s.

Try these questions: How fast would a star be moving toward us if all of its wavelengths shifted to nine-tenths of their rest wavelengths? If a star is moving away from us at 100 km/s, what is the change, $\Delta\lambda/\lambda_0$? If a star is moving toward us at 0.1 c, what is the change, $\Delta\lambda/\lambda_0$, in its wavelengths?

(Answers appear at the end of the book.)

Modern physics was born when Newton set out to understand the motions of the planets. Two and a half centuries later, scientists—including Maxwell, Planck, Einstein, Rutherford, and Doppler, among many others—discovered the basic properties of electromagnetic radiation and the structures of atoms. As we will see in the following chapters, the fruits of their labors have important implications for astronomy even today.

R I V U X G

WEB LINK 4.3 **FIGURE 4-17** Proper Motion of Barnard's Star Two images of Barnard's star, taken a year apart in 2000 and 2001, show the proper motion of the star during that time. In addition to having the largest known proper motion (10.3″ per year), Barnard's star is one of the closest stars to Earth.

4-8 Frontiers yet to be discovered

Spectra provide much of what we know about the cosmos. Spectra of objects in space continue to yield new insights into the chemical composition of the planets, moons, local space debris, stars, interstellar gas and dust, and galaxies. Furthermore, as we obtain more accurate Doppler measurements of objects in our Galaxy, in other nearby galaxies, and, indeed, of entire galaxies, we are better able to determine the rate at which the objects in the universe are moving relative to one another. The more we understand of the chemistries and motions of objects in space, the more accurately we are able to explain the evolution of the solar system, stars, galaxies, and the universe.

SUMMARY OF KEY IDEAS

• By studying the wavelengths of electromagnetic radiation emitted and absorbed by an astronomical object, astronomers can learn about its temperature, chemical composition, rotation rate, companion objects, and movement through space.

Blackbody Radiation
• A blackbody is a hypothetical object that perfectly absorbs electromagnetic radiation at all wavelengths. The relative intensities of radiation that it emits at different wavelengths depend only on its temperature. Stars closely approximate blackbodies.

• Wien's law states that the peak wavelength of radiation emitted by a blackbody is inversely proportional to its temperature—the higher its temperature, the shorter the peak wavelength. The intensities of radiation emitted at various wavelengths by a blackbody at a given temperature are shown as a blackbody curve.

• The Stefan-Boltzmann law shows that a hotter blackbody emits more radiation at every wavelength than does a cooler blackbody.

• The motion of an object toward or away from an observer causes the observer to see all of the colors from the object blueshifted or redshifted, respectively. This effect is generically called a Doppler shift.

Discovering Spectra
• Spectroscopy—the study of electromagnetic spectra—provides important information about the chemical composition of remote astronomical objects.

• Kirchhoff's three laws of spectral analysis describe the conditions under which absorption lines, emission lines, and a continuous spectrum can be observed.

• Spectral lines serve as distinctive "fingerprints" that identify the chemical elements and compounds comprising a light source.

Atoms and Spectra
• An atom consists of a small, dense nucleus (composed of protons and neutrons) surrounded by electrons. Atoms of different elements have different numbers of protons, while different isotopes have different numbers of neutrons.

• Quantum mechanics describes the behavior of particles and shows that electrons can only be in certain allowed orbits around the nucleus.

• The nuclei of some atoms are stable, while others (radioactive ones) spontaneously split into pieces.

• The spectral lines of atoms of a particular element correspond to the various electron transitions between allowed orbits of that element with different energy levels of those atoms. When an electron shifts from one energy level to another, a photon of the appropriate energy (and hence a specific wavelength) is absorbed or emitted by the atom.

• The spectrum of hydrogen at visible wavelengths consists of part of the Balmer series, which arises from electron transitions between the second energy level of the hydrogen atom and higher levels.

• Every different element, isotope, and molecule has a different set of spectral lines.

• When a neutral atom loses or gains one or more electrons it is said to be charged. The atom loses an electron when the electron absorbs a sufficiently energetic photon, which rips it away from the nucleus.

• The equation that describes the Doppler effect states that the size of a wavelength shift is proportional to the radial velocity between the light source and the observer.

WHAT DID YOU THINK?

1 *Which is hotter, a "red-hot" or a "blue-hot" object?* Of all objects that glow visibly from heat generated or energy stored inside them, those that glow red are the coolest.

2 *What color does the Sun emit most brightly?* The Sun emits all wavelengths of electromagnetic radiation. The colors it emits most intensely are in the blue-green part of the spectrum. Because the human eye is less sensitive to blue-green than to yellow, and because Earth's atmosphere scatters blue-green wavelengths more readily than longer wavelengths, we normally see the Sun as yellow.

3 *How can we determine the age of space debris found on Earth?* We measure how much the long-lived radioactive elements, such as ^{238}U, have decayed in the object. Carbon dating is only reliable for things that formed within the past 100,000 years. It cannot be used for determining the age of debris found on Earth, which was all formed more than 4.5 billion years ago.

Key Terms for Review

absorption line, 109
absorption line spectrum, 113
atom, 113
atomic number, 114
blackbody, 105
blackbody curve, 105
blueshift, 118
continuous spectrum (continuum), 113
diffraction grating, 112
Doppler shift, 118
electromagnetic force, 114
electron, 114
element, 110
emission line, 110
emission line spectrum, 113
energy flux, 107
excited state, 116
ground state, 116
ion, 116
ionization, 116
isotope, 115

Kirchhoff's laws, 113
luminosity, 107
molecule, 116
neutron, 114
nucleus (of an atom), 114
periodic table, 111
Planck's law, 109
proper motion, 118
proton, 114
quantum mechanics, 116
radial velocity, 118
radioactive, 114
redshift, 118
spectral analysis, 110
spectrograph, 111
spectroscope, 110
Stefan-Boltzmann law, 106
strong nuclear force, 114
transition (of an electron), 116
transverse velocity, 118
weak nuclear force, 114
Wien's law, 106

Review Questions

1. A blackbody glowing with which of the following colors is hottest? **a.** yellow, **b.** red, **c.** orange, **d.** violet, **e.** blue

2. Of the following photons, which has the lowest energy? **a.** infrared, **b.** gamma rays, **c.** visible light, **d.** ultraviolet, **e.** X ray

3. The spectrum of which of the following objects will show a blueshift? **a.** an object moving just eastward on the celestial sphere, **b.** an object moving just northward on the celestial sphere, **c.** an object moving directly toward Earth, **d.** an object moving directly away from Earth, **e.** an object that is not moving relative to Earth

4. What is a blackbody? What does it mean to say that a star appears almost like a blackbody? If stars appear to be like blackbodies, why are they not black?

5. What is Wien's law? How could you use it to determine the temperature of a star's surface?

6. What is the Stefan-Boltzmann law? How do astronomers use it?

7. Using Wien's law and the Stefan-Boltzmann law, state the changes in color and intensity that are observed as the temperature of a hot, glowing object increases.

8. What color will an interstellar gas cloud composed of hydrogen glow, and why?

9. What is an element? List the names of five different elements, and briefly explain what makes them different from each other.

10. How are the three isotopes of hydrogen different from each other?

11. Explain how the spectrum of hydrogen is related to the structure of the hydrogen atom.

12. Why do different elements have different patterns of lines in their spectra?

13. Explain why the Doppler shift tells us only about the motion directly along the line of sight between a light source and an observer, but not about motion across the celestial sphere.

14. What is the Doppler shift, and why is it important to astronomers?

Advanced Questions

The answers to all computational problems, which are preceded by an asterisk (*), appear at the end of the book.

***15.** Approximately how many times around the world could a beam of light travel in 1 s?

***16.** The bright star Regulus in the constellation of Leo (the Lion) has a surface temperature of 12,200 K. Approximately what is the dominant wavelength (λ_{max}) of the light that it emits?

***17.** The bright star Procyon in the constellation of Canis Minor (the Little Dog) emits the greatest intensity of radiation at a wavelength $\lambda_{max} = 445$ nm. Approximately what is the surface temperature of the star in Kelvins?

***18.** As observed from Earth, the wavelength of H_β in the spectrum of the star Megrez in the Big Dipper is 486.112 nm. Laboratory measurements demonstrate that the normal wavelength of this spectral line is 486.133 nm. Is the star coming toward us or moving away from us? At what speed?

***19.** In the spectrum of the bright star Rigel, H_α is observed to have a wavelength of 656.331 nm. Is the star coming toward us or moving away from us? How fast?

***20.** Imagine driving down a street toward a traffic light. How fast would you have to go so that the red light (700 nm) would appear green (500 nm)? What fraction is this of the speed of light, c ($c = 300,000$ km/s)?

Discussion Questions

21. Compare the technique of identifying chemicals by their spectral line patterns with that of identifying people by their fingerprints.

22. Suppose you look up at the night sky and observe some of the brightest stars with your naked eye. Is there any way of telling which stars are hotter and which are cooler? Explain.

23. How could we exclude Earth's atmosphere as the source of the iron absorption lines in Figure 4-6?

What If...

***24.** The Sun were twice its actual diameter but still had the same surface temperature? At what wavelength would that new Sun emit its radiation most intensely? How many times brighter would the new Sun be than the present Sun? How might things on Earth be different under the bigger Sun?

25. All of the nearby stars were observed to have red-shifted spectra. What conclusions could we draw? What other measurement would be useful to have for each star, and why?

26. No stars had any Doppler shift. What would that say about stellar motions relative to the solar system? Is that possible? *Hint:* Consider Newton's law of gravitation from Chapter 3.

27. The spectrum of the Sun (Figure 4-4) had several absorption lines missing when taken from above Earth's atmosphere compared to its spectrum taken at Earth's surface. What would this indicate about the chemistries of the Sun and of the Earth's atmosphere?

Web Questions

*28. Measuring Stellar Temperatures Access the Active Integrated Media Module (AIMM) "Blackbody Curves" in Chapter 4 of the *Discovering the Universe: From the Stars to the Planets* Web site. (a) Use the module to determine the range of temperatures over which a star's peak wavelength is in the visible spectrum. (b) Determine whether any of the following stars have a peak wavelength in the visible spectrum: Rigel, $T = 14,000$ K; Deneb, $T = 9500$ K; Arcturus, $T = 4500$ K; Vega, $T = 11,500$ K; Betelgeuse, $T = 3100$ K.

Observing Projects

29. Obtain a glass prism or a diffraction grating, available from science museum stores, catalogs, or your physics department. Look through the prism or grating at various light sources, such as an ordinary incandescent lightbulb, a neon sign, and a mercury vapor street lamp. *Do not look at the Sun! Looking directly at the Sun causes blindness.* Do you have any trouble seeing spectra? What do you have to do to see a spectrum? Describe the differences in the spectra of the various light sources you observed.

30. Use *Starry Night Enthusiast*™ to determine which planets, if any, are up at night now. Press **Home**, select the *Find* tab, clear any words in the "Search All Databases" box. Turn off the motion and set the time to when you will observe. The planets that will be up this evening will have positive "Alt" values (altitudes, or angles above the horizon) in the list below that box. Double click on any of the planets from Saturn to Mercury that will be up to see its location in the sky. If a planet is visible, arrange to observe it and make an accurate drawing of its position among the background stars. If no planet is visible, you can perform these observations using the Moon. Observe the planet (or Moon) approximately one week later and again make a drawing. Based on its movement among the stars, is the body undergoing prograde or retrograde motion? Explain using the techniques in *An Astronomer's Toolbox 1-1* or more accurate methods suggested by your instructor, determine the angle across the celestial sphere through which the planet has moved in this time.

31. Use *Starry Night Enthusiast*™ to examine three distant celestial objects. First select **Favorites > Guides > Atlas** from the menu to view the entire sky. Open the **Options** pane, expand the **Deep Space** layer and select only the **Messier Objects** item. Then use the *Find* pane to locate objects (i), (ii), (iii) listed below. Double click on each in turn, when it appears below the "Search All Databases" box as you type. Use the Zoom buttons in the toolbar to magnify the object. Based on its appearance, state whether it is likely to have a continuum spectrum, an absorption line spectrum, or an emission line spectrum, and explain your reasoning. Bear in mind that because nebulae may contain stars and other objects, they can have more than one type of spectrum. (i) Trifid Nebula, (ii) Ring Nebula, (iii) Lagoon Nebula. Your teacher may assign additional objects for you to find and study.

PART II

Understanding the Stars

Interstellar dust illuminated by a pulse of light emitted from the red giant star, V838 Monocerotis, in the center of the image. [NASA, ESA, and H. H. Bond ([STScI])

AN ASTRONOMER'S ALMANAC

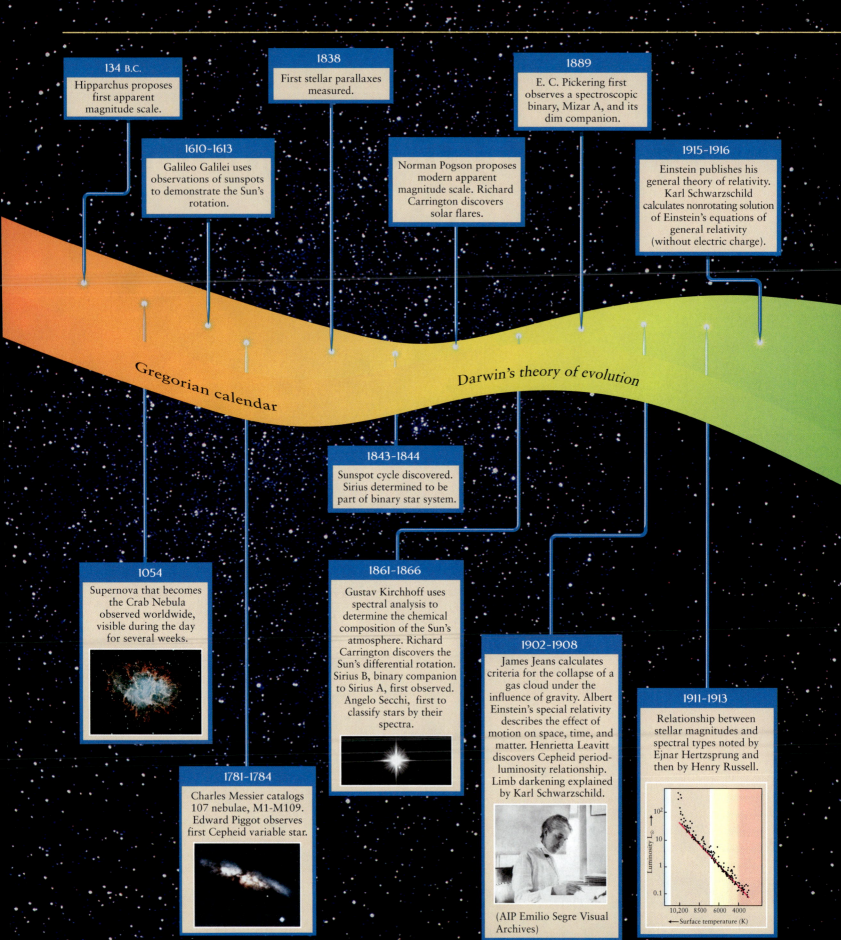

134 B.C.
Hipparchus proposes first apparent magnitude scale.

1838
First stellar parallaxes measured.

1889
E. C. Pickering first observes a spectroscopic binary, Mizar A, and its dim companion.

1610-1613
Galileo Galilei uses observations of sunspots to demonstrate the Sun's rotation.

1838 Norman Pogson proposes modern apparent magnitude scale. Richard Carrington discovers solar flares.

1915-1916
Einstein publishes his general theory of relativity. Karl Schwarzschild calculates nonrotating solution of Einstein's equations of general relativity (without electric charge).

Gregorian calendar

Darwin's theory of evolution

1843-1844
Sunspot cycle discovered. Sirius determined to be part of binary star system.

1054
Supernova that becomes the Crab Nebula observed worldwide, visible during the day for several weeks.

1861-1866
Gustav Kirchhoff uses spectral analysis to determine the chemical composition of the Sun's atmosphere. Richard Carrington discovers the Sun's differential rotation. Sirius B, binary companion to Sirius A, first observed. Angelo Secchi, first to classify stars by their spectra.

1902-1908
James Jeans calculates criteria for the collapse of a gas cloud under the influence of gravity. Albert Einstein's special relativity describes the effect of motion on space, time, and matter. Henrietta Leavitt discovers Cepheid period-luminosity relationship. Limb darkening explained by Karl Schwarzschild.

(AIP Emilio Segre Visual Archives)

1911-1913
Relationship between stellar magnitudes and spectral types noted by Ejnar Hertzsprung and then by Henry Russell.

1781-1784
Charles Messier catalogs 107 nebulae, M1-M109. Edward Piggot observes first Cepheid variable star.

THE STARS

1918-1924
Annie Cannon classifies hundreds of thousands of stellar spectra, leading to OBAFGKM spectral classification scheme for stars.

1920-1924
Cecilia Payne and Meghnad Saha explain relationship between Balmer lines and surface temperatures of stars. Arthur Eddington recognizes relationship of stellar mass to stellar evolution rate; formulates mass-luminosity relationship for main sequence stars.

1950-1959
Solar wind discovered by Eugene Parker. Distinction made between Cepheid types I and II by Walter Baade.

1994-1997
Hubble Space Telescope observations reveal a supermassive black hole in the center of the galaxy M87. Existence of brown dwarfs confirmed. First isolated, non-pulsar neutron star discovered; first observational evidence of rotating gas in a black hole's ergoregion.

1925-1929
Wolfgang Pauli proves that two identical particles cannot be in the same place at the same time. George Gamow proposes hydrogen fusion as energy source for stars.

1941
Mechanism for Cepheid variability explained by Arthur Eddington.

1987
Supernova 1987A observed as it brightened and then as it faded.

Until 2006, astronomers thought 2/3 of nearby stars are in binary systems, 1/3 single stars. Charles Lada discovers many dim stars in our neighborhood, changing the fractions to 1/3 binary, 2/3 single.

Quantum mechanics

Structure of DNA

Cloning

1930-1938
Subrahmanyan Chandrasekhar calculates maximum mass of a white dwarf. James Chadwick discovers the neutron. Fritz Zwicky and Walter Baade propose existence of neutron stars. J. Robert Oppenheimer and George Volkov calculate maximum mass of a neutron star.

1938-1939
Fusion shown to be source of the Sun's energy. Hans Bethe and Carl von Weizscker detail the process of main sequence fusion. J. Robert Oppenheimer and Hartland Snyder demonstrate that collapsing, massive stars form black holes.

1963-1968
Roy Kerr calculates rotating solution of Einstein's equations of general relativity (without electric charge). Jocelyn Bell and Anthony Hewish discover the first pulsar.

(Robin Scagell/ Galaxy)

1971-1978
First black hole (Cygnus X-1) identified. Stephen Hawking proposes primordial black holes; shows that black holes evaporate.

2001
Hubble makes first direct observation of atmosphere around an extrasolar planet.

CHAPTER 5

The Sun: Our Extraordinary Ordinary Star

Magnetic fields carry gases above the Sun's surface in loops that can extend hundreds of thousands of kilometers (NASA)

R I V U X G

WHAT DO YOU THINK?

1 What percentage of the solar system's mass is in the Sun?

2 Does the Sun have a solid and liquid interior like Earth?

3 What is the surface of the Sun like?

4 Does the Sun rotate? If so, how fast?

5 What makes the Sun shine?

Answers to these questions appear in the text beside the corresponding numbers in the margins and at the end of the chapter.

Dawn. Light and heat from the Sun begin to arouse our senses and signal the start of daily activity. Earliest societies, realizing that the Sun is essential to the existence and maintenance of life on Earth, revered it. The same respect for the Suns awesome power motivated astronomers in the nineteenth and twentieth centuries to figure out how it shines.

A quick glance (never longer) at the Sun shows a brilliant disk of light, apparently no larger than the Moon. Despite appearances, the Sun is a body of monumental size and mass compared to any object we have studied so far. If the Sun's surface were as close to Earth as the Moon's surface is now, the Sun would spread two-thirds of the way across the sky, not to mention that it would be so hot on Earth that our planet would quickly vaporize. The Sun contains as much mass as 333,000 Earths. Indeed, all of the planets and other objects orbiting the Sun *combined* have only 0.15% as much mass as the Sun.

The Sun's tremendous mass not only holds the planets in orbit, but also explains why it shines. The Sun's mass creates a crushing force at its center that literally fuses atoms together—a process that generates the energy that eventually leaves the Sun and lights up our days. Some of the energy rushing into space (namely the ultraviolet radiation, visible light, and infrared radiation emitted by the Sun) is essential to maintaining Earth's habitability.

The Sun is the closest star to Earth. By studying it, astronomers have come to learn how most stars throughout the universe work. As powerful and majestic as it is in our sky, we are going to discover in the next few chapters that the Sun is an ordinary star, neither among the most massive nor among the least massive. Similarly, it is neither among the brightest nor among the dimmest of stars. Figure 5-1 lists the Sun's properties.

In this chapter you will discover

- why the Sun is a typical star
- how today's technology has led to new understanding of solar phenomena, from sunspots to the powerful ejections of matter that sometimes enter our atmosphere
- that some features of the Sun generated by its varying magnetic field occur in cycles
- how the Sun generates the energy that makes it shine
- new insights into the nature of matter from solar neutrinos

THE SUN'S ATMOSPHERE

Although astronomers often speak of the solar "surface," the Sun is so hot that it has neither liquid nor solid matter anywhere inside it. Moving down through the Sun, we continually encounter ever denser and hotter gases.

5-1 The photosphere is the visible layer of the Sun

The Sun appears to have a surface only because most of its visible light comes from one specific gas layer (see Figure 5-1). This region, which is about 400-km thick, is appropriately called the **photosphere** ("sphere of light"). The density of the photosphere's gas is low by Earth standards, about 0.01% as thick as the air we breathe. The photosphere has a blackbody spectrum (recall Chapter 4) that corresponds to an average temperature of 5800 K (review Figure 4-3).

The photosphere is the innermost of the three layers that comprise the Sun's atmosphere. Because the upper two layers (discussed in the following two sections) are transparent to most wavelengths of visible light, we see through them down to the photosphere. We cannot, however, see through the shimmering gases of the photosphere, so everything below the photosphere is the Sun's interior.

As you can see in Figure 5-1, the photosphere appears darkest toward the edge, or **limb**, of the solar disk, a phenomenon, called **limb darkening**. This occurs because we see regions of different temperatures at different depths in the photosphere. Here is how it works: Wherever we look on the Sun, we see light through roughly equal amounts of the Sun's atmosphere. Because the Sun is spherical, the light we see leaving the Sun at different places actually comes from different levels of the photosphere (Figure 5-2). Looking from Earth at the center of the Sun's disk (*bottom black line*, Figure 5-2), we see farther into the Sun's atmosphere than when we look toward the limb (*top black line*, Figure 5-2). The photosphere is hottest and brightest at its base, and so the center of the Sun's disk, where we see deepest into it, looks brighter and yellower than does its limb.

Under good observing conditions and using a telescope and special dark filters for protection, you can see

The Sun: Vital Statistics

Mass ($1\ M_\odot$):	1.989×10^{30} kg $= 3.33 \times 10^5$ Earth masses	Sun symbol: $\odot$
Visual radius ($1\ R_\odot$):	6.960×10^5 km $= 10^9$ Earth radii	
Luminosity ($1\ L_\odot$):	3.827×10^{26} W	
Mean angular diameter:	32 arcmin	
Rotation periods:	Equatorial: 25 days	
	Polar: 35 days	
Mean density:	1408 kg/m^3	
Distances from Earth:	Mean (1 AU): 1.496×10^8 km	
	Maximum: 1.520×10^8 km	
	Minimum: 1.470×10^8 km	
Mean light travel time to Earth:	8.32 min	
Mean temperatures:	Surface: 5800 K	
	Center: 1.55×10^7 K	
Composition (by mass):	74% hydrogen, 25% helium, 1% other elements	
Composition (by number of atoms):	92.1% hydrogen, 7.8% helium, 0.1% other elements	
Distance from center of Galaxy:	26,000 ly $=$ 8000 pc	
Orbital period around center of Galaxy:	220 million years	
Orbital speed around center of Galaxy:	220 km/s	

R I V U X G

WEB LINK 5-1

FIGURE 5-1 Our Star, the Sun The Sun emits most of its visible light from a thin layer of gas, called the photosphere, as shown. Although the Sun has no solid or even liquid region, we see the photos-phere as its "surface." Astronomers always take great care when viewing the Sun by using extremely dark filters or by projecting the Sun's image onto a screen. (Celestron International)

a blotchy pattern, called *granulation,* on same like the photosphere (Figure 5-3). The lightly colored **granules** measure about 1000 km across and are surrounded by relatively dark boundaries. Time-lapse photography shows that granules form, disappear, and then re-form in cycles that last several minutes. At any single moment, several million granules cover the solar surface.

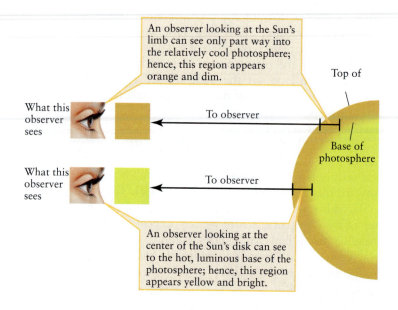

An observer looking at the Sun's limb can see only part way into the relatively cool photosphere; hence, this region appears orange and dim.

Top of

What this observer sees

To observer

Base of photosphere

What this observer sees

To observer

An observer looking at the center of the Sun's disk can see to the hot, luminous base of the photosphere; hence, this region appears yellow and bright.

Like liquid in a pot of simmering soup, the gases in granules rise and fall. In Section 4-7 we learned that radial motion of a light source affects the wavelengths of its spectral lines through the Doppler effect (see Figures 4-15 and 4-16 and *An Astronomer's Toolbox 4-3*). By carefully measuring the Doppler shifts of spectral lines in various parts of individual solar granules, astronomers have determined that hot gases move upward in the center of each granule and cooler gases cascade downward around its edges. This motion is caused by **convection** in which gas or liquid, heated from below, expands and rises like a bubble. Reaching the surface, the hot gas radiates its energy, cools, and descends. Once the Sun's hot gases arrive at the photosphere, they radiate energy

FIGURE 5-2 Limb Darkening The Sun's edge, or limb, appears distinctly darker and more orange than does its center, as seen from Earth (see Figure 5-1). This occurs because we look through the same amount of solar atmosphere at all places. As a consequence, we see higher in the Sun's photosphere near its limb than when we look at its central regions. The higher photosphere is cooler and, because it is a blackbody, darker and more orange than the lower, hotter region of the photosphere.

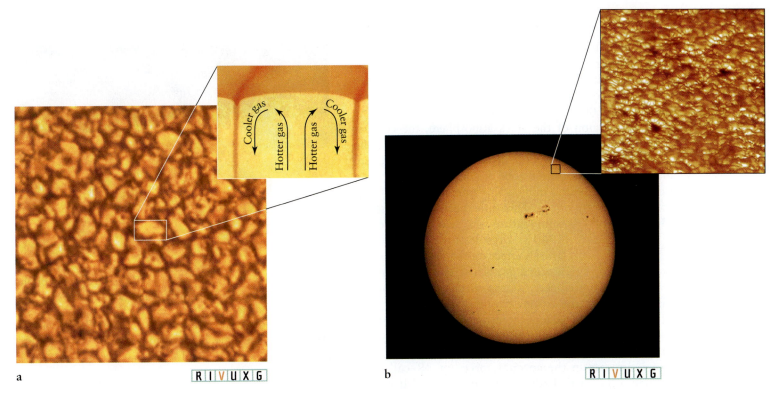

a R I V U X G

b R I V U X G

  FIGURE 5-3 Solar Granulation
(a) High-resolution photographs of the Sun's surface reveal a blotchy pattern, called granulation. Granules, which measure about 1000 km across, are convection cells in the Sun's photosphere. *(Inset)* Gas rising upward produces the bright granules. Cooler gas sinks downward along the darker, cooler boundaries between granules. This convective motion transports energy from the Sun's interior outward to the solar atmosphere. (b) At lower resolution, the Sun's surface appears relatively smooth (the dark regions will be discussed shortly). *(Inset)* Seen near the Sun's limb, granules bulge upward at their centers as a result of the convection that creates them. (a: MSFC/NASA; inset: Goran Scharmer, Lund Observatory; b: Celestron International; inset: G. Scharmer [ISP, RSAS] et al., Lockheed-Martin Solar & Astrophysics)

out into space. We see this energy as visible light and other electromagnetic radiation. Upon radiating, the gases cool, spill over the edges of the granule, and plunge back down into the Sun along the boundaries between granules (see Figure 5-3 inset).

According to the Stefan-Boltzmann law (see Section 4-2), hotter regions emit more photons per square meter than do cooler regions. Note in Figure 5-3a that the centers of the granules are brighter than their edges. Observations indicate that a granule's center is typically 100 K hotter than its edge, thus explaining the observed brightness difference.

5-2 The chromosphere is characterized by spikes of gas, called *spicules*

Immediately above the photosphere is a dim layer of less dense stellar gas, called the **chromosphere** ("sphere of color"). This unfortunate name suggests that it is the layer we normally see, but for centuries it was visible only when the photosphere was blocked during a total solar eclipse.

During an eclipse, the chromosphere is visible as a pinkish strip some 2000-km thick around the edge of the dark Moon. Today, astronomers can also study the chromosphere through filters that pass light with specific wavelengths strongly emitted by it—but not by the photosphere—or through telescopes sensitive to nonvisible wavelengths that the chromosphere emits intensely (Figure 5-4).

High-resolution images of the chromosphere reveal numerous spikes, which are jets of gas called **spicules** (Figures 5-4 and 5-5a). A typical spicule rises for several minutes at the rate of 72,000 km/h (45,000 mph) to a height of nearly 10,000 km (Figure 5-5b). Then it collapses and fades away. At any one time, roughly a third of a million spicules cover a few percent of the Sun's chromosphere.

Which part of the disk of the Sun appears brighter, the center or the edge? Why?

Spicules are generally located on the boundaries of enormous regions of rising and falling chromospheric gas called **supergranules** (Figure 5-5a). A typical supergranule has a diameter slightly larger than Earth's and contains about 900 granules.

5-3 Temperatures increase higher in the Sun's atmosphere

It seems plausible that the temperature should decrease as you rise through the Sun's atmosphere. After all, by moving upward, you move farther from the Sun's internal heat source. Indeed, starting in the photosphere at 5800 K, the temperature drops to around 4000 K in the lower chromosphere. Surprisingly, however, the temperature then begins to increase much more as you ascend, reaching about 10,000 K at the top of the chromosphere. The outermost region of the Sun's atmosphere, the **corona**, extends several million kilometers from the top of the chromosphere (Figure 5-6). Between the chromosphere and corona is a **transition zone** in which the temperature skyrockets to about 1 million K (Figure 5-6c).

> Which part of the Sun do we normally see?

This unexpected increase in temperature was discovered around 1940 as a result of the high temperature's effect on the spectrum of the Sun's corona. The hotter a gas is, the more it is ionized (electrons are stripped off the atoms). Astronomers discovered that the corona contains the emission lines of several highly ionized (therefore, very hot) elements. For example, a prominent green line caused by the presence of Fe XIV, an iron atom stripped of 13 electrons, appears in the coronal spectrum. (Recall from Section 4-5 that Fe I is a neutral iron atom.) It is now

RIVUXG

FIGURE 5-4 The Chromosphere This photograph of the chromosphere was taken by the *Hinode* (Japanese for "sunrise") satellite. The dark bumps are the tops of granules, and the light regions are hotter gases in spicules. The spicules on the edge or limb of the Sun give a sense of the height of these gas jets. (Hinode JAXA/NASA)

known that coronal temperatures are typically in the range of 1 to 2 million K, although some regions are even hotter.

Although its temperature is extremely high, the density of gas in the corona is very low, about 10 trillion times less dense than the air at sea level on Earth. The low

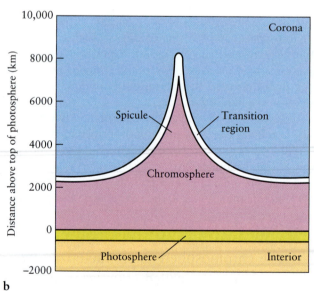

a   RIVUXG   b

FIGURE 5-5 Spicules and Supergranules (a) Spicules appear in this photograph of the Sun's chromosphere. The Sun appears rose-colored in this image because it was taken through an H_α filter that passes red light from hydrogen and effectively blocks most of the photosphere's light. Surrounded by spicules, supergranules are regions of rising and falling gas in the chromosphere. Each supergranule spans hundreds of granules in the photosphere below. Inset: A view of spicules from above. (b) The spicules are jets of gas that surge upward into the Sun's outer atmosphere. This schematic diagram shows a spicule and its relationship to the solar atmosphere's layers. The photosphere is about 400-km thick. The chromosphere above it extends to an altitude of about 2000 km, with spicules jutting up to nearly 10,000 km above the photosphere. The outermost layer, the corona (discussed in Section 5-3), extends millions of kilometers above the photosphere. (a: NASA; inset: Swedish Solar Telescope/Royal Swedish Academy of Sciences, La Palma, Spain, by Bert De Pontieu, Lockheed-Martin Solar and Astrophysics Lab)

a

In this narrow transition region between the chromosphere and the corona, the temperature rises abruptly by about a factor of 100.

c

b

FIGURE 5-6 The Solar Corona (a) This visible-light photograph was taken during the total solar eclipse of July 11, 1991. Numerous streamers are visible, extending millions of kilometers above the solar surface. (b) This X-ray image of the Sun's corona, taken by the *Yohkoh* satellite in 1999, provides hints of the complex activity taking place on and in the Sun. The million-degree gases in the corona emit the X rays visible here. (c) This graph shows how temperature varies with altitude in the Sun's chromosphere and corona and in the transition region between them (white). Note that both the height and temperature scales are nonlinear. (a: R. Christen and M. Christen, Astro-Physics Inc.; b: Greg Slater; c: adapted from A. Gabriel)

density partly accounts for the dimness of the corona, which otherwise would outshine the photosphere. Astronomers have mounting evidence that the corona is heated by energy carried aloft and released there by the Sun's complex magnetic fields, discussed in Sections 5-5 and 5-6.

The total amount of visible light that we receive from the solar corona is comparable to the brightness of the full Moon—or only about one-millionth as bright as the photosphere. As with the chromosphere, the corona can be seen only when the photosphere is blocked out or through special filters or at nonvisible wavelengths (such as ultraviolet and X ray), at which the corona is especially bright compared to the photosphere. The photosphere is blocked naturally during a total eclipse or artificially with a specially designed telescope, called a *coronagraph*. Figure 5-6a is a photograph of the corona taken during a total eclipse (see also Figure 1-26). Figure 5-6b shows a stunning X-ray image of the corona.

Just as Earth's gravity prevents most of our atmosphere from escaping into space, so, too, does the Sun's

gravity keep most of its outer layers from leaving. However, some of the gas in the corona is moving fast enough—around a million kilometers per hour—to escape the Sun's gravity forever and race into space. This outflow of particles is called the **solar wind**. We will see its effects on various parts of the solar system in the last section of this book. The *heliosphere* is a bubble in space created by the solar wind and contains the Sun and planets. This outflow of gases from the Sun prevents most of the gases flowing in space from other stars from entering our solar system.

The Sun ejects around a million tons of matter each second as the solar wind. Even at this rate of emission, the mass loss due to the solar wind will amount to only a few tenths of a percent of the Sun's total mass throughout its lifetime. Although electrons and hydrogen and helium nuclei comprise 99.9% of the solar wind, silicon, sulfur, calcium, chromium, nickel, neon, and argon ions have also been detected in it. The solar wind particles reach speeds up to 2.9×10^6 km/h (1.8×10^6 mph). The wind achieves these high speeds, in part, by being accelerated by the Sun's magnetic field, discussed in Section 5-5. By the time the solar wind travels 1 AU and arrives in the vicinity of Earth, it has spread out so much that there are typically just 5 particles per cubic centimeter (cm³),

> Which of the Sun's three atmospheric layers is coolest? Which is densest?

and they are moving around 1.1 to 1.4 × 10⁶ km/h (6.8 to 8.5 × 10⁵ mi/h). For comparison, there are 6 × 10¹⁹ molecules/cm³ in the air we breathe, and the winds at Earth's surface are moving less than 240 km/h (150 mi/h).

Astronomers think that the Sun's surface chemical composition today is nearly identical to the composition of the gas and dust from which it formed, called the solar nebula. To better understand the chemistry of the early solar system, the *Genesis* spacecraft was launched to collect pristine solar wind particles far from the Earth and to return them here. By studying the amounts of the different chemicals in the solar wind, astronomers hope to better understand how the Sun works, as well as to learn the chemistries of the earlier generation of stars that exploded and whose gases eventually recondensed and formed the Sun and the other objects in the solar system. *Genesis*'s parachute failed and the spacecraft crashed into Earth at 320 km/h (about 200 mph), creating a small crater. Nevertheless, the collected samples have been salvaged and are being studied.

An important result from the *Genesis* mission concerns the chemistry of neon in the Sun. The amounts of different neon isotopes (atoms of neon that have different numbers of neutrons in their nuclei) in the Moon rocks brought back by Apollo astronauts suggested that the Sun was more active (hotter) in the past than it is today. The amounts of neon isotopes collected by *Genesis* were different from those in the Moon rocks. Astronomers deduce that the neon at different depths of the Moon was chemically altered by different amounts due to the radiation from space over billions of years. Taking this into account, they conclude that the Sun has maintained a more nearly constant temperature than previously believed.

THE ACTIVE SUN

Granules, supergranules, spicules, and the solar wind occur continuously. But the Sun's atmosphere is periodically disrupted by magnetic fields that stir things up, creating a group of phenomena, known collectively as the *active* Sun. The Sun's most obvious transient features are **sunspots**, regions of the photosphere that appear dark because they are cooler than the rest of the Sun's lower atmosphere. Sometimes sunspots occur in isolation (Figure 5-7a), but often they arise in clusters, called *sunspot groups* (Figure 5-7b).

INSIGHT INTO SCIENCE

Perception versus Reality Our senses (and often our technology) are limited, so we must be careful about interpreting what we perceive. For example, the sunspots in Figure 5-7 certainly look like black spots. However, they appear black only in contrast to the bright light around them. As we will discuss shortly, sunspots are actually red and orange.

a

b

RIVUXG

FIGURE 5-7 Sunspots (a) This dark region on the Sun is a typical isolated sunspot. Granulation is visible in the surrounding, undisturbed photosphere. (b) This high-resolution photograph shows a sunspot group in which several sunspots overlap and others occur nearby. (a: Royal Swedish Academy of Science; b: National Solar Observatory)

a

b R I V U X G

c R I V U X G

FIGURE 5-8 The Sunspot Cycle (a) The number of sunspots on the Sun varies with a period of about 11 years. The most recent sunspot maximum occurred in 2001, and the most recent sunspot minimum occurred in 2007. (b) The active Sun has many sunspots (this photo was taken in 1979). (c) The Sun has many fewer sunspots when it is not active (this photo, showing a time with no sunspots, was taken in 1989). (b and c: NOAO)

5-4 Sunspots reveal the solar cycle and the Sun's rotation

Like other transient features of the active Sun, the average number and location of sunspots vary in fairly predictable cycles because of the changes in magnetic fields that produce them (as we will explore in Section 5-5). As shown in Figure 5-8a, the average sunspot cycle lasts approximately 11 years. Within an 11-year cycle, most sunspots appear at a **sunspot maximum** (Figure 5-8b). Sunspot maxima occurred most recently in 1979, 1989, and 2001. During a **sunspot minimum**, the Sun is often devoid of sunspots, as it was in 1986, 1996, and 2007 (Figure 5-8c). The next trough in the cycle is projected to occur in 2018.

A typical sunspot is 10,000 km across and lasts between a few hours and a few months. Each sunspot has two parts: a dark, central region, called the *umbra*, and a brighter ring that surrounds the umbra, called the *penumbra*, visible in Figure 5-7a, b. Although these are the same names as the blocked areas of an eclipse (see Section 1-11), we will see in the next section that their causes are completely different—another good example of how certain astronomical terms can have different meanings in different contexts.

Seen without the surrounding brilliant granules that outshine it, a sunspot's umbra appears red and its penumbra is orange. From Wien's law (see Section 4-2), these colors indicate that the umbra is typically 4300 K and the penumbra is 5000 K, both cooler than the normal photosphere.

On rare occasions, a sunspot group is so large that it can be seen with the unaided eye. Chinese astronomers recorded such sightings 2000 years ago, and a huge sunspot group visible to the naked eye was seen in 2001. *Always use special dark filters or other means to protect your eyes when viewing the Sun. Looking directly at the Sun for more than a few moments can cause eye damage!* Of course, a telescope gives a much better view, so it was not until Galileo did so that anyone examined sunspots in detail.

By following sunspots as they moved across the solar disk (Figure 5-9), Galileo discovered that *the Sun rotates once in about 4 weeks*. Galileo initially observed the Sun directly, undoubtedly contributing to his blindness. Eventually his protégé, Benedetto Castelli, developed the technique of projecting sunlight onto a screen to safely observe sunspots. Because a typical sunspot group lasts about 2 months, it can be followed for two solar rotations. Sunspot activity reveals that different latitudes of the Sun rotate at different rates, called **differential rotation**. The equatorial regions rotate more rapidly than the polar regions. A sunspot near the solar equator takes 25 days to go once around the Sun, whereas a sunspot at 30° north or south of the equator takes about 27 days. The rotation period at 75° north or south of the equator is about 33 days, and near the poles it is as long as 35 days.

The average latitude at which new sunspots appear changes throughout the sunspot cycle. At the beginning of each cycle, the sunspots appear mostly at about 30° north and south latitudes. Those that form later in the cycle typically occur closer to the equator. Figure 5-10 shows that the range of latitudes over which sunspots are located and the average latitude at which they are found vary at the same 11-year rate as does the number of sunspots. Satellite measurements reveal that the Sun emits about 0.1% more energy at the peak of the sunspot cycle than at its minimum.

> Which area of the photosphere is hottest, a sunspot's umbra, its penumbra, or a granule?

5-5 The Sun's magnetic fields create sunspots

In 1908, the American astronomer George Ellery Hale discovered that sunspots are directly linked to intense magnetic fields on the Sun. When Hale focused a spectroscope on sunlight coming from a sunspot (Figure 5-11a), he found that each spectral line in the normal solar spectrum

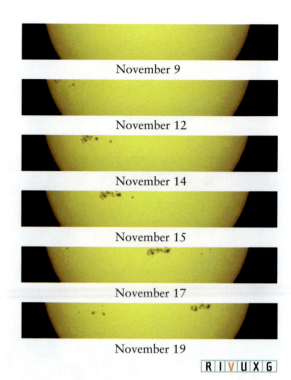

November 9

November 12

November 14

November 15

November 17

November 19

R I V U X G

FIGURE 5-9 The Sun's Rotation This series of photographs taken in 1999 shows the same sunspot group over one third of a solar rotation. Note how the sunspot groups have changed over this time. By observing a group of sunspots from one day to the next in this same manner, Galileo found that the Sun rotates once in about 4 weeks. Sunspot activity also reveals the Sun's differential rotation: The equatorial regions rotate faster than the polar regions. (Carnegie Observatories)

a

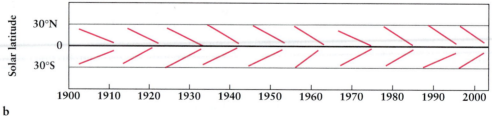

b

FIGURE 5-10 Locations of Sunspots throughout the Sunspot Cycle (a) This "butterfly" diagram of sunspot locations shows that sunspots occur at changing latitudes throughout each cycle. From most common locations to least, the diagram is color-coded yellow, red, green, and white. (b) This graph shows that at the beginning of each sunspot cycle, the average sunspot is located at moderate latitudes, around 30° north or south. Sunspots arising later in each cycle typically form closer and closer to the Sun's equator. (NASA/MSFC/NSSTC/ HATHAWAY 2007/09)

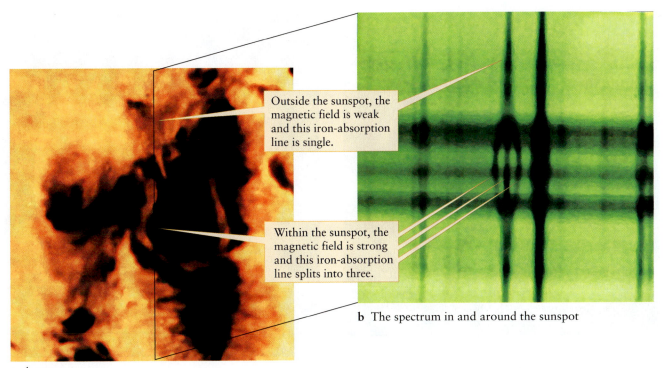

Outside the sunspot, the magnetic field is weak and this iron-absorption line is single.

Within the sunspot, the magnetic field is strong and this iron-absorption line splits into three.

b The spectrum in and around the sunspot

a A sunspot

R I V U X G

FIGURE 5-11 Zeeman Splitting by a Sunspot's Magnetic Field (a) The black line drawn across the sunspot indicates the location toward which the slit of the spectroscope was aimed. (b) In the resulting spectrogram, one line in the middle of the normal solar spectrum is split into three components by the Sun's magnetic field. The amount of splitting between the three lines is used to determine the magnetic field's strength. Typical sunspots have magnetic fields some 5000 times stronger than Earth's magnetic field. (a and b: NOAO)

is flanked by additional, closely spaced spectral lines not usually observed (Figure 5-11b). This "splitting" of a single spectral line into two or more lines is called the **Zeeman effect**. The Dutch physicist Pieter Zeeman, who first observed it in the laboratory in 1896, showed that an intense magnetic field splits the spectral lines of a light source inside the field. The more intense the magnetic field, the more the split lines are separated.

Sunspots are areas where concentrated north or south magnetic fields project through the hot gases of the photosphere. But how do the fields create the spots? The answer lies in the interaction between the magnetic fields and the photosphere's gases. Because of the photosphere's high temperature, many atoms in it are *ionized*: One or more of their electrons have been stripped off by high-energy photons there. As a result, the photosphere is a mixture of electrically charged ions and electrons, called a **plasma**.

Plasmas are extremely good conductors of electricity and are repelled from regions of high magnetic field. Therefore, the magnetic field protruding through the photosphere prevents hot, ionized gases inside the Sun from rising to the surface as they normally do. Thus, such regions of the photosphere are left relatively devoid of hot gas and are therefore cooler and darker than the surrounding solar surface. We observe these darker, cooler regions as sunspots.

The fact that magnetic fields create sunspots presented a problem for astronomers studying the Sun. If you have ever played with toy magnets, you may have discovered that two north poles or two south poles repel each other. How is it then that bundles of same-pole magnetic fields can stay together where they pierce the Sun's photosphere? The answer lies under the Sun's visible layer.

To learn more about the Sun's interior, astronomers record its vibrations—a study called **helioseismology**—just as geologists use earthquakes to study Earth's interior structure. Although there are no true sunquakes, the Sun does vibrate at a variety of frequencies, somewhat like a ringing bell. These vibrations, first noted in 1960, can be detected by using sensitive Doppler shift measurements. They have shown, for example, that portions of the Sun's surface move up and down by about 10 km every 5 min (Figure 5-12a).

Slower vibrations with periods ranging from 20 min to nearly an hour were discovered in the 1970s. Oscillations lasting several days have been detected, as have pulses 16 months long. One important discovery from helioseismology is that deep inside, the Sun rotates

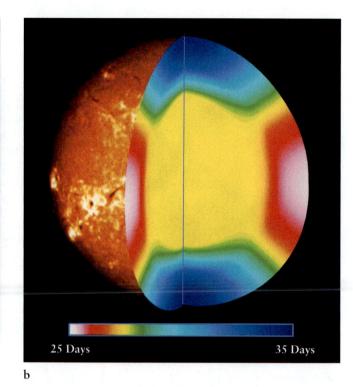

25 Days   35 Days

a

b

  FIGURE 5-12   Helioseismology (a) This computer-generated image shows one of the millions of ways in which the Sun vibrates because of sound waves resonating in its interior. The regions that are moving outward are blue; those moving inward are red. The cutaway shows how deep these oscillations are believed to extend. (b) This cutaway picture of the Sun shows how the rate of solar rotation varies with depth and latitude. Red and yellow denote faster-than-average motion; blue regions move more slowly than average. The pattern of differential surface rotation, which varies from 25 days at the equator to 35 days near the poles, persists at least 19,000 km down into the Sun's convective layer. Sunspots preferentially occur on the boundaries between different rotating regions. Earthlike jet streams and other wind patterns have also been discovered in the Sun's atmosphere.

like a rigid body, rather than with the differential rotation we see on its surface (Figure 5-12b).

Observing the vibrations of the Sun around sunspots, astronomers in 2001 apparently found the answer to the question of how sunspots persist for months despite the repulsion of the magnetic fields inside them. They discovered that below the photosphere, the gases that surround each sunspot whip around like a hurricane as large across as Earth's diameter. The circulation of charged gases around the magnetic fields holds them in place.

Magnetic field lines form complete loops, with each magnetic field having a north pole and a south pole. Likewise, when the Sun's magnetic field emerges through one sunspot or sunspot group, it forms a loop that reenters the Sun at another sunspot or sunspot group (Figure 5-13; *insets*). We associate the name *north pole* or *south pole* with each sunspot or sunspot group, depending on whether the magnetic field there is pointing outward (a south pole) or pointing inward (a north pole). Sunspots and sunspot groups are connected in pairs—one where the magnetic field points out of the Sun, the other where it points into the Sun.

During his observations, Hale found that on one hemisphere of the Sun, sunspots with a magnetic north pole always come into view before the corresponding sunspots with a magnetic south pole. At the same time on the other hemisphere, the order is reversed (see Figure 5-13; *insets*). Hale also found that this pattern reverses itself about every 11 years. The hemisphere where north magnetic poles come first during one 11-year cycle has south magnetic poles coming first during the next. Astronomers, therefore, speak of the 22-year **solar cycle**, the time it takes the magnetic fields to return to the original orientation.

Based on these observations, we can now explain the Sun's magnetic field. This field is created as a result of the Sun's rotation and the resulting motion of the ionized particles found throughout it. This explanation was proposed in 1960 by another American astronomer, Horace Babcock, as the **magnetic dynamo** model, in an effort to explain the 22-year solar cycle. When the Sun is quiet, its magnetic field lies just below the surface in the highly conducting plasma located there, unlike Earth's field, which passes through our planet's center.

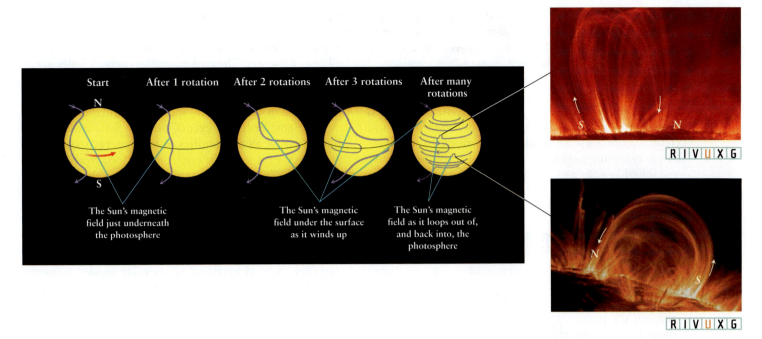

Start   After 1 rotation   After 2 rotations   After 3 rotations   After many rotations

The Sun's magnetic field just underneath the photosphere

The Sun's magnetic field under the surface as it winds up

The Sun's magnetic field as it loops out of, and back into, the photosphere

R I V U X G

R I V U X G

FIGURE 5-13 Babcock's Magnetic Dynamo In a plausible partial explanation for the sunspot cycle, differential rotation wraps a magnetic field around the Sun just under its surface. Convection under the photosphere tangles the field, which becomes buoyant and rises through the photosphere, creating sunspots and sunspot groups. (Insets) In each group, the sunspot that appears first as the Sun rotates (on the right side of each loop as you view them) has the same polarity as the Sun's magnetic pole in that hemisphere (N in the upper hemisphere and S in the lower hemisphere). To understand why, note that magnetic fields make complete loops: If a north pole enters the Sun, it leads to a south pole emerging, and vice versa. Follow the magnetic field in the northern hemisphere from where it enters the Sun (near the top) in counterclockwise loops until you encounter the bump in that hemisphere. Where the field emerges in the bump is a south pole and where it reenters the Sun is a north pole, as drawn.

Conversely, following the fields in the southern hemisphere, starting at the south pole, you encounter the southern bump from the left, emerging as a north pole and reentering the Sun as a south pole, as drawn. The Sun's magnetic fields are revealed by the radiation emitted from the gas they trap. These ultraviolet images show coronal loops up to 160,000 km (100,000 mi) high, with gases moving along the magnetic field lines at speeds of 100 km/s (60 mi/s). (Top inset: TRACE, Stanford-Lockheed Institute for Space Research, and NASA Small Explorer Program; bottom inset: NASA)

As shown in Figure 5-13, the Sun's differential rotation causes its magnetic field to become increasingly stretched. Like stretching a rubber band, stretching the field this way causes it to store energy. At the same time, the magnetic field becomes tangled like the jumble of computer wires behind my desk as convection of the gases under the photosphere also causes the fields to rise and fall. Unlike a rubber band, however, magnetic field lines cannot break to release the energy stored in them. Rather, the fields must untangle themselves.

This untangling process begins as the mixed-up regions of magnetic field trap gases. This gas expands, thereby becoming buoyant and floating up through the solar surface, carrying the magnetic fields with them. Sunspots, with magnetic fields typically 5000 times stronger than Earth's magnetic field, form where loops or tangles of solar magnetic field leave and reenter the Sun. The gas bottled up in the loops of magnetic field eventually leaks out, and the fields untangle, interact with other parts of the Sun's magnetic field, and

gradually settle back under the photosphere, at which point the sunspots associated with them disappear. This process of magnetic fields piercing the Sun's surface begins at high latitudes, and over the next 11 years it occurs closer and closer to the Sun's equator (see Figure 5-10).

Because of how these fields interact and vanish, every 11 years the Sun's entire magnetic field is reversed—the Sun's north magnetic pole becomes its south magnetic pole, and vice versa. After another 11-year cycle, the field is back to its original orientation. This is why the solar cycle is 22 years long. The most recent reversal of the Sun's magnetic field occurred in 2001.

Notable irregularities occur in the solar cycle. For example, the overall reversal of the Sun's magnetic field is often piecemeal and haphazard. More intriguing still is the strong historical evidence that all traces of sunspots and the sunspot cycle have vanished for decades at a time. For example, in 1893 the British astronomer E. Walter Maunder used historical observations to conclude that virtually no sunspots occurred from 1645 through

1715 (see Figure 5-8a). This period, called the *Maunder minimum*, coincides with a period of cold so extreme in Europe that it was called the *Little Ice Age*. At the same time, western North America was subject to severe drought.

Similar sunspot-free periods apparently occurred at irregular intervals in earlier times as well. Conversely, periods of increased sunspot activity in the eleventh and twelfth centuries coincided with periods of warmer-than-average temperatures. It remains to be seen, however, whether scientists can find all of the links between the number of sunspots and periods of extreme temperatures on Earth.

INSIGHT INTO SCIENCE

Cause and Effect While extremes of sunspot activity often occur at the same times as weather extremes on Earth, models of these phenomena must also take into account the possibility of purely terrestrial causes for these temperature changes. Remember, just because one event follows another does not mean that the first *causes* the second.

5-6 Solar magnetic fields also create other atmospheric phenomena

Figure 5-14 shows the active Sun's chromosphere and corona. The bright areas in this photograph are called **plages** (pronounced *plahzh*, from the French word for "beaches"). By studying the light emitted by calcium or hydrogen atoms in the plages, we know that they are hotter, and therefore brighter, than the surrounding chromosphere. Plages, which often appear just before nearby sunspots form, are thought to be created by the magnetic fields under the photosphere crowding upward just before they emerge through the photosphere. In pushing upward, the fields compress the gases of the upper Sun, causing this gas to become hotter and therefore to glow more brightly.

The dark streaks in Figure 5-14 are features in the corona, called **filaments**. These are huge volumes of gas lofted upward from the photosphere by the Sun's magnetic field. When viewed from the side rather than from above, filaments form gigantic loops or arches, called **prominences** (see Figures 5-14 and 5-15a). The temperature of the gas in prominences can reach 50,000 K. These features are almost always associated with sunspots. Some prominences last for only a few hours, whereas others persist for months. The gas in the most energetic prominences escape the magnetic fields that confine them and surge out into space (Figure 5-15b).

X-ray photographs also reveal numerous coronal bright and dark spots that are hotter or colder, respec-

> What holds the gas in loops above the Sun's surface?

tively, than the surrounding corona (Figure 5-16). Temperatures in the bright regions occasionally reach 4 million K. Many of the bright coronal hot spots are located over sunspots. The darker, cooler **coronal holes** act as conduits for gases to flow out of the Sun. Therefore, when a coronal hole on the rotating Sun faces Earth, the solar wind in our direction increases dramatically.

Violent, eruptive events on the Sun, called **solar flares**, release vast quantities of high-energy particles, as well as X rays and ultraviolet radiation, from the Sun (Figure 5-17). Flares are so powerful that they leave the region of the surface of the Sun in their vicinity quaking for an hour or more. At the maximum of the sunspot cycle, about 1100 flares occur per year. Most of them last for less than an hour, but, during that time, temperatures soar to 5 million K. By following the paths of particles emitted by flares as they are guided outward by the Sun's magnetic field, astronomers have recently begun mapping that field as it extends out beyond Earth (Figure 5-18).

Astronomers have also observed huge, balloon-shaped volumes of high-energy gas being ejected from the corona. These **coronal mass ejections** typically expel 2 trillion tons of matter at 1.4×10^6 km/h, and

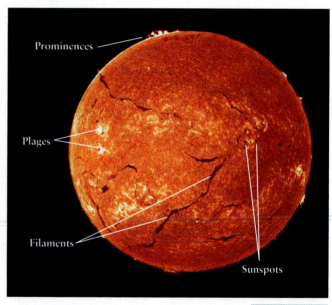

R I V U X G

FIGURE 5-14 Active Sun in H$_\alpha$ This photograph shows the chromosphere and corona during a solar maximum, when sunspots are abundant. The image was taken through a filter that allowed only light from H$_\alpha$ emission to pass through. The hot, upper layers of the Sun's atmosphere are strong emitters of H$_\alpha$ photons. (See Section 4-6 for details of the Balmer series, H$_\alpha$–H$_\infty$.) A few large sunspots are evident. Most notable are features that do not appear at the solar minimum, such as the snakelike features shown here, called filaments, bright areas, called plages, and prominences (filaments seen edge-on) observed at the solar limb. (NASA)

a

R I V U X G

Approximate size
of Earth for
comparison

b

R I V U X G

FIGURE 5-15 Prominences (a) A huge prominence arches above the solar surface in this *SOHO* image taken in 2001. The radiation that exposed this picture is from singly ionized helium at a wavelength of 30.4 nm, corresponding to a temperature of about 50,000 K. (b) The gas in some promi-

nences is so energetic that it breaks free from the magnetic fields that shape and confine it. This eruptive prominence occurred in 1999 and did not strike Earth *(shown for size)*. (a: ESA/NASA; b: Joseph B. Gurman, Solar Data Analysis Center, NASA)

Coronal hole

R I V U X G

R I V U X G

WEB LINK 5.6

FIGURE 5-16 A Coronal Hole This X-ray picture of the Sun's corona was taken by the *SOHO* satellite. A huge coronal hole dominates the lower right side of the corona. The bright regions are emissions from sunspot groups. (*SOHO/EIT/ESA/NASA*)

WEB LINK 5.7

FIGURE 5-17 A Flare Solar flares, which are associated with sunspot groups, produce energetic emissions of particles from the Sun. This image, taken in 2000 by *SOHO*, shows a twisted flare in which the Sun's magnetic field lines are still threaded through the region of emerging particles. (NASA) each one lasts for up to a few hours (Figure 5-19). Coronal mass ejections have enough energy to break through the Sun's magnetic fields that normally contain them or even to carry fields outward. Flares have been observed to create shock waves that initiate some of the coronal mass ejections. The origins of others are still under investigation. The numbers of plages, prominences, flares, and coronal mass ejections vary with the same 11-year cycle as the number of sunspots. Coronal mass ejections, the major source of hazardous particles from the Sun, occur with varying frequency throughout the sunspot cycle, but they never completely cease.

Some coronal mass ejections, solar flares, and filaments head toward Earth. It takes about 8 min for their electromagnetic radiation to get here; their high-energy particles arrive a few days later. At times when these surges of particles are *not* coming to us, the normal solar wind particles are trapped by Earth's magnetic fields in regions called the Van Allen belts (see Section 15-4 for more details on them). The additional particles from a coronal mass ejection or other solar event that *do* come our way overwhelm the Van Allen belts, enabling matter in them to cascade Earthward. One of the most spectacular of such events in recent years occurred on and around October 31, 2003 (Halloween in the United States), filling the night skies with aurorae in many places around the world. Perhaps you saw them.

Using the technology described in Chapter 3, several spacecraft now monitor the Sun and the space

FIGURE 5-18 A Snapshot of the Sun's Global Magnetic Field By following the paths of particles emitted by a solar flare, astronomers have begun mapping the solar magnetic field outside the Sun. The field guides the outflowing particles, which, in turn, emit radio waves that indicate the position of the field. These data were collected by the *Ulysses* spacecraft in 1994. *Ulysses* was the first spacecraft to explore interplanetary space from high above the plane of the ecliptic.

between it and Earth. These include the *Solar and Heliospheric Observatory (SOHO)* launched in 1995 by the European Space Agency (*ESA*) and NASA, the *Transition Region and Coronal Explorer (TRACE)* launched in 1998, the *Reuven Ramaty High Energy Solar Spectroscopic Imager (RHESSI)* launched in 2002, and the *Solar Radiation and Climate Experiment (SORCE)* launched in 2003. These spacecraft observe the Sun in gamma ray, X ray, ultraviolet, and visible parts of the spectrum, and they provide both scientific and space weather information.

a—A coronal mass ejection R I V U X G b Two to 4 days later

FIGURE 5-19 A Coronal Mass Ejection *(Left)* X-ray image of a coronal mass ejection from the Sun taken by *SOHO*. *(Right)* Two to 4 days later, the highest-energy gases from the ejection reach 1 AU. If they come our way, most particles are deflected by Earth's magnetic field *(blue)*. However, as shown, some particles leak Earthward, causing aurorae, disrupting radio communications and electric power transmission, damaging satellites, and ejecting some of Earth's atmosphere into interplanetary space. (*SOHO/EIT/LASCO/ESA/NASA*)

Space weather can profoundly affect humans in space, as well as our assets out there and on the ground. Strong surges of high-energy particles from the Sun can potentially damage satellites, disrupt radio communications, short out power grids here, produce intense aurorae in Earth's atmosphere, cause some of Earth's atmosphere to gush into interplanetary space, and harm or kill people in space. The satellites mentioned previously provide data that allow astronomers to accurately forecast several days in advance the arrival of dangerous levels of solar particles. In the past few years, modeling the Sun's surface using helioseismology has become so sophisticated that we can even deduce what is happening on the side of the Sun facing away from Earth. As the Sun rotates, any hazardous activity there will sweep into view. By knowing what is coming even weeks ahead of time, astronauts and spacecraft can be even better protected than ever before.

THE SUN'S INTERIOR

During the 1800s, geologists and biologists found convincing evidence that Earth must have existed in more or less its present form for at least hundreds of millions of years. This fact posed severe problems for astrophysicists, because at that time it seemed impossible to explain how the Sun could continue to shine for so long, radiating immense amounts of energy into space. If the Sun were shining by burning coal or hydrogen gas as we could burn these substances on Earth, the Sun would be ablaze for only 5000 years before consuming all of its fuel.

Everyday experience tells us that the Sun is the source of an enormous amount of energy. We have just seen that observations reveal hot gas, intense magnetic fields, and a variety of continuous and transient features on the Sun's surface. But the energy does not come from the surface gas or the magnetic fields—they have no mechanisms to create it. The origin of the Sun's electromagnetic radiation is deep within it. To understand why the Sun shines, we must understand where its energy originates and how that energy is transported to its surface.

5-7 Thermonuclear reactions in the core of the Sun produce its energy

In 1905, Albert Einstein provided an important clue to the source of the Sun's energy with his special theory of relativity. One of the implications of this theory is that matter and energy are related by the simple equation:

$$E = mc^2$$

In other words, a mass (m) can be converted into an amount of energy (E) equivalent to mc^2, where c is the speed of light. Because $c^2 (= c \times c)$ is huge, namely 9×10^{10} km^2/s^2, a small amount of matter can be converted into an awesome amount of energy.

Inspired by Einstein's work, physicists discovered **5** that the Sun's energy output comes from the conversion of matter into energy. In the 1920s, the British astrophysicist Arthur Eddington proposed that the temperature at the center of the Sun, its **core**, is much greater than had ever been imagined. Calculations eventually revealed that the temperature there is about 15.5×10^6 K. Physicists also showed that at temperatures above about 10×10^6 K, hydrogen fuses into the element helium. In each such reaction, a tiny amount of mass is converted into energy in the form of gamma rays. Enough fusion occurs in its core to account for all of the energy emitted by the Sun.

> What two features on the Sun are actually the same thing, seen from different angles?

The process of fusing nuclei at such extreme temperatures is called **thermonuclear fusion** (see *An Astronomer's Toolbox 5-1*). In particular, conversion of hydrogen into helium is called **hydrogen fusion**. The same process provides the devastating energy released by a hydrogen bomb. (We will encounter the thermonuclear fusion of helium and other elements in later chapters.) Some of the energy generated by hydrogen fusion in the Sun's core eventually escapes through the photosphere into space. That energy makes the Sun shine.

Hydrogen fusion is also called *hydrogen burning*, even though nothing is burned in the conventional sense. The ordinary burning of wood, coal, or any

© Sidney Harris

flammable substance is a chemical process involving only the electrons orbiting the nuclei of the atoms. Thermonuclear fusion is a far more energetic process that involves violent collisions that change the atomic nuclei themselves. *An Astronomer's Toolbox 5-1* shows you just how energetic.

> Where in the Sun is the energy that we see generated?

You may have heard statements like "matter is always conserved" or "that energy is always conserved." We know that both of these concepts are incorrect, because mass can be converted into energy, and vice versa. What is true, however, is that the total amount of energy plus mass (multiplied by c^2) is conserved. So, the destruction of mass and the creation of energy by the Sun do not violate any laws of nature.

AN ASTRONOMER'S TOOLBOX 5-1

Thermonuclear Fusion

What drives the thermonuclear fusion that powers the Sun? For nuclei to fuse, they must be brought together at incredibly high temperatures and pressures. That is exactly what occurs in the Sun's core, where the entire mass of the Sun compresses inward. The core's temperature is 15.5 million K, its pressure is about 3.4×10^{11} atm, and its density is 160 times greater than that of water.

Normally, nuclei cannot contact one another because the positive electric charge on each proton prevents nearby protons from coming too closely. (Remember that like charges repel each other.) But in the extreme heat and pressure of the Sun's center, the protons move so fast in such close proximity that they can stick, or fuse, together.

The nuclear transformations inside the Sun follow several routes, but each begins with the simplest atom, hydrogen (H). Most hydrogen nuclei consist of a single proton. The final outcome of fusion is creation of the nucleus of the next simplest atom, helium (He), consisting of two protons and two neutrons. The fusion of hydrogen into helium takes several steps.

The diagram on page 145 shows the most common path for hydrogen fusion in the Sun. This particular sequence is called the *proton-proton*, or *PP, chain*.

Note that proton-proton fusion releases positively charged electrons, e^+, called **positrons**, and neutral, nearly massless particles, called **neutrinos**, v. When these positrons encounter regular electrons in the Sun's core, both particles are annihilated, and their mass is converted into energy in the form of gamma-ray photons. The final fusion, in which the helium forms, also returns two protons to the core, which are then available to fuse again.

Because the PP chain produces neutrinos and the Sun's energy, we can summarize hydrogen fusion this way:

$$4\,^1\text{H} \rightarrow 1\,^4\text{He} + \text{neutrinos} + \text{energy}$$

Example: From our summary equation, we can calculate the energy released during a fusion reaction. We simply look at how much mass is converted into energy:

Mass of 4 hydrogen atoms = 6.693×10^{-27} kg
− Mass of 1 helium atom = 6.645×10^{-27} kg
Mass lost = 0.048×10^{-27} kg

Thus, a small fraction (0.7%) of the mass of the hydrogen going into the nuclear reactions does not show up in the mass of the helium. Ignoring the relatively small mass of the neutrino, this lost mass is converted into energy, as predicted by Einstein's famous equation:

$$E = mc^2$$
$$= (0.048 \times 10^{-27}\text{ kg}) \times (3 \times 10^8\text{ m/s})^2$$
$$= 4.3 \times 10^{-12}\text{ J}$$

Compare! The energy released from the formation of a single helium atom would light a 10-watt lightbulb for almost one-half a trillionth of a second.

Now let us add up the energy emitted from the entire Sun. As noted earlier in this chapter, the Sun's mass, usually designated 1 $M_\odot$, is equal to 333,000 Earth masses. Its total energy output per second, called the **solar luminosity** and denoted 1 $L_\odot$, is 3.9×10^{26} W. To produce this luminosity, the Sun converts 600 million metric tons of hydrogen into helium within its core each second. This is twice the weight of the Empire State Building in New York City. This enormous rate is possible because the Sun contains a vast supply of hydrogen—enough to continue the present rate of energy output for another five billion years.

Try these questions: How many helium atoms would have to be created from hydrogen to release enough energy to light a 60-watt bulb for 12 h? How many joules of energy would be released if a U.S. penny (2.5×10^{-3} kg) were converted entirely into energy? (Answers appear at the end of the book.)

Hydrogen fusion in the Sun usually takes place in a sequence of steps called the proton-proton chain. Each of these steps releases energy that heats the Sun and gives it its luminosity.

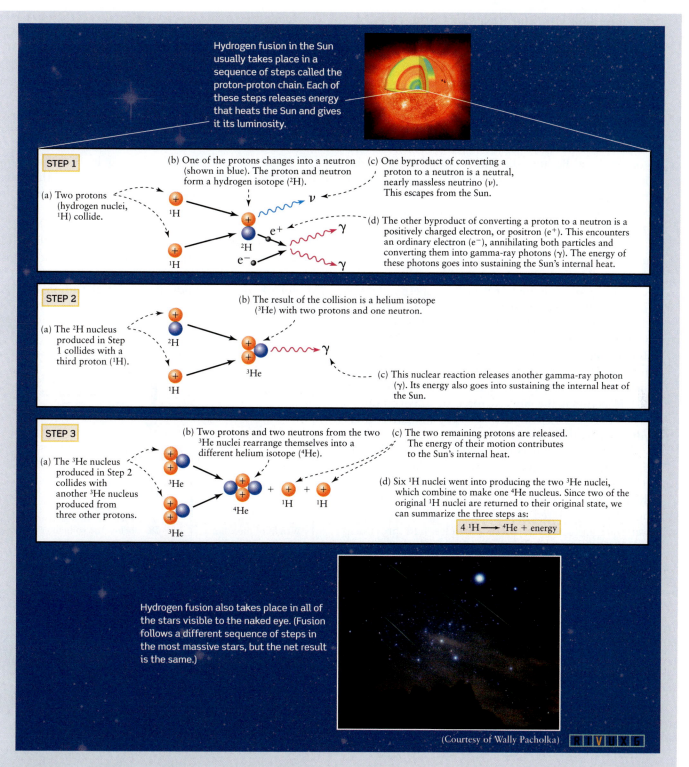

STEP 1

(a) Two protons (hydrogen nuclei, ^{1}H) collide.

(b) One of the protons changes into a neutron (shown in blue). The proton and neutron form a hydrogen isotope (^{2}H).

(c) One byproduct of converting a proton to a neutron is a neutral, nearly massless neutrino (ν). This escapes from the Sun.

(d) The other byproduct of converting a proton to a neutron is a positively charged electron, or positron (e$^+$). This encounters an ordinary electron (e$^-$), annihilating both particles and converting them into gamma-ray photons (γ). The energy of these photons goes into sustaining the Sun's internal heat.

STEP 2

(a) The ^{2}H nucleus produced in Step 1 collides with a third proton (^{1}H).

(b) The result of the collision is a helium isotope (^{3}He) with two protons and one neutron.

(c) This nuclear reaction releases another gamma-ray photon (γ). Its energy also goes into sustaining the internal heat of the Sun.

STEP 3

(a) The ^{3}He nucleus produced in Step 2 collides with another ^{3}He nucleus produced from three other protons.

(b) Two protons and two neutrons from the two ^{3}He nuclei rearrange themselves into a different helium isotope (^{4}He).

(c) The two remaining protons are released. The energy of their motion contributes to the Sun's internal heat.

(d) Six ^{1}H nuclei went into producing the two ^{3}He nuclei, which combine to make one ^{4}He nucleus. Since two of the original ^{1}H nuclei are returned to their original state, we can summarize the three steps as:

$$4\ ^1\text{H} \longrightarrow\ ^4\text{He} + \text{energy}$$

Hydrogen fusion also takes place in all of the stars visible to the naked eye. (Fusion follows a different sequence of steps in the most massive stars, but the net result is the same.)

(Courtesy of Wally Pacholka)   R I V U X G

TOOLBOX 5-1 Steps to Fuse Hydrogen into Helium by the Proton–Proton Chain

5-8 The solar model describes how energy escapes from the Sun's core

A scientific theory describing the Sun's interior, called the **solar model**, explains how the energy from nuclear fusion in the Sun's core gets to its photosphere. The model begins with the inward force due to the Sun's gravity. This force increases the pressure and temperature in the Sun's core, thereby causing hydrogen fusion to occur there. Because the Sun is not shrinking today, however, there must be an outward force throughout its interior that counters the inward force of gravity. That outward force is created by the gamma-ray photons generated during fusion.

To begin their journeys outward, newly-created gamma rays slam into nearby ions and electrons in the solar core. The ions and electrons absorb the energy of the photons and rebound at very high speeds. Because they are densely packed together, they do not travel far before striking other particles. In each collision, the ions and electrons exert forces on one another. They then rebound, emitting photons and colliding with other particles. The result of these frequent interactions is an outward force sufficient to counterbalance gravity. The balance between the inward force of gravity and the outward force from the motion of the hot gas is called **hydrostatic equilibrium** (Figure 5-20).

Maintaining the Sun's hydrostatic equilibrium requires that the photons created in the Sun's core give up some of their energy. Photons collide with particles, which collide with other particles and thereby emit new photons. The emitted photons have slightly less energy than the photons that these particles had previously absorbed. The energy the photons lose provides the outward force keeping the Sun in equilibrium. Photon energies slowly diminish as the photons travel outward from the Sun's core to the photosphere, an odyssey that typically takes 170,000 years.

The outward movement of energy by photons hitting particles, which then bounce off other particles and thereby reemit photons, is called *radiative transport*, because individual photons are responsible for carrying energy from collision to collision. Calculations show that radiative transport is the dominant means of outward energy flow in the **radiative zone**, extending from the core to about 70% of the way out to the photosphere.

Near the photosphere, energy is carried the remaining distance to the surface by the bulk motion of the hot gas, rather than by the flying, energetic photons. As we discussed earlier, this circulation of gas blobs is called *convection*. We thus say that the Sun has a **convective zone**. Hot gas travels up to the top of this zone by convection and from there radiates photons into space. These photons are what we see as sunlight, and you can see that the photosphere, which is at the bottom of the solar atmosphere, is also at the top of the convective zone. Once the gases in the convective zone lose energy by emitting photons into space, they cool and settle back into the Sun. As noted earlier, this convective flow is also the origin of solar granules, and the departing energy is the visible light and other electromagnetic radiation that the Sun emits into space. Figure 5-21a is a sketch of our model of the internal structure of the Sun.

Different gamma rays created in the Sun's core lose different amounts of energy as they and their successors travel upward through the Sun. Therefore, the photons emitted from the photosphere have a wide range of energies and, hence, wavelengths. The most intense emission is in the visible part of the electromagnetic spectrum. This process of energy loss by photons traveling up from the Sun's core is the origin of the blackbody nature of the photosphere's spectrum.

a

b

FIGURE 5-20 Hydrostatic Equilibrium (a) Matter deep inside the Sun is in hydrostatic equilibrium, meaning that upward and downward forces (or equivalently, pressures) on the gases are balanced. (b) When the forces on the divers in water are in hydrostatic equilibrium, the divers neither sink nor rise. (JUPITER-IMAGES/ Comstock Images/Alamy)

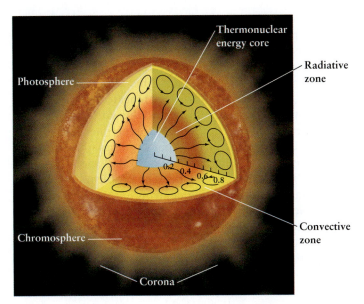

Photosphere

Chromosphere

Corona

Thermonuclear energy core

Radiative zone

Convective zone

a

FIGURE 5-21 The Solar Model (a) Thermonuclear reactions occur in the Sun's core, which extends to a distance of 0.25 solar radius from the center. In this model, energy from the core radiates outward to a distance of 0.7 solar radius. Convection is responsible for energy transport in the Sun's outer layers. (b) The Sun's internal structure is displayed here with graphs that show how the luminosity, mass, temperature, and density vary with the distance from the Sun's center. A solar radius (the distance from the Sun's center to the photosphere) equals 696,000 km. (c) Ten most common elements in the Sun, by the numbers of atoms of each and by the percentage of the Sun's total mass they each comprise. (a: NASA)

b

Element	Number of atoms (percent)	Percent of total mass
Hydrogen	91.2	71.0
Helium	8.7	27.1
Oxygen	0.078	0.97
Carbon	0.043	0.40
Nitrogen	0.0088	0.096
Silicon	0.0045	0.099
Magnesium	0.0038	0.076
Neon	0.0035	0.058
Iron	0.030	0.014
Sulfur	0.015	0.040

c

Our model of the Sun's interior can be expressed as a set of mathematical equations, called the *equations of stellar structure*. The model quantitatively describes the Sun's internal characteristics, such as its pressure, temperature, and density at various depths. These equations also describe the conditions inside other stars as they evolve. Because the equations are so complex, astrophysicists today use computers to solve them. Figure 5-21b presents their results graphically. The four graphs show how the Sun's luminosity, mass, temperature, and density vary from the Sun's center to its photosphere. For example, the upper left graph gives the percentage of the Sun's luminosity created within that radius. The luminosity increases to 100% at about one-quarter of the way from the Sun's center to its photosphere. This tells us that all of the Sun's energy is produced within a volume extending out to $0.25 R_\odot$, where $R_\odot$ is the radius of the Sun.

The mass curve rises to nearly 100% at about 0.6 $R_\odot$ from the Sun's center. Almost all of the Sun's mass is

Why does the Sun not collapse under the influence of its own enormous gravitational attraction?

therefore confined to a volume extending only 60% of the distance from the Sun's center to its visible surface. This helps explain why the density of the gas in the photosphere is 10^4 times lower than the density of the air we breathe.

5-9 The mystery of the missing neutrinos inspired research into the fundamental nature of matter

As explained in *An Astronomer's Toolbox 5-1*, for every proton that changes into a neutron during thermonuclear fusion, a *neutrino* is released. Neutrinos have no electric charge, and they are extraordinarily difficult to detect because they rarely interact with ordinary matter. The Sun is largely transparent to neutrinos, allowing these particles to stream outward, unimpeded, from its core. Likewise, most of the neutrinos from the Sun that arrive at Earth pass right through it.

According to our model of fusion in the Sun, nearly 10^{38} *solar neutrinos* per second are produced at the Sun's center. This output is so huge that here on Earth roughly 100 billion solar neutrinos per second pass through every square centimeter of your body!

Occasionally, however, solar neutrinos strike neutrons and convert them into protons. If astronomers could detect even a few of these converted protons, it might be possible to build a "neutrino telescope" that could be used to detect the thermonuclear inferno in the Sun's core that is hidden from the view of telescopes that collect photons.

 Inspired by such possibilities, American chemist Raymond Davis designed and built a large neutrino detector. This device consisted of a huge tank that contained 100,000 gallons of perchloroethylene (C_2Cl_4, the fluid your local dry cleaner uses) located deep in the Homestake gold mine in Lead, South Dakota. All neutrino experiments are performed underground to help prevent them from being contaminated by other sources of energy, like high-energy protons from space. Because matter is virtually transparent to neutrinos, most of the solar neutrinos passed right through Davis's tank. On rare occasions, however, a solar neutrino hit the nucleus of one of the chlorine atoms in the cleaning fluid and converted one of its neutrons into a proton, creating a radioactive atom of argon. The rate at which argon was produced was therefore correlated with the number of solar neutrinos arriving at Earth.

On average, solar neutrinos created one radioactive argon atom every 3 days in Davis's tank. To the consternation of astronomers, this rate corresponds to only one-third of the neutrinos predicted to be created by the fusion in the Sun's core. There were three possible explanations of this unexpected result: The experiment could be faulty; the Sun might not be fusing at the expected rate; or our understanding of the properties of neutrinos could be in error. This experiment, which began in the mid-1960s, was repeated with extreme care by other researchers around the world who got the same results, suggesting that the experiment was not at fault. Calculations of how much energy the Sun must generate to shine as it does confirmed the number of neutrinos created per second. The problem had to lie in our understanding of the neutrino.

Neutrinos, like the planet Neptune, were discovered because they had to exist. Back in the first few decades of the twentieth century, physicists observed neutrons in nuclei spontaneously transforming into protons and electrons. The transformation is a result of the *weak nuclear force*, briefly introduced in Section 4-5. When these scientists calculated the energy and momenta of the neutron and the particles into which it changed, the numbers didn't match—conservation of energy and momentum (see Section 2-7) appeared to be violated. If this were indeed the case, it would have required a fundamental reworking of many of the laws of nature. A simpler solution was proposed in 1930 by physicist Wolfgang Pauli, who posited the existence of a hard-to-detect particle that carried away the "lost" energy and momentum (it was named *neutrino* in 1931 by physicist Enrico Fermi).

Among the properties neutrinos had to have are very small masses compared to any then-known particles that have mass and very little interaction with other matter. This latter property followed because trillions upon trillions of them are being created in the Sun every second, with trillions passing through every cubic meter of Earth every second, and there were (and are) virtually no interactions between them and us. Indeed, neutrinos were not detected for 26 years, until 1956. Because electrons or positrons (identical to electrons, except with opposite electric charge) always accompany the formation of the neutrinos created in the Sun, they are often called *electron neutrinos*.

Two other kinds of neutrinos were subsequently proposed. One, the *muon neutrino*, is emitted in reactions in which elementary particles, called muons (or antimuons), are released, while the other, the *tau neutrino*, accompanies the formation of other elementary particles, the tau and antitau particles. The Sun emits neither muon nor tau neutrinos, and the first generation of detectors could only detect electron neutrinos. The way out of the neutrino dilemma lay in the existence of muon and tau neutrinos.

Initially, the three so-called "flavors" of neutrinos (electron, muon, and tau) were thought to be massless, and, if so, the equations revealed that they could not change from one flavor into another. *But,* if the neutrinos have even the slightest mass, then they can transform from one to another. The mixing of massive neutrinos would occur spontaneously, meaning that if an electron neutrino left the Sun's core and sped toward Earth, there was a finite probability (about 65%, actually) that by the time it got here, it would have transformed into a tau or muon neutrino. So, if neutrinos have mass, we should see only one-third of them (which we do), because the rest would have changed flavor before they got to us.

In 1998, the super Kamiokande neutrino detector in Japan observed the other two types of neutrinos. The observations were repeated in 2002 at the Sudbury Neutrino Observatory (Figure 5-22) in Ontario, Canada. At least a dozen other neutrino detectors have begun operations or are being built to further study solar neutrinos.

To understand how the present generation of neutrino detectors work, let us consider one of the several interactions that occur between neutrinos and other matter. The Sudbury Neutrino Observatory is filled with water that contains a rare type of hydrogen nucleus, called *deuterium*, which consists of a proton and a neutron. When a solar neutrino is absorbed by a deuterium nucleus, the nucleus breaks apart into two protons and an electron. As this electron rushes through the water, it emits a flash of light, called **Cerenkov radiation**. As the Russian physicist Pavel A. Cerenkov (pronounced *Che-ren-kov*) first observed, the flash occurs whenever a particle moves through a medium, such as water, faster than light can. Such motion does not violate the tenet that the speed of light *in a vacuum* (3×10^5 km/s) is the ultimate speed limit in the universe. Light is slowed considerably as it passes through water, and high-energy particles can exceed this reduced speed of light in a medium without violating the laws of nature.

Thus, scientists detect neutrinos by observing Cerenkov radiation flashes with light-sensitive devices, called *photomultipliers*, mounted in the water (see Figure 5-22). The three flavors of neutrinos have different interactions with water, all of which have been detected. This evidence that neutrinos can change *requires* that they have mass, and it explains the earlier low rate of solar neutrino observations. Astronomical research therefore led to a change in our understanding of some of the lowest-mass particles in existence.

R I V U X G

FIGURE 5-22 The Solar Neutrino Experiment Located 2073 m (6800 ft) underground in the Creighton nickel mine in Sudbury, Canada, the Sudbury Neutrino Observatory is centered around a tank that contains 1000 tons of water. Occasionally, a neutrino entering the tank interacts with one or another of the particles already there. Such interactions create flashes of light, called Cerenkov radiation. Some 9600 light detectors sense this light. The numerous silver protrusions are the back sides of the light detectors prior to their being wired and connected to electronics in the lab, seen at the bottom of the photograph. (Ernest Orlando Lawrence/Berkeley National Laboratory)

5-10 Frontiers yet to be discovered

Despite centuries of observations and decades of modeling it, the Sun holds innumerable secrets yet to be uncovered. We have launched an armada of spacecraft specifically designed to observe it in different parts of the spectrum. Astronomers soon hope to understand more about such questions as how the Sun's magnetic field is generated and how it changes throughout the solar cycle; how the Sun's atmosphere heats so dramatically in the transition zone; why the solar luminosity varies with time; just how the Sun's changing output affects the temperatures on Earth; the details of how flares and coronal mass ejections occur; and why the Sun rotates differentially.

Why did the earlier neutrino detectors not detect the predicted number of neutrinos from the Sun?

SUMMARY OF KEY IDEAS

The Sun's Atmosphere

• The thin shell of the Sun's gases we see are from its photosphere, the lowest level of its atmosphere. The gases in this layer shine nearly as a blackbody. The photosphere's base is at the top of the convective zone.

• Convection of gas from below the photosphere produces features called granules.

• Above the photosphere is a layer of hotter, but less dense, gas called the chromosphere. Jets of gas, called spicules, rise up into the chromosphere along the boundaries of supergranules.

• The outermost layer of thin gases in the solar atmosphere, called the corona, extends outward to become the solar wind at great distances from the Sun. The gases of the corona are very hot, but they have extremely low densities.

The Active Sun

• Some surface features on the Sun vary periodically in an 11-year cycle, and the magnetic fields that cause these changes actually varying over a 22-year cycle.

• Sunspots are relatively cool regions produced by local concentrations of the Sun's magnetic field protruding through the photosphere. The average number of sunspots and their average latitude vary in an 11-year cycle.

• A prominence is gas lifted into the Sun's corona by magnetic fields. A solar flare is a brief, but violent, eruption of hot, ionized gases from a sunspot group. Coronal mass ejections send out large quantities of gas from the Sun. Coronal mass ejections and flares that head our way affect satellites, communication, and electric power, and cause aurorae.

• The magnetic dynamo model suggests that many transient features of the solar cycle are caused by the effects of differential rotation and convection on the Sun's magnetic field.

The Sun's Interior

• The Sun's energy is produced by the thermonuclear process called hydrogen fusion, in which four hydrogen nuclei release energy when they fuse to produce a single helium nucleus.

• The energy released in a thermonuclear reaction comes from the conversion of matter into energy, according to Einstein's equation $E = mc^2$.

• The solar model is a theoretical description of the Sun's interior derived from calculations based on the laws of physics. The solar model reveals that hydrogen fusion occurs in a core that extends from the center to about a quarter of the Sun's visible radius.

• Throughout most of the Sun's interior, energy moves outward from the core by radiative diffusion. In the Sun's outer layers, energy is transported to the Sun's surface by convection.

• Neutrinos were originally believed to be massless. The electron neutrinos generated and emitted by the Sun were originally detected at a lower rate than is predicted by our model of thermonuclear fusion. The discrepancy occurred because electron neutrinos have mass, which causes many of them to change into other forms of neutrinos before they reach Earth. These alternative forms are now being detected.

WHAT DID YOU THINK?

1 *What percentage of the solar system's mass is in the Sun?* The Sun contains about 99.85% of the solar system's mass.

2 *Does the Sun have a solid and liquid interior like Earth?* No. The entire Sun is composed of hot gases.

3 *What is the surface of the Sun like?* The Sun has no solid surface. Indeed, it has no solids or liquids anywhere. The level we see, the photosphere, is composed of hot, churning gases.

4 *Does the Sun rotate? If so, how fast?* The Sun's surface rotates differentially, varying between once every 35 days near its poles and once every 25 days at its equator.

5 *What makes the Sun shine?* Thermonuclear fusion in the Sun's core is the source of the Sun's energy.

Key Terms for Review

Cerenkov radiation, 149
chromosphere, 131
convection, 130
convective zone, 146
core (of the Sun), 143
corona, 132
coronal hole, 140
coronal mass ejection, 140
filament, 140
granule, 130
helioseismology, 137
hydrogen fusion, 143
hydrostatic equilibrium, 146
limb (of the Sun), 129
limb darkening, 129
magnetic dynamo, 138
neutrino, 144
photosphere, 129
plages, 140
plasma, 137
positron, 144
prominence, 140
radiative zone, 146

Review Questions

The answers to all computational problems, which are preceded by an asterisk (*), appear at the end of the book.

1. Describe the features of the Sun's atmosphere that are always present.

2. Describe the three main layers of the solar atmosphere and how you would best observe them.

3. Name and describe seven features of the active Sun. Which two are the same, seen from different angles? Which can have a direct impact on Earth? Explain.

4. Describe the three main layers of the Sun's interior.

*5. When will the next sunspot minimum and sunspot maximum occur after the maximum in 2001 and the minimum in 2007? Explain your reasoning.

6. Why is the solar cycle said to have a period of 22 years, even though the sunspot cycle is only 11 years long?

7. How do astronomers detect the presence of a magnetic field in hot gases, such as the field in the solar photosphere?

8. Describe the dangers in attempting to observe the Sun. How have astronomers learned to circumvent these hazards?

9. Give an everyday example of hydrostatic equilibrium not presented in the book.

10. Give some everyday examples of heat transfer by convection and radiative transport.

11. What do astronomers mean by a "model of the Sun"?

12. Why do thermonuclear reactions in the Sun take place only in its core?

13. What is hydrogen fusion? This process is sometimes called "hydrogen burning." How is hydrogen burning fundamentally unlike the burning of a log in a fireplace?

14. Describe the Sun's interior, including the main physical processes that occur at various depths within the Sun.

15. What is a neutrino, and why are astronomers so interested in detecting neutrinos from the Sun?

Advanced Questions

*16. Using the mass and size of the Sun, calculate the Sun's average density. Compare your answer to the average densities of the outer planets. (*Hint*: The volume of a sphere of radius r is $\frac{4}{3}\pi r^3$.)

*17. Assuming that the current rate of hydrogen fusion in the Sun remains constant, what fraction of the Sun's mass will be converted into helium over the next five billion years? How will this affect the chemical composition of the Sun?

*18. Calculate the wavelengths at which the photosphere, chromosphere, and corona emit the most radiation. Explain how the results of your calculations suggest the best way to observe these regions of the solar atmosphere. (*Hint*: Use Wien's law and assume that the average temperatures of the photosphere, chromosphere, and corona are 5800 K, 50,000 K, and 1.5×10^6 K, respectively.)

19. When we are near a sunspot maximum, the Hubble Space Telescope must be put in a higher orbit. Why? (*Hint*: Think about how the increased solar energy affects Earth's atmosphere.)

Discussion Questions

20. Discuss the extent to which cultures around the world have worshiped the Sun as a deity throughout history. Why do you think our star inspires such widespread veneration?

21. Discuss some of the difficulties of correlating solar activity with changes in the terrestrial climate.

22. Describe some advantages and disadvantages of observing the Sun **a.** from space and **b.** from Earth's South Pole. What kinds of phenomena and issues do solar astronomers want to explore from both Earth-orbiting and Antarctic observatories?

What If . . .

23. The Sun were not rotating? What about it would be different?

24. The typical solar wind were much stronger (say 100 times stronger) than it is now? What differences would there be in the solar system?

25. The typical solar wind were much weaker (say 100 times weaker) than it is now? What differences would there be in the solar system?

26. The Sun's brightness and heat output periodically changed, as they do in many stars? What differences would there be in the solar system? Unless your instructor gives you another time frame, assume that the light and heat vary together in a cycle that lasts 1 year.

Web Questions

27. Search the Web for all solar neutrino experiments. Make a list of them and indicate which are on-line and which are still under construction. Summarize the results found by the active detectors.

28. Search the Web for information about features of the solar atmosphere, called *sigmoids*. What are they? What causes them? How do they provide a way to predict coronal mass ejections?

29. Determining the Lifetime of a Solar Granule Access and view the video "Granules on the Sun's Surface" in Chapter 5 of the *Discovering the Universe: From the Stars to the Planets* Web site. You will use it to determine the approximate lifetime of a solar granule. Select an area on the Sun's image and slowly and rhythmically repeat *Start, Stop, Start, Stop* until you can consistently predict the appearance and disappearance of granules. While keeping your rhythm, move to a different area of the video and continue monitoring the appearance and disappearance of granules. When you are confident that you have the timing right, move your eyes (or use a partner) to the clock shown in the video. Using your *Start-Stop* cycle, determine the length of time between the appearance and disappearance of the granules and record your answer.

Observing Projects

30. Use a telescope to view the Sun by using a solar filter or by projecting the Sun's image onto a screen or sheet of white paper. *Do not look directly at the Sun! Looking at the Sun causes blindness.* Do you see any sunspots? If so, sketch their appearance. Can you distinguish between the umbrae and penumbrae of the sunspots? Can you see limb darkening? Can you see granulation?

31. If you have access to an H_α filter attached to a telescope specially designed for viewing the Sun safely, use this instrument to examine the solar surface. How does the filtered appearance of the Sun differ from that in white light (see Observing Project 30)? What do sunspots look like in H_α? Can you see any prominences? Can you see any filaments? Are the filaments in the H_α image near any sunspots seen in white light?

*32. Use *Starry Night Enthusiast*™ software to determine the Sun's rotation rate. Select **Favourites > Solar System > Sun > Flares and Prominences** from the menu. Note the locations of the features you see on the Sun so you can use them for reference as it rotates. Change the **Time Flow Rate** to **1 sidereal day** and use the single step forward button ($|\blacktriangleright$) to observe the Sun's rotation. Step through enough time steps to determine the rotation rate for the Sun. Compare the number of days for one rotation to the rotation information for the Sun presented in the text and state which part of the Sun the program's rotation is portraying. (Note that the program does not show the Sun's differential rotation.)

CHAPTER 6

Characterizing Stars

Stars Come in Many Colors
(Hubble Heritage Team/
AURA/STScI/NASA)

R I V U X G

WHAT DO YOU THINK?

1 How near to us is the closest star other than the Sun?

2 How luminous is the Sun compared with other stars?

3 What colors are stars, and why do they have these colors?

4 Are brighter stars hotter than dimmer stars?

5 Compared to the Sun, what sizes are other stars?

6 Are most stars isolated from other stars, as the Sun is?

Answers to these questions appear in the text beside the corresponding numbers in the margins and at the end of the chapter.

Legend tells of a race of remarkable insects, the Ephemera, that inhabited a great forest. These noble creatures were blessed with great intelligence, but were cursed with tragically short life spans. To the Ephemera, the forest seemed eternal and unchanging. Members of each generation lived out their brief lives without ever noticing any changes in their leafy world. Nevertheless, careful observations and reasoning led some Ephemera to postulate that the forest was not static. They began to suspect that small green shoots grew to become huge trees and that mature trees eventually died, toppled over, and littered the forest with rotting logs, enriching the soil for future trees. Although unable to witness the transformation personally, the Ephemera predicted the existence of life processes stretching over the mind-boggling periods of many years.

To us, the cosmos seems eternal and unchanging; the views that greet us every night are virtually indistinguishable from those seen by our ancestors. This permanence, however, is an illusion. We see the stars because they emit vast amounts of radiation, created as a result of converting one element into another in their interiors. Because all stars contain finite amounts of the elements fueling these emissions, at different times in the life of each star, its various fuels are used up. This causes the stars to change profoundly.

We do have one advantage over the Ephemera, who were so small that they could see only a few trees, leaves, shoots, and rocks. Astronomers are witnesses to literally billions of stars in various stages of evolution. Therefore, although we cannot see any one star go through more than a tiny fraction of the process, we see every stage of **stellar evolution**. By understanding the changes that stars undergo, we gain insight into our place in the cosmos. We begin our journey to the stars by learning the observed properties of stars that astronomers use to develop models of why stars shine and how they evolve.

In this chapter you will discover

- that the distances to many nearby stars can be measured directly, whereas the distances to farther ones are determined indirectly

- the observed properties of stars on which astronomers base their models of stellar evolution

- how astronomers analyze starlight to determine a star's temperature and chemical composition

- how the total energy emitted by stars and their surface temperatures are related

- the different classes of stars

- the variety and importance of binary star systems

- how astronomers calculate stellar masses

6-1 Distances to nearby stars are determined by stellar parallax

Nothing in human experience prepares us for understanding the distances between the stars. On Earth, the greatest distance we read about in our daily lives is perhaps halfway around the globe, 20,000 km, which is about the distance from New York City to Perth, Australia. By observing the Sun, the Moon, and planets, we have some comprehension of hundreds of thousands, or even millions, of kilometers.

How far away are the nearest stars? Ask 10 people and you will probably hear answers that range from millions to billions of kilometers or miles. In fact, the closest star other than the Sun, Proxima Centauri, in the constellation Centaurus, is about 40 *trillion* km (25 trillion mi) away. It takes light over 4 years to get here from there. Most of the stars you see in the night sky are many times farther away. Given such enormous separations between astronomical bodies, how do we measure the distance to a nearby star?

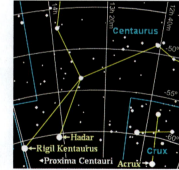

Because Proxima Centauri is closer to us than other stars, its position among the background stars changes as Earth orbits the Sun. This is precisely the same effect that we see when, for example, Mars appears to have retrograde motion as we pass between it and the Sun (see Figures 2-2 and 2-3) or that Tycho Brahe sought for the supernova of 1572 (see Figure 2-7). The apparent motion of nearby stars among the background of more distant stars, due to Earth's motion around the Sun, is called **stellar parallax**.

Parallax is an everyday phenomenon. We experience it when nearby objects appear to shift their positions against a distant background as we move (review Figure 2-6). We also experience it continuously when we are awake, because our eyes change angle when looking at objects at different distances. As you view a tree 10 m away, your eyes cross only slightly (Figure 6-1a). Looking

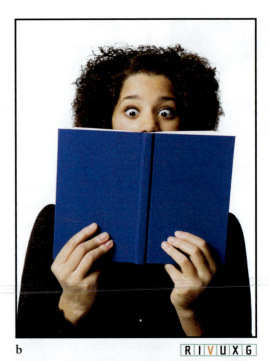

a   b   R I V U X G

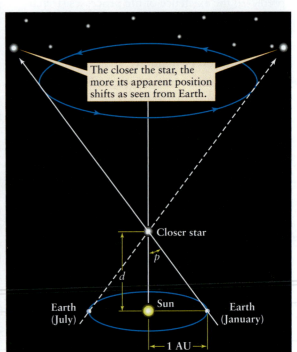

c Parallax of a nearby star   d Parallax of an even closer star

FIGURE 6-1 Using Parallax to Determine Distance (a, b) Our eyes change the angle between their lines of sight as we look at things that are different distances away. Our eyes are adjusting for the parallax of the things we see. This change helps our brains determine the distances to objects and is analogous to how astronomers determine the distance to objects in space. (c) As Earth orbits the Sun, a nearby star appears to shift its position against the background of distant stars. The star's parallax angle (*p*) is equal to the angle between the Sun and Earth, as seen from the star. The stars on the scale of this drawing are shown much closer than they are in reality. If drawn to the correct scale, the closest star, other than the Sun, would be about 5 km (3.2 mi) away. (d) The closer the star is to us, the greater the parallax angle *p*. The distance to the star (in parsecs) is found by taking the inverse of the parallax angle *p* (in arcseconds), $d = 1/p$. (a, b: Mark Andersen/JupiterImages) at something closer, say this book, your eyes cross much more in order for both of them to focus on the same word (Figure 6-1b). The parallax angle formed between your eyes and the tree or the book lets your brain judge just how close these objects are to you. Working out distances of nearby stars is done similarly, but it requires painstaking measurements of tiny angles and a little geometry.

As the Earth moves from one side of its orbit around the Sun to the other, a nearby star's apparent position shifts among the more distant stars. Referring to Figure 6-1c, the parallax angle, p, is half the angle by which Earth shifts positions through the year as seen from that star, measured in arcseconds. The difference in parallax angles for stars at different distances can be seen by comparing Figures 6-1c and 6-1d. An Astronomer's Toolbox 6-1 explores some details of how distances are determined from parallax angles.

The first stellar parallax measurement was made in 1838 by Friedrich Wilhelm Bessel, a German astronomer and mathematician. He found the parallax angle of the star 61 Cygni to be ⅓ arcsec, and so its distance is about 3 pc (see An Astronomer's Toolbox 2-1 to review astronomical distance units). The precision

of stellar parallax measurements is limited by the angular resolution of the telescope, as discussed in Section 3-6. Because parallax angles smaller than about 0.01 arcsec are difficult to measure from Earth-based observatories, the stellar parallax method using Earth-based telescopes gives stellar distances only up to about 100 pc.

Telescopes in space are unhampered by our atmosphere and therefore have higher resolutions than Earth-based telescopes. Parallax measurements made in space thus enable astronomers to determine the distances to stars well beyond the reach of ground-based observations.

 In 1989, the European Space Agency (ESA) launched a satellite, called *Hipparcos* (an acronym for *high precision parallax collecting* satellite, and named for Hipparchus, an astronomer in ancient Greece who created an early classification system for stars). Although the satellite failed to achieve its proper orbit, astronomers used it to measure the distances to over 2.5 million of the nearest stars up to 500 ly (150 pc) away. The success of *Hipparcos* has led

---

AN ASTRONOMER'S TOOLBOX 6-1

Distances to Nearby Stars

Recall from *An Astronomer's Toolbox 2-1* that 1 parsec (1 pc) is the distance at which two objects 1 AU apart appear 1 arcsec apart. This distance is 3.09×10^{13} km, or 206,265 AU. The word *parsec* (from *parallax second*) originated in the use of parallax to measure distance. Using parsecs, we can write down an especially simple equation for the distances to stars:

$$\text{distance to a star in parsecs} = \frac{1}{\text{parallax angle of that star in arcseconds}}$$

or

$$d = \frac{1}{p}$$

where d is the distance to the star and p is the parallax angle of that star.

The equation is only this simple in these units, which is one of the main reasons why many astronomers discuss cosmic distances in parsecs rather than light-years. We will continue to primarily use light-years (ly) throughout this book, however, as they are more intuitive. In light-years, this same equation becomes approximately

$$\text{distance to a star in light-years} \approx \frac{3.26}{\text{parallax angle of that star in arcseconds}}$$

or

$$d_{ly} \approx \frac{3.26}{p}$$

where d_{ly} is the distance to a star in light-years.

Example: The nearest star, Proxima Centauri, has a parallax angle of 0.77 arcsec, and so its distance is 1/0.77, or approximately 1.3 pc. Equivalently, Proxima Centauri is 4.24 ly away. The parallax of Proxima Centauri is comparable to the angular diameter of a dime seen from a distance of 3 km. The parallax angles of the 25 nearest stars are listed in Appendix Table E–5.

Try these questions: What is the average distance from Earth to the Sun in parsecs? What is the parallax angle of a star 100 ly away? How far away is a star with a parallax angle of 0.3850? Rigel is 773 ly away; how far away is it in parsecs? What is its parallax angle in arcseconds? *(Answers appear at the end of the book.)*

to plans for better satellites to collect parallax data from stars even farther away.

Despite the information gained from stellar parallax, astronomers need to know the distances to more remote stars for which parallax cannot yet be measured. Several methods of determining ever-greater distances will be introduced later in this chapter and in Chapters 7, 8, and 12.

Having established that different stars are at different distances from Earth, we now consider the brightnesses that stars appear to have as seen from our planet. Combining the distances and the varied brightness we observe will enable us to calculate how much light stars actually emit, and thereby explore their evolution.

> For which object—the Moon, Mars, or the star Sirius—is the parallax angle smallest from Earth?

MAGNITUDE SCALES

Greek astronomers, from Hipparchus in the second century B.C. to Ptolemy in the second century A.D., undertook the classification of stars strictly by evaluating how bright they appear to be compared to each other. This made sense because back then astronomers assumed the stars were all at the same distance from us, and, therefore, differences in brightness due to different stellar distances from Earth were not expected. Classification regardless of distance is still useful to help us navigate around the night sky, and so we will study it now. In Section 6-3 we will add the effects of different distances on how bright stars appear from Earth.

6-2 Apparent magnitude measures the brightness of stars as seen from Earth

The brightnesses of stars measured without regard to their distances from Earth are called **apparent magnitudes**, denoted by lowercase m. The brightest stars were originally said to be of first magnitude, and their apparent magnitudes were designated $m = +1$. Those stars that appeared to be about half as bright as first-magnitude stars were said to be second-magnitude stars (designated $m = +2$), and so forth, down to sixth-magnitude stars, the dimmest ones visible to the unaided eye. (Greek astronomers did not try to classify the Sun's dazzling brightness in this scheme.) Because stars do not appear with discrete levels of brightness, this system has noninteger magnitudes as well, such as $+3.5$ or $+4.8$.

More quantitative methods of classifying stars were developed in the mid-nineteenth century (see *An Astronomer's Toolbox 6-2*). While maintaining the basic idea that brighter objects have smaller numbers than dimmer ones, the scale used today gives the star Vega an apparent magnitude of 0.0. Because some bodies are brighter than Vega, astronomers assign *negative* numbers

---

AN ASTRONOMER'S TOOLBOX 6-2

Details of the Magnitude Scales

The magnitude scales were created before accurate measurements of the relative brightnesses of stars could be made, and they have since been refined. Specifically, careful measurements reveal that the original first-magnitude stars were about 100 times brighter than the original sixth-magnitude stars. Therefore, astronomers chose the brightness factor of exactly 100 to *define* the range of brightness between modern first- and sixth-magnitude stars. In other words, it takes 100 stars of apparent magnitude $m = +6$ to provide as much light as we receive from a single star of apparent magnitude $m = +1$.

To find out how much brighter each magnitude is from the next dimmer one, we note that there are five integer magnitudes between first and sixth magnitude. Going from $m = +6$ to $m = +5$ increases (multiplies) the brightness we see by the same factor as going from $m = +5$ to $m = +4$, and so on. Going from $m = +6$ to $m = +1$ requires multiplying the same brightness factor from one magnitude to the next 5 times. The number we must multiply 5 times to get the range of brightness of 100 is $100^{1/5} \approx 2.512$, or, put another way, $2.512 \times 2.512 \times 2.512 \times 2.512 \times 2.512 \approx 100$. This means that *each successively brighter magnitude is approximately 2.512 times brighter than the preceding magnitude.*

Example: An $m = +3$ star is approximately 2.512 times brighter than an $m = +4$ star. Equivalently, it takes 2.512 fourth-magnitude stars to provide as much light as we receive from a single third-magnitude star.

Try these questions after reading Section 6-3: How much brighter is an $m = 0$ star than an $m = +4$ star? How much brighter is an $M = -2$ star than an $M = +5$ star? If one star is 7.93 times brighter than another and the brighter star has an absolute magnitude of $M = +3$, what is the absolute magnitude of the dimmer star? (*Hint for last question:* Recall that magnitudes need not be integers.)

(Answers appear at the end of the book.)

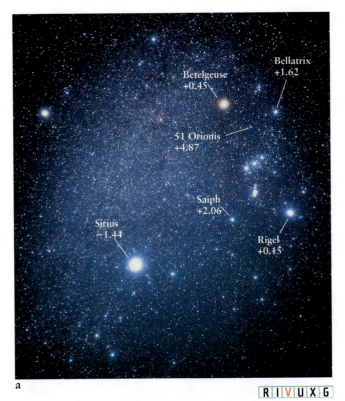

The magnitude scale shows (from top to bottom):
Sun (−26.7)
Full moon (−12.6)
Venus (at brightest) (−4.4)
Sirius (brightest star) (−1.44)
Naked eye limit (+6.0)
Binocular limit (+10.0)
Pluto (+15.1)
Hubble Space Telescope and large Earth-based telescopes (+30.0)

a

b

R I V U X G

FIGURE 6-2 Apparent Magnitude Scale (a) Several stars in and around the constellation Orion labeled with their names and apparent magnitudes. For a discussion of star names, see *Guided Discovery: Star Names*. (b) Astronomers denote the brightnesses of objects in the sky by their apparent magnitudes. Stars visible to the naked eye have magnitudes between $m = -1.44$ (Sirius) and about $m = +6.0$. However, CCD (charge-coupled device) photography through the Hubble Space Telescope or a large Earth-based telescope can reveal stars and other objects nearly as faint as magnitude $m = +30$. (a: Okiro Fujii, L'Astronomie)

to the apparent magnitudes of the very *brightest* objects. Sirius, the brightest star in the night sky, has an apparent magnitude of $m = -1.44$. Figure 6-2a shows it along with the apparent magnitudes of some of the stars in Orion. Using this convention, we can describe other bright objects in the sky, such as the Sun, Moon, comets, and planets. At its brightest, Venus shines with an apparent magnitude of $m = -4.4$, the full Moon has an apparent magnitude of $m = -12.6$, and the Sun has an apparent magnitude of $m = -26.7$. *Remember:* Objects with negative apparent magnitudes appear brighter than those with positive apparent magnitudes—the more negative, the brighter.

Astronomers also extended the magnitude scale so that dimmer stars, visible only through telescopes, could be included in the magnitude system. For example, the dimmest stars visible through a pair of powerful binoculars have an apparent magnitude of about $m = +10$. Time-exposure photographs through telescopes reveal even dimmer stars. The Keck telescopes and the Hubble Space Telescope, among others, image stars nearly as dim as magnitude $m = +30$. Figure 6-2b illustrates the modern apparent magnitude scale. Similarly, apparent magnitudes from entire groups of stars, such as distant galaxies, can be measured.

Knowing, as we now do, that stars are at different distances from us, the apparent magnitudes do not directly reveal fundamental stellar properties. All other things being equal, the closer of two identical stars appears brighter to us (has a smaller apparent magnitude) than the farther star. We take the different distances into account with either of two measures: absolute magnitude and luminosity.

INSIGHT INTO SCIENCE

Bigger Is Not Necessarily Brighter Numbering as well as naming schemes may be counterintuitive in science. You might expect brighter stars to have larger, more positive numbers than dimmer stars, but the apparent magnitude scheme is just the opposite. Similarly, we will see shortly that on the standard plot of stars used by astronomers, the Hertzsprung-Russell diagram, the hottest stars fall on the *left* and the coolest stars on the *right*.

What might cause the closer of two identical stars to appear dimmer than the farther one?

Star Names

Many prominent stars received names from early Arabic or Greek stargazers. For example, Aldebaran is from the Arabic for "The Follower" because it follows the prominent star cluster Pleiades around the sky. Betelgeuse, located in the shoulder of the constellation Orion, takes its name from the Arabic for "Armpit of the Central One." The name Sirius comes from the Greek for "sparkling," while Arcturus is from the Greek for "bear guard," because it is the brightest star in the constellation Bootes, the herdsman next to the Great Bear (Ursa Major).

Unlike Betelgeuse or Aldebaran, most stars never received fanciful names. Indeed, most are so dim that they have been observed only through telescopes in the last 2 centuries. To study the stars yourself, you must be able to keep track of them. Astronomers have created a system of labels for all of them.

Up to the 24 most prominent stars in each constellation are assigned Greek lowercase letters:

α alpha	ι iota	ρ rho
β beta	κ kappa	σ sigma
γ gamma	λ lambda	τ tau
δ delta	μ mu	υ upsilon
ε epsilon	ν nu	φ phi
ζ zeta	ξ xi	χ chi
η eta	o omicron	ψ psi
θ theta	π pi	ω omega

A bright star's name is a Greek letter together with the Latin possessive (technically, genitive) form of the constellation name. For example, the possessive forms of the zodiac constellations are

Constellation	Possessive
Aries	Arietis
Taurus	Tauri
Gemini	Geminorum
Cancer	Cancri
Leo	Leonis
Virgo	Virginis
Libra	Librae
Scorpius	Scorpii
Ophiuchus	Ophiuchi
Sagittarius	Sagittarii
Capricornus	Capricorni
Aquarius	Aquarii
Pisces	Piscium

In most cases, the brightest star in the constellation is α, the second brightest is β, the third is γ, and so on. For example, the brightest star in the constellation Leo is called α Leonis. This name is more informative than its common name, Regulus.

For the millions of stars extending beyond the 24 brightest in a constellation, a variety of catalogues lists the stars numerically. For example, HDE 226868 is a bright, blue star, the 226,868th star in the *Henry Draper Extended Catalogue* of stars.

 WEB LINK 6.3

Caveat emptor: The naming of stars is the responsibility of the International Astronomical Union, which does not ever recognize the commercial sale of star names. Companies that offer to name a star for a price do not have any official standing or recognition in the astronomical community, and neither do the names they promulgate. You can pay them to name a star for you, but the name is absolutely not official.

6-3 Absolute magnitudes and luminosities do not depend on distance

To determine the total energy emitted by each star in space, astronomers need to take into account the fact that different stars are at different distances from Earth. They do this by calculating the brightness each one would have if they were all at the same distance from here. Knowing how far away a star really is and how bright it appears (its apparent magnitude), we can calculate how bright it would be at any distance. **Absolute magnitude, M,** is the brightness each star would have at a distance of 10 pc. Unfortunately, absolute magnitudes have the same counterintuitive numeric scale as apparent magnitudes have.

To understand the relationship between the apparent magnitude and the absolute magnitude, we need to know how the brightness of an object changes with distance. Suppose we observe two identical stars, one twice as far away as the other. How much dimmer will the farther one appear to be, as seen from Earth? As light moves outward from a source, it spreads out over increasingly larger areas of space and its brightness decreases. Thus, the farther away a source of light is, the dimmer it appears. The **inverse-square law** provides the rule for just how quickly the brightness of an object changes with distance.

Imagine light moving out from a star (Figure 6-3a). Start with a small square area of light that has moved out a distance $d = 1$. The light in that square has a certain brightness. When the same square of light has gone twice as far ($d = 2$), you can see on the figure that it has become 4 times larger. The light in each of the same-sized squares at $d = 2$ contains one-quarter of the photons that were in the single square at $d = 1$. Therefore, each small square at $d = 2$ is one-quarter as bright as the same-sized square at $d = 1$. Similarly, when the square of light has moved to $d = 3$, there are now nine squares the same size as the original, each one-ninth as bright as the original square at $d = 1$. For example, the Sun emits 3.83×10^{26} W (watts) of power (compared to 100 W for a bright home lightbulb). The Sun's power is spread over a wider area as it travels through space. When it passes Mercury, sunlight provides 9140 watts per square meter (W/m^2) of space. That same energy has spread out so much that by the time it passes us at 1 AU from the Sun, it provides only 1370 W/m^2. You see the same effect every day when you look at a car's headlights at different distances (Figure 6-3b).

This inverse-square law can be summarized mathematically as follows:

Apparent brightness decreases inversely with the square of the distance between the source and the observer.

Returning to our two identical stars, one twice as far away as the other, the farther one's brightness decreases to $(\frac{1}{2})^2 = \frac{1}{4}$ the brightness of its closer twin by the time the starlight gets to us.

Absolute magnitudes are usually determined by calculations based on apparent magnitudes. Of course, we cannot just move a star to 10 pc distance and then remeasure its apparent magnitude. However, we can calculate the absolute magnitude of a nearby star: We first measure its apparent magnitude and then we find its distance by measuring its parallax angle. Combining these numbers, as shown in *An Astronomer's Toolbox 6-3* gives the star's absolute magnitude.

Knowing the Sun's true distance and its apparent magnitude, we can use the inverse-square law to determine how bright it would appear at 10 pc. At that distance, it would have an apparent magnitude of $m = +4.8$. Therefore, the absolute magnitude of the Sun is $M = +4.8$.

Because absolute magnitudes tell astronomers how bright stars are compared with one another, this information can be used to evaluate models of stellar evolution, which we discuss in the next few chapters. Absolute magnitudes range from roughly $M = -10$ for the brightest stars to $M = +17$ for the dimmest. Although absolute magnitudes give us comparisons between the energy outputs of stars, we also need to know the total energy they release. The total amount of electromagnetic power (energy emitted each second) is called a star's **luminosity**. We saw in Chapter 4 that the greater its luminosity, the brighter the object. Therefore, the smaller or more negative a star's absolute magnitude, the greater its luminosity. For convenience, stellar luminosities are expressed in multiples of the Sun's luminosity, denoted $L_\odot$. As we saw

With greater distance from the star, its light is spread over a larger area and its apparent brightness is less.

Star

$d = 1$
$d = 2$
$d = 3$

a

30 m away

20 m away

b   10 m away   R I V U X G

ANIMATION 6.2

FIGURE 6-3 The Inverse-Square Law (a) The same amount of radiation from a light source must illuminate an ever-increasing area as the distance from the light source increases. The decrease in brightness follows the inverse-square law, which means, for example, that tripling the distance decreases the brightness by a factor of 9. (b) The car is seen at distances of 10 m, 20 m, and 30 m, showing the effect described in part (a). (b: Royalty Free/CORBIS)

The Distance-Magnitude Relationship

The closer a star, the brighter it appears. The inverse-square law leads to a simple equation for absolute magnitude, M. Suppose a star's apparent magnitude is m and its distance from Earth is d (measured in parsecs). Then

$$M = m - 5 \log (d/10)$$

where log stands for the base-10 logarithm. This distance-magnitude relation can be rewritten as

$$m - M = 5 \log d - 5$$

Example: Consider Proxima Centauri, the nearest star to Earth, other than the Sun. By measuring its parallax angle, we know this star is at a distance from Earth of $d = 1.3$ pc. Its apparent magnitude is $m = +11.1$. Therefore, its absolute magnitude is

$$M = 11.1 - 5 \log (1.3/10) = 11.1 - (-4.4) = +15.5$$

Compare! The Sun is an average star with M = +4.8, so Proxima Centauri is an intrinsically dim star. If you know any two of d, m, and M, you can calculate the third variable. For example, if we know a star's absolute and apparent magnitudes, the equation can be used to determine its distance.

Try these questions: A star is observed to have an apparent magnitude $m = +0.268$ and an absolute magnitude M = −0.01. How far from Earth is the star in parsecs and light-years? A star is observed to have an apparent magnitude $m = +1.17$ and is at a distance of 25.1 ly from Earth. What is its absolute magnitude? *(Remember to convert to parsecs first.)* A star is at a distance from Earth of 8.61 ly and has an absolute magnitude of $m = +1.45$. What is its apparent magnitude? You can check your results and identify the stars by referring to Appendix Table E-5 and Table E-6.
(Answers appear at the end of the book.)

---

earlier, this value is about 3.83×10^{26} W. The intrinsically brightest stars (M = −10) have luminosities of 10^6 $L_\odot$. In other words, each of these stars has the energy output of a million Suns. The dimmest stars (M = +17) have luminosities of 10^{-5} $L_\odot$. We will provide both luminosity and absolute magnitude data about stars in the chapters that follow.

> Why can the total energy output of a star be specified equally by its absolute magnitude and its luminosity?

THE TEMPERATURES OF STARS

Armed with knowledge of stellar luminosities, astronomers early in the twentieth century searched for ways to use this information to determine how stars evolve. A major step in this process was to find a graph on which physically similar stars are located near each other, because such groupings often provide insight into how objects work. Astronomers found this graph when they plotted either the luminosity or, equivalently, the absolute magnitude versus the stars' surface temperatures. The resulting graph led to our models of both stellar activity and evolution. To find the surface temperatures, we start with a fact that is easily overlooked: Stars are not all the same color.

6-4 A star's color reveals its surface temperature

If you look carefully at the stars with your naked eyes, you can see that they have different colors, most commonly red, orange, yellow, white, and blue-white (see Figures 6-2 and 6-4a). As we saw for the Sun, stars behave very nearly like blackbodies. Therefore, their colors are determined by their surface temperatures. Recall from Figure 4-2 that different blackbody curves are associated with blackbodies of different temperatures. These curves differ in both their intensities and the locations of their peaks, as described by Wien's law.

Three typical blackbody curves are presented in Figure 6-4. The intensity of light from a relatively cool star peaks at long wavelengths, so the star looks red. The intensity of light from a very hot star peaks at shorter wavelengths, making it look blue. The maximum intensity of a star of intermediate temperature, such as the Sun, is found near the middle of the visible spectrum.

To accurately determine the peaks of stars' blackbody spectra and, hence, their surface temperatures, astronomers need to know which blackbody curve most accurately describes each star. **Photometry** provides this information. In this process, a telescope collects starlight that is then passed through one of a set of col-

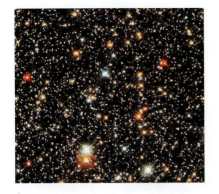

FIGURE 6-4 Temperature and Color (a) This beautiful Hubble Space Telescope image shows the variety of colors of stars. (b) These diagrams show the relationship between the color of a star and its surface temperature. The intensity of light emitted by three stars is plotted against wavelength (compare with Figure 4-2). The range of visible wavelengths is indicated. The location of the peak of each star's intensity curve, relative to the visible-light band, determines the apparent color of its visible light. The insets show stars of about these surface temperatures. Ultraviolet (uv) extends to 10 nm. See Figure 3-6 for more on wavelengths of the spectrum. (a: Hubble Heritage Team/AURA/STScI/NASA; left inset: Andrea Dupree/Harvard-Smithsonian CFA, Ronald Gilliland/STScI, NASA and ESA; center inset: NSO/AURA/NSF; right inset: Till Credner, Allthesky.com)

a

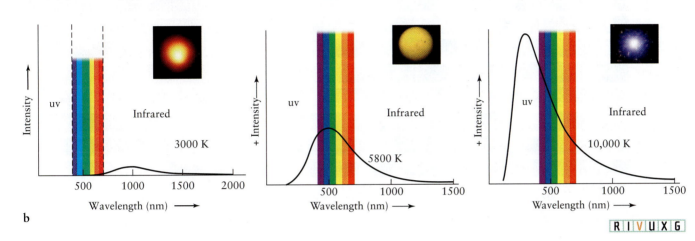

b

R I V U X G

ored filters and recorded on a CCD. At least three photometric images are taken of each star through filters that pass different wavelengths of light. The intensity of a star's image is different at different wavelengths. By plotting the three data points and finding which unique blackbody curve they fit on, we can determine that star's temperature.

For example, if a star's surface is very hot, perhaps 10,000 K (e.g., the hottest star in Figure 6-4b), its radiation is skewed toward the ultraviolet, which makes the star bright as detected through an ultraviolet-passing filter, dimmer through a blue filter, and dimmer still through a yellow filter. Regulus, in Leo, is such a star. If a star is cool, say, 3000 K (e.g., the coolest star in Figure 6-4b), its radiation peaks at long visible or even infrared wavelengths, making the star bright through a red filter, dimmer through a yellow filter, and dimmer still through a blue filter. Betelgeuse (Figure 6-2a) and Aldebaran are examples of such cool stars. Humans are blackbody radiators with peak temperatures around 300 K. Our peak is in the infrared.

6-5 A star's spectrum also reveals its surface temperature

Another way to determine a star's surface temperature is from a study of its spectrum, a technique called stellar spectroscopy (see Chapter 4). Recall that spectral lines result from the absorption and scattering of starlight by gases in the star's atmosphere, in interstellar space, and in Earth's atmosphere. Astronomers begin by taking the spectrum of a star and then identifying and discarding spectral lines due to interstellar gas and Earth's atmosphere. What is left are spectral lines created in the star's atmosphere. These lines provide us with knowledge of the chemical compositions of the outer layers of stars.

At first glance, there seems to be a bewildering variety of stellar spectra, some of which are shown in Figure 6-5. Some stellar spectra show prominent absorption lines of hydrogen. Others exhibit many absorption lines of calcium and iron. Still others are dominated by broad absorption lines created by molecules, such as titanium oxide.

Fortunately, these spectra all hold clues about the nature of stars and, in particular, their surface temperatures. Consider the abundant hydrogen gas found in nearly every star's atmosphere. Although hydrogen accounts for about three-quarters of the mass of a typical star, strong (meaning dark) hydrogen absorption

A star of which temperature has the most intense blue radiation: 3000 K, 5200 K, 6800 K?

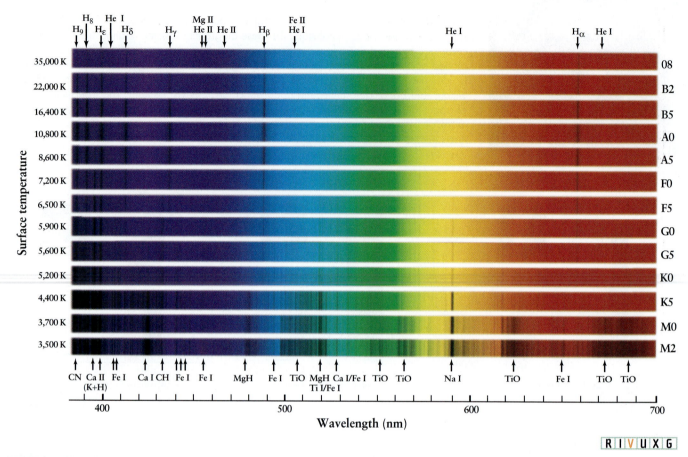

FIGURE 6-5 The Spectra of Stars with Different Surface Temperatures. The corresponding spectral types are indicated on the right side of each spectrum. (Note that stars of each spectral type have a range of temperature.) The hydrogen Balmer lines are strongest in stars with surface temperatures of about 10,000 K (called *A-type* stars). Cooler stars (*G-* and *K-type* stars) exhibit numerous atomic lines caused by various elements, indicating temperatures from 4000 to 6000 K. Several of the broad, dark bands in the spectrum of the coolest stars (*M-type* stars) are caused by titanium oxide (TiO) molecules, which can exist only if the temperature is below about 3500 K. Recall from Section 4-5 that the Roman numeral I after a chemical symbol means that the absorption line is caused by a neutral atom; a numeral II means that the absorption is caused by atoms that have each lost one electron. (R. Bell, University of Maryland, and M. Briley, University of Wisconsin at Oshkosh)

lines do not show up in every star's spectrum. The strength of the absorption lines depends on the star's temperature. In the late 1920s, Harvard astronomer Cecilia Payne and physicist Meghnad Saha succeeded in explaining why a star's visible hydrogen spectrum is affected by its surface temperature.

Niels Bohr's model of the hydrogen atom (recall Figure 4-12) shows why. As Bohr proposed in the early 1900s, the visible hydrogen lines, called *Balmer lines,* are produced when photons excite electrons in the second, $n = 2$, energy level of hydrogen to a higher energy level. Recall from Section 4-6 that these are labeled H_α, H_β, H_γ, etc. At 10,000 K, most hydrogen electrons absorb enough energy from photons to normally reside in the $n = 2$ level. When these electrons gain further energy, they move to higher energy levels. Therefore, a star with that surface temperature produces the strongest Balmer lines.

Why do hotter or cooler stars produce weaker (less dark) Balmer absorption lines? Suppose first that the star is much hotter than 10,000 K. High-energy photons that stream through the photosphere completely strip away (ionize) electrons from most of the hydrogen atoms there. Because an ionized hydrogen atom cannot produce spectral lines, a very hot star has very dim hydrogen Balmer lines, even though it contains great quantities of hydrogen. Conversely, if a star is much cooler than 10,000 K, most of the photons that escape from it possess too little energy to keep many electrons in the $n = 2$ (first excited state) of the hydrogen atoms, or to boost electrons in the $n = 2$ state to higher energy states. Therefore, these stars also produce dim Balmer lines. To produce strong Balmer lines, a star must be hot enough to excite electrons out of the $n = 2$ state but not hot enough to ionize a significant fraction of the atoms.

As you can see in Figure 6-5, at temperatures much cooler or hotter than 10,000 K, the spectral lines of other elements dominate a star's spectrum. For example, the spectral lines of neutral helium, while dim at around 10,000 K, are relatively strong at around 25,000 K because photons have enough energy to excite helium atoms without tearing away the electrons. Conversely, the spectral lines of neutral iron are especially pronounced at around 3500 K. By surveying the relative strengths of a variety of absorption lines, astronomers can determine a star's surface temperature to high precision.

6-6 Stars are classified by their spectra

We have seen that astronomers can determine a star's surface temperature from either the peak of its blackbody or the strength of its various spectral lines. To make use of the diverse stellar spectra, astronomers since the middle of the nineteenth century have grouped similar spectra into classes, or **spectral types**. According to one classification scheme popular in the late 1800s, a star was assigned a letter between A and P, depending on the strength of the Balmer hydrogen lines in the star's spectrum. It was assumed that the strength of these lines was uniquely related to the star's surface temperature, but as we have seen in the preceding section, this belief is unjustified.

In the early 1900s, William Pickering and Williamina Fleming, followed by Annie Jump Cannon, and their colleagues at Harvard Observatory (Figure 6-6) set up the spectral classification scheme we use today. Many of the early A through P categories were dropped because the Balmer lines in a star's spectrum can be weak whether the star is very cool or very hot. The remaining Balmer-based spectral types were thus reordered by stellar surface temperature into the **OBAFGKM sequence**. This sequence is most easily learned with the aid of a mnemonic, such as "**O**h, **B**e **A** **F**ine **G**uy (or **G**irl), **K**iss **M**e!"

> Why do very hot stars have weak hydrogen lines, even though they are composed primarily of hydrogen?

The hottest stars have surface temperatures of 30,000 to 50,000 K and are classified as O type; their spectra are dominated by He II and Si IV (triply ionized silicon). The coolest stars have surface temperatures of 2500 to 3000 K and are classified as M type. Table 6-1 includes representative examples of each spectral type.

The wide range of temperatures covered by each spectral type prompted astronomers to further subdivide the OBAFGKM temperature sequence. Each spectral type is now broken up into 10 temperature subranges. These 10 finer steps are indicated by adding an integer from 0 (hottest) through 9 (coolest). Thus, an A8 star is hotter than an A9 star, which is hotter than an F0 star, which is hotter than an F1 star, and so on. Test yourself on this: What class of star is just slightly cooler than a K9? (The caption for Figure 6-7 has the answer.) The Sun, whose spectrum is dominated by singly ionized metals (especially Fe II and Ca II), is a G2 star.

> Which star is hottest: F5, B6, or M3?

a

b

R | I | V | U | X | G

FIGURE 6-6 *Classifying the Spectra of Stars* The modern classification scheme for stars, based on their spectra, was developed at the Harvard College Observatory in the late nineteenth century. Female astronomers, initially led by Edward C. Pickering *(not shown)* and (a) Williamina Fleming, standing, and then by (b) Annie Jump Cannon, analyzed hundreds of thousands of spectra. Social conventions of the time prevented most female astronomers from using research telescopes or receiving salaries comparable to those of men. (a: Harvard College Observatory; b: © Bettmann/CORBIS)

WEB LINK 6.6

TABLE 6-1 The Spectral Sequence

Spectral class	Color	Temperature (K)	Spectral lines	Examples
O	Blue-violet	50,000–30,000	Ionized atoms, especially helium	Naos (ζ Puppis), Mintaka (δ Orionis)
B	Blue-white	30,000–11,000	Neutral helium, some hydrogen	Spica (α Virginis), Rigel (β Orionis)
A	White	11,000–7500	Strong hydrogen, some ionized metals	Sirius (α Canis Majoris), Vega (α Lyrae)
F	Yellow-white	7500–5900	Hydrogen and ionized metals, such as calcium and iron	Canopus (α Carinae), Procyon (α Canis Minoris)
G	Yellow	5900–5200	Both neutral and ionized metals, especially ionized calcium	Sun, Capella (α Aurigae)
K	Orange	5200–3900	Neutral metals	Arcturus (α Boötis), Aldebaran (α Tauri)
M	Red-orange	3900–2500	Strong titanium oxide and some neutral calcium	Antares (α Scorpii), Betelgeuse (α Orionis)

INSIGHT INTO SCIENCE

Tolerate Idiosyncrasies In astronomy, the word *metal* applies to all elements other than hydrogen or helium. In chemistry, elements are classified as metals or nonmetals according to their physical and chemical properties. To a chemist, sodium and iron are metals, whereas carbon and oxygen are not.

TYPES OF STARS

Keeping in mind that patterns of objects often reveal information about them, early twentieth-century astronomers began searching for relationships between different properties of stars. The relationship between luminosities and surface temperatures turned out to be a key to understanding types of stars. As often happens in such scientific quests, the significance of this relationship was discovered nearly simultaneously by two independent researchers.

6-7 The Hertzsprung-Russell diagram identifies distinct groups of stars

Around 1911, the Danish astronomer Ejnar Hertzsprung noticed that patterns emerge when the luminosities of stars (or their equivalent absolute magnitudes) are plotted against their surface temperatures (or their equivalent spectral types). Luminosity and absolute magnitude indicate the total energy emitted by a star or other body. They refer to what we loosely call "brightness." Within 2

years, the American astronomer Henry Norris Russell independently discovered the same result. Graphs of stellar luminosity or absolute magnitude against surface temperature or spectral type are now known as **Hertzsprung-Russell diagrams**, or **H-R diagrams**.

The H-R diagram is valuable because it shows that stars do not have random surface temperatures and luminosities; the two factors are correlated. Figure 6-7 is a typical H-R diagram. Each dot represents a star whose luminosity and spectral type have been determined. The surface temperatures are plotted along the top of this figure and the absolute magnitudes along the right side. You can therefore see the equivalence of the temperature and spectral type and the equivalence of the luminosity and absolute magnitude.

Bright stars are near the top of the diagram; dim stars are near the bottom. Contrary to intuition, hot (O and B) stars are toward the left side and cool (M) stars are toward the right. Hertzsprung and Russell made this choice because of the standard sequence OBAFGKM.

INSIGHT INTO SCIENCE

From Patterns to Models Scientists look at patterns of behavior in related objects as valuable clues to underlying properties and their causes. Guided by this data, they then create mathematical models, which are used to make fresh predictions. For example, astronomers analyze the relationship between the luminosities and surface temperatures of stars as displayed on an H-R diagram to gain insight into the internal activities of stars.

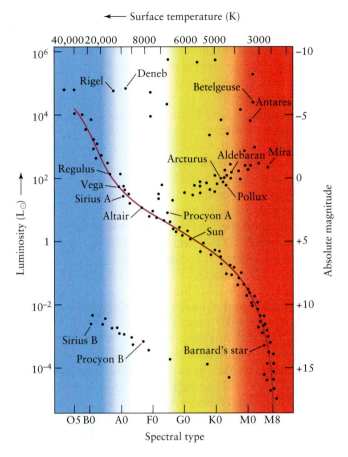

4

To the right of the main sequence on the H-R diagram is a second major grouping of stars. These stars are bright but cool. From the Stefan-Boltzmann law (see *An Astronomer's Toolbox 4-1*), we know that a cool object radiates much less light from each unit of surface area than does a hot object. To be so luminous, these cool stars must therefore be huge compared to main-sequence stars of the same temperature, so they are called **giant stars**.

Giants are typically 10 to 100 times the radius of the Sun and have surface temperatures between 2000 and 20,000 K. The cooler members of this class of stars (those with surface temperatures between 2000 and 4500 K) are often called **red giants** because they appear reddish in the nighttime sky. Aldebaran in the constellation Taurus and Arcturus in Boötes are examples of red giants that you can easily see with the naked eye.

A few rare stars are considerably bigger and more luminous than typical giants. Located along the top of the H-R diagram, these superluminous stars are appropriately called **supergiants**. They can extend in radius up to about 1000 R☉. Betelgeuse in Orion and Antares in Scorpius are two examples that are visible in the nighttime sky. Together, giants and supergiants comprise less than 1% of the stars in our vicinity.

The remaining 8% of stars in our neighborhood of space fall in a final grouping toward the lower left and bottom of the H-R diagram. As their placement on the diagram shows, these stars are hot, dim, and tiny compared to the Sun. Called **white dwarfs**, we will see that they are actually remnants of stars. White dwarfs are roughly the same size as Earth, and because of their low luminosities, they can be seen only with the aid of a telescope.

These results are summarized in Figure 6-8, where the dashed lines indicate the radii of stars. Notice that most main-sequence stars are roughly the same size as the Sun. It is important to keep in mind that, although the H-R diagram is invaluable in helping astronomers *organize* stars by their physical properties (temperature and absolute magnitude), the diagrams do not *explain* the physical mechanisms that produce these characteristics. As we will see shortly, the locations of stars on the H-R diagram provided crucial clues that astrophysicists used in developing and testing their theories of how stars shine.

> For a given spectral class (i.e., temperature), a star of which luminosity class—main sequence, giant, or supergiant—is brightest?

FIGURE 6-7 A Hertzsprung-Russell Diagram On an H-R diagram, the luminosities of stars are plotted against their spectral types. Each dot on this graph represents a star whose luminosity and spectral type have been determined. Some well-known stars are identified. The data points are grouped in just a few regions of the diagram, revealing that luminosity and spectral type are correlated: Main-sequence stars fall along the red curve, giants are to the right, supergiants are on the top, and white dwarfs are below the main sequence. The absolute magnitudes and surface temperatures are listed at the right and top of the graph, respectively. These are sometimes used on H-R diagrams instead of luminosities and spectral types. (*Answer to text question:* An M0 star is the next coolest after a K9.)

The band of stars in Figure 6-7 stretching diagonally across the H-R diagram and on which a red curve is superimposed represents most of the stars we see in the nighttime sky. Called the **main sequence**, it extends from the hot, bright, bluish stars in the upper left corner of the diagram down to the cool dim stars (called *red dwarfs*) in the lower right corner. Each star on this band is a **main-sequence star**. Just over 91% of the stars surrounding the solar system fall into this category.

Observations reveal that the number of main-sequence stars decreases with increasing surface temperature. Therefore, along the main sequence, the cooler M, K, and G stars are the most common ones and the hot O stars are the rarest. The Sun (spectral type G2, absolute magnitude +4.8) is a main-sequence star.

6-8 Luminosity classes set the stage for understanding stellar evolution

We have seen that a star's surface temperature largely determines which lines are prominent in its spectrum. Therefore, classifying stars by spectral type is the same

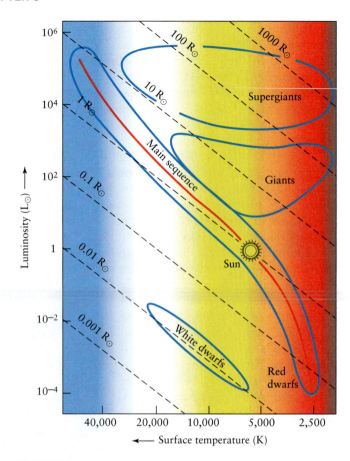

FIGURE 6-8 The Types of Stars and Their Sizes On this H-R diagram, stellar luminosities are plotted against the surface temperatures of stars. The dashed diagonal lines indicate stellar radii. For stars of the same radius, hotter stars (corresponding to moving from right to left on the H-R diagram) glow more intensely and are more luminous (corresponding to moving upward on the diagram) than cooler stars. While individual stars are not plotted, we show the regions of the diagram in which main-sequence, giant, supergiant, and white dwarf stars are found. Note that the Sun is intermediate in luminosity, surface temperature, and radius; it is very much a middle-of-the-road star.

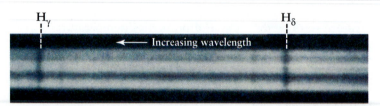

a   A supergiant star has a low-density, low-pressure atmosphere: its spectrum has narrow absorption lines

b   A main-sequence star has a denser, higher-pressure atmosphere: its spectrum has broad absorption lines

R I V U X G

as categorizing them according to surface temperature. However, as you can see in Figure 6-8, stars of the same surface temperature can have different luminosities. As an example, a star with a surface temperature of 5800 K could be a white dwarf, a main-sequence star, a giant, or a supergiant.

Astronomers have found subcategories in which to place stars by studying their spectral absorption lines in detail. This refined classification is possible primarily because absorption lines are affected by the density and pressure of the gas in a star's atmosphere, both of which depend on whether the star is a white dwarf, main-sequence star, giant, or supergiant (Figure 6-9).

Based upon the differences in stellar spectra, W. W. Morgan and P. C. Keenan of the Yerkes Observatory developed a system of **luminosity classes** in the 1930s. Luminosity classes Ia and Ib encompass all of the supergiants; giants of various luminosity are assigned classes II, III, and IV; and main-sequence stars are luminosity class V. Figure 6-10 summarizes these results. We will see in Chapter 7 that the different luminosity classes correspond to different stages of stellar evolution. White dwarfs are not given a luminosity class because, as we will see in Chapter 8, they are stellar remnants that do not create energy by fusion like the Sun and other stars in the five luminosity classes.

Plotting luminosity classes on the H-R diagram (see Figure 6-10) provides a useful subdivision of star types. In fact, astronomers commonly describe a star by both its spectral type and its luminosity class. The Sun, for example, is a G2 V star. This notation supplies a great deal of information about the star, because its spectral type is correlated with its surface temperature. Thus, an astronomer knows immediately that a G2 V star is a main-sequence star with a luminosity of 1 $L_\odot$ and a surface temperature of around 5800 K. Similarly, knowing that Aldebaran is a K5 III star tells an astronomer that it is a red giant with a luminosity of around 370 $L_\odot$ and a surface temperature of about 4400 K (see Figure 6-7).

The H-R diagram also provides us with enough data to determine a star's radius. Recall from An Astronomer's Toolbox 4-1 that there is a relationship between a star's luminosity, L, temperature, T, and radius, r: $L = \sigma T^4 4\pi r^2$, where σ (sigma) is a constant. Knowing two properties, the luminosity and temperature, we can solve this equation for the third, namely the radius. Lines of constant stellar radii are drawn on Figure 6-8.

FIGURE 6-9 Stellar Size and Spectra These spectra are from two stars of the same spectral type (B8) and, hence, the same surface temperature (13,400 K) but different radii and luminosities: (a) the B8 supergiant Rigel (luminosity 58,000 $L_\odot$) in Orion and (b) the B8 main-sequence star Algol (luminosity 100 $L_\odot$) in Perseus. (From W. W. Morgan, P. C. Keenan, and E. Kellman, An Atlas of Stellar Spectra)

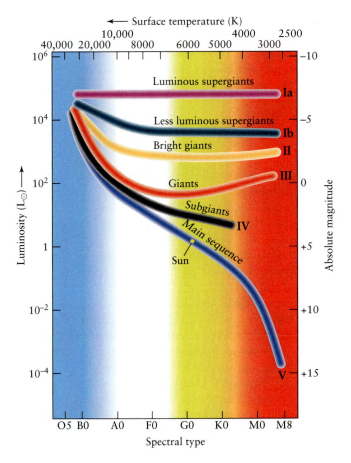

FIGURE 6-10 Luminosity Classes Dividing the H-R diagram into regions, called luminosity classes, permits finer distinctions between giants and supergiants. Luminosity classes Ia and Ib encompass the supergiants. Luminosity classes II, III, and IV indicate giants of different brightness. Luminosity class V indicates main-sequence stars. White dwarfs do not have their own luminosity class.

6-9 A star's spectral type and luminosity class provide a second distance-measuring technique

A star's spectral type and luminosity class, combined with information from the H-R diagram, give astronomers information necessary for determining the distances to stars millions of light-years away, far beyond the maximum distance that can be measured using stellar parallax. The process works like this:

1. Astronomers observe a distant star's apparent magnitude and spectrum.

2. From its spectrum, they determine what spectral class the star belongs to (or, equivalently, its surface temperature).

3. The spectrum also reveals the star's luminosity class, that is, whether it is a supergiant, giant, or main-sequence star. Combining the temperature and the luminosity class determines where it lies on the H-R diagram (see Figure 6-10). From this position, the star's approximate absolute magnitude can be read off the diagram.

> A main-sequence star of which spectral type—F5, A8, or K0—is largest?

4. Using the apparent and absolute magnitudes in the distance-magnitude relationship (see *An Astronomer's Toolbox 6-3*), the star's distance can then be calculated.

Consider, for example, the star Regulus in the constellation Leo (see margin chart for Leo on page 163). Its spectrum reveals Regulus to be a B7 V star (a hot, blue, main-sequence star). Placing it on the H-R diagram (see Figure 6-7), we see that its luminosity is 140 $L_\odot$ and its absolute magnitude is −0.52. Given the star's apparent magnitude, we can use the distance-magnitude relationship to determine its distance from Earth. Because both the spectral type and luminosity class are obtained spectroscopically, this method of determining distances is called **spectroscopic parallax**. (This name is misleading, however, because no parallax angle is involved.)

Spectroscopic parallax is limited in accuracy because of the spread of stars in each luminosity class—the stars in each class do not fall on a single, narrow line (hence the use of the word *approximate* in the previous numbered list). It is also limited in that spectra of distant stars become increasingly hard to determine. As a result, errors of 10% in distance are common using spectroscopic parallax. Accepting such errors, the power of this method is that it can be used for stars at much greater distances than those determined by stellar parallax. Indeed, it even provides distances to stars in other galaxies tens of millions of light-years away.

> What observations must be made to determine a star's distance using spectroscopic parallax?

STELLAR MASSES

The final clue that astronomers needed to understand how stars work is their mass—how much matter they have. If the masses were related to the stellar luminosities and locations on the H-R diagram (and they are), then astronomers would have enough information to start making scientific models of why stars shine. The mass, it turns out, determines the amount of energy each star has available to manufacture light and other electromagnetic radiation. Models of how stars generate that radiation, developed in the twentieth century and still being refined, provide us with further insights into how nature works. With the knowledge gained about thermonuclear fusion in stars, physicists are trying to control this fusion to generate electricity.

The problem with finding stellar masses is that it cannot be done directly by examining isolated stars. The mass of a star can be determined only by its gravitational effects on other bodies, using Newton's law of gravity. When Newton derived the formula for Kepler's third law, he discovered that for any two objects in orbit, such as a planet around the Sun, a moon around a planet, or two stars around each other, the period of their orbits is related to the sum of their masses (see *An Astronomer's Toolbox 6-4*). You have seen in Section 2-5 that for objects orbiting the Sun, we can ignore the mass of the smaller body, thereby getting Kepler's original third law, as stated in Chapter 2. For stars orbiting each other, we use Newton's full equation, which includes the masses of both bodies.

Fortunately for astronomers, one-fifth of the stars near our solar system are members of star systems in which two stars orbit each other. Put another way, *one fifth of the objects we see as single stars are actually pairs of stars in orbit around each other.* They are so distant that we cannot resolve the individual stars in the pairs and so they appear to us as single objects. However, using telescopes to observe the periods of the orbits and the distances between stars, astronomers can determine the sum of the stellar masses of the pair. Often, they can determine individual stellar masses as well.

6-10 Binary stars provide information about stellar masses

A pair of stars located at nearly the same position in the night sky is called a *double star*. Between 1782 and 1838, William Herschel, his sister Caroline, and his son John catalogued thousands of them.

Some double stars are not physically close together and do not orbit each other. These **optical doubles** (sometimes called *apparent binaries*) just happen to lie in the same direction, as seen from Earth. For example, δ Herculis, visible with the naked eye, is an optical double with a dim star. Through a telescope these stars appear close together, but that is an optical illusion.

Other double stars are true **binary stars**—pairs in which two stars orbit each other. In the case of **visual binaries**, both stars can be seen, using a telescope if necessary (Figure 6-11). Astronomers can plot the orbit of one star around the other in a visual binary.

To see how we determine the masses of stars, consider a visual binary. Newton rewrote Kepler's third law as a relation between the masses of the stars (in solar masses), their orbital period around each other in Earth years (this period is the same for both, so it does not matter which star we assume orbits the other), and the average separation between the stars in AU (which equals the semimajor axis of the elliptical orbit):

$$\text{The sum of the masses} = \frac{\text{The cube of the semimajor axis}}{\text{The square of the period}}$$

Thus, by observing the separation between a pair of stars and how long one of them takes to complete its orbit, we can calculate the sum of the masses of the two stars, presented mathematically in *An Astronomer's Toolbox 6-4*.

In many cases, we can also determine the individual masses of the stars. To do this, we need to know the distances of the stars from the **center of mass** of the pair. This is the point about which both stars in a binary system move in elliptical orbits. The center of mass is determined by the masses and separation of the stars and can be understood by analogy to a pair of connected masses whirling and sliding on a frictionless surface. While the two masses go around each other, their center of mass moves in a straight line. One point somewhere between their centers moves in a straight line. This point is its center of mass. Just as the two masses orbit around their center of mass, the two stars in a binary system orbit around their center of mass under the influence of their mutual gravitational attraction (Figure 6-12). The more massive star is closer to the center of mass, just as a heavier child must be closer to the pivot point or fulcrum of a seesaw for it to balance (Figure 6-12b).

We can locate the center of mass of a visual binary by plotting the separate orbits of the two stars, as in Figure 6-11, using the background stars as reference points. The center of mass lies at the common focus of the two elliptical orbits. Comparing the relative sizes of the two orbits around the center of mass yields the ratio of the two stars' masses, M_1/M_2. To get the relative sizes of the orbits, however, we must know the plane of orbit of the two stars. This is not always possible, which is why we cannot always determi1ne the individual masses of the stars. However, when the stars in a binary system periodically eclipse each other, as seen from Earth, we know that the plane of their orbit must be nearly parallel to our line of sight to them (Figure 6-13a, b). Such **eclipsing binaries** can be detected even when the stars cannot be resolved as two distinct images in a telescope. The apparent magnitude of the image of an eclipsing binary dims each time one star blocks out part of the other. An astronomer can measure light intensity from binaries very accurately as a function of time. From these **light curves**, such as those shown in Figure 6-13, we can see at a glance whether the eclipse is partial, creating a V-shaped trough in the

AN ASTRONOMER'S TOOLBOX 6-4

Kepler's Third Law and Stellar Masses

The same gravitational force that holds Earth in orbit around the Sun also keeps pairs of stars in orbit around each other. For any pair of orbiting bodies, the orbits are ellipses, and Kepler's third law can be written as

$$M_1 + M_2 = a^3/P^2$$

Here, M_1 and M_2 are the two masses (expressed in solar masses), a is the length of the semimajor axis of the ellipse (in astronomical units), and P is the orbital period (in years). Note that a is also the average separation between the two bodies. Thus, the sum of the masses can be found once we know the orbital period and average separation.

Example: Suppose two stars make up a double system, and one star follows an ellipse with a semimajor axis of 4 AU. We find that it takes 2.5 years to complete one orbit. Then the sum of the stars' masses is

$$M_1 + M_2 = (4)^3/(2.5)^2 = 10.2\ M_\odot$$

The total mass of the system is 10.2 $M_\odot$.

Try these questions: Why can Kepler's third law for a planet orbiting the Sun be written as $P^2 = a^3$? (*Hint:* For the Sun and a planet, the combined mass is extremely close to the Sun's mass alone.) Two identical stars are observed to orbit each other with a period of 2 years at a separation of 2.32 AU. What are their combined masses? How long would it take a 1 $M_\odot$ and a ½ $M_\odot$ star to orbit each other if their separation were 20 AU?
(Answers appear at the end of the book.)

May 2, 1996	May 3, 1996	May 4, 1996
May 24, 1996	June 1, 1996	June 4, 1996

R I V U X G

FIGURE 6-11 A Binary Star System About one-fifth of the objects we see as "stars" in our region of the Milky Way galaxy are actually double stars. Mizar in Ursa Major is a binary system with stars separated by only about 0.01 arcsec. The images and plots show the relative positions of the two stars over nearly half of their orbital period. The orbital motion of the two binary stars around each other is evident. Either star can be considered fixed in making such plots. (Navy Prototype Optical Interferometer, Flagstaff, AZ. Courtesy of Dr. Christian A. Hummel)

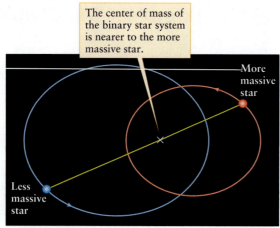

The center of mass of the binary star system is nearer to the more massive star.

More massive star

Less massive star

a A binary star system

The center of mass of the system of two children is nearer to the more massive child.

Fulcrum

b A "binary system" of two children

FIGURE 6-12 Center of Mass of a Binary Star System (a) Two stars move in elliptical orbits around a common center of mass. Although the orbits cross each other, the two stars are always on opposite sides of the center of mass and thus never collide. (b) A see- saw balances if the center of mass of the two children is at the fulcrum. When balanced, the heavier child is always closer to the fulcrum, just as the more massive star is closer to the center of mass of a binary star system.

light curve (Figure 6-13a), or total, creating a flat-bottomed trough (Figure 6-13b). These data provide details about how close the plane of their orbit is to being perpendicular to our line of sight.

When we can determine M_1/M_2, we can find the individual masses, because we already know the sum

$M_1 + M_2$ from Kepler's third law (i.e., two equations and two unknowns yield the values for both unknowns). The individual masses are determined by combining the ratio of the masses and the sum of the masses. The range of stellar masses thus determined extends from 0.08 $M_\odot$ to about 150 $M_\odot$.

a Partial eclipse

Light

Orbital period

Time

b Total eclipse

Light

Time to cross disk of large star

Orbital period

Time

Eclipse begins   Eclipse ends

Eclipsing binary star

c

R I V U X G

FIGURE 6-13 Representative Light Curves of Eclipsing Binaries The shape of the light curve (blue) reveals that the pairs of stars have orbits in planes nearly perpendicular to our line of sight. It also provides details about the two stars that make up an eclipsing binary. Illustrated here are (a) a partial eclipse and (b) a total eclipse. (c) The binary star NN Serpens, indicated by the arrow, undergoes a total eclipse. The telescope was moved during the exposure so that the sky drifted slowly from left to right. During the 10.5-min eclipse, the dimmer, but larger, star in the binary system (an M6 V star) passed in front of the more luminous but smaller star (a white dwarf). The binary became so dim that it almost disappeared. (European Southern Observatory)

Light curves can also reveal information about stellar atmospheres. Suppose that one star of a binary is a tiny white dwarf (a stellar remnant the size of Earth) and the other is a giant (a late stage of stellar evolution, in which the star has greatly expanded from its previous size). By observing exactly how the light from the white dwarf is gradually cut off as it begins to move behind the edge of the giant during an eclipse, astronomers can infer the pressure and density in the upper atmosphere of the giant. Such information is invaluable in testing models of stellar structure.

Many binary stars are separated by several AU or more. Other than orbiting each other, these stars behave as though they are isolated; that is, the models we develop in the following chapters for the evolution of isolated stars apply to them as well. However, there are also **close binary** systems, with only a few stellar diameters separating the stars. Such stars are so close together that the gravity of each one dramatically affects the appearance and evolution of the other. If one member of a close binary is a giant, some of the gas of its outer lay-

ers is pulled onto its more compact companion—mass is transferred from one star to the other. We will explore more about such systems in Chapter 7.

To determine the sum of the masses of two stars in a binary star system, what observations of them must be made?

6-11 Main-sequence stars have a relationship between mass and luminosity

As binary star data accumulated, an important trend began to emerge. On the main sequence, the more luminous the star, the more massive it is. This mass-luminosity relation can be conveniently displayed graphically (Figure 6-14a); note that the axes of this figure are logarithmic. The Sun's mass lies in the middle of both logarithmic ranges. The **mass-luminosity relation** demonstrates that *the main sequence on the H-R diagram is a progression in mass as well as in luminosity and surface temperature.* The hot, bright, bluish stars in the upper

a

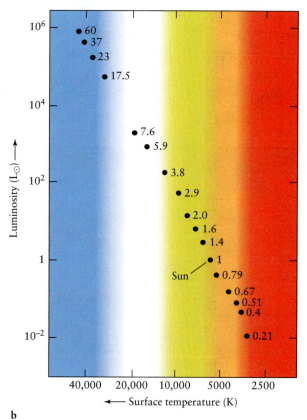

b

FIGURE 6-14 The Mass-Luminosity Relation (a) For main-sequence stars, mass and luminosity are directly correlated—the more massive a star, the more luminous it is. A main-sequence star of mass 10 M$_\odot$ has roughly 3000 times the Sun's luminosity (3000 L$_\odot$); one with 0.1 M$_\odot$ has a luminosity of only about 0.001 L$_\odot$. To fit them on the page, the luminosities and masses are plotted using log-

arithmic scales. (b) On this H-R diagram, each dot represents a main-sequence star. The number next to each dot is the mass of that star in solar masses (M$_\odot$). As you move up the main sequence from the lower right to the upper left, the mass, luminosity, and surface temperature of main-sequence stars all increase.

left corner of the H-R diagram (see Figure 6-14b) are the most massive main-sequence stars. As we move down the sequence, stellar masses decrease until we reach the dim, cool, reddish stars in the lower right corner of the H-R diagram. The correlation of stellar luminosities with mass suggests (correctly, as it turns out)

that a star's energy production is closely linked to its mass. Indeed, the mass is *the* crucial factor in explaining why stars are as bright as they are and how they evolve.

6-12 The orbital motion of binary stars affects the wavelengths of their spectral lines

Many binary stars are scattered throughout our Milky Way Galaxy, but only those that are nearby or are wide-

ly separated can be distinguished as visual binaries. A remote binary often presents the appearance of a single star because even our best telescopes cannot resolve the images of its individual stars. These binary systems can still be detected, however, through spectroscopy.

Spectral analysis yields incongruous spectral lines for some stars. For example, the spectrum of what appears at first to be a single star may include strong absorption lines for both hydrogen (indicating a hot, type A star) and titanium oxide (indicating a cool, type M star). Because a single star cannot display both types of absorption lines prominently, such an observation must be revealing a binary system.

Spectroscopy can also detect the movements of stars orbiting each other because of the Doppler shift in spectral lines. As we saw in Section 4-7, an approaching source of light has shorter wavelengths than if the same source were stationary, and a receding source has longer wavelengths. The amount of the shift is proportional to

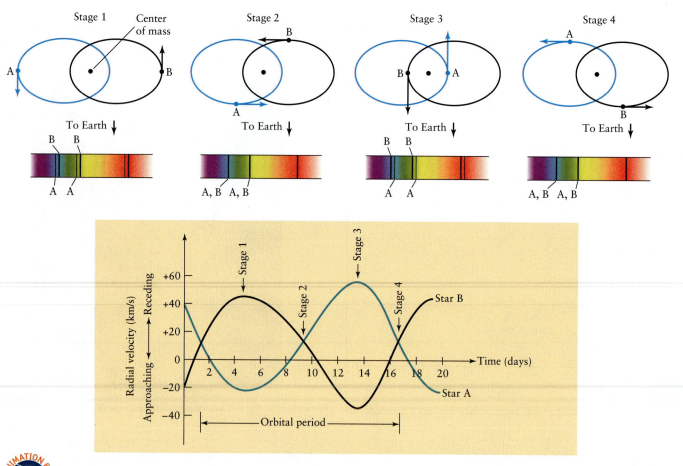

FIGURE 6-15 Spectral-Line Motion in Binary Star Systems (a) The diagrams at the top indicate the positions and motions of the stars, labeled A and B, relative to Earth. Below each diagram is the spectrum we would observe for these two stars at each stage. The changes in colors (wavelengths) of the spectral lines are due to changes in the stars' Doppler shifts, as seen from Earth. (b) This graph displays the radial-velocity curves of the binary HD 171978. (The HD means that this is a star from the Henry Draper Catalogue of stars.) The entire binary is moving away from us at 12 km/s, which is why the pattern of radial velocity curves is displaced upward from the zero-velocity line.

the speed of the light source: The greater the speed, the greater the shift.

If the two stars in a binary are orbiting at more than a few kilometers per second, they produce spectral lines that shift back and forth regularly. Such stars are called **spectroscopic binaries**. The motions of the stars revolving around their center of mass produce the periodic shifting of the spectral lines. In many spectroscopic binaries, one of the stars is so dim that its spectral lines cannot be detected. Instead, a single set of spectral lines from the star that we can see shifts regularly back and forth. Such a *single-line spectroscopic binary* yields less information about its two stars than does a *double-line spectroscopic binary*, in which the lines from both stars are visible. In particular, a double-line binary, in which one star eclipses the other, yields the individual masses of its two stars, whereas a single-line binary cannot provide that information.

Figure 6-15a shows four spectra of a spectroscopic binary system. At Stage 1, star A is moving toward Earth; hence its spectrum is blueshifted. Conversely, star B is moving away, so its spectrum is redshifted. In the spectrum below the drawing, you can see two sets of spectral lines, each slightly offset from each other. A few days later, at Stage 2, the stars have progressed along their orbits so that star A is moving toward the right and star B toward the left, as seen from Earth. Because neither star is moving toward or away from us, their spectral lines return to their normal positions. Stage 3 shows the opposite spectra from Stage 1 (star A is now redshifted and star B is blueshifted), while Stage 4 shows the same spectrum as Stage 2. Figure 6-16 shows spectra taken of the binary star system κ (kappa) Arietis, demonstrating these shifts.

The shifts in spectral lines can yield significant information about the orbital velocities of the stars in a spectroscopic binary. This information is best displayed as a **radial-velocity curve**, in which radial velocity is graphed over time (see Figure 6-15b). In Figure 6-15b, the wavy pattern repeats with a period of about 15 days, which is the orbital period of the binary. This pattern is displaced upward from the zero-velocity line by about 12 km/s, which is the overall motion of the binary system away from Earth. Superimposed on this overall recessional motion are the periodic approaches and recessions of the two stars as they orbit around each other.

Combining radial-velocity curves and light curves, we can obtain other useful information from eclipsing binaries. If an eclipsing binary is also a double-line spectroscopic binary, astronomers can calculate the masses, diameters, varied brightness, speeds, and stellar separation of each star. These stars are rare, however, because the orbital planes of most spectroscopic binaries are tilted so that eclipses do not occur, as seen from Earth.

While most of the stars we see in the night sky are in binary systems or are single objects, some are in groups of three or more. For example, the Pleiades in the constellation Taurus is a cluster of hundreds of stars, of which seven

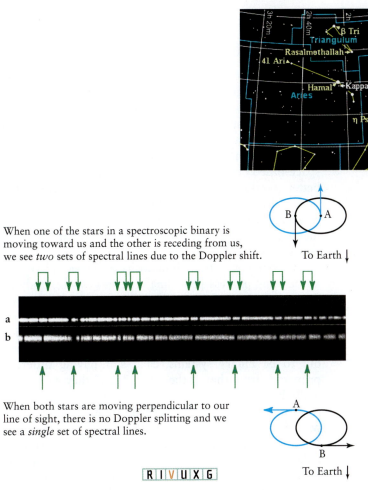

When one of the stars in a spectroscopic binary is moving toward us and the other is receding from us, we see *two* sets of spectral lines due to the Doppler shift.

To Earth ↓

a
b

When both stars are moving perpendicular to our line of sight, there is no Doppler splitting and we see a *single* set of spectral lines.

R I V U X G

To Earth ↓

FIGURE 6-16 A Double-Line Spectroscopic Binary The spectrum of the double-line spectroscopic binary κ (kappa) Arietis has spectral lines that shift back and forth as the two stars revolve around each other. (a) The stars are moving parallel to the line of sight, with one star approaching Earth, the other star receding, as in Stage 1 or 3 of Figure 6-15a. These motions produce two sets of shifted spectral lines. (b) Both stars are moving perpendicular to our line of sight, as in Stage 2 or 4 of Figure 6-15a. As a result, the spectral lines of the two stars have merged. (Lick Observatory)

(the "seven sisters") are often visible to the naked eye. Indeed, astronomers observe that many stars throughout our Milky Way Galaxy and elsewhere are found in even larger groups, which we will study in the next few chapters.

This chapter has provided us with observational data that serve, in large measure, as the basis for the models of stellar activity and evolution presented in the next three chapters. The models make predictions that lead to new observations and refined models. As we proceed through these models, we will take up some further observations driven by these refinements.

What do the spectral lines of a spectroscopic binary do, as observed from Earth?

6-13 Frontiers yet to be discovered

The legacy of *Hipparcos* is a better understanding of the size of our region of the cosmos. As noted earlier, stellar parallax is the most accurate means of measuring distances to the stars, requiring the fewest assumptions about how stars work and the nature of the material between us and the stars. Therefore, a major goal of space-based observational astronomy is to get more and better parallax measurements for stars at greater distances. To this end, satellites will be sent far from Earth to increase the baseline used to make these measurements (see Figure 6-1) and hence the angular accuracy that is possible.

Classifying stars by luminosity and spectral type is an ongoing process providing an ever-growing body of data about stellar properties. This information yields continual surprises. For example, there is a growing number of individual "stars" whose properties astronomers thought they understood but which turned out to be binary systems, each star having different properties from that of the single "star" they were originally believed to be.

Another significant piece of information astronomers need to understand star formation better is how many stars of a given mass are formed. Observations strongly indicate that very high-mass stars form quite infrequently, but is the same true of very low-mass stars? Or do the numbers of stars formed increase with decreasing mass? Or perhaps there is a peak of star formation at some mass with decreasing numbers of stars that have higher and lower masses. The number of stars formed with different masses is expressed as the **initial mass function**. Although there are several theoretic initial mass functions, in the end, observations will determine which is correct.

SUMMARY OF KEY IDEAS

• Stars differ in size, luminosity, temperature, color, mass, and chemical composition—facts that help astronomers understand stellar structure and evolution.

Magnitude Scales

• Determining stellar distances from Earth is the first step to understanding the nature of the stars. Distances to the nearer stars can be determined by stellar parallax, which is the apparent shift of a star's location against the background stars while Earth moves along its orbit around the Sun. The distances to more remote stars are determined using spectroscopic parallax.

• The apparent magnitude of a star, denoted m, is a measure of how bright the star appears to Earth-based observers. The absolute magnitude of a star, denoted M, is a measure of the star's true brightness and is directly related to the star's energy output, or luminosity.

• The absolute magnitude of a star is the apparent magnitude it would have if viewed from a distance of 10 pc. Absolute magnitudes can be calculated from the star's apparent magnitude and distance from Earth.

• The luminosity of a star is the amount of energy emitted by it each second.

The Temperatures of Stars

• Stellar temperatures can be determined from stars' colors or stellar spectra.

• Stars are classified into spectral types (O, B, A, F, G, K, and M) based on their spectra or, equivalently, their surface temperatures.

Types of Stars

• The Hertzsprung-Russell (H-R) diagram is a graph on which luminosities of stars are plotted against their spectral types (or, equivalently, their absolute magnitudes are plotted against surface temperatures).

• The H-R diagram reveals the existence of four major groupings of stars: main-sequence stars, giants, supergiants, and white dwarfs.

• The mass-luminosity relation expresses a direct correlation between a main-sequence star's mass and the total energy it emits.

• Distances to stars can be determined using their spectral types and luminosity classes.

Stellar Masses

• Binary stars are surprisingly common. Those that can be resolved into two distinct star images (even if it takes a telescope to do this) are called visual binaries.

• The masses of the two stars in a binary system can be computed from measurements of the orbital period and orbital dimensions of the system.

• Some binaries can be detected and analyzed, even though the system may be so distant (or the two stars so close together) that the two star images cannot be resolved with a telescope.

• A spectroscopic binary is a system detected from the periodic shift of its spectral lines. This shift is caused by the Doppler effect as the orbits of the stars carry them alternately toward and away from Earth.

• An eclipsing binary is a system whose orbits are viewed nearly edge-on from Earth, so that one star periodically eclipses the other. Detailed information about the stars in an eclipsing binary can be obtained by studying the binary's light curve.

• Mass transfer occurs between binary stars that are close together.

WHAT DID YOU THINK?

1 *How near to us is the closest star other than the Sun?* Proxima Centauri is about 40 trillion km (25 trillion mi) away. Light from there will take about 4 years to reach Earth.

2 *How luminous is the Sun compared with other stars?* The most luminous stars are about a million times brighter, and the least luminous stars are about a hundred thousand times dimmer than the Sun.

3 *What colors are stars, and why do they have these colors?* Stars are found in a wide range of colors, from red through violet as well as white. They have these colors because they have different surface temperatures.

4 *Are brighter stars hotter than dimmer stars?* Not necessarily. Many brighter stars, such as red giants, are cooler but larger, than hotter, dimmer stars, such as white dwarfs.

5 *Compared to the Sun, what sizes are other stars?* Stars range from more than 1000 times the Sun's diameter to less than 1/100 the Sun's diameter.

6 *Are most stars isolated from other stars, as the Sun is?* No. In the vicinity of the Sun, one-third of the stars are found in pairs or larger groups.

Key Terms for Review

absolute magnitude, 160
apparent magnitude, 158
binary star, 170
center of mass, 170
close binary, 173
eclipsing binary, 170
giant star, 167
Hertzsprung-Russell
 (H-R) diagram, 166
initial mass function, 176
inverse-square law, 160
light curve, 170
luminosity, 161
luminosity class, 168
main sequence, 167
main-sequence star, 167
mass-luminosity relation,
 173

OBAFGKM sequence,
 165
optical double, 170
photometry, 162
radial-velocity curve, 175
red giant, 167
spectral types, 165
spectroscopic binary, 175
spectroscopic parallax,
 169
stellar evolution, 155
stellar parallax, 155
stellar spectroscopy, 163
supergiant, 167
visual binary, 170
white dwarf, 167

Review Questions

The answers to all computational problems, which are preceded by an asterisk (*), appear at the end of the book.

1. Stellar parallax measurements are used in astronomy to determine which of the following properties of stars?

a. speeds, b. rotation rates, c. distances, d. colors, e. temperatures

2. The brightness a star would have if it were at 10 pc from Earth is called its a. absolute magnitude, b. apparent magnitude, c. luminosity, d. spectral type, e. center of mass

3. Measurements of a binary star system are required to determine what property of stars? a. luminosity, b. apparent magnitude, c. distance, d. mass, e. temperature

4. A star with which of the following apparent magnitudes appears brightest from Earth? a. 6.8, b. 3.2, c. 0.41, d. −0.44, e. −1.5

5. A star of what spectral class has the strongest (darkest) H_α line? (Hint: See Figure 6-5): a. B2, b. A0, c. A5, d. G5, e. M0

6. Describe how the parallax method of finding a star's distance is similar to the binocular (two-eye) vision of animals.

7. What is stellar parallax?

8. How do astronomers use stellar parallax to measure the distances to stars?

9. Why do stellar parallax measurements work only with relatively nearby stars?

10. What is the difference between apparent magnitude and absolute magnitude?

11. Briefly describe how you would determine the absolute magnitude of a nearby star.

12. What does a star's luminosity measure?

13. Why is the magnitude scale "backward" from what common sense dictates?

14. Does the star Betelgeuse, whose apparent magnitude is $m = +0.5$, look brighter or dimmer than the star Pollux, whose apparent magnitude is $m = +1.1$?

*15. Consider two identical stars, with one star 5 times farther away than the other. How much brighter will the closer star appear than the more distant one?

16. How and why is the spectrum of a star related to its surface temperature?

17. What is the primary chemical component of most stars?

18. A star of which spectral type has the strongest Na I absorption lines? At approximately what wavelength is this line normally found? (Hint: See Figure 6-5).

19. Why does a G2 star have many more absorption lines than a B0 star?

20. Draw an H-R diagram and sketch the regions occupied by main-sequence stars, giants, supergiants, and white dwarfs. Briefly discuss the different ways you could have labeled the axes of your graph.

21. To test your understanding of the H-R diagram, do Interactive Exercise 6-1. You can print out your answers, if required.

22. How can observations of a visual binary lead to information about the masses of its stars?

23. What is a radial-velocity curve? What kinds of stellar systems exhibit such curves?

24. What is the difference between a single-line and a double-line spectroscopic binary?

25. What is meant by the light curve of an eclipsing binary? What sorts of information can be determined from such a light curve?

26. What is the mass-luminosity relation? To what kind of stars does it apply?

27. Refer to Figure 6-7: **a.** What are the hottest and coolest named stars on the diagram?, **b.** What are the brightest and dimmest named stars on the diagram?, **c.** What are the hottest and coolest named main-sequence stars on the diagram?, **d.** What named stars are white dwarfs? giants? supergiants?

Advanced Questions

28. What is the inverse-square law? Use it to explain why a headlight on a car can appear brighter than a star, even though the headlight emits far less light energy per second.

*29. Van Maanen's star, named after the Dutch astronomer who discovered it, has a parallax angle of 0.232 arcsec. How far away is the star?

30. Explain how sailors on a ship traveling parallel to a coastline at a known speed can use parallax angle measurements to determine the distance to the shore.

*31. Suppose that a dim star were located 2 million AU from the Sun. Find **a.** the distance to the star in parsecs and **b.** the parallax angle of the star.

*32. How many times brighter is a star of apparent magnitude $m = -1$ than a star of apparent magnitude $m = +7$?

33. Sketch the radial-velocity curve of a binary whose stars are moving in nearly circular orbits that are **a.** perpendicular and **b.** parallel to our line of sight.

34. Sketch the light curve of an eclipsing binary whose stars are moving along highly elongated orbits with the major axes of the orbits **a.** pointed toward Earth and **b.** perpendicular to our line of sight.

*35. a. What is the approximate mass of a main-sequence star that is 10,000 times as luminous as the Sun? **b.** What is the approximate luminosity of a main-sequence star whose mass is one-tenth that of the Sun?

*36. What is the approximate surface temperature of a main-sequence star with a luminosity 100 times as bright as the Sun?

Discussion Questions

37. Discuss the advantages and disadvantages of measuring stellar parallax from a space telescope in a large solar orbit, say, at the distance of Jupiter from the Sun.

38. How does a star's rotation affect the appearance of its spectral lines? (Hint: Assume we are looking at the star from above its equator. Then, at every instant, half of the spinning star is approaching Earth, while the other half is receding. Consider the resulting Doppler shift in the spectral lines.)

What If...

39. All of the stars in our Milky Way Galaxy actually were the same distance from Earth, say, 10 pc? Describe what the night sky might look like. How about the daytime sky? (Hint: If you are doing this quantitatively, assume for simplicity that all of the stars have the same absolute magnitudes as the Sun and that there are roughly 200 billion stars in the Galaxy).

40. All stellar parallax angles, p, were observed to be increasing? What would that imply about the motions of stars? Is there another observational technique that could be used to confirm this motion?

41. A star's parallax were observed to oscillate—regularly increase in angle and then decrease in angle—over a period of 100 years? What would that imply about the star?

42. The Sun were part of a binary star system of two main-sequence stars and Earth orbited around one of the stars, while the other star just slightly changed Earth's distance of orbit, compared to its orbit today? You might want to discuss such things as climate, tides, impacts from meteoroids and asteroids, and habitability.

43. The Sun were an M-type star, rather than a G-type star? Assuming that Earth orbiting the M-type star had the same composition and orbit distance as it does today, what would be different here?

44. The Sun were a B-type star, rather than a G-type star? Assuming that Earth orbiting the B-type star had the same composition and orbit distance as it does today, what would be different here? We will pick up this question again in Chapter 8 for further insights.

Web Questions

*45. Distances to Stars Using Parallax Access the Active Integrated Media Module "Using Parallax to Determine Distance in Characterizing Stars" of the *Discovering the Universe: From the Stars to the Planets* Web site. Use this to determine the distance in parsecs and in light-years to each of the following stars:

a. Betelgeuse (parallax $p = 0.007630$); b. Vega ($p = 0.1290$); c. Antares ($p = 0.005400$); and d. Sirius ($p = 0.3790$).

46. Search the Web for the periods of 10 binary star systems. Plot these on a graph of time (on the horizontal axis) versus number of systems. If possible, combine this information with similar data from your classmates. Do you see any patterns in the periods of these star systems?

47. To explore the range of periods of binary star systems, locate on the Web binary star systems with periods of a. less than 1 week, b. between 1 day and 1 week, c. between 1 and 2 years, d. between 40 and 50 years; and e. more than 400 years.

Observing Projects

48. Use your *Starry Night Enthusiast*™ software to locate the bright stars listed in Appendix Table E-6 visible from your location. Click the **Home** button in the toolbar and then click the **Stop** button in the toolbar to stop time flow. Set the time in the program to your proposed observing time tonight. Open the **Options** pane and expand the **Constellations** layer. Display the major features of the constellations by selecting **Boundaries**, **Labels** and **Stick Figures** options. Locate and form a list of a set of the stars in Appendix Table E-6 that are visible to you tonight by opening the **Find** pane, typing the name of each object in turn and pressing the Enter key. Ignore (choose **Cancel**) the stars that the computer says are not visible tonight or are never visible from your location. Include in your list their apparent magnitudes and the constellations in which they can be found, and arrange to observe them. You may find it helpful to print out relevant star charts around these chosen stars, using *Starry Night Enthusiast*™. You can determine how well you can distinguish between stars of different brightness by comparing stars of similar but different magnitude, such as Antares and Spica, for example. You can use your star charts and identify and observe other stars, estimating their apparent magnitudes based on the ones on your list. (As a check on these estimates, you can produce a second reference star chart using *Starry Night Enthusiast*™ that indicates apparent magnitudes of stars in any region. Open the **Options** menu on the top bar of the window and click on **Stars > Stars. . .** to display **Stars Options**, click on **Labels** and set **Magnitude** in the **Show** drop-down list.)

49. Locate the stars Betelgeuse and Rigel in Orion. Observe them both by eye and through a small telescope. Are they the same color? To determine their color(s), it helps to compare them to their neighbors.

50. Locate one or more of the following double star systems: Regulus in Leo, Algieba in Leo, Irak [ε (eta) Boo] in Boötes, Ras Algethi in Hercules, Albireo in Cygnus, Vega in Lyra, Polaris in Ursa Minor, Rigel in Orion, Antares in Scorpius, Sirius in Canis Major, and Schedar in Cassiopeia. You can find their locations and print star charts, using your *Starry Night Enthusiast*™ software. View them with and without a telescope. What are their colors?

51. Mizar and Alcor In this exercise, you can use *Starry Night Enthusiast*™ ™ to explore two stars in the Big Dipper. Click the **Home** button in the toolbar. Open the **Find** pane and type "Mizar" in the search box followed by the **Enter** key. If the program indicates that Mizar is not visible at the current time, click the **Best Time** button. As seen from Earth, Mizar has a companion star, Alcor. Use the **Zoom** button in the toolbar to zoom in until a second star moves away from Mizar. Why does it appear to move away? Use the Heads-up display feature of the **Hand Tool** to identify this star as Alcor. Where is the smaller star Alcor located with reference to Mizar? Continue zooming in until the field of view is about 2° wide and then use the angular measurement feature of the **Hand Tool** to measure the angular separation and actual physical distance between Mizar and Alcor. The closest star to Earth is Proxima Centauri, about 4.3 ly away. Comparing the data collected on Mizar and Alcor with this information, consider whether you think that Mizar and Alcor are a visible binary, a spectroscopic binary, or an optical double, and explain your reasoning to support this conclusion.

CHAPTER 7

The Lives of Stars from Birth Through Middle Age

Cosmic Mountains of Creation A Spitzer Space Telescope image of the star-forming region W5, 7000 light-years away, in the constellation Cassiopeia. Young stars can be seen developing in its clouds of gas and dust. (Lori Allen [Harvard-Smithsonian CfA] et al., JPL-Caltech, NASA)

R I V U X G

WHAT DO YOU THINK?

1 How do stars form?

2 Are stars still forming today? If so, where?

3 Do more massive stars shine longer than less massive ones? How is this reasoned?

Answers to these questions appear in the text beside the corresponding numbers in the margins and at the end of the chapter.

Time changes every-thing. We grow up, we grow older, and we die. We are most familiar with changes brought about by aging in our own lives and the lives of other people, but everything else in the universe alters with time, too. These changes are due to the forces in nature that cause interactions between matter and energy. For example, over time, exposure to the effects of weather and pollution causes even solid stone to crumble (Figure 7-1).

Gravity provides the energy that enables stars to shine. As we saw in studying the Sun (Chapter 5), gravity compresses stars, thereby heating them and causing fusion to occur. As a result, stars emit huge amounts of radiation. Over time, their chemical compositions, masses, and brightnesses all vary—they age. Major stages in the life of each star can last from days to hundreds of billions of years. Cumulative lifetimes of stars range from millions to hundreds of billions of years. We will begin exploring their evolution in this and the next chapter.

INSIGHT INTO SCIENCE

Beware Poetic License All too often, the names and descriptions we use for nonliving objects seem to give them human qualities. Anthropomorphism—assigning human attributes to nonhuman creatures or even to nonliving objects, such as stars—is descriptive but not scientific. In fact, scientists avoid this practice where possible, because it creates unjustified expectations. Nevertheless, anthropomorphism does creep into astronomy. For lack of better terms, astronomers describe stellar change with words related to living things, such as *evolution, birth, aging, maturing, growing old,* and *dying,* when talking about stars. This is poetic license of a sort: It piques our interest but muddles our science. As you study stellar evolution, always be mindful of that.

In this chapter you will discover

- how stars form
- what a stellar "nursery" looks like
- how astronomers use the physical properties of stars to learn about stellar evolution
- the remarkable transformations of older stars into giants
- how the Hertzsprung-Russell (H-R) diagram is your guide to the stellar life cycle
- how pairs of orbiting stars change each other

PROTOSTARS AND PRE–MAIN-SEQUENCE STARS

The solar system is thought to have formed from a collapsing, rotating cloud of gas and dust some 4.6 billion years ago. This scientific theory requires that interstellar matter existed before then so that the solar system had something from which to form. Taking this reasoning one step further, if such interstellar matter exists today, perhaps star formation continues.

7-1 Gas and dust exist between the stars

Matter does indeed exist between the stars. We can see a relatively small amount of it through visible-light telescopes (Figure 7-2a); however, the bulk of it is too cold to be seen optically and requires the use of infrared (Figure 7-2b) and radio telescopes. We call the material between stars the **interstellar medium**, and it contains at least 10% of all the observed mass in our Galaxy. Observations of the spectra of the interstellar medium reveal that it is composed of gas containing isolated atoms and molecules and tiny pieces of dust. Table 7-1 summarizes the composition of the interstellar medium.

A variety of molecules is found in interstellar space, including molecular hydrogen (H_2), carbon monoxide (CO), carbon dioxide (CO_2), water (H_2O), ammonia (NH_3), formaldehyde (H_2CO), and simple sugars, among many others. We can identify these molecules by their

TABLE 7-1 Composition of the Interstellar Medium

	Particle number (%)	Mass (%)
Hydrogen (atoms and molecules)	90	74
Helium	9	25
Metals*	1	1

*Metals are all elements except hydrogen and helium.

unique spectral emissions, just as we can identify elements from atomic spectra.

The dust found in space also has a variety of structures. Some carbon-based (that is, organic) particles out there are typically 0.005 micrometers (μm) across (10,000 times smaller than the diameter of a typical human hair). Among these are collections of molecules similar to those found in engine exhaust and burnt meat (Figure 7-3), called *polycyclic aromatic hydrocarbons* (PAHs), which are composed of only carbon and hydrogen atoms. Much larger pieces of space dust have cores of carbon or silicon compounds surrounded with mantles of ice and other materials. These grow to more than 0.30 μm in diameter (nearly 200 times smaller than the diameter of a human hair).

Working with the knowledge that new stars form from the gas and dust of the interstellar medium, astronomers map this matter to identify places to look for young stars. When it can be seen in the visible part of the spectrum, the gas and dust in the interstellar medium glow as a result of scattered light from stars in its vicinity. For example, the interstellar medium is dramatically highlighted by stars in the Pleiades star cluster (Figure 7-2a) located in the constellation Taurus. The bluish haze, starlight scattered by the interstellar gas and dust, is called a **reflection nebula**. A **nebula** (plural: *nebulae*) is a dense region of interstellar gas and dust. Nebulae are often embedded in much larger bodies of gas and dust, called **molecular clouds**. The largest molecular clouds, called **giant molecular clouds,** contain upward of a million solar masses of matter and extend up to about 300 light-years across.

Astronomers identify several other types of nebulae. Most important among these are dark and emission nebulae. **Dark nebulae** are sufficiently dense regions of interstellar gas and dust, so they prevent most of the visible light from behind them getting to us. They look like regions of empty space, whereas they are actually among the densest of interstellar nebulae. **Emission nebulae** are regions of interstellar gas and dust that glow from energy they receive from nearby stars, from exploding stars, and from collisions between nebulae.

Finding more distant gas and dust is difficult because the nearby interstellar medium dims, or even blocks, light from behind it, in just the same way that thick clouds or several thin cloud layers in our sky can prevent us from seeing the Sun or stars behind them. This darkening of light by intervening gas and dust in space is called **interstellar extinction**. Dark nebulae are extreme examples of interstellar extinction.

Even when we can see a star or other object at visible wavelengths through the interstellar medium, it appears redder than it actually is, a phenomenon called **interstellar reddening**. This effect occurs because short-

FIGURE 7-1 Everything Ages In less than 60 years (1935 to 1994), corrosive gases in the air caused this statue of George Washington in New York City to erode even more rapidly than normal. Even without the significant pollution that humans add to the air, everything in contact with it decays. (a: 1994 NYC Parks Photo Archive/Fundamental Photographs; b: Copyright 1994 Kristen Brochman/Fundamental Photographs)

wavelength starlight is scattered by dust grains more than is the long-wavelength light. (Violet is scattered most, followed by blue, green, yellow, and orange, with red scattered least.) Therefore, when we observe a star through gas and dust, we are seeing less short-wavelength light from it than we would if the interstellar medium were not there (Figure 7-4a). The more gas and dust between the objects and us, the redder the objects appear (Figure 7-4b). Interstellar reddening is different from reddening due to the Doppler shift (see Section 4-7). The Doppler shift causes all wavelengths of electromagnetic radiation to lengthen equally, whereas interstellar reddening, due to the stronger scattering of shorter wavelengths, does not change the wavelengths of the starlight we receive—only their intensities.

We have been able to discover distant stars and clouds of interstellar gas and dust whose visible light is obscured by the nearby interstellar medium because the distant objects emit radio or infrared photons that are scattered relatively little on their way to us. Likewise, we use radio and infrared telescopes to find nearby interstellar gas and dust that are too cold to emit much visible light.

Stars form from gas and dust that are sufficiently cool so that they can collapse together. Most of this matter is hydrogen in the form of molecules (rather than atoms). However, molecular hydrogen is relatively hard to detect in space. Therefore, radio astronomers often

a

b

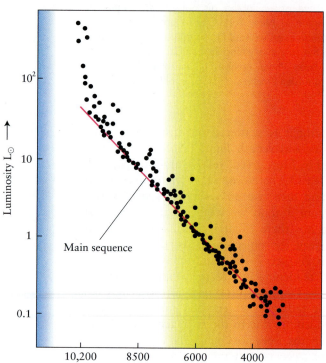

Luminosity $L_\odot$ →

10^2

10

1

0.1

Main sequence

10,200   8500   6000   4000

← Surface temperature (K)

c

FIGURE 7-2 Stars and the Interstellar Medium (a) This open cluster, called the Pleiades, can easily be seen with the naked eye in the constellation Taurus (the Bull). It lies about 375 light-years (116 pc) from Earth. The stars are not shedding mass, unlike the star in Figures 7-14a and 7-22. The blue glow surrounding the stars of the Pleiades is a reflection nebula created as some of the stars' radiation scatters off preexisting dust grains in their vicinity. (b) The same region of the sky in a false-color infrared image taken by the Spitzer Space Telescope. Gases are seen here to exist in more areas than can be detected in visible light. (c) Each dot plotted on this H-R diagram represents a star in the Pleiades, whose luminosity and surface temperature have been determined. Note that most of the cool, low-mass stars have arrived at the main sequence, indicating that hydrogen fusion has begun in their cores. The cluster has a diameter of about 5 light-years, is about 100 million years old, and contains about 500 stars. (a: Anglo-Australian Observatory; b: NASA/JPL-Caltech/J. Stauffer[SSC/Caltech])

In mapping the locations of CO emission (Figure 7-5a), astronomers came to realize that interstellar gas and dust are often concentrated in giant molecular clouds. In some cases, these clouds appear as dark areas silhouetted against a glowing background light, such as Orion's famous Horsehead Nebula (Figure 7-5c). In other cases, the clouds appear as dark nebulae that obscure the background stars (Figure 7-6). Some 6000 of these molecular clouds are estimated to exist in our Milky Way Galaxy and have masses that range from 10^5 to 2×10^6 $M_\odot$ and diameters that range from 50 to 300 light-years. The density inside each of these clouds ranges from 10^2 to 10^5 hydrogen molecules per cubic centimeter—thousands of times greater than the average density of the gas and dust dispersed throughout interstellar space, but some 10^{15} times less dense than the air we breathe. Having located interstellar matter, astronomers turned to explaining why some of this gas and dust collapses to form new stars (and planets).

search instead for carbon monoxide, which emits lots of photons at short radio (microwave) wavelengths of 2.6 and 1.3 mm. Calculations based on the known abundances of elements reveal that there are about 10,000 hydrogen molecules (H_2) for every CO molecule in the interstellar medium. Consequently, wherever astronomers detect strong emission of CO, they deduce that an enormous amount of hydrogen gas must also be present.

FIGURE 7-3 A Connection to Interstellar Space The charred layer created by overcooking this beef contains compounds of carbon and hydrogen, called polycyclic aromatic hydrocarbons. These molecules are also found in interstellar clouds. (R.J. Lerich)

7-2 Supernovae, collisions of interstellar clouds, and starlight trigger new star formation

We will see in detail in Chapter 8 that a supernova is a violent detonation that ends the life cycle of a massive star. The core of the doomed star collapses in a matter of seconds, releasing vast quantities of particles and energy that blow the star apart. The star's outer layers are blasted into space at speeds of several thousand kilometers per second.

Astronomers have found the ashes (more properly, the gas and dust) of many such dead stars scattered across the sky. These **supernova remnants** are another type of nebula. Supernova remnants, such as the Cygnus Loop shown in Figure 7-7, have a distinctly arched appearance, as would be expected for a shell of gas expanding at supersonic speeds. As it passes through the surrounding interstellar medium, the supernova remnant slams into preexisting matter, exciting the electrons in the atoms and molecules there, causing the gases to glow. If the expanding shell of a supernova remnant rams into a giant molecular cloud, it can compress the cloud, thus stimulating star birth in it. There is evidence that such an event happened around the time the solar system formed. For more about seeing nebulae on your own, see *Guided Discovery: Observing the Nebulae* on page 188.

A simple collision between two interstellar clouds can also create regions sufficiently dense and cool to collapse and form new stars. Likewise, radiation from O and B stars, which are especially bright and hot, will ionize (remove electrons from their atomic orbits) the gas that surrounds them, which then moves away and compresses the nearby interstellar medium (Figure 7-8).

Once a giant molecular cloud contracts and cools enough, gravitational attraction causes small regions of gas and dust in it to collapse. As these regions become denser, they block light from behind and inside them, becoming dark **Bok globules** (very small dark nebulae; Figure 7-8), named after astronomer Bart Bok,

> **1**
> The interstellar medium is composed primarily of what kinds of things?

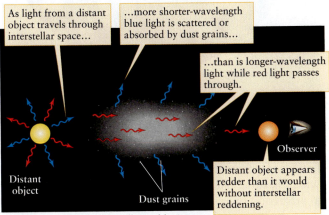

a How dust causes interstellar reddening

As light from a distant object travels through interstellar space...

...more shorter-wavelength blue light is scattered or absorbed by dust grains...

...than is longer-wavelength light while red light passes through.

Observer

Distant object

Dust grains

Distant object appears redder than it would without interstellar reddening.

b Reddening depends on distance

NGC 3576: A closer nebula (9,000 ly from Earth)

NGC 3603: A more distant nebula (20,000 ly from Earth)

R I V U X G

FIGURE 7-4 Interstellar Reddening (a) Dust in interstellar space scatters more short-wavelength (*blue*) light passing through it than longer-wavelength colors. Therefore, stars and other objects seen through interstellar clouds appear redder than they would otherwise.

(b) Light from these two nebulae pass through different amounts of interstellar dust and therefore they appear to have different colors. Because NGC 3603 is farther away, it appears a ruddier shade of red than does NGC 3576. (Anglo-Australian Observatory)

WEB LINK 7.2

FIGURE 7-5 A Gas- and Dust-Rich Region of Orion (a) This color-coded radio map of a large section of the sky shows the extent of giant molecular clouds in Orion and Monoceros as seen in the radio part of the spectrum. The intensity of carbon monoxide (CO) emission is displayed by colors in the order of the rainbow, from violet for the weakest to red for the strongest. Black indicates no detectable emission. The locations of four prominent star-forming nebulae are indicated on the star chart overlay. Note that the Orion and Horsehead nebulae are sites of intense CO emission, indicating that stars are forming in these regions. (b, c) A variety of nebulae appear in the sky around Alnitak, also called ζ (zeta) Orionis, the easternmost star in the belt of Orion. To the left of Alnitak is a bright, red emission nebula, called NGC 2024. The glowing gases in emission nebulae are excited by ultraviolet radiation from young, massive stars. Dust grains obscure part of NGC 2024, giving the appearance of black streaks, while the distinctively shaped dust cloud, called the Horsehead Nebula, blocks the light from the background nebula IC 434. The Horsehead is part of a larger complex of dark interstellar matter, seen in the lower left of this image. Above and to the left of the Horsehead Nebula is the reflection nebula NGC 2023, whose dust grains scatter blue light from stars between us and it more effectively than any other color. All of this nebulosity lies about 1600 light-years from Earth, while the star Alnitak is only 815 light-years away from us. NGC refers to the New General Catalog of stars and IC stands for Index Catalogs, two supplements to the NGC. (a: R. Maddalena, M. Morris, J. Moscowitz, and P. Thaddeus; b: Royal Observatory, Edinburgh; c: R. C. Mitchell, Central Washington University)

Radio image

Visible image

a R I V U X G

b R I V U X G

who studied them. Infrared observations show compact regions of gas and dust, called **dense cores**, inside Bok globules. These cores are destined to become stars.

To collapse, dark cores must be cool, because the hotter a gas, the more its particles collide with one another. The collisions between atoms and molecules create pressure in the cloud. If the temperature is too high, this pressure overcomes the gravitational attraction between the particles, preventing them from drawing close enough together to form stars.

Dense cores have temperatures of around 10 K, which is sufficiently cold so that their own gravitational attraction overwhelms the pressure they feel and thus pulls the gas together to form new stars. The conditions of temperature and density in the gas that ensure this collapse are called a **Jeans instability**, after the British physicist Sir James Jeans, who calculated them in 1902.

Often a giant molecular cloud has several hundred or even thousands of dense cores. In that case, hundreds or thousands of stars form together. Such stellar nurseries will become **open clusters** of stars like the Pleiades

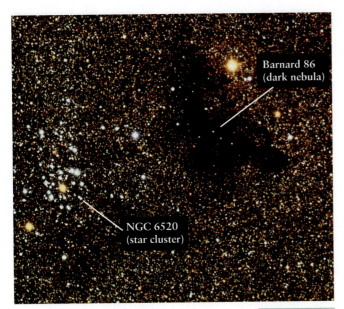

R I V U X G

FIGURE 7-6 A Dark Nebula The dark nebula Barnard 86 is located in Sagittarius. It is visible in this photograph simply because it blocks out light from the stars beyond it. The bluish stars to the left of the dark nebula are members of a star cluster, called NGC 6520. (Anglo-Australian Observatory)

(see Figure 7-2). Open clusters are gravitationally unbound systems, meaning that the stars in them eventually drift apart. It is likely that the solar system formed in an open cluster. In 2004, astronomers detected an open cluster in our Galaxy that contains roughly half a million stars in a volume about 6 light-years across. It includes at least 200 O and B stars. This cluster, Westerlund 1, is the most massive open cluster known to exist in our Galaxy. Called a *super star cluster,* such systems have been observed in distant galaxies, but never before in the Milky Way.

At first, a collapsing dense core is just a cool, dusty region thousands of times larger than our solar system. The dense core actually collapses from the inside out. The inner region falls in rapidly, leaving the outer layers of the dense core to drift in at a more leisurely rate. This process of increasing mass in the central region is called *accretion,* and the newly forming object at the center is called a **protostar** (Figure 7-9). Although fusion has not begun, a protostar emits energy, some of which comes from the compression and heating of its interior caused by the gravitational force from its growing mass of hot gas. However, most of the energy it releases comes from infalling gases colliding with the surface of the protostar. Figure 7-10 is an infrared image showing the locations of myriad protostars in and around a nebula in Centaurus.

What types of events can initiate the process of star formation?

a

R I V U X G

b

R I V U X G

FIGURE 7-7 A Supernova Remnant (a) X-ray image of the Cygnus Loop, the remnant of a supernova that occurred nearly 20,000 years ago. The expanding spherical shell of gas now has a diameter of about 120 light-years. The entire Cygnus Loop has an angular diameter in our sky 6 times wider than the Moon.

(b) This visible-light Hubble Space Telescope image of part of the Cygnus Loop shows emission from different atoms false color-coded with blue from oxygen, red from sulfur, and green from hydrogen. (a: Nancy Levenson/NASA; b: Jeff Hester, Arizona State University and NASA)

WEB LINK 7.3

GUIDED DISCOVERY

Observing the Nebulae

Distant nebulae—clusters of stars and glowing gases—are among the most impressive objects in the night sky. Binoculars and the naked eye are enough to let you "get your hands dirty" exploring them.

You can observe the Great Nebula of Orion (M42) during winter in the northern hemisphere with the naked eye. In all likelihood, you have seen it dozens of times without knowing it. To locate the Great Nebula, find the constellation Orion (using, for example, the star charts at the end of the book). Locate Orion's belt. Due south of the belt are three stars in a row that make up Orion's sword. Examine the sword very carefully with your naked eye. Do any of the stars in it look at all odd? Now look at them through a pair of binoculars. Which one is different from the others? That one is the Great Nebula in Orion, not a star at all! How does what you see compare with Figure 7-16?

The North America Nebula and the Pelican Nebula in Cygnus are best spotted in autumn. Pick a dark, Moonless night and use binoculars rather than a telescope. Higher magnification reveals too small a region of the sky for you to see the entirety of these vast, dim nebulae. To find them, first locate the bright star Deneb on the tail of Cygnus (using, for example, the star chart for Cygnus in this chapter, *Starry Night Enthusiast™*, or the star charts at the end of the book). The North America Nebula is located 3° east of Deneb, whereas the Pelican Nebula is located 2° southeast of it. These are both small angles, so sweep around the sky east of Deneb. If your binoculars are powerful enough, you should be able to see the outlines that give these nebulae their names.

The constellations of Orion and Monoceros encompass one of the most accessible regions of the sky for studying star formation and the interaction of young stars with the interstellar medium. Figure 7-5 shows a map of this region made with a radio telescope tuned to a wavelength of 2.6 mm, which is emitted by CO. Note the extensive areas covered by giant molecular clouds. Such comprehensive maps of CO emission help astronomers understand how the large-scale structure of the interstellar medium is related to the formation of stars.

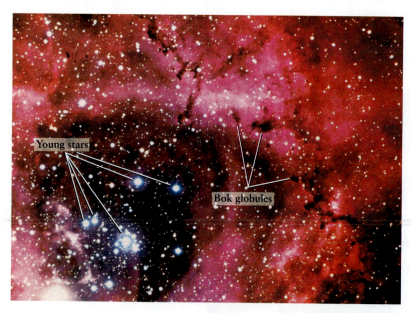

R I V U X G

FIGURE 7-8 The Core of the Rosette Nebula The large, circular Rosette Nebula (NGC 2237) is near one end of a sprawling giant molecular cloud in the constellation Monoceros (the Unicorn). Radiation from young, hot stars has blown gas away from the center of this nebula. Some of this gas has become clumped in Bok globules that appear silhouetted against the glowing background gases. New star formation is taking place within these globules. The entire Rosette Nebula has an angular diameter on the sky nearly 3 times that of the Moon, and it lies some 3000 light-years from Earth. (Anglo-Australian Observatory)

If a dense core is not spinning, it collapses into a sphere, which, ultimately, becomes an isolated star. If it is spinning, it collapses into a disk, which may then condense into two or three stars. Or, if the disk has a low enough mass, it may become a single star with orbiting protoplanets.

7-3 When a protostar ceases to accumulate mass, it becomes a pre–main-sequence star

Protostars are physically larger than the main-sequence stars into which they are evolving. A protostar of 1 $M_\odot$, for example, has about 5 times the diameter of the Sun. Because of their large sizes, protostars emit great quantities of radiation and gas (analogous to the solar wind), and they can be observed as sources of infrared radiation. At this point, they cannot be seen in visible light because they are enshrouded by their outer layer of gas and dust (see Figure 7-9). Much matter is still slowly falling inward from the dense core's outer shell. Eventually, the radiation and particles that flow off the protostar exert enough outward force to halt the infall of this gas and dust. As a result, mass accretion stops, and the protostar becomes a **pre–main-sequence star**.

A pre–main-sequence star contracts slowly, unlike the rapid collapse of a protostar. When the temperature at its core reaches 10^7 K, hydrogen fusion begins. As we saw in Chapter 5, this thermonuclear process releases

a A dark nebula R I V U X G

b A hidden protostar within the dark nebula R I V U X G

FIGURE 7-9 Protostar in a Bok Globule **(a)** This visible-light image shows a small dark nebula (equivalently, Bok globule) called L1014 located in the constellation Cygnus. **(b)** When viewed in the infrared, a protostar is visible within the nebula. (a: Deep Sky Survey; b: NASA/JPL-Caltech/N. Evans, University of Texas at Austin)

enormous amounts of energy. The outpouring of energy from hydrogen fusion creates enough pressure inside the pre–main-sequence star to stop its contraction. In the final stages of pre–main-sequence evolution, the outer shell of gas and dust finally dissipates (Figure 7-11; see also Figure 7-8). For the first time, the star is revealed via visible light to the outside universe.

7-4 The evolutionary track of a pre–main-sequence star depends on its mass

The more massive a pre–main-sequence star is, the more rapidly it begins hydrogen fusion in its core. For example, calculations indicate that a 5-$M_\odot$ pre–main-sequence star starts fusing less than a million years after it first forms from a protostar, whereas a 1-$M_\odot$ pre–main-sequence star takes a few tens of millions of years to begin fusion.

Astrophysicists use computers and the equations of stellar structure (described in Section 5-8) to model the evolution of a pre–main-sequence star. By calculating changes in the energy that the contracting star emits, computer simulations can follow its changing position on a Hertzsprung-Russell (H-R) diagram (Figure 7-12). Keep in mind that such an **evolutionary track** represents changes in a star's temperature and luminosity, not its motion in space.

Protostars transform into pre–main-sequence stars as they cross a curve, called the **birth line** (see the blue line on Figure 7-12). A star's exact location on this curve depends primarily on its mass and, to a much smaller extent, on the amount of metal it contains. (Recall from Section 6-6 that all elements other than hydrogen and helium are considered metals by astronomers.)

Spectroscopic observations of pre–main-sequence stars show that many are vigorously ejecting gas just before they reach the main sequence. Gas-ejecting stars in spectral classes G and cooler (that is, G, K, and M) are

R I V U X G

FIGURE 7-10 A Cluster of Protostars Over 300 protostars (yellow circles) were observed in the infrared by the Spitzer Space Telescope. This cluster of newly forming stars is 13,700 light-years away in the constellation Centaurus. The nebula, some of whose gas is being converted into stars, is called RCW 49 and contains more than 2200 stars and protostars. Most of the interior of this nebula is hidden from our eyes by the dust it contains. (NASA/JPL-Caltech/E. Churchwell, University of Wisconsin)

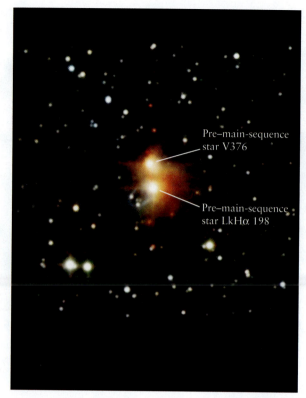

RIVUXG

FIGURE 7-11 Pre–Main-Sequence Stars Seen in infrared, the two large bright objects in the center of this image are pre–main-sequence stars. They have recently shed their cocoons of gas and dust, but still have strong stellar winds that create their irregular shapes. The two stars are an optical double, that is, they are not orbiting each other. (Atlas Image courtesy of 2MASS/UMASS/IPAC-Caltech/NASA/NSF)

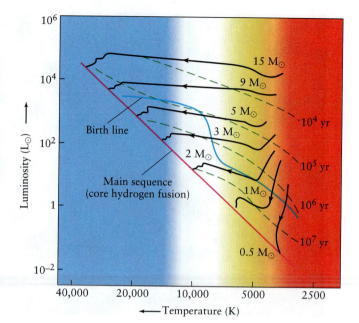

FIGURE 7-12 Pre–Main-Sequence Evolutionary Tracks This H-R diagram shows evolutionary tracks based on models of seven stars having different masses. The *dashed lines* indicate the stage reached after the indicated number of years of evolution. The birth line, shown in *blue,* is the location where each protostar stops accreting matter and becomes a pre–main-sequence star. Note that all tracks terminate on the main sequence at points that agree with the mass-luminosity relation (see Figure 6-14a).

Why can't protostars be observed with visible-light telescopes?

called **T Tauri stars,** after the first example discovered in the constellation of Taurus. Some astronomers propose that the onset of hydrogen fusion is preceded by vigorous chromospheric activity marked by enormous spicules (see Section 5-2) and flares (see Section 5-6) that propel the star's outermost layers back into space. In fact, an infant star going through its T Tauri stage can lose as much as 0.4 $M_\odot$ of matter and also shed its cocoon while still a pre–main-sequence star. In light of all this activity, it is not surprising that observations reveal T Tauri stars to be variable, meaning that they change brightness much more than, say, the Sun does today. We will discuss variable stars further in Sections 7-12 and 7-13.

Pre–main-sequence stars more massive than 2 $M_\odot$ become hotter without much change in overall lumi-

How are T Tauri stars different from main-sequence stars, like the Sun?

nosity. The evolutionary tracks of these pre–main-sequence stars thus traverse the H-R diagram nearly horizontally, from right to left. A star more massive than about 7 $M_\odot$

has no pre–main-sequence phase at all. Its gravitational compression is so great that it begins to fuse hydrogen in its protostellar phase.

Calculations reveal that the minimum temperature required to start normal hydrogen fusion (see *An Astronomer's Toolbox 5-1*) in the core of a star is 10 million K. However, pre–main-sequence stars less massive than 0.08 $M_\odot$ do not have enough gravitational force compressing and heating their cores to ever get this hot. As a result, these small bodies contract to become planet-like orbs of hydrogen and helium, called **brown dwarfs** (Figure 7-13). See *Guided Discovery: Extrasolar Planets and Brown Dwarfs* for further discussion of these important, albeit nonstellar, bodies.

Until 2004, the upper limit to main-sequence stellar mass was thought to be around 120 $M_\odot$. This number was based on calculations that show that above this mass, protostars rapidly develop extremely high fusion rates in their cores, which lead to extremely high surface temperatures—temperatures so great that their outer layers are superheated and thereby expelled into interstellar space. This, in turn, decreases their masses and their temperatures. An example of a very massive star in the process of shedding mass as it settles down onto the main sequence is the Pistol Star (Figure 7-14a). One of the most luminous stars in our Galaxy, the Pistol Star may have started as a protostar with 200 $M_\odot$. Observations

GUIDED DISCOVERY

Extrasolar Planets and Brown Dwarfs

The lowest mass that an object can have and still maintain the fusion of normal hydrogen into helium, as occurs in the Sun (see *An Astronomer's Toolbox 5-1*), is 0.08 M☉, or about 75 times the mass of Jupiter. Astronomers have discovered hundreds of objects in our Galaxy with less than this mass. Like Jupiter, they are primarily composed of hydrogen and helium, with traces of other elements. Many of them are found in orbit around stars, while some are found as free-floating masses that apparently formed without ever orbiting a star. An intriguing question has arisen: *What should they be called?*

Although normal hydrogen fusion does not occur in them, bodies with between 13 and 75 times Jupiter's mass do fuse deuterium (a rare form of hydrogen) into helium and those with between 60 and 75 times Jupiter's mass also fuse lithium (three protons and four neutrons) into helium. Both of these types of fusion occur very briefly in cosmic terms because of the limited supplies of deuterium and lithium in any known object in space. All objects between 13 and 75 times Jupiter's mass are called brown dwarfs. Objects with less than 13 times Jupiter's mass that are orbiting stars are *extrasolar planets*, while free-floating bodies with less than 13 times Jupiter's mass are often called *sub-brown dwarfs*. Bear in mind that these definitions are still undergoing discussion and revision in the astronomical community.

Because they emit relatively little energy compared to stars, extrasolar planets and brown dwarfs are dim and therefore very challenging to observe. Those in orbit are detected by their gravitational or eclipsing effects on the stars they orbit. The first brown dwarf was discovered only in 1994. Named Gliese 229B, it is located in orbit around a star, Gliese 229A (see Figure 7-13), in the constellation Lepus, about 18 ly (6 pc) from Earth. Since then, hundreds more brown dwarfs have been found, along with more than a dozen sub-brown dwarfs. Many of these are found in active star-forming regions, such as the Orion Nebula (see Figure 7-16) and the rho Ophiuchi cloud (see accompanying figure). Astronomers have also found more than 260 extrasolar planets. In 2002, astronomers observed clouds and storms on a brown dwarf similar to, but probably much larger than, the storms observed on the giant planets in our solar system.

Brown dwarfs of larger mass have the interesting feature that when they fuse deuterium or lithium, the helium they create moves upward, out of the core where it is formed. This helium is replaced with fresh deuterium or lithium fuel to fuse. The upward motion of the helium and downward motion of deuterium and lithium-rich hydrogen are due to convection, and, as a result of this motion, eventually all the deuterium and lithium are consumed. We say that these brown dwarfs are fully convective. This is different behavior than we find in the Sun (see Chapter 5), which has a separate core, convective zone, and radiative zone that do not share atoms. Flares have been observed from brown dwarfs. By analogy to the Sun's flares caused by magnetic fields emerging from its surface, astronomers believe that some brown dwarfs rotate and have magnetic fields.

Based on the numbers and locations of the known brown dwarfs and sub-brown dwarfs, astronomers estimate that there may be as many of these bodies in our Milky Way Galaxy as there are stars. Even in these numbers, brown dwarfs do not contribute a substantial amount of mass or gravitational force in the Galaxy because they have such small individual masses.

A Stellar Nursery Full of Brown Dwarfs Besides containing more than 100 young stars, the rho Ophiuchi cloud, located 540 light-years away in the constellation Ophiuchus, contains at least 30 brown dwarfs. By studying these objects, astronomers expect to learn more about early stellar evolution. This infrared image is color coded, with *red* indicating 7.7-μm radiation and *blue* indicating 14.5-μm radiation. (Infrared Space Observatory, NASA)

R I V U X G

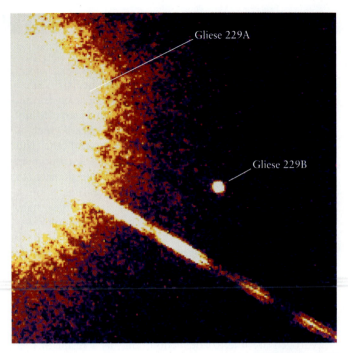

Gliese 229A

Gliese 229B

R I V U X G

WEB LINK 7.5 **FIGURE 7-13** A Brown Dwarf Located 18 light-years (6 pc) from Earth in the constellation Lepus (the Hare), Gliese 229B was the first confirmed brown dwarf ever observed. With a surface temperature of about 1000 K, its spectrum is similar to that of Jupiter. Gliese 229B is in orbit around a star. The overexposed image of part of its companion, Gliese 229A, appears on the left. The two bodies are separated by about 43 AU. Gliese 229B has from 20 to 50 times the mass of Jupiter, but the brown dwarf is compressed to the same size as Jupiter. The spike of light was produced when Gliese 229A overloaded part of the Hubble Space Telescope's electronics. (S. Kularni, California Institute of Technology; D. Golimowski, Johns Hopkins University; NASA)

(Figure 7-14b). Astronomers are trying to reconcile the theory of massive star formation with this observation.

7-5 H II regions harbor young star clusters

We can detect young open clusters of stars from the magnificent glows they create in the nebulae in which they form. Figures 7-5 and 7-15 show examples of these emission nebulae. Because these nebulae are predominantly ionized hydrogen, they are also called **H II regions**. To see why H II regions occur, remember that the most massive pre–main-sequence stars, those of spectral types O and B,

reveal that every few thousand years it expels shells of gas, and it may have less than 10 $M_\odot$ left when the expulsion of matter stops. The entire process of mass loss by very massive stars takes only a few million years.

In 2004, astronomers observed a main-sequence star that apparently has between 130 and 150 $M_\odot$

R I V U X G   a   R I V U X G   b   R I V U X G

FIGURE 7-14 Mass Loss from a Supermassive Star (a) The Quintuplet Cluster is 25,000 light-years from Earth. *(Inset)* Within the cluster is one of the brightest known stars, called the *Pistol*. Astronomers calculate that the Pistol formed nearly 3 million years ago and originally had 100–200 $M_\odot$. The structure of the gas cloud suggests the star ejected the gas we see in two episodes, 6000 and 4000 years ago. The gas from any previous ejections is so thinly spread now that we cannot see it. The nebula shown in the *inset* is more than 4 ly (1.25 pc) across—it would stretch from the Sun nearly to the closest star, Proxima Centauri. The name *Pistol* was given to the star based on early, low-resolution radio images of its gas, which initially looked like an old-fashioned pistol aimed to the left near the top of the *inset*. (b) The largest, most massive known star, LBV 1806-20, is 5 million times brighter and apparently some 150 times more massive than the Sun. This drawing shows the star's color and its size compared to the Sun. (a: D. Filger, NASA; b: Image by Dr. Stephen Eikenberry, Meghan Kennedy/University of Florida)

1. This emission nebula (about 2200 pc away and about 20 pc across) surrounds the star cluster M16.

2. Star formation is still taking place within this dark, dusty nebula.

3. Hot, luminous stars (beyond the upper edge of this image) emit ultraviolet radiation: This makes the dark nebula evaporate, leaving these pillars.

4. At the tip of each of these pillars is a nebula containing a young star.

5. Eventually, the nebulae evaporate, revealing the stars.

R I V U X G

FIGURE 7-15 An H II Region This emission nebula, M16, called the *Eagle Nebula* because of its shape, surrounds a star cluster. It is so numbered because it was the sixteenth object in the Messier Catalogue of astronomical objects. Star formation is presently occurring in M16, which is located 7000 light-years from Earth in the constellation of Serpens Cauda (the Serpent's Tail). Several bright, hot O and B stars are responsible for the ionizing radiation that causes the gases to glow. *(Inset)* Star formation is occurring inside these dark pillars of gas and dust. Intense ultraviolet radiation from existing massive stars off to the right of this image is evaporating the dense cores in the pillars, thereby prematurely terminating star formation there. Newly revealed stars are visible at the tips of the columns. (Anglo-Australian Observatory; J. Hester and P. Scowen, Arizona State University; NASA)

are exceptionally hot. Their surface temperatures are typically 15,000 to 35,000 K, causing them to emit vast quantities of ultraviolet radiation. This energetic radiation easily ionizes any surrounding hydrogen gas, thereby creating an H II region. Photons from an O5 star can ionize hydrogen atoms up to 500 light-years away.

H II denotes ionized hydrogen, which has no electrons in orbit; therefore, how do we observe it? While some hydrogen atoms in the H II regions are being knocked apart by ultraviolet photons, some of the free protons and electrons manage to get back together. As these new hydrogen atoms assemble, their electrons return to their ground states ($n = 1$). This downward cascade through each atom's energy levels, releasing photons with each jump between levels, is what makes the nebula glow. Particularly prominent is the transition from $n = 3$ to $n = 2$, which produces H_α photons at 656 nm in the red portion of the visible spectrum (review the emission line spectrum in Figure 4-8c, d). Thus, the nebula around a newborn star cluster often shines with a distinctive reddish hue (see Figure 7-15). Nebulae that have oxygen often look green when seen with the eye through a telescope because this gas has a green emission line at 501 nm. Because the eye is more sensitive to green light than red, the dimmer oxygen-emission line appears brighter in our brains than does the H_α line, which shows up well on CCD images, which usually appear red.

An H II region is a small, bright "hot spot" in a giant molecular cloud. The collection of hot, bright O and B stars that produces the ionizing ultraviolet radiation is called an **OB association**. The famous Orion Nebula

(Figure 7-16) is an example. Four O and B stars in an open cluster called the Trapezium, at the heart of the Orion Nebula, are the primary sources of the ionizing radiation that causes the surrounding gases to glow. The Orion Nebula is embedded in a giant molecular cloud whose mass is estimated at 500,000 $M_\odot$.

The OB association that creates an H II region also affects the rest of the giant molecular cloud in

> What are groups of high-mass stars called?

which it is imbedded (see Figure 7-16 insets). Detailed models indicate that vigorous stellar winds, along with ionizing ultraviolet radiation from these stars, carve out a cavity in the cloud. Where this outflow is supersonic, it creates a shock wave, such as the sonic boom created by fast-flying aircraft (see the shock waves shown in the right inset of Figure 7-16). The shock wave forms along the outer edge of the expanding H II region, compressing hydrogen gas as it passes and thereby stimulating a new round of star birth. As more O and B stars form, they power the expansion of the H II region still farther into the giant molecular cloud. Meanwhile, the older O and B stars left behind begin to disperse (Figure 7-17). In this way, an OB association "eats into" a giant molecular cloud, creating stars in its wake. The inset in Figure 7-17 shows star formation around a single O star.

7-6 Plotting a star cluster on an H-R diagram reveals its age

As noted in Section 7-2, stars are often observed to form in open clusters, such as seen in the nearby Orion Nebula (see Figure 7-16). Numerous other star-forming regions have been identified, and the young open clusters in them offer astronomers a rich source of information about stars in their infancy. By measuring each

RIVUXG

Trapezium stars

Young stars

Shock waves

RIVUXG   RIVUXG

FIGURE 7-16 The Orion Nebula The middle "star" in Orion's sword is actually the Orion Nebula, part of a huge system of interstellar gas and dust in which new stars are now forming. The Orion Nebula is a region visible to the naked eye. It is 1600 ly (490 pc) from Earth and has a diameter of roughly 16 ly (5 pc). This nebula's mass is about 300 $M_\odot$. (Left inset) This view at visible wavelengths shows the inner regions of the Orion Nebula. At the lower left are four massive stars, the brightest members of the Trapezium star cluster, which cause the nebula to glow. (Right inset) This view shows that infrared radiation penetrates interstellar dust that absorbs visible photons. Numerous infrared objects, many of which are stars in the early stages of formation, can be seen, along with shock waves caused by matter flowing out of protostars faster than the speed of sound waves in the nebula. Shock waves from the Trapezium stars may have helped trigger the formation of the protostars in this view. (ESO—European Southern Observatory; *left inset:* C. R. O'Dell, S. K. Wong, and NASA; *right inset:* R. Thompson, M. Rieke, G. Schneider, S. Stolovy, E. Erickson, D. Axon, and NASA)

ANIMATION 7.1   WEB LINK 7.6

RIVUXG

FIGURE 7-17 The Evolution of an OB Association High-speed particles and ultraviolet radiation from young O and B stars produce a shock wave that compresses gas farther into the molecular cloud, stimulating new star formation deeper into the cloud. Meanwhile, older stars are left behind. (Inset) Stars forming around a massive star 2500 ly (770 pc) away in the constellation Monoceros's Cone Nebula. The stars (small dots on the right side of the inset) arrayed around the bright, massive central star are believed to have formed as a result of the central star compressing surrounding gas with high-speed particles and radiation. The younger stars are just 0.04–0.08 light-years from the central star. (Adapted from C. Lada, L. Blitz, and B. Elmegreen; *inset:* R. Thompson, M. Rieke, G. Schneider, and NASA)

star's apparent magnitude, color, and distance, an astronomer can deduce its luminosity and surface temperature. The data for all the stars in the cluster can then be plotted on an H-R diagram, as shown for the Pleiades in Figure 7-2 or for the cluster NGC 2264 in Figure 7-18.

Because all of the stars in a cluster begin forming at the same time, and stars with different masses arrive on the main sequence at different times, astronomers can use the H-R diagram to determine the age of a cluster. For example, note that the hottest stars in NGC 2264 lie on the main sequence. These hot stars have surface temperatures around 20,000 K and are extremely bright and massive. Their radiation also causes the surrounding gases to glow. Most of the stars cooler than about 10,000 K have not yet arrived at the main sequence. These less massive stars, which are in the final stages of pre–main-sequence contraction, are just now beginning to ignite thermonuclear reactions at their centers.

From the H-R diagram of a young cluster we can see which are the lowest-mass stars that have already entered the main sequence. Using this information, along with theories of stellar evolution, we can determine the cluster's age. The cluster NGC 2264, for example, is roughly 2 million years old. In contrast, nearly all of the stars in the Pleiades (see Figure 7-2c) have completed their pre–main-sequence stage. That cluster's age is calculated to be about 100 million years, which is how long it takes for the least massive stars to finally begin hydrogen fusion in their cores.

As noted earlier, open clusters, such as the Pleiades and NGC 2264, possess barely enough mass to hold themselves together. A star moving faster than the average speed for the cluster occasionally escapes. This lowers the total gravitational force of the cluster, making it easier for other stars to leave. Astronomers predict that a few hundred million years after they begin forming, the stars in an open cluster have separated from each other and mixed with the rest of the stars in the Galaxy. At that time, most open clusters have completely dispersed.

Which star arrives on the main sequence first, one that is 0.5 M$_\odot$ or one that is 2 M$_\odot$?

MAIN-SEQUENCE AND GIANT STARS

In the first part of this chapter, we saw how stars form from dense cores of gas and dust inside giant molecular

NGC 2264

R I V U X G

a

b

FIGURE 7-18 Plotting the Ages of Stars (a) This photograph shows a region of ionized hydrogen and the young star cluster NGC 2264 in the constellation Monoceros. The red nebulosity is located about 2600 light-years from Earth and contains numerous stars that are about to begin hydrogen fusion in their cores. (b) Each dot plotted on this H-R diagram repre- sents a star in this cluster whose luminosity and surface temperature have been measured. Note that most of the cool, low-mass stars have not yet arrived at the main sequence. Calculations of stellar evolution indicate that this star cluster started forming about two million years ago. (a: David Malin/Anglo-Australian Observatory)

clouds. As these cores collapse, the gas and dust in their centers quickly form protostars. Eventually, protostars have pulled in (accreted) most of the mass available to them and they become pre–main-sequence stars that slowly contract. As a pre–main-sequence star evolves, fusion begins in its core.

Recall from Chapter 5 that in our Sun, thermal pressure created by fusion in the core pushes outward, everywhere balancing the inward force of gravity so the Sun neither expands nor collapses. The Sun is said to be in **hydrostatic equilibrium** (see Figure 5-20). When each layer of a pre–main-sequence star can finally support all of the layers above it (meaning that collapse ceases), that star also comes into hydrostatic equilibrium, and a main-sequence star is born (see Figure 7-12). *Main-sequence stars are those stars in hydrostatic equilibrium, in which nuclear reactions fuse hydrogen into helium in their cores at nearly constant rates.* Figure 7-19 summarizes the star-formation process.

7-7 Stars spend most of their lives on the main sequence

The **zero-age main sequence (ZAMS)** is the set of locations on the H-R diagram where pre–main-sequence stars of different masses first become stable objects, neither shrinking nor expanding. This is the solid red line that appears on several of the H-R diagrams in this chapter. Note that the theoretical evolutionary tracks in Figure 7-12 end at locations along the main sequence that agree with the observed mass-luminosity relation (recall Figure 6-14): The most massive main-sequence stars are the most luminous, whereas the least massive stars are the least luminous.

The energy emitted by stars is primarily generated by thermonuclear fusion (see *An Astronomer's Toolbox 5-1*). More luminous (brighter) stars generate and radiate more energy each second than do dimmer stars. Put another way, higher-mass stars are consuming more fuel than are lower-mass stars. This

3

In this cold, dark nebula, gas atoms and dust particles move so slowly that gravity can draw them together.

Gas and dust begin to condense into clumps, forming the cores of protostars.

Protostellar cores within the dark nebula

As the cores condense, their density and temperature both increase.

As the protostars continue to heat up and accrete matter from the nebula, they begin to glow due to their increasing temperature. If the core is rotating, some gas and dust in it forms a disk around the protostar.

In the T Tauri stage, the young star ejects mass into space in opposite directions. A stellar wind blows away the remaining parts of the nebula that surround the star, exposing the star to space.

When accretion stops, the protostar becomes a pre–main-sequence star. Its core heats to 10^7 K and fusion begins there. The mass that is continuing to fall onto the star forms an accretion disk.

The ejected mass can induce a shock wave in the surrounding interstellar material, triggering the formation of additional stars.

Processes that cause the star to lose or gain mass come to an end, and the star stabilizes as a main-sequence star in hydrostatic equilibrium. A system of planets often forms around the star in the disk of gas and dust.

ANIMATION 7.2

FIGURE 7-19 A Summary of the Star-Formation Process This set of photographs and drawings takes you through the star-formation process for stars with less than about 1.5 M$_\odot$.

TABLE 7-2 Main-Sequence Lifetimes

Mass (M$_\odot$)	Surface temperature (K)	Luminosity (L$_\odot$)	Time on main sequence (10^6 years)	Spectral class
25	35,000	80,000	3	O
15	30,000	10,000	15	B
3	11,000	60	500	A
1.5	7000	5	3000	F
1.0 (Sun)	6000	1	10,000	G
0.75	5000	0.5	15,000	K
0.50	4000	0.03	200,000	M

raises the interesting question of whether higher-mass stars have enough extra nuclear fuel to shine longer than lower-mass stars. The answer is *no*. The equations of nuclear fusion reveal that higher-mass stars consume fuel so quickly that these stars actually spend *less* time at each stage of stellar evolution than do lower-mass stars. For example, the tremendous internal pressure created by gravity inside O and B stars forces them to consume all of their core hydrogen in only a few million years, as shown in Table 7-2. Conversely, stars of very low mass take hundreds of billions of years to convert their cores from hydrogen into helium. Because, as we will see, the universe is roughly 14 billion years old, no main-sequence star with mass less than about 0.75 M$_\odot$ has yet moved into the next stage of stellar evolution.

The equations further reveal that the conversion of hydrogen into helium in every star's core takes a long time compared to any other stage of its stellar evolution. (These stages are presented in the following sections and the next two chapters.) That is why the vast majority of stars represented on an H-R diagram are located on the main sequence. This raises another intriguing question: Since most stars are on the main sequence, why are most of the stars visible to the naked eye other kinds of stars? The answer will become apparent shortly.

About how long ago was the Sun a ZAMS star?

EVOLUTION OF STARS WITH MASSES BETWEEN 0.08 AND 0.4 M$_\odot$

Because stars with mass less than 0.4 M$_\odot$ and stars with mass more than 0.4 M$_\odot$ evolve beyond the ZAMS so differently, we explore their evolutionary stages separately.

7-8 Red dwarfs convert essentially their entire mass into helium

The lowest-mass main-sequence stars, called **red dwarfs**, have masses between 0.08 M$_\odot$ and 0.4 M$_\odot$. They have the lowest core temperatures of all stars. Because fusion rates depend critically on temperature, these stars produce the least amount of energy and therefore emit the least energy; they are the dimmest of all main-sequence stars.

Red dwarfs are also different from higher-mass main-sequence stars in that the lower-mass stars transport the helium that they create out of their cores. This hot helium moves upward by convection, while cooler, denser gases from the hydrogen-rich outer layers move downward and into the core (Figure 7-20). This hydrogen is fused into helium, too. As a result, these stars eventually convert their entire mass into helium.

Fusion in the cores of red dwarfs is so slow that they remain on the main sequence for hundreds of billions of years. (In comparison, the theory of solar evolution predicts that the Sun's total lifetime on the main sequence will be about 10 billion years.) When a red dwarf has converted its entire mass into helium, fusion ceases inside it because the star does not have enough pressure in its core to heat and fuse that helium into anything else. These helium bodies will then radiate the heat they generated while on the main sequence, becoming cooler and dimmer. They will move down and to the right on the H-R diagram.

We saw in Section 6-11 how there is a relationship between the masses and luminosities of main-sequence stars (e.g., see Figure 6-14). Besides being the lowest mass, dimmest main-sequence stars, red dwarfs are also the most common stars in our Galaxy, representing some 85% of all its stars. However, even the nearest red dwarfs are so dim that astronomers continue to discover more of them. Red dwarfs typically form as isolated bodies surrounded with planets, unlike more massive

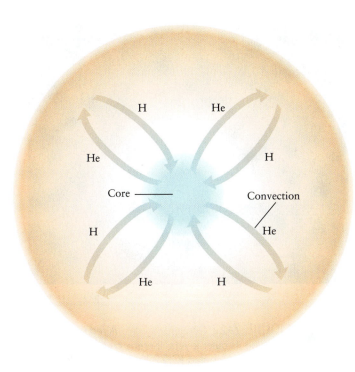

FIGURE 7-20 Fully Convective Star This drawing shows how the helium created in the cores of red dwarfs rises into the outer layers of the star by convection, while the hydrogen from the outer layers descend into the core. This process continues until the entire star is composed of helium.

stars, which are apt to form in binary star systems. Because red dwarfs are the most common stars and are usually single, astronomers have recently concluded that most stars in our galaxy are isolated, rather than binary star systems, as they thought until just a few years ago.

EARLY AND MIDDLE EVOLUTION OF STARS WITH MORE THAN 0.4 M$_\odot$

The changes in stars with more than $0.4 M_\odot$ are much more complicated. In this chapter, we consider what happens to these stars on the main sequence and in the evolutionary stage after it. In the following two chapters, we explore the ends of their lives.

7-9 When core hydrogen fusion slows down, a main-sequence star with mass greater than 0.4 M$_\odot$ becomes a giant

Recall that the main sequence is defined by the fusion of hydrogen into helium in a star's core. Unlike lower-mass stars, stars with more than $0.4 M_\odot$ do not transport heli-

um out of their cores by convection. Therefore, the chemical composition of their cores differs from the composition of their outer layers throughout their lives.

Is the name *red dwarf* appropriate for the stars so named? Why or why not?

When the hydrogen in the core of a main-sequence star with more than $0.4 M_\odot$ is mostly converted into helium, fusion in the core drastically slows down. The temperature of the core at that time is not high enough to enable its helium to fuse into other elements. However, the star continues to evolve and, contrary to intuition, the end of core hydrogen fusion actually causes these stars to expand in size.

Recall from our discussion of the Sun in Chapter 5 that, while on the main sequence, a star's outer layers are supported by thermal pressure, in which energy is supplied by fusion in the core. As the rate of this fusion decreases, a star can no longer support the crushing weight of its outer layers, and so the hydrogen-rich gas just outside the core is compressed inward and thereby heated enough to begin fusing into helium (Figure 7-21). This process is called **hydrogen shell fusion** because it occurs in a shell a few thousand kilometers thick surrounding the core. Recall from *An Astronomer's Toolbox 5-1* that the proton-proton chain is the major source of energy inside the Sun. This fusion process also predominates in hydrogen shell fusion.

Why, then, does the star expand to become a giant? Essentially, the star begins generating more

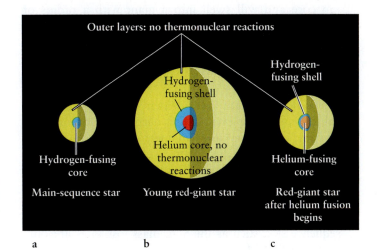

FIGURE 7-21 Evolution of Stars Off the Main Sequence (a) Hydrogen fusion occurs in the cores of main-sequence stars. (b) When the core is converted into helium, fusion there ceases and then begins in a shell that surrounds the core. The star expands into the giant phase. This newly formed helium sinks into the core, which heats up. (c) Eventually, the core reaches 10^8 K, whereupon core helium fusion begins. This causes the core to expand, slowing the hydrogen shell fusion and thereby forcing the outer layers of the star to contract.

energy in its shell than it did when it was in hydrostatic equilibrium as a main-sequence star. The extra energy is absorbed by the gases in the outer layers of the star, which are thereby heated more than they had been by energy from hydrogen core fusion. Because heated gas expands, this increase in internal temperature causes the entire star to swell. The details involved in this transition to the giant phase are still being explored.

INSIGHT INTO SCIENCE

More Than the Sum of Its Parts We have gotten to the point in science where some computer models that numerically solve complex sets of equations can accurately reproduce phenomena without fully explaining them. For example, computer models of the equations that describe how stars evolve show them expanding into the giant phase. This gives us some confidence that the equations are meaningful representations of reality. However, the models are so complex that astrophysicists are unable to identify the specific physical effects that take place that cause the stars to expand.

Far from the fusing shell, the surface gases cool and the temperature of the star's bloated surface decreases between 3000 and 6000 K, depending on the star's total mass. In about 5 billion years, our Sun will have a helium core and will swell into a giant with a radius of about ½ AU, vaporizing Mercury and perhaps causing Venus to spiral into the Sun. Earth will be scorched to a cinder.

Giant stars are so enormous that their outer layers constantly leak gases into space. At times, this mass loss is significant (Figure 7-22). When studied spectroscopically, the escaping gases exhibit narrow absorption lines, and the lines from gases coming toward us are slightly blueshifted, due to the Doppler effect. This small shift corresponds to a speed of 10 km/s, typical of the expansion velocities with which gases leave the tenuous outer layers of giants. A typical rate of mass loss for a giant is roughly 10^{-7} $M_{\odot}$ per year. For comparison, in a main-sequence star, such as the Sun, mass loss rates are only around 10^{-14} $M_{\odot}$ per year.

Although the surface of a giant is cooler than that of the main-sequence star from which it evolved, the giant is more luminous: It can emit more photons each second because it has so much more surface area. As a full-fledged giant (Figure 7-23a), our Sun will shine 2000 times more brightly than it does today.

> Where in the Sun will the energy be generated to expand it into the giant phase?

R I V U X G

FIGURE 7-22 A Mass-Loss Star A red giant star is shedding its outer layers, thereby creating this reflection nebula, labeled IC 2220 and called *Toby Jug*, located in the constellation Carina. The star is embedded inside the nebula and is not visible in this image. (Anglo-Australian Observatory)

7-10 Helium fusion begins at the center of a giant

When a star first becomes a giant, its hydrogen-fusing shell surrounds a small, compact core a few times the diameter of Earth, composed of almost pure helium. At first, no thermonuclear reactions occur in the helium-rich core of a giant because the temperature there is too low to fuse helium nuclei. The hydrogen-fusing shell creates more helium, which settles into the core. The eventual transformation of the core follows two dramatically different routes, depending on the star's total mass. The lower-mass giants, from 0.4 $M_{\odot}$ to about 2 $M_{\odot}$ go down an exceptionally bizarre path.

0.4 to 2 $M_{\odot}$—Helium Flash The helium atoms in cores of young giant stars with masses between 0.4 and about 2 $M_{\odot}$ are squeezed by the gravitational force into a crystal-like solid. At the pressures they are under, these atoms are completely ionized, separating into nuclei and electrons. The electrons, distributed between the nuclei, are so closely crowded together that they come under the influence of a phenomenon, called the **Pauli exclusion principle**, first formulated in 1925 by the Austrian physicist Wolfgang Pauli. According to this principle, nature does not allow two identical particles to exist in the same place at the same time. As the electrons are pressed closer and closer together, the exclusion principle prevents them from freezing in place between the

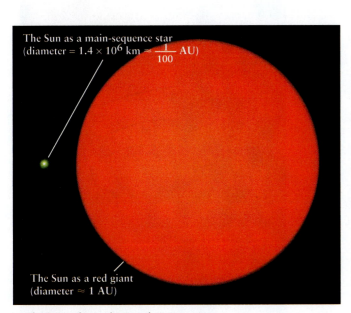

The Sun as a main-sequence star
(diameter = 1.4×10^6 km ≈ $\frac{1}{100}$ AU)

The Sun as a red giant
(diameter ≈ 1 AU)

a The Sun today and as a red giant

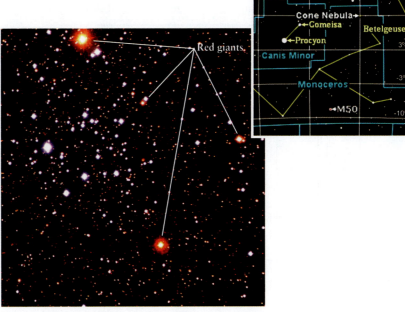

Red giants

Cone Nebula
Comeisa   Betelgeuse
Procyon
Canis Minor
Monoceros
M50

b Red giant stars in the star cluster M50

R I V U X G

FIGURE 7-23 The Sun Today and as a Giant
(a) In about 5 billion years, when the Sun
expands to become a giant, its diameter will
increase a hundredfold from what it is now, while its core
becomes more compact. Today, the Sun's energy is produced in a hydrogen-fusing core whose diameter is about
200,000 km. When the Sun becomes a giant, it will draw
its energy from a hydrogen-fusing shell that surrounds a
compact helium-rich core. The helium core will have a diameter of only 30,000 km. The Sun's diameter will be about
100 times larger, and it will be about 2000 times more luminous as a giant than it is today. (b) This composite of visible
and infrared images shows red giant stars in the open
cluster M50 in the constellation of Monoceros (the Unicorn).
(T. Credner and S. Kohle, Astronomical Institutes of the University
of Bonn)

nuclei. Instead, many of them must vibrate faster and
faster so that they do not become "identical," here
meaning being in the same place and moving with the
same speeds as adjacent electrons.

Astronomers call the electrons in this situation
degenerate. Degeneracy therefore keeps the electrons
apart, provides a pressure in the core, and prevents the
core from collapsing. We say that the helium-rich core
of a low-mass giant is supported by **electron degeneracy pressure**. This state provides a greater outward pressure than did the normal pressure in the star before
degeneracy developed in its core. The solid core just
grows from in-falling helium and heats up under the
gravitational influence of the star's mass.

Here is the weirdest part. The equations predict that,
unlike the pressure of an ordinary gas, the pressure inside
a degenerate core does not change with temperature. In a
normal gas, such as the air we breathe, when the temperature increases, the pressure also increases, causing the
gas to expand and cool. When a degenerate core's temperature increases the pressure there does not increase.
Without the "safety valve" of increasing pressure, the
star's core cannot expand and cool when overheated.

Eventually, the core temperature reaches about 100
million K, at which time **core helium fusion** begins (see
Figure 7-21c). This fusion creates energy, so the core's
temperature increases but in its degenerate state, the
core's pressure is unable to increase. Therefore, the core
does not immediately expand, and for several hours
both the core temperature and the fusion rate increase
dramatically. This situation is called the **helium flash**.

When the core temperature reaches around 3.5×10^8 K, the degenerate helium is forced to transform back
into an ordinary gas. Suddenly, the usual safety valve
operates, namely that the high temperature increases the
core pressure, causing the core to expand and cool, mod-
erating the rate of fusion occurring there within a few
hours. The expanding core, in turn, enlarges and cools
the hydrogen-fusing shell around it. Fusion in the shell
therefore decreases, so the flow of photons from the shell
to the outer layers of the star drops, causing the outer
regions to cool. As they cool, they contract—the star
shrinks until it reaches a new hydrostatic equilibrium.

Keep in mind that the helium flash occurs in and
acts upon the star's core. It does not cause a sudden
flash in brightness. A 0.4-to-2 $M_\odot$ helium-fusing giant

a

b

FIGURE 7-24 Post–Main-Sequence Evolution (a) The luminosity of the Sun changes as our star evolves. It began as a protostar with decreasing luminosity. On the main sequence today, it gradually brightens. Giant phase evolution occurs more rapidly, with faster and larger changes of luminosity. Note the change in scale of the horizontal axis at 12 billion years. (b) Model-based evolutionary tracks of five stars are shown on this H-R diagram. In the high-mass stars, core helium fusion ignites smoothly where the evolutionary tracks make a sharp turn upward into the giant region of the diagram.

is left smaller, dimmer, and hotter than it was before it began fusing helium. Figure 7-24a shows the expected changing brightness of the Sun when going through this process. Figure 7-24b includes evolutionary tracks of two stars in this mass range as they leave the main sequence, move into the giant phase, go through the helium flash, and settle back down. We will follow their evolution through the rest of the giant phase after briefly exploring how stars with more than 2 $M_\odot$ proceed off the main sequence.

More than 2 $M_\odot$—No Helium Flash The pressure created by the gravitational force of a star with more than 2 solar masses is enough to smoothly compress and heat its core until it reaches the fusion temperature of helium. In other words, the cores of giants in this mass range never become degenerate. When an ordinary gas, such as the gas that exists in the cores of these stars, is compressed, it heats up; when it expands, it cools down. If the energy production created by helium fusion overheats its core, the core expands, cooling its gases and slowing the rate of thermonuclear reactions. Conversely, if too little energy is being created to support the star's overlying layers, they move inward, compressing the core. The resulting increase in temperature speeds up the thermonuclear reactions and thus increases the energy output, which stops the contraction. Either way, the "safety valve" keeps it from collapsing or exploding.

Figure 7-24b shows the evolution off the main sequence of three stars in this mass range, which do not undergo helium flash. These stars move up and to the right across the giant region of the H-R diagram, getting larger until core fusion begins, whereupon they contract slightly—less than stars that undergo the helium flash.

7-11 Life in the giant phase has its ups and downs

The dominant process for fusing helium in the cores of giants is called the *triple-alpha process*, in which three helium atoms fuse to become a carbon atom. (Early in the twentieth century, particles emitted by radioactive nuclei were called alpha particles. They were later identified as helium nuclei.) The triple alpha process works in two steps. First, two helium nuclei fuse to create beryllium. Within 10^{-8} s, a third helium nucleus collides with the short-lived beryllium to create carbon.

Will the Sun undergo a helium flash? Why or why not?

This process also releases energy and can be summarized as follows:

$$^4He + {}^4He + {}^4He \rightarrow {}^{12}C + \gamma,$$

where γ denotes energy emitted as photons. The fusion in giants does not stop there. Some of the carbon created can then fuse with another helium nucleus to produce oxygen:

$$^{12}C + {}^4He \rightarrow {}^{16}O + \gamma$$

As in main-sequence stars, energy is released in the form of gamma rays. Giant stars fuse helium in their cores for about 10% as long as they spend fusing hydrogen while on the main sequence. For example, the Sun will spend a total of about 10 billion years on the main sequence, after which it will be a giant for about 1 billion years. While helium fusion is occurring in a giant's core, further gravitational contraction of the star's core ceases.

Despite the fact that all stars spend most of their lives on the main sequence, most of the stars visible to the naked eye are giants! This occurs because the relatively few giants in our neighborhood are so luminous that they outshine their neighbors in the night sky.

VARIABLE STARS

After core helium fusion begins, the evolutionary tracks of mature stars move part way back toward the main sequence (right to left on an H-R diagram; Figure 7-24). Their pressure-temperature safety valves are not perfect, so as they shrink, they overheat and core fusion causes them to expand to larger sizes than they would be at equilibrium. Becoming too large, their pressure and temperature decrease and they collapse, overcompressing themselves and overheating themselves again. This is analogous to dropping a tennis ball, which bounces repeatedly after striking the ground. In other words, a star under these conditions becomes unstable and pulsates. The region on the H-R diagram between the main sequence and the right side of the giant branch, where this is going on, is called the **instability strip** (Figure 7-25). These so-called **variable stars** can be easily identified by their changes in brightness amid a field of stars of constant luminosity.

Lower-mass, post–helium-flash stars pass through the lower end of the instability strip as they move horizontally on the H-R diagram along their evolutionary tracks. These stars become **RR Lyrae variables**, named after the prototype in the constellation of Lyra (the Lyre). RR Lyrae variables all have periods shorter than 1 day. Higher-mass stars pass back and forth through the upper end of the instability strip on the H-R diagram. These stars become Cepheid variables, often simply called *Cepheids*.

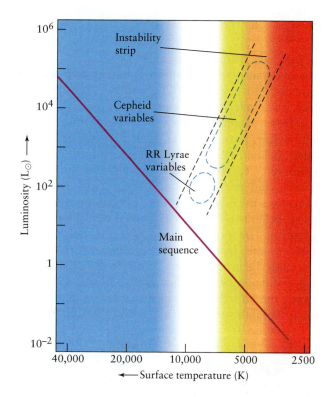

FIGURE 7-25 The Instability Strip The instability strip occupies a region between the main sequence and the giant branch on the H-R diagram. A star passing through this region along its evolutionary track becomes unstable and pulsates.

7-12 A Cepheid pulsates because it is alternately expanding and contracting

How many helium atoms does it take to make one oxygen atom?

A **Cepheid variable** is characterized by the way in which its light output varies—rapid brightening followed by gradual dimming. A Cepheid variable brightens and fades because the star's outer layers cyclically expand and contract. Lines in the spectrum of δ (delta) Cephei shift back and forth with the same 5.4-day period that characterizes its variations in magnitude. According to the Doppler effect measured from it, these shifts mean that the star's surface is alternately approaching and receding from us.

When a Cepheid variable pulsates, its gases alternately heat up and cool down; the surface temperature cycles between about 6300 K and 5000 K. Thus, the characteristic light curve (luminosity versus time curve, as in Figure 6-13) of a Cepheid variable results from changes in both size and surface temperature.

Just as a bouncing ball eventually comes to rest, a pulsating star would soon stop pulsating without some mechanism to maintain its oscillations. The means by which a variable star keeps going is to provide an extra push from inside compared to stars that are stable. Variable stars have a layer of gas rich in partially ionized helium (helium with one electron stripped off). This gas, absorbing a lot of photons leaving the core, heats, expands, and pushes the outer layers outward, just as steam raises the lid of a pot (Figure 7-26). Eventually, the ionized helium layer is spread so thin that photons pass through it. The ionized helium cools and contracts, and the star's outer layers collapse, compressing the interior gases until the process repeats itself. In our analogy with the pot of boiling water, when enough of the steam inside escapes, the lid falls back and the steam inside builds up again. Variability only occurs when the conditions of temperature and pressure inside the star are appropriate. As a result, giants are variable only when the stars are in the instability strip of an H-R diagram.

> **How do astronomers observe that Cepheids are changing size?**

7-13 Cepheids enable astronomers to estimate vast distances

Cepheids are important to astronomers because there is a direct relationship between a Cephid period of pulsation and its average luminosity. This correlation is called, appropriately enough, the **period-luminosity relation**. We can determine the distance to a Cepheid in four steps. First, we observe its period. Second, we use the period-luminosity relation to determine its luminosity and, hence, its absolute magnitude (recall that the luminosity and absolute magnitudes are directly related to each other). Third, we observe its apparent magnitude. Finally, we use the distance-magnitude relationship in *An Astronomer's Toolbox 6-3* to calculate its distance.

Dim Cepheid variables pulsate rapidly, with periods of 1 to 2 days, and have average brightness variations of a few hundred times our Sun's luminosity. The most luminous Cepheids have the longest periods of all Cepheids, with variations occurring over 100 days and average brightness variations equaling 10,000 $L_\odot$. Being so bright, we can see Cepheids far beyond the boundaries of our Milky Way Galaxy. Because the changes in brightness of Cepheids can be seen at distances where other techniques for measuring distance, such as stellar parallax, fail, the period-luminosity relation plays an important role in determining the overall size and structure of the universe. We will explore this application further in Chapter 11.

The details of a Cepheid's pulsation depend on the abundance of heavy elements in its atmosphere. The average luminosity of metal-rich Cepheids is roughly 4 times greater than the average luminosity of metal-poor Cepheids that have the same period. Thus, there are two classes: **Type I Cepheids** (also called δ *Cephei stars*), which are the brighter, metal-rich stars, and **Type II Cepheids** (also called *W Virginis stars*), which are the dimmer, metal-poor stars. The period-luminosity relations for both types of variables are shown in Figure 7-27.

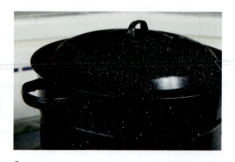

a   b   c

R I V U X G

d

FIGURE 7-26 Analogy for Cepheid Variability (a) As pressure builds up in this pot, the force on the lid (Cepheid's outer layers) increases. (b) When the pressure inside the pot is sufficient, it lifts the lid off (expands the star's outer layers) and thereby allows some of the energy inside to escape. This process cycles (two are shown), as do the luminosity and temperature of Cepheid stars. (a–d: Janet Horton)

Firm Foundations? To determine things far from our everyday size and time scales, scientists often must base their results on the predictions of complex theories, even before all of the testing and refinement of the theories are complete. An error anywhere along the chain of ideas leads to incorrect results. For example, using the Cepheid variable stars to measure distances requires combining several intermediate concepts, including believing that the period-luminosity relationship applies well to all Cepheids, believing that there are just two different types of Cepheids, and assuming that the peak luminosity depends only on the Cepheid's distance from us (not correct if there are gas and dust between us and the star, as there often are). If any of these or myriad other assumptions is wrong, the calculated distances will be incorrect.

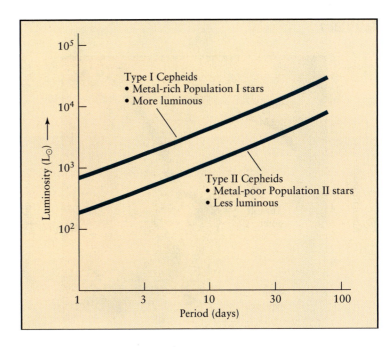

FIGURE 7-27 The Period-Luminosity Relation The period of a Cepheid variable is directly related to its average luminosity: The more luminous the Cepheid, the longer its period and the slower its pulsations. Type I Cepheids (δ Cephei stars) are metal-rich stars. They are brighter than the Type II Cepheids (W Virginis stars), which are metal-poor stars. (Adapted from H. C. Arp)

7-14 Globular clusters are bound groups of old stars

Many giant stars are found in groups, called **globular clusters**, so named because of their spherical shapes. A typical globular cluster, like the one shown in Figure 7-28, may contain up to a million stars in a volume 200 light-years across. Like open clusters, the stars in globular clusters all formed at about the same time. Unlike open clusters of young stars, however, globular clusters are gravitationally bound groups of old stars from which few stars escape. Globular clusters in our Galaxy (other galaxies have them, too) are distributed in a sphere around the center of the Milky Way.

Astronomers know that globular clusters are old because they contain no high-mass main-sequence stars. If you measure the luminosity and surface temperature of many stars in a globular cluster and plot the data on a color-magnitude diagram (Figure 7-29), you will find that the upper half of the main sequence is missing. All of the high-mass main-sequence stars evolved long ago

into giants, leaving behind only lower-mass, slowly evolving stars still undergoing core hydrogen fusion.

The H-R diagram of a globular cluster typically shows a horizontal grouping of stars to the left of the center portion of the diagram (see Figure 7-29). These hot stars, called **horizontal branch stars**, undergo helium fusion in their cores, and they have luminosities of about 50 L⊙ (see Figure 7-29). Eventually, these stars will move back toward the giant region as core helium fusion and hydrogen shell fusion devour their fuel.

> What are typical periods of Cepheid variable stars?

Horizontal-branch stars

Red giants

FIGURE 7-28 A Globular Cluster This cluster, M10, is about 85 light-years across and is located in the constellation Ophiuchus (the Serpent Holder), roughly 16,000 light-years from Earth. Most of the stars here are either red giants or blue, horizontal-branch stars with both core helium fusion and hydrogen shell fusion. (T. Credner and S. Kohle, Astronomical Institutes of the University of Bonn)

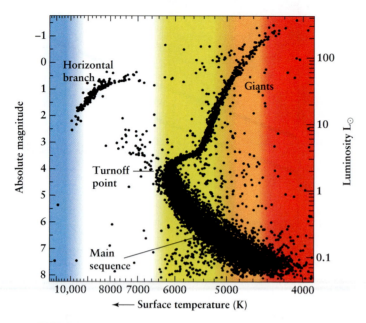

FIGURE 7-29 An H-R Diagram of a Globular Cluster Each dot on this graph represents the absolute magnitude and surface temperature of a star in the globular cluster M55. Note that the upper half of the main sequence is missing. The horizontal branch stars are stars that recently experienced the helium flash in their cores and now exhibit core helium fusion and hydrogen shell fusion.

As noted earlier, an H-R diagram of a cluster can also be used to determine the age of the cluster assuming that all of the stars in the cluster formed at the same time. In the H-R diagram for a young open cluster (review Figure 7-18b), almost all of the hot massive stars are still on the main sequence. As a cluster gets older, like the Pleiades, stars begin to leave the main sequence (see Figure 7-2c). The high-mass, high-luminosity stars are the first to become giants.

Over time, the main sequence in a cluster gets shorter and shorter. The top of the surviving portion of the main sequence is called the **turnoff point** (see Figure 7-29). Stars at the turnoff point are just beginning to exhaust the hydrogen in their cores. The theory of stellar evolution enables us to predict how long stars with different masses last on the main sequence before moving toward the giant stage. Therefore, the age of the stars currently at the turnoff point is the age of the cluster (recall Table 7-2).

Consider the plot for the globular cluster M55, shown in Figure 7-29. In M55, $0.8\text{-}M_\odot$ stars are just leaving the main sequence. Therefore, according to Table 7-2, the cluster's age is approximately 13.5 billion years.

Using data collected from the *Hipparcos* satellite, Figure 7-30a is a composite diagram of isolated and cluster stars, reflecting the overall structure of the H-R diagram. Using data from stellar evolution theory as

FIGURE 7-30 Structure of the H-R Diagram (a) Data taken by the Hipparcos satellite placed 41,453 stars more precisely on the H-R diagram than any previous observations. This figure shows the overall structure of the H-R diagram. The thickness of the main sequence is due in large part to stars of different ages turning off the main sequence at differ-

ent places, as shown in (b). (b) The black bands indicate where data from various star clusters fall on the H-R diagram. The ages of turnoff points (in years) are listed in red alongside the main sequence. The age of a cluster can be estimated from the location of the turnoff point, where the cluster's most massive stars are just now leaving the main sequence.

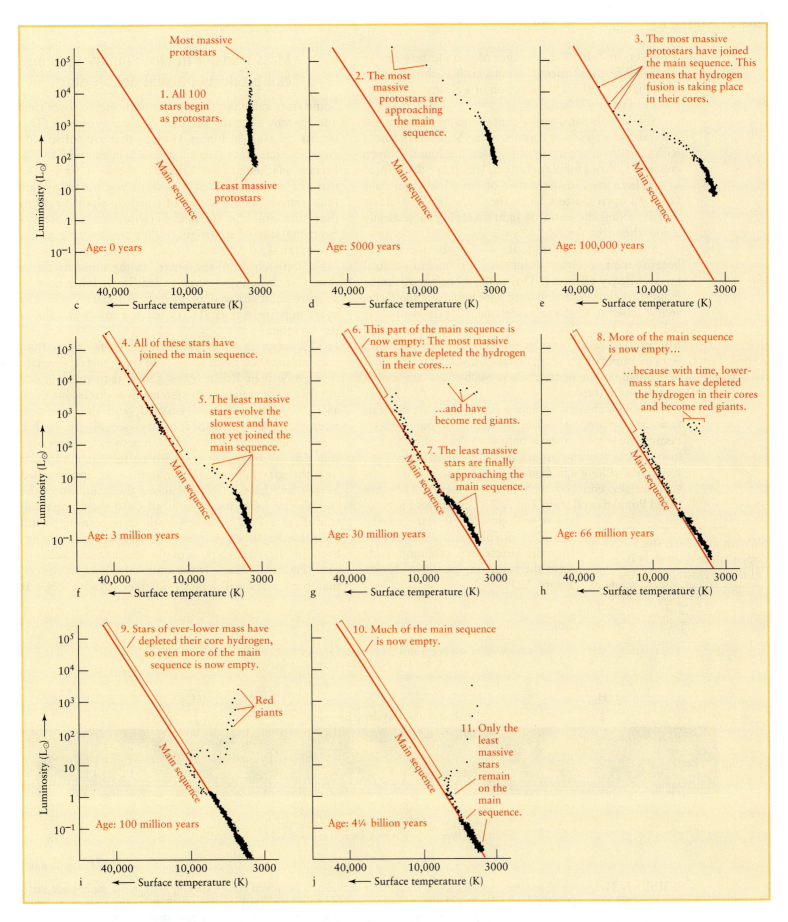

FIGURE 7-30 *(continued)*. (c–j) Summarizing the evolution of a theoretical cluster of 100 stars, as shown by their locations on the H-R diagram. (In principle, each star's evolution could be fol-lowed separately.) After a star passes through the red giant phase, it is deleted from the diagram.

presented in Table 7-2, the ages of star clusters can be estimated from the turnoff points, such as those depicted in Figure 7-30b. The evolution of a cluster is summarized in Figure 7-30c–j.

While globular clusters contain many of the oldest stars, the youngest star clusters in the Milky Way (those with their main sequences still intact) are found in open clusters. Unlike globular clusters, which are distributed in a sphere around the center of the Milky Way, our Galaxy's open clusters lie in the plane of the Galaxy, that is, along the band of light sweeping majestically across the night sky.

Stars in open clusters are said to be metal-rich, because their spectra contain many prominent spectral lines of heavy elements. This material originally came from stars that exploded long ago, enriching the interstellar gases with the heavy elements formed in their cores. The young, open clusters are therefore formed from the debris of older generations of stars. The Sun is an example of a young, metal-rich star that was likely formed in an open cluster. Such stars are called **Population I stars.**

Globular clusters, containing older stars, are generally located above or below the plane of our Galaxy. Because their spectra show only weak lines of heavy elements, these ancient stars are said to be metal-poor. They were created long ago from gases that had not yet been substantially enriched with heavy elements. They are called **Population II stars.** Figure 7-31 compares the spectra of a Population II star and the Sun. The difference in chemistry of the stars affects their evolution, such as in the variability of Cepheids (see Figure 7-27).

> Which are older, stars in open clusters or stars in globular clusters?

7-15 Mass transfer in close binary systems can produce unusual double stars

Some stars substantially change their masses over time and thereby alter the patterns of their evolution. This occurs for stars in binary systems that are sufficiently close to each other (see Chapter 6). In the mid-1800s, French mathematician Édouard Roche pointed out that the atmospheres of two stars in a binary system must remain within a pair of teardrop-shaped regions that surround the stars. Otherwise, the gas escapes from the star of its origin, either transferring to the other star, where the teardrops make contact, or escaping completely from the binary in the opposite direction. Cut through, these so-called **Roche lobes** take on a figure-eight shape (*dashed lines* in Figure 7-32). The more massive star is always located inside the larger Roche lobe.

The stars in many binaries are so far apart that, even during their giant stages, the stars' surfaces remain well inside their Roche lobes. Other than orbiting one another, each star in such systems lives out its life cycle as if it were single and isolated, and the system is referred to as a **detached binary** (see Figure 7-32a). If two stars are relatively close together, however, one star may fill or overflow its Roche lobe as it expands into the giant phase. This system is a **semidetached binary**, and gases then flow across the point where the two Roche lobes touch and fall onto the companion star (Figure 7-32b). When both stars completely fill their Roche lobes, the system is a **contact binary**, because the two stars actually touch and exchange gas (Figure 7-32c). It is quite unlikely, however, that both stars will exactly fill their Roche lobes at

The spectrum of this Population II star shows absorption lines of hydrogen (such as H_γ and H_δ) but only very weak absorption lines of metals; such a star is **metal-poor.**

H_δ Increasing wavelength ⟶ H_γ

a

b

The spectrum of this Population I star has stronger absorption lines of metals; such a star is **metal-rich.**

R I V U X G

FIGURE 7-31 Spectra of a Metal-Poor and a Metal-Rich Star These spectra compare (a) a metal-poor (Population II) and (b) a metal-rich (Population I) star (the Sun) of the same surface temperature. Numerous spectral lines prominent in the solar spectrum are caused by elements heavier than hydrogen and helium. Note that corresponding lines in the metal-poor star's spectrum are weak or absent. Both spectra cover a wavelength range that includes two strong hydrogen absorption lines, labeled H_γ (410 nm) and H_δ (434 nm). (Lick Observatory)

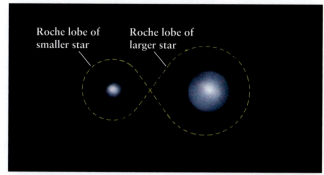

a Detached binary: Neither star fills its Roche lobe.

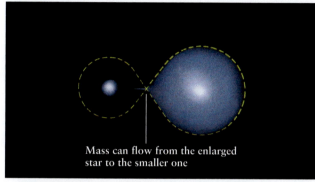

Mass can flow from the enlarged
star to the smaller one

b Semidetached binary: One star fills its Roche lobe.

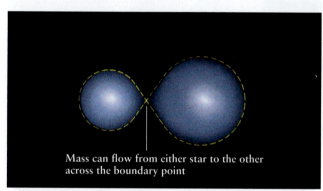

Mass can flow from either star to the other
across the boundary point

c Contact binary: Both stars fill their Roche lobes.

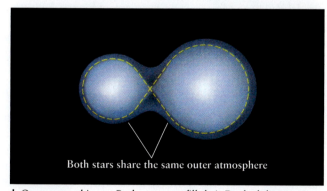

Both stars share the same outer atmosphere

d Overcontact binary: Both stars overfill their Roche lobes.

WEB LINK 7.12 **FIGURE 7-32** Detached, Semidetached, Contact, and Overcontact Binaries (a) In a detached binary, neither star fills its Roche lobe. (b) If one star fills its Roche lobe, the binary is semidetached. Mass transfer is often observed in semidetached binaries. (c) In a contact binary, both stars fill their Roche lobes. (d) The two stars in an overcontact binary both overfill their Roche lobes. The two stars actually share the same outer atmosphere.

the same time. It is more likely that they overflow their lobes, giving rise to a common atmospheric envelope. Such a system is an **overcontact binary** (Figure 7-32d).

Semidetached and contact binaries are easiest to detect if they are also eclipsing binaries (recall Figure 6-13). Their light curves have a distinctly rounded appearance, caused by these tidally distorted egg-shaped stars. The eclipsing binary β (beta) Persei, commonly called Algol (from the Arabic term for "demon"), is a semidetached binary that can easily be seen with the naked eye in the constellation Perseus. From Algol's light curve (Figure 7-33a), astronomers have determined that the binary contains a star that fills its Roche lobe. Sometime in the past, as it expanded and became a giant, this star dumped a significant amount of gas onto its companion.

Mass transfer is still occurring in a semidetached eclipsing binary, called β Lyrae, in the constellation Lyra. Like β Persei, β Lyrae contains a giant that fills its Roche lobe (Figure 7-33b). For many years, astronomers were puzzled that the detached companion star in β Lyrae is severely underluminous, contributing virtually no light at all to the visible radiation coming from the system. Furthermore, the spectrum of β Lyrae contains unusual features, some of which are consistent with gas flowing between the stars and around the system as a whole.

The β Lyrae system was explained in 1963, when Su-Shu Huang proposed that the underluminous star in β Lyrae is enveloped in a huge **accretion disk** of gas captured from its bloated companion. The disk is so large and thick that it completely shrouds the secondary star, making it impossible to observe at visible wavelengths. The primary star is overflowing its Roche lobe, with gases streaming onto the disk at the rate of 1 $M_\odot$ per hundred thousand years.

The fate of a semidetached system like β Persei or β Lyrae depends primarily on how fast its stars evolve. If the detached star expands to fill its Roche lobe while the companion star fills its own Roche lobe, then the result is an overcontact binary. An example is W Ursae Majoris, in which two stars share the same photosphere (Figure 7-33c).

Theories of the evolution of binary systems that transfer mass reveal that some remarkable transformations are possible, such as the case of φ (phi) Persei. As shown in Figure 7-34, the more massive star in a binary can actually be transformed into the lower-mass star, and vice versa! When stars in such systems gain mass, their increased gravitational force

In which type of binary system are the two stars most actively exchanging mass?

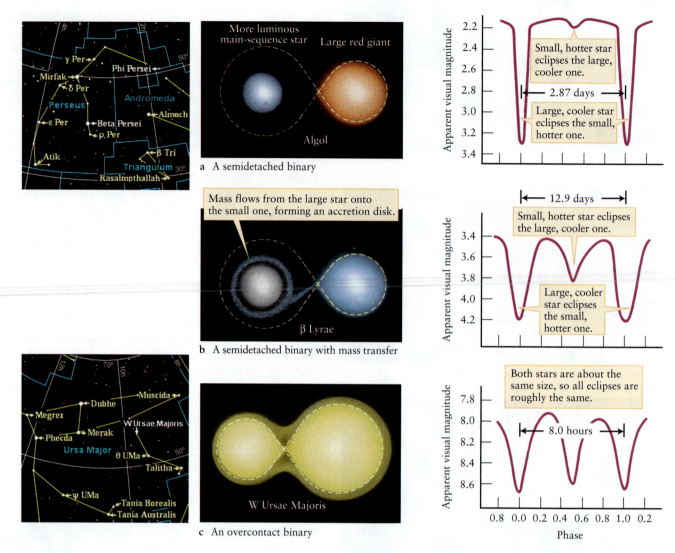

FIGURE 7-33 Three Close Binaries Sketches of and light curves for three eclipsing binaries are shown. The phase denotes the fraction of the orbital period from one primary minimum to the next. (a) Algol, also known as β Persei, is a semidetached binary. The deep eclipse occurs when the giant star *(right)* blocks the light from the smaller, but more luminous, main-sequence star. (b) β Lyrae is a semidetached binary in which mass transfer has produced an accretion disk that surrounds the detached star. This disk is so thick and opaque that it renders the secondary star almost invisible. (c) W Ursae Majoris is an overcontact binary. Both stars therefore share their outer atmospheres. The short, 8-h period of this binary indicates that the stars are very close to each other.

increases their rate of fusion, which increases the rate at which they evolve. Conversely, the stars that lose mass begin evolving more slowly than before their mass loss. In Chapters 8 and 9 we will see that mass transfer produces some of the most extraordinary objects in the sky.

7-16 Frontiers yet to be discovered

Stars, whether isolated, in pairs, or in much larger groups, are our primary source of information about the universe. Details of each step of stellar evolution remain to be explored; the pre–main-sequence evolution of stars is being revealed in sources like the Eagle Nebula (see Figure 7-15). These observations provide data with which to compare the theories of early star formation. Over the past two decades, observations have revealed that more stars lose mass in the form of stellar winds while on the main sequence than previously expected. Consequently, we have a lot to learn about why such stars shed mass.

As noted, we still need to understand the details of why stars swell into the red giant phase. Also, the mechanisms that cause stars to be variable are still under investigation.

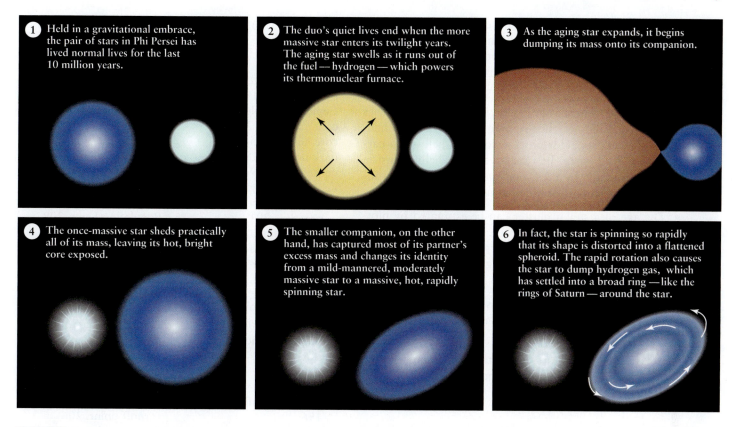

1. Held in a gravitational embrace, the pair of stars in *Phi Persei* has lived normal lives for the last 10 million years.

2. The duo's quiet lives end when the more massive star enters its twilight years. The aging star swells as it runs out of the fuel — hydrogen — which powers its thermonuclear furnace.

3. As the aging star expands, it begins dumping its mass onto its companion.

4. The once-massive star sheds practically all of its mass, leaving its hot, bright core exposed.

5. The smaller companion, on the other hand, has captured most of its partner's excess mass and changes its identity from a mild-mannered, moderately massive star to a massive, hot, rapidly spinning star.

6. In fact, the star is spinning so rapidly that its shape is distorted into a flattened spheroid. The rapid rotation also causes the star to dump hydrogen gas, which has settled into a broad ring — like the rings of Saturn — around the star.

FIGURE 7-34 Mass Exchange Between Close Binary Stars This sequence of drawings shows how close binary stars can initially be isolated but, as they age, grow and exchange mass. Such mass exchange leads to different fates than if the same stars had evolved in isolation.

SUMMARY OF KEY IDEAS

Protostars and Pre–Main-Sequence Stars

• Enormous, cold clouds of gas and dust, called giant molecular clouds, are scattered about the disk of the Galaxy.

• Star formation begins when gravitational attraction causes clumps of gas and dust, called protostars, to coalesce in Bok globules within a giant molecular cloud. As a protostar contracts, its matter begins to heat and glow. When the contraction slows down, the protostar becomes a pre–main-sequence star. When the pre–main-sequence star's core temperature becomes high enough to begin hydrogen fusion and stop contracting, it becomes a main-sequence star.

• The most massive pre–main-sequence stars take the shortest time to become main-sequence stars (O and B stars).

• In the final stages of pre–main-sequence contraction, when hydrogen fusion is about to begin in the core, the pre–main-sequence star may undergo vigorous chromospheric activity that ejects large amounts of matter into space. G, K, and M stars at this stage are called T Tauri stars.

• A collection of a few hundred or a few thousand newborn stars formed in the plane of the Galaxy is called an open cluster. Stars escape from open clusters, most of which eventually dissipate.

Main-Sequence and Giant Stars

• The Sun has been a main-sequence star for 4.6 billion years and should remain so for about another 5 billion years. Less massive stars than the Sun evolve more slowly and have longer main-sequence lifetimes. More massive stars than the Sun evolve more rapidly and have shorter main-sequence lifetimes.

• Main-sequence stars with mass between 0.08 and $0.4 M_\odot$ convert all of their mass into helium and then stop fusing. Their lifetimes last hundreds of billions of years, so none of these stars has yet left the main sequence.

• Core hydrogen fusion ceases when hydrogen in the core of a main-sequence star with $M > 0.4 M_\odot$ is gone, leaving a core of nearly pure helium surrounded by a shell where hydrogen fusion continues. Hydrogen shell fusion adds more helium to the star's core, which contracts and becomes hotter. The outer atmosphere expands considerably, and the star becomes a giant.

• When the central temperature of a giant reaches about 100 million K, the thermonuclear process of helium fusion begins. This process converts helium to carbon, then to oxygen. In a massive giant, helium fusion begins gradually. In a less massive giant, it begins suddenly in a process called the helium flash.

• The age of a stellar cluster can be estimated by plotting its stars on an H-R diagram. The upper portion of the main sequence disappears first, because more massive main-sequence stars become giants before low-mass stars do.

• Giants undergo extensive mass loss, sometimes producing shells of ejected material that surround the entire star.

• Relatively young stars are metal-rich (Population I); ancient stars are metal-poor (Population II).

Variable Stars

• When a star's evolutionary track carries it through a region, called the instability strip in the H-R diagram, the star becomes unstable and begins to pulsate.

• RR Lyrae variables are low-mass, pulsating variables with short periods. Cepheid variables are high-mass, pulsating variables exhibiting a regular relationship between the period of pulsation and luminosity.

• Mass can be transferred from one star to another in close binary systems. When this occurs, the evolution of the two stars changes.

WHAT DID YOU THINK?

1   *How do stars form?* Stars form from the mutual gravitational attraction between gas and dust inside giant molecular clouds.

2   *Are stars still forming today? If so, where?* Yes. Astronomers have seen stars that have just arrived on the main sequence, as well as infrared images of gas and dust clouds in the process of forming stars. This star formation occurs in interstellar clouds of gas and dust.

3   *Do more massive stars shine longer than less massive ones? How is this reasoned?* No. Lower-mass stars last longer because the lower gravitational force inside them causes fusion to take place at slower rates compared to the fusion inside higher-mass stars.

Key Terms for Review

accretion disk, 209
birth line, 189
Bok globule, 185
brown dwarf, 190
Cepheid variable, 203
contact binary, 208
core helium fusion, 201
dark nebulae, 183
dense core, 186
detached binary, 208
electron degeneracy
    pressure, 201
emission nebula, 183
evolutionary track, 189
giant molecular
    cloud, 183
globular cluster, 205
H II regions, 192
helium flash, 201
horizontal branch
    star, 205
hydrogen shell
    fusion, 199
hydrostatic equilibrium,
    196
instability strip, 203
interstellar
    extinction, 183
interstellar medium, 182
interstellar reddening, 183

Jeans instability, 186
molecular cloud, 183
nebula (*plural* nebulae),
    183
OB association, 193
open cluster, 186
overcontact binary, 209
Pauli exclusion
    principle, 200
period-luminosity
    relation, 204
Population I star, 208
Population II star, 208
pre–main-sequence
    star, 188
protostar, 187
red dwarf, 198
reflection nebula, 183
Roche lobe, 208
RR Lyrae variable, 203
semidetached binary,
    208
supernova remnant, 185
T Tauri stars, 190
turnoff point, 206
Type I Cepheid, 204
Type II Cepheid, 204
variable stars, 203
zero-age main sequence
    (ZAMS), 196

Review Questions

1. Consider a star behind a cloud of interstellar gas and dust as seen from our perspective. Which of the following would you see? **a.** The star appears brighter than it would if the cloud were not present, **b.** The star appears to be moving toward us, **c.** The star appears redder than it would if the cloud were not present, **d.** The star would be invisible at all wavelengths, **e.** The star would always appear green.

2. What is the lowest mass that a star can have on the main sequence? **a.** There is no lower limit, **b.** 0.003 $M_\odot$ **c.** 0.08 $M_\odot$, **d.** 0.4 $M_\odot$ **e.** 2.0 $M_\odot$

3. What is the source of energy that enables a main-sequence star to shine? **a.** Friction between its atoms, **b.** Fusion of hydrogen in a shell that surrounds the core, **c.** Fusion of helium in its core, **d.** Fusion of hydrogen in its core **e.** Burning of gases on its surface

4. What are giant molecular clouds, and what role do they play in star formation?

5. Why are low temperatures necessary for dense cores to form and contract into protostars?

6. Why do thermonuclear reactions not occur on the surface of a main-sequence star?

7. What is an evolutionary track, and how do such tracks help us interpret the H-R diagram?

8. Draw the pre–main-sequence evolutionary track of the Sun on an H-R diagram. Briefly describe what was occurring throughout the solar system at various stages along this track. (You may find it useful to review Chapter 6.)

9. On what grounds are astronomers able to say that the Sun has about 5 billion years remaining in its main-sequence stage?

10. What will happen inside the Sun 5 billion years from now when it begins to evolve into a giant?

11. How is the evolution of a main-sequence star with less than 0.4 $M_\odot$ fundamentally different from that of a main-sequence star with more than 0.4 $M_\odot$?

12. Draw the post–main-sequence evolutionary track of the Sun on an H-R diagram up to the point when the Sun becomes a helium-fusing giant. Briefly describe what might occur throughout the solar system as the Sun undergoes this transition.

13. What does it mean when an astronomer says that a star "moves" from one place to another on an H-R diagram?

14. What is the helium flash and what causes it?

15. Explain how and why the turnoff point on the H-R diagram of a cluster is related to the cluster's age.

16. Why do astronomers believe that globular clusters are made of old stars?

17. What are Cepheid variables, and how are they related to the instability strip?

18. What occurs in Cepheid stars that is analogous to the vapor raising the lid on a pot of boiling water?

19. What are RR Lyrae variables, and how are they related to the instability strip?

20. What is a Roche lobe, and what is its significance in close binary systems?

21. What are the differences between detached, semidetached, contact, and overcontact binaries?

Advanced Questions

The answers to all computational problems, which are preceded by an asterisk (*), appear at the end of the book.

22. How is a degenerate gas different from an ordinary gas?

23. If you took a spectrum of a reflection nebula, would you see absorption lines, emission lines, or no lines? Explain your answer.

24. Why is it useful to plot the *apparent* magnitudes of stars in a single cluster on an H-R diagram?

25. What might happen to the massive outer planets when the Sun becomes a giant?

*26. How many 1.5-$M_\odot$ main-sequence stars would it take to equal the luminosity of one 15-$M_\odot$ star?

*27. How many times longer does a 1.5-$M_\odot$ star fuse hydrogen in its core than does a 15-$M_\odot$ star?

28. Why does a shock wave from a supernova produce relatively few high-mass O and B stars compared to the number of low-mass A, F, G, K, and M stars produced?

29. How would you distinguish a newly formed protostar from a giant, given that they occupy the same location on the H-R diagram?

30. What observational consequences would we find in H-R diagrams for star clusters as a result of the universe having a finite age? Could we use these consequences to establish constraints on the possible age of the universe? Explain.

Discussion Questions

31. Discuss the possibility of life-forms and biological processes occuring in giant molecular clouds. In what ways might conditions favor or hinder biological evolution?

32. Is there any evidence that Earth has ever passed through a star-forming region in space?

What If . . .

33. The solar system passed through a giant molecular cloud? How would this encounter affect Earth and life on it?

34. The Sun began entering its giant phase today? What would happen to Earth and to life on it?

35. Earth were orbiting a 0.5-$M_\odot$ star at a distance of 1 AU? What would be different for Earth and life on it? What effects would moving Earth closer to the lower-mass Sun have?

36. The Sun were a variable star? How would this change life on Earth, and, assuming we could live in orbit around such a star, how might it change our perspective of the cosmos?

Web Questions

37. To test your understanding of where stars are formed, do Interactive Exercise 7-1 on the Web. You can print out your results, if required.

38. To test your understanding of variable stars, do Interactive Exercise 7-2 on the Web. You can print out your results, if required.

39. To test your understanding of close binary star systems, do Interactive Exercise 7-3 on the Web. You can print out your results, if required.

Observing Projects

40. Use a telescope to observe at least two of the following interstellar gas clouds: M42 (Orion), M43, M20 (Trifid), M8 (Lagoon), M17 (Omega). You can easily locate them with the aid of your *Starry Night Enthusiast*™ program, star charts published in such magazines as *Astronomy* and *Sky & Telescope*, or the list of coordinates below. In each case, can you identify the stars responsible for the ionizing radiation that causes the nebula to glow? Draw a picture of what you see through the telescope and compare it with a photograph of the object. Which portions of the nebula are not visible through your telescope? Why are they not visible?

Nebula	Right ascension	Declination
M42 (Orion)	5h 35.4m	−5° 27′
M43	5h 35.6m	−5° 16′
M20 (Trifid)	18h 02.6m	−23° 02′
M8 (Lagoon)	18h 03.8m	−24° 23′
M17 (Omega)	18h 20.8m	−16° 11′

41. Several star clusters can be seen quite well with a good pair of binoculars. You can easily locate many of them with the aid of your *Starry Night Enthusiast*™ program, star charts published during the summer months in such magazines as *Astronomy* and *Sky & Telescope*, or the list of coordinates below. Observe as many of these clusters as you can. Look at them through a telescope, if available. Note the overall distribution of stars in each cluster. Can you see any of these clusters with the naked eye? What difference do you note between binocular and telescopic images of individual clusters?

Star cluster	Right ascension	Declination
M45 (Pleiades)	3h 47.0m	+24° 07′
Hyades	4h 27.0m	+16° 00′
Praesepe	8h 40.1m	+19° 59′
Coma	12h 25.0m	+26° 00′
M11	18h 51.1m	−06° 16′

42. Use the *Starry Night Enthusiast*™ program to examine a star-forming region. Select **Favourites > Deep Space > Clusters and Nebulas > Nebulas in Sagittarius** from the menu. The view is centered on the constellation Sagittarius with several prominent clusters and nebulae labeled. Time is passing at 1 sidereal day per frame. You can temporarily stop motion by left-clicking (or clicking on a Mac) on the screen, as necessary. **a.** In which month is M20, the Trifid Nebula, highest in the sky at noon? Explain how you determined this. **b.** In which month is M20 highest in the sky at midnight, so that it is best placed for observing with a telescope? Explain how you determined this. **c.** Click the right mouse button (Ctrl+Click on a Mac) with the cursor over M20, the Trifid Nebula, and select **Magnify** from the Object Contextual menu. Explain why different areas of the nebula have the colors that they have.

WHAT IF... Earth Orbited a 1.5-$M_\odot$ Sun?

Earth is at a perfect distance from a wonderful star. The Sun provides just enough heat so that liquid water, necessary to sustain life, can exist on Earth. Would our planet still be suitable for the evolution of life if the Sun were 1.5 $M_\odot$ rather than 1.0 $M_\odot$? To begin with, we need to know how far from the new Sun, let's call it Sol II, to place Earth.

A Very Sunny Day Sol II's surface temperature would be 8400 K and it would appear blue-white in our sky. Sol II would give off 7 times as much energy per second as our present Sun, due to the combination of a higher temperature and a 20% larger radius than the Sun. The effect of Sol II's increased energy emission would require that Earth be located much farther away from Sol II than our present distance from the Sun. To understand why, consider the increase in infrared (heat) output from Sol II. That extra heat would have the initial effect of increasing the average global temperature on Earth at our present distance by about 10 K (about 20°F). This does not seem like a lot, but that slight increase would quickly boost the atmospheric temperature much higher.

The extra heat from Sol II would also cause more ocean water to evaporate into the atmosphere. Because water is a greenhouse gas (that is, it traps infrared radiation), the air temperature would increase, causing even more water to evaporate from the oceans, which, in turn, would cause the air to heat even more. This vicious cycle, called *runaway greenhouse effect*, would make Earth's surface so hot and dry that it would be uninhabitable.

By moving Earth about 2.6 times farther away from Sol II, the temperature would become suitable for life. At that distance, the year would be 1249 days long. While moving away from the heat is one thing, moving away from the ultraviolet radiation emitted by Sol II is something else altogether.

Ultraviolet Excess Just by increasing the Sun's mass by 50%, the ultraviolet radiation emitted would be several thousand times stronger. This is because the energy output of stars with different surface temperatures varies with wavelength. While the output of visible light would change only slightly, the output of ultraviolet radiation would be vastly greater. Therefore, even though Earth's surface temperature would be suitable farther from Sol II, the flood of ultraviolet radiation would be so strong that the ozone layer would be overwhelmed, and the level of ultraviolet radiation at Earth's surface would be much higher than it is today. This would be so even with the greater concentration of ozone created by the increased ultraviolet from Sol II.

Life would have to evolve greater protection from ultraviolet radiation than it has today. If this were not a great enough challenge, suppose intelligent life-forms were evolving on Earth and orbiting Sol II 4.6 billion years after the solar system formed. They would discover that their star had evolved so rapidly that it was just about to expand into the giant phase!

R I V U X G

(John Elk III/Bruce Coleman)

CHAPTER 8

The Deaths of Stars

Tycho's Supernova First seen as a visible light object by Tycho Brahe in 1572, this X-ray image of Tycho's supernova was taken in 2003 by the Chandra X-ray telescope. Multimillion-degree gas and dust (*shown in red and green*) are expanding outward at about 10 million km/h following a shell of high-energy electrons (*blue*). (NASA/CXC/Rutgers/J. Warren & J. Hughes et al.)

R I V U X G

WHAT DO YOU THINK?

1 Will the Sun someday cease to shine brightly? If so, how will this occur?

2 What is a nova? How does it differ from a supernova?

3 What are the origins of the carbon, silicon, oxygen, iron, uranium, and other heavy elements on Earth?

4 What are cosmic rays? Where do they come from?

5 What is a pulsar?

Answers to these questions appear in the text beside the corresponding numbers in the margins and at the end of the chapter.

Most video and computer games have one thing in common: Players move through different levels, each progressively more difficult than the last. The biggest challenge comes at the final level, when you have to face the most powerful or evil entity. Stellar evolution is similar, with stars passing through different stages of stellar activity before they can move on to the next stage. There is one big difference—a game player can win the final level, but a star at that stage of evolution always ceases to shine with the vigor that it had previously. In the end, stars eject vast quantities of gas and dust into interstellar space. In human terms, they die. In this chapter and in Chapter 9, we learn how the later stages of stellar evolution are significantly different for stars with different masses. Some of them stop evolving by relatively mild emissions of their outer layers, while others have spectacular finales.

In this chapter you will discover

- what happens to stars when core helium fusion ceases
- how heavy elements are created
- the characteristics of the end of stellar evolution
- why some stars go out relatively gently, and others go out with a bang
- the incredible density of the matter in neutron stars and how these objects are observed

LOW-MASS STARS AND PLANETARY NEBULAE

We have seen that red dwarf stars, with masses less than $0.4 M_\odot$, never get to the giant phase of stellar evolution. Rather, they convert all their mass to helium, stop fusing, and thereafter they just cool off. We will now explore the fates of stars greater than $0.4 M_\odot$ as they proceed along through the giant phase and beyond.

In Chapter 7, we discovered that when hydrogen fusion in the shell surrounding the core first begins, the energy from it causes a star to expand and become a giant. During this process, low-mass stars move over to,

and ascend, the giant branch on the H-R diagram for the first time (Figure 8-1a; also, review Figure 7-24b). As this happens, mass is expelled into space in the form of stellar winds, which reduce the masses of these stars. Then core helium fusion begins, with stars of less than about 2 $M_\odot$ undergoing a core helium flash in the giant stage (Figure 8-1a). Stars with more than 2 $M_\odot$ begin helium fusion more gradually. In either case, after core fusion begins again, stars shrink and move onto the *horizontal branch* (Figure 8-1b). Their cores are eventually converted into carbon and oxygen, helium fusion ceases, and these stars undergo another stage that closely parallels the end of core hydrogen fusion (Figure 8-1c). The destinies of stars depend on their masses, and we have two mass ranges to consider: 0.4 to 8 $M_\odot$ (hereafter, *low-mass* stars), and more than 8 $M_\odot$ (hereafter, *high-mass* stars).

8-1 Low-mass stars become supergiants before expanding into planetary nebulae

Calculations reveal that carbon and oxygen atoms require a temperature of at least 600 million K to fuse. Because the core of a low-mass giant on the horizontal branch only reaches about 200 million K, fusion of these elements in the core does not occur. Photon production in the core therefore decreases when its helium fuel is nearly used up, and the inner regions of the star again contract, compressing and heating the shell of helium-rich gas just outside the core. As a result, **helium shell fusion** begins outside of the core; this shell is itself surrounded by a hydrogen-fusing shell (Figure 8-2). All of this takes place within a volume roughly the size of Earth.

Once helium shell fusion commences, the new outpouring of energy again pushes the outer envelope of the star out. A low-mass star thus leaves the horizontal branch and ascends the H-R diagram for a second, and final, time. Powered by the fusion from two shells, such an **asymptotic giant branch star**, or **AGB star**, becomes brighter than ever before. When the Sun moves along the asymptotic giant branch, it will expand to a radius of about 1 AU, enveloping Earth in its tenuous outer layers (Figure 8-2). An 8 $M_\odot$ star will have a diameter as big as the orbit of Mars and shine with the luminosity of $10^4 L_\odot$. At the top of the asymptotic giant branch, these stars are now so bright that they are classified as low-temperature, red supergiants (see Figure 8-1c).

AGB stars are destined to self-destruct. Giant stars of all spectral types emit strong stellar winds. Low-mass stars expel their outer layers more slowly than higher-mass stars, but they all reduce their masses significantly from what they were on the main sequence. During their initial ascent up the giant branch, stars can lose as much as 30% of their

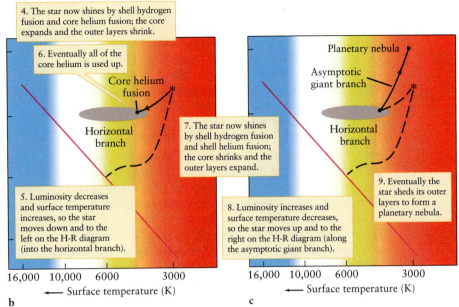

1. The star shines by shell hydrogen fusion: The inert core shrinks and the outer layers expand.

3. Core helium fusion begins with the helium flash (*).

Helium flash *

Main sequence

Horizontal branch

Giant branch

2. Luminosity increases and surface temperature decreases, so the star moves up and to the right on the H-R diagram (along the giant branch).

a

4. The star now shines by shell hydrogen fusion and core helium fusion; the core expands and the outer layers shrink.

6. Eventually all of the core helium is used up.

Core helium fusion

Horizontal branch

7. The star now shines by shell hydrogen fusion and shell helium fusion; the core shrinks and the outer layers expand.

5. Luminosity decreases and surface temperature increases, so the star moves down and to the left on the H-R diagram (into the horizontal branch).

b

Planetary nebula

Asymptotic giant branch

Horizontal branch

9. Eventually the star sheds its outer layers to form a planetary nebula.

8. Luminosity increases and surface temperature decreases, so the star moves up and to the right on the H-R diagram (along the asymptotic giant branch).

c

ANIMATION 8.1

FIGURE 8-1 *Post–Main-Sequence Evolution of Low-Mass Stars* (a) A typical evolutionary track on the H-R diagram as a star makes the transition from the main sequence to the giant phase. The asterisk (*) shows the helium flash occurring in a low-mass star. (b) After the helium flash, the star converts its helium core into carbon and oxygen. While doing so, its core re-expands, decreasing shell fusion. As a result, the star's outer layers recontract. (c) After the helium core is completely transformed into carbon and oxygen, the core recollapses, and the outer layers re-expand, powered up the asymptotic giant branch by hydrogen shell fusion and helium shell fusion.

masses. On the asymptotic giant branch they lose even more, often surrounding themselves with thickening cocoons of gas and dust. A star of the Sun's mass loses about 10^{-5} M$_\odot$ per year at this stage, eventually dumping half of its mass back into space.

This mass loss limits the force of gravity available to compress the star's core and the regions of shell fusion. A fine balance is struck: Our low-mass AGB star is compressed until its core of carbon and oxygen becomes degenerate, meaning that the electrons provide a growing repulsive force that stops its contraction (as discussed in Section 7-10). However, even for the most massive of these low-mass stars (now much less than their original 8 M$_\odot$ due to their stellar winds), the core temperature cannot reach the 600 million K necessary to fuse carbon or oxygen. Therefore, no further core fusion occurs.

The final stage through which a low-mass star passes begins with a "thermal runaway" (meaning a rapid rise in temperature) in the helium shell, just like the helium flash in its core earlier in its life (described in section 7-10). This occurs because the triple alpha process (see Section 7-11) is extremely sensitive to temperature, and, as the temperature goes up slightly, the fusion rate skyrockets. The increase in energy output from the thin helium-fusing shell, called a **helium shell flash**, expands the star, thereby briefly decreasing the temperature in the shell and slowing the rate of fusion.

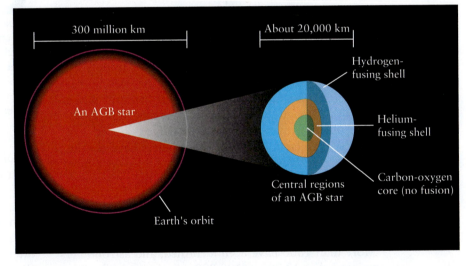

300 million km

About 20,000 km

An AGB star

Hydrogen-fusing shell

Helium-fusing shell

Carbon-oxygen core (no fusion)

Central regions of an AGB star

Earth's orbit

FIGURE 8-2 *The Structure of an Old Low-Mass Star* Near the end of its life, a low-mass star, like the Sun, travels up the AGB and becomes a supergiant. (The Sun will be about as large as the diameter of Earth's orbit.) The star's inert core, the hydrogen-fusing shell, and the helium-fusing shell are contained within a volume roughly the size of Earth. The inner layers are not shown to scale here.

A star undergoes several helium shell flashes as its helium shell thickens. During each flash, the helium shell's energy output jumps a thousandfold. These brief outbursts are separated by relatively quiet intervals lasting about 100,000 years, during which the helium shell gradually becomes thicker. The energy from the increased fusion during the helium shell flashes causes the star's outer layers to expand and, therefore, cool further. Eventually, the outer gases are sufficiently cool so that electrons and ions in them can recombine. This is exactly the opposite of ionization (discussed in Section 4-5). Whereas ionization requires an electron to absorb a photon, recombination forces an electron to emit a photon.

Just as the collisions of particles create pressure (force acting over an area), the collisions of photons with matter also create pressure. The impacts of photons emitted during recombination, along with the photons from the helium flashes and the photons normally flowing outward from fusion, generate enough pressure to eject more and more of the star's outer layers into space. Eventually, the mass of the star decreases so much that there is not enough pressure to sustain any shell fusion. As the ejected material expands and cools, some of it condenses to form dust grains. When enough of the gas has left the star for the core to be visible, the expanding dust and gases are considered to be a **planetary nebula** (see Figure 8-1c). As the outer layers of the star are shed, an increasingly hot interior is revealed, and the star moves to the left across the H-R diagram (Figure 8-3). Including the mass lost prior to the planetary nebula phase, low-mass stars lose as much as 80% of their masses by shedding their outer layers.

INSIGHT INTO SCIENCE

Names Often Don't Change as Understanding Improves Planetary nebulae have *nothing* to do with planets. This term was coined by the astronomer William Herschel in the eighteenth century. When viewed through the telescopes of the day, these glowing objects, often green in appearance, looked like planets similar to Uranus. Astronomers have continued to use this term, but do not be fooled—we now know that planetary nebulae indicate the presence of dying stars, not planets.

1 Planetary nebulae are quite common in our Galaxy. More than 1800 have been identified, and astronomers estimate that 20,000 to 100,000 exist in the Milky Way. Indeed, the fate of the Sun is to become a planetary nebula. The rate at which the gases are shed in a planetary nebula is so slow that they really are not explosive, compared either to our common conception of an explosion on Earth or to the supernova explosions that we will examine shortly.

Because stellar winds from red giants and subsequent planetary nebulae are so plentiful, astronomers estimate that they return a total of about 5 $M_\odot$ per year of gas and dust to the disk of our Galaxy. This mass amounts to about 85% of all matter expelled by all types of stars. This material goes into the formation of new, metal-rich Population I stars and associated planets. Thus, planetary

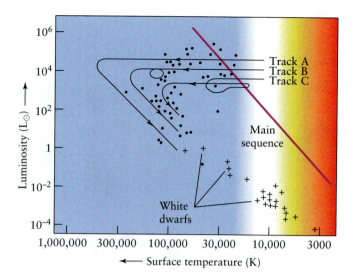

Evolutionary track	Mass ($M_\odot$)		
	Red supergiant	Ejected nebula	White dwarf
A	3.0	1.8	1.2
B	1.5	0.7	0.8
C	0.8	0.2	0.6

FIGURE 8-3 *Evolution from Supergiants to White Dwarfs* The evolutionary tracks of three low-mass supergiants are shown as they eject planetary nebulae. The table gives their masses as supergiants, the amount of mass they lose as planetary nebulae, and their remaining (white dwarf) masses. The loops indicate periods of instability and adjustment for the white dwarfs. The dots on this graph represent the central stars of planetary nebulae whose surface temperatures and luminosities have been determined. The crosses are white dwarfs for which similar data exist.

nebulae play an important role in the chemical and physical evolution of the Galaxy.

The outflowing gases ejected in a planetary nebula take a breathtaking variety of shapes when they interact with gases surrounding their stars, with companion stars, and with the stars' magnetic fields, which are often 10 to 100 times stronger than the Sun's field (Figures 8-4 and 8-5). The Hourglass Nebula (Figure 8-5c) appears initially to have shed mass in a doughnut shape around itself. In the star's final death throes, this gas and dust forced the outflow to go in two directions perpendicular to the plane of the doughnut, creating what is called a *bipolar planetary nebula*.

Spectroscopic observations of planetary nebulae show bright emission lines of hydrogen, carbon, neon, magnesium, oxygen, and nitrogen. From the Doppler shifts of these lines, we conclude that the gas and dust are moving outward from the dying low-mass star at speeds of 10 to 30 km/s. A typical planetary nebula moving outward for 10,000 years has a diameter of a few light-years.

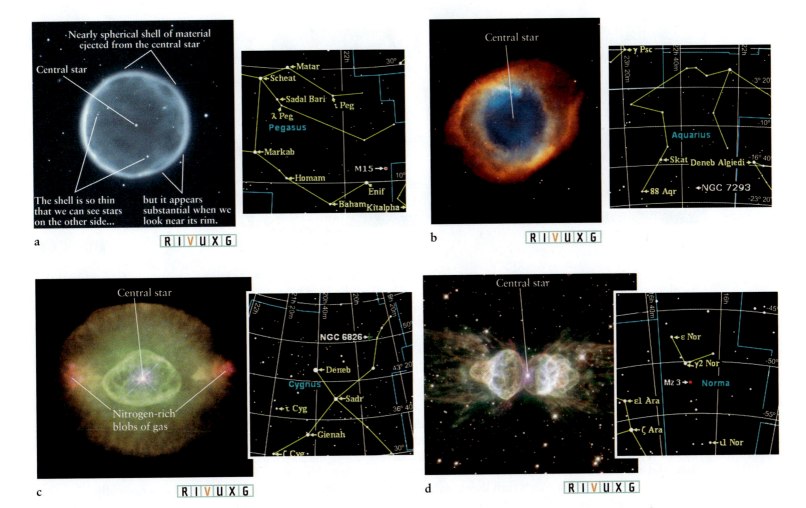

FIGURE 8-4 *Some Shapes of Planetary Nebulae*
The outer shells of dying low-mass stars are ejected in a wonderful variety of patterns. (a) An exceptionally spherical remnant, this shell of expanding gas, in the globular cluster M15 in the constellation Pegasus, is about 7000 ly (2150 pc) away from Earth. (b) The Helix Nebula, NGC 7293, located in the constellation Aquarius, about 700 ly (215 pc) from Earth, has an angular diameter equal to about half that of the full Moon. Red gas is mostly hydrogen and nitrogen, whereas the blue gas is rich in oxygen. (c) NGC 6826 shows, among other features, lobes of nitrogen-rich gas (red). The process by which they were ejected is as yet unknown. This planetary nebula is located in Cygnus. (d) MZ 3 (Menzel 3), in the constellation Norma (the Carpenter's Square), is 3000 ly (900 pc) from Earth. The dying star, creating these bubbles of gas, may be part of a binary system. (a: WIYN/NOAL/NSF; b–d: NASA, NOAO, ESA, The Hubble Helix Nebula Team, M. Meixner/STScI, and T. A. Rector/ NRAO; b: Howard Bons, STScI/Robin Ciardullo, Pennsylvania State University/ NASA; c: AURA/STScI/NASA)

By astronomical standards, a planetary nebula is short-lived. After about 50,000 years, its gases spread over distances so far from the cooling central star that its nebulosity simply fades from view. The gases then mingle and mix with the surrounding interstellar medium, enriching it with moderately heavy elements.

8-2 The burned-out core of a low-mass star becomes a white dwarf

The exposed carbon-oxygen cores of low-mass stars are stable objects supported by electron degeneracy pressure. Such burned-out hulks are called **white dwarfs,** and this is the fate of the Sun's core. Typical white dwarfs are roughly the size of Earth and covered with a thin coating of hydrogen and helium. The densities of these stellar "corpses" are typically 10^9 kg/m^3. In other words, a teaspoonful of white dwarf matter brought to Earth would weigh 5 tons.

Theoretical evolutionary tracks of three burned-out stellar cores are shown in Figure 8-3. These particular white dwarfs evolved from main-sequence stars of between 0.8 and 3.0 M$_\odot$. During the ejection phase, the appearance of these remnants changes rapidly. They appear to race along their evolutionary tracks on the

What luminosity class will the Sun be in just prior to when it becomes a planetary nebula?

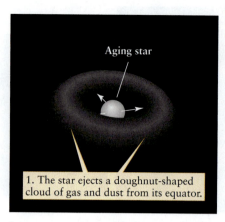

Aging star

1. The star ejects a doughnut-shaped cloud of gas and dust from its equator.

a

2. The star then ejects gas from its entire surface.

3. The doughnut channels the ejected gas into two oppositely directed streams.

b

Early, doughnut-shaped ejecta

Star

Later, more uniform, gas ejecta

c

R I V U X G

FIGURE 8-5 Formation of a Bipolar Planetary Nebula Bipolar planetary nebulae may form in two steps. Astronomers hypothesize that (a) first, a doughnut-shaped cloud of gas and dust is emitted from the star's equator, (b) followed by outflow that is channeled by the original gas to squirt out perpendicularly to the plane of the doughnut. (c) The Hourglass Nebula appears to be a textbook example of such a system. The bright ring is believed to be the doughnut-shaped region of gas lit by energy from the planetary nebula. The Hourglass is located about 8000 ly (2500 pc) from Earth. (c: R. Sahai and J. Trauer, JPL; WFPC-2 Science Team; and NASA)

H-R diagram, sometimes executing loops called thermal pulses as they temporarily reheat (tracks B and C in Figure 8-3). Finally, as the ejected planetary nebulae fade and the stellar cores cool, the evolutionary tracks of the dying stars take a sharp turn downward toward the white dwarf region of the diagram.

As billions of years pass, an isolated white dwarf radiates its stored energy into space, cooling and decreasing in luminosity. It moves down and to the right on the H-R diagram. The Sun will become a cold, dark, dense sphere of degenerate gases rich in oxygen and carbon, about the size of Earth. We can actually follow the theory a little further. It predicts that after billions of years of radiating away their heat, the interior temperatures of white dwarfs will decrease to about 4000 K, and the carbon and oxygen in them will solidify, transforming them into giant crystals.

Many white dwarfs are found in our solar neighborhood, but all are too faint to be seen with the naked eye. The first white dwarf to be discovered is a companion to the bright star Sirius. The binary nature of Sirius was first deduced in 1844 by the German astronomer Friedrich Bessel, who noticed that the star was moving back and forth, as if orbited by an unseen object. This companion, Sirius B (Figure 8-6), was first glimpsed in 1862. Recent satellite observations at ultraviolet wavelengths—where white dwarfs emit most of their light—demonstrate that the surface temperature of Sirius B is about 30,000 K.

Of what two elements is a white dwarf composed?

8-3   White dwarfs in close binary systems can create powerful explosions

Occasionally, a star in the sky suddenly becomes between 10^4 and 10^6 times brighter than it is normally. This phenomenon is called a **nova**. Its abrupt rise in brightness is followed by a gradual decline that may stretch over several months or more (Figures 8-7 and 8-8). Astronomers typically observe two or three novae in our Galaxy each year. Novae are fairly common, with about 20 estimated to occur in the Milky Way Galaxy annually. We do not see all of them because of the interstellar gas and dust that obscures many parts of the Galaxy.

Painstaking observations strongly suggest that novae occur in close binary systems that contain a white dwarf. The ordinary companion star fills its Roche lobe (recall Figures 7-32, 7-33, and 7-34), so it gradually deposits fresh hydrogen onto the white dwarf. This new mass becomes a dense layer covering the hot surface of the white dwarf. As more gas is deposited and compressed, the temperature in the hydrogen layer continues to increase. Finally, at about 10^7 K, hydrogen fusion ignites throughout the layer, blowing it into interstellar space. This explosion is the nova. After a nova, fusion ceases on the white dwarf. The companion star, however, may retain enough mass to supply a new layer of surface hydrogen, enabling some novae to reoccur. White dwarfs are not damaged by novae initiated on their surfaces.

Recent visible observations by the Hubble Space Telescope and X-ray observations by *Chandra* reveal

a   R I V U X G   b   R I V U X G

FIGURE 8-6 Sirius and Its White Dwarf Companion (a) Sirius, the brightest-appearing star in the night sky, is actually a double star. The smaller star, Sirius B, is a white dwarf, seen here at the five o'clock position in the glare of Sirius. The spikes and rays around the bright star, Sirius A, are created by optical effects within the telescope. (b) Since Sirius A (11,000 K) and Sirius B (30,000 K) are hot blackbodies, they are strong emitters of X rays. (a: R. B. Minton; b: NASA/SAO/CXC)

that novae are much more complex than astronomers had thought. Rather than smooth shells of expanding gas, they are composed of thousands of clumps of gas. These observations are stimulating astrophysicists to reexamine the details of the nova mechanism described above to try to explain these newly observed effects.

 Theoretical models predict that degenerate matter, such as the electrons in a white dwarf, can withstand only a finite amount of pressure, above which the degenerate matter cannot exert enough outward force to prevent the core from collapsing further. This limits the mass of a white dwarf to less than 1.4 $M_\odot$. This mass is called the **Chandrasekhar limit**, after the astrophysicist Subrahmanyan Chandrasekhar, who received a Nobel prize in 1990 for his pioneering theoretical studies of white dwarfs a half-century earlier.

> Why are novae not associated with isolated white dwarfs?

a   b   R I V U X G

FIGURE 8-7 Nova Herculis 1934 These two pictures show a nova (a) shortly after peak brightness as a magnitude −3 star and (b) 2 months later, when it had faded to magnitude +12. Novae are named after the constellation and year in which they appear. (UCO/Lick Observatory)

1. Material from a star accretes onto a companion white dwarf.

2. When enough accreted material builds up, thermonuclear reactions occur on the white dwarf's surface, creating a burst of visible light.

3. The nova fades over several weeks.

FIGURE 8-8 The Light Curve of a Nova This graph shows the history of Nova Cygni 1975, a nova that was observed to blaze forth in the constellation of Cygnus in September 1975. The rapid rise in magnitude followed by a gradual decline is characteristic of many novae, although some oscillate in intensity as they become dimmer.

8-4   Accreting white dwarfs in close binary systems can also explode as Type Ia supernovae

Whereas a nova just removes hydrogen and helium that build up on the surface of a white dwarf, some of these stellar remnants actually blow apart completely. This event is called a **Type Ia supernova** that occurs with a white dwarf in a semidetached binary system. To trigger a Type Ia supernova, a swollen giant companion star dumps gas onto a white dwarf that is close to the Chandrasekhar limit. When the added gas causes the white dwarf's mass to cross that limit, the increased pressure deep inside enables carbon fusion to begin in the core. In a catastrophic runaway process reminiscent of the helium flash, the rate of carbon fusion skyrockets, and, with no outer layers to absorb the energy, the star blows up.

A Type Ia supernova is powered by nuclear energy. What we see is the fallout from a gigantic thermonuclear explosion, which produces a wide array of radioactive isotopes. Especially abundant is an unstable isotope of nickel that decays into a radioactive isotope of cobalt. Most of the electromagnetic display of a Type Ia supernova results directly from the radioactive decay of this nickel into cobalt.

A Type Ia supernova's spectrum lacks hydrogen lines because the star from which it came was virtually free of this element. The light curve of the supernova, which shows its change of brightness over time, begins with a sudden rise (Figure 8-9). It typically reaches an absolute magnitude of -19 at peak brightness and then declines gradually for more than a year.

Because these supernovae are so bright, they can be seen far beyond the bounds of the Milky Way. Whereas our Galaxy is about 10^5 light-years in diameter, Type Ia supernovae can be seen in other galaxies at distances of over 10^9 light-years away. Because they all peak at the same absolute magnitude, these supernovae can be used to determine distances to myriad galaxies throughout the universe. This is done in two steps. First, the apparent magnitude of a Type Ia supernova at its peak brightness is observed. Second, because we know its absolute magnitude, we use the distance-magnitude relationship (see *An Astronomer's Toolbox 6-3*) to calculate its distance.

HIGH-MASS STARS AND TYPE-II SUPERNOVAE

In main-sequence stars of at least 8 $M_\odot$, sufficient gravitational force compresses and heats the core enough to enable its carbon and oxygen to fuse. This leads to the creation of a host of other elements that are eventually ejected into interstellar space.

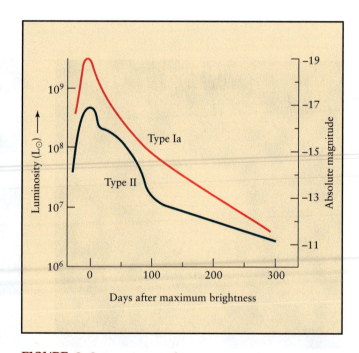

FIGURE 8-9 Supernova Light Curves A Type Ia supernova, which gradually declines in brightness, is caused by an exploding white dwarf in a close binary system. A Type II supernova is caused by the explosive death of a massive star and usually has alternating intervals of steep and gradual declines in brightness.

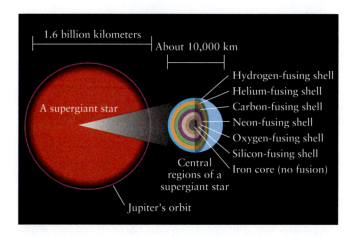

FIGURE 8-10 The Structure of an Old High-Mass Star Near the end of its life, a high-mass star becomes a supergiant with a diameter almost as wide as the orbit of Jupiter. The star's energy comes from six concentric fusing shells, all contained within a volume roughly the same size as Earth.

8-5 A series of fusion reactions in high-mass stars leads to luminous supergiants

When helium fusion ends in the core of a star more massive than 8 $M_{\odot}$, gravitational compression collapses the carbon-oxygen core, driving the star's central temperature above 600 million K. Now, carbon fusion begins, producing such elements as neon and magnesium. When a star compresses itself enough to elevate its central temperature to 1.2 billion K, neon fusion occurs, creating oxygen and magnesium. When the central temperature of the star then reaches 1.5 billion K, oxygen fusion begins, producing sulfur, silicon and phosphorus. The process of converting lower-mass elements into higher-mass ones is called **nucleosynthesis**.

As the core's temperature increases, the core's fusion rate soars. Also, each succeeding core stage has fewer and fewer nuclei than the previous one. For example, because each helium atom is made up of four hydrogen atoms, only one quarter as many helium atoms exist in a star's core as there were hydrogen atoms. Combining the higher fusion rate and fewer fusible atoms causes each cycle of core thermonuclear fusion to occur more rapidly than the previous stage. For example, detailed calculations for a star that was initially 25 $M_{\odot}$ on the main sequence reveal that carbon fusion occurs for 600 years, neon fusion for 1 year, and oxygen fusion for only 6 months. After half a year of core oxygen fusion, gravitational compression forces the central temperature up to 2.7 billion K, and silicon fusion begins. This thermonuclear process proceeds so furiously that the entire core supply of silicon in a 25-$M_{\odot}$ star is converted into iron in 1 day.

Each stage of fusion adds a new shell of matter outside of the core (Figure 8-10), creating a structure that resembles the layers of an enormous onion. Together, the tremendous numbers of fusion-generated photons created in all of these shells, as well as in the core, push the outer layers of the star farther and farther outward. The star expands to become a luminous supergiant, almost as wide across as Jupiter's orbit around the Sun.

Why do type Ia supernovae not have any hydrogen lines in their spectra?

High-mass main-sequence stars evolve into luminous supergiant stars that are brighter than 10^5 $L_{\odot}$, emit winds throughout most of their existence, and have mass-loss rates that exceed those of giants (see Section 7-9). Figure 8-11 shows a supergiant star losing mass. Betelgeuse (see Figure 6-2a), 470 light-years away in the constellation of Orion, is another good example of a supergiant experiencing mass loss. Spectroscopic observations show that Betelgeuse is losing mass at the rate of 1.7×10^{-7} $M_{\odot}$ per year and is surrounded by ejected gas, in a *circumstellar shell*. This huge shell is expanding at 10 km/s, and escaping gases have been detected at distances of 10,000 AU from the star. The expanding circumstellar shell has an overall diameter of one-third of a light-year.

Where in massive stars are the higher-mass elements formed?

Silicon fusion in high-mass stars involves many types of nuclear reactions, but its major final product is iron. Eventually, the core is converted entirely into iron,

R I V U X G

FIGURE 8-11 Mass Loss by a Supermassive Star Gas and dust ejected by the massive star HD 148937 in the constellation Norma. Orange is H_{α}, blue is oxygen, and red is sulfur. Located about 4200 ly (1300 pc) away, this star has 40 $M_{\odot}$ and is more than half way through its 6-million-year lifespan. (Gemini Observatory/AURA) which is surrounded by layers of shell fusion that consume the star's remaining reserves of fuel (see Figure 8-10). Whereas the star's enormously bloated atmosphere is nearly as big as the orbit of Jupiter, its entire energy-producing region is again contained in a volume the size of Earth. The buildup of an inert core of iron nuclei and electrons signals the impending violent death of a massive star.

8-6   High-mass stars blow apart in violent supernova explosions

Unlike lighter elements, iron cannot fuel further thermonuclear reactions. The protons and neutrons inside iron nuclei are already so tightly bound together that no further energy can be extracted by fusing still more particles with them. The sequence of fusion stages in the cores of high-mass stars therefore ends.

Because iron atoms do not fuse and emit energy, the electrons in the core must now support the star's outer layers by the strength of electron degeneracy pressure alone. Soon, however, the continued deposition of fresh iron from the silicon-fusing shell causes the core's mass to exceed the Chandrasekhar limit. Electron degeneracy suddenly fails to support the star's enormous weight, and the core collapses.

Any isolated main-sequence star of more than 8 $M_\odot$ develops an iron core as a luminous supergiant. When the core collapses, a rapid series of cataclysms is triggered that tears the star apart in a few seconds. Let us see how this happens in the death of a 25-$M_\odot$ star, according to computer simulations (Table 8-1).

In a 25-$M_\odot$ star, electron degeneracy pressure fails when the density is sufficiently high inside the iron core. The core, only some 3000 km in diameter, then collapses immediately. In roughly $\frac{1}{10}$ s, the central temperature exceeds 5 billion K. Gamma-ray photons associated with this intense heat have so much energy that they begin to break apart the iron nuclei, a process called **photodisintegration**. Although millions of years passed from the time the star arrived on the main sequence with a hydrogen- and helium-filled core until its core became iron, it takes less than a second to convert the core back into elemental protons, neutrons, and electrons!

Within another $\frac{1}{10}$ s, as the density continues to climb, the electrons in the core are forced to combine with the core's protons to produce neutrons, and the process releases a flood of neutrinos. As we saw in Section 5-9, neutrinos have no electric charge and very little mass, and most pass through Earth or the Sun without interaction. Nevertheless, the matter deep inside a collapsing high-mass star is so fantastically dense that its newly created neutrinos cannot immediately escape from the star's core. Slamming into nearby particles, the neutrinos provide a pressure pushing outward from the core.

According to the calculations, about $\frac{1}{4}$ s after the collapse begins, the density of the entire core reaches 4×10^{17} kg/m^3, which is *nuclear density*, the density at which neutrons and protons are normally packed together inside atomic nuclei. (Compare this with the density of water, 10^3 kg/m^3.) When the neutron-rich material of the core reaches nuclear density, it suddenly stiffens. This means that it resists the collapse much more than when it was less dense. The collapse of the core halts so abruptly that it rebounds and begins rushing back out in a process called *core bounce*.

 The model predicts that, during this critical stage, the star's unsupported layers of shell-fusing matter are plunging inward at up to 15% of the speed of light. The outward-flowing neutrinos and rebounding core slam into this matter (the *red regions* in Figure 8-12a). The impact stops the core's rebound while causing the infalling matter to reverse

TABLE 8-1   Evolutionary Stages of a 25-$M_\odot$ Star

Stage	Central temperature (K)	Central density (kg/m^3)	Duration of stage
Hydrogen fusion	4×10^7	5×10^3	7×10^6 years
Helium fusion	2×10^8	7×10^5	5×10^5 years
Carbon fusion	6×10^8	2×10^8	600 years
Neon fusion	1.2×10^9	4×10^9	1 year
Oxygen fusion	1.5×10^9	1×10^{10}	6 months
Silicon fusion	2.7×10^9	3×10^{10}	1 day
Core collapse	5.4×10^9	3×10^{12}	0.2 seconds
Core bounce	2.3×10^{10}	4×10^{17}	milliseconds
Supernova explosion	about 10^9	varies	hours

course. In just a fraction of a second, a tremendous volume of matter begins to move back up toward the star's surface. This matter accelerates rapidly as it encounters less and less resistance, and soon it forms an outgoing shock wave. After a few hours, this shock wave reaches the star's surface, lifting the star's outer layers away from the core in a mighty blast. The star becomes a **Type II supernova.**

As the supernova expands, the star's luminosity, which was already some 10^4 to 10^5 $L_\odot$, suddenly increases by another factor of between 10^3 and 10^6. The brightest supernova can therefore emit nearly as much energy as a hundred billion stars as luminous as our Sun! In other words, a supernova is often 10% as bright as all of the other stars in its galaxy combined.

As the layers of a massive dying star are blasted into space, they are compressed so much by the neutrinos

and the shock wave that fusion actually occurs during the explosion, creating a broad assortment of heavy elements and larger particles of interstellar dust. Indeed, the products of shell fusion and this final burst of fusion during supernovae are where many of the elements on Earth and other terrestrial planets come from, including many of the atoms in our bodies. We really are made of stardust.

Computer simulations indicate that the onion skins of different elements do not just move out of a star going supernova in lockstep (Figures 8-12a, b). Therefore, the ejected gases are not expected to be uniform shells of matter. Indeed, observations reveal that the material emerging in a supernova comes out quite irregularly. Figures 8-12c, d, and e show the distributions of some of the elements in the supernova Cassiopeia A (Cas A). Other observations show that some of the dense ele-

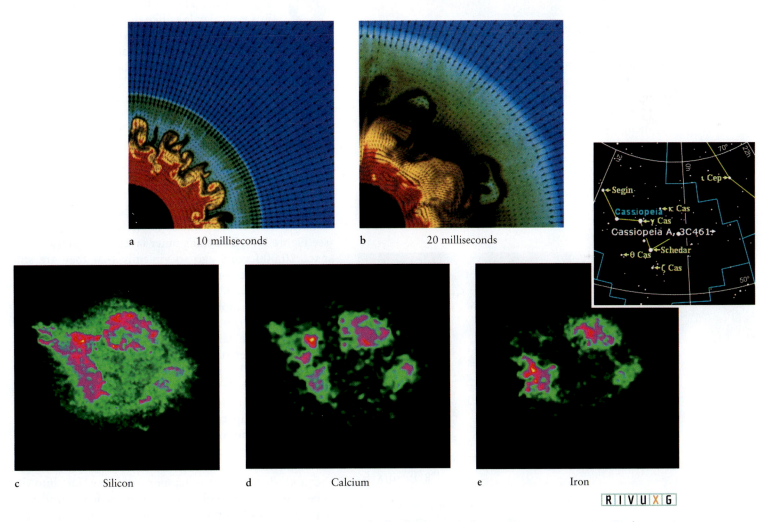

a   10 milliseconds   b   20 milliseconds

c   Silicon   d   Calcium   e   Iron

R I V U X G

FIGURE 8-12 Supernovae Proceed Irregularly Images (a) and (b) are computer simulations showing the chaotic flow of gas deep inside the star as it begins to explode as a supernova. This helps account for the globs of iron and other heavy elements emitted from deep inside, as well as the lopsided distribution of all elements in the supernova remnant, as shown in (c), (d), and (e). These three pictures are X-ray images of supernova remnant Cassiopeia A taken by *Chandra* at different wavelengths. (a and b: Courtesy of Adam Burrows, University of Arizona and Bruce Fryxell, NASA/GSFC; c, d, and e: U. Hwang et al., NASA/GSFC) ments, like iron, ejected in supernovae are actually created deep inside the innermost silicon and oxygen layers of the supernovae and then shot out like bullets or geysers. Other elements are created throughout the expanding stars. However, calculations and observations reveal that supernovae do not create enough of the heaviest elements, like gold, silver, and platinum, to account for the amounts of these elements found on Earth. We will explore their likely origin in Section 8-14.

Over its lifetime, our model 25-$M_\odot$ star ejects more than 20 $M_\odot$ of its mass back into space by stellar winds, by gas ejected during unstable periods, and during its supernova phase. Less massive stars return proportionately less mass to the interstellar medium.

What role do neutrinos play in Type II supernovae?

INSIGHT INTO SCIENCE

The Process of Science The best scientific theories and models also provide explanations about matters that they were not initially designed to study. In modeling stellar evolution and supernovae, the processes that create and eject metals (that is, elements heavier than hydrogen and helium) from stars provide the elements necessary for life to exist.

R I V U X G

FIGURE 8-13 The Gum Nebula The Gum Nebula is the largest known supernova remnant. It spans 60° across the sky and is centered roughly on the southern constellation of Vela. The nearest portions of this expanding nebula are only 300 light-years from Earth. The supernova explosion occurred about 11,000 years ago, and the remnant now has a diameter of about 2300 light-years. Only the central regions of the nebula are shown here. (Royal Observatory, Edinburgh)

The spectrum and light curve of a Type II supernova are both significantly different from those of a Type Ia supernova. Whereas hydrogen lines are absent from a Type Ia's spectrum, they are strong in the spectra of Type II supernovae. Both types begin with a sudden rise in brightness (see Figure 13-9). However, a Type II supernova usually peaks at −17, a hundred times dimmer than a Type Ia. Whereas Type Ia supernovae decline gradually for more than a year, Type II supernovae alternate between periods of steep and gradual declines in brightness. Type II light curves, therefore, have a steplike appearance.

8-7 Supernova remnants are observed in many places

Astronomers have seen more than 1000 supernovae. These observations suggest that in a typical galaxy like the Milky Way, Type Ia supernovae occur approximately once every 36 years, whereas Type II supernovae occur about once every 44 years. Thus, there should be about five supernovae exploding in our Milky Way Galaxy each century. Vigorous stellar evolution in our Milky Way occurs primarily in the Galaxy's disk, because stars there continue to form the giant molecular clouds. The disk (which we are in and which we see edge-on as the milky-white band of stars and glowing gas and dust across the sky) is also where supernovae explode.

Astronomers find the debris of local supernova explosions in the Milky Way scattered across the Galaxy's disk. A beautiful example is the Cygnus Loop (see Figure 7-7). The star's outer layers were blasted into space 20,000 years ago so violently that they are still traveling at supersonic speeds. As this expanding shell of gas plows through the interstellar medium, it collides with atoms and molecules, making the gases glow.

Many supernova remnants cover sizable fractions of the sky. The largest is the Gum Nebula, with an angular diameter of 60° (its central region is shown in Figure 8-13). This nebula looks so big because it is so close: Its near side is only about 300 light-years from Earth. Studies of the Gum Nebula's expansion rate suggest that the supernova exploded about 11,000 years ago. At maximum brilliance, the exploding star was probably as bright as the Moon at first quarter.

The Milky Way's disk is so filled with interstellar gas and dust that we cannot see very far into it. Supernovae are thought to erupt every few decades in remote parts of our Galaxy, but their detonations are hidden from optical telescopes by the interstellar medium. Many supernova remnants in our Galaxy can be detected only at nonvisible wavelengths, ranging from X rays through radio waves. For example, Figure 8-14 shows images of the supernova remnant Cassiopeia A as seen in X ray (Figure 8-14a) and radio (Figure 8-14b).

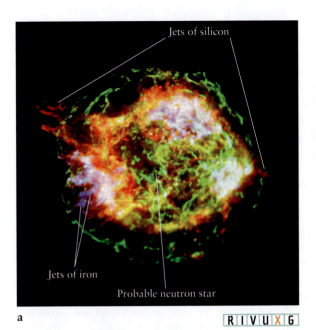

Jets of silicon

Jets of iron

Probable neutron star

a

RIVUXG

b

RIVUXG

ANIMATION 8.4   WEB LINK 8.6

FIGURE 8-14 Cassiopeia A Supernova remnants, such as Cassiopeia A, are typically strong sources of X rays and radio waves. (a) An X-ray picture of Cassiopeia A taken by *Chandra*. The opposing jets of silicon, probably guided by powerful magnetic fields, were ejected early in the supernova, before the iron-rich jets were released. (b) A corresponding radio image produced by the Very Large Array (VLA). Radiation from the supernova that produced this nebula first reached Earth 300 years ago. The explosion occurred about 10,000 light years from here. (a: NASA/ CXC/ GSFC/U. Hwang et al.; b: Very Large Array)

Visible-light photographs of this part of the sky reveal only a few small, faint wisps. Thus, radio searches for supernova remnants are more fruitful than visual searches. Only 24 supernova remnants have been found in visible-light photographs, but more than 3100 remnants in our Galaxy and in others have been discovered by radio astronomers.

From the expansion rate of the nebulosity in Cassiopeia A, astronomers calculate that the light from that supernova explosion first reached Earth about 300 years ago. Although telescopes were in wide use by the late 1600s, no record of the event is known. In fact, the last supernova of a star near its maximum brightness in our Galaxy, seen by the naked eye, was observed by Johannes Kepler in 1604. In 1572, Tycho Brahe also recorded the sudden appearance of an exceptionally bright star in the sky.

In 2002, astronomers discovered that 2 million years ago a group of bright O and B stars, called the Scorpius-Centaurus OB association, passed within 150 light-years of Earth, and that one or more supernovae in this group may have occurred at that time. Such a close explosion could have damaged Earth's ozone layer and caused the extinction of some ocean life at the end of the Pliocene epoch.

Besides the gas and dust enriched with heavy elements, supernovae leave a heritage long after their remnants have dissipated. Much of this gas and dust is recycled into new stars and planets, such as our solar system.

> Why might we not be able to see all the supernovae in our Galaxy?

8-8 Cosmic rays are not rays at all

In observations connected with supernovae, astronomers have detected particles flying through space at speeds that exceed 90% of the speed of light. These high-speed particles are called **cosmic rays** or **primary cosmic rays**. They were named *rays* before their true identity as particles was known, and, as often happens, this misleading name has stuck. Cosmic rays are primarily hydrogen nuclei, along with less than 1% of the nuclei of all the more massive elements and a few percent of electrons and positrons (positively charged electrons).

The origins of cosmic rays are just beginning to become clear. At first, many astronomers thought that moderate-energy cosmic rays were emitted directly by supernovae. This made sense because these explosions are the best-known source of the energy necessary to give such cosmic rays their tremendous speeds. However, observations by the *Advanced Composition Explorer* satellite reveal that the isotopes of nickel coming to Earth in the form of cosmic rays are not the isotopes emitted by supernovae. Instead, evidence is mounting that many cosmic rays are created when supernova debris slams into gas already in space, causing some of this preexisting matter to accelerate and become cosmic rays.

Much less common than cosmic rays from supernova remnants are ultrahigh-energy cosmic rays that come from outside of our Galaxy. Each of these particles has as much energy as a baseball thrown by a major league pitcher. Observations of their paths suggest that many of them come from galaxies containing supermassive black holes, which we will discuss in Section 9-8. Whether these black holes themselves generate the ultrahigh-energy cosmic rays is still under investigation.

Most primary cosmic rays coming in our direction collide with gas in the atmosphere some 15 km above Earth's surface. This is fortunate, because each cosmic ray packs an enormous amount of energy that potentially could harm living tissue. The collision in the air divides a cosmic ray's energy among several gas particles, which are shoved Earthward. These, in turn, often hit other gas atoms, creating a cascade of lower-energy particles that eventually reach Earth's surface as a **cosmic ray shower** (Figure 8-15). These cosmic rays created from atoms in our atmosphere are called **secondary cosmic rays**.

> What are cosmic rays?

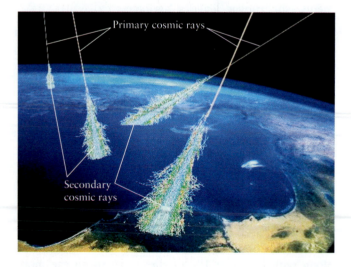

FIGURE 8-15 Cosmic Ray Shower Cosmic rays from space slam into particles in the atmosphere, breaking them up and sending them Earthward. These debris are called secondary cosmic rays. This process of impact and breaking up continues as secondary cosmic rays travel downward, creating a cosmic ray shower, as depicted in this artist's conception of four such events. (Simon Swordy [U. Chicago], NASA)

8-9   Supernova 1987A offered a detailed look at a massive star's death

Some supernovae in other galaxies can be seen without a telescope. In 1885, a supernova in the Andromeda Galaxy was just barely visible to the unaided eye. On February 23, 1987, a supernova was observed in the Large Magellanic Cloud, a galaxy near our own Milky Way. The supernova, designated SN 1987A, was the first to be observed that year (Figure 8-16), and it was so bright that it could easily be seen with the naked eye. What made SN 1987A such an exciting discovery was that it gave astronomers a rare opportunity to study the death of a nearby massive star using modern equipment that provided data with which to verify the theory of supernovae described earlier.

At first, SN 1987A reached only a tenth of the luminosity predicted for an exploding massive star. For the next 85 days, it gradually brightened. Then it began dimming, as is characteristic of an ordinary supernova. Fortunately, the star had been observed before it became a supernova (Figure 8-16 *insets*). The Large Magellanic Cloud is about 160,000 light-years from Earth—near enough to us that many of its stars have been individually observed and catalogued. The star had been identified as a B3 I supergiant. The theory of the evolution of such stars provided an explanation for SN 1987A's unusually slow brightening.

When this star was on the main sequence, its mass was about 20 $M_\odot$, although by the time it exploded, it had shed many solar masses. The evolutionary track for an aging 20-$M_\odot$ star wanders back and forth across the top of the H-R diagram, so the star alternates between being a hot (blue) supergiant and a cool (red) supergiant. The star's size changes significantly as its surface temperature changes. A blue supergiant is only 10 times larger in diameter than the Sun, but a red supergiant of the same luminosity is 1000 times larger. Because the doomed star was a relatively small blue supergiant when it exploded, it reached only a tenth of the brightness that it would have attained had it been a red supergiant at that time.

Ordinary telescopic observations of a supernova explosion can only show us the expanding outer layers of the dying star. No photons created in the core can penetrate through the supernova gases directly to our telescopes. As the outer layers of the supernova thin, however, neutrinos escape directly from the core into space. By detecting these neutrinos and measuring their properties, astronomers can learn many details about the star's collapsing core, especially about core bounce. Several neutrino detectors were operating when the neutrinos from SN 1987A reached Earth.

Nearly a day before SN 1987A was first observed in the sky, teams of scientists at neutrino detectors in Japan and the United States reported observing Cerenkov flashes (see Section 5-9) from a burst of neutrinos. The Kamiokande II detector in Japan detected 12 neutrinos at about the same time that 8 were found by the IMB (Irvine-Michigan-Brookhaven) detector in a salt mine under Lake Erie. Neutrinos preceded the visible supernova outburst because they escaped from the dying star before the shock wave from the collapsing core reached the star's surface. They were detected in Earth's northern hemisphere, where the supernova is always below the horizon, after having passed through Earth. The discovery of these neutrinos provided strong support for the theory that supernovae are caused by the collapse of the star's core, its subsequent bounce, and the flow of neutrinos that help cause the supernova, as described in Section 8-6.

About three and a half years after SN 1987A's detonation was first seen from Earth, astronomers using the Hubble Space Telescope obtained a picture showing several rings of glowing gas around the exploded star (Figure 8-17). This gas had been emitted by the star 20,000 years before the supernova, when a hydrogen-rich stellar envelope was ejected by stellar winds from the doomed star, as discussed earlier in this chapter. This gas expanded in an hourglass shape, because it was blocked from expanding around the star's equator, either by a preexisting ring of gas or by the orbit of an as-yet-unseen companion star.

While emitting these gases, the doomed star was a red supergiant. It then shrank and in its final blue giant phase, the same star emitted higher-speed gas that compressed the hourglass gas into three rings—a narrow one around the equator and wider ones at the top and bottom of the hourglass. This relic gas is now being illuminated by photons from the supernova. In 1998, astronomers began to observe gas ejected from the supernova striking the preexisting circumstellar gas. This collision has caused the ring of gas to brighten, and it should eventually illuminate more of the earlier gas that had been ejected.

SN 1987A provided astronomers with invaluable information, in part, because the doomed star had been studied before it exploded and its distance from Earth was known. The supernova was also located in an unobscured part of the sky, and neutrino detectors happened to be operating at the time of the outburst.

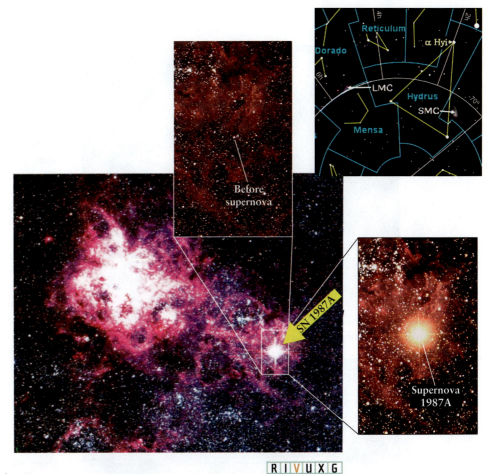

R I V U X G

FIGURE 8-16 Supernova 1987A A supernova was discovered in a nearby galaxy called the Large Magellanic Cloud (LMC) in 1987. This photograph shows a portion of the LMC that includes the supernova and a huge H II region, called the Tarantula Nebula. At its maximum brightness, observers at southern latitudes saw the supernova without a telescope. (Insets) The star before and after it exploded. (European Southern Observatory; insets: Anglo-Australian Observatory/ David Malin Images)

WEB LINK 8.7

Astronomers will be monitoring the progress of this supernova for years to come.

Why were neutrinos from SN 1987A observed before the light from this event?

NEUTRON STARS AND PULSARS

When stars that initially had between 8 and 25 $M_\odot$ explode as supernovae, their expanding cores are stopped by the outer layers into which they slam. The cores recontract and remain intact as highly compressed clumps of neutrons, called **neutron stars**. Isolated neutron stars, resulting from these explosions, are stable stellar remnants with masses between 1.4 $M_\odot$ and about 3 $M_\odot$.

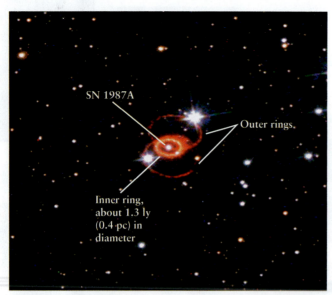

a Supernova 1987A seen in 1996   R I V U X G

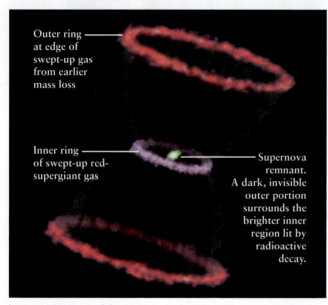

b An explanation of the rings

FIGURE 8-17 Shells of Gas Around SN 1987A (a) Intense radiation from the supernova explosion caused three rings of gas surrounding SN 1987A to glow in this Hubble Space Telescope image. This gas was ejected from the star 20,000 years before the star detonated. All three rings lie in parallel planes. The inner ring is about 1.3 light-years across. The white and colored spots are unrelated stars. (b) When the progenitor star of SN 1987A was still a red supergiant, a slowly moving wind from the star filled the surrounding space with a thin gas. When the star contracted into a blue supergiant, it produced a faster-moving stellar wind. The interaction between the fast and slow winds somehow caused gases to pile up along an hourglass-shaped shell surrounding the star. The burst of ultraviolet radiation from the supernova ionized the gas in the rings, causing the rings to glow. The supernova itself, at the center of the hourglass, glows because of energy released from radioactive decay. (Robert P. Kirshner and Peter Challis, Harvard-Smithsonian Center for Astrophysics; STScI)

8-10   The cores of many Type II supernovae become neutron stars

The neutron was discovered during laboratory experiments in 1932. Inspired by the realization that white dwarfs are supported by electron degeneracy pressure, Fritz Zwicky and Walter Baade proposed, within a year, that a similar repulsion between neutrons could support remnant neutron cores. They used the fact that the Pauli exclusion principle also prevents neutrons with identical properties from packing too closely together. When neutrons are pressed together, many of them have to move rapidly (so as not to be identical to their neighbors). This motion provides a pressure, called **neutron degeneracy pressure**, that equations predict is even greater than electron degeneracy pressure (see Section 7-10) and thus allows stellar remnants with masses beyond the Chandrasekhar limit to be stable.

Most scientists ignored Zwicky and Baade's theory for years. After all, a neutron star seemed to be a rather unlikely object. To transform protons and electrons into neutrons, the density in the star would have to be equal to nuclear density, about 10^{17} kg/m³. Thus, a teaspoon full of neutron star matter brought back to Earth would weigh about 1 billion tons. Put another way, neutron star matter is 200 million times denser than that found in a white dwarf. Furthermore, an object compacted to nuclear density would be very small. A 2-$M_\odot$ neutron star would have a diameter of only 20 km (12 mi) and would fit inside any major city on Earth. The surface gravity on one of these neutron stars would be so strong that the escape velocity would equal one-half the speed of light. All of these conditions appeared to be so outrageous that few astronomers paid any serious attention to the subject of neutron stars—until 1968.

In that year, Jocelyn Bell, then a graduate student in radio astronomy at Cambridge University, noticed that the radio telescope she was using was detecting regular pulses from one particular location in the sky. Careful repetition of the observations demonstrated that the radio pulses were arriving with a regular period of 1.3373011s. The regularity of this pulsating radio source was so striking that the Cambridge team suspected that they might be detecting signals from an advanced alien civilization, which accounts for the name first assigned to it: LGM1 (short for Little Green Men 1). This possibility was soon discarded as several more of these pulsating radio sources, which soon came to be known as **pulsars**, were discovered across the sky. The broad distribution of pulsars would require incredibly widespread civilizations, inconsistent with what we have found in our searches for alien intelligence (see Chapter 19). In all cases, the periods were extremely regular, ranging between 0.2 and 1.5 s (Figure 8-18).

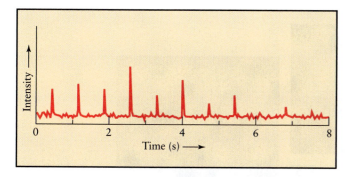

WEB LINK 8.9 **FIGURE 8-18** A Recording of a Pulsar This chart recording shows the intensity of radio emissions from one of the first pulsars to be discovered, PSR 0329+54. Note that some of the pulses are weak and others are strong. Nevertheless, the spacing between pulses is so regular (0.714 s) that it is more precise than most clocks on Earth.

When the discovery of pulsars was officially announced in early 1968, astronomers around the world began proposing all sorts of explanations. Many of these theories were bizarre, and arguments raged for months. We have already ruled out alien civilizations. Astronomers therefore looked for some recurring event in a star's life. We know that a star sheds matter as it swells to a giant. Perhaps some stars alternately expand and contract, emitting energy as they pulsate. However, this explanation also fails because any star expanding and contracting as fast as a pulsar emits pulses would explode.

Late in 1968, all controversy was laid to rest with the discovery of a pulsar in the middle of the Crab Nebula. In A.D. 1054, Native American, Chinese, and possibly other peoples saw and recorded the appearance of a "guest star" in the constellation of Taurus. When we turn a telescope toward this location, we find the Crab Nebula, shown in Figure 8-19. It looks like the residue of an explosion and is, in fact, a supernova remnant. At the center of the Crab Nebula is the Crab pulsar (Figure 8-19b).

With expansion and contraction ruled out, the likely scenario is that pulsars are spinning objects. If the Crab pulsar is spinning, it is spinning too fast to be a white dwarf. Its period is 0.033 s, which means that it rotates 30 times each second. At that speed, something as wide as a white dwarf would immediately fly apart. Strengthening the case against pulsars being white dwarfs, astronomers discovered numerous other pulsars rotating even faster, including PSR 1937+21 (PSR for pulsar and 1937+21 for the object's right approximate ascension and declination), a stellar remnant that pulses 640 times each second. Staying with the belief that they are spinning, astronomers concluded that pulsars must be incredibly compact. Calculations

revealed that, if neutron stars exist, they have a sufficiently small diameter to remain intact while rotating as fast as pulsars pulsate. Indeed, this was the *only* explanation that withstood the process of scientific scrutiny, and, as a result, the theory of neutron stars described earlier was quickly developed.

How big is a neutron star?

8-11   A rotating magnetic field explains the pulses from a neutron star

Why are neutron stars rotating so fast, and how do they emit pulses of electromagnetic energy? Based on observations of the rotation of the Sun, Betelgeuse, and other nearby stars, astronomers theorize that many, perhaps most, stars on the main sequence have some angular momentum (rotation). The Sun, for example, takes nearly a full month to rotate once around its axis. But the rotation rate of collapsing stars increases, just as pirouetting ice skaters speed up when they pull in their arms, as shown in Figure 2-12. Recall from Section 2-7 that this occurs because the total amount of angular momentum in an isolated system always remains constant. To conserve angular momentum, the core of a high-mass main-sequence star rotating once a month spins faster than once a second when it collapses to the size of a neutron star.

In addition to rapid rotation, most neutron stars are believed to have intense magnetic fields. In an average star like our Sun, the magnetic field is spread out over millions upon millions of square kilometers just under the star's surface (see Chapter 5). However, when a star of solar dimensions collapses down to a neutron star, its magnetic field becomes very concentrated, in the same way as stalks of wheat in a field become concentrated when you gather them in a bundle (Figure 8-20). The strength of a collapsing star's magnetic field increases to as much as 10^{15} times Earth's magnetic field. A magnet that strong located halfway to the Moon would lift iron rods off Earth.

The axis of rotation of a typical neutron star is not the same as the axis connecting its north and south magnetic poles (Figure 8-21), just as the magnetic and rotation axes of the Sun and planets (except Saturn) are different from each other. As the neutron star rotates, its powerful magnetic field therefore rapidly changes direction. Like a giant electric generator, the star creates intense electric fields, which act on protons and electrons near its surface. The powerful electric fields channel these charged particles, causing them to flow out from the neutron star's polar regions, as sketched in Figure 8-21. As the particles stream along the field, they accelerate and emit energy. The result is two very thin beams of radiation that pour out of the neutron star's north and south magnetic polar regions and sweep through space—a pulsar.

5

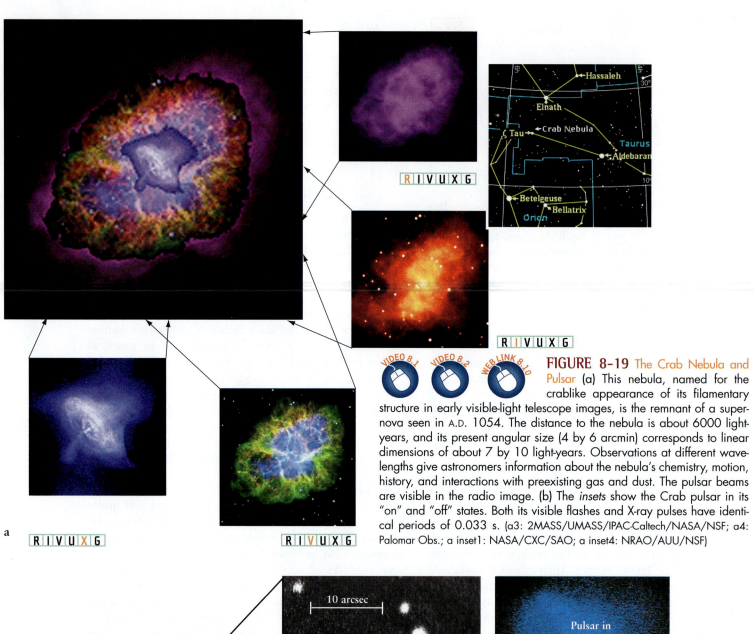

FIGURE 8-19 The Crab Nebula and Pulsar (a) This nebula, named for the crablike appearance of its filamentary structure in early visible-light telescope images, is the remnant of a supernova seen in A.D. 1054. The distance to the nebula is about 6000 light-years, and its present angular size (4 by 6 arcmin) corresponds to linear dimensions of about 7 by 10 light-years. Observations at different wavelengths give astronomers information about the nebula's chemistry, motion, history, and interactions with preexisting gas and dust. The pulsar beams are visible in the radio image. (b) The *insets* show the Crab pulsar in its "on" and "off" states. Both its visible flashes and X-ray pulses have identical periods of 0.033 s. (a3: 2MASS/UMASS/IPAC-Caltech/NASA/NSF; a4: Palomar Obs.; a inset1: NASA/CXC/SAO; a inset4: NRAO/AUU/NSF)

a

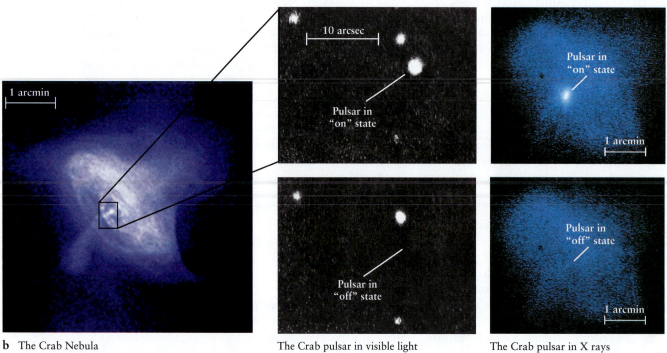

b The Crab Nebula   The Crab pulsar in visible light   The Crab pulsar in X rays

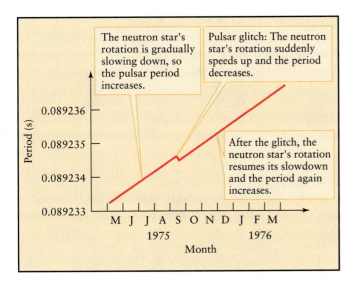

The neutron star's rotation is gradually slowing down, so the pulsar period increases.

Pulsar glitch: The neutron star's rotation suddenly speeds up and the period decreases.

After the glitch, the neutron star's rotation resumes its slowdown and the period again increases.

FIGURE 8-23 A Glitch Interrupts the Vela Pulsar's Spindown Rate An isolated pulsar radiates energy, which causes it to slow down. This "spin down" is not always smooth. As it slows down, it becomes more spherical, and so its spinning, solid surface must readjust its shape. Because the surface is brittle, this readjustment is often sudden, like the cracking of glass, which causes the angular momentum of the pulsar to suddenly jump. Such an event, called a glitch and shown here for the Vela pulsar in 1975, changes the pulsar's rotation period.

one of them. The absorption lines are believed to be due to scattering of radiation from the neutron star by gases in its atmosphere, just as scattering in the Sun's atmosphere causes absorption lines in its spectrum. In 2005, astronomers found indications that some isolated neutron stars expel energy long after they form. In particular, the Cassiopeia A remnant has regions glowing from radiation deposited on them as recently as 50 years ago (some 275 years after the supernova itself). The cause of this activity is still being determined.

Just as Earth's crust has earthquakes, calculations show that the crusts of rotating neutron stars have equivalent events. When a rotating neutron star (pulsar) undergoes such a quake, its rotation rate suddenly changes to conserve its angular momentum. This creates a **glitch** in the rate at which the pulsar is slowing down. Such events have been observed in a number of pulsars (Figure 8-23).

While most pulsars appear to be isolated bodies, more than 90 have been discovered with companions (not necessarily other neutron stars). These pulsars are called *binary pulsars* and are of interest for several reasons. First, many binary pulsars pull mass onto themselves from their companions. This infall causes the pulsar to speed up (like the skater in Figure 2-12

pulling in her arms). As a result of this effect, some pulsars spin nearly a thousand times per second and are called *millisecond pulsars*. Second, some pulsars, like the one labeled PSR 1916+13, have compact massive companions, which are often also neutron stars. Observations of binary pulsars with neutron star companions reveal that the two bodies are spiraling toward each other, as predicted by Einstein's gravitational theory, called *general relativity* (see Sections 2-1 and 14-4 for details).

In 2004, astronomers observed, for the first time, a pair of pulsars in orbit around each other (Figure 8-24). This discovery provides an exciting opportunity to further test the predictions of general relativity. For example, this theory predicts that these two pulsars should be precessing (the directions that their rotation axes point should be changing; see Section 1-8) and that, like the binary pulsars just discussed, they should be spiraling in toward each other. Observations of this system in the next few years will test these predictions.

> Why do glitches change the rotation rates of pulsars?

8-14 Colliding neutron stars may provide some of the heavy elements in the universe

In Sections 8-5 and 8-6 we discussed nucleosynthesis, the process of creating the heavier elements. We saw how a variety of elements up to iron are created inside

FIGURE 8-24 Double Pulsar This artist's conception shows two pulsars that orbit their center of mass. The double pulsar they represent is called PSR J0737-3039, which is about 1500 light-years from Earth in the constellation Puppis. One pulsar has a 23-ms period and the other has a 2.8-s period. The two orbit once every 2.4 h. (Michael Kramer/Jodrell Bank Observatory, University of Manchester)

massive stars before the supernova occurs and are then ejected into space during the explosion. We also saw that even heavier elements are created and ejected during the supernovae themselves. However, models indicate that supernovae do not create enough of the elements heavier than iron to account for the amounts of these elements found in the universe.

Computer simulations of what happens during collisions of two neutron stars predict that, in these impacts, some of the neutrons are splashed into space, where many of them decay back into protons and help form various elements. The interactions that follow create a variety of heavy elements in amounts consistent with observations.

> Why does most of the iron on Earth not come from the iron cores of high-mass stars that explode as supernovae?

8-15   Binary neutron stars create pulsating X-ray sources

Neutron stars in binary systems may also hold the key to another regular pulse from the sky—pulses from X-ray sources. During the 1960s, astronomers obtained tantalizing X-ray views of the sky during short rocket and balloon flights that briefly lifted X-ray detectors above Earth's atmosphere. Several strong X-ray sources were discovered, and each was named after the constellation in which it was located.

Astronomers were so intrigued by these preliminary discoveries that NASA built and launched *Explorer 42*, an X-ray-detecting satellite that could make observations 24 h a day. The satellite was launched from Kenya (to place it in an orbit above Earth's equator) in 1970, on the seventh anniversary of Kenyan independence. To commemorate the occasion, *Explorer 42* was renamed *Uhuru*, which means "freedom" in Swahili.

Uhuru gave us our first comprehensive look at the X-ray sky. As the satellite slowly rotated, its X-ray detectors swept across the heavens. Each time an X-ray source came into view, signals were transmitted to receiving stations on the ground. Before its battery and transmitter failed in early 1973, *Uhuru* had succeeded in locating 339 X-ray sources.

The discovery of pulsars was still fresh in everyone's mind when the *Uhuru* team discovered X-ray pulses coming from Centaurus X-3 in early 1971. Figure 8-25 shows data from one sweep of *Uhuru's* detectors across Centaurus X-3. The pulses have a regular period of 4.84 s. A few months later, similar pulses were discovered coming from a source called Hercules X-1, which has a period of 1.24 s. Because the periods of these two X-ray sources are so short, astronomers suspected that they had found rapidly rotating neutron stars.

It soon became clear, however, that systems such as Centaurus X-3 and Hercules X-1 are not ordinary pulsars like the Crab or Vela pulsars. Every 2087 days, Centaurus X-3 completely turns off for almost 12 h. This fact suggests that Centaurus X-3 is an eclipsing binary (see Figure 6-13), and that the X-ray source takes nearly 12 h to pass behind its companion star.

The case for the binary nature of Hercules X-1 is even more compelling. It has an off state corresponding to a 6-h eclipse every 1.7 days, and careful timing of the X-ray pulses shows a periodic Doppler shift every 1.7 days. This information provides direct evidence of orbital motion around a companion star. When the X-ray source is approaching us, its pulses are separated by slightly less than 1.24 s. When the source is receding from us, slightly more than 1.24 s elapse between the pulses.

Astronomers now realize that systems such as Centaurus X-3 and Hercules X-1 are examples of binary systems in which one object is a neutron star. Because all of these binaries have very short orbital periods, the distance between the ordinary star and the neutron star must be small. This proximity enables the neutron star to capture gas escaping from the companion star (Figure 8-26).

To explain the pulsations of X-ray sources such as Centaurus X-3 or Hercules X-1, astronomers propose that the ordinary star either fills or nearly fills its Roche lobe. Either way, matter escapes from the star. If the star fills its lobe, as in the case of Hercules X-1, mass loss results from direct overflow through the Roche lobe (Figure 8-26). If the star's surface lies just inside its lobe, as with Centaurus X-3, a stellar wind carries off the mass. A typical rate of mass loss for the

FIGURE 8-25 X-Ray Pulses from Centaurus X-3 This graph shows the intensity of X rays detected by *Uhuru* as Centaurus X-3 moved across the satellite's field of view. The variation in the height of the pulses was a result of the changing orientation of *Uhuru's* X-ray detectors toward the source as the satellite rotated. The short pulse period suggests that the source is a rotating neutron star.

ordinary star in such a pair is approximately 10^{-9} $M_\odot$ per year.

The neutron star in a pulsating X-ray source, like an ordinary pulsar, rotates rapidly and has a powerful magnetic field inclined to the axis of rotation (recall Figure 8-21). As the gas from the companion star falls toward the neutron star, the magnetic field funnels the incoming matter down onto its magnetic polar regions. The neutron star's gravity is so strong that the gas is traveling at nearly half the speed of light by the time it crashes onto the star's surface. This violent impact creates hot spots at both poles, with temperatures of about 10^8 K, that emit abundant X rays with a luminosity nearly 100,000 times brighter than the Sun. Observations of three such X-ray sources in 2005 revealed the impact sites on the neutron stars to range in size from 100 m to 4 km across. As the neutron star rotates, two beams of X rays from the polar caps sweep around the sky. If Earth happens to be in the path of one of the beams, we can observe this pulsating X-ray source. The pulse period is thus equal to the neutron star's rotation period. For example, the neutron star in Hercules X-1 completes one rotation every 1.24 s.

8-16 Neutron stars in binary systems can also emit powerful isolated bursts of X rays

Neutron stars can acquire additional mass from a companion star, which leads to the flaring up of the neutron stars. Beginning in late 1975, astronomers analyzing data from X-ray satellites detected sudden, powerful bursts of radiation. The record of a typical burst is shown in Figure 8-27. The source, called an **X-ray burster**, emits X rays at a constant low level; then, suddenly and without warning, an abrupt increase occurs, followed by a gradual decline. A typical burst lasts for 20 s. Several dozen X-ray bursters have been located, most lying in the plane of the Milky Way Galaxy.

X-ray bursters, like novae, are believed to arise from mass transfer in binary star systems. With a burster, however, the stellar remnant is a neutron star rather than a white dwarf. Gases escaping from the ordinary companion star fall onto the neutron star. The energy released as this gas crashes down onto the neutron star's surface produces the low-level X rays that are continuously emitted by the burster.

Most of the gas falling onto the neutron star is hydrogen, which becomes compressed against the hot surface of the star by the star's powerful surface gravity. In fact, temperatures and pressures in this accreting layer are so high that the arriving hydrogen is promptly converted by fusion into helium. Constant hydrogen fusion soon pro-

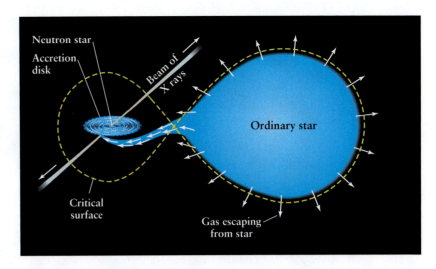

FIGURE 8-26 A Model of a Pulsating X-Ray Source Gas transfers from an ordinary star to the neutron star. The infalling gas is funneled down onto the neutron star's magnetic poles, where it strikes the star with enough energy to create two X-ray-emitting hot spots. As the neutron star spins, beams of X rays from the hot spots sweep around the sky.

duces a layer of helium that covers the entire neutron star. When the helium layer is about 1-m thick, helium fusion ignites explosively, and we observe a sudden burst of X rays. In other words, whereas explosive hydrogen fusion on a white dwarf produces a nova, explosive helium fusion on a neutron star produces an X-ray burster.

Is an X-ray burster similar to a supernova in the evolution of its source?

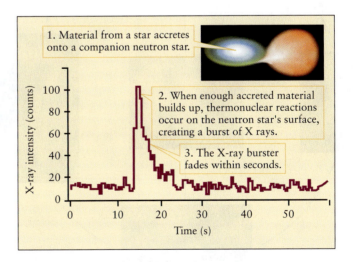

FIGURE 8-27 X Rays from an X-Ray Burster A burster emits X rays with a constant low intensity interspersed with occasional powerful bursts. This burst was recorded in September 1975 by an X-ray telescope that was pointed toward the globular cluster NGC 6624.

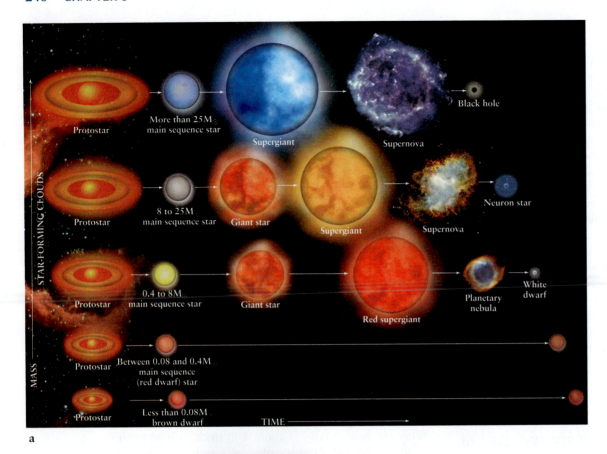

Protostar
More than 25M☉ main sequence star
Supergiant
Supernova
Black hole

Protostar
8 to 25M☉ main sequence star
Giant star
Supergiant
Supernova
Neuron star

Protostar
0.4 to 8M☉ main sequence star
Giant star
Red supergiant
Planetary nebula
White dwarf

Protostar
Between 0.08 and 0.4M☉ main sequence (red dwarf) star

Protostar
Less than 0.08M☉ brown dwarf

STAR-FORMING CLOUDS

MASS

TIME

a

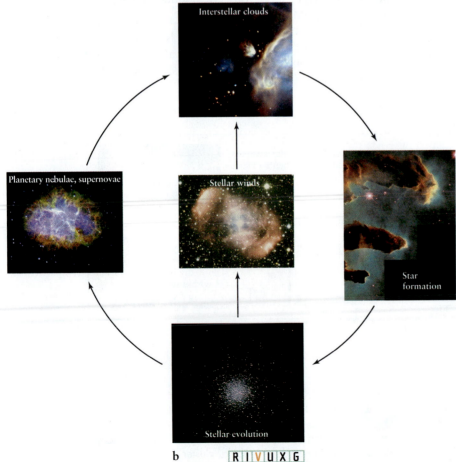

Interstellar clouds

Planetary nebulae, supernovae

Stellar winds

Star formation

Stellar evolution

b

R I V U X G

FIGURE 8-28 A Summary of Stellar Evolution (a) The evolution of isolated stars depends primarily on their masses. The higher the mass, the shorter the lifetime. Stars less massive than about 8 M☉ can eject enough mass to become white dwarfs. High-mass stars can produce Type II supernovae and become neutron stars or black holes. The horizontal *(time)* axis is not to scale, but the relative lifetimes are accurate. (b) The cycle of stellar evolution is summarized in this figure. (*Top inset:* Infrared Space Observatory, NASA; *right inset:* Anglo-Australian Observatory/J. Hester and P. Scowen, Arizona State University/NASA; *bottom inset:* NASA; *left inset:* NASA; *middle inset:* Anglo-Australian Observatory)

8-17 Smaller, more exotic stellar remnants composed of quarks may exist

In death as well as in life, the mass of a star determines its fate. Just as there is an upper limit to the mass of a white dwarf (the Chandrasekhar limit), there is also an upper limit to the mass of a neutron star, called the *Oppenheimer-Volkov limit*. Above this limit, neutron degeneracy pressure cannot support the overbearing weight of the star's matter pressing in from all sides. The Chandrasekhar limit for a white dwarf is 1.4 $M_\odot$, and the Oppenheimer-Volkov limit for a neutron star is about 3 $M_\odot$.

According to calculations first done in 1964 and experiments verifying the theory done in 1968, many elementary particles, such as neutrons (and protons, but not electrons), are actually composed of smaller particles, called **quarks**. Normally, quarks cannot exist as separate objects—in the cases of neutrons and protons, they clump together in groups of three. However, when the pressure on them is great enough, neutrons can become so compressed that they dissolve into their constituent quarks. As free particles, quarks obey the Pauli exclusion principle, meaning that identical quarks cannot occupy the same place at the same time. As with electrons and neutrons, quarks therefore create a pressure that resists collapse, thereby creating quark stars. The upper limit for the mass of quark stars is unknown, but it could not be much greater than the Oppenheimer-Volkov limit.

If the remnant core from a supernova is above the mass limit of neutron stars (and quark stars, if their existence is verified), then there is another fate in store for it. As noted earlier, the gravitational attraction near a neutron star is so strong that the particles escaping from its surface must be moving at roughly half the speed of light. If the stellar corpse has slightly greater than 3 $M_\odot$, so much matter becomes crushed into such a small volume that the repulsion created by the Pauli exclusion principle on its neutrons is quickly overcome. The remnant therefore collapses further and the escape velocity exceeds the speed of light. Because nothing can travel faster than light, nothing—not even light—can leave such a collapsed object. The discovery of neutron stars and the calculations regarding the limit to the Pauli exclusion principle inspired astrophysicists to consider seriously one of the most bizarre and fantastic objects ever predicted by modern science—the black hole, which we examine in Chapter 9.

Figure 8-28 summarizes the evolutionary history of stars, as discussed in Chapters 7 and 8.

8-18 Frontiers yet to be discovered

The incredible telescopes now in existence or scheduled to go on-line soon are revolutionizing our understanding of the final stages of star formation. In coming years, astronomers hope to pin down the details of how planetary nebulae, novae, supernovae, and stellar remnants behave. Among specific questions to be answered—What are the physical processes that cause planetary nebulae? Why do planetary nebulae have such a variety of shapes? What causes the hourglass shape of so many stellar remnants? Why are novae composed of many blobs of gas? Do quark stars really exist, and, if so, what are their properties?

> What is the Oppenheimer-Volkov limit?

SUMMARY OF KEY IDEAS

- Stars with higher masses fuse more elements than do stars with lower masses.

- Stars lose mass via stellar winds throughout their lives.

Low-Mass Stars and Planetary Nebulae
- A low-mass (below 8 $M_\odot$) main-sequence star becomes a giant when hydrogen shell fusion begins. It becomes a horizontal-branch star when core helium fusion begins. It enters the asymptotic giant branch and becomes a supergiant when helium shell fusion starts.

- Stellar winds during the thermal pulse phase eject mass from the star's outer layers.

- The burned-out core of a low-mass star becomes a dense carbon-oxygen body, called a white dwarf, with about the same diameter as that of Earth. The maximum mass of a white dwarf (the Chandrasekhar limit) is 1.4 $M_\odot$.

- Explosive hydrogen fusion may occur in the surface layer of a white dwarf in some close binary systems, producing sudden increases in luminosity that we call novae.

- An accreting white dwarf in a close binary system can also become a Type Ia supernova when carbon fusion ignites explosively throughout such a degenerate star.

High-Mass Stars and Supernovae
- After exhausting its central supply of hydrogen and helium, the core of a high-mass (above 8 $M_\odot$) star undergoes a sequence of other thermonuclear reactions. These include carbon fusion, neon fusion, oxygen fusion, and silicon fusion. This last fusion eventually produces an iron core.

- A high-mass star dies in a supernova explosion that ejects most of the star's matter into space at very high speeds. This Type II supernova is triggered by the gravitational collapse and subsequent bounce of the doomed star's core.

- Neutrinos were detected from Supernova 1987A, which was visible to the naked eye. Its development supported theories of Type II supernovae.

Neutron Stars, Pulsars, and (perhaps) Quark Stars

• The core of a high-mass main sequence star containing between 8 and 25 $M_\odot$ becomes a neutron star. The cores of slightly more massive stars may become quark stars. A neutron star is a very dense stellar corpse consisting of closely packed neutrons in a sphere roughly 20 km in diameter. The maximum mass of a neutron star, called the Oppenheimer-Volkov limit, is about 3 $M_\odot$.

• A pulsar is a rapidly rotating neutron star with a powerful magnetic field that makes it a source of periodic radio and other electromagnetic pulses. Energy pours out of the polar regions of the neutron star in intense beams that sweep across the sky.

• Some X-ray sources exhibit regular pulses. These objects are believed to be neutron stars in close binary systems with ordinary stars.

• Explosive helium fusion may occur in the surface layer of a companion neutron star, producing the sudden increase in X-ray radiation, called an X-ray burster.

WHAT DID YOU THINK?

1   *Will the Sun someday cease to shine brightly? If so, how will this occur?* Yes. The Sun will shed matter as a planetary nebula in about 6 billion years and then cease nuclear fusion. Its remnant white dwarf will dim over the succeeding billions of years.

2   *What is a nova? How does it differ from a supernova?* A nova is a relatively gentle explosion of hydrogen gas on the surface of a white dwarf in a binary star system. Supernovae, on the other hand, are explosions that cause the nearly complete destruction of massive stars.

3   *What are the origins of the carbon, silicon, oxygen, iron, uranium, and other heavy elements on Earth?* These elements are created during stellar evolution, by supernovae, and by colliding neutron stars.

4   *What are cosmic rays? Where do they come from?* Cosmic rays are high-speed particles (mostly hydrogen and other atomic nuclei) in space. Many of them are thought to have been created as a result of supernovae.

5   *What is a pulsar?* A pulsar is a rotating neutron star in which the magnetic field's axis does not coincide with the rotation axis. The beam of radiation it emits periodically sweeps across our region of space.

Key Terms for Review

Review Questions

1. Will the Sun shed most of its mass, and, if so, what is that event called? **a.** yes, as a planetary nebula, **b.** yes, as a supernova, **c.** yes, as a white dwarf, **d.** yes, as a neutron blast, **e.** no

2. A white dwarf is composed of primarily **a.** neutrons, **b.** hydrogen and helium, **c.** iron, **d.** cosmic rays, **e.** carbon and oxygen.

3. What prevents a neutron star from collapsing? **a.** hydrogen fusion, **b.** friction, **c.** electron degeneracy pressure, **d.** neutron degeneracy pressure, **e.** helium fusion

4. What is the difference between a giant star and a supergiant star?

5. Why is knowing the temperature in a star's core so important in determining which nuclear reactions can occur there?

6. What determines the temperature in the core of a star?

7. What is a planetary nebula, and how does it form?

8. What is the Chandrasekhar limit?

9. What is a neutron star?

10. Compare a white dwarf and a neutron star. Which of these stellar corpses is more common? Why?

 11. To test your understanding of the stages of stellar evolution, do Interactive Exercise 8-1. You can print out your results, if required.

12. What is the Oppenheimer-Volkov limit?

13. On an H-R diagram, sketch the evolutionary track that the Sun will follow between the time it leaves the main sequence and when it becomes a white dwarf. Approximately how much mass will the Sun have when it becomes a white dwarf? Where will the rest of the mass have gone?

14. Why have searches for supernova remnants at visible wavelengths been less fruitful than searches at other wavelengths?

 15. To test your understanding of how stars "die," do Interactive Exercise 8-2. You can print out your results, if required.

 16. To test your understanding of neutron stars, do Interactive Exercise 8-3. You can print out your results, if required.

17. Why do astronomers believe that pulsars are rapidly rotating neutron stars?

 18. To test your understanding of rotating neutron stars, pulsars, and novae, do Interactive Exercise 8-4. You can print out your results, if required.

19. What is the difference between Type Ia and Type II supernovae?

20. Compare a nova with a Type Ia supernova. What do they have in common? How are they different?

21. Compare a nova and an X-ray burster. What do they have in common? How are they different?

22. Describe what X-ray pulsars, pulsating X-ray sources, and X-ray bursters have in common. How are they different manifestations of the same type of astronomical object?

Advanced Questions

The answers to all computational problems, which are preceded by an asterisk (*), appear at the end of the book.

23. What prevents thermonuclear reactions from occurring at the center of a white dwarf? If no thermonuclear reactions take place in its core, why doesn't the star collapse?

24. Suppose you wanted to determine the age of a planetary nebula. What observations would you make, and how would you use the resulting data?

25. Why is the rate of expansion of the gas shell in a planetary nebula often not uniform in all directions?

26. What kinds of stars would you monitor if you wished to observe a supernova explosion from its very beginning? Look up the tabulated lists of the nearest and brightness stars in Appendix Tables E-5 and E-6. Which, if any, of these stars are possible supernova candidates? Explain.

27. To determine the period of a pulsar accurately, astronomers must take Earth's orbital motion around the Sun into account. Explain why.

*28. The distance to the Crab Nebula is about 2000 pc. When did it actually explode?

 29. To test your overall understanding of stellar evolution, do Interactive Exercise 8-5. You can print out your results, if required.

Discussion Questions

30. Suppose that you discover a small glowing disk of light while searching the sky with a telescope. How would you determine whether this object was a planetary nebula? What else could this object be?

31. Immediately after the first pulsar was discovered, one explanation offered was that the pulses were signals from an extraterrestrial civilization. Why did astronomers discard this idea?

32. Describe how astronomers can determine whether a supernova at a known distance is Type Ia or Type II, assuming that they can see the supernova from the time it begins to brighten. There are at least two valid answers to this question.

33. Describe how astronomers can determine whether a supernova at an unknown distance is Type Ia or Type II.

What If . . .

34. Earth passed through an old supernova remnant? What would happen to Earth and life on it?

35. The Sun were a B-type star, rather than a G-type star? Assuming that Earth orbiting the B-type star had the same composition and orbital distance that it has today, what would be different on Earth? If you answered this question in Chapter 6, you might want to see how the material in this chapter enhanced your previous answer.

36. The Sun were an M-type star, rather than a G-type star? Assuming that Earth orbiting the M-type star had the same composition and orbital distance that it has today, what would be different on Earth?

37. The Sun were expanding into a giant star today? How might we cope with this change?

Web Questions

38. It has been claimed that the Dogon tribe in Western Africa has known for thousands of years that Sirius is a binary star. Search the Web for information about these claims. What is their basis? Why are scientists skeptical, and how do they refute these claims?

39. Search the Web for recent information about SN 1987A. Sketch the shape of the supernova remnant. Has a

pulsar yet been detected in the center of this supernova remnant? If so, how fast is it spinning? Has the supernova debris thinned out enough to give a clear view of the neutron star?

40. Search the Web for information about SN 1994I, a supernova that occurred in the galaxy M51 (NGC 5194). Why was this supernova unusual? Was it bright enough to have been seen by amateur astronomers?

Observing Projects

41. Planetary nebulae are found all across the sky. Some of the brightest are listed in Chapter 8 of the *Discovering the Universe: From the Stars to the Planets* Web site. With the help of star maps or *Starry Night Enthusiast*™ software, locate and observe as many planetary nebulae as possible on clear, Moonless nights, using the largest telescope at your disposal. Some of the more notable planetary nebulae include: Little Dumbbell (M76), NGC 1535, Eskimo, Ghost of Jupiter, Owl (M97), Ring (M57), Blinking Planetary, Dumbbell, Saturn Nebula, and NGC 7662. Note and compare the various shapes of the different nebulae. In which cases can you see the central star?

42. Two supernova remnants can be seen through modest telescopes, one in the winter sky and the other in the summer sky. Both are quite faint, however, so you should schedule your observations for a Moonless night. The winter sky contains the Crab Nebula, which is discussed in detail in this chapter. It is located near the star marking the eastern horn of Taurus (the Bull) and can be found using *Starry Night Enthusiast*™ software. Its coordinates are R.A. 5^h 34.5^m and Dec. = $+22°\ 00'$. Whereas the entire Crab Nebula easily fits into the field of view of an eyepiece, the Veil (or Cirrus) Nebula in the summer sky is so vast that you can see only a small fraction of it at a time. The easiest way to find the Veil Nebula is to aim the telescope at the star 52 Cygni (R.A. = $20^h\ 45.7^m$ and Dec. = $+30°\ 43'$), which lies on one of the brightest portions of the nebula. This star is also accessible in *Starry Night Enthusiast*™. If you then move the telescope slightly north or south until 52 Cygni is just out of the field of view, you should see giant wisps of glowing gas.

43. Use a telescope to observe the triple star system 40 Eridani, whose coordinates are R.A. = $4^h\ 15.3^m$ and Dec. = $-7°\ 39'$. The primary, a 4.4-magnitude yellowish star like the Sun, has a 9.6-magnitude white dwarf companion, the most easily seen white dwarf in the sky. On a clear, dark night with a moderately large telescope, you should also see that the white dwarf has an 11th-magnitude companion, which completes this interesting trio.

WHAT IF... A Supernova Exploded Near Earth?

The Flash What would happen on Earth if a supernova occurred only 50 light-years away? (Considering the titanic forces that supernovae release, it should come as no surprise that the high-energy electromagnetic radiation from such an explosion detonating much closer than this distance would immediately kill virtually all life on Earth.) Neutrinos would foreshadow by a few hours the visible flash and pending flood of X rays and gamma rays from the supernova. The doomed star would then grow tremendously luminous, 50 times brighter than the Moon and only 8000 times less bright than the Sun. It would be brighter than the light from all other stars in the night sky combined.

The lethal X rays and gamma rays would be the first causes of death on Earth. Within days, organisms from virtually all species of plants and animals would begin dying of radiation poisoning. Entire radiation-sensitive species might then be annihilated. The radiation would also cause many cancers and other internal diseases over succeeding years, leading to many more plant and animal deaths. At the same time, ultraviolet radiation reaching Earth's surface would cause an astronomical jump in the rates of skin cancer and cataracts.

The first blast of ultraviolet radiation from the supernova would destroy Earth's ozone layer within a matter of days, transforming the ozone primarily into atomic oxygen. With the removal of this protective barrier, ultraviolet radiation from both the supernova and the Sun would saturate Earth's surface. After the intensity of ultraviolet radiation from the supernova diminished, sunlight would begin repairing the ozone layer, eventually returning it to normal levels.

The brightness and emission of all the electromagnetic radiation from the supernova would peak after a month. It would then fade, but the supernova remnant would be visible for millennia as an expanding cloud of gas and dust in the night sky. Fortunately for the life that survived the onslaught of high-energy radiation from the supernova, the remnant cloud would primarily emit harmless visible light.

The Aftermath The most energetic particles from the supernova and its environs, the so-called *cosmic rays,* would also affect Earth's surface. Moving at 90% of the speed of light, cosmic rays would arrive here only 5 to 10 years after the photons. Cosmic ray energies are much higher than those of any particles normally existing on Earth. For example, a typical cosmic ray has enough energy to light a small lightbulb for 1 s; it takes over 600,000 trillion electrons flowing through wiring in your house to do the same thing.

Some cosmic rays would slam into nitrogen or oxygen molecules in the air, shattering the air molecules and creating cosmic ray showers. The highest-energy cosmic rays from the supernova would reach Earth intact, as do the highest-energy cosmic rays from other sources today. These would break up atoms of the objects they strike on Earth's surface, enhancing the earlier biological damage caused by the supernova's electromagnetic radiation.

The bulk of the matter ejected into space by the supernova would travel at speeds of 16,000 km/s (60 million km/h), nearly 20 times slower than the emitted photons. Thus, the bulk of the supernova remnant would take at least 1000 years longer to reach us than did its initial radiation. By the time the bulk of the blast wave reaches the solar system, it would have become so thin and diffuse that it would probably do little damage to Earth's atmosphere. However, we could expect this material to deposit a thin but exotic mix of elements into the upper atmosphere. All of this material would eventually fall to Earth and alter the chemistry of both the ocean and the soil.

The Outcome The damage done to life by both the electromagnetic radiation and cosmic ray particles would disrupt the global food chain. In the oceans, large quantities of plankton and other microscopic organisms at the foundation of the chain would die off. As a result, many of the larger aquatic species that feed on these smaller organisms would starve. Similarly, on the surface, most plant life would wither, and many herbivorous animals would starve as a result. This would, of course, lead to the death and dislocation of animals throughout the food chain. Surviving plants and animals would undergo genetic mutations due to the supernova's radiation. Most of these genetic changes would lead to immediate death of the altered plant or animal, or their offspring, but a few changes would be beneficial and enable the mutated organisms to thrive.

How would the human race fare after the supernova? Our long-term survival would depend, in part, on whether surviving humans could coexist with surviving plants and animals, and with each other.

CHAPTER 9

Black Holes: Matters of Gravity

Black-hole Accretion Disk Artist's conception of gases spiraling toward a black hole. Some of the gas is so hot that it expands and jets outward perpendicular to the accretion disk, guided by magnetic fields. (A. Hobart, CXC)

R I V U X G

WHAT DO YOU THINK?

1 Are black holes empty holes in space? If not, what are they?

2 Does a black hole have a solid surface? If not, what is at its surface?

3 What power or force enables black holes to draw things into them?

4 How close to a black hole do you have to be for its special effects to be apparent?

5 Can you use black holes to travel to different places in the universe?

6 Do black holes last forever? If not, what happens to them?

Answers to these questions appear in the text beside the corresponding numbers in the margins and at the end of the chapter.

Stars are formed and held together by gravity. At each stage that we have considered, they are prevented from collapsing further by the outward forces generated by fusion and electromagnetism. In this chapter we explore the fate of stellar remnants with more than 3 $M_\odot$ in which gravitation "wins" the battle, but eventually loses the war between the inward and outward forces.

When the gravitational force of an object is so great that it overcomes all opposing repulsive forces or pressures (like neutron degeneracy pressure), the object collapses in on itself. Its gravitational attraction then becomes so strong that nothing—not even light—can escape from it. When this happens, the matter and the space around it become a *black hole*.

Black holes inspire awe, fear, and uncertainty. Many people harbor the mistaken belief that black holes are giant vacuum cleaners destined to "suck up" all matter in the universe. Happily, the equations that describe them reveal that black holes are more benign than that, but they are still truly strange.

The idea that gravity could prevent matter from escaping was first put forward in the eighteenth century, but it was not until the twentieth century, when Albert Einstein presented his ideas of *relativity* (how matter and motion affect mass, distance, and time), that the modern concept of a black hole developed. We must first explore the relativity theories in order to understand these exotic objects.

In this chapter you will discover

- that Einstein's theory of general relativity predicts the existence of regions of space and time that are severely distorted by the extremely dense matter they contain
- that space and time are not separate entities
- how black holes arise
- the surprisingly simple theoretical properties of black holes
- that X rays and jets of gas are created near many black holes
- the fate of black holes
- the unsurpassed energy emitted by gamma-ray bursts

THE RELATIVITY THEORIES

It seems reasonable that if I watch you traveling at, say, 100 km/h (60 mph), my measurements of your mass, the rate your watch runs, and the length of your car would be the same as the measurements you make of these things. In fact, they will not be the same. At the beginning of the twentieth century, Einstein began a revolution in science by disregarding his common sense, making a few physical assumptions about nature, and using mathematics to explore their consequences. The resulting equations revealed profound, albeit highly counterintuitive, insights into how nature works. Armed with the equations that Einstein derived, scientists have been exploring and using the consequences of his theories to develop new technology and a deeper understanding of how matter, energy, space, and time interact.

9-1 Special relativity changes our conception of space and time

 In 1905, Albert Einstein derived his **theory of special relativity**, a description of how motion affects our measurements of distance, time, and mass. He was guided by two innovative ideas. Although the implications of the theory proved revolutionary, the first notion seems simple:

Your description of physical reality is the same regardless of the (constant) velocity at which you move.

In other words, as long as you are moving in a straight line at a constant speed, you experience the same laws of physics as anyone else moving at any other constant speed and in any other direction. To illustrate this first idea, suppose you were inside a closed boxcar moving smoothly in a straight line at 100 km/h and you dropped a pen from a height of 2 m. You time how long it takes the pen to reach the floor (about 0.64 s). Then the train is stopped and you repeat the same experiment in the station. The time it would take the pen to fall would be identical.

The second idea seems more bizarre:

Regardless of your speed or direction, you always measure the speed of light to be the same.

Suppose that you are in a (very powerful) car moving toward a distant street lamp at 150,000 km/s (93,000 mi/s). In simpler terms, this is 50% the speed of light, denoted $0.5c$ (Figure 9-1a). A friend of yours leaning on the lamppost sees you coming and also sees light from the lamp heading toward you at c. How fast do *you* see the photons of light coming at you? It would seem reasonable

Light photons from lamp as seen from ground have velocity = 1c

Velocity = 0.5c

View of light and car from the ground

a

Light photons from lamp as seen from car have velocity = 1c

View of light and ground from the car

b

FIGURE 9-1 The Speed of Light Is Constant (a) As seen from the ground, photons of light from the lamp are traveling toward the car with a speed $v = c$ (ignoring the slight decrease in speed due to the presence of the air). (b) As seen from the car, moving toward the lamp at $v = 0.5c$, the photons are also traveling at the speed $v = c$. The difference in color between the light in the two figures is due to the Doppler shift, as explained in *An Astronomer's Toolbox 4-3*.

that you clock them at $c + 0.5c$, which is the speed of the photons toward you plus your speed toward them, but this is incorrect. You will, in fact, see the photons coming toward you with the same speed, c, that your friend sees them leaving the lamp (Figure 9-1b). You would, however, see that they have a different color than your friend sees. That difference is due to the Doppler shift, as explained in *An Astronomer's Toolbox 4-3*.

Einstein's theory of special relativity expresses these two assumptions mathematically, and the results of these assumptions have been confirmed in innumerable experiments. Perhaps the most well-known result of special relativity is the relationship between matter and energy, $E = mc^2$. The fact that mass is multiplied by the square of a very large number reveals that a little mass creates a lot of energy. This equation implicitly states that, contrary to most daily experience (and what most people are taught as children), matter *can* be converted into energy and energy *can* be converted into matter. This conversion takes place when stars fuse elements together, like the hydrogen fusion that occurs in the Sun today. Three other fundamental results of special relativity also defy our everyday experience:

1. **The length of an object decreases as its speed increases.** In other words, if a train moves toward or away from you, you would measure its length to be shorter than you would measure its length to be when it is stopped at the station (Figure 9-2). (This result also explains why the car in Figure 9-1 appears too short.) This result is called *length contraction*. However, if you measure the length of the moving train while you are inside it and moving with it, your measurement of its length will be the same as you measured on the ground when it was at rest. The word *relativity* emphasizes the importance of the relative speed between the observer and the measured object.

2. **Clocks that you see as moving run more slowly than do clocks you see at rest.** This result is called *time*

dilation. Indeed, the faster a clock moves relative to you, the slower it appears to tick from your perspective. For example, air travelers actually age more slowly than they would if they did not fly (although because of the relatively low speeds of aircraft travel compared to the speed of light, this difference is imperceptibly small). People flying in aircraft do not feel time moving more slowly, however, because their biological activities slow down at the same rate as the clocks around them. Only an observer moving more slowly sees that the clock and the activities of the high-speed travelers have slowed. These connections between motion and clocks mean that space and time

This train is at rest relative to you.

The same train is now moving at very high speed relative to you.

FIGURE 9-2 Movement and Space According to the theory of special relativity, the faster an object moves, the shorter it becomes in its direction of motion as observed by someone not moving with the object. It becomes infinitesimally short as its speed approaches the speed of light. The dimensions perpendicular to the object's motion are unchanged.

a   b   c

FIGURE 9-3 Curved Spacetime (a) This flat surface represents two dimensions in spacetime. In the absence of any matter in the spacetime, straight lines are straight in our intuitive sense (the sheet is flat). (b) In the presence of matter, spacetime curves, as shown by the curvature of the sheet when mass is laid on it. Straight lines, defined by the paths that light rays take, are no longer straight in the "usual" sense. Besides changing the path of photons, this curvature also creates gravity, which (c) pulls the two masses toward each other.

cannot be considered as two separate concepts. Relativity requires us to combine them in a single entity, thus creating the concept of **spacetime**.

3.   The mass of an object increases as it moves faster. The concept of mass is discussed in Section 2-7. The equations of special relativity reveal that an object moving at the speed of light would have an infinite amount of mass. It is impossible for any object with mass to reach this speed, much less go faster than the speed of light, because an infinite amount of mass is more than the total mass that exists in the universe. Furthermore, the equation $F = ma$ (F is the force exerted on an object, m is its mass, and a is the acceleration it undergoes; see Section 2-7) reveals that if an object's mass is infinite, the force necessary to accelerate this object must also be infinite. Because the total force available to move matter in the universe is finite, no massive object can be sped up to the speed of light: The speed of light is the universal speed limit.

How much mass, m, would have to be destroyed to create an amount of energy, E?

9-2   General relativity explains how matter warps spacetime, creating gravitational attraction

WEB LINK 9.2   While special relativity tells how the mass of an object is related to its (constant) speed, this theory is special (in the sense of limited) because it does not account for the effects of acceleration and gravitation. They were added by Einstein in 1915 in a more general theory, called the **theory of general relativity** (or, more simply, general relativity). It describes how spacetime changes shape in the presence of matter (Figure 9-3a, b). The greater the mass, the more the distortion or *curvature*. Furthermore, the curvature of spacetime creates attraction between all pieces of matter in the universe (Figure 9-3b)—this attraction is what we call the *gravitational force*.

Another result of general relativity is that time slows down in the presence of matter. The greater the concentration of matter, the slower clocks tick (Figure 9-4).

Two identical clocks synchronized in space far from other masses

Time moves more slowly near a larger mass than near a smaller mass.

FIGURE 9-4 Time Slows Down Near Matter (a) Two clocks in space set at exactly the same time are (b) brought to Earth and the Moon. From a vantage point far from Earth and the Moon, the clock on Earth is ticking more slowly than the clock on the Moon. This occurs because mass slows down the flow of time, and Earth has more mass (and a higher density, which adds to the effect) than the Moon.

a   b

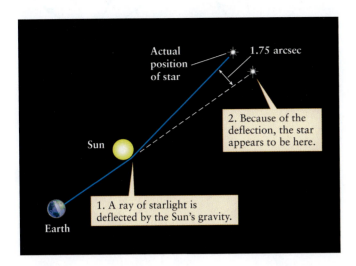

FIGURE 9-5 *Curved Spacetime and the Path of Light* The warping of space by matter causes light to be deflected. This was the first prediction of general relativity to be confirmed, in 1919. This confirmation came when stars behind the Sun were observed during an eclipse. The star in this drawing was not observed where it was supposed to be, as a result of the Sun's gravity changing the path of its light.

For example, time passes more slowly for us here on Earth's surface than it would for astronauts on the Moon, which has less mass than Earth.

Newton's laws of motion and his universal law of gravitation are accurate only for objects with relatively small masses, slow velocities compared to the speed of light, and low densities (such as those for objects found on Earth). Newton's laws are also limited to motion sufficiently far from large masses (such as the Sun) or high-density objects (such as neutron stars). We can use Newton's laws to accurately predict the paths of projectiles on Earth and the motion of our planet around the Sun, but they do not describe very well the orbit of Mercury around the Sun (because that planet is so close to our star) or two neutron stars that orbit each other. In such cases, general relativity correctly predicts how objects move and how time passes. Closer to home, the accurate determination of location using GPS satellites orbiting the Earth requires the use of the equations of general relativity.

9-3 Spacetime affects the behavior of light

Besides affecting the behavior of objects, the curvature of spacetime changes the path and wavelength of light that passes near any matter. (These behaviors are not predicted by Newton's laws.) Light travels along trajectories in space, called *geodesics.* You can get an idea of how geodesics work by imagining that you are flying on the shortest possible route from one city to another. Your path would be analogous to a geodesic for a photon, and like that particle in the presence of matter in space, you would actually be following the curve of Earth's surface,

rather than going in a "straight line." The first experimental verification of general relativity was its prediction that geodesics are sufficiently curved to cause light from stars behind the Sun to arc around it (Figure 9-5).

Photons that leave the vicinity of a star also lose energy in climbing out of the star's gravitational field. They do not slow down like a bullet fired upward. Rather, they shift to longer wavelengths (Figure 9-6), an effect we see in the spectra of some white dwarfs, whose light appears redder than it would if this effect did not occur. This shift of wavelengths leaving the vicinity of a massive object is called **gravitational redshift.**

General relativity has been confirmed again and again, as seen in these observations:

- Light is measurably deflected by the gravitational curving of space due to the presence of matter like stars or entire galaxies containing billions of stars (see Figure 9-5).

- The perihelion position of Mercury as seen from the Sun shifts, or precesses, by 43 arcsec per year more

> What property of matter does general relativity address that is not included in special relativity?

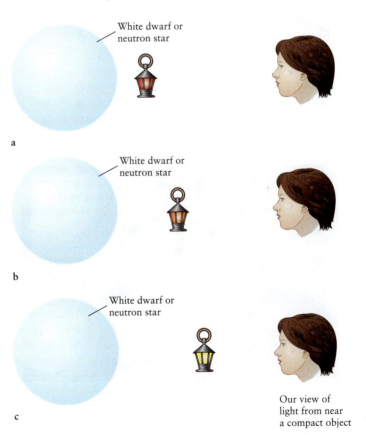

FIGURE 9-6 *Gravitational Redshift* The color of light from the same object located at different distances from a mass appears different as seen from far away. The photons that leave the vicinity of the massive object lose energy and therefore redshift. The closer the light source is to the mass, the redder the light appears, and hence the name gravitational redshift. The same argument applies to light leaving the surfaces of different stars.

than is predicted by Newtonian gravitational theory (Figure 9-7).

- The orbits of stars in binary systems follow the paths predicted by general relativity, rather than those predicted by Newtonian gravitation.

Does light change direction as it goes past Earth? Why or why not?

- The spectra of stars are observed to have the gravitational redshifts predicted by general relativity (see Figure 9-6).

9-4 General relativity predicts the fate of massive star cores—black holes

Let us now return to stellar evolution and consider what happens when high-mass main-sequence stars (with more than 25 $M_\odot$) explode as supernovae. As with the stars that contain between 8 $M_\odot$ and 25 $M_\odot$, high-mass stars create neutron stars. The neutron-star remnants of these high-mass stars, however, have more than 3 $M_\odot$. This is problematic because above this mass, the gravitational force of the neutron star overcomes the neutron degeneracy pressure created by the interactions among its particles. As a result, such remnant stars collapse further, and no force in nature can stop their infall.

We can use the equations of general relativity to understand the fate of collapsing neutron stars. We just saw that all matter warps the space around itself (see Figures 9-3 and 9-5). When matter gets sufficiently dense, it actually causes space near it to curve so much that it closes in on itself (Figure 9-8). Photons flying

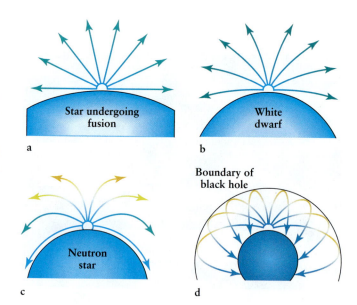

FIGURE 9-8 Trapping of Light by a Black Hole (a) The paths and color of light rays departing from a main-sequence, giant, or supergiant star are affected very little by the star's gravitational force. (b) Light leaving the vicinity of a white dwarf curves and redshifts more, whereas (c) near a neutron star, some of the photons actually return to the star's surface. (d) Inside a black hole, all light remains trapped. Most photons curve back in, except those that fly straight upward, which become infinitely redshifted, thereby disappearing.

outward at an angle from such a collapsing star arc back inward. Photons flying straight upward redshift so much that they lose all their energy (see Section 3-3) and hence they cease to exist.

If light, the fastest moving of all known things, cannot escape from the vicinity of such dense matter, then nothing can! Such regions out of which no matter or any form of electromagnetic radiation can leave are called **black holes**. Plummeting in on itself, a collapsing neutron star becomes so dense that it ceases to consist of neutrons. General relativity predicts that in creating a black hole, matter compresses to infinite density (and zero volume), a state called a **singularity**. However, we know that general relativity and our other theories of physics are invalid in such a situation. A more comprehensive theory of nature must be developed to explain the state of matter in a black hole's singularity. Efforts are under way to develop such a theory. The best-known nascent theory is called *superstrings*, but its mathematics is daunting and much work remains to be done. All we can say with confidence is that the matter in black holes becomes incredibly dense and compact.

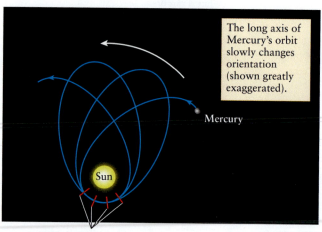

FIGURE 9-7 Mercury's Orbit Explained by General Relativity The location of Mercury's perihelion (its position closest to the Sun) and its long axis of orbit change, or precess, with each orbit. This occurs because of the gravitational influences of the other planets plus the curvature of space as predicted by Einstein's theory of general relativity. The amount of precession is inconsistent with the prediction of the orbit made by Newton's law of gravity alone.

The long axis of Mercury's orbit slowly changes orientation (shown greatly exaggerated).

Mercury

Sun

Perihelia of Mercury's orbit

INSIDE A BLACK HOLE

The formation of a black hole is complicated, but its nature is surprisingly simple: It contains matter at its center or in a ring, it has a boundary shaped like a sphere, it either has a net electric charge or not, and it either rotates or does not rotate.

9-5 Matter in a black hole becomes much simpler than elsewhere in the universe

2 A black hole is separated from the rest of the universe by a boundary, called its **event horizon**. The event horizon is not like the surface of a solid or liquid body. No matter exists at this location except for the instant it takes infalling mass to cross the event horizon and enter the black hole.

We cannot look inside a black hole because no electromagnetic radiation escapes from it. Our understanding of its structure comes from the equations of general relativity. According to Einstein's theory, the event horizon is a sphere. The distance from the center of the black hole to its event horizon is called the **Schwarzschild radius** (abbreviated R_{Sch}), after the German physicist Karl Schwarzschild, who first determined its properties. *An Astronomer's Toolbox 9-1* shows how to calculate this distance, which depends only on the black hole's mass. The more massive the black hole, the larger its event horizon.

According to the equations of general relativity, when a stellar remnant collapses to a black hole, it loses its internal magnetic field. The field's energy radiates away in the form of electromagnetic radiation and **gravitational radiation**. Emitted as **gravitational waves**, this energy travels as ripples in the very fabric of spacetime. Gravitational waves are also created when neutron stars or black holes collide or when stars and stellar remnants are in close orbits around each other. (Actually, gravitational radiation is emitted whenever any two things move around each other, such as a pair of dancers. When the moving bodies are less massive than stars or stellar remnants, however, we have no hope of detecting their very weak gravitational radiation with either present or projected technology.)

Gravitational radiation has not yet been directly detected, unlike, say, visible light that we "see" by its effects on our eyes or on a CCD. However, gravitational radiation has been observed indirectly by its effects on the orbits of some binary star systems. In particular, a pair of neutron stars orbiting each other emit so much energy as gravitational waves that the two bodies spiral toward each other. The change in their orbit is correctly predicted by general relativity. This agreement earned a Nobel Prize in Physics in 1993 for Joseph Taylor and Russell Hulse, who discovered the first binary pulsar in 1974 and measured the changes in its neutron stars' orbits.

The ripples in spacetime created by gravitational waves from stars or stellar remnants are incredibly tiny. On Earth, each meter-wide volume of space changes by less than 10^{-25} m as the waves pass by. Astronomers around the world are building *gravitational wave detectors* to measure these small changes (Figure 9-9). They have been in operation since 2002, and astronomers expect that they will eventually provide a way of directly observing high-energy gravitational activity in the cosmos.

Characteristics of Black Holes Besides losing its internal magnetic field, matter within a black hole loses almost all traces of its composition and origin. Indeed, it retains only three properties that it had before entering the black hole: its *mass*, its *angular momentum*, and its *electric charge*. Familiar concepts, such as

> Why do neutron stars of more than about 3 $M_\odot$ collapse to form black holes?

AN ASTRONOMER'S TOOLBOX 9-1

The Sizes of Black Holes

In 1918, Karl Schwarzschild, a German physicist, discovered the first solution of Einstein's equations. His solution describes the nature of nonrotating black holes. According to Einstein's general theory of relativity, the Schwarzschild radius R_{Sch} of any black hole can be found from its mass, M:

$$R_{Sch} = \frac{2GM}{c^2},$$

where R_{Sch} is measured in meters; M is the black hole's mass in kilograms; c is the speed of light, 3×10^8 m/s; and G is the gravitational constant, 6.67×10^{-11} m^3/kg·s^2. Expressing M in terms of solar masses, this equation can be conveniently approximated by

$$R_{Sch} \approx 3 \, M_{BH},$$

where M_{BH} is the black hole's mass in solar masses and R_{Sch} is in kilometers.

Example: A 5-$M_\odot$ main-sequence star has a radius of 3×10^6 km, whereas the previous equation reveals that a black hole with the same mass has a 15-km Schwarzschild radius.

Try these questions: A primordial black hole with the mass of Mount Everest would have a Schwarzschild radius of just 1.5×10^{-15} m! What is Mount Everest's mass? What is the Schwarzschild radius of a galactic black hole that contains 3 billion solar masses? A black hole with a Schwarzschild radius equal to the radius of Earth has how much mass?

(Answers appear at the end of the book.)

FIGURE 9-9 Laser Interferometer Gravitational Wave Observatory (LIGO) Located in Hanford, Washington, this is one of several gravitational wave (colloquially, gravity wave) detectors around the world. It has two perpendicular arms, each 4 km long. Gravity waves that pass the detector cause unequal changes in the lengths of the arms. These changes are detected by lasers inside each arm. (LIGO Laboratory)

proton, neutron, electron, atom, and molecule, no longer apply. In addition, because few large bodies appear to have a net charge, it is also doubtful that black holes do. We therefore predict that there are only two different types of black holes: those that rotate and those that do not.

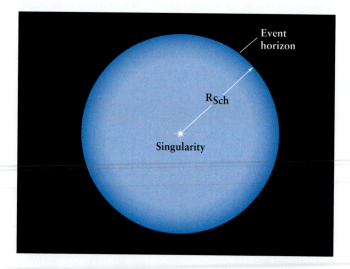

FIGURE 9-10 Structure of a Schwarzschild (Nonrotating) Black Hole A nonrotating black hole has only two notable features: its singularity and its boundary. Its mass, called a singularity because it is so dense, collects at its center. The spherical boundary between the black hole and the outside universe is called the event horizon. The distance from the center to the event horizon is the Schwarzschild radius, R_{Sch}. There is no solid, liquid, or gas surface at the event horizon. In fact, except for its location at the boundary of the black hole, an event horizon lacks any features at all.

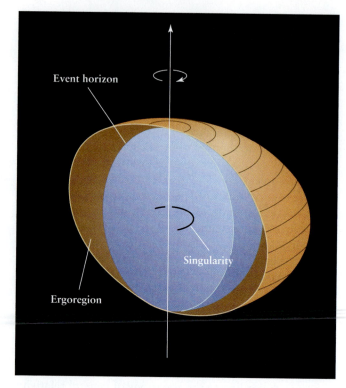

FIGURE 9-11 Structure of a Kerr (Rotating) Black Hole Rotating black holes are only slightly more complex than non-rotating ones. The singularity of a Kerr black hole is located in an infinitely thin ring around the center of the hole. It appears as an arc in this cutaway drawing. The event horizon is again a spherical surface. There is also a doughnut-shaped region, called the ergoregion, just outside the event horizon, in which nothing can remain at rest. Space in the ergoregion is being curved or pulled around by the rotating black hole.

Types of Black Holes If the mass creating a black hole is not rotating, the black hole that is formed does not rotate either. We call nonrotating black holes **Schwarzschild black holes** (Figure 9-10). General relativity predicts that all the mass in such a black hole collapses to a point of infinite density at its center, the singularity mentioned earlier. The rest of the volume from the event horizon to the singularity of a Schwarzschild black hole is empty space.

When the matter that creates a black hole possesses angular momentum, that matter collapses to a ring-shaped singularity located inside the black hole between its center and the event horizon (Figure 9-11). Such rotating black holes are called **Kerr black holes**, in honor of the New Zealand mathematician Roy Kerr, who first calculated their structure in 1963. Once again, the black hole is empty except for the singularity. Most Kerr black holes should be spinning thousands of times every second, even faster than the pulsars that we studied in Chapter 8. The black hole with the highest known rotation rate, discovered in 2006, rotates a thousand times per second, consistent with this theory.

Unlike Schwarzschild black holes, the equations indicate that Kerr black holes possess a doughnut-shaped

Ergoregion

Black hole

Event horizon

R I V U X G

FIGURE 9-12 Swirling Space in an Ergoregion Just as the chocolate in this blender is being dragged by the spinning blade, so, too, is space dragged by a spinning (Kerr) black hole. (Jack Andersen/Food Pix)

region directly *outside* of their event horizons in which objects cannot remain at rest without falling into the black hole. Called **ergoregions**, they are volumes of space-time that the rotating black hole drags around, like so much batter in a blender (Figure 9-12). In 1997, matter orbiting a black hole was observed to behave in a fashion consistent with the existence of an ergoregion surrounding the hole. If it is moving fast enough, an object that enters the ergoregion can again fly out of it; however, if it stops in the ergoregion, it must fall into the black hole.

9-6 Falling into a black hole is an infinite voyage

Imagine being in a spacecraft orbiting only 1000 **3** Schwarzschild radii (15,000 km) from an isolated 5-$M_\odot$ black hole (see *An Astronomer's Toolbox 9-1* for the equation of the Schwarzschild radius). You are held in orbit by the black hole's gravitational force. Even at that short distance, the only effect the black hole has on you is its gravitational attraction. Only when you get very, very close to the event horizon do bizarre things begin to happen. To investigate these changes, you send a cube-shaped probe with a clock toward the black hole, with the same side of the cube always facing "downward" toward the black hole. The probe emits a blue glow so that you can follow its progress. What happens to the cube as it approaches the black hole?

> Where is the singularity of a rotating black hole?

From the time you eject it until it reaches about 100 **4** Schwarzschild radii (1500 km), you see the probe descend as if it were falling toward a planet or moon (Figure 9-13a). Time measured by the probe's clock slows down only slightly. At 100 Schwarzschild radii, however, the probe begins to respond significantly to the powerful tidal effect from the black hole. The face of the probe closest to the event horizon receives perceptibly more gravitational pull from the hole than its farther-away parts, and it begins to stretch apart.

By the time the probe comes within a few Schwarzschild radii of the event horizon, the tidal forces on it are so great that it violently elongates. The part of the probe closest to the black hole accelerates downward and away from the rest of the probe. Furthermore, the sides of the probe are drawn together: They are falling in straight lines toward a common center. The net gravitational effect of moving close to the event

Clock on a spaceship
far from the black hole

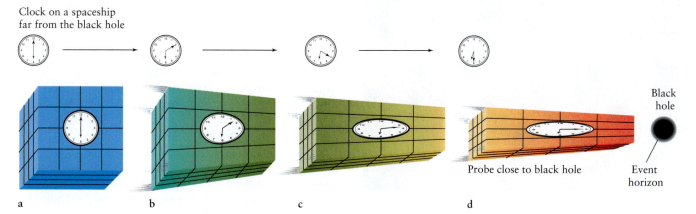

Black hole

Probe close to black hole

Event horizon

a   b   c   d

WEB LINK 9-4

FIGURE 9-13 Effect of a Black Hole's Tidal Force on Infalling Matter (a) A cube-shaped probe 1500 km from a 5-$M_\odot$ black hole. (b, c, d) Near the Schwarzschild radius, the probe is pulled long and thin by the difference in the gravitational forces felt by its different

sides. This tidal effect is a greatly magnified version of the Moon's gravitational force on Earth. The probe changes color as its photons undergo extreme gravitational redshift and time slows down on the probe, as seen from far away.

horizon is for the probe to be pulled long and thin. From a practical perspective, this means that the probe would be violently torn apart, because it is not composed of perfectly elastic material.

As the probe nears the black hole, the blue photons leaving it must give up more and more energy to escape the increasing gravitational force. However, unlike a projectile fired upward, photons cannot slow down. Rather, they lose energy by increasing their wavelengths (see *An Astronomer's Toolbox 3-1*). This is another example of the gravitational redshift predicted by general relativity. The closer the probe gets to the event horizon, the more its light is redshifted—first to green, then yellow, then orange, then red, then infrared, and, finally, to radio waves (see Figure 9-13b, c, d).

Stranger still is the black hole's effect on time. General relativity predicts that when the probe approaches within a few Schwarzschild radii of the black hole, its infall rate will slow down as seen from far away. Also, signals from the probe show you that its clocks are running much more slowly than they did when it left your spacecraft. Time dilation becomes so great near the event horizon that the probe will appear to hover above it and its clocks will stop.

> How would a 1M⊙ black hole 1 AU from Earth affect our planet?

Anyone in the probe would observe something else altogether: They see the probe actually cross the event horizon and continue falling toward the black hole's singularity in a normal period of time, according to their own watch. Pulled apart by tidal effects, the probe disintegrates as it falls inward. Contrary to the science fiction concept of traveling great distances quickly by passing through a black hole, calculations indicate that objects entering them could not survive passage through, even if there were a way to come out somewhere else.

Could a black hole be connected to another part of spacetime or even some other universe? General relativity predicts such connections, called **wormholes**, for Kerr black holes, but astrophysicists are skeptical that the equations are correct in this regard. Their conviction is called **cosmic censorship**: Nothing can leave a local region of space that contains a singularity (that is, a black hole).

R I V U X G

FIGURE 9-14 Formation of an Accretion Disk Just as the water in this photograph swirls around waiting to get down the drain, the matter pulled toward a black hole spirals inward. Angular momentum of the infalling gas and dust causes them to form an accretion disk around the hole. (Chris Collins/Corbis)

EVIDENCE FOR BLACK HOLES

9-7 Several binary star systems contain black holes

Black holes are more than fine points of relativity theory: They are real. Their presence has been observed by their effects on the orbits of other stars and on gas and dust near them, and more of them are being located all of the time. To find evidence for black holes, we look first to binary star systems.

The technique for detecting black holes formed from collapsing stars is based on the interaction between the black hole and its binary companion. When one star in a close binary becomes a black hole, its gravitational attraction pulls off some of its companion's atmosphere. However, such black holes have diameters of only a few kilometers, so there is not enough room for all that gas to fall straight in. Rather, the infalling gas swirls into the black hole like water going down a bathtub drain (Figure 9-14). The gas waiting to fall in forms an **accretion disk**, a disk of gas and dust spiraling in toward the black hole. Magnetic fields in the disk help pull the debris inward. Calculations reveal that this gas is compressed and heated so much from the collisions of its particles that it gives off X rays (Figure 9-15). Thus, if a visible star has a sufficiently tiny, sufficiently massive X-ray-emitting companion, we have located a black hole. See *Guided Discovery: Identifying Stellar-Remnant Black Holes* for details on how this is done.

To date, in the Milky Way Galaxy alone, at least 10 stellar-remnant black holes in binary star systems have

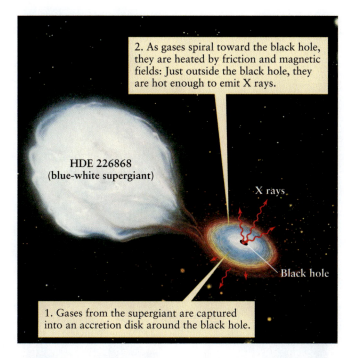

2. As gases spiral toward the black hole, they are heated by friction and magnetic fields: Just outside the black hole, they are hot enough to emit X rays.

HDE 226868
(blue-white supergiant)

X rays

Black hole

1. Gases from the supergiant are captured into an accretion disk around the black hole.

FIGURE 9-15 X Rays Generated by Accretion of Matter Near a Black Hole Stellar-remnant black holes, such as Cygnus X-1, LMC X-3, V404 Cygni, and probably A0620-00, are detected in close binary star systems. This drawing (of the Cygnus X-1 system) shows how gas from the 30 $M_\odot$ companion star, HDE 226868, transfers to the black hole, which has at least 11 $M_\odot$. This process creates an accretion disk. As the gas spirals inward, friction and compression heat it so much that the gas emits X rays, which astronomers can detect. (Courtesy of D. Norton, Science Graphics)

been identified (Table 9-1). There is also growing evidence that black holes can collide with each other or with different types of objects, such as neutron stars.

9-8 Other black holes range in mass up to billions of solar masses

Based on the equations of general relativity, the fate of massive neutron stars led as early as 1939 to the idea of black holes. Since then, calculations have supported at least three other types of black holes:

1. Early in the life of the universe, black holes could have formed from the condensation of vast amounts of gas, as well as the collisions of stars, during the process of galaxy formation, thereby creating **supermassive black holes**, each with millions or billions of solar masses.

2. We saw in Section 7-2 that stars form in clusters. If a cluster forms with enough stars concentrated in a sufficiently small volume of space, collisions between stars in the center of the cluster can lead to the formation of a black hole with between a few hundred and

a few thousand solar masses. These are called **intermediate-mass black holes**.

Why is the gas in an accretion disk heated?

3. Black holes could have formed during the explosive beginning of the universe as tiny amounts of matter were compressed sufficiently to form **primordial black holes**. These bodies would have masses from grams to the mass of a planet.

Supermassive Black Holes In May 1994, the Hubble Space Telescope obtained compelling evidence for a black hole at the center of the galaxy M87. In the nucleus of M87 is a tiny, bright source of light. Spectra showed that nearby gas and stars are orbiting it extremely rapidly. They can be held in place only if the bright object contains some three *billion* solar masses (Figure 9-16). Given that its size is only slightly larger than the solar system, the source can only be a black hole.

Since 1994, black holes in the centers of many galaxies have been identified by their X-ray emissions and gravitational effects on surrounding gas and stars. For example, a distinct, frisbee-shaped accretion disk around the supermassive black hole in NGC 7052 was seen by the Hubble Space Telescope in 1998 (Figure 9-17). As we will see in the next chapter, a black hole containing several million solar masses has even been found at the center of our Milky Way, only 26,000 light-years from Earth. Indeed, evidence for central, supermassive black holes has been found in most of the several dozen nearest galaxies.

TABLE 9-1 Stellar Black Holes in the Milky Way

X-ray source name	Mass* of companion	Mass* of black hole
Cygnus X-1	24–42	11–21
V404 Cygni	~0.6	10–15
V461 Sgr	~10	5.5–8
GS 2000+25	~0.7	6–14
H 1705-250	0.3–0.6	6.4–6.9
GRO J1655-40	2.34	7.02
A 0620-00	0.2–0.7	5–10
GS 1124-T68	0.5–0.8	4.2–6.5
GRO J042+32	~0.3	6–14
4U 1543-47	~2.5	2.7–7.5

*Solar masses.
"Revisiting the Black Hole." R. Blandford and N. Gehrels. *Physics Today,* June 1999.

Identifying Stellar–Remnant Black Holes

Shortly after the *Uhuru* X-ray satellite was launched in the early 1970s, astronomers found a promising black-hole candidate—an X-ray source, called Cygnus X-1. This source is highly variable and irregular. Its strong X-ray emission flickers on time scales as short as a hundredth of a second. If different parts of an X-ray source grew bright at different times, its emission would be a continuous stream. For Cygnus X-1 to flicker, the entire star must brighten and dim as a unit. Therefore, light must have time to travel across Cygnus X-1 between pulses. Because light travels 3×10^8 km/s, it travels 3000 km in the "flicker time" of a hundredth of a second. This means that for all of Cygnus X-1 to brighten and darken simultaneously, it must be about 3000 km across, smaller in diameter than Earth.

Cygnus X-1 occasionally emits radio radiation, and, in 1971, radio astronomers succeeded in associating Cygnus X-1 with the visible star HDE 226868 (see accompanying figure), a B0 supergiant with a surface temperature of about 31,000 K. Because such stars do not emit significant X rays, HDE 226868 alone cannot be the Cygnus X-1 X-ray source. Spectroscopic observations soon showed that the lines in the spectrum of HDE 226868 shift back and forth within a period of 5.6 days. This behavior is characteristic of a single-line spectroscopic binary (see Section 6-12), and HDE 226868's companion is too dim to produce its own set of spectral lines. The clear implication is that HDE 226868 and Cygnus X-1 are the two components of a binary star system.

The B0 supergiant HDE 226868's mass is estimated at about 30 $M_\odot$, like other B0 supergiants. As a result, Cygnus X-1 must have at least 11 $M_\odot$; otherwise, it would not exert enough gravitational pull to make the B0 star wobble by the amount deduced from the periodic Doppler shift of its spectral lines. Cygnus X-1 cannot be a white dwarf or a neutron star, because its mass is too large for either of these objects. The only remaining possibility is that it must be a fully collapsed star—a black hole.

In the early 1980s, a similar binary system was identified in the nearby galaxy called the *Large Magellanic Cloud*. The X-ray source, called LMC X-3, exhibits rapid fluctuations, just like those of Cygnus X-1. LMC X-3 orbits a B3 main-sequence star every 1.7 days. From its orbital data, astronomers conclude that the mass of LMC X-3 is about 10 $M_\odot$, which would make it a black hole.

R I V U X G

HDE 226868 This star is the visual companion of the X-ray source Cygnus X-1. This binary system is located about 8000 light-years from Earth and contains a black hole of at least 11 $M_\odot$ in orbit with HDE 226868, a B0 blue supergiant star. The photograph was taken with the 200-in. telescope at Palomar Observatory on Palomar Mountain, north of San Diego. The slightly dimmer star above is an optical double that is not part of the binary system. (J. Kristian, Carnegie Observatories)

WEB LINK 9.6

Another black-hole candidate is a spectroscopic binary in the constellation Monoceros that contains the flickering X-ray source A0620-00. The visible companion of A0620-00 is an orange-colored dwarf star. It is a low-mass, main-sequence star of spectral type K that orbits the X-ray source every 7.75 h. From orbital data, astronomers conclude that the mass of A0620-00 must be greater than 5 $M_\odot$, probably about 9 $M_\odot$.

Earlier in this book we studied how tidal effects of one body on another occur in several situations, including in the Earth-Moon system and in the heating of Io and Europa by Jupiter. In 2004, astronomers observed several supermassive black holes in other galaxies absorbing parts of passing stars. These black holes created such high tides on those stars that the stars were pulled apart. Some of each star's mass was then drawn into the black hole. These discoveries were made because infalling gas from the stars was rapidly heated to millions of degrees Kelvin, causing the gas to emit a burst of X rays that was observed by orbiting telescopes.

As we will explore in more detail in Chapter 13, galaxies were created from condensing gas and stars in the early universe. Our technology is now providing us with observations of that epoch of the universe, and the formation process of supermassive black holes in the centers of galaxies is now being studied. These black holes apparently formed, in part, from the gas that was condensing to create galaxies and, in part, from the collisions of stars and black holes that came into existence shortly after the beginning of time. There is also observational evidence that the formation of supermassive black holes is still ongoing. We will explore this further in Chapter 11.

R I V U X G

FIGURE 9-16 Supermassive Black Hole The bright region in the center of galaxy M87 has stars and gas held in tight orbits by a black hole. M87's bright nucleus (center of the region in the white box) is only about the size of the solar system but it pulls on the nearby stars with so much force that astronomers calculate that it is a 3×10^9 $M_\odot$ black hole. (Holland Ford, STScI/Johns Hopkins University; Richard Harms, Applied Research Corp.; Zlatan Tsvetanov, Arthur Davidsen, and Gerard Kriss at Johns Hopkins University; Ralph Bohlin and George Hartig at STScI; Linda Dressel and Ajay K. Kochhar, Applied Research Corp., Landover, MD; and Bruce Margon, University of Washington, Seattle)

a

R I V U X G

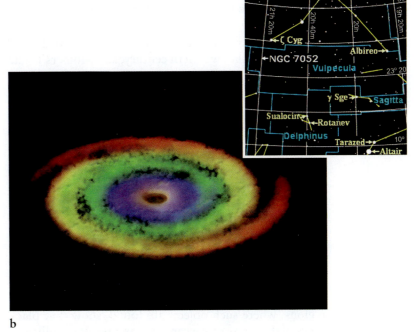

b

FIGURE 9-17 Accretion Disk Around a Supermassive Black Hole (a) Swirling around a 3×10^8 $M_\odot$ black hole in the center of the galaxy NGC 7052, this disk of gas and dust is 3700 ly across. The gas is cascading into the black hole, which will consume it all over the next few billion years. The black hole appears bright because of light emitted by the hot, accreting gas outside its event horizon. NGC 7052 is 191 million ly from Earth in the constellation Vulpecula. (b) This drawing shows how the gases spiraling inward in an accretion disk heat up as they approach the black hole. Color coding follows Wien's displacement law: red (coolest), followed by orange, yellow, green, blue, and violet (hottest). (R. P. van der Marel, STScI/F. C. van den Bosch, University of Washington/NASA; b: NASA/CXC/SAO)

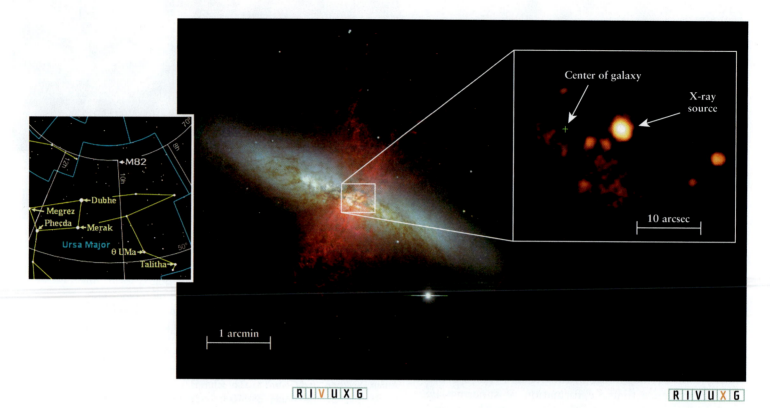

R I V U X G   R I V U X G

FIGURE 9-18 An "Intermediate-Mass" Black Hole M82 is an unusual galaxy in the constellation Ursa Major. The *inset* shows an image of the central region of M82 from the *Chandra* X-ray Observatory. The bright, compact X-ray source shown varies in its light output over a period of months. The properties of this source suggest that it is a black hole of roughly 500 $M_\odot$. (Subaru Telescope, National Astronomical Observatory of Japan; *inset:* NASA/SAO/CXC)

Intermediate-Mass Black Holes At the end of the twentieth century, astronomers began discovering objects that appear to be black holes with between 10^2 and 10^5 $M_\odot$. Again, these objects were identified as potential black holes by the intensity and spectra of the X rays that they emit and by the orbits of stars around them. For example, near the center of the galaxy M82, astronomers found what appears to be a 500–1000 $M_\odot$ black hole (Figure 9-18), while in the galaxy NGC 1313, astronomers have observed what appear to be two intermediate-mass black holes, each with 200–500 $M_\odot$. Most intermediate mass black holes are observed in globular clusters.

In support of the belief that these objects are black holes, computer simulations of stars in the crowded regions, where such objects are found, show that black holes in this mass range can form as a result of collisions and close tidal interactions between stars. Do not fear—such a black hole will not develop near Earth. There have to be at least a million times more stars per volume than there are in our stellar neighborhood for such frequent stellar collisions or near misses to occur.

WEB LINK 9.8

Where are supermassive black holes located?

VIDEO 9.1

Primordial Black Holes Even more exotic black holes may have formed along with the universe itself. The British astrophysicist Stephen Hawking has proposed that the Big Bang explosion from which most astronomers believe the universe emerged may have been chaotic and powerful enough to have compressed tiny knots of matter into primordial black holes. Their masses may have ranged from a few grams to greater than the mass of Earth. Astronomers have not yet observed evidence of primordial black holes, although that does not mean they do not necessarily exist—just that we cannot yet detect them.

9-9 Black holes and neutron stars in binary systems often create jets of gas

We saw in Section 8-11 how beams of radiation are emitted by neutron stars. Similarly, many neutron stars and black holes emit pairs of gas jets shooting out in opposite directions.

Consider a black hole (neutron stars behave analogously) in a binary system. If the companion star is still fusing material, then its outer layers can be pulled off, as discussed in Section 7-15. The infall of this matter toward the compact companion is too rapid for all of the mass to enter the event horizon immediately. The resulting accretion disk around the black hole is the key to the jets.

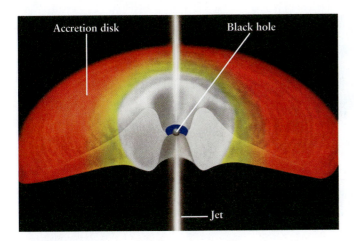

FIGURE 9-19 *Jets Created by a Black Hole in a Binary System* Some of the matter spiraling inward in the accretion disk around a black hole is superheated and redirected outward to produce two powerful jets of particles that travel at close to the speed of light. The companion star is off to one side of this drawing.

As the disk mass spirals down toward the event horizon, this gas is being compressed into a smaller volume. Such compression heats the gas, which, in turn, causes its pressure to increase. As a result of the increased pressure, the gas expands, forming a doughnut-shaped region around the black hole (Figure 9-19). As the gas starts the final plunge toward the black hole, its temperature skyrockets to tens or even hundreds of millions of degrees Kelvin. At such temperatures, the pressure is so great that much of the infalling gas expands and, finding little resistance perpendicular to the plane of the accretion disk, it squirts out as two jets (Figure 9-19). These jets are prevented from spreading out by a magnetic field created as the hot gases orbit in the accretion disk and by pressure from surrounding gas.

Whereas neutron stars and stellar-mass black holes create jets as a result of being in binary star systems, supermassive black holes have so much gravitational attraction that they pull huge quantities of nearby interstellar gas and dust into orbit around themselves without needing a binary companion. Figure 9-17a shows such a disk, and Figure 9-16 shows one of the jets emitted by the supermassive black hole in M87. We will explore more about jets created by black holes in Chapter 12.

GAMMA-RAY BURSTS

Some black holes, neutron stars, and the supernovae that accompany their formation are apparently the sources of the most energetic and enigmatic energy emissions known in the universe. Called **gamma-ray bursts**, these pulses of gamma rays (the shortest-wavelength electromagnetic radiation) emit more energy than supernovae and from even smaller volumes of space.

9-10 Gamma-ray bursts are the most powerful explosions in the known universe

Because atomic bomb blasts emit pulses of gamma rays, the U.S. military put the gamma-ray-detecting Vela satellites in orbit around Earth to monitor illegal nuclear explosions in the 1960s. In 1973, astronomers were told that since 1967, these satellites had been detecting bursts of gamma rays from objects in space. This news intrigued the astronomical community because, at that time, there was no known mechanism to emit such radiation.

> Why doesn't all the gas in a black hole's accretion disk enter the black hole?

Gamma-ray bursts each last between a few milliseconds and about 1000 s (just over 16 min). There appear to be at least three types of such bursts. One group typically lasts for a few tenths of a second. The second group typically lasts for about 40 s. The third group appears to have properties of both of the first two groups, hence it is called a *hybrid gamma-ray burst*. Unlike X-ray bursts (see Section 8-16), each gamma-ray burst occurs only once.

More than 4000 gamma-ray bursts have been observed, and more are being discovered at a rate of about 1 per day. Plotting their locations on the celestial sphere (Figure 9-20), astronomers have discovered that the shorter-lived bursts occur slightly more commonly in the plane of the Milky Way Galaxy than elsewhere, while the longer-lived ones occur randomly over the sky. In 2004, astronomers detected the first remnant of a gamma-ray burst that occurred in the Milky Way.

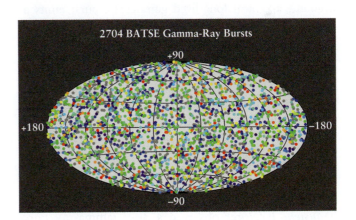

FIGURE 9-20 *The Most Powerful Known Bursts* Gamma-ray bursts have been observed everywhere in the sky, indicating that, unlike X-ray bursters, most do not originate in the disk of the Milky Way Galaxy. This map of the entire sky "unfolded" onto the page shows 2704 bursts detected by the Burst and Transient Source Experiment (BATSE) aboard the Compton Gamma-Ray Observatory. The colors indicate the brightness of the bursts; they are coded brightest in *red*, dimmest in *violet*. *Gray dots* indicate incomplete information about the burst strength. (NASA)

22 seconds

48 seconds

73 seconds

R I V U X G

FIGURE 9-21 Gamma-Ray Bursts In 1999, astronomers photographed the visible-light counterpart of a 100-s gamma-ray burst some 9 billion light-years away in the constellation Boötes. The times indicated are after the burst began. (Carl Akerlof, University of Michigan, Los Alamos National Laboratory, Lawrence Livermore National Laboratory)

Since 1997, optical and infrared "afterglows" of many gamma-ray bursts have been detected (Figure 9-21). These often persist for days, which allow astronomers time to study their spectra in detail. Such observations reveal that the light has passed through intergalactic gas clouds. Combining this with observations of the bursts' spectra and redshifts reveals that the longer-lasting bursts typically originate more than 6 billion light-years away, far outside our Galaxy. The most distant gamma-ray burst that has been observed occurred 12.8 billion light-years away. Many of the shorter bursts occur in more nearby galaxies. Taking their distance into account, a typical long-lived gamma-ray burst emits as much energy in 100 s as the Sun will emit over the 10 billion years that it will be on the main sequence.

Deep sky observations of some gamma-ray bursts have begun to show galaxies in the same places. As Figure 9-22 reveals, the bursts do not appear in the centers of their host galaxies. Likewise, the bursts in the plane of the Milky Way are not located near the galactic nucleus. These observations indicate that bursts are not associated with the supermassive black holes in the galactic cores. Research suggests that gamma-ray bursts are instead associated with stellar cataclysms.

In 2003, observations by NASA's *High-Energy Transient Explorer* observed a 30-s gamma-ray burst that was seen to occur at the same place as a particularly powerful supernova, called a *hypernova*, that occurs in **Wolf-Rayet stars**. These are rotating stars of at least 20 $M_\odot$ with strong magnetic fields and stellar winds. The bursts associated with these stars occur when gamma rays and a lot of gas are funneled out along the rotation axis of the star. If we are looking down the axis from Earth, the gamma rays create the burst, and the gas, traveling more slowly, creates the subsequent afterglow. Observations reveal that hypernovae and the associated gamma-ray bursts are extremely messy affairs, with powerful aftershocks, rather than a single, smooth outburst. The details of the explosions and the mechanism by which they generate gamma-ray bursts are still under investigation.

Other gamma-ray bursts are thought to occur when black holes collide; when a black hole swallows a neutron star; or when a pair of neutron stars collide and form a black hole. Calculations confirm that the collision of such dense objects should indeed lead to a rapid, intense burst of gamma rays.

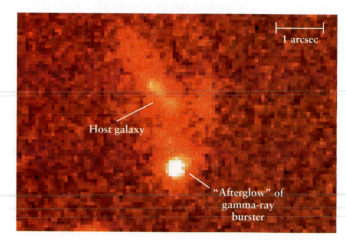

R I V U X G

FIGURE 9-22 The Host Galaxy of a Gamma-Ray Burst The Hubble Space Telescope recorded this visible-light image in 1999, 16 days after a gamma-ray burst was observed at this location. The image shows a faint galaxy that is presumed to be the home of the gamma-ray burst. The galaxy has a bright blue color, indicating the presence of many recently formed stars. (Andrew Fruchter, STScI; and NASA)

To better understand gamma-ray bursts, astronomers have constructed telescopes both on Earth and in orbit dedicated to rapidly shifting to look in the direction in which these events occur in space. These telescopes, including the *Swift Gamma-Ray Burst Mission* and the *High Energy Transient Explorer-2*, see these bursts and provide new insights into their origins and dynamics. *Swift* carries not only a gamma-ray telescope to locate the bursts, but also separate X-ray and optical telescopes to study their afterglows.

9-11 Black holes evaporate

In exploring the fate of black holes, astrophysicists find that, once again, these objects confound common sense—they evaporate! With a black hole cloaked behind the event horizon and its mass collapsed into a singularity, it must seem that there is no way of getting mass from the black hole back out into the universe again. No way, that is, until you recall that mass and energy are two sides of the same coin. What if there was a way of converting the black hole's mass into a form of energy that *could* get out, such as gravitational energy? The transformation does occur, according to an idea first proposed by Stephen Hawking.

Black holes convert their mass into energy by a process called **virtual particle** production. If things in this chapter aren't weird enough, consider this: The laws of nature allow pairs of particles, called virtual particles, to spontaneously appear. Each pair consists of a particle and its antiparticle—such as a proton and an antiproton, an electron and an antielectron (positron), or a pair of photons. The particles in each pair are identical, except for having opposite electric charges, and they annihilate each other and normally disappear without a trace within a very short period of time, typically 10^{-21} s.

We know that virtual particles actually flash into existence spontaneously throughout the universe because some have been made into real particles in the laboratory. This is done using high-speed particles in particle accelerators that slam into the pair of virtual particles and push them apart before they annihilate each other. The mass or energy given to the virtual particles in making them real comes from the laboratory particles, which lose an equal amount of energy. Calculations reveal that billions and billions of pairs of virtual particles are seething in and out of existence in your body every second.

Now, suppose two virtual particles form *just outside* the event horizon of a black hole. They would experience an exceptionally powerful gravitational field. If one particle is created slightly farther from the hole than its companion, the two virtual particles feel a tidal force from the black hole's tremendous gravitational pull. If the gravitational tidal force is strong enough, it can pull the two virtual particles apart before they can annihilate each other. This separation makes them real (Figure 9-23).

To conserve momentum, at least one of the newly formed real particles always falls into the black hole. But the other particle will sometimes have enough energy to escape from the vicinity of the black hole. This latter particle flies free into space, and the black hole has effectively emitted energy equal to $E = mc^2$, where m is the mass of the freed particle and c is the speed of light. Alternatively, the black hole may lose the energy due to one of two photons created, $E = hc/\lambda$, where λ is the wavelength of the photon and h is Planck's constant.

How long do gamma-ray bursts last?

Where the virtual particles become real is a temporary void in space outside the event horizon. The black hole fills the void with energy that comes from its mass. This energy is transmitted outside the event horizon as gravitational radiation to replace the energy taken away by the newly formed particles. This is called the **Hawking process,** and the particles leaving the vicinity of the black hole are called *Hawking radiation* (even though many of the particles are not electromagnetic radiation).

The time it takes a black hole to completely evaporate by the Hawking process increases with the mass of the hole. More massive black holes have lower tidal forces at their event horizons than do less massive and, therefore, smaller-diameter black holes. Therefore, the *rate* at which virtual particles are converted into real particles around bigger black holes is actually slower than it is around smaller ones. The faster the real particles are produced, the faster they wear down the mass of the black hole; thus, lower-mass black holes evaporate more

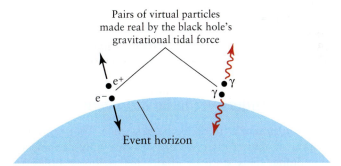

FIGURE 9-23 Evaporation of a Black Hole Throughout the universe, pairs of virtual particles spontaneously appear and disappear so quickly that they do not violate any laws of nature. The tidal force just outside of the event horizon of a black hole is strong enough to tear apart two virtual particles that appear there before they destroy each other. The gravitational energy that goes into separating them makes them real and, therefore, permanent. At least one of each pair of newly created particles falls into the black hole. Sometimes the other particle escapes into the universe. Because the gravitational energy used to create the particles came from the black hole, the hole loses mass and shrinks, eventually evaporating completely. Here we see just a few particles in the making: an electron (e$^-$) and a positron (e$^+$) and a pair of photons (γ).

R I V U X G

FIGURE 9-24 Mount Everest Primordial black holes, formed at the beginning of time, may have had masses similar to that of Mount Everest, as shown here. Calculations indicate that the universe is old enough for such black holes to have evaporated. Their final particle production rate is so high that they should look as though they are exploding. (Galen Rowell/Corbis)

rapidly. Indeed, calculations indicate that in its final moments of evaporation, a black hole should create particles so quickly that it appears to explode.

A stellar black hole of 5 $M_\odot$ would take more than 10^{62} years to evaporate (much longer than the present age of the universe), whereas a 10^{10}-kg primordial black hole (equivalent to the mass of Mount Everest; Figure 9-24) would take only about 15 billion years to evaporate. Depending on the precise age of the universe, a black hole of this size and age should be in the final throes of evaporating. Astronomers are trying to identify events in space that correspond to the violent, particle-generating deaths of primordial black holes.

> Why does at least one particle in the Hawking process always fall into the black hole?

9-12 Frontiers yet to be discovered

Black holes take us to the limits of science. They represent a state of matter that we are not yet fully capable of explaining. This state of affairs makes black holes an especially intriguing field of study. Understanding the nature of black hole singularities is a major research goal. Numerous other frontiers exist in this realm. For example, by directly observing gravitational radiation, we will have a new tool for exploring violent activity in the universe. We still need to understand the details of black hole formation, both from stars and by other means. We have yet to understand in detail why mid-mass and high-mass black holes have the masses they have.

If black holes merge, as is now expected, the details of these events and their effects on the rest of the universe have still to be understood. It also remains to be shown whether or not the cosmic censorship theorem is

correct. If not, the implications of the existence of wormholes need to be examined in more depth. Among the more intriguing issues surrounding black holes is whether primordial black holes exist. Although the evaporation process has not yet been seen, we should soon have the technology to discover it, if it is occurring.

Gamma-ray bursts still require much explanation. How does so much energy get transformed so quickly into electromagnetic radiation in these events? Why do some supernovae emit most of their energy as gamma rays, while others emit most as less energetic electromagnetic photons?

SUMMARY OF KEY IDEAS

- If a stellar corpse is more massive than about 3 $M_\odot$, gravitational compression overcomes neutron degeneracy (or the theorized quark degeneracy pressure) and forces it to collapse further and become a black hole.

- A black hole is an object so dense that the escape velocity from it exceeds the speed of light.

The Relativity Theories
- Special relativity reveals that space and time are intimately connected and change with an observer's motion.

- As seen by observers moving more slowly, the faster an object moves, the slower time passes for it (time dilation) and the shorter it becomes (length contraction).

- According to general relativity, mass causes space to curve and time to slow down. These effects are significant only near large masses or compact objects.

- Observations indicate that some binary star systems harbor black holes. In such systems, gases captured by the black hole from the companion star heat up and emit detectable X rays and jets of gas.

- Supermassive black holes exist in the centers of many galaxies. Intermediate-mass black holes exist in clusters of stars. Very low mass (primordial) black holes may have formed at the beginning of the universe.

Inside a Black Hole
- The event horizon of a black hole is a spherical boundary where the escape velocity equals the speed of light. No matter or electromagnetic radiation can escape from inside the event horizon. The distance from the center of the black hole to the event horizon is called the Schwarzschild radius.

- The matter inside a black hole collapses to a singularity. The singularity for nonrotating matter is a point at the center of the black hole. For rotating matter, the singularity is a ring inside the event horizon.

- Matter inside a black hole has only three physical properties: mass, angular momentum, and electric charge.

- Nonrotating black holes are called Schwarzschild black holes. Rotating black holes are called Kerr black holes. The event horizon of a Kerr black hole is surrounded by an ergoregion in which all matter must constantly move to avoid being pulled into the black hole.

- Matter that approaches a black hole's event horizon is stretched and torn by the extreme tidal forces generated by the black hole; light from the matter is redshifted; and time slows down.

- Black holes can evaporate by the Hawking process, in which virtual particles near the black hole become real. These transformations of virtual particles into real ones decrease the mass of a black hole until, eventually, it disappears.

Gamma-Ray Bursts
- Gamma-ray bursts are events believed to be caused by some supernovae and by the collisions of dense objects, such as neutron stars or black holes. Some occur in the Milky Way and nearby galaxies, whereas many occur billions of light-years away from Earth.

- Typical gamma-ray bursts occur for a few tens of seconds and emit more energy than the Sun will radiate over its entire 10-billion-year lifetime.

WHAT DID YOU THINK?

1 *Are black holes empty holes in space? If not, what are they?* No, black holes contain highly compressed matter—they are not empty.

2 *Does a black hole have a solid surface? If not, what is at its surface?* No. The surface of a black hole, called the *event horizon*, is empty space. No stationary matter exists there.

3 *What power or force enables black holes to draw things into them?* The only force that pulls things in is the gravitational attraction of the matter in the black hole.

4 *How close to a black hole do you have to be for its special effects to be apparent?* About 100 times the Schwarzschild radius.

5 *Can you use black holes to travel to different places in the universe?* No, most astronomers believe that the wormholes predicted by general relativity do not exist.

6 *Do black holes last forever? If not, what happens to them?* No, black holes evaporate.

Key Terms for Review

accretion disk, 256
black hole, 252
cosmic censorship, 256
ergoregion, 255
event horizon, 253
gamma-ray burst, 261
gravitational radiation, 253
gravitational redshift, 251
gravitational wave, 253
Hawking process, 263
intermediate-mass black hole, 257
Kerr black hole, 254
primordial black hole, 257
Schwarzschild black hole, 254
Schwarzschild radius, 253
singularity, 252
spacetime, 250
supermassive black hole, 257
theory of general relativity, 250
theory of special relativity, 248
virtual particle, 263
wormhole, 256
Wolf-Rayet stars, 262

Review Questions

1. Which property, if any, of normal matter ceases to exist in a black hole? a. mass, b. chemical composition, c. angular momentum, d. electric charge, e. all of these properties exist in a black hole

2. Supermassive black holes are found in which of the following locations? a. in the centers of galaxies, b. in globular clusters, c. in open (or galactic) clusters, d. between galaxies, e. in orbit with a single star

3. Which feature is found with Kerr black holes but not Schwarzschild black holes? a. a singularity, b. an event horizon, c. gravitational redshift of photons outside of the black hole, d. an ergoregion, e. warping of nearby spacetime

4. Under what conditions do all outward pressures on a collapsing star fail to stop its inward motion?

5. In what way is a black hole blacker than black ink or a black piece of paper?

6. If the Sun suddenly became a black hole, how would Earth's orbit be affected?

7. What is cosmic censorship?

8. What are the differences between rotating and nonrotating black holes?

9. Why are all of the observed stellar-remnant black-hole candidates members of close binary systems?

10. If light cannot escape from a black hole, how can we detect X rays from such objects?

Advanced Questions

The answers to all computational problems, which are preceded by an asterisk (*), appear at the end of the book.

*11. What is the Schwarzschild radius of a black hole, measured in kilometers, containing 3 $M_\odot$? 30 $M_\odot$?

12. If more massive stars evolve and die before less massive ones, why do some black-hole candidates have lower masses than their stellar companions?

13. Under what circumstances might a neutron star in a binary star system become a black hole?

14. Which type of black hole, nonrotating or rotating, do science fiction writers use (implicitly) in sending spaceships from one place to another through the hole? Why would the other type not be suitable?

*15. You are standing in a train car at rest in the station, as in Figure 9-2. You measure the car's length to be 10 meters long. The train speeds up to the considerable speed of 0.95c with you still inside the car. How long will you measure the car to be now? Explain how you got your answer.

What If . . .

16. A black hole of 5 $M_\odot$ passed by Earth at, say, Neptune's distance from the Sun? What would happen to our planet's orbit and to life here? Hint: The Sun's gravitational force is about 180 times greater than the maximum effect that black hole would have on us. The Sun's

force on Earth is about 17,000 times greater than Jupiter's effect on us, for comparison.

17. A primordial black hole with the mass of our Moon approached, passed through, and exited Earth? What might happen to our planet and to life here? Hints: You may want to calculate the black hole's Schwarzschild radius. Also, specify whether this is a "high speed" or "low speed" event.

18. A primordial black hole exploded nearby? What would astronomers observe?

Web Questions

19. To test your understanding of black hole structure, do Interactive Exercise 9-1 on the Web. You can print out your answers, if required.

20. Search the Web for information about the stellar-mass black-hole candidate, named V4641 Sgr. In what ways does it resemble other black-hole candidates, such as Cygnus X-1 and V404 Cygni? In what ways is it different and more dramatic? How do astronomers currently explain why V4641 Sgr is different?

PART III

Understanding the Universe

R I V U X G

Combined infrared and ultraviolet image of the Andromeda Galaxy,
2.5 million light-years away and 260,000 light-years in diameter.
Red indicates cool, dusty regions; blue are hot, young stars; older stars
are depicted in green; and pinkish regions are where old and young
stars coexist. (NASA/JPL-Caltech/K. Gordon [Univ. of Ariz.] & GALEX Science)

AN ASTRONOMER'S ALMANAC

1755
Milky Way proposed by Immanuel Kant to be only one among many galaxies.

1914-1922
Redshift observed in spectra of galaxies taken by Vesto Slipher.

1927-1929
Bertil Lindblad calculates the solar system's motion through the Milky Way. BL Lac discovered.

1943
First Seyfert galaxy detected by Carl Seyfert.

1785
William Herschel calculates the location of solar system within the Milky Way, but incorrectly.

1917-1918
Harlow Shapley observes that Milky Way is surrounded by a spherical distribution of globular clusters not centered on Earth; determines, essentially correctly, the location of the solar system in the Milky Way and the location of Galactic center.

Photography, Radio

Television, Radar

Transistors

1517-1519
Ferdinand Magellan, first European to record existence of Large and Small Magellanic Cloud galaxies.

1922
Alexander Friedmann discovers expanding universe solution to Einstein's equations of general relativity.

1930-1936
Robert Trumpler establishes presence of interstellar dust. Karl Jansky detects intense radio waves from direction of Galactic center. Grote Reber builds first radio telescope. Georges Lemaitre proposes that Big Bang occurred.

964
Andromeda Galaxy discovered by Abd al-Rahman al-Sufi.

1912-1915
Henrietta Leavitt discovers the period–luminosity relationship for Cepheid Type I stars. Einstein's general theory of relativity predicts a dynamic universe.

1924-1929
Edwin Hubble uses Cepheid variable stars to prove that nebulae are galaxies external to the Milky Way; proposes "tuning fork" classification scheme for galaxies based on their shapes; determines the relationship between the distance to superclusters of galaxies and their speed away from the Milky Way.

1948
Ralph Alpher, Hans Bethe, and George Gamow predict the cosmic microwave background radiation.

1845
William Parsons (Lord Rosse) observes and sketches spiral-shaped nebulae later shown to be external galaxies.

THE UNIVERSE

1951

Walter Baade and Rudolph Minkowski discover optical counterpart to Cygnus A.
Orion and Perseus arms of Milky Way discovered by William Morgan.

1959

Third Cambridge radio catalog published.

1965-1968

Arno Penzias and Robert Wilson discover the cosmic microwave background radiation.
Donald Lynden-Bell proposes black holes as engines for quasars and other active galaxies.

Accretion disk   Supermassive black hole

1990-1992

COBE spacecraft confirms blackbody nature of the cosmic microwave background and discovers anisotropy in the cosmic background radiation.

1980-1981

Alan Guth proposes inflationary phase of the evolution of the universe. Robert Kirshner, August Oemler, Paul Schecter, and Stephen Shectman detect void in the distribution of galaxies in the constellation Boötes.

Lasers, Sputnik

1952

Simple experiment by Stanley Miller and Harold Urey yields first organic molecules created in a laboratory.

Microprocessor, CCD

1960-1961

Frank Drake conducts an early search for extraterrestrial intelligence using a radio telescope; derives an equation to calculate the likely number of extraterrestrial civilizations.

1973-1978

Jerry Ostriker and James Peebles discover that the visible mass of a galaxy is not sufficient to keep it together. Brent Tully and Richard Fisher discover a relationship between galactic luminosity and stellar velocities.
Vera Rubin and Kent Ford show that rotation curves of spiral galaxies are flatter than predicted by Newtonian gravitation.

Internet

1989

Black holes discovered at centers of Andromeda and its companion galaxy, M32.

1998

Type Ia supernova observations reveal that the universe is accelerating outward.

2003

Hubble constant, H_0, determined to be 71 ±0.4 km/s/Mpc.

CHAPTER 10

The Milky Way Galaxy

R I V U X G

The center of the Milky Way Galaxy, as seen by the *Chandra* X-ray telescope. The false colors reveal extremely hot gases that surround black holes, neutron stars, and white dwarves. The image is approximately 300 by 800 light-years. The supermassive black hole at the center of our Galaxy is enshrouded by the hot, white gas cloud in the center of the image. (NASA/UMass/D. Wang et al.).

WHAT DO YOU THINK?

1 What is the shape of the Milky Way Galaxy?

2 Where is our solar system located in the Milky Way Galaxy?

3 Is the Sun moving through the Milky Way Galaxy and, if so, about how fast?

Answers to these questions appear in the text beside the corresponding numbers in the margins and at the end of the chapter.

By the beginning of the twentieth century, astronomers knew that we live in a forest of stars and interstellar matter. Just as nearby trees and leaves hide what lies behind them, the stars and interstellar clouds we can see block light from more distant objects. A discussion ensued back then as to whether there is only one forest. This culminated in the inconclusive Shapley-Curtis debate on the subject of whether the Milky Way is the entire universe. Observations presented in 1924 finally enabled astronomers to step outside our forest and see clearly that there are others, separated from ours by vast regions of relatively empty space. Our forest is called the **Milky Way Galaxy.** Observations have since revealed that the other forests, also called *galaxies,* number in the billions. By the end of the twentieth century, astronomers knew that the galaxies that fill the universe are clustered together, rather than randomly dotting the cosmic landscape. They had also discovered that the clusters of galaxies are gathered together in "superclusters."

The more stars, nebulae, galaxies, and other features that astronomers observed in the cosmos, the more evident it became that something else dwelled in each galaxy that they could not see. Wraith-like, this "dark matter" interacted with everything else without showing itself. As the twenty-first century dawned, astronomers started getting glimpses of these wraiths, revealing that there are several kinds of them. Nevertheless, other than that it interacts gravitationally, we still do not know the nature of most of this dark matter.

In this part of the book, we explore the galaxies, clusters of galaxies, superclusters of galaxies, and the nature of the space that connects them, along with what little we know about the dark matter and other entities that populate the universe. We begin by studying our Milky Way. For those fortunate enough to live away from bright outdoor lights and air pollution, the Milky Way appears as a veiled band overlaid with the glow of individual stars. Centuries of observations have firmly established that our solar system is part of an enormous assemblage of hundreds of billions of stars, along with gas, dust, and other matter, all held together by their mutual gravitational attraction. Most of the stars in our Galaxy are located in a disk that looks from its edge like a flying saucer in an old science fiction movie (Figure 10-1a). The inner part of the disk is filled with stars, some of which form a bar, as shown (Figure 10-1b). Spiral arms swirl out from the ends of the bar, and, within these arms, new stars form from the debris of earlier generations of stars. The Galaxy's remaining stars are located in a spherical halo that surrounds the disk. (Note: only Earth's Galaxy is capitalized.)

In the following two chapters, we will examine the rest of the galaxies and the supermassive black holes that lie at the hearts of most of them. Then, like biologists who study the origins and evolution of life, we consider what astronomers know about the formation, evolution, and fate of the universe. The book ends with a discussion of astrobiology: the interrelationship between life and astronomy.

In this chapter you will discover

- the Milky Way Galaxy—billions of stars, along with gas and dust, bound together by mutual gravitational attraction

- the structure of our Milky Way Galaxy

- Earth's location in the Milky Way

- how interstellar gas and dust enable star formation to continue

- that observations reveal the presence of significant mass in the Milky Way that astronomers have yet to identify

- that there is a massive black hole at the center of our Galaxy

DEFINING THE MILKY WAY

Prior to the twentieth century, astronomers did not know the large-scale distribution of stars and other matter in the universe. Throughout history, most people, including many astronomers, believed that the Milky Way contains all of the stars in the cosmos. In other words, they thought that the "Galaxy" and the "universe" were the same thing. We begin the study of galaxies by learning how that belief changed.

a

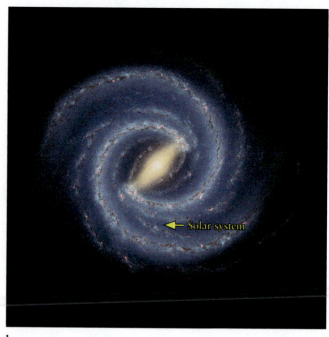

b

FIGURE 10-1 Schematic Diagrams of the Milky Way (a) This edge-on view shows the Milky Way's disk, containing most of its stars, gas, and dust, and its halo, containing many old stars. Individual stars in the halo are too dim to show, so the bright regions in the halo represent clusters of stars. (b) Our Galaxy has two major arms and several shorter arm segments, all spiraling out from the ends of a bar of stars and gas that passes through the Galaxy's center. The bar's existence and the presence of two major arms were both confirmed by the *Spitzer Space Telescope*. (b: NASA/JPL-Caltech/R. Hurt [SSC])

10-1 Studies of Cepheid variable stars revealed that the Milky Way is only one of many galaxies

The belief that our Galaxy is but one of many was put forth in 1755, when the German philosopher Immanuel Kant suggested that vast collections of stars lie far beyond the confines of the Milky Way. Less than a century later, the Irish astronomer William Parsons observed the structure of some of those "island universes" proposed by Kant. Parsons was the third Earl of Rosse in Ireland. He was rich, he liked machines, and he was fascinated by astronomy. Accordingly, he set about building gigantic telescopes. In February 1845, his pièce de résistance was finished. This telescope's massive mirror measured 1.8 m (6 ft) in diameter and was mounted at one end of an 18-m (60-ft) tube controlled by cables, straps, pulleys, and cranes (Figure 10-2a). For many years, this triumph of nineteenth-century engineering enjoyed distinction as the largest telescope in the world.

With this new telescope, Lord Rosse examined many of the glowing interstellar clouds previously discovered and catalogued by the Herschels. William Herschel, his sister Caroline, and his son John, among others, discovered and recorded details of many fuzzy-looking astronomical objects, called **nebulae** (singular: **nebula**). With the high resolution provided by his telescope, Lord Rosse observed that some of these nebulae have a distinct spiral structure. A particularly good example is M51, also called the *Whirlpool Galaxy* or *NGC 5194*.

Lord Rosse had no photographic equipment in 1845, so he made drawings of what he observed. Figure 10-2b is his drawing of M51. Views like this inspired him to echo Kant's proposal of island universes. Figure 10-2c shows a modern photograph of M51. It is interesting to note the differences between the perception and interpretation of astronomical objects and a camera's recording of them.

Most astronomers of Rosse's day did not agree with the notion of island universes outside of our Galaxy. They thought that the Milky Way contained all stars in the universe—that the Milky Way *was* the universe. In April 1920, a formal discussion, now known as the **Shapley–Curtis debate**, was held at the National Academy of Sciences in Washington, D.C. Harlow Shapley argued that the spiral nebulae are relatively small, nearby objects scattered around our Galaxy. Heber D. Curtis championed the *island universe* theory, arguing that each of these spiral nebulae is a separate rotating system of stars, much like our own Galaxy. While the Shapley–Curtis debate focused scientific attention on the size of the universe, nothing was decided, because no one had any firm evidence to demonstrate exactly how far away the spiral nebulae were. Astronomers desperately needed to devise a way to measure the distances to them. A young teacher and former basketball player from Kentucky who moved to Chicago to study astronomy finally achieved this goal. His name was Edwin Hubble.

a

R I V U X G

b

M51 (NGC 5194)   NGC 5195

c

R I V U X G

WEB LINK 10.2

FIGURE 10-2 High-Tech Telescope of the Mid-Nineteenth Century (a) Built in 1845, this structure houses a 1.8-m-diameter telescope, the largest of its day. The improved resolution it provided over other telescopes was similar to the improvement that the Hubble Space Telescope provided over Earthbound optical instruments when it was launched. The telescope, as shown here, was restored to its original state in the three years from 1996–1998. (b) Using his telescope, Lord Rosse made this sketch of the spiral structure of the galaxy M51 and its companion galaxy NGC 5195. (c) A modern photograph of M51 (also called NGC 5194) and NGC 5195. The spiral galaxy M51 in the constellation of Canes Venatici is known as the Whirlpool Galaxy because of its distinctive appearance. The two galaxies are about 20 million ly from Earth. (a: Birr Castle Demesne; b: Lund Humphries; c: NOAO)

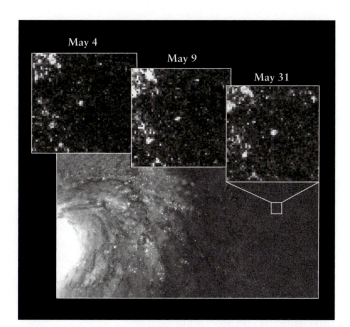

May 4

May 9

May 31

R I V U X G

In 1923, Hubble took an historic photograph of M31, then called the Andromeda Nebula. It was one of the spiral nebulae around which controversy raged. On the photographic plate he discovered what first appeared to be a nova. Referring to previous plates of that region, he soon realized that the object was actually a Cepheid variable star. As we saw in Section 7-12, these pulsating stars vary in brightness periodically. Further scrutiny over the next several months revealed many other Cepheids. Figure 10-3 shows a Cepheid in the galaxy M 100, at different stages of brightness.

VIDEO 10.1   WEB LINK 10.3

FIGURE 10-3 A Cepheid Variable Star in Galaxy M100 One of the most reliable ways to determine the distance to moderately remote galaxies is to locate Cepheid variable stars in them, as discussed in the text. At 56 million ly (17 Mpc) from Earth, the galaxy M100 in the constellation Coma Berenices is among the most remote objects whose distances have been determined using Cepheids. Insets: The Cepheid in this view, one of 20 located to date in M100, is shown at different stages in its brightness cycle, which recurs over several weeks. (Dr. Wendy L. Freedman, Observatories of the Carnegie Institution of Washington; NASA)

Only a decade before, in 1912, the American astronomer Henrietta Leavitt had published an important study of Cepheid variables. Leavitt examined many of them in the Small Magellanic Cloud, then also believed to be a nebula, but now known to be a galaxy very close to the Milky Way. Leavitt's study led her to the period-luminosity relation for Type I Cepheids (see Section 7-13). Leavitt established that a direct relationship exists between a Cepheid's luminosity (or absolute magnitude) and its period of oscillation, which we saw in Figure 7-27 and which is presented in more detail in Figure 10-4. By observing the star's period and apparent magnitude, its distance can be calculated, as described in *An Astronomer's Toolbox 10-1*.

Thanks to this work, Hubble knew that he could use the characteristics of the fluctuating light of Cepheid variable stars to help him calculate the distance to M31 and put the Shapley–Curtis debate to rest once and for all. His observations of Type I Cepheids led him to determine that M31 is some 2.2 million light-years *beyond* the Milky Way. This proves that M31 is not an open or globular cluster in our Galaxy, but rather an enormous separate stellar system—a separate galaxy. M31, now called the *Andromeda Galaxy*, is the most distant object in the universe that can be seen with the naked eye. Similar calculations have been done for all galaxies in which Cepheids can be observed. The distances to even more remote galaxies have been determined from observations of Type Ia supernovae in them. (see *An Astronomer's Toolbox 10-1*).

> Referring to Figure 10-4, how would the brightness of a Cepheid variable with peak luminosity of 1000 $L_\odot$ change if it were observed every 5 days?

INSIGHT INTO SCIENCE

Room for Debate Lacking definitive data, competent scientists can develop and believe strikingly different explanations for the same observations. At the time of the Shapley–Curtis debate, the observations allowed both points of view. It was only with the advent of the distance measurement technique used by Hubble that the spiral nebulae were definitely shown to lie outside of the Galaxy.

Hubble's results, which he presented at the end of 1924, did settle the Shapley–Curtis debate. The universe was recognized to be far larger and populated with far bigger objects than most astronomers had imagined. Hubble had discovered the realm of the galaxies. Today, we know that the universe contains myriad galaxies, of which the Milky Way is just one. Like the Milky Way, each **galaxy** is a grouping of millions, billions, or even

FIGURE 10-4 The Period-Luminosity Relation This graph shows the relationship between the periods and average luminosities of classical (Type I) Cepheid variables and the closely related RR Lyrae stars (discussed in Chapter 7). Each dot represents a Cepheid or RR Lyrae whose luminosity and period have been measured.

trillions of stars, along with gas, dust, and matter in other forms, all gravitationally bound together.

We now apply Hubble's method to find Earth's place in the Milky Way Galaxy. Recall that the Sun's proximity to us makes it our best-understood star. It might seem that the nearness of the stars and clouds in the Milky Way would make it the best-understood galaxy. However, the clouds of gas and dust that surround the solar system make it very challenging for astronomers to survey completely the distant parts of the Galaxy, which are only now coming into focus.

THE STRUCTURE OF OUR GALAXY

Because the band of the Milky Way completely encircles us, astronomers long ago suspected that the Sun and all of the stars that we see are part of it. In the 1780s, William Herschel took the first steps toward mapping its structure. He attempted to deduce the Sun's location in the Galaxy by counting the number of stars in 683 regions of the sky. He reasoned that the greatest density of stars should be seen toward the Galaxy's center and a lesser density seen toward the edge. However, Herschel found roughly the same density of stars all along the Milky Way. He therefore concluded that we are at the center of the Galaxy.

Herschel was wrong: Earth has no privileged place in the Milky Way. The Sun is about 26,000 ly (8000 pc) from the Galaxy's center, the **galactic nucleus**. Herschel's physical understanding of the cosmos was incomplete, so he misinterpreted his observations and thus came to an incorrect conclusion.

Cepheids and Supernovae as Indicators of Distance

Because their periods are directly linked to their luminosities, Cepheid variables are among the most reliable tools that astronomers have for determining the distances to relatively nearby galaxies. To this day, astronomers use this link—much as Hubble did back in the 1920s—to measure intergalactic distances. More recently, they have begun to use Type Ia supernovae, which are far more luminous and thus can be seen much farther away, to determine the distances to more remote galaxies.

Example: In 1992, a team of astronomers used Cepheid variables in a galaxy, called IC 4182, to deduce that galaxy's distance from Earth. They used the Hubble Space Telescope on 20 separate occasions to record images of the stars in IC 4182. By comparing these images, the astronomers could pick out which stars vary in brightness. In this way, they discovered 27 Cepheids in IC 4182. Using their observations, they determined the (changing) apparent magnitudes of the Cepheid variables and plotted their light curves. The Hubble Space Telescope is particularly well suited for studies of this kind, because its extraordinary angular resolution makes it possible to pick out individual stars at great distances. One such Cepheid has a period of 42.0 days and an average apparent magnitude (m) of +22.0. (See Section 6-2 for an explanation of the apparent magnitude scale.) By comparison, the dimmest star you can see with the naked eye has $m = +6$; this Cepheid in IC 4182 appears less than one one-millionth as bright. The star's spectrum shows that it is a metal-rich Type I Cepheid variable.

According to the period-luminosity relation shown in Figure 10-4, such a Type I Cepheid with a period of 42.0 days has an average luminosity of 33,000 $L_\odot$. In other words, this Cepheid has an average absolute magnitude (M) of −6.5 (compared to M = +4.8 for the Sun). Hence, the difference between the Cepheid's apparent and absolute magnitudes, called its **distance modulus**, is

$$m - M = (+22.0) - (-6.5) = 22.0 + 6.5 = 28.5$$

From *An Astronomer's Toolbox 6-3*, we see that the distance modulus of a star is related to its distance in parsecs (d) by

$$m - M = 5 \log d - 5,$$

where "log" is the base 10 logarithm. This can be rewritten as

$$d = 10^{(m - M + 5)/5} \text{ parsecs}$$

Inserting the value for the distance modulus in this equation, we get the distance to the Cepheid variable and, hence, the distance to the galaxy of which that star is part:

$$d = 10^{(28.5 + 5)/5} \text{ parsecs} = 10^{6.7} \text{ parsecs}$$
$$= 5.0 \times 10^6 \text{ parsecs}$$

The galaxy is 5 Mpc (1 Mpc = 1 megaparsec = 10^6 parsecs), or 16 million (1.6×10^7) light-years, from Earth.

Astronomers are interested in IC 4182 because a Type Ia supernova was observed there in 1937. Type Ia supernovae are exploding white dwarfs that all reach the same maximum brightness at the peak of their outburst (see Section 8-4). Once astronomers knew the peak absolute magnitudes of Type Ia supernovae, they could use these supernovae as distance indicators. Because the distance to IC 4182 is known from its Cepheids, the 1937 observations of the supernova in that galaxy allow us to calibrate Type Ia supernovae as distance indicators. At maximum brightness, the 1937 supernova reached an apparent magnitude of $m = +8.6$. Because the distance modulus of the galaxy ($m - M$) is 28.5, we see that when a Type Ia supernova is at maximum brightness, its absolute magnitude is

$$M = m - (m - M) = 8.6 - 28.5 = -19.9$$

Whenever astronomers find a Type Ia supernova in a remote galaxy, they can combine this absolute magnitude with the observed maximum apparent magnitude to get the galaxy's distance modulus, from which the galaxy's distance can be easily calculated (just as we did above for the Cepheids in IC 4182). This technique has been used to determine the distances to galaxies hundreds of millions of parsecs away.

Try these questions: At what distance in parsecs is a star with distance modulus 20? Epsilon (ϵ) Indi has an apparent magnitude of +4.7 and distance of 3.6 pc. What is its absolute magnitude? What would the Sun's apparent magnitude and distance modulus be as seen from 100 pc away? Its absolute magnitude is +4.83.
(Answers appear at the end of the book.)

10-2 Cepheid variables help us locate our Galaxy's center

While studying star clusters in the 1930s, R. J. Trumpler discovered the reason for Herschel's mistake. Herschel did not know about interstellar gas and dust, which affected his counts of the stars. Trumpler noticed that remote star clusters appear dimmer than would be expected just from their distance alone. Something must be blocking starlight on its way toward Earth. He correctly concluded that interstellar space is not a perfect vacuum. Instead, it contains dust that absorbs light from

R I V U X G

FIGURE 10-5 Our Galaxy This wide-angle photograph spans half the Milky Way. The Northern Cross is at the left, and the Southern Cross is at the right. The center of the Galaxy is in the constellation Sagittarius, in the middle of this photograph. The dark lines and blotches are caused by hundreds of interstellar clouds of gas and dust that obscure the light from background stars, rather than by a lack of stars. (Dirk Hoppe)

distant stars. Great patches of this dust are clearly visible in wide-angle photographs (Figure 10-5). Like the stars, this dust is concentrated in the plane of the Galaxy.

This interstellar dust almost completely obscures from view visible light emanating from the center of our

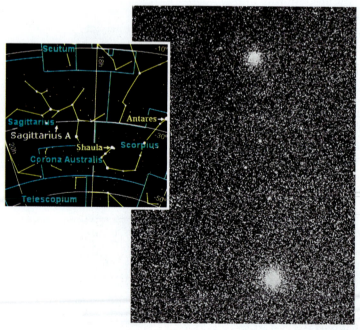

R I V U X G

FIGURE 10-6 A View Toward the Galactic Center More than a million stars in the disk of our Galaxy fill this view, which covers a relatively clear window just 4° south of the galactic nucleus in Sagittarius. Beyond the disk stars you can see two prominent globular clusters. Although most regions of the sky toward Sagittarius are thick with dust, very little obscuring matter appears in this tiny section of the sky. (Harvard Observatory)

Galaxy. Visual photons from there are mostly absorbed or scattered before they reach us. Therefore, Herschel was seeing only nearby stars, and he measured apparent magnitudes that were dimmer than they would have been had there been no interstellar dust. Without adjusting for its effects, he concluded that the stars were farther away than they really are. He also had no idea of the true size of the Galaxy and could not see the vast number of stars located in the general direction of the galactic center that are hidden by the dust.

INSIGHT INTO SCIENCE

A Little Knowledge Incomplete information often leads to incorrect interpretation of data and, therefore, to incorrect conclusions. Herschel's lack of knowledge about the matter in interstellar space prevented him from correctly interpreting the distribution of stars that surround Earth and, thus, led to his inaccurate conclusion about the position of the Sun within the Galaxy.

Because interstellar dust is concentrated in the plane of the Galaxy, the absorption of starlight is strongest in those parts of the sky covered by the Milky Way. Above or below the plane of the Galaxy, our view is relatively unobscured. Knowledge of our true position in the Galaxy eventually came from observations of globular clusters (see Section 7-14). Shapley used the period-luminosity relationship for variable stars to determine the distances to the then-known 93 globular clusters in the sky. (More than 150 are known today.) From their directions and distances, he mapped out the distribution of these clusters in three-dimensional space. By 1917, Shapley had discovered that the globular clusters are located in a spherical distribution centered not on Earth but on a point in the Milky Way toward the constellation Sagittarius. Figure 10-6 shows two globular clusters in a relatively clear part of the sky in that direction. Shapley then made a bold conjecture: *The globular clusters orbit the center of the Milky Way, which is located in Sagittarius.* His pioneering research has since been observationally verified. Earth is not at the center of the Galaxy.

10-3 Nonvisible observations help map the galactic disk

To see into the dust-filled plane of the Milky Way, astronomers use radio wave, infrared, X-ray, and gamma-ray telescopes (see Figure 3-37). These wavelengths are scattered much less by the interstellar gas and dust located throughout the Galaxy's disk than are visible or ultraviolet wavelengths. Observations of the distant parts of the Galaxy were first made using radio telescopes. Radio waves penetrate Earth's atmosphere, so we can observe them anywhere that we can build a radio telescope. (Recall that infrared observations must

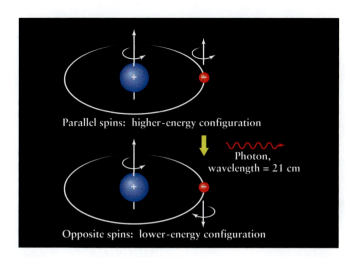

FIGURE 10-7 Electron Spin and the Hydrogen Atom Due to their spins, electrons and protons both act as tiny magnets. When an electron and the proton it orbits are spinning in the same direction, their energy is higher than when they are spinning in opposite directions. When the electron flips from the higher-energy to the lower-energy configuration, the atom loses a tiny amount of energy that is radiated as a radio photon with a wavelength of 21 cm.

be made at high altitudes or in space, and that X-ray and gamma-ray observations are almost always made from space.)

Detecting the radio emission directly from interstellar hydrogen—by far the most abundant element in the universe—is a primary means of mapping the Galaxy. Unfortunately, the major transitions of electrons in the hydrogen atom (see Figure 4-11) produce photons at ultraviolet and visible wavelengths that do not penetrate the interstellar medium. How, then, can radio telescopes directly detect all of this hydrogen? The answer lies in atomic physics.

In addition to mass and charge, particles such as protons and electrons possess a tiny amount of angular momentum, commonly called **spin**. According to the laws of quantum mechanics, the electron and proton in a hydrogen atom can spin only in either parallel or opposite directions (Figure 10-7); they can have no other spin orientations. If the electron in a hydrogen atom flips from one orientation to the other, the atom must gain or lose a tiny amount of energy. In particular, when flipping from parallel to opposite spins, the atom simultaneously emits a low-energy radio photon whose wavelength is 21 cm. This flip happens rarely in each atom, so it is only because the Galaxy has vast quantities of interstellar hydrogen gas that it can be detected at all. In 1951, a team of astronomers first succeeded in detecting the faint hiss of 21-cm radio static from spin flips.

The detection of **21-cm radio radiation** was a major breakthrough in mapping the **disk** of the Galaxy. To see why, suppose that you aim your radio telescope across

the Galaxy, as sketched in Figure 10-8. Your radio receiver picks up 21-cm emission from hydrogen clouds at points 1, 2, 3, and 4. However, the radio waves from these various clouds are Doppler shifted (see *An Astronomer's Toolbox 4-3*) by slightly different amounts, because they have different radial velocities (motion toward or away from Earth). Because these radio waves from gas clouds in different parts of the Galaxy arrive at our radio telescopes, with slightly different wavelengths as a result of being Doppler shifted, it is possible to identify which radio signals come from which gas clouds and thus to produce an initial map of the Galaxy, such as that shown in Figure 10-9a.

What are the two reasons that the gas and dust at points 1, 2, 3, and 4 (Figure 10-8) have different radial velocities, as measured from Earth?

Our radio map reveals numerous arched lanes of neutral hydrogen gas. If this were the overall structure of the Galaxy, then the Milky Way would appear unlike any other observed galaxy (see Chapter 11 for typical images of other galaxies). Indeed, the other disk-shaped galaxies we observe have spiral arms. We need different observations to improve our understanding of the Galaxy's disk features. Note that photographs of a barred spiral galaxy (Figure 10-10) show arms outlined by "spiral tracers"—bright, Population I stars and emission nebulae. As we saw in Chapter 7, these features indicate active star formation. If the Milky Way is spiral, then another useful way to further chart its structure and

FIGURE 10-8 A Technique for Mapping the Galaxy Hydrogen clouds at different locations along our line of sight are moving around the center of the Galaxy at different speeds. Radio waves from the various gas clouds, therefore, exhibit slightly different Doppler shifts, permitting astronomers to sort out the gas clouds and map the Galaxy. (NASA/JPL-Caltech/R. Hurt [SSC])

a

R I V U X G

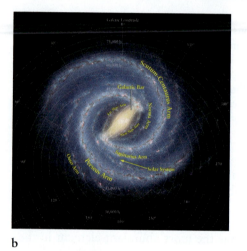

b

FIGURE 10-9 A Map of the Galaxy (a) This map, based on radio telescope surveys of 21-cm radiation, shows the distribution of hydrogen gas in a face-on view of the Galaxy. This view just hints at spiral structure. The galactic nucleus is marked with a *dot* surrounded by a circle. Details in the large, blank, wedge-shaped region toward the *upper left* of the map are unknown, because gas in this part of the sky is moving perpendicular to our line of sight and thus does not exhibit a detectable Doppler shift. *Inset:* This drawing, based on visible-light data, shows that our solar system lies between two arms of the Milky Way Galaxy. (b) This drawing labels two major and several minor spiral arms in the Milky Way. (a: Courtesy of G. Westerhout; *inset:* National Geographic; b: NASA/JPL-Caltech/R. Hart [SSC])

show that it has **spiral arms** is to map the locations of star-forming complexes filled with H II regions, giant molecular clouds, and massive, hot, young stars in groups, called *OB associations*.

Dust absorption limits the range of visual observations in the plane of the Galaxy to less than 10,000 light-years from Earth. However, astronomers can use visible observations of nearby, bright OB associations and associated H II regions to plot the spiral arms near the Sun. Radio observations of hydrogen and carbon monoxide molecules (discussed in Section 7-1) and infrared images of stars taken by the *Spitzer Space Telescope* have been used to chart more remote star-forming regions of the Galaxy. Taken together, all of these observations indicate that our Galaxy has about 200 billion stars located in and between two major spiral arms and several short arm segments (see Figure 10-9b). (Each arm is named after the constellation in which it is centered, as seen from Earth.) Recent observations by the *Spitzer Space Telescope* also confirm previous observations that a bar of stars and gas crosses the center of the Galaxy. We will explore the origin of the spiral arms in Chapter 11, when we present more evidence about the

What situation(s) on Earth are analogous to the obscuration caused by interstellar gas and dust?

cause of spiral structure from observations of other galaxies.

The observable disk of our Galaxy is about 100,000 light-years in diameter and about 2000 light-years thick (Figure 10-11a). Two major spiral arms border either side of the Sun's position. On the side toward the galactic center is the Sagittarius arm, which stargazers in the northern hemisphere see in the summer when they look at the portion of the Milky Way stretching across Scorpius and Sagittarius (see Figures 10-5 and 10-9b). Directed away from the galactic center is the Perseus arm, one of the two major arms of the Galaxy. The Perseus arm is visible in the northern hemisphere in the winter. The other major spiral arm is the Scutum-Centaurus arm, which cannot be seen at visible wavelengths because of obscuring dust in the interstellar medium.

Observations reveal that the Galaxy's arms spiral out from a flattened sphere of stars, called the **central bulge** (sometimes called the nuclear bulge), that is about 20,000 light-years in diameter. This feature is also seen in Figure 10-12, a wide-angle infrared image of the Galaxy taken by the *COBE* satellite. The central bulge is centered on the galactic nucleus 26,000 light-years away from us. The bulge is pierced by a bar of stars, gas, and dust that move one way down the bar and then back the other way (see Figure 10-9b).

H II regions (red) in the spiral arms

Hot luminous, young stars (blue) in the spiral arms

a Visible-light view of M83

R I V U X G

Neutral hydrogen (H I) is concentrated in the spiral arms

b 21-cm radio view of M83

R I V U X G

M83

Menkent

Centaurus

ε Cen

FIGURE 10-10 Two Views of a Barred Spiral Galaxy The galaxy M83 is in the southern constellation of Centaurus, about 12 million light-years from Earth. (a) At visible wavelengths, spiral arms are clearly illuminated by young stars and glowing H II regions. (b) A radio view at 21-cm wavelength shows the emission from neutral hydrogen gas. Note that the spiral arms are more clearly demarcated by visible stars and H II regions than by 21-cm radio emission. (a: S. Van Dyk/IPAC; b: VLA, NRAO)

100,000 ly

26,000 ly

Sun

Central bulge

Galactic nucleus

2,000 ly

Disk

Halo

Globular clusters

Dark matter

Visible matter

10-4 The galactic nucleus is an active, crowded place

If you lived on a planet near the center of the Galaxy, which is called the galactic nucleus, you would see a million stars as bright as Betelgeuse. The total intensity of starlight from all those nearby stars would be equivalent to 200 of our full Moons. Night would never really fall. Stranger still, the Galaxy around you would be filled with intense activity.

Figure 10-13 shows three infrared views that look toward the nucleus of the Galaxy. Figure 10-13a is a wide-angle view covering a 50° segment of the Milky Way through Sagittarius and Scorpius. The prominent band across this image is a thin layer of dust in the plane of the Galaxy. The numerous knots and blobs along the dust layer are interstellar clouds heated by young O and B stars. Figure 10-13b is an *IRAS* (Infrared Astronomical Satellite) view of the galactic center. Numerous streamers of dust *(blue)* surround it. The strongest infrared emission *(white)* comes from **Sagittarius A** (often abbreviated Sgr A),

FIGURE 10-11 Our Galaxy: As seen from the side, three major visible components of our Galaxy are a thin disk, a central bulge, and a halo. As noted earlier, there is also a central bar. The visible Galaxy's diameter is about 100,000 light-years, and the Sun is about 26,000 light-years from the galactic center. The disk contains gas and dust along with Population I (young, metal-rich) stars. The halo is composed almost exclusively of Population II (old, metal-poor) stars. *Inset:* The visible matter in our Galaxy fills only a small volume compared to the distribution of dark matter, whose composition is presently unknown. Its presence is felt by its gravitational effect on visible matter.

FIGURE 10-12 Infrared View of the Milky Way Taken by the *COBE* satellite in 1997, this infrared image shows the disk and central bulge of our Galaxy. Most of the sources scattered above and below the disk are nearby stars. Stars appear white, whereas interstellar dust appears orange. Note that the dust that obscures light from more distant stars in Figure 10-5 is quite bright in this infrared image. (The COBE Project, DIRBE, NASA)

a

Sagittarius (center of Milky Way galaxy)

Solar system (view of Milky Way from Earth)

b

R I V U X G

which is also a grouping of several powerful sources of radio waves. One of these sources, called Sagittarius A* (pronounced "A-star"), is believed to be the galactic nucleus. Figure 10-13c shows stars within 1

light-year of Sagittarius A*, with resolution of 0.02 light-year.

Radio observations add more to the picture of the galactic center. Some of the most detailed radio images of

Galactic center

Sagittarius A*

1 light-year

a A wide-angle (50°) infrared view

R I V U X G

b A close-up view shows a more luminous region at the galactic center.

R I V U X G

c An extreme close-up view centered on Sagittarius A* a radio source at the very center of the Milky Way Galaxy, shows hundreds of stars within 1 ly (0.3 pc).

R I V U X G

WEB LINK 10.5 **FIGURE 10-13** The Galactic Center (a) This wide-angle view at infrared wavelengths shows a 50° segment of the Milky Way centered on the nucleus of the Galaxy. Black represents the dimmest regions of infrared emission, with blue the next strongest, followed by yellow and red; white represents the strongest emission. The prominent band diagonally across this photograph is a layer of dust in the plane of the Galaxy. Numerous knots and blobs along the plane of the Galaxy are interstellar clouds of gas

and dust heated by nearby stars. (b) This close-up infrared view of the galactic center covers the area outlined by the white rectangle in *(a)*. (c) This infrared image shows about 300 of the brightest stars less than 1 ly from Sagittarius A*, which is at the center of the picture. The distribution of stars and their observed motions around the galactic center imply a very high density (about a million solar masses per cubic light-year) of less luminous stars. (a, b: NASA; c: R. Schödel et al. MPE/ESO)

Sgr B2
Sgr B1
Arc
Sgr A
(Galactic center)
SNR 0.3+0.0
Threads
The Cane
Background galaxy
Threads
The Pelican
Sgr C
Snake
Mouse
Sgr E
SNR 359.0-00.9
SNR 359.1-00.5

a

0.2"

N
E

S0-1
S0-2
S0-4
S0-5
S0-16
S0-19
S0-20

Keck/UCLA
Galactic Center Group

1995-2006

b

FIGURE 10-14 Two Views of the Galactic Nucleus
(a) A radio image taken at the VLA of the galactic nucleus and environs. This image covers an area of the sky eight times wider than the Moon. SNR means supernova remnant. The numbers following each SNR are its right ascension and declination. The Sgr (Sagittarius) features are radio-bright objects. (b) The *colored dots* superimposed on this infrared image show the motion of seven stars in the vicinity of the unseen massive object (denoted by the yellow star) at the position of the radio source Sagittarius A*, part of Sgr A in (a). The orbits were measured over an 11-year period. This plot indicates that the stars are held in orbit by a 4×10^6 M$_\odot$ black hole. (a: Naval Research Laboratory produced by N. E. Kassim, D. S. Briggs, T. J. W. Lazio, T. N. LaRosa, J. Imamura & S. D. Hyman. Originally from the NRAO Very Large Array. Courtesy of A. Pedlar, K. Anantharamiah, M. Gross & R. Ekers; b: Keck/UCLA Galactic Center Group)

it come from the Very Large Array (VLA). Figure 10-14a is a wide-angle view of Sagittarius A and surrounding features, including arcs of gas, at least three supernovae remnants, and localized radio sources. Huge filaments, such as the one labeled "Arc," lie perpendicular to the plane of the Galaxy and stretch 200 light-years northward of the galactic disk, then abruptly arch southward toward Sagittarius A. The orderly arrangement of these filaments suggests that a magnetic field may be controlling the distribution and flow of ionized gas, just as magnetic fields on the Sun funnel such gas to create solar prominences. This resulting radio emission, produced by high-speed electrons that spiral around these magnetic fields, is called **synchrotron radiation**. Despite its small size, Sagittarius A is one of the brightest sources of synchrotron radiation in the entire sky.

X-ray observations taken by the *Chandra* telescope in 2004 reveal that the galactic nucleus is also bathed in ultrahot gas with a temperature of 100 million K. If this gas is more than a single outburst from an as-yet-unknown source, it must continually be replenished to remain as hot as it is. Astronomers are still trying to account for the source of this energetic gas.

If the center of our Galaxy is not active and bizarre enough, recent gamma-ray observations reveal positrons being ejected from that region. (Recall that positrons have positive electric charges but are otherwise identical to electrons.) The source of these positrons is still unknown. Positrons can be detected because when a positron and an electron collide, they annihilate each other. Their mass is converted into energy as gamma rays with well-defined wavelengths. These special gamma rays have also been observed emanating from the galactic center.

Infrared observations also reveal stars and gas in very rapid orbits around Sagittarius A* (Figure 10-14b). Something massive must be holding this high-speed matter in such tight orbits around the galactic nucleus. Using Kepler's third law, astronomers calculate that 4×10^6 M$_\odot$ is needed to prevent the stars and gas from flying off into interstellar space. The observed broadening of spectral lines further suggests that an object with the mass of 4 million Suns in a volume only the size of the solar system is concentrated at Sagittarius A*. Astronomers believe that the object is a supermassive black hole. In 2001, the *Chandra* X-ray telescope observed a burst of X rays from this black hole. The X rays were generated by gas heating up as it fell into it. The bigger a black hole, the longer a burst of X rays can last. The duration of the observed blast indicated that the black hole is no wider than 1 AU across, consistent with theoretical calculations.

The presence of supernova remnants at the center of our Galaxy implies what other activity is occurring in that region?

As we saw in Chapter 9, extraordinary activity is also occurring in the nuclei of many other galaxies, implying the presence of supermassive black holes at their centers as well. Astronomers are actively studying these regions in an effort to understand the complex, intriguing events that are happening there.

10-5 Our Galaxy's disk is surrounded by a spherical halo of stars and other matter

As mentioned at the beginning of this chapter, stars have been observed in a spherical distribution, called the **halo,** centered on the galactic nucleus and extended out far beyond the disk (see Figure 10-11). The halo was originally discovered because of the globular clusters that it contains. Looking out of the plane of the Milky Way's disk and between globular clusters, we see apparently unobstructed views of distant galaxies, such as the Whirlpool Galaxy observed by Lord Rosse (see Section 10-1). Appearances can be deceiving. Although the brightness of the globular clusters suggests that they contain most of the stars in the halo, it turns out that about 99% of the halo's stars are isolated *halo field stars* spread all through the halo. The localized concentrations of stars in globular clusters account for only 1% of the halo's stars. It is worth noting, however, that some globular clusters are observed to contain intermediate-mass black holes, which contribute significantly to their total mass.

The various components of the Milky Way Galaxy overlap and interpenetrate each other. For example, the disk slices through the central bulge, and globular clusters and halo field stars periodically pass through the plane of the disk. Figure 10-15 shows the shapes of the orbits of typical central bulge, disk, and halo stars and clusters. Our solar system is presently moving at an angle of 25° from the plane of the Galaxy's disk, as illustrated by the red path in the figure. We will move above the plane about 115 light-years and then return, crossing the plane every 33 million years.

Since 1994, astronomers have observed a dozen small galaxies that orbit in our Galaxy's halo. The Sagittarius Dwarf and the Canis Major Dwarf (Figure 10-16) are named after the constellations in which their central regions lie. They have both spread widely through the halo and periodically pass through the disk of the Milky Way. These two galaxies are much smaller in size and mass than the Milky Way and are losing stars due to the disrupting influence of our Galaxy's strong gravitational attraction. Within the next 100 million years, they will cease to exist, their stars becoming part of the Milky Way. This process of a bigger galaxy consuming a smaller one is called **galactic cannibalism.** The most recently discovered dwarf, called the Ursa Major dwarf spheroidal galaxy, was found in 2005.

10-6 The Galaxy is rotating

Our solar system is moving at 828,000 km (half a million miles) per hour around the center of our Galaxy. Just as the orbital motion of the planets keeps them from falling into the Sun, the motion of the stars and interstellar clouds around the galactic center keeps these bodies apart. If the stars and clouds in our Galaxy were not in orbit, their mutual gravitational forces would have caused them to fall together and form one massive black hole billions of years ago. (Indeed, such infall of gas clouds early in the life of the universe is believed to be the origin of the supermassive black holes located at the centers of the Milky Way and other galaxies.) Just as detecting the positions of the stars and clouds has been difficult, so, too, has been measuring their orbital motions.

Radio observations of 21-cm radiation from hydrogen gas provide important clues about our Galaxy's rotation. By measuring Doppler shifts, astronomers can determine the speed of objects toward or away from us across the Galaxy. These observations clearly indicate that our Galaxy does not rotate like a rigid body (Figure 10-17a) but rather exhibits *differential rotation*, meaning that stars at different distances from the galactic center orbit the Galaxy at different rates (Figure 10-17b).

Observations show that the Sun's distance from the center of the Galaxy varies little throughout its orbit. About 80% of the stars in our neighborhood

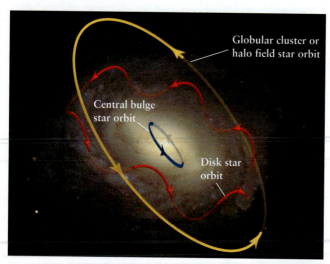

R I V U X G

FIGURE 10-15 Orbits of Stars in Our Galaxy This disk galaxy, M58, looks very similar to what the Milky Way Galaxy would look like from far away. The *colored arrows* show typical orbits of stars in the central bulge *(blue)*, disk *(red)*, and halo *(yellow)*. Interstellar clouds, clusters, and other objects in the various components have similar orbits. (NOAO/AURA/NSF)

a

R I V U X G

b

R I V U X G

VIDEO 10.3 **FIGURE 10-16** The Nearest Galaxy (a) The Canis Major Dwarf is a dwarf elliptical galaxy that lies some 25,000 light-years from the Milky Way. This infrared radiation-based image shows the Milky Way's spiral arms, as well as the distribution of stars being stripped from the Canis Major Dwarf by our Galaxy's gravitational tidal force. Containing only about one billion stars, the Canis Major Dwarf will be completely pulled apart within the next 100 million years or so by the Milky Way. (b) View from Earth of the Canis Major Galaxy and its path of debris. (a: R. Ibata/Strasbourg Observatory, ULP/et al., 2MASS, NASA b: Nicola Martin & Rodrigo Ibata, Observatoire de Strasbourg, 2003)

have similar-shaped orbits. The other 20% have orbits that are either taking them outward from the center or inward toward it, including at least one star moving outward with enough speed to leave the Galaxy forever. To understand differential rotation, we focus on those stars with nearly circular orbits. Because of the stars' differential rotation in the Galaxy, the Sun is like a car on a circular freeway with the fast lane on one side and the slow lane on the other. As sketched in Figure 10-17b, stars in the fast lane (closer to the center of the Galaxy) are passing the Sun and thus appear from our vantage point to be moving in one direction, while stars in the slow lane (farther from the center of the Galaxy than our solar system) are being overtaken by the Sun and therefore appear to be moving in the opposite direction. This is like the retrograde motion of the planets discussed in Section 2-2.

Unfortunately, like the 21-cm observations, studying the motion of nearby stars and gas reveals only how fast they are moving relative to the Sun. To get a complete picture of the Galaxy's rotation, we must find out how fast the Sun itself is orbiting the center of the Galaxy. The Swedish astronomer Bertil Lindblad proposed a method of computing this speed. He noted that globular clusters do not move in the orderly pattern shown in Figure 10-17. Different globular clusters in the halo of our Galaxy orbit in different planes, and so they do not participate in the organized rotation of the objects in the Galaxy's disk. However, the combined velocities of the globular clusters around the center of the Galaxy must average to zero or else they would drift, *en masse*, relative to the rest of the Galaxy. Using the motions of globular clusters as a reference, astronomers calculated the speed of the Sun's orbit around the galactic center given earlier.

Knowing the Sun's speed and its distance from the galactic center, astronomers can calculate the Sun's orbital period. Traveling at 828,000 km/h, our Sun takes about 230 million years to complete one trip around the Galaxy. When the solar system last passed our present location in the Milky Way, early dinosaurs of the Triassic period roamed Earth.

By combining the true speed of the Sun with the relative speed of the stars around us, as measured by radio astronomers, we can also determine the actual orbital speeds of the stars. This computation gives us the **rotation curve** of the Galaxy, a graph that shows the orbital speeds of stars and interstellar clouds at various distances from the center of the Milky Way (Figure 10-18).

Knowing the Sun's velocity around the Galaxy from the rotation curve, we can use Kepler's third law to calculate that the mass of the Galaxy that lies between us and the galactic nucleus is about 110 billion solar masses.

> What effect, which also occurs in the solar system, is responsible for the removal of many stars from smaller galaxies orbiting the Milky Way?

> The complete trip of the solar system around the Galaxy is analogous to what period in Earth's motion in the solar system?

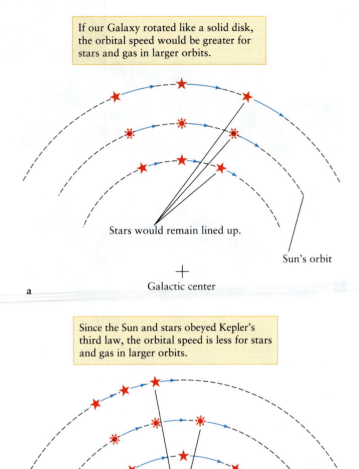

If our Galaxy rotated like a solid disk, the orbital speed would be greater for stars and gas in larger orbits.

Stars would remain lined up.

Sun's orbit

+
Galactic center

a

Since the Sun and stars obeyed Kepler's third law, the orbital speed is less for stars and gas in larger orbits.

Stars in smaller orbits would rapidly overtake those in larger orbits.

Sun's orbit

+
Galactic center

b

FIGURE 10-17 Differential Rotation of the Galaxy (a) If all stars in the Galaxy had the same angular speed, they would orbit in lockstep. (b) However, stars at different distances from the galactic center have different angular speeds. Stars and clouds farther from the center take longer to go around the Galaxy than do stars closer to the center. As a result, stars closer to the Galaxy's center than the Sun are overtaking the solar system, whereas stars farther from the center are lagging behind us.

MYSTERIES AT THE GALACTIC FRINGES

Because the band of the Milky Way extends all around Earth, we know that stars and other matter exist beyond the Sun's orbit. In recent years, astronomers have been astonished to discover just how much matter lies farther from the center of our Galaxy than we are.

10-7 Most of the matter in the Galaxy has not yet been identified

According to Kepler's third law, the farther a star or cloud is from the center of the Milky Way, the slower it should be moving (see the dashed line in Figure 10-18), just as the orbital speeds of the planets decrease with increasing distance from the Sun. Observations reveal, however, that galactic orbital speeds continue to *climb* well beyond the visible edge of the galactic disk. This means there must be more gravitational force from the Galaxy acting on the distant stars and clouds than we can see or have taken into account.

These observations led astronomers to calculate that a surprising amount of matter must therefore lie beyond the Sun's orbit in the Galaxy. They found that nearly 90% of the mass of our Galaxy has yet to be located. If we include the mass of the as-yet-unidentified matter, the Milky Way's total mass exceeds 1×10^{12} M$_\odot$.

To make matters even more mysterious, much of this mass is in the form of **dark matter**: Whatever it is, very little of it shows up on images at any wavelength. Given its effects on the stars and gas in the Galaxy that we can detect, astronomers deduce that this dark matter is spherically distributed all around the Galaxy in a massive halo extending at least 12 times farther from the galactic nucleus than we are, far beyond the halo defined by the visible globular clusters and halo field stars (see *inset* in Figure 10-11). Small fractions of the dark matter are composed of neutrinos and white dwarfs, but the nature of the rest of it—whether black holes, gas,

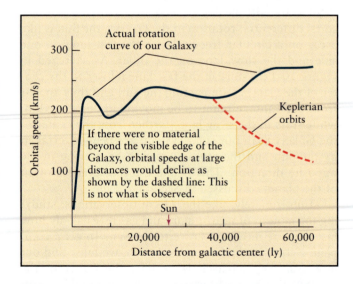

FIGURE 10-18 The Galaxy's Rotation Curve The blue curve shows the orbital speeds of stars and gas in the Galaxy, and the *dashed* red curve shows Keplerian orbits that would be caused by the gravitational force from known objects. Because the data (blue curve) do not show any such decline, there is, apparently, an abundance of dark matter that extends to great distances from the galactic center. This additional mass gives the outer stars higher speeds than they would have otherwise.

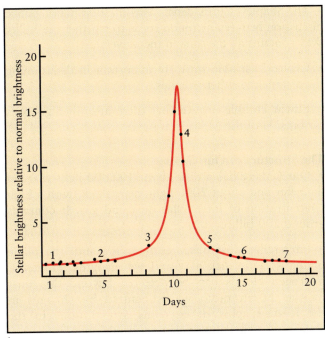

a

b

FIGURE 10-19 Microlensing by Dark Matter in the Galactic Halo (a) Gravitational fields cause light to change direction. A white dwarf, brown dwarf, or black hole in the Galaxy's halo passing between Earth and a more distant star will focus the starlight in our direction, making distant objects appear brighter than they are normally. (b) The light curve of the gravitational microlensing of light from a star in the Galaxy's central bulge by an intervening object.

Jupiter-like bodies, brown dwarfs, or something far more exotic—has yet to be determined. The unidentified matter in our Galaxy is sometimes called the **missing mass**. This is a poor name because the matter is not missing; we just do not yet know its location or nature.

Some searches for dark matter rely on the principle from general relativity that matter causes passing light to change direction (see Figure 9-5). The Galaxy's dark matter may contain relatively massive objects, like brown dwarfs or black holes. If so, these objects will cause the light from more distant stars to change direction as the dark objects pass between those stars and Earth (Figure 10-19a). Such objects are called *MACHOs* (Massive Compact Halo Objects). Indeed, as a dark object moves directly between a more distant star and Earth, the star's light is focused toward us, causing the star to brighten. Called **microlensing**, this brightening of distant stars for several weeks has been observed (Figure 10-19b). Microlensing observations indicate that only a small fraction of the Galaxy's dark matter is in the form of these MACHOs.

10-8 Frontiers yet to be discovered

Despite our being part of it, our knowledge of the Milky Way is limited by the difficulty in making observations through its gas and dust. We have yet to locate all of the globular clusters. Even more challenging will be locating and categorizing all of the stars and clouds of gas and

dust in the disk of the Galaxy. Are there any more small galaxies that the Milky Way is consuming? We also have to determine fully and accurately the structure and extent of the spiral arms and the bar of gas, dust, and stars running through the central bulge.

> Microlensing is similar to the physics of which: reflecting telescopes or refracting telescopes?

Our observations of the Galaxy's nucleus hint at its remarkable activity that has yet to be explained. What is heating the 100-million-K gas? What is creating the positrons there, whose resulting gamma rays we observe? In its farthest reaches, what else is in the Galaxy's halo that creates all of the gravitational force we detect, but whose source is not yet observed? In other words, what is the dark matter? Another major unsolved mystery is the formation process of our Galaxy. Did stars form first? Did the supermassive black hole form and then pull gas and dust into orbit, which then formed stars? Did smaller galaxies form and then coalesce into the Milky Way as it is today? Astronomers are presently working to answer all of these questions.

SUMMARY OF KEY IDEAS

Discovering the Milky Way

• A century ago, astronomers were divided on whether all stars and nebulae are part of the Milky Way Galaxy.

- The Shapley–Curtis debate was the first major public discussion between astronomers as to whether the Milky Way contains all the stars in the universe.

- Cepheid variable stars are important in determining the distance to other galaxies.

- Edwin Hubble proved that there are other galaxies far outside of the Milky Way.

The Structure of Our Galaxy

- Our Galaxy has a disk about 100,000 light-years in diameter and about 2000 light-years thick, with a high concentration of interstellar dust and gas. It contains around 200 billion stars.

- Interstellar dust obscures our view into the plane of the galactic disk at visual wavelengths. However, hydrogen clouds can be detected beyond this dust by the 21-cm radio waves emitted by changes in the relative spins of electrons and protons in the clouds, as well as by other non-visible emissions.

- The center, or galactic nucleus, has been studied at gamma-ray, X-ray, infrared, and radio wavelengths, which pass readily through intervening interstellar dust and H II regions that illuminate the spiral arms. These observations have revealed the dynamic nature of the galactic nucleus, but much about it remains unexplained.

- A supermassive black hole of about 4×10^6 $M_\odot$ exists in the galactic nucleus.

- The galactic nucleus of the Milky Way is surrounded by a flattened sphere of stars, called the central bulge, through which a bar of stars and gas extend. The entire Galaxy is surrounded by a halo of matter that includes a spherical distribution of globular clusters and field stars, as well as large amounts of dark matter.

- A disk with two bright and several dimmer arms of stars, gas, and dust spirals out from the ends of the bar in the galactic central bulge.

- Young OB associations, H II regions, and molecular clouds in the galactic disk outline huge spiral arms where stars are forming.

- The Sun is located about 26,000 light-years from the galactic nucleus, between two major spiral arms. The Sun moves in its orbit at a speed of about 828,000 km/h and takes about 230 million years to complete one orbit around the center of the Galaxy.

Mysteries at the Galactic Fringes

- From studies of the rotation of the Galaxy, astronomers estimate that its total mass is about 1×10^{12} $M_\odot$. Much of this mass is still undetectable.

WHAT DID YOU THINK?

1   *What is the shape of the Milky Way Galaxy?* The Milky Way is a barred spiral galaxy. A bar of stars, gas, and dust runs through its central region. It has at least four spiral arms and is surrounded by a spherical halo of stars and dark matter.

2   *Where is our solar system located in the Milky Way Galaxy?* The solar system is between the Sagittarius and Perseus spiral arms, about 26,000 light-years from the center of the Galaxy.

3   *Is the Sun moving through the Milky Way Galaxy and, if so, about how fast?* Yes. The Sun orbits the center of the Milky Way Galaxy at a speed of 828,000 km/h.

Key Terms for Review

dark matter (missing mass), 284
disk (of a galaxy), 277
distance modulus, 275
galactic cannibalism, 282
galactic nucleus, 274
galaxy, 274
halo (of a galaxy), 282
microlensing, 285
Milky Way Galaxy, 271
missing mass, 285
nebula (*plural* nebulae), 272

central bulge, 278
rotation curve (of a galaxy), 283
Sagittarius A, 279
Shapley–Curtis debate, 272
spin (of an electron or proton), 277
spiral arm, 278
synchrotron radiation, 281
21-cm radio radiation, 277

Review Questions

The answers to computational problems, which are preceded by an asterisk (*), appear at the back of the book.

1. Where in the Galaxy is the solar system located? **a.** in the nucleus **b.** in the halo **c.** in a spiral arm **d.** between two spiral arms **e.** in the central bulge

2. What is located in the nucleus of the Galaxy? **a.** a globular cluster **b.** a spiral arm **c.** a black hole **d.** the solar system **e.** a MACHO

3. Which statement about the Milky Way Galaxy is correct? **a.** Our Galaxy is but one of many galaxies. **b.** Our Galaxy contains all stars in the universe. **c.** All stars in our Galaxy take the same time to complete one orbit. **d.** Most stars in our Galaxy are in the central bulge. **e.** None of the stars in our Galaxy move.

4. What was the Shapley–Curtis debate all about? Was a winner declared at the end of the debate? Whose ideas turned out to be correct?

5. How did Edwin Hubble prove that M31 is not a nebula in our Milky Way Galaxy?

CHAPTER 11

Galaxies

The Starburst Galaxy M82 shows a frenzy of star formation in its central regions and jets of gas perpendicular to the galaxy. Both activities are believed to have been initiated by passage of the galaxy near its neighbor, M81. (Courtesy Mark Westmoquette, University College London; Jay Gallagher, UW-Madison; Linda Smith, University College London; WIYN/NSF; NASA/ESA)

R I V U X G

WHAT DO YOU THINK?

1 Are most of the stars in spiral galaxies located in their spiral arms?

2 Do all galaxies have spiral arms?

3 Are galaxies isolated objects?

4 Is the universe contracting, unchanging in size, or expanding?

Answers to these questions appear in the text beside the corresponding numbers in the margins and at the end of the chapter.

We live in a galactic suburb, far from the center of the action. The Sun is just 1 of about 200 billion stars in the Milky Way, and our Galaxy is but 1 of between 50 billion and 1 trillion galaxies in the visible universe. In studying galaxies, we have entered the realm of truly cosmic systems of stars, gas, and dust. Although nearby stars are separated from us by light-years (trillions of miles), neighboring galaxies are hundreds of thousands of times farther away, and the most distant galaxies are at least 13.2 billion light-years from Earth.

Finding patterns is a powerful tool built into the human psyche. Early in the twentieth century, astronomers observed that galaxies appear in surprisingly few shapes. The classification scheme Edwin Hubble developed is where we begin our study of the large-scale structure of the universe—but that is only the beginning. By the mid-twentieth century, clusters of galaxies were discovered, followed by clusters of clusters (now called *superclusters*) of galaxies. Between these ensembles of stars, gas, and dust are vast regions of space relatively devoid of matter. Having established these patterns, astronomers now know the large-scale structure of visible matter in the universe, which we also present in this chapter.

In this chapter you will discover

- how galaxies are categorized by their shapes
- the processes that produce galaxies of different shapes
- that galaxies are found in clusters that contain huge amounts of dark matter
- why clusters of galaxies form in superclusters
- how some galaxies merge and others devour their neighbors
- that the universe is changing size

TYPES OF GALAXIES

Despite the incredible number of galaxies, there is surprising consistency in their overall shapes. Edwin Hubble began cataloguing their appearance in the 1920s. The Hubble classification of galaxies—spirals, barred spirals, ellipticals, irregulars, and their subclasses—is still used today.

11-1 The winding of a spiral galaxy's arms is correlated to the size of its central bulge

Spiral galaxies without central bars (often called *normal spiral galaxies*) are characterized by a central bulge and arched lanes of stars and glowing interstellar clouds, which appear as arms that spiral out from the bulge. Observations of many spiral galaxies reveal that the more tightly wound the spiral, the larger the central bulge.

As shown in Figure 11-1, spirals with tightly wound spiral arms (and fat central bulges) are called *Sa* (for spiral type *a*) *galaxies*. Those with moderately wound spiral arms (and a moderate central bulge) are *Sb galaxies*. Loosely wound spirals (with small central bulges) are *Sc galaxies*. A typical spiral galaxy contains about 100 billion stars and measures nearly 10^5 light-years in diameter.

The closest spiral galaxy to the Milky Way is the Andromeda Galaxy, M31 (Figure 11-2). M31 is visible to the naked eye as a fuzzy blob in the constellation of Andromeda. The disk of this galaxy is tilted as seen from Earth, typical of many spiral galaxies. Some are tilted so much that their spiral arms are not evident. We can still classify these galaxies as specific spiral types by observing the sizes of their central bulges. For example, M104 (Figure 11-3a) has a huge central bulge. It must therefore be an Sa galaxy with tightly wound arms. An Sb galaxy (Figure 11-3b) has a smaller central bulge. The tiny central bulge of an Sc galaxy (Figure 11-3c) is hardly noticeable at all in this edge-on galaxy.

Besides the degree of winding, the overall appearance of individual spiral arms varies from galaxy to galaxy. In some galaxies, called *flocculent spirals* (from the word meaning "fleecy"), the spiral arms are broad, fuzzy, chaotic, and poorly defined (Figure 11-4a). Other galaxies, called *grand-design spirals*, exhibit beautiful arching arms outlined by brilliant H II regions and OB associations. In these galaxies, the spiral arms are thin, delicate, graceful, and well-defined (Figure 11-4b).

At first glance, it may seem that the shapes of spiral galaxies are created by the orbital motions of strings of stars, with the stars closer to the center of the galaxy orbiting more rapidly and thereby leading the outer edges of the spirals. But, in this case, common sense leads us astray. Although the stars closer to the center of the galaxy do orbit faster than those farther away, spiral arms cannot be orbiting strings of stars. If they were, the spiral arms should "wind up" from the inside out (Figure 11-5), wrapping themselves tightly around the nucleus. This action is not observed.

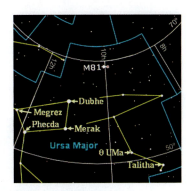

a NGC 1357: an Sa galaxy

R I V U X G

b M81: an Sb galaxy

R I V U X G

c NGC 4321: an Sc galaxy

R I V U X G

FIGURE 11-1 Spiral Galaxies (Nearly Face-on Views) Edwin Hubble classified spiral galaxies according to the tightness of their spiral arms and the sizes of their central bulges. Sa galaxies have the largest central bulges and the most tightly wound spiral arms, whereas Sc galaxies have the smallest central bulges and the least tightly wound arms. The images are different colors because they were taken through filters that pass different colors. (a: NASA/Hubble Space Institute; b: Robert Gendler; c: Anglo-Australian Observatory)

If star motions caused spirals, then after a few galactic rotations, the spiral structure should disappear altogether. Why, then, do we observe spiral arms in our Galaxy and many others? This *winding dilemma* led astronomers to two explanations, one for flocculent spirals and the other for grand-design spirals.

11-2 Explosions create flocculent spirals, and waves create grand-design spirals

Flocculent Spiral Galaxies Consider a disk galaxy so young that it has not yet formed spiral arms. Stars begin forming in one of its giant molecular clouds. The radiation and winds from these stars compress the nearby interstellar gas, triggering the formation of additional stars. Moreover, massive stars quickly explode as supernovae and produce shock waves, which further compress surrounding gases. Shock waves are very powerful compressions of the gas, such as the sonic booms created by lightning, the snap of a whip, or a supersonic aircraft. Thus, as new stars trigger the birth of still other stars, the star-forming region grows, a process called *self-propagating star formation*.

The continuing birth of stars accounts for spiral arms in flocculent spiral galaxies. The galaxy's differential rotation (inner regions orbit faster than outer regions) drags the inner edge of each young region ahead of its outer edge, spreading the star-forming region into a spiral arm. This region of the galaxy is bright because it is highlighted by its brilliant O and B stars, and by the nebulae that these stars cause to glow. However, high-mass O and B stars explode relatively quickly, and each spiral arm dims before it winds up (that is, there is no winding dilemma here). As old arms fade, new star-forming regions appear and create new spiral arms.

Where self-propagating star formation is the only process creating spiral arms in a galaxy, bits and pieces of these arms appear constantly, only to disappear as the bright, massive stars in them die. This process creates chaotic spiral arms, such as those observed in flocculent galaxies (see Figure 11-4a) but not the smooth spirals of grand-design

> Comparing it to the galaxies in Figure 11-1, what is the spiral classification of M 74, shown in Figure 11-4?

FIGURE 11-2 Andromeda (M31) Andromeda is a beautiful spiral galaxy and the only galaxy visible to the naked eye from Earth's northern hemisphere. It has dim, red giant stars (not visible here) extending half a million light-years from its nucleus. Without a telescope, Andromeda appears to be a fuzzy blob in the constellation of the same name. Located only 2.5 Mly (0.77 Mpc) from us, Andromeda is gravitationally bound to the Milky Way, and it covers an area in the sky roughly 5 times as large as the full Moon. Two other galaxies, M32 and M110, are also labeled on this photograph. The points of light that pepper the image are stars in our Galaxy. (Bill and Sally Fletcher/Tom Stack and Associates)

R I V U X G

a M104: an Sa galaxy

R I V U X G

b NGC 891: an Sb galaxy

R I V U X G

c NGC 5907: an Sc galaxy

R I V U X G

FIGURE 11-3 Spiral Galaxies Seen Nearly Edge-on from the Milky Way (a) Because of its large nuclear bulge, this galaxy (called the Sombrero Galaxy) is classified as an Sa. If we could see it face-on, the spiral arms would be tightly wound around a voluminous bulge. (b) Note the smaller nuclear bulge in this Sb galaxy. (c) At visible wavelengths, interstellar dust obscures the relatively insignificant nuclear bulge of this Sc galaxy. (a: European Southern Observatory; b: © Malin/IAC/RGO; c: Brand Ehrhorn/Adam Block/NOAO/AURA/NSF)

galaxies. Thus, self-propagating star formation cannot be the whole story.

Grand-Design Spiral Galaxies Our search for the mechanism that generates grand-design spirals begins in a pond. If you throw a rock into the water, you create ripples. As you know from experience, the ripples move outward in concentric rings from where the rock struck (Figure 11-6a). But suppose the pond water is rotating when you throw in a rock. Now what shape do the ripples have? According to calculations done by astronomer Bertil Lindblad in the 1920s, ripples created in a rotating disk-shaped system of liquid (the pond) or gas and dust (a disk galaxy) will be spiral. These ripples are called **spiral density waves.**

Russian scientists in the 1970s tested Lindblad's model. Lacking the computer power to create simulations of galaxies, they threw pebbles into pie pans of

a M33: a spiral galaxy with flocculent spiral arms

R I V U X G

Grand design spiral arms

b M74: a grand design spiral galaxy

R I V U X G

FIGURE 11-4 Variety in Spiral Arms The differences in spiral galaxies suggest that at least two mechanisms create spiral arms. (a) This flocculent spiral galaxy has fuzzy, poorly defined

spiral arms. (b) This grand-design spiral galaxy has well-defined spiral arms. (a: NASA; b: Gemini Observatory/AURA)

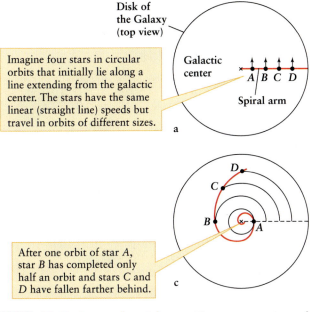

Disk of the Galaxy (top view)

Imagine four stars in circular orbits that initially lie along a line extending from the galactic center. The stars have the same linear (straight line) speeds but travel in orbits of different sizes.

Galactic center

Spiral arm

a

When star A has completed half of an orbit, stars B, C, and D have only completed a quarter or less of an orbit.

b

After one orbit of star A, star B has completed only half an orbit and stars C and D have fallen farther behind.

c

As star A completes its second orbit, the spiral continues to wind tighter.

d

FIGURE 11-5 The Winding Dilemma The rotation curve of the disk stars in our Galaxy indicates that most of them have the same linear (straight-line) speed. Those farther from the center take longer to go around because they have a greater distance to travel at the same speed than stars closer to the center of the Galaxy, which orbit in smaller circles. All four dots in these drawings have circular orbits at the same linear speed. Think of them as (a) a few stars initially in a straight line. As time goes on, the outer stars are left behind, creating (b-d) a spiral shape that becomes more and more tightly wound. Such tightening is not observed in our Galaxy or in other galaxies.

water rotating on phonograph turntables. The resulting water wave patterns were indeed spirals, as drawn in Figure 11-6b. Furthermore, spiral density waves never wind up. In the case of the gas and dust in a galaxy, these waves orbit in a rigid spiral pattern at about half the average speed of the interstellar medium through which they move.

In the mid-1960s, two American astrophysicists looked carefully at how galactic spiral density waves traveling through the disk of a galaxy lead to spiral arms. C. C. Lin and Frank Shu argued that density waves cause interstellar gas and dust to pile up temporarily. This occurs because, unlike water waves, which move up and down as they travel along, spiral density waves in the gas and dust of a galaxy are compressional waves, just like sound. As a result, they cause interstellar gas and dust to become denser as this material passes through the waves.

Lin and Shu demonstrated that a spiral arm is a galactic traffic jam. Imagine workers painting a line down a busy freeway. The cars normally cruise at 100 km/h (about 65 mph), but the crew of painters causes a bottleneck. The cars slow down temporarily to avoid hitting other cars and the slowly moving paint truck. As seen from the air, there is a noticeable congestion of cars around the painters (Figure 11-7). An individual car spends only a few moments in the moving traffic jam before resuming its usual speed, but the traffic jam itself lasts all day long.

The vehicles in the traffic jam represent the interstellar gas and dust entering a spiral arm. The paint truck represents a small part of the spiral density wave. Like the cars catching up to and passing the paint truck, the interstellar medium sweeps through the more slowly moving spiral density waves. The waves compress some of this interstellar gas and dust, creating new clouds, which then collapse to form new, metal-rich stars. We see spiral arms because they contain bright O and B stars and copious quantities of dust and gas that these stars illuminate. The sprawling dust lanes in Figure 11-4b attest to the recent passage of that material through a shock wave. While all of this is going on, the density wave maintains its shape, and, overall, the spiral arms in these galaxies are better defined than in the flocculent galaxies discussed earlier.

A few high-mass stars form in most open clusters, such as those created by spiral density waves. So why are spiral galaxies eventually not filled with enough of these bright stars to wash out the spiral arms and to make the disk uniformly bright? The answer is that the massive stars highlighting each density wave are short-lived and explode before they finish passing through it. The remaining, longer-lived, lower-mass stars, like our Sun, fill the space between the spiral arms without emitting as much light (see Figure 11-4b). As pronounced as spiral arms may appear, they contain only 5% more stars than are found between them.

It takes an enormous amount of energy to compress interstellar gas and dust. After a billion years or so, even spiral density waves would begin to fade away. Some driving mechanisms must keep them going, analogous to a   RIVUXG   b

FIGURE 11-6 Ripples in Water (a) The usual circular ripples expanding from the place where a rock was thrown into the water. (b) Ripples in rotating water creating spiral arms, as do ripples in the gas and dust of a disk galaxy. (a: © Royalty Free/Corbis)

throwing more rocks in the pond. We are not yet completely certain what those mechanisms are, but one likely explanation for reinvigorating the spiral structure is the passage of a nearby companion galaxy. As this nearby galaxy periodically passes close by, its gravitational attraction pulls on the gas, stars, and dust of the spiral galaxy to generate new density waves. Indeed, grand-design galaxies are usually found in the presence of a companion galaxy.

Although most spiral galaxies have two arms, a sizable minority has more. The Milky Way, for example, has at least four spiral arms. Astronomers have not yet established why the numbers of arms vary.

Because new stars are formed from interstellar gas and dust, it is plausible that galaxies with different amounts of this material will have different star formation rates and therefore different overall structures. Infrared and radio observations reveal that Sa galaxies typically contain about 4% gas and dust, whereas Sb galaxies contain 8%, and Sc galaxies contain 25%. Continually improving computer simulations of galaxies reveals more and more of the details of how various concentrations of gas and dust affect galactic structure. Figure 11-8 summarizes the activity in a spiral galaxy.

> Fill in the blanks: If analogous events occurred on Earth, we would call self-propagating star formation a _____ reaction or a _____ effect.

Traffic flow →

Paint truck represents part of a spiral density wave moving slowly through a galaxy.

Cars represent faster-moving gas and dust overtaking, and being slowed by, spiral density wave.

FIGURE 11-7 Compression Wave in Traffic Flow When normal traffic flow is slowed down, cars bunch together. In a grand-design galaxy, a density wave moves through the stars and gas. The wave is merely a region of slightly denser matter, which, in turn, creates more gravitational force. This force compresses the gas and enhances star formation, which highlights the spiral density wave.

ANIMATION 11.1

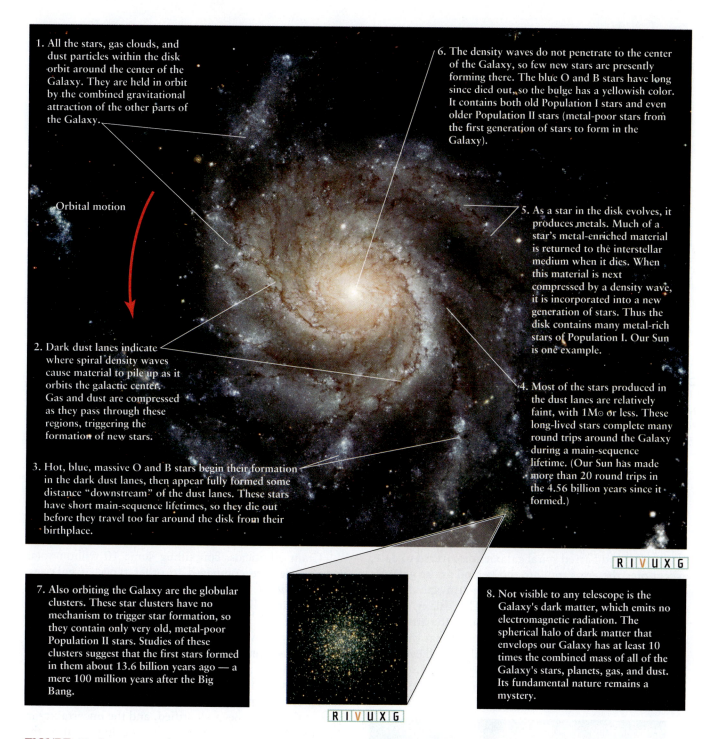

1. All the stars, gas clouds, and dust particles within the disk orbit around the center of the Galaxy. They are held in orbit by the combined gravitational attraction of the other parts of the Galaxy.

Orbital motion

2. Dark dust lanes indicate where spiral density waves cause material to pile up as it orbits the galactic center. Gas and dust are compressed as they pass through these regions, triggering the formation of new stars.

3. Hot, blue, massive O and B stars begin their formation in the dark dust lanes, then appear fully formed some distance "downstream" of the dust lanes. These stars have short main-sequence lifetimes, so they die out before they travel too far around the disk from their birthplace.

6. The density waves do not penetrate to the center of the Galaxy, so few new stars are presently forming there. The blue O and B stars have long since died out, so the bulge has a yellowish color. It contains both old Population I stars and even older Population II stars (metal-poor stars from the first generation of stars to form in the Galaxy).

5. As a star in the disk evolves, it produces metals. Much of a star's metal-enriched material is returned to the interstellar medium when it dies. When this material is next compressed by a density wave, it is incorporated into a new generation of stars. Thus the disk contains many metal-rich stars of Population I. Our Sun is one example.

4. Most of the stars produced in the dust lanes are relatively faint, with 1M☉ or less. These long-lived stars complete many round trips around the Galaxy during a main-sequence lifetime. (Our Sun has made more than 20 round trips in the 4.56 billion years since it formed.)

R I V U X G

7. Also orbiting the Galaxy are the globular clusters. These star clusters have no mechanism to trigger star formation, so they contain only very old, metal-poor Population II stars. Studies of these clusters suggest that the first stars formed in them about 13.6 billion years ago — a mere 100 million years after the Big Bang.

8. Not visible to any telescope is the Galaxy's dark matter, which emits no electromagnetic radiation. The spherical halo of dark matter that envelops our Galaxy has at least 10 times the combined mass of all of the Galaxy's stars, planets, gas, and dust. Its fundamental nature remains a mystery.

R I V U X G

FIGURE 11-8 Dynamics of a Grand-Design Spiral Galaxy This figure summarizes the activities taking place in a grand-design spiral galaxy. (Image of spiral galaxy M101: NASA and ESA; image of globular cluster M3: S. Kafka and K. Honeycutt, Indiana University/WIYN/NOAO/NSF)

11-3 Bars of stars run through the central bulges of barred spiral galaxies

The Milky Way is a **barred spiral**, a spiral galaxy with a bar of stars crossing through the central bulge. Such bars are composed of stars and gas that basically flow first one way down the bar and then turn around and flow the other way. Bars form in spiral galaxies that have less total mass than the normal or "unbarred" spirals described earlier. Indeed, the higher total mass of normal spirals prevents bars from forming. About one-third of all spiral galaxies are barred spirals, the remainder being the normal spirals. The two main arms in barred spirals typically extend from the ends of the bar rather than from the central bulge itself.

Edwin Hubble found that the winding of the spiral arms in barred spirals correlates with the size of the central bulge (Figure 11-9), just as for normal spirals. An *SBa* (for spiral, barred, type *a*) galaxy has a large central bulge (and tightly wound spiral arms). Likewise, a barred

a M58: an SBa galaxy

b M83: an SBb galaxy

c NGC 1365: an SBc galaxy

FIGURE 11-9 Barred Spiral Galaxies As with spiral galaxies, Edwin Hubble classified barred spirals according to the tightness of their spiral arms (which correlates with the sizes of their central bulges). SBa galaxies have the most tightly wound spirals and largest central bulges, SBb galaxies have moderately tight spirals and medium-sized central bulges, and SBc galaxies have the least tightly wound spirals and the smallest central bulges. (a: Johan H. Knapen and Nik Szymanek, University of Hertfordshire; b: ESO, European Southern Observatory; c: Jean-Charles Cuillandre/CFHT/Photo Researchers, Inc.)

> Why do you think that spiral arms do not extend farther inward than the bar in barred spiral galaxies?

spiral with moderately wound spiral arms (and a moderate central bulge) is an *SBb* galaxy, and an *SBc* galaxy has loosely wound spiral arms (and a tiny central bulge). The Milky Way is classified as an SBab, which has properties between those of an SBa and an SBb. Sa and SBa are sometimes called early type galaxies, while Sc and SBc are called late type galaxies.

The motions of the stars in many nearby galaxies have been measured from Doppler shifts of their spectra. In all spiral and barred spiral galaxies discovered to date, except one, the arms trail around behind as the galaxies rotate. They are thus called **trailing-arm spirals**. The exception is the galaxy NGC 4622. This galaxy has the points of its spirals leading the motion of the arms. It is a *leading-arm spiral*. Calculations predict that the leading arms are unstable and that this galaxy will eventually switch and become a trailing-arm spiral.

11-4 Elliptical galaxies display a wide variety of sizes and masses

Elliptical galaxies, named for their distinctive shapes, have no spiral arms. They range tremendously in size and mass—from the biggest to the smallest galaxies in the universe. Figure 11-10 includes two *giant elliptical galaxies*, M84 and M86, that form part of a cluster of galaxies in the constellation Virgo. These enormous giant elliptical galaxies are each about 2 million light-years in diameter, 20 times the diameter of the Milky Way Galaxy.

Giant ellipticals, containing some 10 trillion solar masses, are rare compared to other types of galaxies, whereas *dwarf elliptical galaxies* are extremely common. Dwarf ellipticals are only a fraction of the size of an average elliptical galaxy and contain so few stars—only a few million—that we see these galaxies as nearly transparent. Because you can actually see straight through the center of a dwarf elliptical galaxy and out the other side (Figure 11-11), distant ones are hard to detect. Hence, many more dwarf ellipticals undoubtedly exist than have been identified, and the uncertainty in their number is a major reason that we do not know the total number of galaxies in the visible universe.

Hubble subdivided elliptical galaxies according to how round or oval they look. The roundest elliptical galaxies are called *E0 galaxies*, whereas the most elongated are *E7 galaxies*. These latter are about three times longer than they are wide. Elliptical galaxies of intermediate elongation are numbered E1 to E6 (Figure 11-12).

FIGURE 11-10 Giant Elliptical Galaxies The Virgo cluster is a rich, sprawling collection of more than 2000 galaxies about 50 million light-years from Earth. Only the center of this huge cluster appears in this photograph. The two largest galaxies in the cluster are the giant elliptical galaxies M84 and M86. (Royal Observatory, Edinburgh)

RIVUXG

The Hubble scheme classifies galaxies solely by their appearance from Earth. But whenever we observe anything in the sky, we are seeing a two-dimensional view of a three-dimensional object. In the case of elliptical galaxies, we observe length and width, but we have no way of knowing anything about the third dimension of depth. What looks like an E0 galaxy (basically circular) might actually be egg-shaped when viewed from another angle. Conversely, an elongated E7 galaxy might look circular when viewed face-on.

Elliptical galaxies look far less dramatic than their spiral and barred spiral cousins because they contain relatively little interstellar gas and dust. Because stars form in interstellar clouds, relatively few stars should be forming in ellipticals compared to the numbers forming in spirals. Observations of spectra confirm the hypothesis that these galaxies contain primarily Population II, low-mass, long-lived stars. However, billions of years ago, many ellipticals contained bright, massive stars that have since exploded as supernovae. We know this because the *Chandra* X-ray telescope has discovered large numbers of remnant black holes and neutron stars in several of these galaxies.

11-5 Galaxies without global structure are called irregular

Hubble found some galaxies that cannot be classified as spirals, barred spirals, or ellipticals. He called these **irregular galaxies**. They are generally rich in interstellar gas, dust, and both young and old stars. Irregular galaxies with numerous OB associations and only hints of organized structure are denoted *Irr I*. The nearby Large Magellanic Cloud (LMC) (Figure 11-13a) and the Small Magellanic Cloud (SMC) are examples of Irr I. Both can be seen with the naked eye from southern latitudes.

Occasionally, irregulars are observed that appear highly distorted and completely asymmetrical, as though created by collisions between galaxies or by violent activ-

Leo 1: an E4 galaxy   R I V U X G

FIGURE 11-11 A Dwarf Elliptical Galaxy This nearby E4 galaxy, called Leo I, is about 600,000 light-year from Earth. It is only 3000 light-year in diameter and is so sparsely populated with stars that you can see right through its center. It is a satellite of the Milky Way. (NASA)

ity in their nuclei. These are denoted *Irr II,* as shown by NGC 4485 (Figure 11-13b). Irregular galaxies are typically smaller and less massive than spirals, containing between 10^8 and about 3×10^{10} M$_\odot$.

Besides being an E0, what other classification could M105 in Figure 11-12 have?

11-6 Hubble presented spiral and elliptical galaxies in a tuning fork-shaped diagram

Edwin Hubble connected the three regularly shaped types of galaxies—spirals, barred spirals, and ellipticals—in a diagram shaped like a tuning fork (Figure 11-14). According to his scheme, S0 or SB0 galaxies, called **lenticulars** (lens-shaped), are an intermediate type between ellipticals and the two kinds of spirals. Although

a M105: an E0 galaxy

b M49: an E4 galaxy

c NGC 4526: an E7 galaxy   R I V U X G

FIGURE 11-12 Elliptical Galaxies Hubble classified elliptical galaxies according to how round or elongated they appear. An E0 galaxy is round; a very elongated elliptical galaxy is an E7. Three examples are shown. (a: J. D. Wray, McDonald Observatory; b, c: 1999 Princeton University Press/Zsolt Frei and James E. Gunn) a Large Magellanic Cloud, an Irr I galaxy

b NGC 4485 (Irr II) and NGC 4490 (Sc) galaxies

R I V U X G

FIGURE 11-13 Irregular Galaxies (a) At a distance of only 179,000 light-years, the Large Magellanic Cloud (LMC), an Irr I irregular galaxy, is the third closest known companion of our Milky Way Galaxy. (The Milky Way's closest known companion, the Canis Major Dwarf, is shown in Figure 10-16.) About 62,000 light-years across, the LMC spans 22° across the sky, about 44 times the angular size of the full Moon. Note the huge H II region (called the Tarantula Nebula or 30 Doradus). Its diameter of 800 light-years and mass of 5 million solar masses make it the largest known H II region. (b) The small irregular (Irr II) galaxy NGC 4485 interacts with the highly distorted Sc galaxy NGC 4490, also called the *Cocoon Galaxy*. This pair is located in the constellation Canes Venatici. (a: Anglo-Australian Observatory; b: M. Altmann/Hoher Observatory)

> The Large Magellanic Cloud is visible to the naked eye in the southern hemisphere. From Figure 11-13, what two things do you think it could be mistaken for?

they often look somewhat like ellipticals, lenticular galaxies have both a central bulge and a disk, like spiral galaxies, but they lack spiral arms. For want of any better scheme, the irregular galaxies are sometimes placed between the ends of the tuning fork tines of the Hubble diagram.

The idea that one type of galaxy may change into another type has been in and out of favor with astronomers for decades. Indeed, such transformations were part of Hubble's motivation in relating the different types of galaxies as he did. As we will see later in this chapter and in Chapter 13, extensive observations by the Hubble Space Telescope and by ground-based telescopes reveal that interactions between galaxies sometimes do

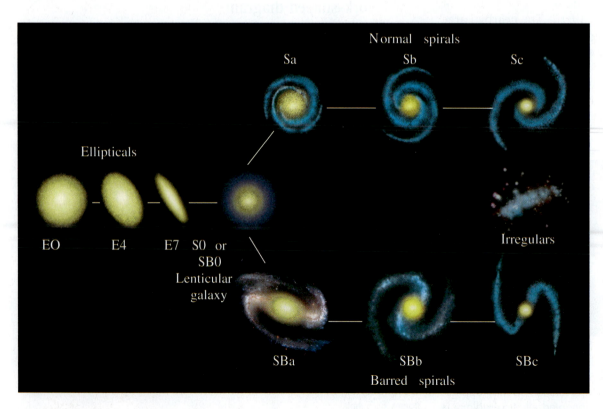

FIGURE 11-14 Hubble's Tuning Fork Diagram Hubble summarized his classification scheme for galaxies with this tuning fork diagram. Elliptical galaxies are classified by how oval they appear, whereas spirals and barred spirals are classified by the sizes of their central bulges and the correlated winding of their spiral arms. An S0 or SB0 galaxy, also called a lenticular galaxy, is an intermediate type between ellipticals and spirals. It has a disk but no spiral arms.

TABLE 11-1 Some Properties of Galaxies

	Spiral (S) and barred spiral (SB) galaxies	Elliptical galaxies (E)	Irregular galaxies (Irr)
Mass ($M_\odot$)	10^9 to 4×10^{11}	10^5 to 10^{13}	10^8 to 3×10^{10}
Luminosity ($L_\odot$)	10^8 to 2×10^{10}	3×10^5 to 10^{11}	10^7 to 10^9
Diameter (ly)	1.6×10^4 to 8×10^5	3×10^3 to 6.5×10^5	3×10^3 to 3×10^4
Stellar populations	Disk: young Population I central bulge; halo: Population II and old Population I	Population II and old Population I	Mostly Population I
Percentage of observed galaxies	77%	*20%	3%

*This percentage does not include dwarf elliptical galaxies that are as yet too dim and distant to detect. Hence, the actual percentage of galaxies that are ellipticals is likely to be higher than shown here.

lead to changes in their structures. For example, when two disk galaxies merge they often morph into a giant elliptical galaxy. However, it appears that most galaxies remain unchanged in overall shape once they are fully formed. The properties of the different types of galaxies are summarized in Table 11-1.

11-7 Galaxies built up in size over time

In Chapter 13, we will explore the observational evidence for the formation and early evolution of galaxies. Nevertheless, it is worth summarizing here what astronomers know about why the galaxies have the structures we observe. With the exception of dwarf ellipticals, galaxies formed from smaller ensembles of stars, gas, and dust. The underlying dark matter that clumped together in the young universe drew this matter together, often assisted by a supermassive black hole formed early in the life of the universe. This black hole became the nucleus of the galaxy. The more massive the black hole, the larger a disk galaxy's central bulge. Once formed, about three quarters of all galaxies have maintained their structures, whereas the remaining quarter have changed (for example, from spiral to elliptical).

Normal (unbarred) spiral galaxies formed when swirling eddies of gas-rich matter collided and formed a disk centered on an especially massive black hole and lots of dark matter. When the amount of dark matter and the mass of the central black hole were too low, the matter in the central regions of the disk had bar-shaped orbits, creating barred-spiral galaxies. Disk galaxies formed when the gases that created them had time to interact with each other without forming too many stars. If the collision of the gases led to especially rapid star formation, the resulting galaxies were elliptical or irregular. Giant elliptical galaxies were (and still are) formed from the collisions of two spiral galaxies (either type).

CLUSTERS AND SUPERCLUSTERS

As we consider the vastness of galaxies, it is hard to imagine that structures this huge orbit each other in groups. But they do.

11-8 Galaxies occur in clusters, which occur in larger clumps called superclusters ▪3

Galaxies are not scattered randomly throughout the universe, but rather they are grouped together in **clusters**. The Fornax cluster (Figure 11-15), so called because it is located in the con-

> What physical property of matter discussed in Chapter 2, besides gravitation, is essential in forming all spiral galaxies?

R I V U X G

FIGURE 11-15 A Cluster of Galaxies This group of galaxies, called the Fornax cluster, is about 60 million light-years from Earth. Both elliptical and spiral galaxies are easily identified. The barred spiral galaxy at the lower left is NGC 1365, the largest and most impressive member of the cluster. For a closer view of NGC 1365, see Figure 11-9. (Anglo-Australian Observatory/Royal Observatory, Edinburgh)

FIGURE 11-16 Superclusters in Our Neighborhood This diagram shows the distances and relative positions of superclusters within 950 million light-years of Earth. *Note* also the labeling of some of the voids, which are large, relatively empty regions between superclusters. (Kirk Korista)

cluster orbit each other and, occasionally, even collide.

Clusters of galaxies are themselves grouped together in huge associations, called **superclusters**. A typical supercluster contains dozens of individual clusters spread over a volume 150 million light-years across. Figure 11-16 shows the distribution of superclusters in our vicinity of the universe. The nearer ones, out to the Virgo cluster, are members of our *Local Supercluster*. Observations indicate that most superclusters are not gravitationally bound units—most clusters in each supercluster are drifting away from most of the other clusters in that same supercluster. Furthermore, the superclusters are all moving away from one another.

 How superclusters are arranged in space became clearer, beginning in the early 1980s, when astronomers began systematic surveys of the positions of observable objects in the nearby universe. Completed in 2005, the first Sloan Digital Sky Survey observed nearly 200 million galaxies, stars, and other celestial objects, and measured the redshifts of over a million of them. This survey, made in the northern hemisphere, covered about one-quarter of the sky and gives us a three-dimensional map through 100 times the volume of space as we had mapped before. Another set of observations, called the Two Degree Field (2dF) Survey, studied regions of both the northern and southern hemispheres.

These data reveal that most superclusters are located on the boundaries between enormous bubble-like *voids* in which few galaxies are observed (Figure 11-17). These voids are roughly spherical and measure between 100 million and 400 million light-years in diameter. They are not completely empty, however. Observations reveal hydrogen clouds in some of them, while others may be subdivided by strings of dim galaxies.

The distribution of clusters of galaxies throughout the universe appears to have the structure of a sponge (Figure 11-18) or soap bubbles, with the galaxies locat-

Astronomers observe very hot (10^7 K) gas between clusters of galaxies. Why does this gas not appear in Figure 11-15?

stellation of Fornax (the Furnace), is typical. Observations of galactic motion indicate that members of a cluster of galaxies are gravitationally bound together: The galaxies in a

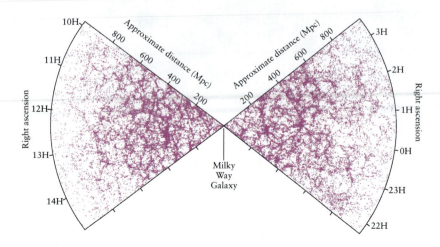

The 2dF Galaxy Survey

FIGURE 11-17 Structure in the Universe This map shows the distribution of 62,559 galaxies in two wedges extending in opposite directions from Earth out to distances of 3.5 billion light-years. For an explanation of right ascension (r.a.), see Section 1-3. Note the prominent voids surrounded by thin areas full of galaxies. (Courtesy of the 2dF Galaxy Redshift Survey Team)

FIGURE 11-18 Foamy Structure of the Universe A sponge that recreates the distribution of bright clusters of galaxies throughout the universe. The empty spaces in the foam are analogous to the voids found throughout the universe. The spongy regions are analogous to the locations of most of the galaxies. (Image Source/Super Stock)

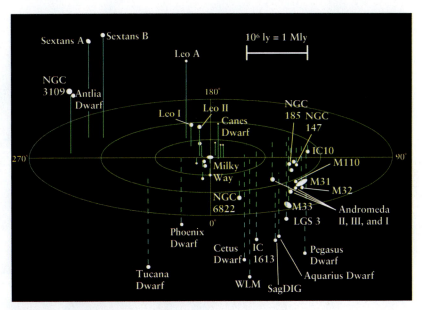

FIGURE 11-19 The Local Group Our Galaxy belongs to a poor, irregular cluster that consists of about 40 galaxies, called the *Local Group*. This map shows the distribution of about three-quarters of the galaxies. The Andromeda Galaxy (M31) is the largest and most massive galaxy in the Local Group. The second largest is the Milky Way itself. M31 and the Milky Way are each surrounded by a dozen satellite galaxies. The recently discovered Canis Major Dwarf Galaxy is the Milky Way's nearest known neighbor.

ed where the sponge or soapy material is, surrounded by voids. Astronomers believe that this spongy pattern contains important clues about conditions in the early universe that have not yet been fully fathomed.

11-9 Clusters of galaxies may appear densely or sparsely populated and regular or irregular in shape

A cluster of galaxies is said to be either a **poor cluster** or a **rich cluster**, depending on whether it contains less than or more than a thousand galaxies. For example, the Milky Way Galaxy, the Andromeda Galaxy, and the Large and Small Magellanic Clouds belong to a poor cluster, called the **Local Group**. The Local Group contains roughly 40 galaxies, over a third of which are dwarf ellipticals. Figure 11-19 shows a map covering most of the Local Group. The Virgo cluster is the nearest rich cluster, with over 2000 galaxies.

New galaxies in the Local Group continue to be found. This happens primarily because astronomers are developing techniques to observe objects that lie in the same plane as, but beyond, the Milky Way. In addition to the 1994 discovery of the Sagittarius Dwarf Galaxy, the dwarf galaxy Antlia (Figure 11-20) was first observed in 1997 and the Canis Major Dwarf (see Figure 10-16) was found in 2003. The Canis Major Dwarf is only about 25,000 light-years from the Milky Way.

Astronomers also categorize clusters of galaxies according to their shapes. A **regular cluster** is distinctly spherical, with a marked concentration of galaxies at its center. Numerous gravitational interactions over billions

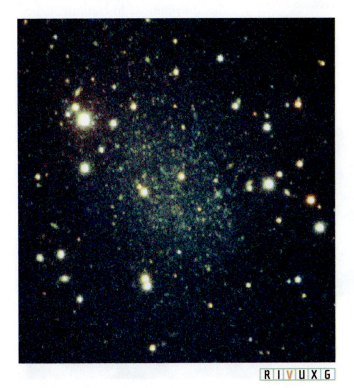

R I V U X G

FIGURE 11-20 A Recently Discovered Member of the Local Group The galaxy Antlia was first detected in 1997. It lies about 3 million light-years away, outside the region depicted in Figure 11-19. This galaxy contains only about a million stars. (M. J. Irwin, Royal Greenwich Observatory, and A. B. Whiting and G. K. T. Hau, Institute of Astronomy, Cambridge University)

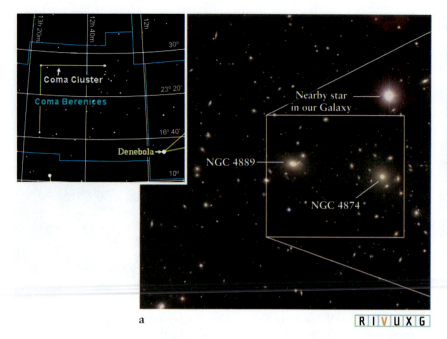

a ☐R☐I☐V☐U☐X☐G☐

b ☐R☐I☐V☐U☐X☐G☐

FIGURE 11-21 The Coma Cluster (a) This rich, regular cluster, containing thousands of galaxies, is about 300 million light-years from Earth. (b) Regular clusters are composed mostly of elliptical and lenticular galaxies and are sources of X rays. This *Chandra* image shows Coma's central region, which is 1.5 million light-years across. The gas cloud emitting most of these X rays has a temperature of 100 million K. (a: O. Lopez-Cruz and I. Shelton, NOAO/AURA/NSF; b: NASA/CXC/SAO/A. Vikhlinin, et al.)

> What are two reasons that new galaxies are still being discovered in the Local Group?

of years have spread the galaxies into this distribution. In contrast, galaxies in an **irregular cluster** are more randomly scattered about a sprawling region of the sky.

The nearest example of a rich, regular cluster is the Coma cluster, located about 300 million light-years from us in the constellation of Coma Berenices (Figure 11-21). Despite its great distance, more than 1000 bright galaxies within it are easily visible from Earth. It is likely that the Coma cluster contains many thousands of dwarf ellipticals soon to be detected—maybe as many as 10,000 galaxies overall.

Rich, regular clusters, such as the Coma cluster, contain mostly elliptical and lenticular galaxies. Only 15% of the Coma cluster's galaxies are spirals and irregulars. Irregular clusters, such as the Virgo cluster and the Hercules cluster (Figure 11-22), display a more even mixture of galaxy types. Two-thirds of the 200 brightest galaxies in the Hercules cluster are spirals, nearly a fifth are ellipticals, and the rest are irregular.

11-10 Galaxies in a cluster can collide and combine

Occasionally, two or more galaxies in a cluster or two or more galaxies from different, nearby clusters collide.

Sometimes they pass through one another (Figure 11-23) and sometimes they merge. In 2002, the Hubble Space Telescope observed four galaxies colliding simultaneously. There is so much space between stars that the probability of two stars crashing into each other during galactic collisions is extremely small. On the other hand, the galaxies' huge clouds of interstellar gas and dust are so large that they do collide, slamming

☐R☐I☐V☐U☐X☐G☐

FIGURE 11-22 The Hercules Cluster This irregular cluster, which is about 700 million light-years from Earth, contains a high proportion of spiral galaxies, often associated in pairs and small groups. (NOAO)

Cartwheel Galaxy   R I V U X G

FIGURE 11-23 A Composite Image of the Cartwheel Galaxy
This ring-shaped assemblage 500 million light-years from Earth
is likely the result of one galaxy, probably the blue-white one
below it at the eight o'clock position, having passed through the
middle of the larger one. Astronomers suspect that the passage
created a circular density wave in the Cartwheel that stimulated
a burst of star formation, creating many bright blue and white
stars. Ultraviolet is shown in blue, visible light in green, infrared
in red, and X-ray in violet. (NASA/JPL-Caltech/P. Appleton
[SSC/Caltech])

into each other and producing strong shock waves. The
colliding interstellar clouds are stopped in their tracks.

A violent collision can strip both galaxies of their
interstellar gas and dust. The energy associated with
the collision heats this gas and dust to extremely high
temperatures. This process is a major source of the hot
intergalactic gas often observed in rich, regular clusters. The *Chandra* X-ray Observatory has observed
intergalactic gas with temperatures between 300,000
and 100 million K.

Another source of intergalactic gas is gas escaping
from dwarf galaxies. This gas is ejected by supernovae
of stars created in a burst of star formation when the
galaxies formed. Whereas our Galaxy's gravitational
force traps supernova remnants in the disk, this gas
escapes into intergalactic space from the relatively weak
gravitational pull of dwarf galaxies.

Less violent collisions and near misses between
galaxies allow the compressed interstellar gas and dust to
cool sufficiently to have new stars form from it. These
collisions can also stimulate galaxywide star formation,
which may account for the **starburst galaxies** that blaze

with the light of numerous newborn stars. These galaxies
are characterized by bright centers surrounded by clouds
of warm interstellar dust, indicating a recent, vigorous
episode of star birth (Figure 11-24). Their warm dust is
so abundant that starburst galaxies are among the most
luminous objects in the universe at infrared wavelengths.

Observational evidence indicates that globular clusters are formed during collisions of interstellar gas and
dust. Unlike the globular clusters in our Galaxy, some
galaxies contain very young globular clusters. Whether
those in our Galaxy formed as a result of an early collision of the Milky Way with another galaxy has yet to be
determined.

The Irr II, starburst galaxy M82 is one member of
a nearby cluster of about a dozen galaxies that include
the beautiful spiral galaxy M81 and a fainter companion, called NGC 3077 (Figure 11-25a). A radio survey
of that region of the sky revealed enormous streams of
hydrogen gas connecting these and other, smaller galaxies in that cluster (Figure 11-25b). The loops and twists
in these streamers suggest that the three galaxies have
had several close encounters over the ages, most recently about 600 million years ago. Gravitational interactions between colliding galaxies can also result in thousands of stars being hurled out into intergalactic space
along huge, arching streams, as seen in Figure 11-26a.

In rich clusters, astronomers observe near misses
between galaxies (Figure 11-26b). If galaxies are surrounded by extended halos of dim stars, these near
misses strip the galaxies of these outlying stars. In this
way, a loosely dispersed sea of dim stars populates the
space between galaxies in a cluster.

Similarly, after galaxies collide, some interior
stars are flung far and wide, scattering material into intergalactic space. Such isolated
stars have been observed by the Hubble Space Telescope.
However, other stars slow down, often causing the remnants of the galaxies to merge. Several dramatic examples of **galactic mergers** have been discovered, such as
the galaxy known as NGC 6240 (Figure 11-27). The
supermassive black holes in the nuclei of the original two
galaxies (see Section 9-8) have not yet merged (see inset:
Figure 11-27). They are expected to do so within a few
hundred million years.

As mentioned in Chapter 10, with regard to the
Milky Way tearing apart the Canis Major Dwarf Galaxy,
astronomers also speak of galactic cannibalism, which
occurs when a large galaxy captures and "devours" a
smaller one. Cannibalism differs from mergers in that the
"dining" galaxy is significantly bigger than its "dinner,"
whereas merging galaxies are about the same size.

Many astronomers suspect that giant ellipticals are
the product of galactic mergers and cannibalism, including spiral-spiral collisions. Observations reveal that
giant galaxies typically occupy the centers of rich clusters. Once formed, giant ellipticals can continue to grow
because smaller galaxies are often located around them

WEB LINK 11.10

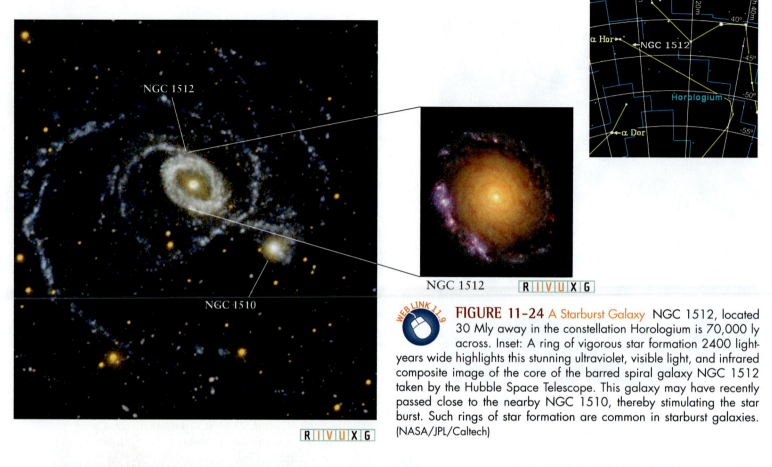

NGC 1512

FIGURE 11-24 A Starburst Galaxy NGC 1512, located 30 Mly away in the constellation Horologium is 70,000 ly across. Inset: A ring of vigorous star formation 2400 light-years wide highlights this stunning ultraviolet, visible light, and infrared composite image of the core of the barred spiral galaxy NGC 1512 taken by the Hubble Space Telescope. This galaxy may have recently passed close to the nearby NGC 1510, thereby stimulating the star burst. Such rings of star formation are common in starburst galaxies. (NASA/JPL/Caltech)

FIGURE 11-25 The M81 Group The Irr II, starburst galaxy M82 (see inset) is in a nearby cluster of about a dozen galaxies, including the spectacular spiral M81. Several of the galaxies in this cluster are connected by streamers of hydrogen gas. (a) Three brightest galaxies at visual wavelengths.

(b) This radio image, created from data taken by the Very Large Array, shows the streamers of hydrogen gas that connect the bright galaxies and also several dim ones, seen as regions of bright orange here. (a: Palomar Sky Survey; b: M. S. Yun, VLA, and Harvard)

Interaction between the galaxies
has produced new, blue stars

"Tail" of stars and
gas pulled out of the
interacting galaxies

Galaxies with material
flowing between them

 a NGC 4676

R I V U X G

 b IC 2163 *(left)* and NGC 2207

R I V U X G

FIGURE 11-26 Interacting and Colliding Galaxies (a) Pairs of colliding galaxies often exhibit long "antennae" of stars ejected by the collision. This particular system is known as NGC 4676 or "the Mice" (because of its tails of stars and gas). It is 300 million light-years from Earth in the constellation Coma Berenices. The collision has stimulated a firestorm of new star formation, as can be seen in the bright blue regions. Mass can also be seen flowing between the two galaxies, which will eventually merge. (b) These two galaxies, NGC 2207 (right) and IC 2163, are orbiting and tidally distorting each other. Their most recent close encounter occurred 40 million years ago when the two were perpendicular to each other and about 1 galactic diameter apart. Computer simulations indicate that they should eventually coalesce. (a: NASA, H. Ford/JHU, G. Illingworth/UCSC/Lick, M. Clampin/STScI, G. Hartig/STScI, The ACS Science Team, and ESA; b: NASA)

(see Figure 11-10). As they pass through the extended halo of a giant elliptical, these smaller galaxies slow down and are eventually consumed by the larger galaxy.

Figure 11-28, a computer simulation, shows a large, disk-shaped galaxy devouring a small satellite galaxy. The large galaxy consists of 90% stars (blue) and 10% gas (white) by mass. It is surrounded by a halo of dark matter whose mass is about 3.3 times that of the disk.

The satellite galaxy, which has a tenth of the mass of the large galaxy, contains only stars (yellow). Initially, the satellite is in circular orbit about the large galaxy. Note that spiral arms appear in the large galaxy as the collision proceeds. Two billion years elapse as the satellite spirals in toward the core of the large galaxy. Although much material is stripped from the satellite, most of its stars plunge into the central region of the large galaxy.

Hubble Optical

NGC 6240

R I V U X G

Chandra X ray

R I V U X G

Supermassive black holes

FIGURE 11-27 *Merging Galaxies* This contorted object, NGC 6240, in the constellation Ophiuchus is the result of two spiral galaxies in the process of merging. The widespread blue-green area reveals that the collision between the two galaxies has triggered an immense burst of star formation. Inset: The *Chandra* X ray telescope shows that at the heart of this system are two supermassive black holes. Within a few hundred million years, these black holes are expected to merge into a single, more massive black hole. (R. P. van der Marel/J. Gersgen/STScI/NASA; *inset*: S. Komossa/ G. Hasinger/MPE, et al., CXC/NASA)

11-11 Dark matter helps hold together clusters of galaxies

What keeps the rapidly moving galaxies together in clusters? There must be sufficient matter to provide the necessary gravitational force to bind galaxies together in clusters and, in some cases, clusters into superclusters. However, *no cluster of galaxies contains enough visible matter to stay bound together.* Most of the matter binding clusters of galaxies is dark matter (matter

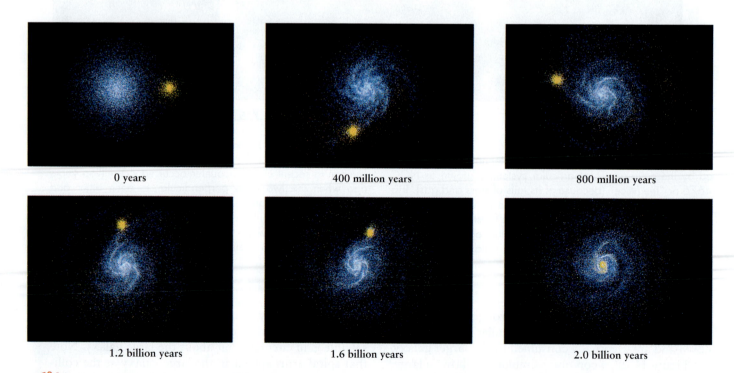

0 years

400 million years

800 million years

1.2 billion years

1.6 billion years

2.0 billion years

FIGURE 11-28 *Simulated Galactic Cannibalism* This computer simulation shows a small galaxy (yellow stars) being devoured by a larger, disk-shaped one (blue stars, white gas). Note how spiral arms are gener- ated in the disk galaxy by its interaction with the satellite galaxy. (Lars Hernquist, Institute for Advanced Study with simulations performed at the Pittsburgh Supercomputing Center)

VIDEO 11.2

FIGURE 11-29 The Rotation Curves of Four Spiral Galaxies
This graph shows how the orbital speed of material in the disks of four spiral galaxies varies with the distance from the center of each galaxy. If most of each galaxy's mass were concentrated near its center these curves would fall off at large distances. But these and many other galaxies have flat rotation curves that do not fall off. This indicates the presence of extended halos of dark matter. See Figure 10-18 to compare these to the Milky Way's rotation curve.

that we cannot yet directly detect). This must be true because, otherwise, the galaxies would have long ago wandered apart in random directions, and the clusters would no longer exist. Recall that a similar problem arose in Chapter 10 regarding dark matter in individual galaxies. Analyses of the total visible mass demonstrate that the total mass needed to bind together a typical rich cluster is 10 times greater than the mass of material that shows up on visible-light images. Furthermore, observations of the distribution of galaxies in clusters have determined that the dark matter usually has the same distribution as the galaxies in the cluster, and that each galaxy corresponds to a concentration of dark matter.

Quite a bit of evidence has accumulated to support the idea of extended halos surrounding galaxies. Many galaxies have rotation curves similar to those of our Milky Way (recall Figure 10-18). These rotation curves remain remarkably flat out to surprisingly great distances from the galaxies' centers. For example, Figure 11-29 shows the rotation curves of four spiral galaxies. In all cases, the orbital speed is fairly constant. According to Kepler's third law, we should see a decline in orbital speed toward the outer portions of a galaxy. In most cases, we do not, which implies that we still have not detected the true edge of these and many similar galaxies. As with the Milky Way Galaxy, astronomers conclude that a considerable amount of dark matter must extend well beyond the visible portions of most galaxies.

Astronomers using X-ray telescopes have discovered the nature of some of the dark matter (which is therefore no longer considered dark). Satellite observations have revealed X rays pouring from the space between galaxies in rich clusters. This radiation is emitted by substantial amounts of hot intergalactic gas at temperatures between 10 million and 100 million K. Figure 11-21b shows this hot gas in the center of the Coma cluster. Observations by the *Chandra* X-ray Observatory indicate that this intergalactic gas contains twice as much mass as exists in all of the galaxies.

Nature also provides astronomers with assistance in visualizing dark matter by the way its mass gravitationally warps space, as described by Einstein's general theory of relativity (see Section 9-2). We saw in Chapter 10 how the same effect occurs on small scales here in the Milky Way. The local distortion is called *microlensing*, and when dark matter and the visible matter in entire galaxies or clusters of galaxies focus light from more distant galaxies, the effect is called **gravitational lensing** (Figure 11-30a). Figure 11-30b shows the gravitationally lensed images of distant galaxies caused by dark matter and closer galaxies.

In 2000, astronomers observed an intergalactic cloud of gas the size of a galaxy. Observations revealed that for that gas to remain bound together, it must contain about 100 billion solar masses of matter. Because no galaxy exists in that space, astronomers conclude that the gas is held together by that much dark matter. In 2007, astronomers saw distant galaxies that undergo gravitational lensing without the presence of an intervening visible galaxy. They deduced that the effect was caused by just the presence of dark matter, whose distribution is shown as a *blue haze* in Figure 11-30c.

The amount of gravitational lensing by dark matter alone, or by dark matter and a galaxy, or by a cluster of galaxies provides astronomers with more than just enlarged, distorted images of distant objects. It tells us how much mass is contained in the dark matter and galaxy or cluster that does the lensing. The greater the mass doing the lensing, the greater the effect on the distant object. The visible matter in the galaxies accounts for only about 10% to 20% of the gravitational lensing that occurs. This means that dark matter accounts for 80% to 90% of the mass in the universe.

> If the Milky Way collided with another spiral galaxy, in what type of galaxy would our descendants likely live?

SUPERCLUSTERS IN MOTION

Whenever an astronomer finds an object in the sky, one of the first tasks is to determine its composition. As we saw in Chapters 3 and 4, that means attaching a spectrograph to a telescope and recording the object's

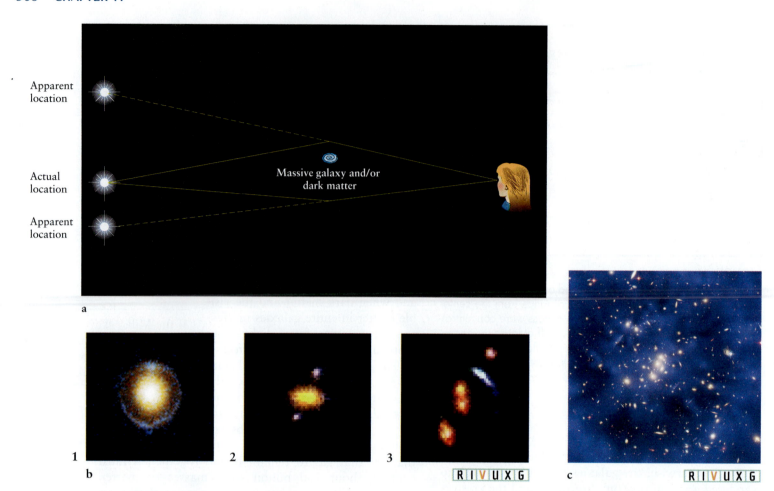

a

b

1   2   3

RIVUXG   c   RIVUXG

FIGURE 11-30 Gravitational Lensing of Extremely Distant Galaxies (a) Schematic of how a gravitational lens works. Light from the distant object changes direction due to the gravitational attraction of the intervening galaxy and underlying dark matter. The more distant galaxy appears in different places than it actually is. (b) Three examples of gravitational lensing: (1) The bluer arc is a galaxy that has been lensed by the redder elliptical galaxy. (2) A pair of bluish images of the same object lensed symmetrically by the brighter, redder galaxy between them. (3) The lensed object appears as a blue arc under the gravitational influence of the group of four galaxies. (c) Superimposed in blue on this image of the galaxy cluster CL 0024+17 is the location of dark matter that is gravitationally lensing the galaxies behind it. (b: NASA, ESA, A. Bloton [Harvard-Smithsonian CfA] and the SLACS Team; c: NASA, ESA, and M.J. Jee [Johns Hopkins University])

If the dark matter in Figure 11-30c suddenly vanished, what would we see of the clusters of galaxies behind it?

spectrum. As long ago as 1914, V. M. Slipher, working at the Lowell Observatory in Arizona, took spectra of "spiral nebulae," now known to be spiral galaxies. He was surprised to discover that the spectral lines of 11 of the 15 spiral nebulae that he studied were substantially redshifted. These redshifts indicate that they are moving away from the Milky Way at significant speeds. This left scientists with a major puzzle: Is the entire universe expanding?

11-12 The redshifts of superclusters indicate that the universe is expanding

During the 1920s, Edwin Hubble and Milton Humason recorded the spectra of many galaxies with the 100-in telescope on Mount Wilson, confirming that most galaxies are rapidly receding from the Milky Way. This implies that the universe *is* expanding, although, as we will see in a moment, it does not imply that the Milky Way is at the center of it!

Using the Doppler effect (recall Figure 3-6 and *An Astronomer's Toolbox 4-3*), Hubble calculated the speed at which the galaxies are moving away from us. Using techniques, such as the brightness of Cepheid variables (see Section 10-1), he also estimated the distances to several of these galaxies. He found a direct correlation between the distance to a galaxy and the size of its redshift: *Galaxies in distant clusters and superclusters are moving away from us more rapidly than galaxies in nearby clusters and superclusters.* Figure 11-31 shows this correlation for five elliptical galaxies.

GALAXIES in	REDSHIFTS

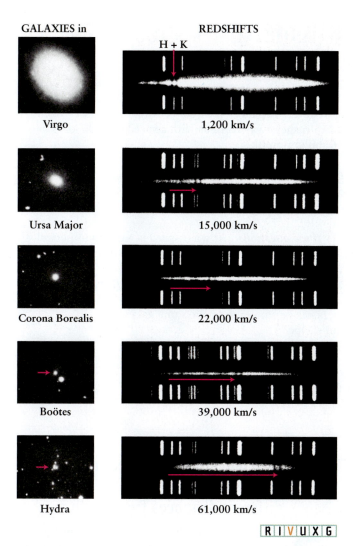

Virgo — 1,200 km/s

Ursa Major — 15,000 km/s

Corona Borealis — 22,000 km/s

Boötes — 39,000 km/s

Hydra — 61,000 km/s

R I V U X G

FIGURE 11-31 Five Galaxies and Their Spectra The photographs of these five elliptical galaxies were all taken at the same magnification. They are labeled according to the constellation in which each galaxy is located. The spectrum of each galaxy is the hazy band between the comparison spectra at the top and bottom of each plate. In all five cases, the so-called H and K lines of calcium are seen. The recessional velocity (calculated from the Doppler shifts of the H and K lines) appears below each spectrum. Note that the fainter—and thus more distant—a galaxy is, the greater is its redshift. (Carnegie Observatories)

This recessional motion pervades the universe and is now called the **Hubble flow.**

The separation of clusters in the same supercluster is slower than the general Hubble flow of the universe because these clusters are relatively close to one another, so their gravitational attractions slow each other down. Keep in mind that the Hubble flow does not occur for galaxies in any given cluster, because all of those galaxies are gravitationally bound to each anoth-

er. To recap: The Hubble flow applies to the separation of superclusters from one another and many clusters of galaxies from one another.

When Hubble plotted the redshift data on a graph of distance versus velocity, he found that the points lie nearly along a straight line. Figure 11-32 shows the relationship between the distances to galaxies and their recessional motion, including many of whose distances were measured using the apparent magnitude of Type Ia supernovae. The information in this figure can be stated as a formula, called the **Hubble law:**

$$\text{Recessional velocity} = H_0 \times \text{distance}$$

This is simply the equation of a straight line, with the slope of the line denoted by the constant H_0, commonly called the **Hubble constant.** *An Astronomer's Toolbox 11-1* shows how the Hubble law works.

The relationship between the distances to galaxies and their redshifts was one of the most important astronomical discoveries of the twentieth century. It tells us that we are living in an expanding universe, and the Hubble law reveals the speed of that expansion. It also shows why we are not in the center of the universe, even though the other superclusters are all receding from us, as discussed in *Guided Discovery: The Expanding Universe.*

The discovery that the universe is expanding raised the question of whether the universe will expand forever

FIGURE 11-32 The Hubble Law The distances and recessional velocities of distant galaxies are plotted on this graph. The straight line is the "best fit" for the data. This linear relationship between distance and speed is called the Hubble law. For historical reasons, distances between galaxies, clusters of galaxies, and superclusters of galaxies are usually given in megaparsecs, Mpc, rather than millions of light-years.

AN ASTRONOMER'S TOOLBOX 11-1

The Hubble Law

The Hubble law describes our expanding universe. It is a simple formula:

$$v = H_0 \times d,$$

where v is a galaxy's recessional velocity, d is its distance from Earth, and H_0 is the Hubble constant. It is the formula for the straight line displayed in Figure 11-32, and the Hubble constant is the slope of this line.

As Figure 11-32 shows, we usually measure a galaxy's velocity in kilometers per second and its distance in megaparsecs. You will see in Section 11-13 that astronomers have determined Hubble's constant to be $H_0 \approx 71$ km/s/Mpc. If you have distances in millions of light-years (Mly), you can convert H_0 using 1.00 Mpc = 3.26 Mly.

Example: A galaxy 1 Mpc from us is moving away at a rate of

$$v = (71 \text{ km/s/Mpc}) \times (1 \text{ Mpc}) = 71 \text{ km/s}$$

A galaxy 2 Mpc away is therefore receding at 142 km/s, and so on. A galaxy located 100 million parsecs from Earth rushes away from us with a speed of 7100 km/s or 16 million mph!

Compare! While a supercluster of galaxies 10 Mpc away recedes at 710 km/s, the solar system orbits the center of our Galaxy at a comparable 230 km/s.

Try these questions: At what speed is a galaxy observed to be 8 billion light-years away moving from us? How far away from us is a galaxy observed to be moving away at 216 km/s? If H_0 were 50 km/s/Mpc, would a galaxy 4 billion light-years away be moving away from us faster or slower than the same galaxy with $H_0 = 71$ km/s/Mpc?

(Answers appear at the end of the book.)

---

or whether the gravitational attraction between all of its parts is enough to cause it someday to stop and recollapse. This is like trying to escape forever from Earth. A normal jump might raise you up ⅓ m or so, but if you had a sufficiently powerful cannon to shoot you upward fast enough, you could escape Earth's gravitational attraction completely and enter interplanetary space. To determine whether the superclusters are expanding away from one another fast enough to escape their mutual gravitation forever, astronomers have worked hard to determine the exact value of H_0. We explore that here and pick up the question of the fate of the universe in Chapter 13.

If the Hubble constant were three times its present value, how much slower or faster would superclusters be moving apart?

Power in Numbers Much of the power of science comes from expressing systematic data in mathematical form. By themselves, the spectra of galaxies give a qualitative picture of galactic motion, and the pattern of their recessional velocities suggests that a large-scale property of the universe is in effect here. When the data are converted into the equation, called the *Hubble law,* it becomes clear that the universe is expanding, and we can begin to look for the reason behind this expansion.

11-13 Different techniques determine the expansion of the universe at different distances from Earth

WEB LINK 11.12

To determine the Hubble constant, astronomers must measure the redshifts and distances to many galaxies. Although redshift measurements from spectra can be quite precise, it is very difficult to measure accurately the distances to remote galaxies. We cannot use the method of stellar parallax at the distances of galaxies. Recall from Section 10-1 that only the distances to the stars in our Galaxy within about 150 pc (500 ly) can be determined very precisely this way.

Distances to other galaxies are determined from observations of the apparent magnitudes of objects in them or apparent magnitudes of entire galaxies. These observations are compared to known absolute magnitudes for similar objects. Having the apparent magnitude and the absolute magnitude enables astronomers to calculate distances, as discussed in *An Astronomer's Toolbox 6-3*. Astronomers use the term **standard candle** to denote any object whose absolute magnitude is known. The various techniques used to determine large-scale distances are presented in *Guided Discovery: The Tully-Fisher Relation and Other Distance-Measuring Techniques*. Figure 11-33 summarizes the distances for which all of the measuring techniques described here are valid. Including data from all of the methods yields a

GUIDED DISCOVERY

The Tully-Fisher Relation and Other Distance-Measuring Techniques

Spectroscopic parallax (see Section 6-9) provides distances to stars up to 10 kpc (33 kly, where kly denotes 1000 light-years). This distance is still within the Milky Way. To determine distances to other galaxies, we need sources in them that become very bright compared to the luminosity of normal stars, and whose absolute magnitudes are well known. Using the brightnesses of RR Lyrae variable stars (see Sections 7-13 and 7-15), astronomers can measure distances to the Large and Small Magellanic Clouds or about 100 kpc (330 kly). To measure the Hubble flow, we must determine distances to much more distant galaxies.

Cepheid variable stars (see Section 10-2 and Figure 10-4) can now be seen out to 30 Mpc (100 Mly, where Mly denotes a million light-years) from Earth. The distances to galaxies in this nearby volume of space can thus be determined from the Cepheid period-luminosity law. Plotting the distances to galaxies versus their recessional velocity for the small volume of space extending out 30 Mpc from Earth (shown on the left side of the graph in Figure 11-32) enabled astronomers to get a first estimate for H_0 from the slope of the line through these data. This gave $H_0 \approx 75$ km/s/Mpc. Errors in both distance and recessional velocities led to inaccuracies in the slope of this curve.

Beyond 30 Mpc, even the brightest Cepheid variables, which have absolute magnitudes of about −6, are not visible with current technology. In the 1970s, the astronomers Brent Tully and Richard Fisher developed another method for determining distances. They discovered that the width of the hydrogen 21-cm emission line of a spiral galaxy (discussed in Chapter 10) is related to the galaxy's absolute magnitude: *The broader the line, the brighter the galaxy.* The connection goes like this: Interstellar gas orbits in galaxies, as we have seen. Different hydrogen gas clouds in different places in the same galaxy have different speeds toward or away from us and so we observe a variety of slightly different wavelengths from this gas, centered around 21 cm. This combination of emissions makes the 21-cm emission line appear "broad."

The **Tully-Fisher relation** says that the greater the galaxy's mass, the greater its luminosity and the more rapidly the stars and gas orbit in it. The faster it rotates, the greater the range of speeds and hence the greater the Doppler shifts we see from it. Measuring the width of the 21-cm line, therefore, tells us the galaxy's mass. Assuming that stars with different masses, and hence luminosities, occur with the same frequencies in each galaxy, the galaxy's mass tells us its *absolute* magnitude. Combining this with observations of the galaxy's *apparent* magnitude allows us to calculate its distance from us using the distance-magnitude relationship (see *An Astronomer's Toolbox 6-3*).

Because line widths can be measured quite accurately, astronomers can use the Tully-Fisher relation to determine the luminosities (absolute magnitudes) of many spiral galaxies and thus their distances.

The key method for determining the distances to the most remote galaxies is to measure the brightness of type Ia supernova explosions within them. These supernovae all reach an absolute magnitude of −19 at the peak of their outbursts (see accompanying figure). Using their observed apparent magnitudes and the distance modulus equation (see *An Astronomer's Toolbox 10-1*), their distances can be calculated. With this method, astronomers in 2001 measured a supernova nearly 3 billion parsecs (10 billion light-years) away, and more distant type Ia supernovae have been measured since then. Measurements of distances to galaxies using the Tully Fisher relation and type Ia supernovae added data to Figure 11-32 and yield a more accurate Hubble constant of 71 km/s/Mpc.

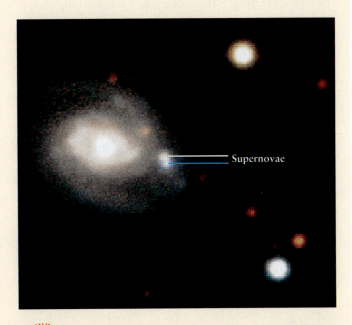

Two Supernovae in NGC 664 In 1997 the rare occurrence of two supernovae in the same galaxy at the same time was observed in the spiral galaxy NGC 664, located about 300 Mly (90 Mpc) from Earth. Supernovae observed in remote galaxies are important standard candles used by astronomers to determine the distances to these faraway objects. The two supernovae overlap each other, as shown. The upper, yellow-orange supernova was observed to occur 2 months before the hotter, blue one, which was observed to occur less than 2 weeks before this image was made and had not yet achieved maximum brightness. (Perry Berlind and Peter Garnavich, Harvard Smithsonian Center for Astrophysics)

What are a few of the techniques that we use to measure distances on Earth?

distance-velocity graph, as shown in Figure 11-32. Using this diagram, astronomers have calculated that the Hubble constant is

$$H_0 = 71 \pm 4 \text{ km/s/Mpc}$$

The plus or minus 4 km/s/Mpc indicates the possible range for H_0, when taking into account all of the errors that may still remain in the observations and calculations. A significant source of error is the effect of intergalactic gas that causes the standard candles to appear dimmer than they would if the gas were not present. Astronomers must determine the amounts of both intergalactic and interstellar gas and take their effects into account, something that is often hard to do.

11-14 Astronomers are looking back to a time when galaxies were first forming

The Hubble Space Telescope has been looking deeper and deeper into the universe. In 1998, working in concert with the Keck I telescope, it observed galaxies 8 Bly (where Bly denotes a billion light-years) away, many of them in the process of merging (Figure 11-34). The same year it saw galaxies nearly 12 Bly away (Figure 11-34b). In 2004, galaxies up to 13.2 Bly away were observed. We are seeing these

In what way are telescopes time machines?

latter objects as they were less than a billion years after the universe came into existence. Although many of the early galaxies are not as well formed as nearby galaxies, the existence of galaxy-sized collections of stars so far back in time is an important clue to the processes of star and galaxy formation that occurred in the early universe.

The observational data now being accumulated from the period of time when the universe was less than a few billion years old are helping astronomers develop and test theories of the evolution of the early universe. We will explore this point further in Chapter 13.

11-15 Frontiers yet to be discovered

The farther out in the universe that we explore, the more there is to discover. In the realm of the galaxies, we have yet to understand fully how the different types of galaxies form. Are nuclear bulges formed as a result of collisions between galaxies? Are globular clusters the remnants of previous galaxies that have been consumed by bigger galaxies? How long does it take interacting galaxies to combine? Do supermassive black holes form before, simultaneously with, or after galaxies come into existence? Moving into the realm of the global structure of the universe, what is the overall distribution of clusters and superclusters of galaxies in the universe? Are most of them really found on spongelike or soap-bubble-like surfaces and, if so, what is that structure telling us about the underlying distribution of dark matter and other properties of the cosmos?

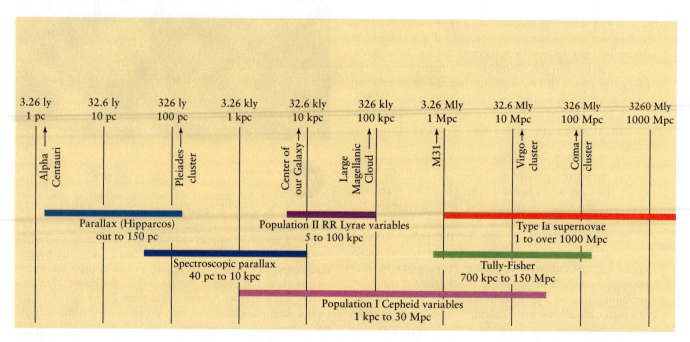

FIGURE 11-33 Techniques for Measuring Cosmological Distances Astronomers use different methods to determine different distances in the universe. All of the methods shown here are discussed in the text.

a

RIVUXG

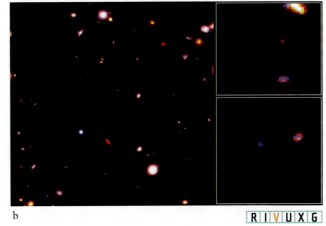

b

RIVUXG

FIGURE 11-34 Distant Galaxies (a) The young cluster of galaxies MS1054-03, shown on the left, contains many orbiting pairs of galaxies, as well as remnants of recent galaxy collisions. Several of these systems are shown at the right. This cluster is located 8 billion light-years away from Earth. (b) This image of more than 300 spiral, elliptical, and irregular galaxies contains several galaxies that are an estimated 12 billion light-years from Earth. Two of the most distant galaxies are shown in the images on the right, in red, at the centers of the pictures. (a, b: P. Van Dokkum, Uner of Granengen, ESA, and NASA)

The Expanding Universe

To understand the universe, scientists build models—mathematical representations of the world around us. In developing these pictures, it often helps to begin with a simple analogy. Because the expansion of the universe is hard to visualize, imagine for a moment the batter for a chocolate chip cake floating in an oven in the International Space Station. As the cake bakes, it expands and the chocolate chips move farther apart. They do not get larger and do not move through the batter. Each chocolate chip remains at rest in its own little bit of cake and is carried along as the cake spreads out.

Now, think of each chocolate chip as a supercluster of galaxies and the batter between the chocolate chips as the rest of spacetime. As the universe expands, the distance between widely separated superclusters of galaxies grows larger and larger. The expansion of the universe is the expansion of spacetime.

Suppose you were on one particular chocolate chip, say the chocolate chip labeled 3 in the oven at the left in the figure. You would see chocolate chips 2 and 4 moving away from you with equal velocities. Every 10 min, chocolate chips 2 and 4 get 1 cm farther away as the cake expands, as shown in the oven on the right. In that same time interval, chocolate chips 1 and 5 move twice as far away from you as chocolate chips 2 and 4. To go twice as far away in the same time, chocolate chips 1 and 5 must be moving twice as fast as the closer chocolate chips 2 and 4. This is exactly what Hubble's law of cosmological expansion (recessional velocity = H_0 × distance) says: *Double the distance and you double the velocity.*

To see why the Earth need not occupy the center of the universe, put yourself on chocolate chip 4. You see chocolate chips 3 and 5 moving away at the same rate—in fact, they recede at the same rate as you saw chocolate chips 2 and 4 move away when you were on chocolate chip 3. From chocolate chip 4, chocolate chips 2 and 6 are moving away twice as fast as chocolate chips 3 and 5. From any chocolate chip, you will see all of the other chocolate chips moving away. Similarly, no matter where we are in the universe, we see all of the other superclusters receding from us in the same manner.

The Expanding Chocolate Chip Cake Analogy The expanding universe can be compared to a chocolate chip cake baking and expanding in the International Space Station. Just as all of the chocolate chips move apart as the cake rises, all of the superclusters of galaxies recede from each other as the universe expands.

SUMMARY OF KEY IDEAS

Types of Galaxies

• The Hubble classification system groups galaxies into four major types: spiral, barred spiral, elliptical, and irregular.

• The arms of spiral and barred spiral galaxies are sites of active star formation.

• According to the theory of self-propagating star formation, spiral arms of flocculent galaxies are caused by the births and deaths of stars over extended regions of a galaxy. Differential rotation of a galaxy stretches the star-forming regions into elongated arches of stars and nebulae that we see as spiral arms.

• According to the spiral density wave theory, spiral arms of grand-design galaxies are caused by spiral density waves. The gravitational field of a spiral density wave compresses the interstellar clouds that pass through it, thereby triggering the formation of stars, including OB associations, which highlight the arms.

• Elliptical galaxies contain much less interstellar gas and dust than do spiral galaxies; little star formation occurs in elliptical galaxies.

Clusters and Superclusters

• Galaxies group into clusters rather than being randomly scattered through the universe.

• A rich cluster contains at least a thousand galaxies; a poor cluster may contain only a few dozen up to a thousand galaxies. A regular cluster has a nearly spherical shape with a central concentration of galaxies; in an irregular cluster, the distribution of galaxies is asymmetrical.

• Our Galaxy is a member of a poor, irregular cluster, called the Local Group.

• Rich, regular clusters contain mostly elliptical and lenticular galaxies; irregular clusters contain more spiral and irregular galaxies. Giant elliptical galaxies are often found near the centers of rich clusters.

• No cluster of galaxies has an observable mass large enough to account for the observed motions of its galaxies; a large amount of unobserved mass must be present between the galaxies.

• Hot intergalactic gases emit X rays in rich clusters.

• When two galaxies collide, their stars initially pass each other, but their interstellar gas and dust collide violently, either stripping the gas and dust from the galaxies or triggering prolific star formation. The gravitational effects of a galactic collision can cast stars out of their galaxies into intergalactic space.

• Galactic mergers occur. Also a large galaxy in a rich cluster may grow steadily through galactic cannibalism.

Superclusters in Motion

• A simple linear relationship exists between the distance from Earth to galaxies in other superclusters and the redshifts of those galaxies (a measure of the speed at which they are receding from us). This relationship is the Hubble law: Recessional velocity = $H_0 \times$ distance, where H_0 is the Hubble constant.

• Astronomers use standard candles—Cepheid variables, the brightest supergiants, globular clusters, H II regions, supernovae in a galaxy, and the Tully-Fisher relation—to calculate intergalactic distances. Because of difficulties in measuring the distances to remote galaxies, the value of the Hubble constant, H_0, is not known with complete certainty.

WHAT DID YOU THINK?

1 *Are most of the stars in spiral galaxies located in their spiral arms?* No. The spiral arms contain only 5% more stars than the regions between the arms.

2 *Do all galaxies have spiral arms?* No. Galaxies may be either spiral, barred spiral, elliptical, or irregular. Only spirals and barred spirals have arms.

3 *Are galaxies isolated objects?* No. Galaxies are grouped in clusters, and clusters are grouped in superclusters.

4 *Is the universe contracting, unchanging in size, or expanding?* The universe is expanding.

Key Terms for Review

Review Questions

1. What are the most massive galaxies in the universe? a. normal spirals, b. barred spirals, c. giant ellipticals d. dwarf ellipticals, e. irregulars

2. Spiral density waves are directly responsible for which of the following? a. flocculent spirals, b. grand-design spirals, c. supernovae, d. the collisions between galaxies, e. galactic cannibalism

3. Which of the following statements about the motion of galaxies is correct? a. All galaxies are moving apart. b. Superclusters of galaxies are all moving apart. c. Superclusters of galaxies are all moving toward each other. d. The Milky Way Galaxy is at the center of the universe. e. All clusters of galaxies in each supercluster are moving toward each other.

4. What is the common name for the sonic boom created by lightning?

5. What spectral classes of stars are the 5% only found in spiral arms of spiral galaxies?

6. What is the Hubble classification scheme? Which category includes the biggest galaxies? Into which category do the smallest galaxies fall? Which type of galaxy is the most common?

7. In which Hubble types of galaxies are new stars most commonly forming? Describe the observational evidence that supports your answer.

8. What is the difference between a flocculent spiral galaxy and a grand-design spiral galaxy?

9. Briefly describe how the theory of self-propagating star formation accounts for the existence of spiral arms in some spiral galaxies.

10. Briefly describe how the spiral density wave theory accounts for the existence of spiral arms in some spiral galaxies.

11. How is it possible that galaxies in our Local Group still remain to be discovered? In what part of the sky are these galaxies located? What sorts of observations might reveal these galaxies?

12. Can any galaxies besides our own be seen with the naked eye? If so, which?

13. What is the difference between a rich cluster of galaxies and a poor one? What is the difference between a regular cluster of galaxies and an irregular one?

14. How can a collision between galaxies produce a starburst galaxy?

15. Why do astronomers believe that considerable quantities of dark matter must exist in clusters of galaxies?

16. Explain why the dark matter in galaxy clusters cannot be neutral hydrogen.

17. What is the Hubble law?

18. Some galaxies in the Local Group exhibit blueshifted spectral lines. Why are these blueshifts not violations of the Hubble law?

19. What is a standard candle? Why are standard candles important to astronomers trying to measure the Hubble constant?

20. What kinds of stars would you expect to find populating space between galaxies in a cluster?

Advanced Questions

The answers to all computational problems, which are preceded by an asterisk (*), appear at the end of the book.

*21. Suppose a spectrum of a distant galaxy showed that its redshift corresponds to a speed of 22,000 km/s. How far away is the galaxy in Mpc?

*22. A cluster of galaxies in the southern constellation of Pavo (the Peacock) is located 100 Mpc from Earth. How fast, on average, are galaxies in this cluster receding from us? Why do different galaxies in the cluster show different velocities?

23. How would you determine what fraction of a distant galaxy's redshift is caused by the galaxy's orbital motion around the center of mass of its cluster?

Discussion Questions

24. Discuss the advantages and disadvantages (e.g., reliability, repeatability, etc.) of using the various standard candles to determine extragalactic distances.

25. Discuss whether the various Hubble types of galaxies actually represent some sort of evolutionary sequence.

26. Discuss the sorts of phenomena that can occur when galaxies collide. Do you think that such collisions can change the Hubble type of a galaxy? Explain.

27. From what you know about stellar evolution, the interstellar medium, and the spiral density wave theory, explain the appearance and structure of the spiral arms of grand-design spiral galaxies.

What If . . .

28. The solar system was located in an active star-forming region in a spiral arm, rather than on the edge of the Orion arm? How would the solar system be different?

29. The Milky Way collided with the Andromeda Galaxy? What might we experience on Earth? (This merger will occur in the distant future.)

30. No distant galaxies showed any redshift or blueshift? Discuss what that would imply about the universe.

31. All distant galaxies showed a blueshift, rather than a redshift? Discuss what that would imply about the universe.

Web Questions

32. To test your understanding of the global properties of galaxies, do Interactive Exercise 11-1 on the Web. You can print out your answers, if required.

33. To test your understanding of the local group of galaxies, do Interactive Exercise 11-2 on the Web. You can print out your answers, if required.

34. To test your understanding of the various types of galaxies, do Interactive Exercise 11-3 on the Web. You can print out your answers, if required.

35. To test your understanding of star formation in the spiral density wave model, do Interactive Exercise 11-4 on the Web. You can print out your results, if required.

Observing Projects

36. Using a telescope with an aperture of at least 30 cm (12 in.) in a dark location, observe as many of the spiral galaxies as you can, listed in Table E-11, in the back of the book. You can locate them and print out their locations on a star chart, using your *Starry Night Enthusiast*™ software. Many of these galaxies are members of the Virgo cluster, which can best be seen from the northern hemisphere during the spring. Because all galaxies are quite faint, be sure to schedule your observations for a Moonless night. The best view is obtained when a galaxy is high in the sky. While at the eyepiece, make a sketch of what you see. Can you distinguish any spiral structure? After completing your observations, compare your sketches with photographs found in such popular books as *Galaxies* by Timothy Ferris (Sierra Club Books, 1980) and *The Color Atlas of the Galaxies* by James Wray (Cambridge University Press, 1988).

37. Using a telescope with an aperture of at least 30 cm (12 in), observe as many of the interacting galaxies listed in Table E-11, in the back of the book as you can. You can locate them and print out their locations on a star chart, using *Starry Night Enthusiast*™ software. As in Observing Project 36, be sure to schedule your observations for a Moonless night when the galaxies you wish to observe will be near the meridian. While at the eyepiece, make a sketch of what you see. Can you distinguish hints of interplay among the galaxies? After completing your observations, compare your sketches with photographs found in popular books, such as *Galaxies* by Timothy Ferris (Sierra Club Books, 1980).

38. Optical Double Galaxies The galaxies M81 and M82 (Figure 11-25) appear to be near each other, as seen on photographs taken from the Sun. In this exercise, you can use *Starry Night Enthusiast*™ to determine just how close they are in

space. Select **Favourites > Deep Space > Galaxies > M81 and 82** from the menu. Click the **Stop** button to stop time flow. The view, covering a field only 1° wide, is centered on M81, also called Bode's Galaxy. To the right is M82, also called the Cigar Galaxy. Note particularly the difference in appearance of these two galaxies. Use the angular measurement tool to determine the angular distance separating these galaxies, as seen from Earth. Compare this separation to the apparent size of the Full Moon (about 30′ in diameter). To find the actual physical distance separating these two galaxies, select **Favourites > Deep Space > Galaxies > Local Universe** from the menu. This view shows the Milky Way and several nearby galaxies against a background of more distant galaxies. Open the **Find** pane and type "M81" in the search box without pressing the enter key. In the list of objects found, click the "+" button immediately left of the label for M81. Then double-click the bottom entry in the expanded list for M81, labeled Spiral Sab Galaxy under the "Kind" heading, to center this galaxy in the view. Type "M82" in the search box of the Find pane without pressing the Enter key. Expand the list for M82 and click the box to the left of the bottom entry to label this galaxy. Zoom in to a field of view 1° wide. Again use the angular measurement tool to measure the distance between these two galaxies. The actual distance separating the galaxies is indicated in blue below their angular separation as seen from this observing position. Are these two galaxies as close to each other as the Milky Way and the Large Magellanic Cloud, which are about 0.195 Mly apart?

39. Galaxy Orientation Here is an opportunity to compare an image in the text with one in *Starry Night Enthusiast*™. Select **Favourites > Deep Space > Galaxies > Local Universe** from the menu. Open the **Find** pane and type "M74" in the search box without pressing the Enter key. Click the "+" button immediately to the left of the label for M74 in the list of found objects. Click the ∨ menu button to the left of the bottom entry in the list for M74 in the search list and select **Go There**. This will take you to an image of the galaxy shown in Figure 11-4b in the text. Do not move the screen image yet. Describe the appearance of the galaxy in the text and then describe the appearance of the galaxy on the screen. Describe how these two views of the same galaxy (book and program) are similar and how they differ. Write an explanation for the difference in appearance. Now use the **Location Scroller** (hold down the **Shift** key while holding down the mouse button and dragging the mouse) until you see the galaxy on the screen face on. Determine whether or not the galaxy on the screen is the same as the galaxy shown in the text. Explain how you make this assertion.

40. Galaxy Types. To give you experience in interpreting the images of galaxies that you see, you can use *Starry Night Enthusiast*™ to visit a variety of galaxies and determine whether they are spiral, barred spiral, or irregular. Begin by selecting

Favourites > Deep Space > Galaxies > Local Universe from the menu.

Now, use the **Find** pane to visit each of the galaxies listed below in turn. For each object, type its name in the search box of the **Find** pane but do not press the Enter key. Be careful—include spaces, where indicated, when typing in object names. Click the "+" button to the immediate left of the object in the list of items found. Click the menu buttons at the extreme left of each entry in the list and select the **Go There** option when you find it. You can press the spacebar to move immediately to the galaxy. Use the **Zoom** buttons and the **Location Scroller** to adjust the view so that the galaxy is seen face on. Then write down whether you think it is a spiral (S), barred spiral (SB), or irregular (Irr). You do not need to worry about the subclasses in this question.

IC 4182	IC 5152	M33	M58	M74
M81	M82	M83	M94	M109
NGC 1232	NGC 1313	NGC 1365	LMC	SMC

41. Clusters of Galaxies Clusters of galaxies contain different numbers and distributions of galaxies. In this exercise you can use *Starry Night Enthusiast*™ to compare a few of these groupings. Select **Favourites > Deep Space > Galaxies > Virgo Cluster** from the menu. Position the cursor over one of the prominent galaxies in the center of the view and open the object contextual menu (right-click on a PC, Ctrl+Click on a Mac) and select **Highlight "GA Virgo Cluster" Filament**. Use the **Increase current elevation** (∧) button in the toolbar until the complete cluster, highlighted in yellow, is visible in the view. Use the **Location Scroller** to examine the cluster from various viewpoints. What is the general shape of the Virgo cluster? If you see other groupings of galaxies as you move the screen, make a sketch of the screen and circle what you believe are other groups on your sketch. To see how astronomers have grouped these other galaxies, open the **Object Contextual** menu over a galaxy within a suspected group and select the **Highlight** option to see the extent of the group to which this galaxy belongs. Choose three of these clusters, centre and zoom in on them, as necessary, and use the **Location Scroller** to study them. Describe their distributions of galaxies compared to the Virgo Cluster. For example, what are their shapes and relative sizes compared to Virgo and to each other?

42. Great Wall of Galaxies In this exercise you can use *Starry Night Enthusiast*™ to view and interpret the Great Wall of Galaxies. Select **Favourites > Observing Projects > Large Scale Structure of Universe > Great Wall** from the menu. The galaxies in the Great Wall will be highlighted in yellow. Keep in mind that the image is centered on the Milky Way. Use the **Location Scroller** (hold down the **Shift** key while holding down the mouse button and moving the mouse) to study this structure from different viewpoints. You will see the Great Wall of Galaxies move around the Milky Way. Study the distribution of galaxies in the Great Wall as you move it and write down an explanation of where the name "Great Wall" originates. Write down any other appropriate names that come to mind from the shape.

WHAT IF... The Solar System Were Located Closer to the Center of the Galaxy?

It is a clear, Moonless night. Stars by the thousands twinkle serenely against the ebony darkness of space. Overhead the soft white span of the Milky Way catches your attention, and you try to see individual stars in its glowing haze.

Most of the 6000 stars visible to the naked eye throughout the year are within 300 light-years of Earth. The rest of the Galaxy's 200 billion or so stars are too dim or too obscured by interstellar gas and dust to be easily observed. What if the solar system were ⅓ of its present distance from the center of our Galaxy? At that distance, we would still be in the realm of the spiral arms, extremely close to the nuclear bulge, and none of the stars we see now would be visible.

Population Explosion Out in the galactic suburbs, where we live today, the population density is about one star per 300 cubic light-years. The displaced Earth would be surrounded by nearly 5 times as many stars, or one per 60 cubic light-years. Also, the solar system would pass much more frequently through the dust-rich spiral arms of the Galaxy. Scattering starlight, this interstellar matter would glow as diaphanous wisps throughout our nighttime sky. Furthermore, whenever the solar system was actually in a spiral arm, several nearby stars would be of the high-mass, high-luminosity variety. The combined light from these stars and the shimmering clouds would be so great that for millions of years at a time night would never fall.

Evolution Interruption Recall from Chapter 8 that high-mass stars evolve much more rapidly than average-mass stars like the Sun. If we were closer to the center of the Milky Way and frequently passing through the spiral arms of the Galaxy, massive stars would explode near us much more frequently than they do today. These nearby supernovae, occurring within 50 light-years of Earth, would deposit lethal radiation, damage life, and cause mass extinctions. As a result, the direction of evolution would change more frequently.

At Earth's current location, the closest star, Proxima Centauri, is more than 4 light-years away. Would Earth be in danger of colliding with a star if the solar system were closer to the center of the Galaxy? Several stars would certainly be much closer than Proxima Centauri is now. However, stars are so small compared to the vastness of a galaxy that the likelihood of a collision would still approach nil.

Much more likely would be the passage of a star so close to the Sun that Earth's orbit would be disturbed. If Earth's orbit became, say, more elliptical, the change of seasons would be noticeably affected. With the seasonal effect of Earth's tilt compounded by greater changes in distance between Earth and the Sun, one hemisphere of Earth would suffer much more extreme temperatures than it does today, while the other would see less variation. This would affect the evolution of life, of course, and the distribution of life-forms on the planet.

Close Encounters Earth-evolved life at our new location in the Galaxy would be more likely to encounter sentient beings on planets that orbit nearby stars. Today, after a century of broadcasting radio and television signals, a 200-light-year diameter sphere of space centered on Earth is filled with such signals. There are about 6300 stars in that sphere. At our new location there would be 31,500 stars in the same volume and the probability of many more stars with life-supporting planets that orbit them. Perhaps one of the reasons that life on Earth has been able to evolve so long unaffected by other intelligent life is that the solar system lies near the fringe of the Galaxy.

CHAPTER 12

Quasars and Other Active Galaxies

One Quasar, Five Images Gravitational lensing by galaxies in the intervening cluster brings five images of a distant quasar in our direction. (The fifth image is behind the large, orange galaxy.) The cluster of galaxies is 7 billion light-years away, whereas the quasar is 10 billion light-years distant. Three images of the same galaxy are also visible. (ESA, NASA, K. Sharon [Tel Aviv University] and E. Ofek [Caltech])

Gravitationally lensed images of the same galaxy

Gravitationally lensed images of the same quasar

R I V U X G

WHAT DO YOU THINK?

1 What does "quasar" stand for?

2 What do quasars look like?

3 Where do quasars get their energy?

Answers to these questions appear in the text beside the corresponding numbers in the margins and at the end of the chapter.

In this chapter we look at active galaxies—energy sources of almost unimaginable power. While the short-lived outputs of supernovae boggle the mind, they represent miniscule amounts of energy compared to quasars, Seyfert galaxies, radio galaxies, and BL Lac objects. Quasars, for example, emit more energy each second than the Sun does in 200 years, and they continue to do so for millions of years. Evidence shows that most galaxies undergo periods of similar activity. Some truly remarkable activity must occur deep within quasars and other ultrahigh energy sources to make them so luminous.

In this chapter you will discover

- bright and unusual objects, called *active galaxies*
- distant, luminous quasars
- the unusual spectra and small volumes of quasars
- the extremely powerful BL Lac objects
- supermassive black holes that serve as central engines for radio galaxies, quasars, Seyfert galaxies, and BL Lac objects

QUASARS

Our knowledge of quasars began in an amateur astronomer's backyard. Grote Reber built the first radio telescope in 1936 behind his home in Illinois, opening the realm of nonvisual astronomy. By 1944, Reber had detected strong radio emissions from sources in the constellations Sagittarius, Cassiopeia, and Cygnus. Two of these sources, Sagittarius A (Sgr A) and Cassiopeia A (Cas A), are in our Galaxy. The first is the galactic nucleus (see Section 10-4), and the second is a supernova remnant (see Chapter 8). However, Reber's third source, called Cygnus A (Cyg A), proved hard to categorize (Figure 12-1). Others quickly refined Reber's observations, but the mystery only deepened in 1954, when Walter Baade and Rudolph Minkowski, using the 200-in optical telescope on Mount Palomar, discovered a strange-looking galaxy at the position of Cygnus A (Figure 12-1 *inset*).

The galaxy associated with Cyg A is very dim. Nevertheless, Baade and Minkowski managed to photograph its spectrum. They detected a redshift that corresponds to a speed of 14,000 km/s. According to the Hubble law, this speed indicates that Cyg A lies 635 million ly (194 Mpc) from Earth.

Because Cyg A is one of the brightest radio sources in the sky, its enormous distance intrigued astronomers. Although barely visible through the giant optical telescope at Palomar, Cyg A's radio waves can be picked up by amateur astronomers with backyard equipment. Its radio energy output must therefore be colossal. In fact, Cyg A shines with a radio luminosity 10^7 times as bright as that of an entire ordinary galaxy, such as Andromeda. The object that creates the Cyg A radio emissions has to be something extraordinary.

12-1 Quasars look like stars but have huge redshifts

Cygnus A is not the only powerful radio source in the far-distant sky. Starting in the 1950s, radio astronomers created long lists of radio sources. One of the most famous lists, the *Third Cambridge Catalogue*, was published in 1959. (The first two catalogues were filled with inaccuracies.) Even today, astronomers often refer to its 471 radio sources by their "3C numbers." Cyg A, for example, is designated 3C 405, because it is the 405th source in the Cambridge list. Because of the extraordinary luminosity of Cyg A, astronomers were eager to learn whether any other sources in the 3C catalog had similar properties.

One interesting case was 3C 48. In 1960, Allan Sandage used the Palomar telescope to discover a "star" at the location of this radio source (Figure 12-2). Recall that stars are blackbodies whose peak intensity is typically visible light and whose radio emission is much less intense. Because ordinary stars are not strong sources of radio emission, 3C 48 had to be something unusual. Indeed, its spectrum showed a series of emission lines that, initially, no one could identify. Although 3C 48 was clearly an oddball, many astronomers thought it was just another strange star in our Galaxy.

Another such "star," called 3C 273, was discovered in 1962. Like 3C 48, this object (Figure 12-3) emits a series of bright spectral emission lines that no one could then identify. These spectral lines are brighter than the background radiation at other wavelengths, called the *continuum* (recall Kirchhoff's laws in Section 4-4).

A breakthrough finally came in 1963, when Maarten Schmidt at the California Institute of Technology identified four of the brightest spectral lines of 3C 273 as four spectral lines of hydrogen. However, these emission lines

R I V U X G

R I V U X G

FIGURE 12-1 Cygnus A (3C 405) Radio image produced from observations made at the Very Large Array. Most of the radio emissions from Cygnus A come from the radio lobes located on either side of the peculiar galaxy seen in the inset, a Hubble Space Telescope image. Each of the two radio lobes extend about 160,000 light-years from the optical galaxy and contain a brilliant, condensed region of radio emission. *Inset:* At the heart of this system of gas lies a strange-looking galaxy that has a redshift that corresponds to a recessional speed of 5% of the speed of light. According to the Hubble law, Cygnus A is therefore 635 million light-years from Earth. Because Cygnus A is one of the brightest radio sources in the sky, this remote galaxy's energy output must be enormous. (R. A. Perley, J. W. Dreher, J. J. Cowan, NRAO; *inset:* William C. Keel, Robert Fosbury)

R I V U X G

Quasar energy source

All of this haze and surrounding green lines are results of the telescope being "overloaded" with energy from the quasar source.

Jet

3C273

R I V U X G

FIGURE 12-2 Quasar 3C 48 For several years, astronomers erroneously believed that this object is simply a peculiar, nearby star that happens to emit radio waves. Actually, the redshift of this starlike object is so great that, according to the Hubble law, it must be roughly 4 billion light-years away. (Alex G. Smith, Rosemary Hill Observatory, University of Florida)

FIGURE 12-3 Quasar 3C 273 This combined X-ray and infrared view shows the starlike object associated with the radio source 3C 273 and the luminous jet it has created. The jet is also visible in the radio and visible parts of the spectrum. By 1963, astronomers determined that the redshift of this quasar is so great that, according to the Hubble law, it is nearly 2 billion light-years from Earth. (S. Jester, D. E. Harris, H. L. Marshall, K. Meisenheimer, H. J. Roser, and R. Perley) from 3C 273 are found at much longer wavelengths than the usual wavelengths of these lines. Schmidt concluded that the hydrogen lines are subjected to a substantial redshift. Furthermore, the intense emission lines mean that something unusual is heating the gas.

Spectra for stars in our Galaxy exhibit comparatively small Doppler shifts, because these stars cannot move extremely fast relative to the Sun without soon escaping from the Galaxy. Schmidt thus concluded that 3C 273 is not a nearby star after all. Pursuing this conclusion, he promptly found that its redshift corresponds to a speed away from us of almost 16% of the speed of light. According to the Hubble law, this huge redshift implies an impressive distance to 3C 273 of roughly 2 billion light-years (Bly).

Figure 12-4 shows the spectrum of 3C 273, which looks nothing like a star's blackbody spectrum with absorption lines. Remember from Chapter 4 that a spectrograph records energy intensity at different wavelengths. The emission lines appear as peaks, and absorption lines appear as valleys. The emission lines are caused by excited gas atoms that emit radiation at specific wavelengths. Figure 12-5 shows the spectrum of an object so far away that the expansion of the universe (see Section 11-12) gives it a redshift of 92% of the speed of light.

> How are the spectra of quasars different from the spectra of stars?

Inspired by Schmidt's success, astronomers looked again at the spectral lines of 3C 48. Sure enough, the emission lines corresponded to hydrogen but with a redshift that corresponds to a velocity of nearly one-third the speed of light. Therefore, 3C 48 must be nearly twice as far away as 3C 273, or about 4 billion light-years from Earth, assuming that the Hubble constant, H_0, is 71 km/s/Mpc.

FIGURE 12-4 Spectra of 3C 273 The visible and infrared spectra of 3C 273 are dominated by four bright emission lines caused by hydrogen. This radiation is redshifted nearly 16% from its rest wavelengths.

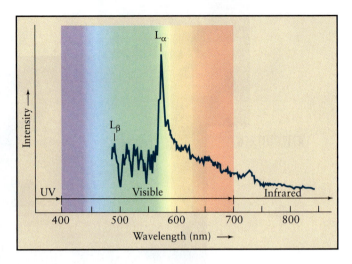

FIGURE 12-5 Spectrum of a High-Redshift Quasar The light from this quasar, known as PKS 2000-330, is so highly redshifted that spectral emission lines normally in the far-ultraviolet (L_α and L_β) are seen at visible wavelengths. Note the many deep absorption lines on the short-wavelength side of L_α. These lines, collectively called Lyman-alpha forest, are believed to be created by remote clouds of gas along our line of sight to the quasar. Hydrogen in these clouds absorbs photons from the quasar at wavelengths less redshifted than the quasar's L_α emission line.

Because of their starlike appearances and strong radio emissions, 3C 48 and 3C 273 were dubbed **quasistellar radio sources**, a term soon shortened to **quasars**. Related objects, called **quasi-stellar objects (QSOs)**, also look like stars but have strong energy outputs in all different parts of the electromagnetic spectrum, with most being strongest in the infrared. More than 100,000 quasars and QSOs have been discovered. In what follows, we use the word *quasars* to cover QSOs, too.

All quasars have redshifts of at least 0.06, or 6%, of the speed of light. The upper limit corresponds to speeds away from Earth of greater than 90% of the speed of light. From the Hubble law, it follows that the distances to these high-redshift quasars are typically in the range of 10 billion to more than 13 billion light-years. Plotting their distances (Figure 12-6), astronomers found that quasars began forming early in the life of the universe, their number peaking about 2 billion years after it came into existence. Their number dropped to near zero about 7 billion years ago. The reason for this evolution will become clear shortly.

12-2 A quasar emits a huge amount of energy from a small volume

Galaxies are big and bright. A typical large galaxy, like our own Milky Way, contains hundreds of billions of stars and shines with

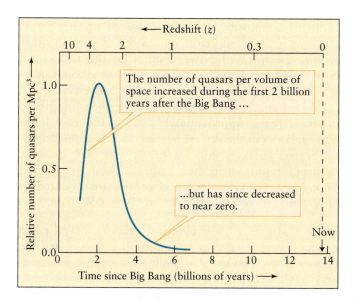

FIGURE 12–6 History of Quasar Formation The greater the redshift of a quasar, the farther it is from Earth and the farther back in time we are seeing it. By observing the number of quasars found at different redshifts, astronomers can calculate how the density of quasars has changed over time.

FIGURE 12-7 Brightness of 3C 279 This graph shows variations in the brightness of the quasar 3C 279. Note the especially large outburst observed in 1937. These data were obtained by carefully examining old photographic plates in the files of the Harvard College Observatory.

the luminosity of 10 billion Suns. The largest and most luminous galaxies, the giant ellipticals, are only 10 times brighter. But beyond 8 billion light-years from Earth, even the brightest galaxies are too faint to be easily detected. Most ordinary galaxies are too dim to be detected at half that distance. If quasars can easily be seen more than 13 billion light-years away, they must be far more luminous than galaxies. Indeed, a typical quasar is 100 times brighter than our Milky Way.

In the mid-1960s, several astronomers discovered that quasars fluctuate in brightness. Going back over old images, they found that some newly identified quasars had actually been photographed in the past but had not been considered as anything special because they looked like stars. One photograph of 3C 273, for example, dates back to 1887. By carefully examining these old images, as well as recent ones, astronomers could see that *quasars occasionally flare up.* Prominent outbursts from quasar 3C 279 reached Earth around 1937 and 1943 (Figure 12-7). During these events, the luminosity of 3C 279 increased by a factor of at least 25. At peak periods, it was emitting at least 10,000 times as much energy as the entire Milky Way.

Fluctuations and their timescales   Suppose that somewhere on a star or other object an event occurred that began making the object more luminous and then returned it to normal brightness with a total variation or cycle time of, for example, one minute. If that fluctuation in luminosity were traveling at the speed of light,

then it would take half a minute to cross the object and complete the process of brightening it, and another half a minute to reverse, so that the object would return to its original brightness one minute after the event began. If the same fluctuation traveled more slowly, then the object would have to be smaller than half a light-minute across for it to return to its original state after one minute. Therefore, the object could not be more than half a light-minute across (ignoring the curve of a sphere). This argument shows that the timescale of fluctuation allows astronomers to calculate the maximum size an object can have.

Many quasars vary in brightness over only a few months, weeks, or days. In fact, X-ray observations reveal large variations in as little as 3 h. This rapid flickering means that the source of the quasar's energy must be quite small by galactic standards. The energy-emitting region of a typical quasar—the "powerhouse" that blazes with the luminosity of 100 galaxies—is less than 1 light-day in diameter. Something must be producing the luminosity of 100 galaxies from a volume with approximately the same diameter as our solar system. Before considering their energy source, let us examine other objects that will turn out to be powered by the same "engine," as it is called by astronomers.

> Which can fluctuate fastest—an object 1 million km in diameter or an object 1 billion km in diameter?

OTHER ACTIVE GALAXIES

Quasars are but one kind of very powerful energy source in space. The others include Seyfert galaxies, radio galaxies, double-radio sources, and BL Lacertae objects. Located at the centers of galaxies, these powerhouses are collectively called **active galactic nuclei,** or **AGN.** The galaxies that contain AGN are called **active galaxies.** Active galaxies have unusually bright, starlike (that is, a pinpoint as seen from Earth) nuclei; their spectra don't look anything like a star's blackbody spectrum; they have strong emission lines; many are highly variable, meaning that their brightness levels change in different parts of the electromagnetic spectrum; they have jets of electrons and protons and beams of radiation emanating from their cores like the emissions observed for 3C 273; and most are more luminous than ordinary galaxies.

12-3 Active galaxies can be either spiral or elliptical

Seyfert Galaxies   Carl Seyfert at the Mount Wilson Observatory in California discovered the first active galaxies in 1943 while surveying spiral galaxies. Now called **Seyfert galaxies,** these spirals reveal exceptionally bright, starlike nuclei and strong emission lines in their spectra. For example, the rich spectrum of NGC 4151 has many prominent emission lines. Some of these lines are produced by iron atoms with a dozen or more electrons stripped away, indicating that it contains some extremely hot gas. Seyfert galaxies also vary in brightness. For example, at times the magnitude of NGC 4151 changes over the course of a few days.

Another example of a Seyfert galaxy is NGC 1566 (Figure 12-8). At infrared wavelengths, this galaxy shines with the brilliance of 10^{11} Suns. This extra-

R I V U X G

FIGURE 12-8 *Seyfert Galaxy NGC 1566*   This Sc galaxy is a Seyfert galaxy some 50 Mly (16 Mpc) from Earth in the southern constellation Dorado (the Goldfish). The nucleus of this galaxy is a strong source of radiation whose spectrum shows emission lines of highly ionized atoms. (Anglo-Australian Observatory)

ordinary luminosity has been observed to vary by as much as 7×10^9 L$_\odot$ over only a few weeks. In other words, the infrared power output of the nucleus of NGC 1566 increases and decreases by an amount nearly equal to the total luminosity of our entire Galaxy.

Many more Seyfert galaxies have been discovered in recent years. Approximately 10% of the most luminous galaxies in the sky are Seyferts. Some of the brightest Seyfert galaxies shine as brightly as faint quasars, with point sources in their centers. These are labeled Type 1 Seyferts. Galaxies whose centers are obscured by large amounts of gas and dust are Type 2 Seyferts.

Radio Galaxies   Some elliptical galaxies, called **radio galaxies** because of their strong radio emission, are like lower-energy quasars. Although their energy source is very small, like that of a quasar, the resulting radio emissions of these active galaxies are spread over very large areas, often hundreds of thousands of light-years across.

Some radio galaxies are accompanied by unusual optical features; namely, the galaxies look as though they are exploding, which they are not. They are therefore called **peculiar galaxies** (denoted *pec*). *Peculiar* is not the classification of a galaxy shape, like *spiral, barred spiral,* or *elliptical.* Any of the Hubble classes of galaxies (see Chapter 11) can be peculiar.

Figure 12-9 is a combined X-ray, optical, and radio image of the peculiar elliptical galaxy NGC 5128 (Centaurus A) in the southern constellation of Centaurus. The dark region that cuts the elliptical galaxy in two is actually a region of very thick dust through which little light from the galaxy behind it can penetrate. The galaxy is ejecting large volumes of hot X-ray–emitting gas, as well as jets of gas in two **radio lobes** shown in this figure. NGC 5128 is classified as an "E0 (pec)." The E0 denotes a circular elliptical galaxy, while (pec) means that this galaxy has the features of a peculiar galaxy.

Detailed observations of Centaurus A reveal that both its radio and X-ray jets (see Figure 12-9 *inset*) originate in the galaxy's nucleus. Particles and energy stream out of the galaxy's nucleus toward the radio lobes at nearly the speed of light. Such galaxies, with two radio lobes, are now called **double-radio sources.** By 1970, radio astronomers had discovered dozens of other double-radio sources, usually in giant elliptical galaxies. Double-radio sources are among the brightest radio objects in the universe.

All double-radio sources appear to have a central engine that ejects electrons, protons and magnetic fields outward along two oppositely-directed jets. After traveling many thousands or even millions of light-years, this material slows down, allowing charged particles and magnetic fields to produce the radio radiation that we detect. This type of radio emission, called *synchrotron radiation* (see Section 10-4), occurs whenever energetic electrons, protons, or ions move in a spiral path within a magnetic field.

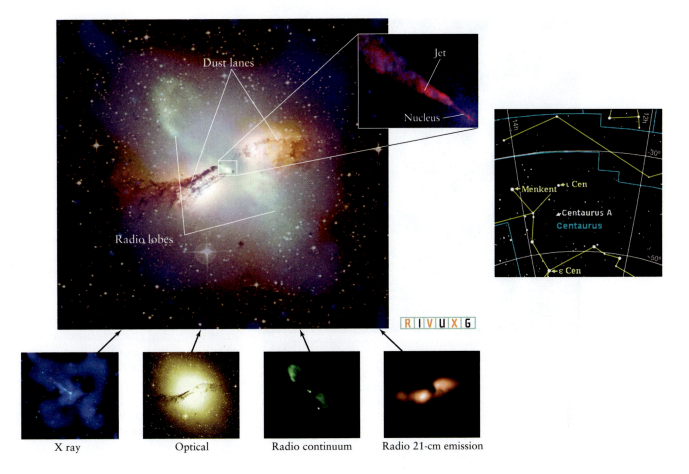

R I V U X G

X ray   Optical   Radio continuum   Radio 21-cm emission

WEB LINK 12.3

FIGURE 12-9 Peculiar Galaxy NGC 5128 (Centaurus A) This extraordinary radio galaxy is located in the constellation Centaurus, 11 million light-years from Earth. At visible wavelengths a dust lane crosses the face of the galaxy. Superimposed on this visible image is a false-color radio image (green) showing that vast quantities of radio radiation pour from matter ejected from the galaxy perpendicular to the dust lane, along with radio emission (rose-colored) along the dust lane, and X-ray emission detected by NASA's *Chandra* X-ray Observatory (blue). The X rays may be from material ejected by the black hole or from the collision of Centaurus A with a smaller galaxy. Inset: This X-ray image from the Einstein Observatory shows that NGC 5128 has a bright X-ray nucleus. An X-ray jet protrudes from the nucleus along a direction perpendicular to the galaxy's dust lane. (*X ray:* NASA/CXC/M. Karovska et al.; *radio 21-cm:* NRAO/VLA/J. Van Gorkom/Schminovich, et al.; *radio continuum:* NRAO/VLA/J. Conden, et al.; *optical:* Digitized Sky Survey U.K. Schmidt Image/STScI; *inset: X ray*—NASA/CXC/Bristol U./M. Hardcastle; *radio*—NRAO/VLA/Bristol U./M. Hardcastle)

The idea that a double-radio source emits jets of particles is supported by the existence of **head-tail sources,** so named because each such source appears to have a region of concentrated radio emission (the head) with two tails of gas that stream from it in opposite directions and then sweep back. NGC 1265, an active elliptical galaxy in the Perseus cluster of galaxies, shows these properties. NGC 1265 is known to be moving at a high speed (2500 km/s) relative to the Perseus cluster as a whole. In a radio map (Figure 12-10), its emission has a distinctly windswept appearance.

Protons and electrons ejected in the two jets from NGC 1265 are deflected by the galaxy's passage through the sparse intergalactic medium. The head-tail source leaves behind a trail of particles, like the trail of smoke that pours from a rapidly moving steam train.

The giant elliptical galaxy M 87 is a relatively weak radio galaxy, emitting only 1% as much energy as the Milky Way. Conversely, the distant, peculiar, elliptical galaxy Cygnus A (Figure 12-1) is a powerful radio galaxy that emits 40 times as much energy as does our Galaxy. Pairs of orbiting black holes and double-radio sources have also been observed, such as 3C 75 (Figure 12-11). Table 12-1 summarizes the relative power emitted by the Sun, the Milky Way Galaxy, quasars, Seyfert galaxies, and radio galaxies.

> What Hubble types of galaxies can be peculiar?

BL Lacertae Objects and Blazars   Although radio galaxies emit less energy from their nuclei than do quasars, there are also objects of the same size that from our vantage point emit even more energy than quasars.

RIVUXG

FIGURE 12-10 Head-Tail Source NGC 1265 This active elliptical galaxy is moving at a high speed through the intergalactic medium. Because of this motion, the two tail jets trail the galaxy at its head, giving this radio source a distinctly windswept appearance. (NRAO)

TABLE 12-1 Galaxy and Quasar Luminosities

Object	Luminosity (watts)
Sun	4×10^{26}
Milky Way Galaxy	10^{37}
Seyfert galaxies	$10^{36} - 10^{38}$
Radio galaxies	$10^{36} - 10^{38}$
Quasars	$10^{38} - 10^{42}$

These are the **BL Lacertae objects.** The name comes from their prototype, BL Lacertae (also called **BL Lac**) in the constellation Lacerta (the Lizard). BL Lac (Figure 12-12) was discovered in 1929, when it was mistaken for a variable star, largely because its brightness varies by a factor of 15 in only a few months. A BL Lac's most intriguing characteristic is an unusually weak set of spectral lines. In other words, its spectrum is not as easily identified as those of stars or interstellar clouds. Nevertheless, the spectra enable astronomers to determine the redshifts of these objects.

BL Lac objects are believed to be elliptical galaxies with bright quasars at their centers, much as Seyfert

RIVUXG

FIGURE 12-11 Binary Head-Tail Source This combined radio and X-ray image of 3C 75 shows the head-tail sources emanating from supermassive black holes in a pair of galaxies that are in the process of merging. The black holes are separated by 25,000 light-years and are 300 million light-years away from Earth. (X-ray: NASA/CXC/D. Hudson, T. Reiprich, et al. [AIFA])

RIVUXG

FIGURE 12-12 BL Lacertae This photograph shows fuzz around BL Lacertae (arrow). The redshift of this fuzz indicates that BL Lacertae is about 900 Mly (280 Mpc) from Earth. BL Lac objects appear to be giant elliptical galaxies with bright quasar-like nuclei, much as Seyfert galaxies are spiral galaxies with quasar-like nuclei. BL Lac objects contain much less gas and dust than Seyfert galaxies. (T. D. Kinman, NOAO/AURA)

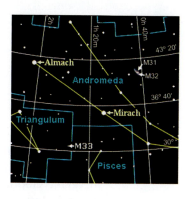

FIGURE 12-13 Elliptical Galaxy M32 This small galaxy is a satellite of M31, a portion of which is seen at the left of this wide-angle photograph. Both galaxies are roughly 2.5 million light-years from Earth. Inset: High-resolution image from the Hubble Space Telescope that shows the center of M32. Note the concentration of stars at the nucleus of the galaxy. The nucleus is only 175 light-years across. (Palomar Observatory; inset: NASA, ESA)

galaxies are spiral galaxies with quasar-like centers. BL Lac objects change intensities—they are variable. Those that have periods of a day or less are also called **blazars.** In 2004, a blazar with 10 billion solar masses located 12.5 billion light-years away was discovered. It is intriguing that this incredibly massive object formed only a billion years after the universe formed.

SUPERMASSIVE ENGINES

Quasars, BL Lac objects, Seyfert galaxies, and radio galaxies are excellent examples of how astronomers use one mechanism to explain an apparently wide variety of objects in space. In 1968, the British astronomer Donald Lynden-Bell suggested that the gravitational field of a supermassive black hole (see Chapter 9) could cause enormous releases of energy for quasars and other active galaxies.

12-4 Supermassive black holes exist at the centers of most galaxies

As we saw in Chapter 9, black holes with a wide range of masses are being discovered throughout the universe. Supermassive black holes (those exceeding a million solar masses or so) have been detected in a rapidly growing number of galaxies, including our Milky Way (see Chapter 11). To emphasize the similarities in active galaxies, we consider four further examples: M31, M32, M87, and M104.

The Andromeda Galaxy (M31) is the largest, most massive galaxy in the Local Group. In the mid-1980s, several astronomers made careful spectroscopic observations of the core of M31. Using measurements of Doppler shifts, they determined that stars within 50 light-years of the galactic core are orbiting the nucleus at exceptionally high speeds, which suggests that a massive object is located at the Galaxy's center. Without the gravity of such an object to keep the stars in their high-speed orbits, they would have escaped from the core region long ago. From such observations, astronomers estimate the mass of the central object in M31 to be about 50 million solar masses. That much matter confined to such a small volume strongly suggests the existence of a supermassive black hole located there.

Near M31, there is a small elliptical galaxy, called M32 (Figure 12-13). High-resolution spectroscopy indicates that stars close to the center of M32 are also orbiting this galaxy's nucleus at unusually high speeds, which is explained by the presence of a supermassive black hole there, too. A picture taken by the Hubble Space Telescope (Figure 12-13 inset) shows that the concentration of stars at the core of M32 is truly remarkable. Its number density of stars is more than 100 million times greater than the density of stars in the Sun's neighborhood. The concentration of stars and their high speeds strongly support the belief that a supermassive black hole also exists at the center of M32.

> What objects that we have studied are individually millions or billions of solar masses?

One early candidate galaxy for study by the Hubble Space Telescope was the giant elliptical galaxy M87 (Figure 12-14). Located some 50 million light-years from Earth, M87 is an active galaxy that has long been recognized as unusual. In 1918, Heber Curtis at the Lick Observatory reported that the center of M87 has "a curious straight ray . . . apparently connected with the nucleus by a thin line of matter." Figure 12-14 shows radio and visible images of M87 to scale. M87 is also a powerful source of X rays. X rays emitted by such small sources are a signature of black holes (as discussed in Section 9-7).

What makes it so hard for us to see deep into elliptical galaxies?

The Hubble Space Telescope image of M87 in Figure 12-14 shows an exceptionally bright, starlike nucleus and a surrounding disk of gas with trailing spiral arms. To produce this fiery glow, stars must be packed so tightly at the center of M87 that their density is at least 300 times greater than that normally found at the centers of giant ellipticals. The motions of the stars in this central clustering support the hypothesis that a black hole with a mass of nearly 3 billion Suns resides at the center of M87.

Even edge-on spiral galaxies are revealing their innermost secrets. John Kormendy used a 3.6-m telescope on Mauna Kea to examine the core of M104 (Figure 12-15) spectroscopically. Once again, high-speed gas orbiting the galaxy's nucleus was found. These observations suggest that the center of this galaxy is dominated by a supermassive black hole that contains a billion solar masses. Similar spectroscopic observations of the edge-on disk galaxy NGC 3115 also reveal evidence of a billion-solar-mass black hole at its core.

12-5 Jets of protons and electrons ejected from around black holes help explain active galaxies

The gravitational energy associated with supermassive black holes at the centers of galaxies creates huge jets of gas comprised of protons and electrons. To see how this works, consider a black hole at the center of a young galaxy filled with gas and dust. Because the centers of galaxies are congested places, a black hole there will capture a massive accretion disk of this debris (Figure 12-16a). According to Kepler's third law, the material in the inner regions of such a disk orbits the most rapidly. The inner matter, therefore, constantly interacts with the more slowly moving gases in the outer regions, and the friction between them heats all of the gas. As energized gases spiral toward the black hole, they are compressed and heated to millions of degrees. This temperature creates great pressure, causing some of the hot gas to expand and eventually squirt out where the resistance to its expansion is lowest—namely, perpendicular to the accretion disk.

R I V U X G

FIGURE 12-14 Giant Elliptical Galaxy M87 M87 is located near the center of the sprawling, rich Virgo cluster, which is about 50 million light-years from Earth. Embedded in this radio image of gas is the galaxy M87 from which it has been ejected. Images at different radio and visible wavelengths reveal a variety of details about the structure of the jets of gas. M87's extraordinarily bright nucleus and the gas jets result from a 3 billion-solar-mass black hole, whose gravity causes huge amounts of gas and an enormous number of stars to crowd around it. (NASA and the Hubble Heritage Team [STScI/AURA]; Frazer Owen [NRAO], John Biretta [STScI] and colleagues)

R I V U X G

FIGURE 12-15 Sombrero Galaxy (M104) This spiral galaxy in Virgo is nearly edge-on to our Earth-based view. Spectroscopic observations indicate that a billion-solar-mass black hole is located at the galaxy's center. You can see the bright region in the galaxy's center created by stars and gas that orbit the black hole. (NASA and the Hubble Heritage Team [STScI/AURA])

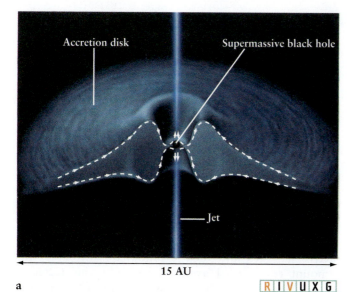

Accretion disk   Supermassive black hole

Jet

15 AU

a   R I V U X G

b   R I V U X G   R I V U X G

FIGURE 12-16
Supermassive Black Holes as Engines for Galactic Activity (a) In the accretion disk around a supermassive black hole, in-swirling gas heats and expands. Pulled inward, compressed, and heated further, some of it is eventually expelled perpendicular to the disk in two jets. (b) The giant elliptical galaxy NGC 4261 is a double-radio source located in the Virgo cluster, about 100 million light-years from Earth. An optical photograph of the galaxy (white) is combined with a radio image (orange and yellow) to show both the visible galaxy, which does not emit much radio energy, and its jets, which do. Inset: This Hubble Space Telescope image of the nucleus of NGC 4261 shows a disk of gas and dust about 800 ly (250 pc) in diameter, orbiting a supermassive black hole. (b: NASA; inset: ESA)

In 1995, the Hubble Space Telescope took a picture of the giant elliptical galaxy NGC 4261 (Figure 12-16b) that astronomers interpret as showing this set of events. A disk of gas and dust about 800 light-years in diameter is seen orbiting a supermassive black hole. The speed of the material in this disk indicates that it is held in orbit by a 1.2-billion-solar-mass object. The *inset* in Figure 12-16b is a Hubble image of the galaxy's nucleus. The dark, horizontal oval in the picture is our oblique view of the accretion disk, and ground-based radio and optical observations show a double-lobed structure (Figure 12-16b) that is bisected by the oval.

What confines the ejected gas to narrow jets? At first, the gas still falling toward the black hole prevents the jets from spreading. To see why, consider air spraying out of the nozzle of a hose. The stream of air broadens and fans out through a wide angle in the air. This would be the case around these black holes if the infalling gas were not there. If the nozzle is placed in a swimming pool, however, the stream of air will not fan out as much at first (Figure 12-17). Similarly, as the two jets of hot gas leave the vicinity of the black hole, they must blast their way through the gas that is still crowding inward. Passage through this material causes the jets to become narrow, concentrated beams. Then, magnetic fields created in the hot, swirling accretion disk spiral around the jets of gas and help keep the ejecta in columns (Figure 12-18).

FIGURE 12-17 Focusing Jets by Pressure
(a) If a high-speed jet of gas or liquid encounters little pressure (from the surrounding air, in this image), then it will spread out. (b) If the jet encounters high pressure, such as occurs when it enters water, then it will maintain longer its shape as a column. (a: Comstock Images/Getty Images; b: Fundamental Photographs, New York)

a   b

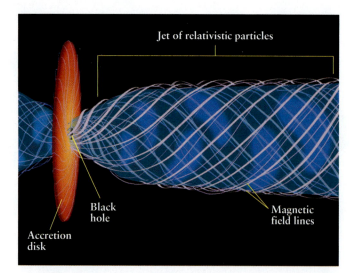

WEB LINK 12.8 **FIGURE 12-18** Focusing Jets by Magnetic Fields
The hot, ionized accretion disk (red-yellow) around
the black hole rotates and creates a magnetic field
that is twisted into spring-shaped spirals above and below the
disk. Some of the accretion disk's gas falling toward the black
hole is overheated and squirted at high speeds into the two
tubes created by the magnetic fields. The fields keep the gas
traveling directly outward from above and below the disk, thus
creating the two jets.

After traveling for hundreds of thousands of years
or more, the jets of gas interact with enough pre-exist-
ing interstellar and intergalactic gas to be stopped,
thereby forming the lobes and other features. In 2003,
astronomers using the *Chandra* X-ray Observatory
observed gas jetting out from a supermassive black hole
in a galaxy at the center of the Perseus cluster of galaxies and striking gas that resides between galaxies. This
intracluster gas became compressed by the jet and generated concentric rings or shells of sound waves 35,000
light-years in wavelength. Put another way, the sound
created by the jet is a B-flat that is 57 octaves below
middle C.

This "rocket nozzle" model explains not only double-radio sources but also the jets and beams we see protruding from other active galaxies. We can now tie
together all of the various objects discussed in this chapter. *The major difference between the different types of
active galaxies is the angle at which we view the central
engine.* As Figure 12-19 shows, an observer sees a radio
galaxy or a double-radio source when the accretion disk
is viewed nearly edge-on, because the jets are nearly in
the plane of the sky. At a steeper angle, a quasar is seen.
If one of the jets is aimed almost directly at Earth, the
galaxy is a BL Lac object. The Seyferts, active spiral
galaxies, are distinguished by whether we can see the
bright core (Type 1) or not (Type 2). Their placements in
Figure 12-19 are still under investigation.

There are many active galaxies that we *should* see
but do not. These have such thick and dusty accretion
disks that little visible light or X rays get through them.
We cannot see the energy sources, for example, of
quasars using visible light or X-ray telescopes in such situations. However, infrared radiation should pass through
the gas and dust. Thus, in 2005, astronomers used the
Spitzer Space Telescope to study regions of galaxies in
which quasars *should* exist. In just one tiny patch of sky
they found 21. Extrapolating to the total volume of space
that contains quasars, they expect "stealth" active galaxies to exist by the millions or more.

The other major factor in the observed properties of
active galaxies and their evolutionary history is the
amount of material available to make the accretion
disk. Overlap in the brightness levels of different types
of active galaxies occurs because central black holes
and their accretion disks have different masses. Their
emissions also evolve with time.

When supermassive black holes are young, they are
surrounded by vast amounts of gas, which create
massive accretion disks. Over time, much of this gas

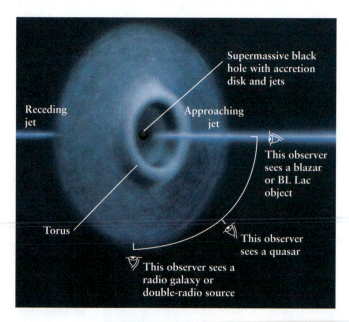

ANIMATION 12.3 **FIGURE 12-19** Orientation of the Central
Engine and Its Jets BL Lacertae objects, quasars,
and double-radio sources appear to be the same
type of object viewed from different directions. If one of the
jets is aimed almost directly at Earth, we see a BL Lac object.
If the jet is somewhat tilted to our line of sight, we see a
quasar. Tilted farther and we see an active galaxy. If the jets
are nearly perpendicular to our line of sight, we see a double-
radio source. The central region of the system is shown in
Figures 12-16a and 12-18.

either enters the black hole or jets away from the disk, as discussed previously. When an accretion disk is depleted, that active galaxy becomes more and more quiet. Because the galaxies we observe close to us are about the same age as the Milky Way, over 13 billion years old, most have small or no accretion disks around the massive black holes in their nuclei. This is why we do not see many active galaxies close to us. However, when galaxies collide, gas can be available to enable quiet galaxies to once again become active.

INSIGHT INTO SCIENCE

Occam's Razor Revisited It is possible to create *separate* theories for quasars, BL Lac objects, and other active galaxies. However, scientists prefer a comprehensive single theory that explains as many things as possible, because it incorporates common properties of the various phenomena. This search for simple, more powerful explanations is the central engine of progress in science.

12-6 Gravity focuses light from quasars

The radiation emitted by all quasars is subject to the gravitational lensing described by Einstein's general theory of relativity (see Sections 9-2 and 11-11). Like the light from stars behind the Sun, when the light from quasars travels toward us, it is deflected as it moves past other galaxies or intergalactic gas and dark matter. As we saw in Section 11-11, this gravitational lensing can lead to our receiving several images of objects. Figure 12-20 shows four images of a quasar in a configuration called an **Einstein cross**. If the background object is exactly behind an intervening quasar, the light is actually focused as a ring, called an **Einstein ring** (see Figure 11-30b), rather than as several separate images.

As you know, light travels at a finite speed. For example, light from the Sun takes about 8 min to reach Earth. We see the Sun as it was 8 min ago. The farther an object is from us, the longer it takes light and other radiation from it to reach Earth. We see a galaxy 1 billion light-years away as it was a billion years ago. When we look at quasars, we are seeing objects as they appeared when the universe was younger still. In the last few years, astronomers using the Sloan digital sky survey have discovered quasars at a distance corresponding to when the universe was only about 800 million years old.

R I V U X G

 FIGURE 12-20 Gravitational Lensing of Quasars Image from the Hubble Space Telescope that shows the gravitational lensing of a quasar in the constellation of Pegasus. The quasar, about 8 billion light-years from Earth, is seen as four separate images that surround a galaxy that is only 400 million light-years away. This pattern is called an Einstein cross. The diffuse image at the center of the Einstein cross is the core of the intervening galaxy. The physical effect that creates these multiple images is the same as that seen for galaxies, as depicted in Figure 11-30. (NASA/ESO)

12-7 Frontiers yet to be discovered

Quasars, active galaxies, BL Lac objects, and double-radio sources are rich areas for observational and theoretical research. Why do different quasars emit most strongly in different parts of the spectrum? Are quasars an early phase of most galaxies, including the Milky Way, as many astronomers are coming to believe? What causes quasar variability? Do we have the proper explanations that blazars, quasars, and radio galaxies are caused by supermassive black holes seen from different orientations? How did supermassive black holes form so quickly in the life of the universe? New discoveries in this realm are being made nearly every week.

> What astronomical objects other than black holes have jets of gas rushing out from them?

> What other astronomical objects have been observed after their light passed through a gravitational lens?

SUMMARY OF KEY IDEAS

• The development of radio astronomy in the late 1940s led to the discovery of very powerful and extremely distant energy sources.

Quasars and Other Active Galaxies

• An active galaxy is an extremely luminous galaxy that has one or more unusual features: an unusually bright, starlike nucleus; strong emission lines in its spectrum; rapid variations in luminosity; and jets or beams of radiation that emanate from its core. Active galaxies include quasars, Seyfert galaxies, radio galaxies, double-radio sources, and BL Lacertae objects.

• A quasar, or quasi-stellar radio source, is an object that looks like a star but has a huge redshift. This redshift corresponds to a distance of billions of light-years from Earth, according to the Hubble law.

• To be seen from Earth, a quasar must be very luminous, typically about 100 times brighter than an ordinary galaxy. Relatively rapid fluctuations in the brightness levels of some quasars indicate that they cannot be much larger than the diameter of our solar system.

• An active spiral galaxy with a bright, starlike nucleus and strong emission lines in its spectrum is categorized as a Seyfert galaxy.

• An active elliptical galaxy is called a radio galaxy. It has a bright nucleus and a pair of radio-bright jets that stream out in opposite directions.

• BL Lacertae (BL Lac) objects (some of which are called blazars) have bright nuclei whose cores show relatively rapid variations in luminosity.

• Double-radio sources contain active galactic nuclei located between two characteristic radio lobes. A head-tail radio source shows evidence of jets of high-speed particles that emerge from an active galaxy.

Supermassive Central Engines

• Many galaxies contain huge concentrations of matter at their centers.

• Some matter that spiral in toward a supermassive black hole is squeezed into two oppositely directed beams that carry particles and energy into intergalactic space.

• The energy sources from quasars, Seyfert galaxies, BL Lac objects, radio galaxies, and double-radio sources are probably matter ejected from the accretion disks that surround supermassive black holes at the centers of galaxies.

WHAT DID YOU THINK?

1 *What does "quasar" stand for?* Quasi-stellar radio source.

2 *What do quasars look like?* They look like stars, but they emit much more energy than any star.

3 *Where do quasars get their energy?* A quasar is thought to be powered by a supermassive black hole with millions or billions of solar masses at the center of a galaxy.

Key Terms for Review

active galaxy (*or* active galactic nuclei *or* AGN), 324
blazar, 327
BL Lacertae (BL Lac) object, 326
double-radio source, 324
Einstein cross, 331
Einstein ring, 331

head-tail source, 325
peculiar galaxy (pec), 324
quasar (quasi-stellar radio source), 322
quasi-stellar object (QSO), 322
radio galaxy, 324
radio lobe, 324
Seyfert galaxy, 324

Review Questions

1. What is a double-radio source seen along one axis of a jet called? **a.** a BL Lac object **b.** a quasar **c.** an active galaxy **d.** a double-radio source **e.** a pulsar

2. What two things does the engine of a quasar contain? **a.** a supermassive black hole and a binary companion **b.** a stellar-mass black hole and a binary companion **c.** a supermassive black hole and an accretion disk **d.** a stellar-mass black hole and an accretion disk **e.** a supergiant star and an accretion disk

3. Suppose you suspected a certain object in the sky to be a quasar. What sort of observations would you perform to confirm your hypothesis?

4. Explain why astronomers do not use any of the standard candles described in Chapter 11 to determine the distances to quasars.

5. Explain how the rate of variability of a source of light can be used to place an upper limit on the size of the source.

6. What is an active galaxy? List the different kinds of active galaxies. How do they differ from one another?

7. Why do astronomers believe that the energy-producing region of a quasar is very small?

8. How is synchrotron radiation produced?

9. What evidence indicates that quasars are extremely distant active galaxies?

10. What is a double-radio source?

11. What is a supermassive black hole? What observational evidence suggests that supermassive black holes might be located at the centers of many galaxies?

12. Why do many astronomers believe that the engine at the center of a quasar is a supermassive black hole surrounded by an accretion disk?

13. How does the orientation of the jets that emanate from the center of a galaxy relative to our line of sight relate to the type of active galaxy that we observe?

Advanced Questions

14. In the 1960s, some astronomers suggested that quasars might be compact objects ejected at high speeds from the centers of nearby ordinary galaxies. Why does the absence of blueshifted quasars disprove this hypothesis?

15. When quasars were first discovered, many astronomers were optimistic that these extremely luminous objects could be used to probe distant regions of the universe. For example, it was hoped that quasars would provide high-redshift data from which the Hubble constant could be more accurately determined. Why have these hopes not been realized?

Discussion Questions

16. Explore the belief that quasars, double-radio sources, and giant elliptical galaxies represent an evolutionary sequence.

17. Some quasars show several sets of absorption lines whose redshifts are less than the redshift of the quasars' emission lines. For example, the quasar PKS 0237-23 has five sets of absorption lines, all with redshifts somewhat less than the redshift of the quasar's emission lines. Propose an explanation for these sets of absorption lines.

What If . . .

18. Earth passed through a jet emitted by a radio galaxy? What would we see and what might happen to Earth?

19. A jet of gas suddenly appeared across the night sky? What might it indicate has recently happened?

20. Sirius suddenly became ring-shaped for a few hours and then returned to normal? What might cause such an event?

Web Question

21. To test your understanding of active galaxies, do Interactive Exercise 12-1 on the Web. You can print out your answers, if required.

CHAPTER 13

Cosmology

Bursts of star formation in the early universe, depicted by science artist Adolf Schaller. (Adolf Schaller, STScl-NASA-K. Lanzetta, SUNY)

WHAT DO YOU THINK?

1 What is the universe?

2 Did the universe have a beginning?

3 Into what is the universe expanding?

4 How strong is gravity compared to the other forces in nature?

5 Will the universe last forever?

Answers to these questions appear in the text beside the corresponding numbers in the margins and at the end of the chapter.

For millennia, our ancestors speculated about whether the universe has existed forever or whether it began some finite time in the past. Did God create the universe, or was its creation an event that required no divine intervention? Science has yet to provide an answer to this question. But astronomers and other scientists *have* accumulated a great deal of knowledge about what has happened to the universe beginning a tiny fraction of a second after it began and what its fate will be. Although the overall theory of cosmic evolution is not complete, its major elements are consistent with observations.

In this chapter we will examine the distribution of matter on the largest scales in the universe and how this matter is moving and changing. Using this information, we will explore the Big Bang model of how the universe has evolved from its earliest moments, when it was smaller than an atom. We will see how it grew and created matter as we know it, forming and evolving large-scale systems, such as galaxies and groupings of galaxies. Finally, we will consider the fate of the universe.

In this chapter you will discover

- cosmology, which seeks to explain how the universe began, how it evolves, and its fate

- the best theory that we have for the evolution of the universe—the Big Bang

- how astronomers trace the emergence of matter and the formation of galaxies

- how astronomers explain the overall structure of the universe

- our understanding of the fate of the universe

THE BIG BANG

1 The **universe** consists of all matter, energy, and spacetime that we can ever detect or that will ever be able to affect us. (We use this definition because there may be matter and energy in other dimensions or matter and energy that are moving away from us so quickly that their influence will never reach us. These ideas are still very preliminary and theoretical, and for this book we will continue to deal with just concepts for which science has experimental evidence.) So far, in this text, we have explored matter on size scales from atoms to superclusters of galaxies. We also learned in Section 11-12 that the superclusters are all moving away from one another, implying that the universe is expanding. In this chapter we take the observational evidence from the rest of the book and use it to explore the Big Bang theory of cosmology. **Cosmology** is the study of the large-scale structure and evolution of the universe.

13-1 General relativity predicts an expanding (or contracting) universe

Modern cosmology almost began in 1915, when Einstein published his theory of general relativity. To his surprise and dismay, the equations predicted that the universe is not static: They indicated that it should be either expanding (which it is) or contracting. But Einstein was not ready for what the equations were telling him.

The prediction of a changing universe flew in the face of the widely accepted belief in an infinite, static universe, a concept promoted by Isaac Newton more than 2 centuries earlier. Newton believed that each star is fixed in place and held under the influence of a uniform gravitational pull from every part of the cosmos. If the stars were not uniformly distributed, he argued, one region would have more mass than another. The denser region's gravity would then attract other stars, causing them to further clump together. Because he did not observe this clumping, Newton concluded that the mass of the universe must be distributed uniformly over an infinite space. At the time that Einstein published the theory of general relativity, the existence of galaxies and clusters and superclusters of galaxies had not yet been established, and the relative motion of superclusters of galaxies had not been observed.

The apparently static universe and the prevailing philosophy that the universe had existed forever made Einstein doubt the implications of his own theory, so he missed the opportunity to propose that we live in a changing universe. Instead, he adjusted his elegant equations to yield a static cosmos. He did this by adding a repulsive (outward-pushing) term, called the **cosmological constant,** to his equations so that gravity's normal attractive force would be counterbalanced and the universe would be static. After observations revealed that the universe is expanding, Einstein said that adding the cosmological constant was the biggest blunder of his career.

Although the value of the cosmological constant that Einstein inserted was wrong, the concept of such a

WEB LINK 13.1

constant may be correct. Observations since 1997 indicate that the universe is not just expanding but actually *accelerating* outward. This means that there must be an outward pressure that more than counteracts the effects of normal gravitation, which is trying to slow the universe's expansion. One of the two current theories that can explain that acceleration is the presence of a cosmological constant that creates the outward pressure. We discuss these two theories further in Section 13-15.

13-2 The expansion of the universe creates a Doppler-like redshift

  Edwin Hubble is credited with discovering that we live in an **expanding universe** (see Section 11-12). The redshifts of clusters and superclusters of galaxies that Hubble found moving away from us appear to come from the Doppler effect, but they actually do not. Recall that the normal Doppler shift is caused by an object moving toward or away from us through *fixed* spacetime (see Section 4-7). However, using Einstein's theory of general relativity, we find that spacetime, the fabric of the universe, is *not fixed* but is actually expanding. This expansion is what is carrying the superclusters away from one another, and, in many cases, it carries clusters in a given supercluster away from each other. (The gravitational force that holds objects such as planets and stars and entire galaxies together is so strong that these bodies and systems are not being pulled apart. The expansion of space just acts to increase the separation between large groups of galaxies.)

The redshift that Hubble observed, caused by the expansion of the universe, is properly called the **cosmological redshift**. In other words, the photons that we observe from galaxies in other superclusters are all redshifted because space is expanding. To understand why, consider a wave drawn on a rubber band (Figure 13-1). The wave has an unstretched wavelength (see Section 3-1) of λ_0. As the rubber band stretches, the wavelength increases ($\lambda > \lambda_0$). Now imagine a photon coming toward us from a distant galaxy. As the photon travels through space, space is expanding, and, like stretching the rubber band, this expansion stretches the photon's wavelength. When the photon reaches our eyes, we see a drawn-out wavelength—the photon is redshifted. The longer the photon's journey, the more its wavelength is stretched by the expansion of the universe. Therefore, astronomers observe larger redshifts in photons from relatively distant galaxies than in photons from relatively nearby galaxies.

As we saw in Section 11-12, Hubble discovered the linear relationship between the distances to galaxies in other superclusters and the redshifts of those galaxies' spectral lines. For example, a galaxy twice as far from Earth as another galaxy has twice the cosmological redshift of the closer one. The normal Doppler shift has the

a A wave drawn on a rubber band …

b … increases in wavelength as the rubber band is stretched.

FIGURE 13-1 Cosmological Redshift Just as the waves drawn on this rubber band are stretched along with the rubber band, so too are the wavelengths of photons stretched as the universe expands.

same relationship. As a result of the two effects being described by the same equations, the normal Doppler shift and the cosmological redshift predict the same relationship between redshift and motion—except for the most distant galaxies and quasars, where effects of relativity must be taken into account. Working with relatively nearby galaxies, Hubble was fully justified in using the Doppler equation to calculate the recession of galaxies.

> If the cosmological constant that Einstein used had the opposite effect—that is, if it were attractive—what would happen to a universe initially at rest, as Einstein had supposed to be true?

> What are the three redshifts presented in this book?

13-3 The Hubble constant is related to the age of the universe

Hubble's law gives us a way to estimate the age of the universe. Imagine watching a movie of two superclusters receding from each other. If we then run the film backward, time runs in reverse, and we observe the superclusters approaching each other. The time it would take for them to collide is the time since they were last combined as parts of the same matter. *Assuming that the universe began expanding when it came into existence and that it has always expanded at a constant rate*, we can use a simple equation to estimate the age of the universe:

$$\text{Time since the Big Bang} = \frac{\text{separation distance}}{\text{recessional velocity}}$$

Recall that Hubble found the relationship

Recessional velocity = H_0 × separation distance,

which we can rewrite as

$$H_0 = \frac{\text{recessional velocity}}{\text{separation distance}}$$

Comparing the first and last equations here (see *An Astronomer's Toolbox 13-1*), we have that Hubble's constant is the reciprocal of the time since the universe began. Using a Hubble constant of 71 km/s/Mpc, we find:

$$1/H_0 = 1/71 \text{ km/s/Mpc} = 13.7 \text{ billion years}$$

The universe has *not* been expanding at a constant rate, and we do not yet know the complete history of its motion. Nevertheless, calculations reveal that the age of the universe given by $1/H_0$ is within ¼ of a billion years of our best determination of the age of the universe, and we use 13.7 billion years as the age of the universe in what follows.

> If the universe were twice as old as it is now, how would the Hubble constant compare to the value it has today?

13-4 Remnants of the Big Bang have been detected

In 1927, the Belgian astrophysicist and Catholic priest Georges Lemaître proposed that, as a consequence of the equations of Einstein's theory of general relativity, the universe is expanding. Using Hubble's subsequent observations of an expanding universe, Lemaître logically concluded that the superclusters must have expanded from a much smaller volume. He proposed that the universe began as an extraordinarily hot and dense *primordial atom* of energy. Just as energy created in an explosion causes debris to expand outward, the primordial energy created at the beginning of time caused the universe to expand. This initial event is now called the **Big Bang**. Space and time (i.e., spacetime; see Chapter 9) did not exist before that moment. Rather than expanding into preexisting spacetime, the Big Bang explosion created spacetime, and that spacetime (in which we live) has been expanding ever since. The Big Bang led to the formation of all spacetime, matter, and energy.

Shortly after World War II, the astrophysicists Ralph Alpher and George Gamow proposed that seconds after the Big Bang, the universe was an incredibly hot blackbody—in excess of 10^{32} K. As heat is a meas-

---

AN ASTRONOMER'S TOOLBOX 13-1

H_0 and the Age of the Universe

The Hubble constant may not look like a measure of time, but it is. All we need to do to see this is a little practice with unit conversion. The procedure is to remove all distance units from H_0 and then invert it. Recall that we are using H_0 = 71 km/s/Mpc, which has different units of distance in the numerator and denominator. As we saw in Chapter 11, this mix of units makes it easy to determine how fast a galaxy is receding once its distance is known.

Example: A galaxy 10 Mpc away is receding at 71 km/s/Mpc × 10 Mpc = 710 km/s.

We know that 1 pc = 3.09×10^{13} km, and so 1 Mpc = 3.09×10^{19} km. Because there are 3.156×10^7 s in a year, we get the conversion factors we need:

$$\frac{1}{H_0} = \frac{1}{71 \text{ km/s/Mpc}} \times (3.09 \times 10^{19} \text{ km/Mpc})$$

$$\times \frac{1}{3.156 \times 10^7 \text{ s/years}} = 13.7 \times 10^9 \text{ years}$$

If the universe had always been expanding at the same speed, then $1/H_0$ would be the age of the universe.

However, for the first several billion years of its existence, the expansion of the universe was slowing down, and it has been speeding up since.

Decades of observations by many astronomers have confirmed that all distant galaxies are moving away from Earth. Not only that, but *the Hubble law relating velocities and distances is the same in all directions.* The Hubble constant, H_0, is the same no matter where you aim your telescope. The fact that the recession rates are the same in all directions is the condition called *isotropy*. The universe is isotropic.

The general expansion away from us seems to suggest that Earth is at the center of the universe. After all, where else could we be if all of the distant galaxies are moving away from us? In fact, the answer is that we could be practically anywhere in the universe! To understand why, see *Guided Discovery: The Expanding Universe* (Chapter 11).

Try these questions: How old would the universe be if Hubble's constant were half its actual value? Twice its actual value?

(Answers appear at the end of the book.)

RIVUXG

FIGURE 13-2 *Bell Labs Horn Antenna* This Bell Laboratories horn antenna at Holmdel, New Jersey, was used by Arno Penzias *(right)* and Robert Wilson in 1965 to detect the cosmic microwave background. (Lucent Technologies, Bell Laboratories)

ure of how many photons there are, and how energetic they are, the early universe must therefore have been filled with high-energy, electromagnetic radiation. As the universe expanded, the cosmological redshift stretched the wavelength of this radiation so that most of the photons left over from this event are now radio waves. Recall from Section 3-3 that photon energies decrease with increasing wavelength. Because the early photons are being redshifted, the energy they have is decreasing. This energy is measured as heat, so the universe is cooling off as it expands.

WEB LINK 13.4

Calculations indicated that if the universe began with a hot Big Bang, the remnants of that energy should still fill all space today, giving the entire universe a temperature only a few Kelvin above absolute zero. This radiation is called the **cosmic microwave background** because the peak of its blackbody spectrum should lie in the microwave part of the radio spectrum. In the early 1960s, Robert Dicke, P. J. E. Peebles, David Wilkinson, and their colleagues at Princeton University began designing a microwave radio telescope to detect it.

Meanwhile, just a few miles from Princeton, two physicists had already detected the cosmic background radiation. Arno Penzias and Robert Wilson of Bell Laboratories were working on a new horn antenna designed to relay telephone calls to Earth-orbiting communications satellites (Figure 13-2). However, Penzias and Wilson were mystified. No matter where in the sky they pointed their antenna, they detected a faint background noise. All efforts to eliminate this

background noise failed, even the careful removal of static noise-generating pigeon droppings from the antenna.

Thanks to a colleague, they learned of the then-theoretical cosmic microwave background and the work of Dicke, Peebles, and Wilkinson in trying to locate it. Communicating with the Princeton astronomers, Penzias and Wilson presented their finding and thus were able to claim the first detection—their annoying noise was, in fact, the remnant energy of the Big Bang. The cosmic microwave background, also commonly called the *3-degree background radiation,* is required by the Big Bang theory but is neither required nor predicted by any other competing cosmological theory. Detection of the cosmic microwave background is a principal reason why the Big Bang is accepted by astronomers as the correct cosmological theory.

> What other objects that we have studied in this book have (nearly) blackbody spectra?

INSIGHT INTO SCIENCE

Consider Plausible Alternatives Until They Are Shown to Be Inconsistent with Observations Sir Fred Hoyle, who aptly named the Big Bang in 1949, was one of the very few astrophysicists who never believed that it occurred. He, Herman Bondi, and Thomas Gold proposed the *steady-state theory* in which the universe has existed forever, is continuously expanding, and, as it expands, new matter is created that takes the place of the receding matter. Where this new matter might come from is unknown, but the laws of physics do not necessarily exclude its creation. Today, astronomers accept the Big Bang theory rather than the steady-state theory because cosmic microwave background observations are consistent with the Big Bang theory but not with the steady-state theory.

Precise measurements of the cosmic microwave background were first made by the *Cosmic Background Explorer (COBE)* satellite, which operated between 1989 and 1994. The radiation has since been measured more precisely by other equipment, such as the *Wilkinson Microwave Anisotropy Probe (WMAP)* and the balloon-carried BOOMERANG (Figure 13-3). The data taken by *COBE* and shown in Figure 13-4 reveal that, as predicted, this ancient radiation has the spectrum of a blackbody with a temperature of approximately 2.73 K. It is called 3-degree background radiation. The observations by *COBE* earned the 2006 Nobel prize in physics for astrophysicists John C. Mather and George F. Smoot.

b

R I V U X G

FIGURE 13-3 In Search of Primordial Photons (a) The *Wilkinson Microwave Anisotropy Probe (WMAP)* satellite, launched in 2001, improved upon the measurements of the spectrum and angular distribution of the cosmic microwave background taken by the *COBE* satellite. (b) The balloon-carried telescope BOOMERANG orbited above Antarctica for 10 days, collecting data used to resolve the cosmic micro-wave background, with 10 times higher resolution than that of *COBE*. All of these experiments found local temperature variations across the sky but no overall deviation from a blackbody spectrum. (a: NASA/WMAP Science Team; b: The BOOMERANG Group, University of California, Santa Barbara)

13-5 The universe has two symmetries—isotropy and homogeneity

Observations show that the cosmic microwave back-ground is almost perfectly isotropic: Its intensity is near-ly the same in every direction in the sky. Figure 13-5 shows a map of the microwave sky. It is very, very slight-ly warmer than average in the direction of the constella-tion Leo and slightly cooler in the opposite direction, toward Aquarius. After subtracting out Earth's rotation and revolution, including the effect of the Moon on our orbit, and the solar system's orbit in our Galaxy, the tiny

FIGURE 13-4 Spectrum of the Cosmic Microwave Background The little squares on this graph are *COBE*'s measurements of the brightness of the cosmic microwave background plotted against wavelength. To a remarkably high degree of accuracy, the data falls along a blackbody curve for 2.73 K. The peak of the curve is at a wavelength of 1.1 mm, in accordance with Wien's law. (Courtesy of E. Cheng; NASA COBE Science Team)

temperature variation depicted in Figure 13-5 results from the Milky Way's motion through the cosmos. Here is how that works: As our Galaxy moves toward Leo, our motion through the cosmic microwave background radi-ation creates a normal Doppler shift that causes photons from that direction to appear to have shorter wave-lengths—radiation from the direction of Leo is blueshift-ed, whereas radiation from the opposite direction, Aquarius, is redshifted because we are moving away from that region (Figure 13-6). A photon's wavelength deter-mines its energy (see Section 3-3). The energy of the pho-tons then determines the temperature that we measure for that part of the sky. Seeing slightly shorter-wavelength photons in the direction of Leo means that space in that direction appears slightly warmer, because of our motion, than space in other directions.

The observed temperature differences shown in Figure 13-5 mean that the solar system is moving toward Leo at a speed of 390 km/s. Taking into account the known velocity of the Sun around the center of our Galaxy, astronomers calculate that the entire Milky Way Galaxy is moving relative to the cosmic microwave background at 600 km/s—some 1.3 million mi/h—in the general direction of the Centaurus cluster. The grav-itational attraction of four nearby clusters of galaxies, most significantly the *Great Attractor*, and an enormous supercluster, called the *Shapley concentration*, is believed to be pulling us in that direction. Subtracting the motion of Earth, the Sun, the Milky Way, and the Local Cluster with respect to the microwave back-ground, the average intensity of radiation is found to be the same in all directions to about 1 part in 100,000.

Although the existence of the cosmic background radiation provides compelling support for the Big Bang theory, its near-isotropy was, until recently, a major problem. The issue was that the original Big Bang theory did not require space to be isotropic—after accounting for Earth's motion through the cosmos, the temperature in one direction *could* have been many degrees different than the temperature in any other direction, but it is not.

Isotropy is not limited to the blackbody radiation observed throughout the universe. It is also found on sufficiently large scales when exploring the numbers of galaxies in different directions. Astronomers have counted the numbers of galaxies at different distances from us and in a variety of directions. The number of galaxies stays roughly constant, allowing, as we saw in Section 11-8, for the distribution of most galaxies on the boundaries of sponge-shaped regions or bubbles throughout the universe. Because of this *uniformity with distance*, we say that the universe is also homogeneous. Isotropy and **homogeneity** must be explained by any viable theory of cosmology, such as the Big Bang.

The need to explain isotropy and homogeneity, among other things, has led to numerous refinements in the Big Bang theory. As a result, this theory now provides an accurate scenario for the evolution of the universe from a tiny fraction of a second after it formed and onward.

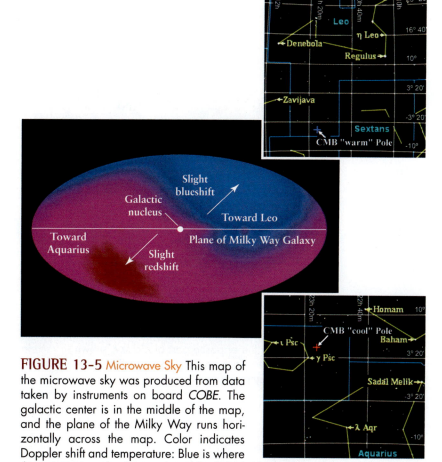

FIGURE 13-5 Microwave Sky This map of the microwave sky was produced from data taken by instruments on board *COBE*. The galactic center is in the middle of the map, and the plane of the Milky Way runs horizontally across the map. Color indicates Doppler shift and temperature: Blue is where the microwave background is blueshifted and appears warmer, while red is where it is redshifted and appears cooler. This Doppler shift across the sky is caused by Earth's motion through the microwave background. The resulting variation in background temperature is quite small, only 0.0033 K above the average radiation temperature of 2.726 K (which we have rounded to 2.73 K in the text). (NASA)

A BRIEF HISTORY OF SPACETIME, MATTER, ENERGY, AND EVERYTHING

The composition of the universe during its first few minutes of existence was profoundly different from what it is today. To understand why, we need to expand briefly on the nature of the four known physical forces: gravity, electromagnetism, and the strong and weak nuclear forces.

13-6 All physical forces in nature were initially unified

Only the effects of electromagnetism and gravity can extend over infinite distances. The electromagnetic force holds electrons in orbit around the nuclei in atoms, which allows normal matter to exist. However, over large volumes of space, the net effects of electromagnetism cancel out because a negative electric charge exists for every positive charge and a south magnetic pole exists for every north magnetic pole.

Although gravity is the weakest of all four forces, it is the only force whose effect is both infinite in extent and attractive for all ordinary matter. This is why gravity determines the evolutionary behavior of stars, the orbits of planets, the existence of galaxies, and many other large-scale phenomena. Indeed, the attractive force of gravity dominates the universe at astronomical distances. (Despite all of this, we will learn shortly that under certain conditions, gravity can be repulsive.)

What object other than Earth could meaningfully have been used at the center of Figure 13-6?

In contrast to gravity and electromagnetism, both strong and the weak nuclear forces act over extremely short ranges. Their influences extend only over atomic nuclei, distances less than about 10^{-15} m. The **strong nuclear force** holds protons and neutrons together. Without this force, nuclei would disintegrate because of the electromagnetic repulsion of their positively charged protons. Thus, the strong nuclear force overpowers the electromagnetic force inside nuclei.

The **weak nuclear force** is at work in certain kinds of radioactive decay, such as the transformation of a neutron into a proton. Protons and neutrons are composed of more basic particles, called **quarks**. A proton is composed of two "up" quarks and one "down" quark,

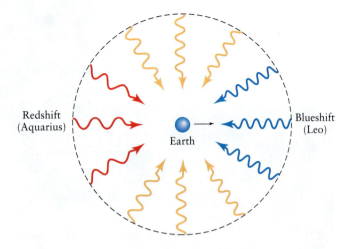

FIGURE 13-6 *Our Motion Through the Microwave Background* Because of the Doppler effect, we detect shorter wavelengths in the microwave background and a higher temperature of radiation in that part of the sky toward which we are moving. This part of the sky is the area shown in blue, in Figure 13-5. In the opposite part of the sky, shown in red, in Figure 13-5, the microwave radiation has longer wavelengths and a cooler temperature.

from the earlier discussion that when the universe was young, it was very hot—hot enough for this unifying of forces to occur. Detailed calculations reveal that in the first 10^{-35} s of existence, the temperature of the universe was more than high enough for the electromagnetic, weak, and strong forces to have the same strength.

The relationship between gravitation and the other three forces is still uncertain. Scientists have not yet been able to include its effects, although it is generally expected that the next generation of theories that explain elementary particles will include it. Efforts to create such a theory are under way, under the general name **Theories of Everything**. Recent mathematical formulations of such theories, called **superstring theories**, are being developed and have begun making predictions that scientists hope to test. Assuming gravity will be joined to the other forces, we include its expected effects here. Figure 13-7 shows that, during the beginning moments of the universe (from 0 to 10^{-43} s, called the **Planck era**), the four forces are believed to have all been unified—only one physical force existed in nature. However, as with GUTs, this remains to be proven.

whereas a neutron is made of two "down" quarks and one "up" quark. The weak nuclear force is at play whenever a quark changes from one variety to another. For example, in one kind of radioactive decay, a neutron transforms into a proton as a result of one of the neutron's "down" quarks changing into an "up" quark.

To examine details of the physical forces, scientists use particle accelerators that hurl high-speed particles at targets. High speed is equivalent to high energy and therefore high temperature. In such experiments, physicists find that *at sufficiently high temperatures, the different forces begin to behave the same way.* For example, at extremely high temperatures, the electromagnetic force, which works over all distances under "normal" circumstances, and the weak force, which only works over very short distances under the same "normal" circumstances, become identical. They are no longer separate forces, but become a single force called the *electroweak force.* The experiments verifying this were done at the CERN particle accelerator in Europe in the 1980s.

Theories have also been developed that describe conditions under which the electromagnetic, weak, and strong forces all have the same strength (i.e., they act as one). Called **Grand Unified Theories** (or **GUTs**), the energies or, equivalently, temperatures at which this unification of forces occurs are greater than can ever be achieved in the laboratory on Earth. But recall

INSIGHT INTO SCIENCE

Different Realms of Science Are Working Together Because physicists who study elementary particles have no hope of building accelerators powerful enough to test the unified theory of the fundamental forces described here, they have joined with astrophysicists and astronomers to use the entire universe as a large laboratory. If there is a way to see what happened during the earliest moments of the universe, they could use those observations to test the theory. Amazingly, such observations can be, and are being, made.

13-7 Equations explain the evolution of the universe, even before matter and energy, as we know them, existed

The earliest time at which our current equations can explain the behavior of the universe is about 10^{-43} s. This is called the **Planck time**, in honor of Max Planck, who helped derive some of the fundamental quantum physics that was used to calculate this time. (The time before this, but still after the Big Bang, is sometimes referred to as the *Planck era.*) At the Planck time, gravity is thought to have become a separate force, leaving electromagnetism and the weak and strong nuclear forces still united as the GUT force. Which present GUT model, if any, is correct, remains to be seen.

The Planck time corresponded to when the currently observable universe was much smaller than the size of an atom today. Keep in mind that the universe was not (and is not) expanding into preexisting spacetime. Rather, the expansion was (and is still) creating spacetime. Therefore, you could not stick a thermometer into the early universe from "outside" to take its temperature. However, if you could somehow measure the total amount of energy in the universe at that early time, it would have had a temperature of about 10^{32} K. There were no separate particles and electromagnetic radiation then, just gravitation and the GUT force, so matter and energy, as we know them, did not exist.

Because disorder is more physically likely to occur than order, astronomers theorize that the matter in the early universe was distributed chaotically. Some places had higher temperatures than other places. *If the universe had continued expanding only at the rate calculated for the Planck time, today we should be able to see different parts of the universe with significantly different temperatures.* However, as we saw in Figure 13-5, the temperatures of every region of the universe are the same to within a tiny fraction of 1° K. This situation is called the **isotropy problem** or **horizon problem**. Realizing how uniform the universe's temperature is, astronomers have hypothesized that very early in its existence, the cosmos underwent a short period of dramatically higher expansion, called the **inflationary epoch**, which smoothed out the universe that we can see.

For the first 10^{-35} s of the universe's existence (the blink of an eye takes about 1/10 of a second), calculations indicate that the universe expanded at the initial rate and that its temperature dropped to 10^{27} K (Figure 13-8). (For comparison, room temperature is about 300 K.) By then, the energy in each volume of the universe had decreased to the point at which the strong nuclear force could no longer remain unified with the electromagnetic and weak nuclear forces. At that instant, the strong nuclear force emerged (see Figure 13-7), distinct from the other two forces, which remained together as the electroweak force, introduced earlier.

Many astronomers hypothesize that before the strong force decoupled from the electroweak force, the universe was in an unstable state, called a *false vacuum*. It was unstable in the same sense as you would be if you

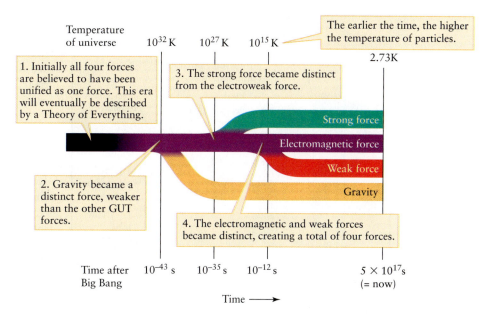

1. Initially all four forces are believed to have been unified as one force. This era will eventually be described by a Theory of Everything.

2. Gravity became a distinct force, weaker than the other GUT forces.

3. The strong force became distinct from the electroweak force.

4. The electromagnetic and weak forces became distinct, creating a total of four forces.

The earlier the time, the higher the temperature of particles.

Temperature of universe 10^{32} K 10^{27} K 10^{15} K   2.73K

Strong force
Electromagnetic force
Weak force
Gravity

Time after Big Bang 10^{-43} s 10^{-35} s 10^{-12} s 5×10^{17} s (= now)

Time →

FIGURE 13-7 Unification of the Four Forces The strength of the four physical forces depends on the speed with which particles interact or, equivalently, their temperature. As shown in this figure, the higher the temperature of the universe, the more the forces resemble each other. Also included here are the ages of the universe at which the various forces were equal.

woke up one day to find yourself on a greased, 10-cm shelf on the face of a cliff thousands of meters above the cliff bottom. The slightest move and you fall. Analogous to the transition you make by falling off the shelf, the young universe is believed to have suddenly made a transition (Figure 13-9) into a stable *true vacuum* state. The universe's transition to the true vacuum state caused it to momentarily accelerate.

Which fundamental force is essential in the process of nuclear fusion?

As the universe began moving toward the true vacuum state, it was already expanding. The transition caused the expansion rate to increase dramatically. From about 10^{-35} s until about 10^{-33} s after the Big Bang, the universe ballooned outward about 10^{50} times larger (Figure 13-10). The currently observable universe expanded to about the size of a soccer ball. This huge rate of expansion of the universe has come to be called **inflation**. The end of the inflationary epoch heralded the first formation of elementary particles: quarks, electrons, photons, and their antiparticles. Every part of the Big Bang became so much larger during inflation that only a tiny fraction of it (we do not know what that fraction is) became the universe that we can observe.

Is the space around you isotropic, as you read this book?

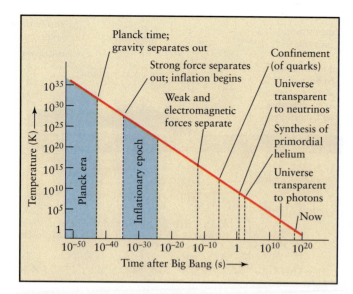

FIGURE 13-8 Early History of the Universe Current theory holds that, as the universe cooled, the four forces separated from their initially unified state. The inflationary epoch lasted from 10^{-35} s to 10^{-24} s after the Big Bang. Quarks became confined together, thereby creating neutrons and protons 10^{-6} s after the Big Bang. The universe became transparent to light (i.e., photons decoupled from matter) when the universe was about 1.6×10^{13} s (379,000 years) old. The physics of the Planck era is presently unknown.

13-8 Homogeneity and isotropy are results of inflation

We can now address the issue of how the Big Bang cosmology explains the universe's isotropy and homogeneity. Most scientists who study the evolution of

the universe believe that it had different temperatures and densities in different places immediately after the Big Bang. Because of inflation, however, we see only a tiny part of it, a region that originally had nearly uniform temperature and density. That uniformity expanded during the inflationary period so much that it created a volume out of which we cannot yet see.

Our view of the cosmos, vast as it appears, is extremely limited. We are limited by the speed of light. Recall that light takes time to travel to Earth, and the farther we look into space, the further back in time we are seeing. We can only see a finite distance into it, namely back to the time of decoupling, about 13.7 billion light-years away. Farther than this distance (or earlier in the life of the universe), the universe was opaque and so we cannot see into it. At a distance of 13.7 billion light-years, we see objects as they were just being formed (still in our nearly uniform region). Therefore, all that we can see from Earth is contained in an enormous sphere (Figure 13-11). The boundary of this sphere is called the **cosmic light horizon.** Light from all of the luminous objects in this volume has had time to reach Earth, but light from objects beyond the cosmic light horizon has not yet arrived here, so we are unaware of objects in these regions. If the speed of light were greater, we could see things farther from us.

Had the universe been perfectly uniform (i.e., everywhere homogeneous and isotropic), it would have been extremely hard for stars, galaxies, and larger groupings of matter to form, pulled together as they were by differences in mass density. Quantum theory played a vital role during the inflationary epoch in creating the clumpiness that we observe today. In 1927, British physicist Paul A. M. Dirac concluded that the laws of nature allow pairs of particles to spontaneously appear, provided that they are identical except for having opposite electric charges, and that they annihilate each other and disappear within a very short period of time, typically 10^{-21} s. As discussed in Section 9-11, these pairs are called *virtual particles*.

Normally, quantum fluctuations (the spontaneous formation of pairs of virtual particles) occur on the size scales of pairs of particles. However, the quantum fluctuations on microscopic scales that were occurring at the beginning of the inflationary period were stretched by inflation to scales that spanned clusters of galaxies and beyond! To picture this, imagine a sponge compressed very tightly so that the spaces and fiber are initially pressed very close together. Released, the sponge

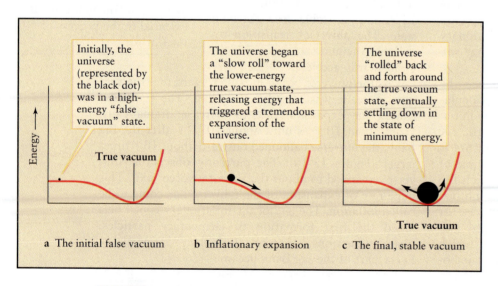

FIGURE 13-9 Cause of Inflation (a) The universe formed in an unstable energy state that (b) began to transition to a stable configuration. (c) This transition provided the energy that caused inflation.

expands, with the relationships between the spaces and the fibers remaining intact, while each part of the sponge grows to a much larger scale. The quantum fluctuations created a spongelike universe, with some regions nearly empty and others relatively full of matter. Because of this internal structure, the universe can only be considered homogeneous on size scales greater than about 1 billion light-years, so that the voids and supercluster regions average out.

Indeed, the theory predicts that tiny quantum fluctuations ultimately became the seeds for the formation of the large-scale structures of the universe that exist today. Because inflation is part of a scientific theory, its occurrence leads to predictable observations. As we will see shortly, telescopes have indeed seen its predicted distributions of matter.

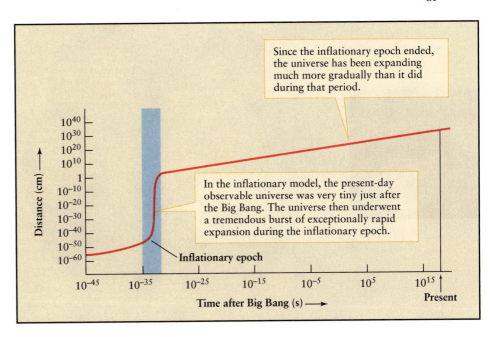

FIGURE 13-10 Observable Universe Before and After Inflation Shortly after the Big Bang, the universe expanded by a factor of about 10^{50} due to inflation. This growth in the size of the presently observable universe occurred in a very brief time (shaded interval).

13-9 During the first second, most of the matter and antimatter in the universe annihilated each other

When inflation ended, the universe resumed the relatively leisurely expansion initiated by the Big Bang (see Figure 13-10). Arriving at the time 10^{-12} s, when the temperature of the universe had dropped to 10^{15} K, the electromagnetic force separated from the weak nuclear force (see Figures 13-7 and 13-8). From that moment on, all four forces interacted with particles essentially as they do today. At 10^{-12} s, the universe's matter was primarily the individual building blocks of protons and neutrons—namely quarks, antiquarks, and the other elementary particles: neutrinos, antineutrinos, electrons, and positrons.

The next significant event is calculated to have occurred at 10^{-6} s, when the universe's temperature dropped to 10^{13} K. Just as it is energetically favorable for boiling water to turn into gas, the temperature of the universe was so high before this time that it was energetically favorable for quarks to exist as isolated particles, rather than to combine in twos and threes to create the particles that exist in the universe today. We know they were initially isolated because similar seas of quarks were recreated under equivalent conditions in the laboratory in 2001 by smashing dense atoms together. After 10^{-6} s, a period appropriately called **confinement**, the universe was sufficiently cool so that quarks could finally stick together to form individual protons, neutrons, and their antiparticles. Figure 13-8 summarizes the connections between particle physics and cosmology.

Equations show that early in the first second, the universe was so hot that virtually all photons were gamma rays possessing incredibly high energies. These energies were high enough so that photons could create matter and antimatter, according to Einstein's equation $E = mc^2$. This process is the reverse of what we saw in Section 5-7, where we discussed how matter is converted into energy in nuclear fusion. To make a particle of mass m, you need a photon with energy E at least as great as mc^2, where c is the speed of light.

> Why is inflation a necessary part of the Big Bang Theory of cosmology?

The creation of matter from energy is routinely observed in laboratory experiments that involve high-energy gamma rays. A highly energetic photon colliding with an atomic nucleus can create a pair of particles (Figure 13-12a). This process, called **pair production**, always creates one ordinary particle and its antiparticle. For example, if one of the particles is an electron, the other is an antielectron, also called a positron. (Recall from Section 9-11 that similar pairs of particles are also believed to be produced near a black hole.)

A positron has the same mass as an electron but the opposite charge. When an electron and a positron meet, they annihilate each other and their energy is converted into photons (Figure 13-12b). Theory predicts that at times earlier than one second in the evolution of the universe, photon collisions created vast numbers of pairs of

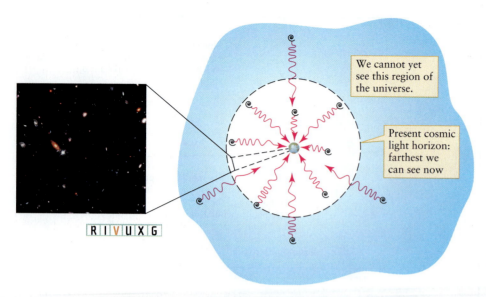

FIGURE 13-11 Observable Universe This diagram shows why we only see part of the entire universe. As time passes, this volume grows, meaning that light from more distant galaxies reaches us. The farthest galaxies we see (inset) are as they were within a few hundred million years after the Big Bang. These galaxies, formed at the same time as the Milky Way, appear young because the light from their beginnings is just now reaching us. While the light from the most distant galaxies we see was traveling toward us, the universe has been growing. Therefore, objects that appear 13 billion light-years away from us are actually about three times farther away today. Put another way, if we could see it all as it is today, rather than as it was when photons from distant objects started their journeys, the visible universe would be nearly 46 billion light-years in radius. *Inset:* This image shows some of the most distant galaxies we have seen. (Robert Williams and the Hubble Deep Field Team, STScI and NASA)

particles. Consequently, shortly after the Big Bang, all space was chock full of protons, neutrons, electrons, and their antiparticles, all immersed in a phenomenally hot bath of high-energy photons.

As the universe expanded, the temperature declined until after the first second, photons could no longer create pairs of particles and antiparticles. The remaining photons still exist and they are the cosmic microwave background that we detect today. Thereafter, particles and antiparticles often collided, annihilating each other and converting their masses back into high-energy photons, and nothing could replenish the dwindling supply of particles and antiparticles.

If all particles had annihilated their antiparticles in the early universe, no matter would be left at all and we would not be here. And yet, here we are and astronomers observe very few antiparticles in the universe. Because particles and antiparticles are always created or destroyed in pairs, why aren't they found in equal numbers in the universe today?

Physicists theorize that the symmetry or equality between the number of particles and the number of antiparticles was broken by the weak force—a *symmetry breaking*. This effect has been observed on Earth in high-energy particle accelerators over the past half century. This means that the number of particles was actually slightly greater than the number of antiparticles. For every billion antiprotons, perhaps a billion plus one protons formed; for every billion antielectrons, a billion plus one electrons formed. Virtually all of the antiparticles created in the first second of time annihilated normal particles shortly thereafter. The remaining particles in the universe were the slight excess of normal matter created as a result of this symmetry breaking.

13-10 The universe changed from being controlled by radiation to being controlled by matter

As the universe expanded and cooled further, the remaining protons and neutrons began colliding and fusing together, a process called **primordial nucleosynthesis**. For the most part, the nuclei with two or more particles were quickly separated again by the gamma rays in which they were bathed. However, during the

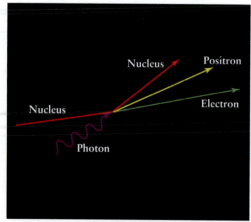

a Nucleus collides with photon and creates positron-electron pair

Time ⟶

b Positron and electron annihilate each other and create two photons

Time ⟶

FIGURE 13-12 Pair Production and Annihilation (a) A particle and an antiparticle can be created when a high-energy photon collides with a nucleus. (b) Conversely, a particle and an antiparticle can annihilate each other and emit energy in the form of gamma rays. (The processes are more complex and are only summarized in these drawings.)

first 3 min, helium and lithium atoms were created and not subsequently split apart to form most of the free helium and at least 10% of the trace element lithium that exists today. A lithium atom has three protons and three neutrons. Hydrogen, helium, and lithium are the three lowest-mass elements. After a few minutes, the universe was too cool to allow fusion to create more massive elements, such as carbon and oxygen. All elements other than the three lowest-mass ones formed later, due to stellar evolution, as discussed in Chapters 5, 7, and 8.

INSIGHT INTO SCIENCE

Share Information Although the evolution of the universe is a truly large-scale event, the activity in the first 3 min was nevertheless intimately connected to the creation and properties of elementary particles, the smallest objects in the universe. Physicists who study elementary particles and astrophysicists who study cosmology now regularly hold meetings to compare notes and connect their fields.

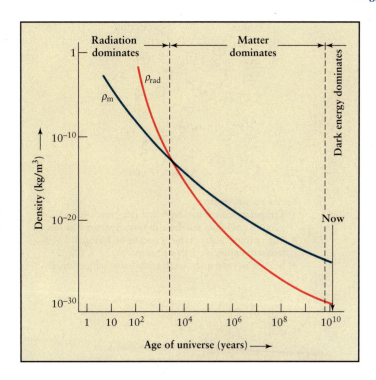

FIGURE 13-13 Evolution of Density For approximately 30,000 years after the Big Bang, the gravitational effects from photons (ρ_{rad}, shown in red) exceeded the attraction of all the matter in the universe (ρ_m, shown in blue). This early period is said to have been radiation-dominated. Later, however, continued expansion of the universe caused ρ_{rad} to become less than ρ_m, at which time the universe became matter-dominated.

Calculations indicate that for tens of thousands of years, photons dominated the behavior of the universe in two important ways. One effect of the photons was to contribute to the gravitational attraction in the universe. This contradicts intuition, because photons are massless and mass is what we normally associate with causing gravitational attraction. Nevertheless, the gravitational effect of photons is a tested prediction of general relativity. For about 30,000 years after the Big Bang, the energy density, and hence the gravitational effect, of photons was greater than the gravitational effect of matter. This situation was called the **radiation-dominated universe.**

The major effect of the radiation-dominated universe was that photons prevented matter from collapsing together and forming stars and galaxies. As the universe expanded, the gravitational effects of both photons and matter decreased as they became more spread out. The effect of photons decreased faster than that of matter (Figure 13-13) because the photons lost energy due to the cosmological redshift, but particles with mass did not lose energy similarly. After about 30,000 years, matter came to dominate the gravitational behavior of the universe, a situation that has since persisted, and thus developed the **matter-dominated universe** in which we live today. It was only after the universe became matter-dominated that gas could respond to gravitational interactions and begin clumping together where the force of gravity was sufficiently strong.

The second way that photons dominated the early universe occurred for about the first 379,000 years. During this time, photons prevented ions and electrons from connecting to form neutral atoms. During that epoch, the universe was completely filled with a shim-

mering expanse of high-energy photons colliding vigorously with protons and electrons. This state of matter, called a *plasma*, prevents any photons from traveling through it. In other words, in a plasma you would see a glow from local photons, but you would not be able to see photons from even a few meters or yards away (Figure 13-14a). The term **primordial fireball** describes the universe during this time.

By around 379,000 years, the universe was so large that the cosmological redshift had stretched photon wavelengths (see Figure 13-1) and therefore decreased photons' energies to the point where they could no longer ionize hydrogen. This stage is called **decoupling,** meaning that the massive particles and the photons were effectively disconnected. Electrons began orbiting nuclei, thereby creating neutral atoms (Figure 13-14b). The space between nuclei became so large that most primordial photons now passed by these particles, rather than colliding with them. Without these collisions to scatter photons, the universe ceased being hazy. It became clear, as it is today, enabling electromagnetic radiation to travel long distances unimpeded and, important from an astronomer's point of view, enabling us to see far into the universe.

How, if at all, would the universe be different, if antiparticles had exceeded particles in the symmetry breaking described here?

Proton   Electron

Photon

a Before recombination:
• Temperatures were so high that electrons and protons could not combine to form hydrogen atoms.
• The universe was opaque: Photons underwent frequent collisions with electrons.
• Matter and radiation were at the same temperature.

Hydrogen atom

b After recombination:
• Temperatures became low enough for hydrogen atoms to form.
• The universe became transparent: Collisions between photons and atoms became infrequent.
• Matter and radiation were no longer at the same temperature.

FIGURE 13-14 Era of Recombination (a) Before recombination, the energies of photons in the cosmic background were high enough to prevent protons and electrons from forming hydrogen atoms. (b) As soon as the energy of the background radiation became too low to ionize hydrogen, neutral atoms came into existence.

This dramatic period, when the universe transformed from being opaque to transparent, took about 100,000 years to complete and is referred to as the **era of recombination,** referring to nuclei and electrons combining. There are many instances, both in space and in the laboratory, where electrons are stripped from atoms. This is the process of ionization mentioned in Section 4-5. After ionization occurs, electrons often recombine with nuclei to re-create neutral atoms. However, the gases created early in the evolution of the universe were never neutral, so the "re" in recombination is misleading. Because the universe was

We say that "most" of the helium in the universe was formed in primordial nucleosynthesis. Where was/is the rest formed?

opaque in its first 379,000 years, we cannot see any further into the past than the era of recombination with telescopes that detect electromagnetic radiation of any wavelength.

 We can use Wien's law to calculate the temperature of the cosmic background radiation at that time. The 3000 K temperature at decoupling corresponds to a peak microwave wavelength of 0.001 mm. The temperature history of the universe is graphed in Figure 13-8.

Since the primordial photons formed, the expansion of the universe has been increasing their wavelengths, so today we see these same photons as the cosmic microwave background. The universe is now about 1000 times larger than it was at decoupling. Although the universe is dominated by matter today, the number of photons left over from the Big Bang is still immense. From the physics of blackbody radiation, astronomers calculate that there are now 400 million cosmic background microwave photons in every cubic meter of space. In contrast, if all visible matter in the universe were uniformly spread throughout space, there would be roughly one hydrogen atom in every three cubic meters of space. In other words, photons continue to outnumber atoms by more than a billion to one. The microwave background contains the most ancient photons we expect ever to be able to observe. This microwave background is a ghostly relic of the former dazzling splendor of the universe.

13-11 Galaxies formed from huge clouds of primordial gas

The Big Bang theory explains the evolution of space-time, matter, and energy since the Planck era. It accounts for the hydrogen, most of the helium, and some of the lithium that exist today. But how do we explain the existence of superclusters of galaxies, clusters of galaxies, and individual galaxies? As noted earlier, these must have formed after the beginning of the matter-dominated era of the universe.

To explain large-scale structures in the early matter-dominated universe, recall the quantum fluctuations that were stretched by inflation. These clumps throughout the inflated universe were initially extremely small differences in density with a wide range of sizes in the microwave background. We know this because the *COBE* and *WMAP* satellites observed the cosmic microwave background and discovered these variations in density back when the universe was only 379,000 years old. Their data show that these variations were only about 1 part in 100,000 denser than the average density of matter in the early universe (Figure 13-15). But these small variations on various size scales are all that is needed to explain the present large-scale structure of the universe.

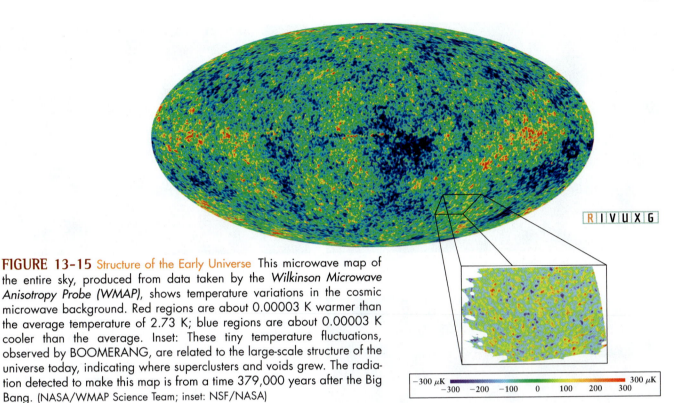

FIGURE 13-15 *Structure of the Early Universe* This microwave map of the entire sky, produced from data taken by the *Wilkinson Microwave Anisotropy Probe (WMAP)*, shows temperature variations in the cosmic microwave background. Red regions are about 0.00003 K warmer than the average temperature of 2.73 K; blue regions are about 0.00003 K cooler than the average. Inset: These tiny temperature fluctuations, observed by BOOMERANG, are related to the large-scale structure of the universe today, indicating where superclusters and voids grew. The radiation detected to make this map is from a time 379,000 years after the Big Bang. (NASA/WMAP Science Team; inset: NSF/NASA)

When the universe was radiation-dominated, the particles in it were moving too fast to clump together and form structure. Then the universe became matter-dominated and particles cooled down sufficiently to clump together. The extra gravitational attraction in regions where the expanded quantum fluctuations created concentrations of matter slightly greater than average caused the particles there to draw together and ultimately form galaxies, clusters, and superclusters.

In the first stage of the clustering process, the contracting matter collapsed inward, heated up, and rebounded outward due to the increased pressure between particles. Gravity drew the particles back inward and the process repeated itself over many cycles. This contracting and re-expanding of vast quantities of gas produced oscillations analogous to sound waves in our air. The gas transferred some of the oscillating energy to the photons that are today the cosmic microwave background, creating regions of slightly hotter and slightly cooler gas. These oscillations in the microwave background have been observed. As we will see, these observations have an important bearing on our understanding of the overall shape of the universe.

Virtually all matter in the early universe consisted of clouds of hydrogen and helium. Between the time of decoupling and the first burst of star formation was a period of little starlight, called the **dark ages.** Note, however, that the dark ages were actually bright due to light from energy created in the Big Bang. There is growing observational evidence that star formation (see Section 7-14) began within 200 million years of the Big Bang (Figure 13-16a). The earliest generation of stars are called *Population III* stars. By 600 million years after the Big Bang, galaxies were beginning to coalesce (Figure 13-16b). We also have observational evidence that galaxies were in clusters within 2 billion years of the Big Bang (Figure 13-16c).

Computer simulations indicate that the first stars were typically 100–500 $M_\odot$, a million times or more brighter than the Sun, and they lived for no more than a few million years before exploding. These stars provided the first metals in the universe (other than the primordial lithium). During the first billion years, it appears that clumps of gas that contained millions or billions of solar masses also collapsed to create supermassive black holes. These would have attracted other matter into orbit around them and thereby served as the seeds for the growth of some galaxies. By observing remote galaxies, astronomers have discovered that most galaxies initially emitted energy we associate with quasars and other active galaxies. This implies that most young galaxies have supermassive black holes at their centers. Recent observations reveal that massive and supermassive black holes typically grow until they have about 0.2% of the mass of a galaxy's nuclear bulge.

From about 600 million years to about 6 billion years after it formed, the universe underwent heavy star formation. Galaxies grew in size during this period, some continuously, others suddenly and quickly. Star formation decreased as the amount of interstellar hydrogen

a Early bursts of star formation

b Arrows indicate galaxies beginning to form 13.4 billion years ago

R I V U X G

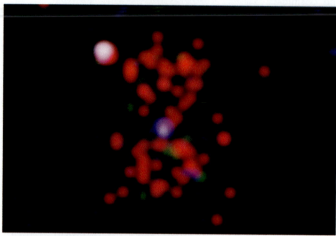

c Massive cluster of galaxies formed 11.2 billion years ago

R I V U X G

FIGURE 13-16 Galaxies Forming by Combining Smaller Units (a) This painting indicates how astronomers visualize the burst of star formation that occurred within a few hundred million years after the Big Bang. The arcs and irregular circles represent interstellar gas illuminated by supernovae. (b) Using the Hubble and Keck telescopes, astronomers discovered two groups of stars (arrows) 13.4 billion light-years away that are believed to be protogalaxies, from which bigger galaxies grew. These protogalaxies were discovered because they were enlarged by the gravitational lensing of an intervening cluster of galaxies. (c) The *Chandra* X-ray telescope imaged gravitationally bound gas around the distant galaxy 3C 294. The X-ray emission from this gas is the signature of an extremely massive cluster of galaxies, in this case, at a distance of about 11.2 billion light-years from us. (a: Adolf Schaller, STScl/NASA/K. Lanzetta, SUNY; b: Richard Ellis (Caltech) and Jean-Paul Kneib (Observatorie Midi-Pyrenees, France), NASA, ESA; c: NASA)

diminished. One observed galaxy, with about eight times the mass in stars as the Milky Way has, reached its full star-forming potential when the universe was only 800 million years old.

Observations also reveal that galaxies were bluer and brighter in the past than they are today. These changes in color and brightness suggest a high abundance of young, bright, hot, massive stars in newly formed galaxies (Figure 13-17a). As galaxies age, these blue O and B stars become supergiants and eventually die. Therefore, galaxies grow somewhat redder and dimmer. This is especially true of those elliptical galaxies that formed early on. They appear to form nearly all of their stars in one vigorous burst of activity that lasts for about a billion years (Figure 13-17b).

In contrast, spiral galaxies have been forming stars for at least the past 10 billion years, although at a gradually decreasing rate. There is still plenty of interstellar hydrogen in the disks of spiral galaxies, like our Milky Way, to fuel star formation today. That is why O and B stars still highlight their spiral arms. Figure 13-17b compares the rates at which spiral and elliptical galaxies form stars.

> If the first elements were hydrogen, helium, and lithium, where did Population II stars in globular clusters and elsewhere get their other metals?

13-12 Star formation activity determines a galaxy's initial structure

Imagine a developing galaxy, called a *protogalaxy*, forming from a cloud of gas. Theory proposes that the rate of star formation determines whether this protogalaxy becomes a spiral or an elliptical galaxy. If stars form slowly enough, then the gas surrounding them has plenty of time to settle by collision with other infalling gas into a flattened disk, just like the early solar system. Star formation continues because the protogalactic disk contains an ample supply of hydrogen, and a spiral or lenticular (disk-shaped but without spiral arms) galaxy is created. If, however, the initial stellar birthrate is high in the protogalaxy, the theory predicts that virtually all

a

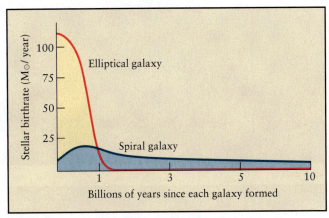

b

FIGURE 13-17 Stellar Birth Rates (a) This figure shows that star formation started quickly and has been tapering off ever since. (b) Most of the stars in an elliptical galaxy are created in a brief burst of star formation when the galaxy is very young. In spiral galaxies, stars form at a more leisurely pace that extends over billions of years.

pregalactic gas is used up in the creation of stars before a disk can form. In this case, an elliptical galaxy is created. Figure 13-18 depicts these contrasting sequences of events. Figure 13-19 summarizes the major evolutionary activities in the early universe.

Not all galaxies maintain their initial structure over the evolution of the universe. As we discussed in Section 11-10, some galaxies collide, and these collisions can change a galaxy's structure from spiral, for example, to elliptical. Indeed, astronomers who use extremely long exposures, called Hubble Deep Field images (see for example, the inset on Figure 13-11), have observed elliptical galaxies during the first few billion years of the universe's existence that are far less uniform in color than are closer ellipticals. They observed blue stars in the young ellipticals consistent with the merger of spirals and a resulting burst of star formation, as well as lots of lower-mass (yellow and red) stars. Our understanding of galactic formation and evolution is far from complete.

Determining the location and nature of the dark matter in the universe (see Sections 10-7 and 11-11) is still of paramount importance in understanding the cosmos. The observable stars, gas, and dust in a galaxy or cluster of galaxies account for only about 10% to 20% of each object's mass. (Recall that this observable matter does not have enough mass to hold galaxies or clusters of galaxies together.) Other than massive neutrinos, we have very little idea of what the remaining 80% to 90% of each galaxy or cluster of galaxies is composed of. However, some progress in understanding the distribution of this dark matter is being made. In 2002,

astronomers who map large numbers of galaxies and use computer simulations of the effects of dark matter determined that on the scales of clusters of galaxies, the locations of the galaxies often coincide with the concentrations of dark matter, while voids between clusters coincide with voids of dark matter. It is likely that the gravitational force of the dark matter caused gas to concentrate in the same regions, thereby stimulating formation of supermassive black holes and galaxies. In 2007, using images of galaxies that were focused by a gravitational lens of dark matter, astronomers were able to plot the locations of this dark matter (Figure 13-20).

If dark matter did not exist, would any gravitational lensing occur in the universe?

THE FATE OF THE UNIVERSE

We now turn to the future. Will the universe last forever? Or will it someday stop expanding and collapse?

13-13 The average density of matter is one factor that determines the future of the universe

As superclusters move apart in the expanding universe, their mutual gravitational attractions act on one another and thereby slow the rate at which they separate. If that were all that mattered in determining the fate of the universe, you could simply locate and add up all of the

Sequence of events ⟶

a Formation of a disk galaxy

b Formation of an elliptical galaxy

FIGURE 13-18 Creation of Spiral and Elliptical Galaxies A galaxy begins as a huge cloud of primordial gas that collapses gravitationally. (a) If the rate of star birth is low, then much of the gas collapses to form a disk, and a spiral galaxy is created. (b) If the rate of star birth is high, then the gas is converted into stars before a disk can form, resulting in an elliptical galaxy.

visible and (presently) dark matter in existence and see if its gravitational force is enough to eventually stop the expansion.

This is analogous to what engineers do in calculating, for example, how high a cannonball shot upward from the surface of Earth will travel. Earth's gravitational force slows the ball's ascent. If the cannonball's speed upward is less than the escape velocity from Earth's gravity (about 11 km/s straight up), it will fall back to Earth. If the ball's speed exceeds the escape velocity, it will leave Earth and continue outward forever, despite the relentless pull of Earth's gravity. On the boundary

between these two scenarios is the situation when the cannonball's speed equals the escape velocity. In that case, the cannonball will just barely escape falling back to Earth, slowing forever, and come to rest an infinite distance away.

By analogy with the cannonball fired from Earth, it would seem that if the universe is moving outward too slowly to overcome the mutual gravitational attraction of its parts, it should stop expanding and someday collapse. If it is moving outward at exactly its escape velocity, it should expand until coming to a stop an infinite time in the future. Or, if it is moving outward fast enough, it should continue to expand forever. However, none of these options is correct!

The laws of physics pertaining to the evolution of the universe, as spelled out in the theory of general relativity and in recent observations of distant objects, reveal that reality is more complex than this simple analogy. Just when astronomers were getting comfortable with the idea that the gravitational force is always attractive, general relativity once again showed that reality ignores our common sense by demonstrating that gravity can be repulsive.

We now know two additional effects that must be considered: The effect of matter on the shape of the universe and the presence of a repulsive gravitational force, called **dark energy** *(not to be confused with dark matter)*. We begin by considering the effect that matter and energy have on the shape of the universe.

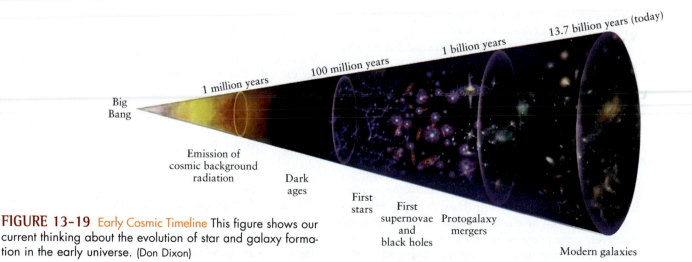

FIGURE 13-19 Early Cosmic Timeline This figure shows our current thinking about the evolution of star and galaxy formation in the early universe. (Don Dixon)

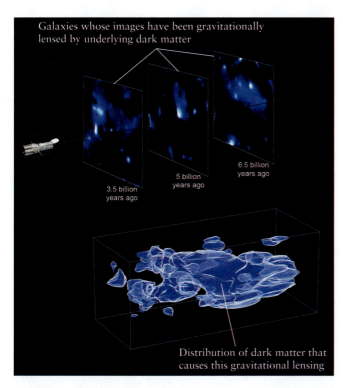

Galaxies whose images have been gravitationally lensed by underlying dark matter

6.5 billion years ago

5 billion years ago

3.5 billion years ago

Distribution of dark matter that causes this gravitational lensing

FIGURE 13-20 Mapping Dark Matter (Top) The Hubble Space Telescope observed that galaxies in the same direction, but at different distances from Earth, undergo different amounts of gravitational lensing. (Bottom) Much of this effect is due to dark matter. By subtracting out the lensing effects of intervening galaxies, the distorted shapes of the galaxies at various distances enable astronomers to determine the distribution of dark matter. (NASA, ESA, and R. Massey [CIT])

13-14 The overall shape of spacetime affects the future of the universe

During the 1920s, Alexandre Friedmann in Russia, Georges Lemaître in Belgium, Willem de Sitter in the Netherlands, and Einstein himself applied the theory of general relativity to the expanding universe. General relativity predicts that the presence of matter curves the fabric of spacetime, as we saw in Chapters 9, 10, and 12, in the form of gravitational lensing and the distortion of spacetime around black holes. Similarly, the presence of energy also curves spacetime—recall that matter and energy are related by $E = mc^2$.

Only three possibilities exist for the overall shape of the universe. They are determined by the amount of mass and energy the universe contains and how fast it is expanding. For example, imagine shining two powerful laser beams out into space. Suppose that we can align these two beams so that they are perfectly parallel as they leave Earth. Suppose, further, that nothing gets in the way of these two beams. We follow them across the spacetime whose shape we wish to determine. The light beams will

begin to diverge due to the expanding universe carrying them apart. This effect happens regardless of any properties of the spacetime, and, because we are not interested in the effect of expansion right now, we compensate for it (i.e., we ignore it) in what follows. The three possibilities for the paths of the laser beams, due to the actual curvature of the universe, are

1. Two beams of light, starting out parallel, gradually get closer and closer together as they move across the universe, eventually intersecting at some enormous distance from Earth (Figure 13-21a). In this case, we say that space is *positively curved*. Analogously, the lines of constant longitude on Earth's surface are parallel at the equator but intersect at the poles. Thus, if the universe has this effect on light, the shape of the universe is like the surface of a sphere: Space is spherical and the universe has positive curvature. This is what would happen if the universe had such high mass and energy densities that the mass and energy literally

> What, if any, objects have we given enough velocity to escape forever from Earth?

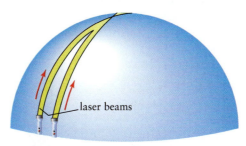

a Parallel light beams converge

b Parallel light beams remain parallel

c Parallel light beams diverge

FIGURE 13-21 Possible Geometries of the Universe The shape of space (represented here as two-dimensional for ease of visualization) is determined by the matter and energy contained in the universe. The curvature is either (a) positive, (b) zero, or (c) negative, depending on whether the average matter and energy density throughout space is greater than, equal to, or less than a critical value. The lines on each curve are initially parallel. They converge, remain parallel, or diverge, depending on the curvature of space.

curved space back in on itself. In the absence of any outward-pushing force, such as would be supplied by a sufficiently large cosmological constant (see Section 13-1), such a universe does not have enough energy to keep expanding forever. It would someday collapse.

It is easiest to understand such a universe by visualizing a two-dimensional version of it as being the surface of a balloon on which you must remain. In such a universe, you could, in principle, keep going in a straight line (which is a curve on the balloon's surface) and eventually end up back where you started. Like the surface of a balloon, a three-dimensional universe with positive curvature has no outside or center and is said to be a **closed universe.**

2. Two beams of light remain parallel regardless of how far they travel (Figure 13-21b). In this case, space is not curved. That is, space is flat and it can extend without limit. This is the structure that meets our common-sense belief.

3. Two initially parallel beams of light gradually diverge farther and farther apart as they move across the universe (Figure 13-21c). This is what would happen if the energy of the Big Bang was sufficiently great to assure that the universe is going to expand forever. In this case, the universe has negative curvature. A horse's saddle is a good example of a negatively curved or *hyperbolic* surface. Initially parallel lines drawn on a saddle always diverge. Thus, in a negatively curved universe, we would describe space as hyperbolic. A hyperbolic universe extends without limit and is called an **open universe.** Both the flat and the hyperbolic universes are open and will therefore grow larger forever.

Most astronomers have a distinct preference for the flat universe. The reason for this expectation that the universe is flat stems from the observations that the universe is homogeneous and isotropic. The only mechanism we have at present to explain these properties of the matter and energy distribution in space is inflation, as discussed earlier. However, the equations predict that, if inflation occurred, the universe must be very nearly flat because the period of inflation stretched the volume of the universe and the matter and energy in it so much.

Telescopic observations strongly support the belief that the universe is flat. This conclusion comes from examining the sizes of the regions of slightly higher and lower temperature in the cosmic microwave background (see Figure 13-15) and comparing them to the sizes predicted for a flat universe. The observations are consistent with the theoretical variations. Furthermore, if the universe had positive curvature, light from the hot regions of the early universe would be curved and thereby focused, creating bigger, brighter images than are observed. Figure 13-22 summarizes the effects of space curvature on observations of the cosmic microwave background.

If the universe is closed, light rays from opposite sides of a hot spot bend toward each other …

If the universe is flat, light rays from opposite sides of a hot spot do not bend at all …

If the universe is open, light rays from opposite sides of a hot spot bend away from each other …

… and as a result, the hot spot appears to us to be larger than it actually is.

… and so the hot spot appears to us with its true size.

… and as a result, the hot spot appears to us to be smaller than it actually is.

FIGURE 13-22 Cosmic Microwave Background and the Curvature of Space Temperature variations in the early universe appear as "hot spots" in the cosmic microwave background. The apparent sizes of these spots depend on the curvature of space. (a) In a closed universe with positive curvature, light rays from opposite sides of a hot spot bend toward each other. Hence, the hot spot appears larger than it actually is (dashed lines). (b) The light rays do not bend in a flat universe. (c) In an open universe, light rays bend apart. The dashed lines show that a hot spot would appear smaller than its actual size. (The BOOMERANG Group, University of California, Santa Barbara)

13-15 Dark energy is causing the universe to accelerate outward

Until the past few years, there was a major inconsistency between the distribution of observed matter and energy in the universe and the flatness of space. Just as there is not enough visible matter to account for galaxies and clusters of galaxies that remain as bound systems, there is not enough observed matter and energy to account for a flat universe. Visible matter accounts for only 4% of the required mass. The cosmic microwave background photons add only 0.005% of the required gravitational effects needed for flatness, and calculations of the mass necessary to keep galaxies and clusters bound, combined with gravitational lensing by dark matter of distant galaxies and quasars, reveal that the dark matter known to exist accounts for only about 23% of the required mass. Therefore, the universe has only about 27% of the required mass and energy necessary to make it flat. Allowing for the errors that still exist in all of these observations, the possible range of mass and energy from all matter and photons in

a

R I V U X G

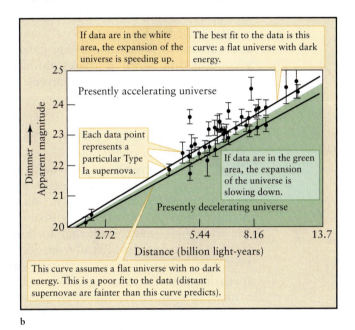

b

FIGURE 13-23 Dimmer Distant Supernova (a) These Hubble Space Telescope images show the galaxy in which the supernova SN 1997ff occurred. This supernova, more than 10 billion light-years away, was dimmer than expected, indicating that the distance to it is greater than the distance it would have if the universe had been continually slowing down since the Big Bang. This supports the notion that an outward (cosmological) force is acting over vast distances in the universe. The arrow on the first inset shows the galaxy in which the supernova was discovered. The bright spot on the second inset shows the supernova by subtracting the constant light emitted by all the other nearby objects. (b) The distances and brightnesses of many very distant supernovae are plotted on this diagram. The locations of the most distant supernovae in the upper region strongly indicate that the universe has been accelerating outward for the past 6 billion years. (a: Adam Riess, Space Telescope Science Institute, NASA)

the universe is still between only 20% and 40% of that required for flatness. Yet flat it certainly appears to be.

By the mid-1990s, combined evidence of the microwave background and large-scale structure had forced astronomers to the conclusion that the remaining energy required to make the universe flat must be a form of dark energy with the remarkable feature that it causes the universe to accelerate outward. Corroborating evidence for the existence of such dark energy was found in the recent observations of Type Ia supernovae in extremely distant galaxies (Figure 13-23).

The light curves (see Section 8-4) for Type Ia supernovae (explosions of white dwarfs in binary star systems) are very well known. Studies show that these supernovae in nearby galaxies behave similarly to those observed in the Milky Way, regardless of their environment. Because they are so bright, such supernovae are excellent standard candles for determining distances to objects billions of light-years away. Assuming that supernovae in distant galaxies also behave similarly to those in our Galaxy, observations first made in 1998 revealed that the supernovae in distant galaxies appear dimmer than they would if the universe had been continually

decelerating or even expanding at a constant speed. In other words, the universe is now expanding faster than it was, so the distant galaxies are farther away than they otherwise have been. Because they are farther away, the supernovae in them appear dimmer than in a continually slowing universe. The universe is *accelerating* outward.

There must be some kind of repulsive force acting to increase the rate at which superclusters separate today. Astronomers hypothesize that this is due to some kind of dark energy that has a repulsive gravitational effect, as mentioned earlier. Confirmation that dark energy exists came in 2003, when astronomers observed light from very distant galaxies passing through more nearby clusters of galaxies on its way toward Earth. As the distant light moved toward a cluster on its way to us, that light was gravitationally blueshifted, meaning the light gained energy as it was pulled toward the cluster. If there were no dark energy in a flat universe, then, as that light left the vicinity of the cluster, it would have been redshifted (lost energy) by exactly the same amount and come to us with the same wavelengths it would have had if it had never passed through the cluster.

However, if dark energy exists, then the universe is accelerating outward (moving outward faster and faster). So the universe would have expanded more during the time that the light was leaving the cluster than when that light was first moving toward the cluster. Under those conditions, the light would have gained more energy from the cluster's gravitational attraction while traveling toward the cluster than the light would have lost to that gravitational attraction as it was moving away from the cluster. In other words, the amount of blueshift that the light underwent falling into the cluster would be greater than the amount of red shift that it underwent leaving the vicinity of the cluster. This was precisely what astronomers discovered: The light from distant galaxies undergoes a net blueshift when it passes through clusters of galaxies on its way to us—the universe is accelerating outward.

Cosmological Constant   There are two theoretical explanations for dark energy. Let us first consider what would happen if the cosmological constant that Einstein introduced and then rejected actually did exist. As Einstein had intended, it would provide a repulsive force to nature. The energy associated with that force would appear in the vacuum of space. Recall from Sections 9-11 and 13-8 that, even in a vacuum, particles are constantly appearing and disappearing. They add energy to the cosmos, so even empty space has energy. The cosmological constant form of energy has the property of contributing a repulsive gravitational force that competes with the attractive gravitational force from matter and radiation, which slows down the expansion. Which gravitational force wins depends on whether there is more vacuum energy or more matter and radiation.

As the universe expands, the average density of matter and energy decreases. In other words, on average there is less matter and energy in each cubic meter of space every second than there was the second before. However, the density of energy created by the cosmological constant is unchanged—as the universe expands the amount of vacuum energy in each cubic meter of space remains constant. If the universe is still expanding when the vacuum energy exceeds the matter and energy per unit volume, then the repulsive gravitational force would dominate and force the universe to start accelerating outward, as is observed.

The concern that some astronomers express about the cosmological constant is that the repulsive force it creates has just the right strength to have allowed the universe to slow down for several billion years and then to slowly cause it to accelerate outward, as seen today. To explain this apparent coincidence would require fine-tuning the ratio of the vacuum energy to the matter and radiation to a remarkably high degree of precision. Scientists do not yet have a good reason why the repulsion created by a cosmological constant is not, say, a hundred times greater, in which case structure throughout the universe, such as stars and galaxies, would not have formed yet, or a billion times less, in which case the repulsive effects of the cosmological constant would be negligible today.

Quintessence   Despite these concerns, observations of distant Type Ia supernovae suggest that the outward force acting on the universe is nearly constant, consistent with the dark energy being created by the physics we characterized with the cosmological constant. However, these observations do not yet rule out the major competing theory of dark energy, called **quintessence.** Scientists are exploring a variety of mathematical descriptions of quintessence. It differs from the vacuum energy associated with the cosmological constant in that the energy of quintessence is not constant but changes slowly with the expansion of the universe and can also be changed by the flow of energy and matter through the universe. Because it can change, quintessence need not have been fine-tuned to a strength that makes the universe flat and creates just the slight acceleration we are seeing. Instead, it could have started out with some arbitrary strength and then been adjusted by physical properties of the universe to match the growth of the universe.

As the universe grew, the energy in quintessence would eventually come to dominate the gravitational energy from matter. If this took place a few billion years ago, quintessence would have begun exerting a negative gravitational force and so the universe would have begun accelerating.

The unexpected discovery that the universe is accelerating outward completely alters astronomers' perspectives about the fate of the cosmos. *It tells us that the universe will expand forever.* At the same time, it neatly solves the mystery of why the universe is flat, as required by the occurrence of inflation in the early universe. Dark energy provides the remaining 73% of the energy needed to account for the universe being flat. The contributions of all of its major components to the mass and energy in the universe are listed in Appendix E-12.

We have seen that our present understanding of the laws of nature is incomplete, especially in that our theories do not explain the nature of the universe during the Planck era (or why the universe began in the first place). Driven by these challenges, scientists have begun to develop superstring theories that may resolve these issues. The following *Guided Discovery: Superstring Theories* is intended to give you a brief introduction to them and what they now predict.

Superstring Theory

To combine gravitation and the other three fundamental forces in nature into one comprehensive Theory of Everything, scientists have had to consider a universe that contains more than the four dimensions we know about today (three of space and one of time). The theories that mathematically describe this new formulation of the universe are called *superstring theories*. There are basically five such theories, which allow all of the particles we have been studying—such as quarks, protons, neutrons, electrons, and photons—to exist. Of course, only one of the superstring theories can be correct, but which, if any, remains to be seen.

Spacetime in superstring theories has 10 dimensions, of which 6 are everywhere rolled up into such tiny volumes that we cannot detect them directly. The other four dimensions are our normal spacetime. Superstring's more general spacetime carries with it properties that allow scientists to combine all four forces in nature into one set of equations.

The difficulty in reconciling quantum mechanics (describing the weak, strong, and electromagnetic forces) and general relativity (describing gravitation) is that the three forces in quantum mechanics are quantized, whereas general relativity is not. In other words, the weak, strong, and electromagnetic forces are transmitted by particles. For example, the quanta of electromagnetism are photons. Gravity, as described by general relativity, is based on a smooth and continuous, rather than quantized, force. Specifically, the distortion of spacetime by matter and energy creates the gravitational force.

Superstring theories begin with a different assumption about all particles and their interactions than do either quantum mechanics or general relativity. The new theories assert that each particle is actually a tiny vibrating string, with different types of particles, each vibrating at a different rates. Gravitation has its own energy-sharing particle, the graviton, analogous to photons for electromagnetism. The interactions between the strings create all of the properties of matter and energy.

The predictions made by superstring theories begin with the assumption that general relativity is the correct "classical" theory for describing the gravitational interaction between matter and energy. This may seem trivial, but, because general relativity today correctly predicts everything in its realm of validity, a more comprehensive (superstring) theory needs to keep that accuracy, or the larger theory is wrong. Some additional predictions of superstring theories include the following:

- The universe cannot have positive curvature (which it does not, as we have seen).

- Some of the clumpy structure that we see as superclusters of galaxies could have been caused by the effects of superstring activity during inflation (with the rest due to the expansion of quantum fluctuations).

- Spacetime may not be entirely smooth. It may have structural defects, like a flawed diamond or ice that has broken into abutting chunks. These defects in spacetime would appear as one-dimensional "cosmic strings" with great density. Their attraction would pull normal matter around them, creating strings of galaxies. These cosmic strings have not yet been found.

- Some of the dark matter may be particles predicted by string theory.

- Most versions of superstring theory include a cosmological constant, but there is no underlying reason yet known for the value of the cosmological constant that may exist today.

- The speed of light is the same for all photons. If different wavelengths of light from the same event arrived at different times, then superstring (and relativity) theories would be wrong.

Superstring theories are, so far, consistent with observations, but it remains to be seen if any one of them will continue to maintain consistency with future observations and to correctly predict observable things that we have not yet seen.

13-16 Frontiers yet to be discovered

Some of the biggest unanswered questions about the cosmos follow from the discoveries presented in this chapter. What caused the Big Bang? What happened during the Planck time before 10^{-43} s? Is our universe the only one, or are there others, perhaps with profoundly different physical properties than ours, that were created at the same time but that are inaccessible to us? What caused the universe to come into existence in a false vacuum? What are the details of the formation processes of galaxies? Why was gas in the early universe not eventually transformed into systems of stars a million times bigger or smaller than galaxies? Did black holes grow with the evolution of the galaxies, as is currently believed? What are the origins of the dark energy? Which, if either, of the current candidates for dark energy—the cosmological constant or quintessence—is correct? One of the things that makes this time in human existence so fascinating is that it is likely we will have answers to most of these questions within our lifetimes.

SUMMARY OF KEY IDEAS

The Big Bang

• Astronomers believe that the universe began as an exceedingly dense cosmic singularity that expanded explosively in an event called the Big Bang. The Hubble law describes the ongoing expansion of the universe and the rate at which superclusters of galaxies move apart.

• The observable universe extends about 13.7 billion light-years in every direction from Earth to what is called the cosmic light horizon. We cannot see any objects that may exist beyond the cosmic light horizon because light from these objects has not had enough time to reach us.

• According to the theory of inflation, early in its existence, the universe expanded super rapidly for a short period, spreading matter that was originally near our location throughout a volume of the universe so large that we cannot yet observe it. The observable universe today is thus a growing volume of space containing matter and radiation that was in close contact with our matter and radiation during the first instant after the Big Bang. This explains the isotropic and homogeneous appearance of the universe.

A Brief History of Spacetime, Matter, Energy, and Everything

• Four basic forces—gravity, electromagnetism, the strong nuclear force, and the weak nuclear force—explain the interactions observed in the universe.

• According to current theory, all four forces were identical just after the Big Bang. At the end of the Planck time (about 10^{-43} s after the Big Bang), gravity became a separate force. A short time later, the strong nuclear force became a distinct force. A final separation created the electromagnetic force and the weak nuclear force.

• Before the Planck time, the universe was so dense that known laws of physics did not describe the behavior of spacetime, matter, and energy back then.

• In its first 30,000 years, the universe was radiation-dominated, during which time photons prevented matter from forming clumps. Then it was matter-dominated, during which time superclusters and smaller clumps of matter formed. Today it is dark-energy-dominated. Dark energy of some sort supplies a repulsive gravitational force that causes superclusters to accelerate away from each other.

• Astronomers think that during the first 379,000 years of the universe, matter and energy formed an opaque plasma, called the primordial fireball. Cosmic microwave background radiation is the greatly redshifted remnant of the universe as it existed about 379,000 years after the Big Bang.

• By 379,000 years after the Big Bang, spacetime expansion caused the temperature of the universe to fall below 3000 K, enabling protons and electrons to combine to form hydrogen atoms. This period is called the era of recombination. The universe became transparent during the era of recombination, meaning that the microwave background radiation contains the oldest photons in the universe.

• Clusters of galaxies and individual galaxies formed from pieces of enormous hydrogen and helium clouds, each of which became a separate supercluster of galaxies.

• All of the superclusters and some of the clusters of galaxies within each supercluster are moving away from one another.

• During the matter-dominated era, structure formed in the universe. As the universe goes farther into the dark-energy-dominated era, the large-scale structure of superclusters of galaxies will fade away.

The Fate of the Universe

• The average density of matter and dark energy in the universe determines the curvature of space and the ultimate fate of the universe.

• Observations show that the universe is flat and that the cosmic microwave background is almost perfectly isotropic, resulting from a brief period of very rapid expansion (the inflationary epoch) in the very early universe.

• The universe is accelerating outward and it will expand forever.

WHAT DID YOU THINK?

1   *What is the universe?* It is all of the matter, energy, and spacetime that will ever be detectable from Earth or that will ever affect us.

2   *Did the universe have a beginning?* Yes, it occurred about 13.7 billion years ago, in an event called the *Big Bang*.

3   *Into what is the universe expanding?* Nothing. The Big Bang created space and time (spacetime), as well as all matter and energy in the universe. Spacetime is expanding to accommodate the expansion of the universe.

4   *How strong is gravity compared to the other forces in nature?* Gravity is by far the weakest force.

5   *Will the universe last forever?* Current observations support the belief that the universe will last forever.

Key Terms for Review

Big Bang, 338
closed universe, 354
confinement, 345
cosmic light horizon, 344
cosmic microwave
    background, 339
cosmological
    constant, 336
cosmological redshift, 337
cosmology, 336
dark ages, 349
dark energy, 514
decoupling, 347
era of recombination, 348
expanding universe, 337
Grand Unified Theory
    (GUT), 342
homogeneity, 341
horizon problem, 343
inflation, 343
inflationary epoch, 343

isotropy, 341
isotropy problem, 343
matter-dominated
    universe, 347
open universe, 354
pair production, 345
Planck time, 342
primordial fireball, 347
primordial
    nucleosynthesis, 346
quark, 341
quintessence, 356
radiation-dominated
    universe, 347
strong nuclear
    force, 341
superstring theories, 342
Theories of
    Everything, 342
universe, 336
weak nuclear force, 341

Review Questions

1. Which force in nature is believed to have formed second? **a.** gravity **b.** electromagnetic force **c.** weak force **d.** strong force **e.** All formed at the same time.

2. Inflation most directly explains which of the following? **a.** why the universe is homogeneous **b.** why the universe is expanding **c.** why the universe is going to last forever **d.** why the universe has a background temperature **e.** why the universe has particles

3. What does it mean when astronomers say that we live in an expanding universe?

4. Explain the difference between a Doppler redshift and a cosmological redshift.

5. In what ways are the fate of the universe, the shape of the universe, and the average density of the universe related?

6. Assuming that the universe will expand forever, what will eventually become of the microwave background radiation?

7. What does it mean to say that the universe is dark-energy-dominated? When was the universe radiation-dominated? When was it matter-dominated? How did radiation domination show itself?

8. Explain the difference between an electron and a positron.

9. Where do astronomers believe most of the photons in the cosmic microwave background originated?

10. Give examples of the actions or roles of each of the four basic physical forces in the universe.

11. What is the observational evidence for the **a.** Big Bang, **b.** inflationary epoch, and **c.** confinement of quarks?

Advanced Questions

12. Explain why the detection of cosmic microwave background radiation was a major blow to the steady-state theory.

13. Some so-called "creation scientists" claim that the universe came into existence 6000 years ago. What is Hubble's constant for such a cosmos? Is this a reasonable number? Explain.

Discussion Questions

14. Discuss the implications that science cannot yet tell us what caused the Big Bang or what, if anything, existed before the Big Bang occurred.

15. Explain why gravitational attraction has dominated the behavior of the universe until recently and why the dark energy determines the fate of the universe.

What If . . .

16. The universe were destined to collapse and the collapse were under way? What would be different in space and on Earth under those conditions?

17. Our solar system formed very early in the evolution of our Galaxy, when the universe was just 2 billion years old? What would be different in space and on Earth?

18. Our solar system formed much later in the evolution of the universe than it actually did? How would observations of stars and galaxies be different than they are now?

19. Our solar system formed with the first generation of stars? What would be different about the solar system? Would Earth exist as an inhabitable world? Why or why not?

Web Questions

20. Temperatures in the Early Universe. Access the Active Integrated Media Module "Black Body Curves" in the *Discovering the Universe: From the Stars to the Planets* Web site. **a.** Use the module to determine by trial-and-error the temperature at which a blackbody spectrum has its peak at a wavelength of 1 μm. **b.** At the time when the temperature of the cosmic background radiation was equal to the value you found in (a), was the universe matter-dominated or radiation-dominated? Explain your answer. Hint: Figure 13-8 might help.

Observing Projects

21. View into the Distant Past The Hubble Space Telescope has been used to explore dark regions of space where no bright stars or galaxies seem to appear. In particular, the Hubble Deep Field and Hubble Ultra Deep Field images were extremely long exposures of these dark regions and revealed that rich fields of very distant galaxies do, in fact, appear in these areas. Light emitted from these galaxies early in the life of the universe is just now reaching us. These images therefore provide us with a glimpse of the universe at a time just one or two billion years after the Big Bang. We will examine these two images. You will need a ruler for this project. In *Starry Night Enthusiast™*, open the **Find** pane, ensure that the **Query (Q)** box is empty and click on the **Down (▼)** arrow in this box to display a list of image sources. a) Click on **Hubble Images** and double-click on **Hubble Deep Field** to center the view on this dark region of space. Note its position with respect to the Big Dipper. (If you cannot identify this region of the northern sky, click on **View > Constellations > Asterisms** and **View > Constellations > Labels**. Remove these indicators after you have identified the region.) Zoom in to a field of view about 3° wide and note that the region still appears to have no bright objects within it. Zoom in again until the **Hubble Deep Field** fills the view (about 7′ × 5′). With your ruler, measure and record the visual angular diameters of five of the largest galaxies. b) Return to the **Find** pane and the list of **Hubble Images** and click on **Hubble Ultra Deep Field** to center the view of this "dark" region of the sky. Press the space bar to go there immediately. If necessary, zoom in to the same angular size as you used to view the **Hubble Deep Field**. Measure the diameters of five of the largest galaxies in this image. Give at least two plausible reasons why the two images have different-sized largest galaxies.

PART IV

Understanding the Solar System

A montage of the planets in our solar system presented in correct relative sizes. The orbits in the background are also drawn to scale. (NASA)

Mercury Earth
Venus Mars
Jupiter
Saturn
Uranus
Neptune

1655-1656

Jupiter's Great Red Spot discovered by Giovanni Cassini. Christiaan Huygens identifies Saturn's rings as rings, and discovers its largest moon, Titan.

1705-1757

Edmund Halley makes first prediction of the period of a comet's orbit, calculating that the comet we now call Halley's would return in 1758. Comet's return toward inner solar system first observed by Johann Palitzsch.

1856-1859

James Clerk Maxwell proves that Saturn's rings are not solid.

1877

Phobos and Deimos discovered by Asaph Hall. Giovanni Schiaparelli observes "*canali*" (Italian for "channels") on Mars.

1821

Alexis Bouvard detects irregularities in Uranus's orbit.

1801

Giuseppi Piazzi discovers the asteroid Ceres.

American Revolution

Telegraph in Use

Telephone

1671

Giovanni Cassini calculates first relatively accurate estimate of distance between Earth and Sun.

1861-1866

Giovanni Schiaparelli determined that meteor showers are caused by Earth passing through comet debris.

1843-1849

Neptune's presence predicted independently by John Adams and Urbain Leverrier. Neptune first observed by Johann Galle. Triton discovered by William Lassell. Edouard Roche shows that Saturn's rings did not form from a preexisting moon.

1610-1613

Galileo Galilei observes phases of Venus and major moons of Jupiter—Io, Europa, Ganymede, and Callisto.

1781-1784

William Herschel first to observe Uranus and to detect clouds around Mars.

THE SOLAR SYSTEM

1930
Clyde Tombaugh discovers Pluto.

1960-1964
Venus's retrograde rotation measured. Mercury's rotation rate determined by Arecibo radio observatory. U.S. missions: *Mariner 4*, first flyby of Mars; *Ranger 7*, first successful U.S. lunar impact.

1969-1972
U.S. *Apollo* missions: *Apollo 11*, first humans to land on the Moon; *Apollo 12, 14, 15, 16,* and *17* landings all successful; failed *Apollo 13* grows legendary.

1977-1978
James Elliot discovers rings of Uranus. James Christy discovers Pluto's moon, Charon.

1990-1998
Magellan spacecraft completes full radar map of Venus. *Clementine* orbiter discovers evidence of water on Moon. *Galileo* spacecraft sends probes into Jupiter's atmosphere.

2006-2009
Pluto, Ceres, Eris, Makemake, and Haumea reclassified as dwarf planets, a new category.

2004
Mars rovers *Spirit* and *Opportunity* land on Mars and find strong indications that liquid water existed there. Spacecraft *Cassini* arrives at Saturn and begins studying that planet and its moons.

in Use

Cold War

Communications Satellites in Use

Berlin Wall Falls

The Internet

1950-1959
Oort Comet cloud proposed by Jan Oort. Kuiper belt of comet bodies proposed by Gerard Kuiper. Soviet *Luna* missions: *Luna 1*, first lunar flyby; *Luna 2*, first lunar impact; *Luna 3* flyby, first photo of lunar far side.

1966-1967
Soviet *Luna 13*, first soft landing on Moon; 2 successful *Lunas, 16* and *17* (robot and rover), followed. U.S. *Suryeyor 1*, first of 4 successful soft landings on the Moon. U.S. *Lunar Orbiter 1*, first of 4 successful orbiters. Soviet *Venera 4* makes first landing on Venus.

1970-1976
First landers on Mars: Soviet *Mars 2* and *3* spacecraft. U.S. *Mariner 10*, first spacecraft to visit Mercury. First U.S. landings on Mars by *Viking 1* and *2* spacecraft.

1979-1989
Voyager I and *II* spacecraft visit outer planets, discovering Jupiter's and Neptune's rings, dozens of their smaller moons, and volcanoes on Io.

1995
A planet is discovered orbiting the star 51 Pegasi, the first of over 250 extra solar planets detected so far.

2001
NEAR Shoemaker spacecraft lands on asteroid Eros.

CHAPTER 14

Formation of the Solar System and Other Planetary Systems

Star and planet systems form from collapsing, rotating clouds of gas and dust. (NASA/JPL-Caltech)

R I V U X G

WHAT DO YOU THINK?

1 Were the Sun and planets among the first generation of objects created in the universe?

2 How long has Earth existed, and how do we know this?

3 What typical shape(s) do moons have, and why?

4 Have any Earthlike planets been discovered orbiting Sunlike stars?

Answers to these questions appear in the text beside the corresponding numbers in the margins and at the end of the chapter.

We have studied all the processes necessary to understand how our **solar system**, the Sun, and everything that orbits the Sun formed. Those processes include the formation of hydrogen and helium in the Big Bang; the formation of the Milky Way Galaxy and the first generation of stars in this Galaxy; the evolution of stars and how the stars transform some of their mass into heavier elements; the processes by which stars return gas and dust to the interstellar medium; and the formation of later generations of stars in the spiral arms of the Galaxy.

Beginning with a brief review of these events, this final section of the book explores the solar system's formation and the properties and evolution of everything in it (other than the Sun, which we covered in Chapter 5). Using rovers and landers on Mars, spacecraft orbiting the Red Planet, Mercury, Saturn, as well as other craft traveling through other regions of the solar system, our knowledge of our nearest neighbors in space is increasing at a breathtaking rate.

In this chapter you will discover

- how the solar system formed

- why the environment of the early solar system was much more violent than it is today

- how astronomers define the various types of objects in the solar system

- how the planets are grouped

- how the moons formed throughout the solar system

- what the debris of the solar system is made of

- that disks of gas and dust, as well as planets, have been observed around a growing number of stars

- that newly forming stars and planetary systems are being observed

THE SOLAR SYSTEM CONTAINS HEAVY ELEMENTS, FORMED FROM AN EARLIER GENERATION OF STARS

How did the solar system form? How were its varied building blocks of rock, metal, ice, and gas created? How has the solar system changed since its formation, and what does its history tell us about the planets we see today? Within the past few decades, telescopes and space probes, along with the theories of modern science, have finally provided some of the answers to these age-old questions.

14-1 Stars transform matter from lighter elements into heavier ones

Our study of the formation of the solar system begins with an inconsistency, namely that Earth, the Moon, and many other objects that orbit the Sun are composed primarily of heavy elements, such as oxygen, silicon, aluminum, iron, carbon, and calcium. Furthermore, most of these bodies contain extremely little hydrogen and helium. However, observations of the spectra of the Sun, other stars, and interstellar clouds reveal that hydrogen and helium are by far the most abundant elements in the universe. These two elements account for about 99.9% of all observed atoms (or, equivalently, 98% of observed mass). All of the other elements combined account for less than 1% of the observed atoms (and hence 2% of the observed mass of the universe). How is it that Earth, the Moon, Mars, Venus, Mercury, and many smaller bodies in the solar system typically contain less than 0.15% hydrogen and helium? Somehow, these familiar objects formed from matter that had been enriched with the heavier elements and depleted in hydrogen and helium.

There is a good reason for the overwhelming abundance of hydrogen and helium throughout the universe. As we discovered in Chapter 13, the universe formed a   R I V U X G   b   R I V U X G   c   R I V U X G

FIGURE 14-1 How Stars Lose Mass (a) The brightest star in Scorpius, Antares, is nearing the end of its existence. Strong winds from its surface are expelling large quantities of gas and dust, creating this nebula reminiscent of an Impressionist painting. The scattering of starlight off this material makes it appear especially bright, even at a distance of 604 light-years. (b) The planetary nebula Abell 39 is 7000 light-years from Earth. In this relatively gentle emission of matter, the central star shed its outer layers of gas and dust in an expanding spherical shell now about 6 light-years across. (c) A supernova is the most powerful known mechanism for a star to shed mass. The Crab Nebula, even though it is about 6000 light-years from Earth, was visible during the day for 3 weeks during 1054. (a: David Malin/Anglo-Australian Observatory; b: George Jacoby [WIYN Obs.] et al., WIYN, AURA, NOAO, NSF; c: Malin/ Pasachoff/Caltech)

about 13.7 billion years ago in a violent event called the *Big Bang*. Only the lightest elements—hydrogen, helium, and a tiny amount of lithium—were created in the process. The first stars, composed only of these three elements, condensed out of this primeval matter, probably within a few hundred million years after the Big Bang.

The difference in chemical composition between the early universe and the solar system shows that the solar system did not form as a *direct* result of the Big Bang. Let us briefly review this chemical transformation, which began deep inside the first stars. Once formed, gravity compresses the matter in their central *cores* of stars so much that the hydrogen there is fused into helium. We have seen that hydrogen fusion has the interesting property that some of the hydrogens mass is converted into electromagnetic energy (photons), a portion of which eventually leaks out through the star's surface, enabling it to shine. The core's helium subsequently fuses to create carbon, also converting some mass into energy. If the star has enough pressure in its core, the carbon transforms into even heavier elements, such as nitrogen, oxygen, neon, silicon, and iron. You can review the details of the formation of heavier elements in Chapters 7 and 8.

Stars also shed matter into space, as we also saw in Chapters 7 and 8 (Figure 14-1). Some of the heavy elements formed inside them are eventually ejected, along with most of the hydrogen and helium remaining in their outer layers. The outer layers of stars are expelled at different rates, ranging from *stellar winds* (for example, the solar wind; also see section 7-9 and Figure 14-1a), to the more energetic planetary nebulae (recall sections 8-1 and 8-2; also see Figure 14-1b), to spectacular supernovae (discussed in sections 8-5 through 8-9; see also Figure 14-1c). Supernovae are such powerful events that fusion occurs during the explosion, creating even denser elements that are also expelled from the star. Planetary nebulae and supernovae leave only tiny stellar cores, remnants of once bright and hot stars.

Of the matter ejected from stars, some of it leaves as small particles of dust, such as the soot formed in a fire or in diesel exhaust. Over time, enough gas and dust were emitted by enough stars to form interstellar clouds rich in hydrogen and helium, but also containing metals. (In a curious alternative use of a common word, astronomers define **metals** to be all of the elements in the universe other than hydrogen and helium.) It is to these clouds that we turn next in understanding the formation of the solar system and the dense planets like Earth.

FIGURE 14-2 *Dusty Regions of Star Formation* (a) These three bright young stars in the constellation Monoceros are still surrounded by much of the gas and dust from which they formed. This is a tiny part of a much larger cloud, known as the *Cone Nebula*. Astronomers hypothesize that the solar system formed from a similar fragment of an interstellar gas and dust cloud. (b) Newly formed stars in the Orion Nebula. Although visible light from many of the stars is blocked by the Nebula, their infrared emission travels through the gas and dust to us. (a: ACS Science & Engineering Team, NASA; b: NASA; K. L. Luhman/Harvard-Smithsonian Center for Astrophysics and G. Schneider, E. Young, G. Rieke, A. Cotera, H. Chen, M. Rieke, R. Thomson/Steward Observatory, University of Arizona, Tucson, AZ)

14-2 Gravity, rotation, and heat shaped the young solar system

The solar system formed from a fragment of a vast cloud of interstellar gas and dust. The region of the cloud from which the solar system developed is called the **solar nebula**. It has a diameter of at least 100 AU and a total mass about 2 to 3 times the mass of the Sun. Matter collapses to form star systems, such as our solar system, for at least four reasons. First, winds from some stars compress nearby gas and dust. Second, the explosive force of a nearby supernova compresses regions of preexisting interstellar clouds. Third, pairs of clouds sometimes collide and compress each other. Fourth, regions of gas and dust radiate away enough heat to cool and condense without outside help. In all of these situations, when a cloud fragment gets sufficiently dense, the mutual gravitational attraction of gas and dust particles in it forces it to collapse, thereby creating new stars and planetary systems (Figure 14-2).

Stellar evolution and star formation are ongoing processes. You can see a cloud in the process of forming stars today with your naked eye, namely the middle "star" in Orion's sword (Figures 1-4 and 3-29), which looks like a fuzzy blob in the night sky. Called the Orion Nebula or Great Nebula of Orion, it is actually part of a gigantic interstellar cloud region in which new stars and planets are being created. Recalling from Chapter 3 the limited information that visible light provides, you can also see in Figures 3-30 and Figure 14-2b that ultraviolet and infrared images reveal even more of this region.

Initially, the solar nebula was a very cold collection of gas and dust—well below the freezing point of water. Although most of it was hydrogen and helium, ice and ice-coated dust grains composed of heavy elements were scattered abundantly across this vast volume. Deep inside the nebula, gravitational attraction caused the gas and dust there to fall rapidly toward its center. (In terms of the physics presented in Chapter 2, the cloud's gravitational potential energy was being converted into kinetic energy—energy of motion.) As a result, the density, pressure, and temperature at the center of the nebula began to increase, producing a concentration of matter called the **protosun**.

As the protosun continued to increase in mass and to contract, atoms in it collided with one another with increasing speed and frequency. Such collisions created heat, causing the protosun's temperature to soar. Therefore, the first heat and light emitted by our Sun came from colliding gas, not from nuclear fusion, as it does today.

Rotation played a key role in the evolution of the solar system. When it first formed, the solar nebula was a very slowly rotating ensemble of particles. This rotation occurred because the formation of interstellar clouds is turbulent, meaning that the clouds are created with many slowly swirling regions, like so much

> Which elements on Earth may have been unchanged since the universe began?

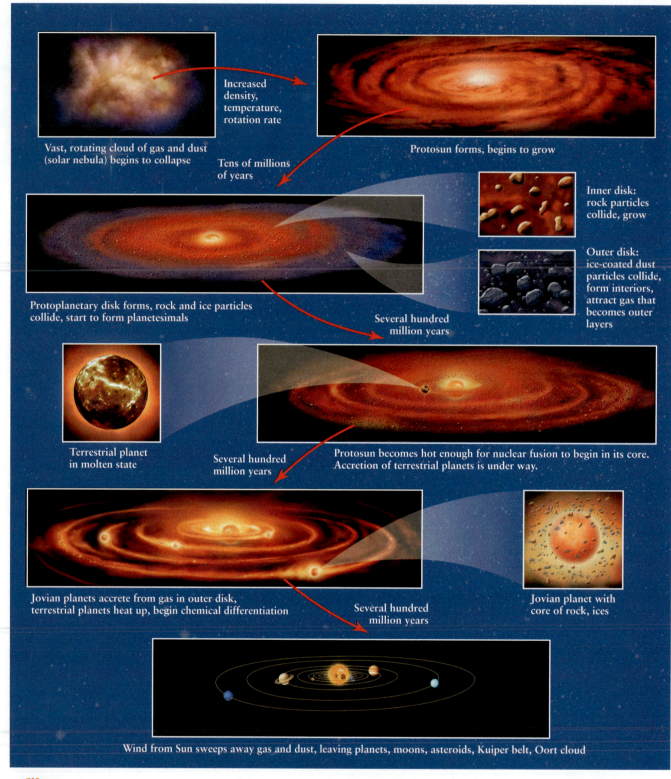

Vast, rotating cloud of gas and dust (solar nebula) begins to collapse

Increased density, temperature, rotation rate

Tens of millions of years

Protosun forms, begins to grow

Protoplanetary disk forms, rock and ice particles collide, start to form planetesimals

Several hundred million years

Inner disk: rock particles collide, grow

Outer disk: ice-coated dust particles collide, form interiors, attract gas that becomes outer layers

Terrestrial planet in molten state

Several hundred million years

Protosun becomes hot enough for nuclear fusion to begin in its core. Accretion of terrestrial planets is under way.

Jovian planets accrete from gas in outer disk, terrestrial planets heat up, begin chemical differentiation

Several hundred million years

Jovian planet with core of rock, ices

Wind from Sun sweeps away gas and dust, leaving planets, moons, asteroids, Kuiper belt, Oort cloud

ANIMATION 14.1

FIGURE 14-3 The Formation of the Solar System This sequence of drawings shows stages in the formation of the solar system.

smoke rising from a fire. As a result of this rotation, the nebula had angular momentum (see *An Astronomer's Toolbox 2-2*), which prevented its outer regions from collapsing into the protosun. A nonswirling collapse would have created a star without any planets or other orbiting matter. Conservation of angular momentum (see Section 2-7) tells us that rotating matter falling inward in the solar nebula revolved faster and faster, like the skater in Figure 2-12 when she pulls her arms in. Even the debris that formed the protosun was orbiting as it fell inward, like water spiraling down a drain. Therefore, the protosun was spinning and the Sun that formed from it should be spinning. It is, as we saw in Chapter 5.

The outer parts of the solar nebula collapsed inward more slowly than the central matter that formed the protosun. This outer region was probably very ragged, but mathematical studies show that the combined effects of gravity, collisions, and rotation transform even an irregular cloud fragment into a rotating disk with a warm center and cold edges (Figure 14-3). You can see an analogous effect of rotation in the way an ice skater's dress is forced into a plane as she spins.

Although we cannot see our solar system as it was before planets formed, astronomers have found what we think are similar disks of gas and dust surrounding other young stars, including those shown in Figure 14-4. Called **protoplanetary disks** or **proplyds**, these systems are undergoing the same initial stage of evolution that we have described here for our solar system.

The protosun's temperature increased as it became denser, and more and more atoms in it collided more often. Radiating more and more heat, the temperature around it began to increase. The increasing temperature vaporized all common icy substances in the inner region of the solar nebula and pushed light gases, like hydrogen and helium, outward. In this inner region, where Earth orbits today, mostly heavier elements remained. These heavy elements eventually came together and formed the four inner planets: Mercury, Venus, Earth, and Mars. The removal of the lighter elements by the protosun from their immediate neighborhood explains why the inner planets are composed mostly of heavy elements. *In other words, we are literally made of star dust!*

14-3 Collisions in the early solar system led to the formation of planets

Although we have used the word *planet* in its everyday usage so far in this book, it is now appropriate to introduce its formal definition. A **planet** is an object that directly orbits a star (rather than orbits another body, like the Moon orbits Earth). A planet also must have

FIGURE 14-4 Young Circumstellar Disks of Matter The heart of the Orion Nebula as seen through the Hubble Space Telescope. The four insets are false-color images of protoplanetary disks within the nebula. A recently formed star is at the center of each disk. The disk in the upper right is seen nearly edge-on. Our solar system is drawn to scale in the lower left image. (C. R. O'Dell and S. K. Wong, Rice University; NASA)

enough mass to pull itself into an essentially spherical shape and enough gravitational force so that it has cleared its orbital neighborhood of most of the smaller debris there. If the planet is rotating, its spherical shape will be modified so that it is wider at its equator than at its poles. Bodies that orbit larger objects, which in turn orbit stars, are called **moons** or **natural satellites**.

The formation of the inner planets was the result of innumerable impacts of small, rocky particles in the solar system's protoplanetary disk. Initially, neighboring dust grains and pebbles in the solar nebula collided and stuck together. Then, over a period of a few million years, these accumulations of dust and pebbles coalesced into larger objects, called **planetesimals,** with diameters of a few kilometers or miles (Figure 14-3).

> What force ensured that once the solar nebula began collapsing, it continued to do so?

The planetesimals collided with anything in their paths. Many were pulverized, many fell into the Sun, but eventually, their mutual gravitational attraction brought a few of them together to form larger objects, called **protoplanets.** Computer simulations of the inner solar system, based on Newton's laws (see Section 2-7), suggest that **accretion**—the coming together of smaller pieces of matter to form larger ones—continued for over 100 million years and should lead to the formation of fewer than

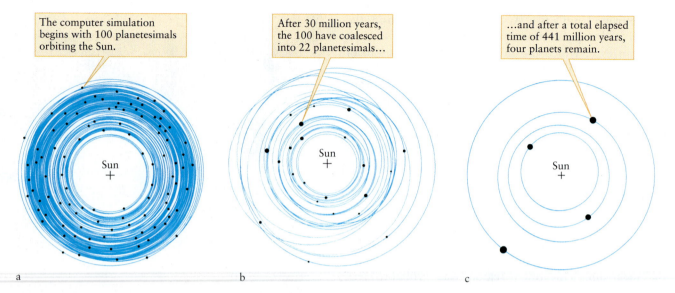

The computer simulation begins with 100 planetesimals orbiting the Sun.

After 30 million years, the 100 have coalesced into 22 planetesimals...

...and after a total elapsed time of 441 million years, four planets remain.

a   b   c

FIGURE 14-5 Accretion of the Inner Planets This computer simulation shows the formation of the inner planets over time.

half a dozen rocky planets (Figure 14-5). The agreement between predictions of the number of planets and their actual number in our inner solar system (four) is encouraging. Observations made in 2004 of star systems that underwent all stages of planet formation indicate that this entire process, often delayed by collisions between large protoplanets, typically takes several hundred million years to complete.

INSIGHT INTO SCIENCE

Computer-Aided Analysis Many equations are so complex that they require computer analysis to enable us to understand their implications. An example is the physics of the formation of the planets and the Sun. Observations then lead to refined computer models.

The traditional theory of giant planet formation, called the **core-accretion model**, posits that while the inner regions of the solar system were heating up, temperatures in the outer regions of the solar nebula remained quite cool. Ice and ice-coated dust, along with hydrogen and helium gas, were able to survive in these cooler regions. The outer planets—Jupiter, Saturn, Uranus, and Neptune—began to form from the accretion of planetesimals. For each, a rocky core even bigger than Earth apparently served as a "seed," and for about a million years the core became coated with additional rock, gas, and ice. When the gas-rich envelope became as massive as the rocky core, the gas and ice began to accumulate at a runaway pace as the growing planet's gravitational attraction pulled more of it onto the surface.

Thereafter, the envelope pulled in most of the gas and ice in its vicinity, creating a shell of water surrounded by a huge volume of hydrogen and helium.

The outer planets contain different amounts of hydrogen, helium, methane, ammonia, and water. Many of their moons are also partially composed of these low-density substances. Discovered in 1930 by American astronomer Clyde Tombaugh, Pluto was originally listed as a planet, a title it lost in 2006. It is now called a *dwarf planet*. The difference between dwarf planets and planets is that dwarf planets do not have enough gravitational force (in other words, mass) to clear their neighborhoods of other orbiting debris. See *Appendix I: Changing Pluto's Status as a Planet* for details of the change in Pluto's classification.

Since the 1980s, a second theory, called the **gravitational instability model** and based on complex computer simulations has presented a completely different view of the formation of the giant planets. These simulations indicate that giant planets condensed into existence within a hundred thousand years without massive terrestrial (Earthlike) cores acting as seeds of their formation.

In systems such as our own solar system, it is likely that the core-accretion model is closer to being correct, but as we will see shortly, many stars have giant planets much closer to them than Mercury is to the Sun. For these systems, the core-accretion model works too slowly, whereas the gravitational instability model may allow such planets to form before the gas near the star has evaporated.

During the millions of years that the inner planets were forming, the Sun was also evolving. Throughout this time, the temperature and pressure at the center of the contracting protosun continued to increase. Finally, its core reached about 10^7 K, which is hot enough to start hydrogen fusion. This process stopped the protosun's collapse, and the Sun was born. Fusion continues today and provides Earth with the energy necessary to sustain life.

Sunlike stars take approximately 100 million years to form from a nebula and settle down, which means that the Sun probably became a full-fledged star before the accretion of the inner planets was completed. Radiation from the Sun heated and disbursed the remaining gases in the solar system, thereby limiting the sizes of the planets and preventing new planetesimal formation.

Because the planets formed in a disk, they should all orbit in more or less the same plane as Earth (that is, nearly in the plane of the ecliptic). Observations bear out this conclusion. The angles of the orbital planes of the other planets with respect to the ecliptic are called their **orbital inclinations**, all of which are 7° or less (Figure 14-6).

Furthermore, according to this model, all of the planets should orbit the Sun in the same "sense" as does Earth. Traveling way above Earth's North Pole and looking down at Earth's orbit, we see that our planet is going counterclockwise around the Sun. All the other planets also orbit counterclockwise as observed from the same vantage point.

To find out more about the solar system's origins, astronomers study the leftover interplanetary debris: asteroids, meteoroids, and comets. Many of these bodies are thought to be virtually unchanged remnants of the solar system's formation. According to radioactive dating of the oldest debris ever discovered from space (see *An Astronomer's Toolbox 4-2*), the solar system was roughly in the form we know today just under 4.6 billion years ago.

Most of the moons in the solar system are planetesimals captured by the gravitational attractions of the planets. The larger ones that orbit Jupiter and Saturn probably formed from disks of gas and dust that orbited those worlds analogous to the disk that orbited the Sun in which the planets coalesced. One moon, ours, was created as the result of a collision between Earth and a Mars-sized body in a highly elliptical orbit (see Chapter 15).

Impacts dominated the early surface development of the planets and moons for hundreds of millions of years. Young Earth, along with every other body that orbited the Sun, was pounded by countless strikes, creating scars, called **craters**. Although most of the craters on our planet's surface have been erased, as we will study in the next chapter, we can still see them on our Moon (Figure 14-7). Indeed, the Moon's virtually airless environment has

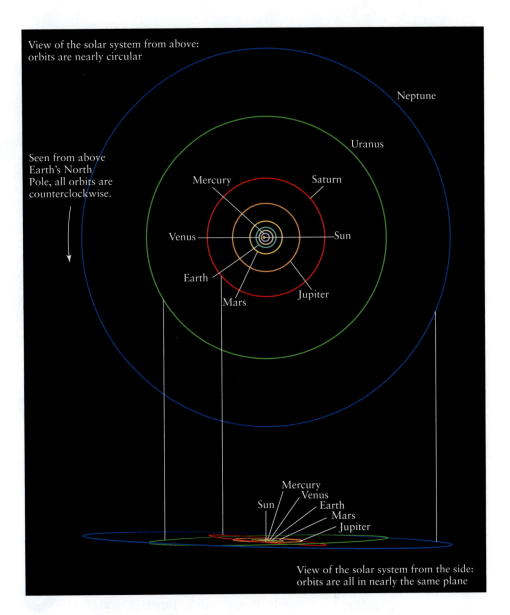

FIGURE 14-6 The Solar System This scale drawing shows the distribution of planetary orbits around the Sun. All orbits are counterclockwise because the view is from above Earth's North Pole. The four terrestrial planets are located close to the Sun; the four giant planets orbit at much greater distances. Seen from above the disk of the solar system, most of the orbits appear nearly circular. Mercury has the most elliptical orbit of any planet.

preserved important information about the early history of the solar system. (*Note* that we capitalize the word *moon* when we are referring only to Earth's Moon.)

Radioactive dating of Moon rocks brought back by the Apollo astronauts indicates that the rate of impacts declined dramatically about 3.8 billion years ago. Since that time, impact cratering has proceeded at a very low level. Thus, most of the craters on the Moon and planets were formed during the first 800 million years of the solar system's history, as the young planets pulled down rocky debris left over from the formative years of the solar nebula.

RIVUXG

FIGURE 14-7 *Our Moon* This photograph, taken by astronauts in 1972, shows thousands of craters produced by impacts of leftover rocky debris from the formation of the solar system. Age-dating of lunar rocks brought back by the astronauts indicates that the Moon is about 4.5 billion years old. Most of the lunar craters were formed during the Moon's first 700 million years of existence, when the rate of bombardment was much greater than it is now. (NASA)

RIVUXG

WEB LINK 14.2

FIGURE 14-8 *An Asteroid* This picture of the asteroid Gaspra was taken in 1991 by the *Galileo* spacecraft on its way to Jupiter. The asteroid measures 12 × 20 × 11 km. Millions of similar chunks of rock orbit the Sun between the orbits of Mars and Jupiter. (NASA)

Conditions developed on the young Earth, especially the presence of liquid water, so that biological evolution could begin to occur there. Strong evidence indicates that liquid water once existed on Mars and may still exist inside it and inside Jupiter's moons Europa and Ganymede, raising the hotly, debated question of whether simple life evolved on those worlds, too.

What two elements comprise most of Saturn's outer layer?

14-4 Minor debris from the formation of the solar system still exists

In addition to the eight planets and their moons, many other objects that still orbit the Sun were created when the solar system formed. Between the orbits of Mars and Jupiter are millions of **asteroids**, bodies composed mostly of rock and metal that are typically less than a kilometer across. This region is called the **asteroid belt**. The largest asteroid, Ceres, has a diameter of about 900 km. The next largest, Pallas and Vesta, are each about 500 km in diameter. Still smaller ones are increasingly numerous. Over one hundred thousand kilometer-sized asteroids have been observed. A close-up picture of the asteroid Gaspra is shown in Figure 14-8.

Other asteroids have highly elliptical orbits that take them across the paths of some of the planets. Some asteroids have their own moons, such as the asteroid Ida, which is orbited by the heavily cratered Dactyl.

WEB LINK 14.3

Pieces of rocky and metallic debris even smaller than asteroids are called **meteoroids**. Typical meteoroids are boulder-sized or smaller and apparently exist throughout the disk of the solar system, although most are in the asteroid belt. Chemical studies show that many meteoroids are small fragments of asteroids that broke off when larger bodies collided.

Beyond the orbit of Neptune lie hundreds of billions of chunks of rock and ice, called **comets**, or more accurately, *comet nuclei*. Many of these comet nuclei orbit in a doughnut-shaped region beyond Neptune's orbit, called the *Kuiper belt*, which is centered on the ecliptic. They are called *Kuiper Belt Objects,* or *KBOs.* Others are believed to be distributed even farther out from the Sun in a spherical distribution, called the *Oort comet cloud.* Many comet nuclei have highly elongated orbits that occasionally bring them close to the Sun. When this happens, the Sun's radiation vaporizes some of their ices, producing the long, flowing tails that we associate with comets (Figure 14-9).

As you are likely to have heard in the media, objects similar to Pluto have been discovered in the Kuiper belt. Like Pluto, they have more elliptical orbits than the planets. It is likely that the KBOs are composed of rock and ice. Collisions probably break up many objects in the

R I V U X G

FIGURE 14-9 A Comet The comet nucleus, the solid part of a comet, is an irregularly-shaped chunk of ice and rocky debris typically 10 km in diameter. When a comet passes near the Sun, solar radiation vaporizes some of the comet's ices, and the resulting gases and dust form one or two tails millions of kilometers long. (Dennis diCicco/CORBIS)

Kuiper belt and Oort cloud, thereby creating the comet nuclei and sending some of them into the inner solar system to become comets. Examples of especially large KBOs that have been observed are Pluto, Eris, Quaoar (pronounced *kwa-o-ar*), and Sedna.

COMPARATIVE PLANETOLOGY

There are two ways of studying objects in the solar system. One is to compare and contrast one feature of all similar objects, such as all of the atmospheres or all of the surfaces of planets, and then move on to the next feature. This has the advantage of showing the "big picture" for each feature as it arises. The other approach is to take an object and explore all of its properties before moving on to the next object. This has the advantage of connecting all of the properties of each object directly to each other, so that you do not, say, connect the atmosphere of Venus with the surface of Mercury. Because both approaches are valuable, we do both.

14-5 Comparisons among the eight planets show distinct similarities and significant differences

The planets that emerged from the accretion of planetesimals in the young solar system were Mercury,

TABLE 14-1 Orbital Characteristics of the Planets

	Average distance from Sun		Orbital period
	(AU)	(10^6 km)	(year)
Mercury	0.39	58	0.24
Venus	0.72	108	0.62
Earth	1.00	150	1.00
Mars	1.52	228	1.88
Jupiter	5.20	778	11.86
Saturn	9.54	1427	29.46
Uranus	19.19	2871	84.01
Neptune	30.06	4497	164.79

Venus, Earth, Mars, Jupiter, Saturn, Uranus, and Neptune. Before exploring the properties of the individual planets in the upcoming chapters, it is instructive to compare a variety of their orbital and physical properties.

Orbits Table 14-1 lists some orbital characteristics of the eight planets. As shown in Figure 14-6, the four inner planets—Mercury, Venus, Earth, and Mars—are crowded close to the Sun. In contrast, the orbits of the four large outer planets—Jupiter, Saturn, Uranus, and Neptune—are widely spaced at greater distances from it. Kepler's laws (see Chapter 2) showed us that all of the planets have elliptical orbits. Even so, most of their orbits are nearly circular. The exception is Mercury, whose orbit is noticeably oval.

> How are asteroids related to many of the meteoroids in the solar system?

Size The planets fall into three size groups (Figure 14-10). The four inner planets form one group, Jupiter and Saturn form the second group, and Uranus and Neptune are the third, intermediate-sized group. Jupiter is the largest planet, with a diameter about 11 times bigger than that of Earth, whereas Mercury, with about a third Earth's diameter, is the smallest planet. Indeed, two moons, Ganymede and Titan, are larger than Mercury. Neptune and Uranus are each about 4 times larger in diameter than Earth, the largest of the four inner planets. The diameters of the planets are given in Table 14-2.

Mass Mass, a measure of the total amount of matter an object contains, is another characteristic that distinguishes the inner planets from the outer planets. The four inner planets have small masses compared to the giant outer ones. Again, first place goes to Jupiter, whose mass is 318 times greater than Earth's (see Table 14-2).

Density Size and mass can be combined in a useful way to provide information about the chemical composition

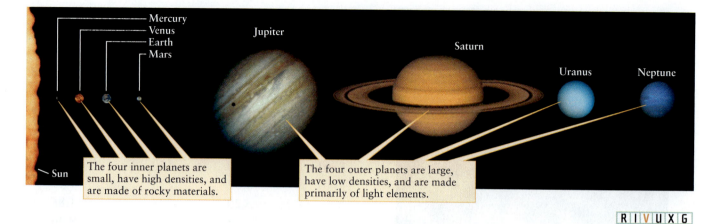

The four inner planets are small, have high densities, and are made of rocky materials.

The four outer planets are large, have low densities, and are made primarily of light elements.

R I V U X G

FIGURE 14-10 The Sun and the Planets This figure shows the eight planets drawn to size scale in order of their distances from the Sun (distances not to scale). The four planets that orbit nearest the Sun (Mercury, Venus, Earth, and Mars) are small and made of rock and metal. The next two planets (Jupiter and Saturn) are large and composed primarily of hydrogen and helium. Uranus and Neptune, also mostly hydrogen and helium, are intermediate in size and contain much water. (Calvin J. Hamilton and NASA/JPL)

of a planet (or any other object). Matter composed of heavy elements, such as iron or lead, has more atoms packed into the same volume than does matter composed of light elements, such as hydrogen, helium, or carbon. Therefore, objects made primarily of heavier elements have a greater average density than objects composed primarily of lighter elements, where average density is given by the equation

$$\text{Average density} = \frac{\text{Total mass}}{\text{Total volume}}$$

The chemical composition (kinds of elements) of any object, therefore, determines its **average density**— how much mass the object has in a unit of volume. For example, a kilogram of copper takes up less space (is denser) than a kilogram of water, even though both have the same mass (Figure 14-11). Average density is

expressed in kilograms per cubic meter. To help us grasp this concept, we often compare the average densities of the planets to the density of something familiar, namely liquid water, which has a density of 1000 kg/m^3.

The four inner planets have high average densities compared to water (see Table 14-2). In particular, the average density of Earth is 5520 kg/m^3. Because the density of typical surface rock is only about 3000 kg/m^3, its high density implies that Earth contains a large amount of material inside it, namely iron and nickel, that is much denser than surface rock. The fact that the inner four planets have densities similar to Earth's (in Latin, *terra*) means that they are composed of similar chemicals. Consequently, these are called **terrestrial planets**. We will explore them in Chapters 15 and 16.

In sharp contrast, Jupiter, Saturn, Uranus, and Neptune have relatively low average densities (see Table 14-2). Indeed, Saturn's average density is less

TABLE 14-2 Physical Characteristics of the Planets

	Diameter		Mass		Average density
	(km)	(Earth = 1)	(kg)	(Earth = 1)	(kg/m^3)
Mercury	4878	0.38	3.3×10^{23}	0.06	5430
Venus	12,100	0.95	4.9×10^{24}	0.81	5250
Earth	12,756	1.00	6.0×10^{24}	1.00	5520
Mars	6786	0.53	6.4×10^{23}	0.11	3950
Jupiter	142,984	11.21	1.9×10^{27}	317.94	1330
Saturn	120,536	9.45	5.7×10^{26}	95.18	690
Uranus	51,118	4.01	8.7×10^{25}	14.53	1290
Neptune	49,528	3.88	1.0×10^{26}	17.14	1640

RIVUXG

FIGURE 14-11 *The Volumes of Objects with Different Densities* All of the objects in this image have the same mass (total number of particles). However, the chemicals from which they form have different densities (number of particles per volume), so they all take up different amounts of space (volume). (2002 Richard Megna/Fundamentals Photographs, NYC)

than that of water. Their low densities support the belief that the giant outer planets contain significant amounts of the lightest elements, hydrogen and helium, as discussed earlier in this chapter. We will see that Jupiter and Saturn share similar compositions, primarily hydrogen and helium, while Uranus and Neptune both contain vast quantities of water as well as much hydrogen and helium. These four are sometimes called *Jovian planets* (the Roman god Jupiter was also called Jove), because they all have relatively low densities and high masses compared to Earth's. However, the differences in the chemistries of Jupiter and Saturn compared to those of Uranus and Neptune make the name *giant planets* (giant compared to Earth) more appropriate in describing these four. We will explore them as two groups in Chapter 17.

Spectra   Further information about the chemical composition of bodies in the solar system is obtained from their spectra. For solar system objects, this radiation is primarily sunlight scattered off the surface or clouds that surround each object. As we saw in Chapter 4, the spectra provide us with details of an object's surface or atmospheric chemical composition. From spectra, we can confirm that the outer layers of the giant planets are primarily hydrogen and helium, and that the surface of Mars is rich in iron oxides, among many other things.

Albedo   The surfaces or the upper cloud layers of the various planets scatter (send in all directions) different amounts of light. The fraction of incoming light returning directly into space is called a body's **albedo.** An object that scatters no light has an albedo of 0.0; for example, powdered charcoal has an albedo of nearly 0.0. An object that scatters all of the light that strikes it (a high-quality mirror comes close) has an albedo of 1.0. The albedo multiplied by 100 gives the percentage of light directly scattered off that body. Three planets (Mercury, Earth, and Mars) have albedos of 0.37 or less. Such albedos result from dark, dry surfaces, or from a mixture of light (water and clouds) and dark (continents) surfaces exposed to space. In contrast, Venus, Jupiter, Saturn, Uranus, and Neptune have albedos of 0.47 or more. High albedos like these imply bright materials exposed to space. The albedos of Venus, Jupiter, Saturn, Uranus, and Neptune are high because they are all completely enshrouded by clouds.

Moons   Every planet, except Mercury and Venus, has moons. There are at least 166 planetary moons known to exist in the solar system (up from 99 known in 2001), and more are still being discovered. Unlike our Moon, most moons are irregularly shaped, more like potatoes than spheres. We will discover that there is as much variety among moons as there is among planets.

> Why is Earth's albedo continually changing?

PLANETS OUTSIDE OUR SOLAR SYSTEM

Understanding the process by which the solar system was formed is being aided by observations of different stages of planet formation occurring around other stars. Although astronomers have been searching for planets outside the solar system for centuries, it was only in the 1980s that observations leading to their discovery were first made.

FIGURE 14-12 A Circumstellar Disk of Matter (a) An edge-on disk of material 225 billion km (140 billion mi) across that orbits the star Beta Pictoris (blocked out in this image) 50 light-years from Earth. Twenty million years old, this disk is believed to be composed primarily of iceberg-like bodies that orbit the star. (b) The central region of the disk is clearly warped in this image. This distortion is believed to be due to the gravitational tug of at least one planet that orbits Beta Pictoris. (c) The solar system drawn to the same scale as the image in (b). (d) Hubble visible light image of the disks of Beta Pictoris. The smaller disk is believed to have been formed by the gravitational pull of a roughly Jupiter-mass planet in that orbit. Because the secondary disk is so dim, the labeling for this image is added in (e). (a: Al Schultz (CSC/STScI); b: Sally Heap (GSFC/NASA); c: NASA; d & e: NASA, ESA, and D. Golimawski)

14-6 Planets that orbit other stars have been discovered

As indicated earlier in the chapter, astronomers have discovered stars in early stages of formation that are surrounded by proplyds: disks of gas and dust. These disks are similar to what the solar nebula was believed to have been like over four and a half billion years ago. Indirect observational evidence for the existence of planets outside the solar system, called *extrasolar planets* or *exoplanets*, first came from the distortion of proplyd disks.

If one or more planets orbit in a proplyd disk around a young star, then the planet's gravitational pull will affect the disk of gas and dust around it, causing the disk to warp or become off-center from its star. An example of serious warping can be seen in the edge-on disk surrounding the star Beta Pictoris (Figure 14-12). This star and the material that orbits it formed only 20 million years ago. In 2006, astronomers resolved (saw the details of) the second disk, labeled in Figure 14-12e). Computer models show that a Jupiter-like planet that orbits out of the plane of the big disk would attract debris from that disk, thereby forming the smaller disk in the plane of its orbit. Beta Pictoris also has millions of comet nuclei, which were discovered in 2001. By studying such systems as Beta Pictoris, we can also begin addressing such questions as whether Earth acquired its water from comets in the young solar system.

The interaction between planets and disks is seen in other systems. The nearby star Fomalhaut has an off-centered disk of gas and dust. This star and its entourage are only some 200 million years old. Observations suggest that this distortion (Figure 14-13) is caused by a planet tugging on the gas as it orbits. The pull of the planet is so great that the center of the disk is displaced 15 AU from the star.

Along with disks of gas and dust and large numbers of comets, asteroids have also been discovered orbiting other stars. The first system discovered with asteroids is Zeta Leporis, a young star just 70 light-years from Earth. That asteroid belt, located about the same distance from its star as is our asteroid belt, is estimated to contain about 200 times as much mass as ours. The pieces of asteroid debris were not seen directly but were inferred by the existence of hot (room-temperature) dust surrounding Zeta Leporis that is being generated by the collisions of asteroids.

Observing extrasolar planets directly as they orbit other stars is extremely challenging because even high-albedo planets like Jupiter are typically very dim compared to the bodies that they orbit. Although no planets have yet been so observed, one object labeled 2M1207b *almost* qualified. This body (Figure 14-14), first observed in 2004, has a mass between 5 and 8 times the mass of Jupiter. It is orbiting a slightly bigger object, which itself is too small to sustain fusion in its core. Hence, neither object shines very brightly. Observations indicate that both objects were formed together, rather than 2M1207b forming from a disk of gas and dust orbiting the bigger one. Bodies that form simultaneously, even if they have masses too small to be stars, are not considered planets!

RIVUXG

FIGURE 14-13 Off-centered Disk The star Fomalhaut, blocked out so that its light does not obscure the disk, is surrounded by gas and dust in a ring whose center is separated from the star by 15 AU, nearly as far as Uranus is from the Sun. This offset is believed to be due to the gravitational effects of a giant planet orbiting Fomalhaut. This system is 25 light years from Earth. The dimmer debris in that system and between us and it scatters light that is considered "noise" in such images. (NASA, ESA, P. Kalas and J. Graham [University of California Berkeley], and M. Clampin)

Larger body 2M1207

distance = 55 AU

Smaller body
2M1207b

R I V U X G

FIGURE 14-14 Image of an Almost-Extrasolar Planet This infrared image, taken at the European Southern Observatory, shows the two bodies 2M1207 and 2M1207b. Neither is quite large nor massive enough to be a star, and evidence suggests that 2M1207b did not form from a disk of gas and dust surrounding the larger body, hence it is not a planet. This system is about 170 ly from the solar system in the constellation Hydra. (ESO/VLT/NACO)

Since 1995, most extrasolar planets have been discovered by their effects on the stars they orbit (Figure 14-15a). Some planets are discovered by measuring changes in the radial velocities of their stars (defined in Section 4-7), which are calculated from the Doppler shifts of the stars' spectra (Figure 14-15b). This motion toward or away from Earth is created by the gravitational pull of the planet. The Doppler shift changes cyclically as the star and planet circle their common center of mass, and the length of time of one cycle is the period of the planet's orbit. Modern spectroscopic techniques are so good that we can detect stars approaching or receding at speeds as low as 4 km/h (2.5 mph).

Other planets are discovered from variations in a star's proper motion, or motion among the background stars (also defined in Section 4-7). A planet's attraction causes its star to deviate from motion in a straight line. This *astrometric* method of discovery looks for just such a wobble (Figure 14-15c).

Yet other planets are located by observing changes in the brightnesses of their stars. As a planet passes or transits between us and its star, it partially eclipses the star (Figure 14-15d). In 2003, astronomers used this *transit method* to detect a planet only one-sixteenth as far from its star as Mercury is from the Sun. That planet orbits once every 28 h, 33 min.

Using this method, astronomers can also study the spectrum of a planet's atmosphere. As a gas giant passed in front of the star HD 209458 (Henry Draper catalog

star number 209458), hydrogen, sodium, oxygen, and carbon in the planet's atmosphere absorbed certain wavelengths of starlight. The Hubble Space Telescope further revealed that the outer layers of this planet are being heated so much that at least 10,000 tons of hydrogen are evaporating into space from it every second. At this lower limit of mass loss, the planet has lost about 0.1% of its mass over its lifetime of 5 billion years.

The transit method has also revealed that some extrasolar planets have much more high-density material in their cores than do our giant planets. The changing intensity of light from the star HD 149026 showed that the planet passing in front of it, which has a mass of 115 Earth masses, is much smaller than other planets of that mass. Calculations reveal that about 70 Earth masses (roughly two thirds) of that planet are rocky material in its core. For comparison, Jupiter's terrestrial core is about 4% of its total mass.

As you can see, many giant planets orbit surprisingly close to their stars (Figure 14-16), much closer than Earth to the Sun. If the core accretion model of how giant planets form is correct, then they are spiraling inward after having formed farther away from their stars than they are now.

In 2004, astronomers began finding extrasolar planets by using a property of space that Albert Einstein discovered. His theory of general relativity, which we explored in detail in Chapter 9, shows that matter warps the surrounding space, causing, among other things, passing light to change direction. This is quite analogous to how light passing through a lens changes direction and is focused—light can also be focused by gravity, an effect called **microlensing** (or *gravitational microlensing*). Figure 14-17 shows how a star with a planet, passing between Earth and a distant star, can focus the light from the distant star, causing it to appear to change brightness. This occurs twice, once as the distant star's light is focused toward us by the closer star and again when the light is focused toward us by the planet.

The vast majority of planets that have been discovered orbiting stars that shine like the Sun have masses ranging between a half and a few times the mass of Jupiter (see, for example, Figure 14-16). The smallest-mass planet known with confidence has about twice the mass of our Moon. However, it, along with at least two other planets, orbits the remnant of a star that exploded, destroying any life that may have existed on its planets. The two other planets in that system each have about 4 times Earth's mass.

In 2005, microlensing was used to detect a terrestrial planet orbiting a shining star. This planet has five and a half Earth's masses. A similar-mass planet, Gliese 581c, was discovered in 2007. While these two worlds are likely to be composed of rock and metal, with water on their surfaces, neither of the stars they orbit are Sunlike. Both stars are so cool compared to the Sun that

4

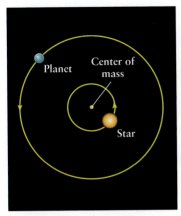

a. A star and its planet

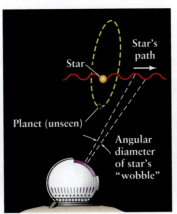

b. The radial velocity method

c. The astrometric method

d. The transient method

FIGURE 14-15 Detecting Planets that Orbit Other Stars (a) A planet and its star both orbit around their common center of mass, always staying on opposite sides of that point. The planet's motion around the center of mass often provides astronomers with the information that a planet is present. (b) As a planet moves toward or away from us, its star moves in the opposite direction. Using spectroscopy, we can measure the Doppler shift of the star's spectrum, which reveals the effects of the unseen planet or planets. (c) If a star and its planet are moving across the sky, the motion of the planet causes the star to orbit its center of mass. This motion appears as a wobbling of the star across the celestial sphere. (d) If a planet happens to move in a plane that takes it across its star (that is, the planet transits the star), as seen from Earth, then the planet will hide some of the starlight, causing the star to dim. This change in brightness will occur periodically and can reveal the presence of a planet.

to be inhabitable, the planets would have to be closer to their stars than Mercury is to the Sun. In those situations, the planets would be in synchronous rotation, like the Moon is around Earth. One side of each planet would be permanently frozen, while the water on the other side would be permanently vaporized.

In 1999, astronomers began observing some stars wobbling in ways too complicated to be caused by a single planet, implying the presence of several planets. In a multiple-planet system, each planet contributes a tug on the star. By combining the effects of two or more planets, the observed pattern of the star's motion (Doppler or astrometric) can be reproduced. The first star discovered with multiple planets was Upsilon Andromedae (Figure 14-18). One of its planets, labeled *Upsilon Andromedae B*, is 10 times closer to its star than Mercury is to the Sun. Using the infrared-sensitive Spitzer Space Telescope, astronomers have determined that the difference in temperature between the daytime and nighttime sides of Upsilon Andromedae B is 1400 K. The star 55 Cancri A, in the constellation Cancer, has at least four planets. So far, 20 stars have been observed with two or more planets that orbit them; several are depicted in Figure 14-16.

A dramatic consequence of the inward spiraling of planets was discovered in 2001, when the remnants of at least one planet were discovered in the atmosphere of a star that still has at least two other planets in orbit around it. This star's atmosphere contained a rare form of lithium that is found in planets, but which is destroyed in stars within 30 million years after they form. The presence of this isotope, ^{6}Li, means that at least one planet spiraled so close to the star that it was vaporized.

14-7 Dust and extrasolar planets orbit a breathtaking variety of stars

While some planets are orbiting stars similar to the Sun, many planets orbit burned-out remnants of stars and even orbit pairs or trios of stars. Most of the planets we have found are around stars in the disk of the Milky Way Galaxy (which is where our solar system resides), but a few have been found in large clumps of stars outside our Galaxy's disk. These clumps, called *globular clusters*, contain very old stars that formed shortly after the universe began. Because it is in a globular cluster, the oldest known planet dates back to the cluster's formation, 13 billion years ago.

The vast majority of the more than 310 extrasolar planets discovered so far have highly elliptical orbits, with eccentricities up to at least to $e = 0.71$ (see Section 2-5). Star systems with massive planets on highly eccentric orbits are unlikely to have life-sustaining Earthlike planets, because the changing location of the massive planet is likely to prevent smaller planets from staying in stable orbits.

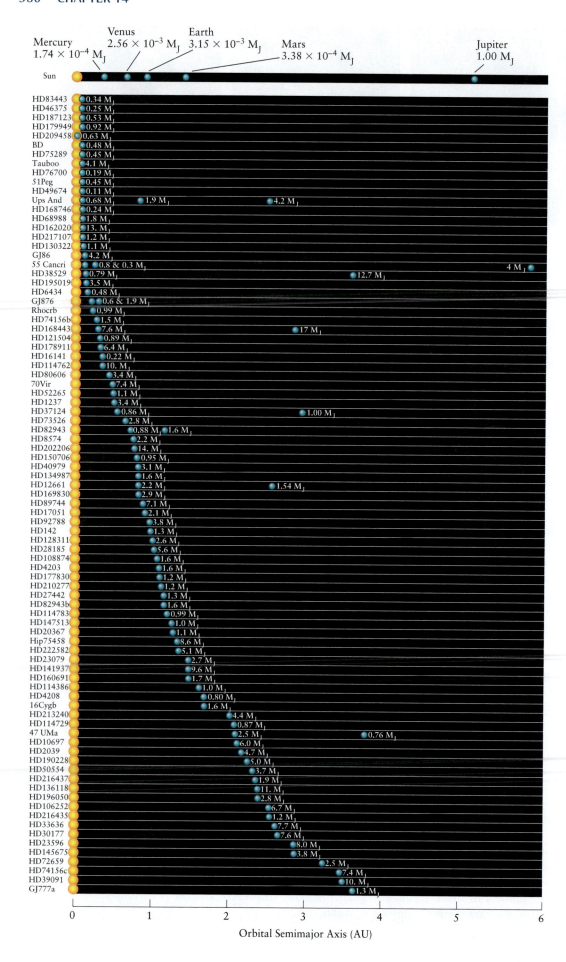

Mercury
1.74×10^{-4} M_J

Venus
2.56×10^{-3} M_J

Earth
3.15×10^{-3} M_J

Mars
3.38×10^{-4} M_J

Jupiter
1.00 M_J

Sun

Star		
HD83443	0.34 M_J	
HD46375	0.25 M_J	
HD187123	0.53 M_J	
HD179949	0.92 M_J	
HD209458	0.63 M_J	
BD	0.48 M_J	
HD75289	0.45 M_J	
Tauboo	4.1 M_J	
HD76700	0.19 M_J	
51Peg	0.45 M_J	
HD49674	0.11 M_J	
Ups And	0.68 M_J 1.9 M_J 4.2 M_J	
HD168746	0.24 M_J	
HD68988	1.8 M_J	
HD162020	$13.$ M_J	
HD217107	1.2 M_J	
HD130322	1.1 M_J	
GJ86	4.2 M_J	
55 Cancri	0.8 & 0.3 M_J 4 M_J	
HD38529	0.79 M_J 12.7 M_J	
HD195019	3.5 M_J	
HD6434	0.48 M_J	
GJ876	0.6 & 1.9 M_J	
Rhocrb	0.99 M_J	
HD74156b	1.5 M_J	
HD168443	7.6 M_J 17 M_J	
HD121504	0.89 M_J	
HD178911	6.4 M_J	
HD16141	0.22 M_J	
HD114762	$10.$ M_J	
HD80606	3.4 M_J	
70Vir	7.4 M_J	
HD52265	1.1 M_J	
HD1237	3.4 M_J	
HD37124	0.86 M_J 1.00 M_J	
HD73526	2.8 M_J	
HD82943	0.88 M_J 1.6 M_J	
HD8574	2.2 M_J	
HD202206	$14.$ M_J	
HD150706	0.95 M_J	
HD40979	3.1 M_J	
HD134987	1.6 M_J	
HD12661	2.2 M_J 1.54 M_J	
HD169830	2.9 M_J	
HD89744	7.1 M_J	
HD17051	2.1 M_J	
HD92788	3.8 M_J	
HD142	1.3 M_J	
HD128311	2.6 M_J	
HD28185	5.6 M_J	
HD108874	1.6 M_J	
HD4203	1.6 M_J	
HD177830	1.2 M_J	
HD210277	1.2 M_J	
HD27442	1.3 M_J	
HD82943b	1.6 M_J	
HD114783	0.99 M_J	
HD147513	1.0 M_J	
HD20367	1.1 M_J	
Hip75458	8.6 M_J	
HD222582	5.1 M_J	
HD23079	2.7 M_J	
HD141937	9.6 M_J	
HD160691	1.7 M_J	
HD114386	1.0 M_J	
HD4208	0.80 M_J	
16Cygb	1.6 M_J	
HD213240	4.4 M_J	
HD114729	0.87 M_J	
47 UMa	2.5 M_J 0.76 M_J	
HD10697	6.0 M_J	
HD2039	4.7 M_J	
HD190228	5.0 M_J	
HD50554	3.7 M_J	
HD216437	1.9 M_J	
HD136118	$11.$ M_J	
HD196050	2.8 M_J	
HD106252	6.7 M_J	
HD216435	1.2 M_J	
HD33636	7.7 M_J	
HD30177	7.6 M_J	
HD23596	8.0 M_J	
HD145675	3.8 M_J	
HD72659	2.5 M_J	
HD74156c	7.4 M_J	
HD39091	$10.$ M_J	
GJ777a	1.3 M_J	

0   1   2   3   4   5   6
Orbital Semimajor Axis (AU)

FIGURE 14-16 Planets and Their Stars This figure shows the separation between extrasolar planets and their stars. The corresponding star names are given on the left of each line. Note that many systems have giant planets that orbit much closer than 1 AU from their stars. (M_J is shorthand for the mass of Jupiter.) For comparison, the solar system is shown at top. (California and Carnegie Planet Search)

FIGURE 14-17 Microlensing Reveals an Extrasolar Planet (a) Gravitational fields cause light to change direction. As a star with a planet passes between Earth and a more distant star, the light from the distant star is focused toward us, making the distant star appear brighter. (b) The focusing of the distant star's light occurs twice, once by the closer star and once by its planet, making the distant star change brightness. For these observations, the closer star and planet are 17,000 light-years away, while the distant star is 24,000 light-years away.

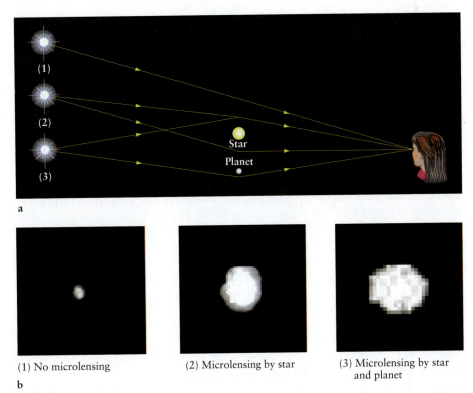

a

(1) No microlensing   (2) Microlensing by star   (3) Microlensing by star and planet

b

Given the growing number of extrasolar planets that are being discovered orbiting a wide range of stars, astronomers are beginning to make estimates for the number of Jupiterlike planets and the number of them orbiting Sunlike stars in our Milky Way Galaxy. These numbers range between a billion and 30 billion planets, of which up to 2 billion are believed to be in orbit around Sunlike stars. All of these numbers will be refined as better estimates are determined for the percentage of star systems with planets.

Although astronomers had expected to find disks of gas and dust orbiting young stars, consistent with the theory explaining the formation of the solar system, they have also begun finding disks of dusty debris orbiting stars as old as the Sun, over 4 billion years. These stars are all known to have planets, and it is most likely that the rubble in them was formed by the collisions of asteroids and comets with each other and perhaps with the planets in these systems.

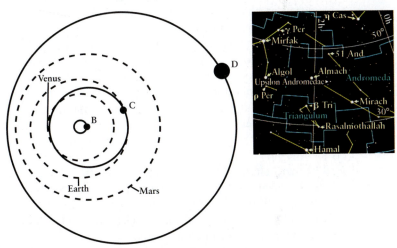

FIGURE 14-18 A Star with Three Planets (a) The star Upsilon Andromedae has at least three planets, discovered by measuring the complex Doppler shift of the star. This star system is located 44 light-years from Earth, and the planets all have masses similar to Jupiter's. (b) The orbital paths of the planets, labeled *B, C,* and *D,* along with the orbits of Venus, Earth, and Mars, drawn for comparison. (Adam Contos, Harvard-Smithsonian Center for Astrophysics)

We should start finding Earthlike planets that orbit Sunlike stars in the coming years. The problem, until recently, has been the sensitivity of our technologies. The difficulty in finding such planets is primarily that they are so dim, small, and low in mass compared to their stars. Their dimness makes it exceedingly hard for current telescopes to see the starlight they scatter compared to the monumental brightness of the stars themselves. When these planets eclipse their stars, the small sizes of the orbiting bodies make the change in the brightness of their star also very hard to detect. Likewise, the small masses of the planets compared to their stars mean that they cause very small Doppler shifts of just a few meters per second, which we are just now able to detect.

When we do start detecting terrestrial planets orbiting other stars, which will be the best candidates for supporting life? We are most likely to discover extraterrestrial life on extrasolar planets orbiting at about 1 AU from stars similar in mass (within about a factor of two) to the Sun. At this distance, such stars provide enough heat to keep water liquid, and these stars shine long enough for life to evolve. We discuss the issues related to other life in the universe in Chapter 19.

14-8 Frontiers yet to be discovered

The formations of the solar system and other planetary systems are so complex that we are continually making refinements to our models of how they started and evolved. Our understanding of the development of the solar system is likely to be greatly advanced by further study of proplyds and planets that orbit other stars. For example, observations of these systems should help us pin down the time scales involved in forming planetary systems. One of the global issues in our solar system that remains to be confirmed is how the different chemical elements became distributed at the distances they are found—why are the inner planets in our solar system terrestrial in composition, while planets even closer to other stars are Jupiter-like gas giants? Is the core-accretion model correct and does it universally explain planet formation, or do some systems form along the lines of the gravitational-instability model?

Another major challenge in solar system astronomy is identifying and determining the properties of the myriad Kuiper belt and Oort comet cloud objects. What are their compositions? Are we correct in classifying Pluto as a large KBO? Although the most poorly understood bodies are the KBOs, every other object in the solar system also has secrets to be uncovered.

The extrasolar planets and their environs present many mysteries, such as how Jupiter-mass planets came to be closer to their planets than any such bodies in our solar system. How common are Earthlike planets, and around what kinds of stars do they typically form?

Although none of the known extrasolar planets is a candidate to sustain life as we know it, we are likely to eventually discover such worlds.

We begin a detailed exploration of the solar system in Chapter 15 by examining the two bodies we know best—Earth and its Moon. By understanding them, we will be better able to make some sense of the remarkably alien neighboring worlds we encounter in the following three chapters.

SUMMARY OF KEY IDEAS

Formation of the Solar System

• Hydrogen, helium, and traces of lithium, the three lightest elements, were formed shortly after the creation of the universe. The heavier elements were produced much later by stars and are cast into space when stars die. By mass, 98% of the observed matter in the universe is hydrogen and helium.

• The solar system formed 4.6 billion years ago from a swirling, disk-shaped cloud of gas, ice, and dust, called the solar nebula.

• The four inner planets formed through the accretion of dust particles into planetesimals and then into larger protoplanets. The four outer planets probably formed through the runaway accretion of gas and ice onto rocky protoplanetary cores over millions of years, but possibly by gravitational collapse in under 100,000 years.

• The Sun formed at the center of the solar nebula. After about 100 million years, the temperature at the protosun's center was high enough to ignite thermonuclear fusion reactions.

• For 800 million years after the Sun formed, impacts of asteroid-like objects on the young planets dominated the history of the solar system.

Comparative Planetology

• The four inner planets of the solar system share many characteristics and are distinctly different from the four giant outer planets.

• The four inner, terrestrial planets are relatively small, have high average densities, and are composed primarily of rock and metal.

• Jupiter and Saturn have large diameters and low densities and are composed primarily of hydrogen and helium. Uranus and Neptune have large quantities of water as well as much hydrogen and helium.

• Pluto, once considered the smallest planet, has a size, density, and composition consistent with other Kuiper belt objects.

• Asteroids are rocky and metallic debris in the solar system, are larger than about a kilometer in diameter, and are found primarily between the orbits of Mars and Jupiter. Meteoroids are smaller pieces of such debris. Comets are debris that contain both ice and rock.

Planets Outside Our Solar System

• Astronomers have observed disks of gas and dust orbiting young stars.

• At least 310 extrasolar planets have been discovered orbiting other stars.

• Most of the extrasolar planets that have been discovered have masses roughly the mass of Jupiter.

• Extrasolar planets are discovered indirectly as a result of their effects on the stars they orbit.

WHAT DID YOU THINK?

1 *Were the Sun and planets among the first generation of objects created in the universe?* No. All matter and energy were created by the Big Bang. However, much of the material that exists in our solar system was processed inside stars that evolved before the solar system existed. The solar system formed billions of years after the Big Bang occurred.

2 *How long has Earth existed, and how do we know this?* Earth formed along with the rest of the solar system, about 4.6 billion years ago. The age is determined from the amount of radioactive decay that has occurred on Earth.

3 *What typical shape(s) do moons have, and why?* Although some moons are spherical, most look roughly like potatoes. Those that are spherical are held together by the force of gravity, pulling down high regions. Those that are potato-shaped are held together by the electromagnetic interaction between atoms, just like rocks. These latter are too small to be reshaped by gravity.

4 *Have any Earthlike planets been discovered orbiting Sunlike stars?* Not really. Most extrasolar planets are Jupiter-like gas giants. The planets similar in mass and size to Earth are either orbiting remnants of stars that exploded or, in the case of Gliese 581, a star much less massive and much cooler than the Sun.

Key Terms for Review

accretion, 369
albedo, 375
asteroid, 372
asteroid belt, 372
average density, 374
comet, 372
core-accretion model, 370

crater, 371
gravitational instability
  model, 370
meteoroid, 372
microlensing, 378
moon (natural
  satellites), 369

orbital inclination, 371
planet, 369
planetesimal, 369
protoplanet, 369
protoplanetary disks
  (proplyds), 369

protosun, 367
solar nebula, 367
solar system, 365
terrestrial planet, 374

Review Questions

1. Which of the following times is closest to the age of the universe when our solar system formed? **a.** 0 years (they formed together), **b.** a million years, **c.** 10 million years, **d.** a billion years, **e.** 10 billion years

2. Pluto is most similar in composition to which of the following objects? **a.** Eris **b.** Jupiter **c.** our Moon **d.** Earth **e.** the Sun

3. To test your understanding of the formation of the solar system, do Interactive Exercise 14-1 on the Web site. Explain what is happening in each figure. You can print out your results, if required.

4. Describe four methods for discovering extrasolar planets.

Advanced Questions

5. What two properties of a planet must be known to determine its average density? How are these properties determined?

6. How can Neptune have more mass than Uranus but a smaller diameter? Hint: See Table 14-2.

7. How would a planet orbiting one of the first-generation stars be different than planets formed today? What chemical elements would that planet contain? Could we humans exist on that world? Why or why not?

What If...

8. Earth had a highly elliptical orbit, like Mercury? What would be different about Earth and life on it?

9. The accretion process of planet formation was still going on in the solar system? How would life on Earth be different, in general, and how might we be different?

10. The solar system passed through a cloud of gas and dust that was beginning to collapse to form a new star and planet system? What might happen to Earth, and how would the passage affect the appearance of the sky?

Web Questions

11. Search the Web for information about recent observations of protoplanetary disks. What insights about the formation of the solar system have astronomers gained from these observations? Explain the evidence astronomers have, from these observations, that planets are forming in these disks.

12. In 2000, extrasolar planets with masses comparable to the mass of Saturn were first detected around the stars HD 16141 (also called 79 Ceti) and HD 46375. Search the Web for information about these planets. How do their masses compare to previously discovered extrasolar planets? Do these planets move around their stars in the same kind of orbit as Saturn follows around the Sun? If not, explain the differences.

13. Search the Web for information about the planet that orbits the star HD 209458. This planet has been detected using two methods. What are they? What have astronomers learned about this planet?

 14. Determining Terrestrial Planet Orbital Periods. Access the animation "Planetary Orbits" in Chapter 14 of the *Discovering the Universe: From the Stars to the Planets* Web site. Focus on the motions of the inner planets during the last half of the animation. Using the stop and start buttons, determine how many days it takes Mars, Venus, and Mercury to orbit the Sun once if it takes Earth approximately 365 days.

15. Search the Web for information about the cool, dim star Gliese 581. What is its mass compared to our Sun's? How bright is it (its luminosity) compared to the Sun? How many planets are known to be orbiting it? How do their masses compare to Earth's? What is special about the planet Gliese 581c?

Observing Projects

 16. You will learn about the solar system in the next five chapters. Meanwhile, the *Starry Night Enthusiast*™ planetarium software can give you an introductory overview. Open the *Starry Night Enthusiast*™ program and select the **Options** tab. Check The **Ecliptic box** in the Guides section and the **Labels** box on the Planets-Moons line of the Solar System section. Stop time advance and move the scene until you are looking westward. Press the **Sunset** button. **a.** Move the sky, examine the ecliptic, and make a list of the planets that are up. **b.** Put the mouse over the clock and increase the time by an hour using the up arrow on your computer. Search along the ecliptic again for planets and add them to your list. **c.** Repeat this procedure until you have reached sunrise. **d.** Set the date for six months from now and repeat **a–c**. How do the lists compare? Explain their similarities and differences.

 17. Many young stars are still embedded in the clouds of gas and dust from which they, and any associated planets, were formed. For example, from the Northern hemisphere, the famous Orion Nebula (Figure 3-29) is visible in the winter evening sky, while the Lagoon, Omega, and Trifid nebulae are visible in the summer night sky. Examine some of these nebulae with a telescope. Describe their appearances. Looking at each nebula, try to determine which stars in your telescope's field of view are actually associated with the nebulosity. Use the **Find** pane in *Starry Night Enthusiast*™ to help you to find these nebulae. The table below gives their coordinates (right ascension and declination; see Section 1-3).

Nebula	Right ascension	Declination
Lagoon	18^h 03.8^m	−24° 23′
Omega	18 20.8	−16 11
Trifid	18 02.3	−23 02
Orion	5 35.4	−5 27

 18. Use the *Starry Night Enthusiast*™ program to observe magnified images of at least four of the planets. Select **Favorites > Guides > Atlas** and click on the **Find** tab to display a list of the planets. If the planets are not listed, delete any words on the Q line. Click on the menu button to the left of a chosen planet's name and select **Magnify**. Describe the planet's appearance. From what you observe, state whether you are looking at the planet's surface or at the top of cloud cover and justify your assertion.

CHAPTER 15

Earth and Moon

Earth and the Moon as seen from space. (NASA)

R I V U X G

WHAT DO YOU THINK?

1 Can Earth's ozone layer, which is now being depleted, be replenished?

2 Who was the first person to walk on the Moon, and when did this event occur?

3 Do we see all parts of the Moon's surface at some time throughout the lunar cycle?

4 Does the Moon rotate and, if so, how fast?

5 What causes the ocean tides?

6 When does the spring tide occur?

Answers to these questions appear in the text beside the corresponding numbers in the margins and at the end of the chapter.

Imagine you are entering our solar system from a distant star system, looking for habitable worlds. Your spacecraft approaches the inner solar system on the opposite side of the Sun from Earth. You encounter Mars, with its thin, chilled, unbreathable atmosphere and barren desert landscapes. Not very promising. Moving much closer to the Sun, the next planet you happen to encounter, Venus, is enshrouded in corrosive clouds hiding a menacingly hot surface. Finally, moving around the Sun and heading back outward you spy Earth, orbiting our star between these two relatively forbidding worlds—one too cold, the other too hot. Ever-changing white clouds pirouette above the browns and blues of Earth's continents and oceans. Its close companion, the Moon, pockmarked by countless craters, causes those oceans to move up and down in a ceaseless rhythm. Is Earth what you are looking for?

In this chapter, we will explore Earth and the Moon, noting what makes Earth suitable for life. Our detailed and rapidly developing knowledge of Earth allows us to present material on topics that range from the magnetic fields surrounding it, to its atmosphere and climate, to the depth and composition of its crust, and the properties extending to its metal-rich core. As you will see, our knowledge about the Moon is much more limited. We will explore in the following two chapters other worlds with many different properties from Earth, leading to vastly different environments from our own.

In this chapter you will discover

- why Earth is such an ideal environment for life

- that Earth is constantly in motion inside and out

- how Earth's magnetic field helps protect us

- what made the craters on the Moon

- how the Sun and the Moon cause Earth's tides

- that both Earth and the Moon have two (different) major types of surface features

- that water ice may have been found at the Moon's poles

EARTH: A DYNAMIC, VITAL WORLD

Looking at Earth from outer space (Figure 15-1), one first notices that its clouds, water, and land combine to scatter, on average, 37% of the sunlight it receives right back into space. That amount of scattered light (an albedo of 0.37) and its variation throughout the year are unqiue among the planets in our solar system. In contrast, cloud-covered worlds, such as Venus, scatter much more light, while bone-dry worlds, such as Mars and our Moon, scatter much less. Furthermore, Earth's albedo varies, consistent with changing cloud and ice layers, as well as with different amounts of land and water facing the Sun. Water, the nearly universal solvent in which life on Earth first formed, covers about 71% of our planet's surface (A solvent is a liquid in which various substances can dissolve.)

Earth is a geologically active, ever-changing world. Earthquakes shake many regions of it. Volcanoes pour huge quantities of molten rock from just under the crust (the outer layer) onto the surface, and gas from within it vents into the oceans and atmosphere. Some mountains are still rising, while others are wearing away. Flowing water erodes topsoil and carves river valleys. Rain and snow help rid the atmosphere of dust particles—and life teems virtually everywhere, making Earth unique in the solar system.

15-1 Earth's atmosphere has evolved over billions of years

Earth's atmosphere is a key part of our planet's dynamic activity and its ability to sustain life. Essential to the existence of complex life is air that contains oxygen molecules. The air we breathe is predominantly a 4-to-1 mixture of nitrogen to oxygen, two gases that are found only in small amounts in the atmospheres of other planets. The theory of planet formation and the models that attempt to reproduce the early history of Earth indicate that this is the third atmosphere to envelop our planet.

Earth's earlier atmospheres contained no free oxygen. Composed of trace remnants of hydrogen and helium left over from the formation of the solar system, the first atmosphere did not last very long. The atoms and molecules in these gases were too light to stay near Earth. Heated by sunlight (meaning that they moved faster and faster), these particles gained enough kinetic energy to overcome Earth's gravitational pull and fly away into space.

The gases of the second atmosphere came from inside Earth, vented through volcanoes and cracks in Earth's surface. That atmosphere was composed primarily of carbon dioxide and water vapor, along with some nitrogen. There was roughly 100 times as much gas in that second atmosphere as there is in the air today. Carbon dioxide and water store a great deal of heat, helping create what is called the *greenhouse effect*, about which we will say more shortly. The high concentration of these gases early in Earth's life stored more heat than the atmosphere does today. Earth back then would have been a serious hothouse had the Sun reached its full intensity at that time. Evidence indicates it did not, and so the high concentrations of carbon dioxide and water in the air prevented the young Earth from freezing.

Oceans began forming when water precipitated out of this atmosphere, as well as from impacts of water-rich space debris, such as the comet nuclei mentioned in Chapter 14. These events occurred within 300 million years of Earth's formation. The rain absorbed and carried down lots of carbon dioxide, and the oceans eventually absorbed about half the carbon dioxide in that second atmosphere. (Water holds a good deal of carbon dioxide, as you can see by the amount of carbon dioxide fizz in a can of soda.) As life evolved in the oceans, much of that dissolved carbon dioxide was transformed into the shells of many organisms and then, as the organisms died, their carbon-rich shells sank to the ocean bottoms. Shells piled up and compressed the shells underneath them into rock, such as limestone.

Early plant life in the oceans and then on the shores removed most of the remaining carbon dioxide in the air by converting it into oxygen plus the nutrients that the plants needed to survive. The most efficient of such conversion mechanisms is photosynthesis, which today helps maintain the balance between carbon dioxide and oxygen in the air. What remained in the air were mostly nitrogen and water vapor.

Oxygen that entered the air early in the life of Earth came from oxygen-rich molecules, such as carbon dioxide, that were broken apart by the Sun's ultraviolet radiation. Then came oxygen released as a by-product of ocean plant life activity. In both cases, this oxygen did not stay in the air for long. Oxygen is highly reactive, and early atmospheric oxygen combined quickly with many elements on Earth's surface, notably iron, to form new compounds, such as iron oxides (otherwise known as rust).

About 2 billion years ago, after all of the surface minerals that could combine with oxygen had done so, the atmosphere began to fill with this gas. With the carbon dioxide gone, the air became the nitrogen-oxygen mixture that we breathe today. Although nitrogen was only a tiny fraction of that earlier atmosphere, it now dominates the thinner air of today.

Because air has mass, it is pulled down toward Earth by gravity, thereby creating a pressure on the surface. We define pressure as a force acting over some area:

$$\text{Pressure} = \frac{\text{Force}}{\text{Area}}$$

The weight of the air pushing down creates a pressure at sea level of 14.7 pounds per square inch, which is commonly denoted as 1 atmosphere (atm). The atmospheric pressure decreases with increasing altitude, falling by roughly half with every gain of 5.5 km of

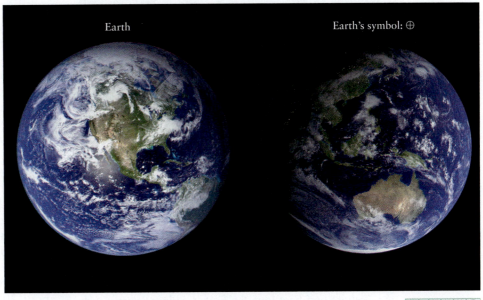

FIGURE 15-1 Views of Earth's Surface An oasis in the forbidding void of space, Earth is a world of unsurpassed beauty and variety. Ever-changing cloud patterns drift through its skies. More than two-thirds of its surface is covered with oceans. This liquid water, in combination with a huge variety of chemicals from its lands, led to the formation and evolution of life over most of the planet's surface. The symbol ⊕ is astronomers' shorthand for Earth. Earth's vital statistics can be found in Appendices E-1 and E-2 (NASA)

Earth

Earth's symbol: ⊕

R I V U X G

altitude. Seventy-five percent of the mass of the atmosphere lies within 11 km (approximately 7 mi, or 36,000 ft) of Earth's surface.

Lower Layers of the Atmosphere   All of Earth's weather—clouds, rain, sleet, and snow—occurs in the lowest layer of the atmosphere, called the **troposphere.** Commercial jets generally fly at the top of this layer to avoid having to plow through the denser air below. If Earth were the size of a typical classroom globe, the troposphere would be only as thick as a sheet of paper wrapped around its surface.

Starting at Earth's surface, atmospheric temperature initially decreases as you go upward, reaching a minimum of 218 K (−67°F) at the top of the troposphere. Above this level is the region called the **stratosphere,** which extends from 11 to 50 km (approximately 7 to 31 mi) above Earth's surface. This is the realm of the **ozone layer.**

INSIGHT INTO SCIENCE

What's in a word? The word "layer" in the term *ozone layer* suggests a narrow region, perhaps a few meters thick. Not so. The ozone layer extends in altitude over some 40 km of the atmosphere!

The Ozone Layer   The ozone layer has been much in the news in recent years. What is it, and how threatened is it? Ozone molecules (three oxygen atoms, O_3, bound together) are created in the stratosphere when short-wavelength ultraviolet radiation from the Sun occasionally splits normal oxygen molecules, O_2, into two separate oxygen atoms. Each resulting oxygen

FIGURE 15-2 Temperature Profile of Earth's Atmosphere The atmospheric temperature changes with altitude because of the way sunlight and heat from the Earth's surface interact with various gases at different heights.

atom then combines with different O_2 molecules to create ozone, O_3. Once created, ozone efficiently absorbs intermediate-wavelength solar ultraviolet rays, thereby heating the air in this layer while preventing much of this lethal radiation from reaching Earth's surface.

The troposphere (lowest part of the atmosphere) is heated primarily by Earth, so the temperature in this part of the atmosphere decreases with altitude—the farther away from the heat source of Earth, the cooler the air gets. However, because the ozone in the stratosphere stores significant heat from the Sun above it, the temperature in the stratosphere actually rises with altitude to about 285 K (50°F) at its top (Figure 15-2).

Not much ozone actually exists in the ozone layer. If all of it were compressed to the density of the air we breathe, it would be a layer only a few millimeters (about ⅛ in) thick. It is enough, however, to protect us from most of the Sun's lethal ultraviolet radiation. Ultraviolet radiation has been a double-edged sword in the history of Earth. Because Earth's early atmosphere lacked ozone, ultraviolet radiation penetrated to the planet's surface, where it provided much of the energy necessary for life to evolve in the oceans. After life developed, however, the formation of the ozone layer by the same ultraviolet radiation was essential in lessening the amount of ultraviolet radiation that reached the surface, thereby allowing life forms to leave the oceans and survive on land.

Today, the ozone layer is still vital, because the energy it absorbs would otherwise cause skin cancer and various eye diseases. In sufficiently high doses, ultraviolet radiation even damages the human immune system by destroying some of the molecules that normally repair cells in our bodies. However, over the past few decades, the ozone layer has been depleted by human-made chemicals, such as chlorofluorocarbons (CFCs). Until the past few years, ozone levels in the mid-latitudes of the northern hemisphere have been decreasing by about 4% per decade. There are some hints that the ozone is decreasing more slowly, but this is still unclear. Much more rapid and dramatic drops in the ozone layer over Earth's poles, especially over the South Pole, have been observed for more than two decades. These seasonal *ozone holes* typically extend over 24 million sq km, or 4.6% of Earth's surface area. There has been some improvement in the past few years, but the holes persist.

Fortunately, as explained earlier, sunlight creates **1** ozone. *If* current international efforts aimed at banning ozone-depleting compounds continue, the ozone layer will naturally replenish itself in a century or so, and the ozone holes will probably stop occurring.

Upper Layers of the Atmosphere   Note in Figure 15-2 that above the stratosphere lies the **mesosphere.** Atmospheric temperature again declines as you move up through the mesosphere, reaching a minimum of about 200 K (−103°F) at an altitude of about 80 km (50 mi). This minimum marks the bottom of the ionosphere, also called the **thermosphere**, above which the Sun's ultraviolet light heats and ionizes atoms, producing charged particles that reflect radio waves. In this era of burgeoning satellite usage and human space travel, the effects of the Sun's electromagnetic radiation, along with the flow of particles from it and from other sources, on Earth's atmosphere and on people in space are increasingly important. As a result, *space weather* satellites are now continually monitoring our neighborhood in space and warning about solar events that could imperil humans and machines out there.

Carbon Dioxide and the Greenhouse Effect Today

Perhaps you have had the experience of parking your car in the sunshine on a warm summer day. You roll up the windows, lock the doors, and go on an errand. After a few hours, you return to discover that the interior of your automobile has become considerably hotter than the outside air temperature (Figure 15-3).

What happened to make your car so warm? First, sunlight entered it through the windows. This radiation was absorbed by the dashboard and the upholstery, increasing their temperatures. Because they become warm, your dashboard and upholstery emit infrared radiation, which you detect as heat and which your car windows do not permit to escape. This energy is therefore trapped inside your car and absorbed by the air and interior surfaces. As more sunlight comes through the windows and is trapped, the temperature continues to increase. The same trapping of sunlight warms a greenhouse and Earth.

The greenhouse effect created by the carbon dioxide (CO_2) in the second atmosphere strongly heated the young Earth. When the carbon dioxide was dramatically reduced, that effect also decreased, but it never went away completely.

The greenhouse effect in our air works like this: All wavelengths of electromagnetic radiation from the Sun approach Earth's atmosphere. Some of this energy is scattered from clouds, land, and water right back into space (Figure 15-4a), while some is absorbed in the atmosphere, thereby heating it. Some visible light, along with some ultraviolet radiation and radio waves, penetrate to Earth's surface and is absorbed by it. This energy helps heat the planet (along with heat from inside) until the surface material can store no further energy. Behaving nearly like a blackbody, Earth radiates the excess energy, primarily as infrared radiation (where its blackbody spectrum peaks).

Unlike most visible light, which passes relatively easily through the gases in our atmosphere, infrared radiation is absorbed by certain gases, especially water, carbon dioxide, and ozone in the air. The energy from these photons further heats the atmosphere. Carbon dioxide levels are increasing in Earth's atmosphere, primarily as a result of human activity, such as the burning of fossil fuels. Increased atmospheric carbon dioxide stores more

FIGURE 15-3 The Greenhouse Effect in a Car The glass windows in this car allow visible light to enter but prevent the infrared radiation released by the car's interior from escaping. The infrared, therefore, heats the air in the car much higher than the outside air. This action also occurs in greenhouses and is called the greenhouse effect. (Tobi Zausner)

heat in the air, and the air becomes warmer. Normally, carbon dioxide is removed from the air (and oxygen added) by the process of photosynthesis in green leaves. With increased deforestation and the expansion of cities, fewer plants and trees are left, so the carbon dioxide in the air remains there longer, overheating the air and creating *global warming*. Figure 15-4b shows how the CO_2 concentration has changed over the past millennium.

The consequences of recent global warming are quite complex and serious. They include the melting of polar ice caps, rising sea levels, changes in ocean heating and circulation patterns, changes in weather patterns world-wide, and increases in undesirable weather events, such as hurricanes, tornadoes, droughts, floods, heat waves, and shoreline erosion. With scientific data on the human effects that add to global warming in hand, many nations of the world are working together to decrease its causes.

> An increase in the vegetation on Earth will have what effect on the carbon dioxide level in the atmosphere?

15-2 Plate tectonics produce major changes on Earth's surface

Large land masses called continents protrude through Earth's oceans. These land masses are regions of Earth's outermost layer, or **crust**, and are composed of relatively low-density rock that floats on denser material below it. By studying the various kinds of rocks at a particular location, geologists can deduce the history of that site.

a

b

FIGURE 15-4 The Greenhouse Effect
(a) Sunlight and heat from Earth's interior warm Earth's surface, which in turn radiates energy, mostly as infrared radiation. Much of this radiation is absorbed by atmospheric carbon dioxide and water, heating the air, which in turn increases Earth's temperature even further. In equilibrium, Earth radiates as much energy as it receives. (b) The amount of carbon dioxide in our atmosphere since 1000 AD has been determined. The increase in carbon dioxide since 1800 due to burning fossil fuels and decreases in forestation have caused a dramatic temperature increase.

For example, whether an area was once covered by an ancient sea or flooded by lava from volcanoes is readily apparent from the kinds of rocks found there at present.

The crust has distinct boundaries (for example, as shown in Figure 15-5), many occurring beneath Earth's oceans. These divisions suggest that the continents are separate bodies. Earthquakes and volcanoes often occur near the boundaries, hinting at activity taking place below Earth's crust.

Anyone who carefully examines the shapes of Earth's continents might conclude that landmasses move. Eastern South America, for example, looks like it would fit snugly against western Africa. Between 1912 and 1915, German scientist Alfred Wegener published his observations on this remarkable fit between landmasses on either side of the Atlantic Ocean, an idea proposed by Isaac Newton two centuries earlier. Wegener was inspired to advocate the theory of **continental drift**—the idea that the continents on either side of the Atlantic Ocean have moved apart.

Wegener argued that a single, gigantic supercontinent called *Pangaea* (a Greek word for "all lands") began to break up and drift apart some 200 million years ago. Initially, most geologists ridiculed Wegener's ideas. Although it was generally accepted that the continents do float on denser rock beneath them, few geologists could accept that entire continents move across Earth's face at speeds as great as several centimeters per year. Wegener and other "continental drifters" could not explain what forces would shove the massive continents around or how less dense continental rock had gotten through the dense rock of the ocean floor.

FIGURE 15-5 The Mid-Atlantic Ridge This artist's rendition of the bottom of the North Atlantic Ocean shows an unusual mountain range in the middle of the ocean floor. Called the Mid-Atlantic Ridge, these mountains are created by lava seeping up from Earth's interior along a rift that extends from Iceland to Antarctica. The black dots indicate locations of earthquakes. (Courtesy of M. Tharp and B. C. Heezen)

Then, in the mid-1950s, geologists discovered long, underwater mountain ranges. The Mid-Atlantic Ridge, for example, stretches from Iceland to Antarctica (see Figure 15-5). Careful examination revealed that molten rock from Earth's interior is being forced upward there, causing the ocean floor to slide apart on either side of the ridge. This **seafloor spreading** is pushing the Americas away from Europe and Africa at a speed of roughly 3 cm per year. The relative motion of the plates worldwide ranges from about 1.5 to 7 cm per year.

INSIGHT INTO SCIENCE

Before they are accepted, scientific theories may undergo years or decades of testing and replication. Newton speculated about continental drift in the late seventeenth century. Wegener had proposed plate tectonics by 1912. His theory was not verified until the 1950s and was not accepted until the late 1960s.

Seafloor spreading provided just the physical mechanism that had been missing from the theory of continental drift. It established Wegener's theory of crustal motion, which was developed into the modern theory of **plate tectonics.**

Earth's surface is made up of about a dozen major thick rafts of rock called *tectonic plates* that move relative to one another. These plates are floating on the slowly moving, denser matter below them. The plates are not always separate, as they are today. Sometimes they merge into a single mass called a *supercontinent*. In recent years, geologists have uncovered evidence that points to a whole succession of supercontinents that broke apart and reassembled. Pangaea is only the most recent supercontinent in this cycle, which repeats on average every 500 million years (Figure 15-6).

a   237 million years ago: the supercontinent Pangaea

b   152 million years ago: the breakup of Pangaea

c   The continents today

FIGURE 15-6 The Supercontinent Pangaea (a) The present continents are pieces of what was once a bigger, united body called Pangaea. (b) Geologists believe that Pangaea must have first split into two smaller supercontinents, which they call Laurasia and Gondwanaland. (c) These bodies later separated into the continents of today. Gondwanaland split into Africa, South America, Australia, and Antarctica, while Laurasia divided to become North America and Eurasia.

The boundaries between plates are the sites of some of the most impressive geological activity on our planet. Earthquakes and volcanoes tend to occur at the boundaries of Earth's crustal plates, where the plates are colliding (convergent boundaries), separating (divergent boundaries), or grinding against each other (slip-strike fault boundaries). If the epicenters of earthquakes are plotted on a map, the boundaries between tectonic plates (see Figure 15-5) stand out clearly. For example, the San Andreas Fault, running along the West Coast of the United States, is the slip-strike fault that occurs where the North American and Pacific plates are rubbing against each other (Figure 15-7b). The Red Sea and Gulf of Suez are expanding as Africa and the Middle East move apart (Figure 15-7c). Great mountain ranges, such as the Himalayas, are thrust up by ongoing collisions between the Indian-Australian and Eurasian plates (Figure 15-7d). During such collisions, one plate moves

FIGURE 15-7 Earth's Major Tectonic Plates (a) Earth's surface is divided into a dozen or so rigid plates that move relative to one another. The boundaries of the plates are the scenes of violent seismic and geologic activity, such as earthquakes, volcanoes, rising mountain ranges, and sinking seafloors. The arrows indicate whether plates are moving apart (←→), together (→←), or sliding past one another ↑↓. (b) Rubbing of Two Plates. The San Andreas fault, running up the west coast of North America, formed because the Pacific Plate is moving northwest along the North American Plate.

(c) The Separation of Two Plates. The plates that carry Egypt and Saudi Arabia are moving apart, leaving the trench that contains the Red Sea. This view, taken by astronauts in 1966, shows the northern Red Sea on the right. (d) The Collision of Two Plates. The plates that carry India and China are colliding. As a result, the Himalayas are being thrust upward. In this photograph, taken by astronauts in 1968, Mount Everest is one of the snow-covered peaks near the center. (a: Digital image by Peter W. Sloss, NOAA-NESDIS-NGDC; b: Craig Aurness/Corbis; c: Gemini 12, NASA; d: *Apollo 7*, NASA) downward (subducts) into Earth under the other. Where old crust subducts, we find deep trenches, such as those off the coasts of Japan and Chile.

Tectonic plate motion helps explain why Earth's oceans have so few craters today: Countless impacts occurred on the oceans, which cover most of our planet. Some of the impacting bodies were slowed or obliterated before cratering the ocean floor, but others were big enough or fast enough to hit the floors. The resulting craters in the oceans have been drawn underground by tectonic activity and thereby erased. (Most craters exposed to the atmosphere were eroded away by wind and water, with the larger ones being drawn underground by plate motion.)

What powers all of this activity? Some enormous energy source has caused shifting among land masses for billions of years. The answer to that question leads us to a scientific model of Earth's interior.

15-3 Earth's interior consists of a rocky mantle and an iron-rich core

Geologists have calculated that Earth was entirely liquid soon after its formation, about 4.6 billion years ago. The violent impact of space debris — along with energy released by the breakup of radioactive elements — heated, melted, and kept the young Earth molten. The process of radioactive atoms breaking apart, called *nuclear fission*, also creates the heat used to generate electricity in nuclear power plants. This differs from the energy generated by *nuclear fusion* in the Sun, wherein lighter atoms are fused together, as mentioned in Chapter 14 and discussed in detail in Chapter 5.

Most of the iron and other dense elements sank toward the center of the young, liquid Earth, just as a rock sinks in a pond. At the same time, most of the less dense materials were forced upward toward the surface (Figure 15-8a). This process, called **planetary differentiation**, produced a layered structure within Earth: a very dense central **core** surrounded by a **mantle** of less dense minerals, which, in turn, was surrounded by a thin crust of relatively light minerals (Figure 15-8b).

Differentiation explains why most of the rocks you find on the ground are composed of lower-density elements such as

FIGURE 15-8 Cutaway Model of Earth Early Earth was initially a homogeneous mixture of elements with no continents or oceans. (a) Molten iron sank to the center and light material floated upward to form a crust. (b) As a result, Earth has a dense iron core and a crust of light rock, with a mantle of intermediate density between them.

silicon and aluminum. Much of the iron, gold, lead, and other denser elements found on Earth today actually had to return to the surface emerging through volcanoes and other lava flows. Despite this return of heavier elements to the surface, the average density of crustal rocks is 3000 kg/m^3, considerably less than the average density of Earth as a whole (5520 kg/m^3).

To understand Earth's interior, geologists have created a mathematical model that includes the effects of differentiation, gravity, radioactivity, chemistry, heat transfer, rotation, and other factors. The model predicts that the temperature in Earth increases steadily from about 290 K on the surface to nearly 5000 K (9000°F) at the center. Radioactivity in Earth's interior helps replenish heat lost through the planet's surface. In addition to temperature, pressure also increases with increasing depth below Earth's surface. The deeper you go, the more mass presses down on you. These factors shape Earth's inner layers.

> What type of boundary exists between Africa and the Middle East?

You might think that a temperature of 5000 K would melt virtually any substance. The mantle, in particular, is composed largely of minerals rich in iron and magnesium, both of which have melting points of only slightly more than 1250 K at Earth's surface. However, the melting point of anything also depends on the pressure to which it is subjected: the higher the pressure, the higher a material's melting point. The high pressures within Earth (1.4 million atm at the bottom of the mantle) keep the mantle solid to a depth of about 2900 km and also allow for a solid iron core in the center of the planet. In other words, the inner core's pressure is so high that it forces the very hot atoms there into a solid.

Geologists have tested their models of Earth's interior by studying the response of the planet to earthquakes.

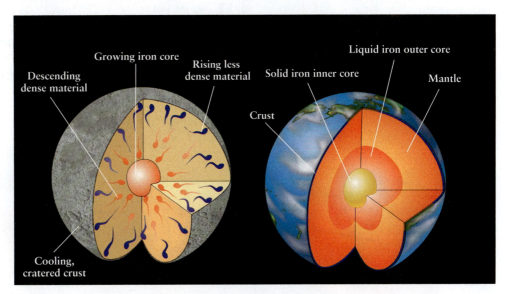

Descending dense material

Growing iron core

Rising less dense material

Cooling, cratered crust

Liquid iron outer core

Solid iron inner core

Crust

Mantle

a

As a result of differentiation, Earth has the layered structure that we see today

b

These events produce a variety of **seismic waves**, vibrations that travel through Earth either as ripples, such as ocean waves, or by compressing matter, such as sound waves. Geologists use sensitive **seismographs** to detect and record these vibratory motions. The varying density and composition of Earth's interior affects the direction of seismic waves traveling through Earth. By studying the deflection of these waves, geologists have been able to determine the structure of the planet's interior.

Analysis of seismic recordings led to the discovery that Earth's solid core has a radius of about 1250 km (775 mi), surrounded by a molten iron core with a thickness of about 2250 km (1375 mi). For comparison, the overall radius of our planet is 6378 km (3963 mi). The cores are composed of roughly 80% iron and 20% lighter elements, such as nickel, oxygen, and sulfur. The interior of our planet therefore has a curious structure: a liquid region sandwiched between a solid inner core and a solid mantle (see Figure 15-8b).

A sophisticated computer model of Earth's interior predicts that the solid inner core is rotating faster than the rest of the planet. Seismic evidence supports the model that the core goes around about 2° more per year than the surface. Refined models and observations like these are revealing more of the secrets of Earth's interior.

The tremendous heat and molten rock inside Earth provide the energy that drives the motion of the tectonic plates in Earth's crust. That heat, from the pressure of the overlying rock and from the decay of radioactive elements, affects Earth's surface today. This heat transport is accomplished by **convection**. In this process a liquid or gas is heated from below and, as a result, it expands, becomes buoyant, and therefore rises, carrying the heat it received with it. Eventually, it releases the heat, cools, condenses, and sinks back down. This motion of fluid causes a circular current, called a *convection current*, to arise. You witness convection every time you see liquid simmering in a pot. Heated from below, blobs of hot fluid move upward. The rising liquid releases its heat at its surface (Figure 15-9a).

Heated from below, the rocks of Earth's upper mantle, although solid, are weak and hot enough to have an oozing, plastic flow (like stretching Silly Putty). As sketched in Figure 15-9b, molten rock, or *magma*, seeps upward along cracks or rifts in the ocean floor, where convection currents of mantle rock are carrying plates apart. Crustal rock sinks back down into Earth where plates collide. The continents ride on top of the plates as they are pushed around by the convection currents circulating beneath them.

1 Convection moves hot water from the bottom to the top...

2 ...where it cools, moves sideways, sinks,...

3 ...warms, and rises again.

a

FIGURE 15-9 Convection and the Mechanism of Plate Tectonics (a) Heat supplied by the heating coil warms the water at the bottom of the pot. The heated water consequently expands and thereby decreases its density. This lower-density water rises (like bubbles in soda) and transfers its heat to the cooler surroundings. When the hot rising water gets to the top of the pot, it loses a lot of heat into the room, becomes denser, and sinks back to the bottom of the pot to repeat the process. (b) Convection currents in Earth's interior are responsible for pushing around rigid plates on its crust. New crust forms in oceanic rifts, where magma oozes upward between separating plates. Mountain ranges and deep oceanic trenches are formed where plates collide and crust sometimes sinks back into the interior. Note that not all tectonic plates move together or apart—some scrape against each other.

Magma   Rift   Divergent boundary   Ocean floor   Convergent boundary   Mountain range   Subduction   Continent

Convection currents   Weak, solid rock

b

Supercontinents like Pangaea apparently sow the seeds of their own destruction by blocking the flow of heat from Earth's interior. As soon as a supercontinent forms, temperatures beneath it increase. As heat accumulates, the supercontinent domes upward and cracks. Overheated molten rock wells up to fill the resulting rifts, which continue to widen as pieces of the fragmenting supercontinent are pushed apart.

Earth's molten, iron-rich interior affects Earth's exterior in another important way, one that has fewer obvious signs than tectonic plate motion. This is the creation of a magnetic field that extends through Earth's surface and out into space. As we will see next, that magnetic field results from the convection of molten metal inside Earth combined with Earth's rotation.

15-4 Earth's magnetic field shields us from the solar wind

If you have ever used a compass, you have seen the effect of Earth's magnetic field. This planetary field is quite similar to the field that surrounds a bar magnet (Figure 15-10a). Magnetic fields are created by electric charges in motion. Thus, the motion of your compass is caused by electric charges moving inside Earth. The electric wires in your house also carry moving charges called electric currents that, in turn, create magnetic fields of their own.

Many geologists believe that convection of molten iron in Earth's outer core, combined with our planet's rotation, creates the electric currents, which, in turn, create Earth's magnetic field. The details of this so-called **dynamo theory** of Earth's magnetic field are still being developed.

Magnetic fields nearly always form complete loops. (The unconnected lines in Figure 15-10b actually are completed outside of the boundaries of the figure.) The magnetic fields near Earth's south rotation pole loop out tens of thousands of kilometers into space and then return near Earth's north rotation pole. The places where the magnetic fields pierce Earth's surface most intensely are called the north and south magnetic poles. These poles move daily. Evidence in solidified lava also reveals that Earth's magnetic field completely reverses or flips (the north magnetic pole becomes the south magnetic pole, and vice versa) on an irregular schedule, ranging upward from a few tens of thousands of years. Each flip takes a few thousand years to complete. The last reversal was about 600,000 years ago. Today, the magnetic and rotation poles of Earth are

> Heated from below, gases also rise. What does this suggest about Earth's atmosphere?

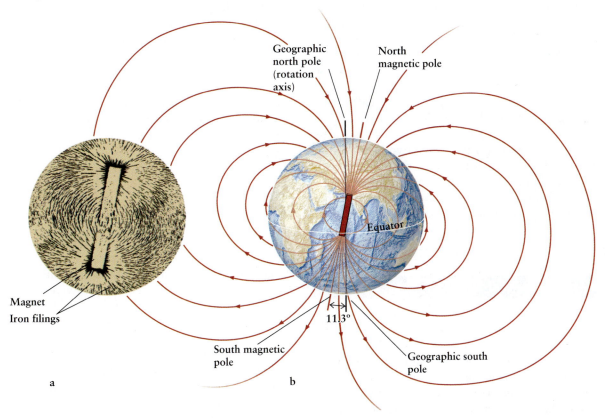

Geographic north pole (rotation axis)

North magnetic pole

Equator

Magnet

Iron filings

11.3°

South magnetic pole

Geographic south pole

a   b

FIGURE 15-10 Earth's Magnetic Field (a) The magnetic field of a bar magnet is revealed by the alignment of iron filings on paper. (b) Although Earth does not contain a bar magnet, its rotation combined with moving electric charges in its core create an equivalent field. Note that the field is not aligned with Earth's rotation axis. By convention, the magnetic pole near Earth's north rotation axis is called the *magnetic north pole* even though it is actually the south pole on a magnet! We will see similar misalignments and flipped magnetic fields when we study other planets. (a: Jules Bucher/Photo Researchers)

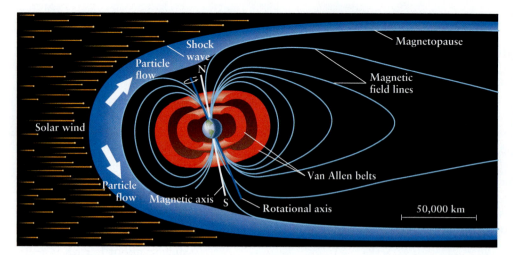

FIGURE 15-11 Earth's Magnetosphere A slice through Earth's magnetic field, which surrounds the entire planet, carves out a cavity in space that excludes charged particles ejected from the Sun, called the solar wind. Most of the particles of the solar wind are deflected around Earth by the fields in a turbulent region colored blue in this drawing. Because of the strength of Earth's magnetic field, our planet traps some charged particles in two huge, doughnut-shaped rings called the Van Allen belts (in red).

about 11.3° apart; some planets have much larger angles between their magnetic and rotation poles. Figure 15-11 is a scale drawing of the magnetic fields around Earth, which comprise Earth's magnetosphere.

Our planet's magnetic field protects Earth's surface from bombardment by particles flowing from the Sun (see Chapter 5) called the **solar wind,** an erratic flow of electrically charged particles ejected from the Sun's upper atmosphere. Near Earth, the particles in the solar wind move at speeds of approximately 400 km/s, or about a million miles per hour.

Magnetic fields can change the direction of moving charged particles. Thus, the solar wind particles heading toward Earth are deflected by our planet's magnetic field. Many particles, therefore, pass around Earth. However, the field also traps some of these charged particles in two huge, doughnut-shaped rings, called the **Van Allen radiation belts,** which surround Earth (see Figure 15-11). The region between the two belts fills up with charged particles, only to be drained of them as a result of lightning on Earth, making that region unstable. These belts, discovered in 1958 during the flight of the first successful U.S. Earth-orbiting satellite, were named after the physicist James Van Allen, who insisted that the satellite carry a Geiger counter to detect charged particles.

> What happens to the Van Allen belts when Earth's magnetic field is flipping?

INSIGHT INTO SCIENCE

Pace of Discovery Discoveries, such as the Van Allen belts, remind us of how much remains to be learned, even about our own planet. Scientists continue to make such fundamental discoveries for many reasons. For example, the technology necessary to make certain observations may just have been developed, or our equations that describe various phenomena may become more complete. Sometimes these equations are so complex that their implications are not immediately clear. Progress in science requires unquenchable curiosity and unfailing open-mindedness.

Sometimes the Van Allen belts overload with particles. The particles then leak through the magnetic fields near the poles and cascade down into Earth's upper atmosphere, usually in a ring-shaped pattern. As these high-speed, charged particles collide with gases in the upper atmosphere, the gases fluoresce (give off light) like the gases in a neon sign. The result is a beautiful, shimmering display, called the **northern lights (aurora borealis)** or the **southern lights (aurora australis),** depending on the hemisphere in which the phenomenon is observed (Figure 15-12).

Occasionally, a violent event on the Sun's surface, called a **coronal mass ejection,** sends a burst of protons and electrons straight through the Van Allen belts and into the atmosphere. These events deplete the ozone layer and cause spectacular auroral displays that can often be seen over a wide range of latitudes and longitudes. Coronal mass ejections also disturb radio transmissions and can damage communications satellites and electric transmission lines.

THE MOON AND TIDES

Earth's only natural satellite provides one of the most dramatic sights in the nighttime sky. The Moon is so large and so close to us that some of its surface features are readily visible to the naked eye. Even without a telescope, you can easily see dark gray and light gray areas that cover vast expanses of the Moon (Figure 15-13). People have dreamed of visiting our nearest neighbor in space for millennia. One of the first movies ever made (1902) was *A Trip to the Moon,* based on Jules Verne's book *From Earth to the Moon.* The wonder of actually sending people there for the first time was one of the few things that has ever caused humanity to stop its quarreling long enough to participate, even vicariously, in a great adventure. To date, 12 people have walked on the Moon. Our exploration of it begins with a study of its surface.

a

b   R I V U X G

c   R I V U X G

FIGURE 15-12 The Northern Lights (Aurora Borealis) A deluge of charged particles from the Sun can overload the Van Allen belts and cascade toward Earth, producing auroras that can be seen over a wide range of latitudes. (a) View of an aurora from the deep space satellite *Dynamics Explorer* 1. The green lines show the locations of land masses under the aurora. Auroras typically occur 100 to 400 km above Earth's surface. (b) View of aurora from space. Part of a Space Shuttle is visible at the bottom left of the picture. (c) Aurora borealis in Alaska. The gorgeous aurora seen here is mostly glowing green due to emission by oxygen atoms in our atmosphere. Some auroras remain stationary for hours, while others shimmer, like curtains blowing in the wind. (a: L. A. Frank and J. D. Crave, University of Iowa; b: MTU/Geological & Mining Engineering & Sciences; c: J. Finch/Photo Researchers, Inc.)

WEB LINK 15.4

R I V U X G

WEB LINK 15.5

FIGURE 15-13 The Moon Our Moon is one of seven large satellites in the solar system. The Moon's diameter of 3476 km (2160 mi) is slightly less than the distance from New York to San Francisco. This photograph is a composite of first-quarter and last-quarter views, in which long shadows enhance the surface features. (NASA)

15-5 The Moon's surface is covered with craters, plains, and mountains

WEB LINK 15.6

Perhaps the most familiar and characteristic features on the Moon are its craters (Figure 15-14a). As discussed in Chapter 14 on the formation of the solar system, all lunar craters that we have examined are the result of bombardment by meteoritic material (rather than volcanic activity). Nearly all of these craters are circular, indicating that they were not merely gouged out by moderate-speed rocks, which would typically create oval craters. Instead, the high-speed (upward of 180,000 km/h or 112,000 mi/h) collisions with the Moon's surface vaporized the rapidly moving debris, often with the power of a large nuclear bomb. The resulting explosions of hot gas and rubble produced the round-rimmed craters that we observe today (Figures 15-14 b, c, and d).

Meteoritic impact causes material from the crater site to be ejected onto the surrounding surface. This pulverized rock is called an **ejecta blanket**. You can see some of the more recent ejecta blankets, which are lighter-colored than the older ones (see Figures 15-13 and 15-14). The ejecta blankets darken with time as their surfaces roughen from impacts by the solar wind and other particles from space. Therefore, the lightness of its ejecta is one clue astronomers use to determine how long ago a crater formed.

Craters larger than about 20 km often form *central peaks* (see Figure 15-14). These occur because the impact

a

R I V U X G

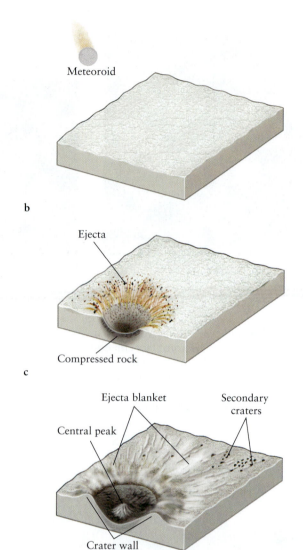

Meteoroid

b

Ejecta

Compressed rock

c

Ejecta blanket   Secondary craters

Central peak

Crater wall

d

FIGURE 15-14 Lunar Craters (a) This photograph, taken from lunar orbit by astronauts, includes the crater Aristillus. Note the crater's central peaks; collapsed, terraced crater wall; and ejecta blanket. Numerous smaller craters resulting from the impact pockmark the surrounding lunar surface. The following three drawings show the crater formation process. (b) An incoming meteoroid, (c) upon impact, is pulverized and the surface explodes outward and downward. (d) After the impact, the ground rebounds, creating the central peak and causing the crater walls to collapse. The lighter region is the ejecta blanket. (a: 2004 Lunar and Planetary Institute/Universities Space Research Association)

compresses the crater floor so much that, afterward, the crater rebounds and pushes the peak upward. As the peak goes up, the crater walls collapse and form *terraces*.

Through an Earth-based telescope, some 30,000 craters, with diameters ranging from 1 km to more than 100 km, are visible. Following a tradition established in the seventeenth century, the most prominent craters are named after astronomers, physicists, mathematicians, and philosophers, such as Kepler, Copernicus, Pythagoras, Plato, and Aristotle. Close-up photographs from lunar orbit reveal millions of craters too small to be seen with Earth-based telescopes. Indeed, extreme close-up photos of the Moon's surface reveal countless microscopic craters (Figure 15-15).

Earth, by comparison, has only about 215 known craters caused by impacts. Many craters on Earth have been pulled inside our planet by plate tectonic motion and obliterated. Many other craters of a few kilometers in diameter and smaller have been worn away (eroded) by weathering effects of wind and water. Earth's atmos-

R I V U X G

FIGURE 15-15 A Microscopic Lunar Crater This photograph made with a microscope shows tiny microcraters less than 1 mm across on a piece of moon rock. (NASA)

a   R I V U X G

FIGURE 15-16 Mare Imbrium and the Surrounding Highlands Mare Imbrium, the largest of the 14 dark plains that dominate Earth-facing side of the Moon, is ringed by lighter-colored highlands strewn with craters and towering mountains. The highlands were created by asteroid impacts pushing land together. (NASA)

b   R I V U X G

phere vaporized countless space rocks that would otherwise have created small craters here. Moreover, those that do hit are slowed down by the atmosphere so that they produce smaller craters than they otherwise would have. We will discuss Earth's craters further in Chapter 18.

Besides its craters, the most obvious characteristic of the Moon visible from Earth is that its surface is various shades of gray (see Figure 15-13). Most prominent are the large, dark gray plains called **maria** (pronounced *mar-ee-uh*). The singular form of this term, **mare** (pronounced *mar-ay*), means "sea" in Latin and was introduced in the seventeenth century when observers who used early telescopes thought that these features were large bodies of water. We know now that no water exists in liquid form on our satellite. Nevertheless, we retain these poetic, fanciful names, including Mare Tranquillitatis (Sea of Tranquility), Mare Nubium (Sea of Clouds), Mare Nectaris (Sea of Nectar), and Mare Serenitatis (Sea of Serenity). As we will discuss shortly, maria are basins on the Moon that were filled in with lava after most cratering ended.

The largest of the maria is Mare Imbrium (Sea of Showers). It is roughly circular and measures 1100 km (700 mi) in diameter (Figure 15-16). Although the maria seem quite smooth in telescopic views from Earth, close-up photographs from lunar orbit and from the surface reveal that they contain small craters and occasional

FIGURE 15-17 Details of a Lunar Mare (a) Close-up views of the lunar surface reveal rilles and numerous small craters on the maria. Astronauts in lunar orbit took this photograph of Mare Tranquillitatis (Sea of Tranquility) in 1969 while searching for potential landing sites for the first human landing. At 1100 km (700 mi) across, this mare is the same size as the distance from London to Rome or from Chicago to Philadelphia. (b) Astronaut David Scott on Hadley's rille during the *Apollo 15* mission to the Moon. (a: NASA; b: Kennedy Space Center/NASA)

cracks, called **rilles** (Figure 15-17). Rilles are the collapsed roofs of underground lava rivers or the beds of long-gone surface lava rivers. When flowing underground lava rivers solidified, the rock they formed shrank and the roofs fell in on the spaces created.

As we will discuss further in Section 15-8, the same side of the Moon always faces Earth. One of the surprises

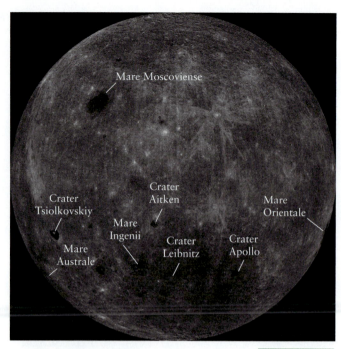

RIVUXG

VIDEO 15.3   WEB LINK 15.7 **FIGURE 15-18** The Far Side of the Moon A composite image from the *Galileo* spacecraft. Mare Orientale lies on the boundary between the near and far sides. (NASA)

stemming from the earliest lunar exploration is that the far side is much more heavily cratered than the near side. A few tiny maria, including Mare Moscoviense, Mare Ingenii, and Mare Australe, plus one moderate-sized mare, Orientale, grace the Moon's far side, the side we never see from Earth (Figure 15-18). Otherwise, craters and mountains cover the entire far side of the Moon.

Detailed measurements by astronauts in lunar orbit demonstrated that the maria on the Moon's Earth-facing side are 2 to 5 km below the average lunar elevation on that side, whereas the ones on the far side are 4 to 5 km above the average lunar elevation. The flat, low-lying, dark gray maria cover only 17% of the entire lunar surface. The remaining 83% is composed of light gray, heavily cratered mountainous regions, called **highlands** (see Figure 15-13). The difference in heights of the maria on the two sides of the Moon apparently stems from the way they were formed. Those on the near side were created by flows of magma that filled low-lying basins (enormous craters) surrounding mountains (highlands) pushed up by large impacts. The small maria on the far side were more likely lava flows that pushed up and out relatively high in the mountains.

Lunar mountain ranges that rim the vast maria basins suggest ancient impacts far more violent than those that produced typical craters. One such mountain range can be seen around the edge of Mare Imbrium (see Figure 15-16). Although they take their names from famous terrestrial ranges, such as the Alps, the Apennines, and the Carpathians, the

Which of the three named craters in Figure 15-14a is the youngest?

lunar mountains were not formed through plate tectonics, as mountains on Earth were. Analyses of Moon rocks and observations from lunar orbit indicate that violent impacts of huge asteroids thrust up the surrounding land to create the lunar highlands.

15-6 Visits to the Moon yielded invaluable information about its history

VIDEO 15.4   WEB LINK 15.8   WEB LINK 15.9 Reaching the Moon became national obsessions for the United States and the former Soviet Union in the early 1960s. A series of successful space missions by the former Soviet Union spurred President John F. Kennedy's goal of landing a person on the Moon by the end of that decade.

The first of six manned lunar landings, *Apollo 11*, set down on Mare Tranquillitatis on July 20, 1969. Astronaut Neil Armstrong was the first human to set foot on the Moon. In the same year, *Apollo 12* also landed on a mare. Human voyages into space encountered a setback when *Apollo 13* experienced a nearly fatal explosion en route to the Moon. Fortunately, a heroic effort by the astronauts and ground personnel brought it safely home. *Apollo 14* through *Apollo 17* took on progressively more challenging lunar terrain. The period of human visitation to our Moon lasted less than 4 years, culminating in 1972, when *Apollo 17* landed amid mountains just east of Mare Serenitatis (Figure 15-19).

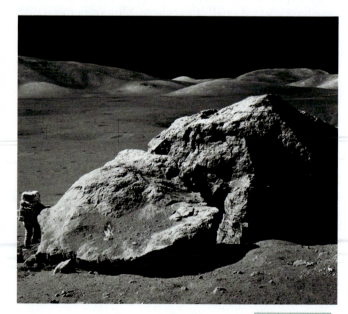

RIVUXG

VIDEO 15.5 **FIGURE 15-19** An Apollo Astronaut on the Moon *Apollo 17* astronaut Harrison Schmitt enters the Taurus-Littrow Valley on the Moon. The enormous boulder seen here slid down a mountain to the right of this image, fracturing on the way. This final Apollo mission landed in the most rugged terrain of any Apollo flight. (NASA)

FIGURE 15-20 The Regolith The Moon's surface is covered by a layer of powdered rock that was created over billions of years as a result of bombardment by space debris. Called *regolith,* this surface material sticks together like wet sand, as illustrated by this *Apollo 11* astronaut's bootprint. (NASA)

Astronauts discovered that the lunar surface is covered with fine powder and rock fragments. This layer, ranging in thickness from 1 to 20 m, is called the **regolith** (from the Greek, meaning "blanket of stone"; Figure 15-20). It formed during 4.5 billion years of relentless micrometeorite bombardment that pulverized the surface rock. We do not refer to this layer as soil, because the term *"soil"* suggests the presence of decayed biological matter, which is not found on the Moon's surface. The rough, powdered regolith absorbs most of the light incident on it. Therefore, the Moon's albedo is only 0.07, meaning that it scatters only 7% of the light striking it.

A major goal of the Apollo program was to study the geology of the Moon and help determine its history. The astronauts brought back a total of 382 kg (842 lb) of lunar rocks, which have provided important information about the early history of the Moon. By carefully measuring trace amounts of radioactive elements, geologists determined that lunar rock from the maria is solidified lava that dates back only 3.1 to 3.8 billion years. This indicates that the maria developed about a billion years after the Moon formed and that the maria were formed from molten rock. Indeed, these Moon rocks are composed mostly of the same minerals that are found in volcanic rocks in Hawaii and Iceland. The rock of these low-lying lunar plains is called **mare basalt** (Figure 15-21). About 17% of the Moon's surface rock is basalt.

In contrast to the dark mare basalt, the lunar highlands are covered with a light-colored rock called

FIGURE 15-21 Mare Basalt This 1.53-kg (3.38-lb) specimen of mare basalt was brought back by *Apollo 15* astronauts in 1971. Small holes that cover about a third of its surface suggest that gas was dissolved in the lava from which this rock solidified. When the lava reached the airless lunar surface, bubbles formed as the pressure dropped and the gas expanded. Some of the bubbles were frozen in place as the rock cooled. (NASA)

anorthosite (Figure 15-22). On Earth, anorthositic rock is found only in very old mountain ranges, such as the Adirondacks in the eastern United States. Compared to the mare basalts, which have more of the heavier elements like iron, manganese, and titanium, anorthosite is rich in calcium and aluminum. Anorthosite is, therefore, less dense than basalt. Many highland rocks brought back to Earth are **impact breccias** (Figure 15-23), which are composites of different rock fused together as a result of meteorite impacts. Typical anorthositic specimens

FIGURE 15-22 Anorthosite The lunar highlands are covered with this ancient type of rock, which is believed to be the material of the original lunar crust. This sample's dimensions are 18 cm × 16 cm × 7 cm. Although this sample is medium gray, other anorthosites retrieved from the Moon have been white, while others are darker gray than this one. This one was brought back by *Apollo 16* astronauts. (NASA)

R I V U X G

FIGURE 15-23 Impact Breccias These rocks are created from shattered debris fused together under high temperature and pressure. Such conditions prevail immediately following impacts of space debris on the Moon's surface. (NASA)

from the highlands are between 4.0 and 4.3 billion years old. These ancient specimens of highland rock probably represent samples of the Moon's original crust (see Figure 15-18).

Four of the five Apollo missions that landed on the Moon set up scientific packages containing seismic detectors (Figure 15-24). These devices are sensitive to moonquakes, motion of the Moon's surface analogous to earthquakes. Between 1969 and 1977, they detected over 12,500 events, including a few quakes strong enough to knock books off shelves (if there had been shelves or books there at the time). Moonquakes occur for any of four reasons: impacts, motion of the lunar surface due to changing tides created by Earth's gravity, expansion and contraction due to heating and cooling of the Moon's surface, and other activity inside the Moon.

With several seismic detectors working simultaneously, geologists were able to deduce information about the Moon's interior from the timing and strength of the vibrations that were detected. The data suggested that the Moon has a small iron core, possibly a layer of hot rock above it, and a thick layer of solid rock above that, surrounded by the crust. The presence of the iron core, amounting to only 3% or 4% of the Moon's total mass, was confirmed by data from the *Lunar Prospector* satellite in 1998. This compares to Earth's core, which has about 30% of our planet's total mass.

Since spacecraft began orbiting the Moon, scientists have noted that their orbits do not follow the expected elliptical paths. Every once in a while, usually over a circular mare, the spacecraft momentarily dip Moonward, as they are pulled by a higher than normal concentration

of mass. These *mass concentrations*, or **mascons**, are local regions of relatively dense rock and metal near the Moon's surface. The mascons found in maria are believed to be magma that filled the bottom of the mare after the basins were formed by impacts of very large bodies. The magma solidified after emerging from inside the Moon. New mascons continue to be discovered, including seven found by the *Lunar Prospector* spacecraft in 1998.

In 1994, the lunar-orbiting *Clementine* spacecraft sent radio signals to the Moon's surface. These signals scattered off the materials there, and then they were received and analyzed on Earth. The results were intriguing, suggesting that frozen water exists in craters near the Moon's south pole. If so, this ice is not in danger of being evaporated by sunlight, because, just as is the case with Earth's poles, the Sun deposits very little energy at the lunar poles (Figure 15-25). The *Lunar Prospector*, using different technology, found evidence for ice at the Moon's north pole; however, when the same spacecraft was intentionally crashed into a crater at the south pole, no water vapor was released. This isolated experiment is by no means definitive. However, in 2003, radio experiments from Earth similar to those run on *Clementine* also failed to see indications of water. Clearly, the issue of whether there are large quantities of water on the Moon is unresolved. If

FIGURE 15-24 *Apollo 11* Landing Site On the Moon's Sea of Tranquility, Astronaut Buzz Aldrin stands next to the package of equipment containing the seismic detector. The corner reflectors are used, even today, to determine the distance from Earth to the Moon. The stereo camera took pairs of images of the Moon's surface. Seeing them through special glasses gives a 3-D close-up view of the Moon's surface. The bottom half of the lander is still on the Moon's surface. The top half brought astronauts Neil Armstrong and Buzz Aldrin back into lunar orbit, where they transferred to the command module to fly home. (NASA)

FIGURE 15-25 The Moon's Interior Based on seismic experiments left on the Moon by Apollo astronauts and the studies done by *Lunar Prospector*, we know that the Moon has a crust, mantle, and core. The lunar crust has an average thickness of about 60 km on Earth-facing side but about 100 km on the far side. The crust and solid upper mantle extend to about 800 km, where the non-rigid inner mantle begins. The Moon's core has a radius of between 300 and 425 km. Although the main features of the Moon's interior are analogous to those of Earth, the proportions and details are quite different. The *Clementine* spacecraft revealed that the south pole has a significant basin. The crust was apparently stripped away by an impact. *Top inset:* A radar image of the Moon's north polar region. The areas computer-colored in white and light gray are regions where the Sun never shines and that may contain water. *Bottom inset:* A radar image of the Moon's south polar region, also showing regions of permanent or near-permanent darkness. (*Insets:* NASA/Galaxy)

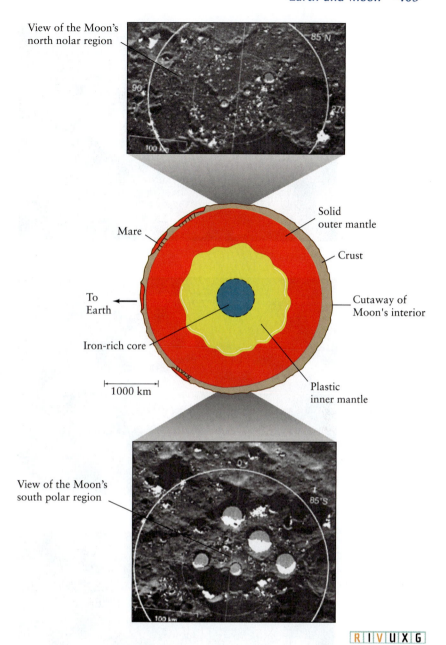

it exists there, the ice was probably deposited on the Moon during comet and asteroid impacts; estimates based on the first observations are that over 3 billion tons of water may be encased there, either mixed in with the regolith or in layers below it.

The presence of water ice would greatly simplify human settlement of the Moon, because ice can provide us not only with liquid water, but also with oxygen and rocket fuel. For example, the Space Shuttle's primary rocket engine is fueled by liquid oxygen and liquid hydrogen. Separating the Moon's water into liquid oxygen and liquid hydrogen would create the same fuel. An existing water reservoir would make establishing a lunar colony much less expensive and more feasible than lugging all that water from Earth or dragging a water-rich comet to the Moon.

Astronauts found no traces of life on the Moon. Life as we know it requires liquid water, of which the Moon has none. The Moon has an atmosphere of sodium gas, so thin as to be considered an excellent vacuum compared to the density of the air we breathe. An atmosphere is held around a planet or moon by that body's gravitational force. If the atmospheric gases are too hot (moving too fast) for the gravitational force to hold them, they leak into space. Our Moon has so little gravitational attraction (about one-sixth what we feel at Earth's surface) that it is unable to retain its tenuous atmosphere. The atmosphere is renewed by sunlight striking the Moon, releasing more gas from its rocks. Earth, in contrast, has enough mass to retain such gases as nitrogen, oxygen, carbon dioxide, water, and argon.

The Moon's slow rotation rate and small iron core suggest that unlike Earth, the Moon should not have a strong magnetic field. Spacecraft orbiting the Moon have confirmed this. However, the *Lunar Prospector*

has discovered local areas where the magnetic fields are strong enough to keep the solar wind away from the Moon's surface. These anomalous regions have yet to be explained.

What type of rock did astronauts from *Apollo 11*, the first flight to land on the Moon, bring back?

15-7 The Moon probably formed from debris cast into space when a huge asteroid struck the young Earth

Before the Apollo program, scientists debated three different theories about the origin of the Moon. The **fission theory** holds that the Moon was pulled out from a rapidly rotating proto-Earth. This theory does not explain why the Moon's overall chemistry is like that of

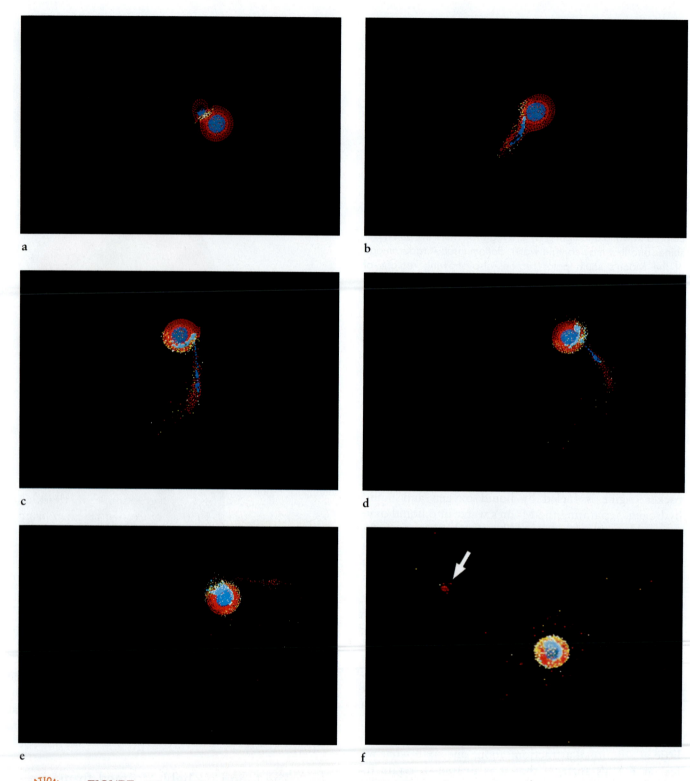

FIGURE 15-26 The Moon's Creation This computer simulation models the creation of the Moon from material ejected by the impact of a large planet-like body with the young Earth. To follow the ejected material as it moves away from Earth, successive views (a–f) pan out to show increasingly larger volumes of space. Blue and green indicate iron from the cores of Earth and the asteroid; red, orange, and yellow indicate rocky mantle material. In this simulation, the impact ejects both mantle and core material, but most of the dense iron falls back onto Earth. The surviving ejected rocky matter coalesces to form the Moon *(indicated by arrow)*. (W. Benz)

ANIMATION 15.5

Earth's outer layers and not its core, and that the Moon has virtually no water incorporated in its rocks, as revealed by the bone-dry samples brought back by Apollo astronauts. The **capture theory** posits that the Moon was formed elsewhere in the solar system and then drawn into orbit around Earth by gravitational forces. However, it is physically very difficult for a planet to capture such a large moon. Furthermore, because bodies formed in different places have different overall chemical compositions, this theory does not explain the similar geochemistries of the Moon and Earth's surface. The **co-creation theory** proposes that Earth and the Moon were formed near each other at the same time. As with the fission theory, this theory fails to explain why, compared to Earth, the Moon has less of the denser elements, such as iron, and why certain types of rocks found on the Moon are not found on Earth.

A fourth theory, now held to be correct by most astronomers, was first proposed in the 1980s. It applies the established fact of age-old collisions in the solar system to the formation of the Moon. Called the **collision-ejection theory**, or *large impact theory,* it proposes that the newly formed Earth was struck at an angle by a Mars-sized asteroid that literally splashed some of Earth's surface layers into orbit around the young planet. Evidence from Moon rocks places this event at around 4.5 billion years ago, within the first 100 million years of Earth's existence. A computer simulation of this cataclysm is shown in Figure 15-26. This orbiting matter became a short-lived ring that eventually condensed into the Moon, just as planetesimals condensed to form Earth.

The collision-ejection theory is consistent with the known facts about the Moon. For example, rock vaporized by the impact would have been depleted of *volatile* (easily evaporated) elements and water, leaving the Moon rocks parched, as we now know they are. Also, the Moon has little heavy iron-rich matter because that material had sunk deep into Earth before the asteroid struck. The material from Earth that splashed into orbit and formed the Moon was mostly the lighter rock floating on Earth's surface. Furthermore, most of the debris from the collision would orbit near the plane of the ecliptic as long as the orbit of the impacting asteroid had been near that plane. Also, the impact of an object large enough to create the Moon could have tipped Earth's axis of rotation and so inaugurated the seasons (Figure 15-27). The fact that this theory is both consistent

with observations and explains two seemingly separate things (the existence of the Moon and the tilt of Earth) makes it especially appealing to scientists.

The surface of the newborn Moon probably remained molten for millennia, due to heat released during the impact of rock fragments falling onto the young satellite and the decay of short-lived radioactive isotopes inside it. As the Moon gradually cooled, low-density lava floating on its surface began to solidify into the anorthositic crust that exists today in the highlands.

By correlating the ages of Moon rocks with the density of craters at the sites where they were collected, geologists have found that the rate of impacts on the Moon changed over the ages. The ancient, heavily cratered lunar highlands are evidence of intense bombardment of asteroids that dominated the Moon's early history. This barrage extended from 4.5 billion years ago, when the Moon's surface first solidified, until about 3.8 billion years ago. This period of impacts was punctuated by a severe pounding 4 billion years ago, part of a solar-system-wide reign of terror.

The frequency of impacts gradually tapered off as meteoroids and planetesimals were swept up by the newly formed planets, as discussed in Chapter 14. Recorded among the final scars at the end of this crater-making era are the impacts of more than a dozen objects, each measuring at least 100 km across. As these huge rocks slammed into the young Moon, they blasted gigantic craters in regions that would later become the maria. Meanwhile, heat from the decay of long-lived radioactive elements, such as uranium and thorium, again began to melt the inside of the Moon. Then, from 3.8 to 3.1 billion years ago, great floods of molten rock gushed up from the lunar interior through the weak

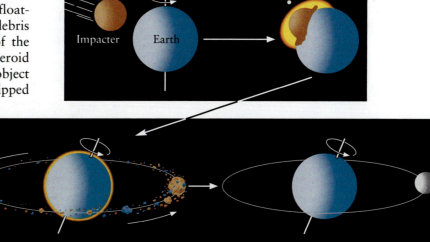

FIGURE 15-27 Tilting Earth's Axis The collision that created the Moon could have also knocked Earth's rotation axis over so that today it has a 23$\frac{1}{2}$° tilt, thereby creating the seasons.

spots in the crust created by these large impacts. Lava filled the impact basins and created the maria we see today. Because most cratering was over by then, maria have few craters.

The reason that the far side of the Moon has only one significant mare apparently stems from the Moon's crust on the side facing Earth being thinner than the crust on the far side. This difference was created when the Moon was young and molten. At that time, the Moon was much closer to Earth (we will explore this shortly). Its rapid orbit and the tides created on it by Earth caused much of the low-density, crustal lava to flow away from Earth to the lunar far side, where it remained when the crust solidified. It was therefore easier for the molten rock inside the Moon to escape through the large craters on the thin crust of our side than through those on the thicker-crusted far side.

Relatively little large-scale activity has happened on the Moon since those ancient times. Although only a few fresh large craters have been formed, the small-sized debris still plentiful in the solar system continues to create small ones. Back in 1953, an impact was observed that has been connected with the formation of a 1.5-km (0.93-mi) diameter crater. Smaller impacts are observed nearly every year when large groups of space debris are known to be crossing the Moon's path. By and large, however, the astronauts visited a world that has remained largely unchanged for more than 3 billion years. Unlike Earth, which is still quite geologically active, any molten rock inside the Moon has apparently stopped

> Why are we fortunate on Earth that the Moon is rarely being struck today?

directly affecting that world's surface. Astronomers hope to learn more about the Moon's interior when modern seismographs are placed on its surface in the not-too-distant future.

15-8 Tides have played several important roles in the history of Earth and the Moon

WEB LINK 15.2

Gravitational attraction keeps Earth and the Moon together, while the angular momentum (see *An Astronomer's Toolbox 2-2*) given to the Moon when it was formed keeps the Moon from being pulled back onto Earth. Contrary to appearances, the Moon does not orbit Earth. Rather, both bodies are in motion around their center of mass, called their *barycenter* (Figure 15-28a), just as a pair of objects tied together and sent spinning across an air table move around a similar common point (Figure 15-28b).

The same surface features on the Moon always face Earth, regardless of what phase the Moon is in (see Figure 1-21). This occurs because the Moon is in **synchronous rotation** around Earth: The Moon rotates on its axis at the same rate that it and Earth orbit their barycenter (Figure 15-29). To see why the Moon must be rotating if it keeps the same side facing us, imagine standing up, holding your arm out in front of you with your palm facing you, and twirling yourself around. You are Earth, while your palm represents the side of the Moon facing Earth. The only way your palm can always face you (be in synchronous rotation) is if your

`3, 4`

a

b

FIGURE 15-28 Motion of Earth-Moon System (a) The paths of Earth and the Moon as their barycenter follows an elliptical orbit around the Sun. (b) Analogously, this time-lapse image shows a pair of disks connected by a piece of wood sliding across a table. Note that its center of mass moves in a straight line, while its other parts follow curved paths. (Copyright Tom Patagnes)

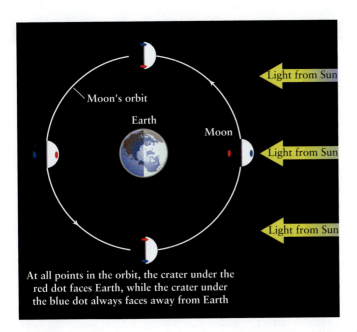

At all points in the orbit, the crater under the red dot faces Earth, while the crater under the blue dot always faces away from Earth

FIGURE 15-29 Synchronous Rotation of the Moon The motion of the Moon around Earth as seen from above Earth's north polar region (ignoring Earth's orbit around Earth-Moon barycenter). For the Moon to keep the same side facing Earth as it orbits our planet, the Moon must rotate on its axis at precisely the same rate that it revolves around Earth.

hand (the Moon) rotates at exactly the same rate that it swings (revolves). So the Moon must be rotating.

We have seen the Moon's "far side" only from spacecraft (see Figure 15-18). The far side of the Moon is often, and incorrectly, called its "dark side." The dark side of the Moon is exactly like the dark side of Earth—it is the side where the Sun is below the horizon. Whenever we view less than a full Moon, we see part of its dark side and some sunlight is then striking part of its far side.

The origin of the Moon's synchronous rotation is in the tidal force that acts between Earth and the Moon. This tidal force also creates ocean tides on Earth. To understand how the Moon came to have synchronous rotation, we investigate the forces acting between Earth and the Moon. To "kill two birds with one stone," the discussion in *Guided Discovery: Tides* focuses on how the Moon creates the tides on Earth. We will use this information to see how Earth once created tides on the Moon that changed the Moon's original rotation rate to synchronous rotation.

5 Two thirds of our tides result from the gravitational tug-of-war between Earth and the Moon as the two bodies orbit their barycenter. The Sun generates most of the remaining third of the tides, with Jupiter and the other planets contributing a tiny fraction.

6 As explained in *Guided Discovery: Tides*, two high tides always occur on opposite sides of Earth (see

Figure 15-30). When the Sun, Moon, and Earth are aligned, the Sun and Moon create pairs of high ocean tides in the same directions, and the resulting combined tides are the highest high tides of the lunar cycle. These **spring tides** (so-called not because they only occur in the spring—they occur in all seasons—but because the term is derived from the German word *springen*, meaning to "spring up") occur at every new and full Moon (see Figure 15-31a). Note that it does not matter whether the Sun and Moon are on the same or opposite sides of Earth when generating the spring tides, because the Sun and Moon both create high tides on opposite sides of Earth. At first quarter and third quarter, the Sun and Moon form a right angle as seen from Earth. The gravitational forces they exert compete with each other, so the tidal distortion is the least pronounced. The especially small tidal shifts on these days are called **neap tides** (Figure 15-31b).

Finally, we can explain the Moon's synchronous rotation: When the Moon was young, it was molten and Earth's gravity created huge tides of molten rock up to 18 m (60 ft) high there. Friction generated by these tides caused the Moon to speed up or slow down (we don't know which) until one high tide there was fixed directly between the Moon's center and the center of Earth. As a result, the Moon's rotation rate became the same as its revolution rate. When it solidified, the Moon was locked into this synchronous rotation.

> Why haven't the ocean tides on Earth put our planet into synchronous rotation with respect to the Moon?

15-9 The Moon is moving away from Earth

A century ago, Sir George Darwin (son of the evolutionist Charles Darwin) proposed that the Moon is moving away from Earth. To understand why, read *Guided Discovery: Tides* and its Figure 15-31, which shows that the high tide nearest the Moon is actually 10° ahead of the Moon in its orbit around Earth. This offset occurs because friction between the ocean and the ocean floor causes the rapidly spinning Earth to drag the tidal bulge ahead of the line between the center of Earth and the center of the Moon. The gravitational force from the water in the high ocean tide nearest the Moon acts back on the Moon, pulling it ahead in its orbit, giving it energy and thereby forcing it to spiral outward. This is similar to the effect you would get if you tied a ball on a string, let it hang down, and then started spinning yourself around. Your hand would act through the string, giving the ball energy and starting it to spin and move farther and farther away from your body.

To test the concept that the Moon is moving away from Earth, *Apollo 11, 14,* and *15* astronauts placed on the Moon sets of reflectors (Figure 15-24), similar to the orange and red ones found on cars. These reflectors are specially designed so that light striking them from any

Tides

Tides result from a combination of Earth's motion around its barycenter with the Moon (or Sun) and the changing strength of the gravitational force from the Moon (or Sun) across the Earth. Ignoring Earth's rotation and focusing on the Moon for the moment, let's first consider the motion of Earth and the Moon around their barycenter (see Figure 15-28a). The barycenter is located 1720 km (1068 mi) below Earth's surface on the line between the centers of Earth and the Moon. As a result of its motion around the barycenter, every point on Earth feels an equally strong force away from the Moon. This force is similar to the outward force that you feel while riding on a merry-go-round. The force away from the Moon is represented at three locations by the force labeled F_{out} in Figure 15-30a.

Opposing this force away from the Moon, Earth feels our satellite's gravitational attraction. Recall from Chapter 2 (*An Astronomer's Toolbox 2-3*) that the gravitational force exerted by one body on another decreases with distance. Therefore, the Moon's gravitational attraction on Earth is greatest at the point on Earth's surface closest to the Moon. This attraction is greater than the force felt at, say, the center of Earth, which in turn is greater than the force felt on the opposite side of Earth from the Moon. The gravitational force in these places is labeled $F_{grav.}$ in Figure 15-30a.

Because forces in the same or opposite directions simply add or subtract, respectively, we can subtract the two forces F_{out} and $F_{grav.}$ at any point on Earth, due to the presence of the Moon. The resulting net force is labeled F_{tide} in Figure 15-30a. This is the tidal force acting at these places. For example, at the point on Earth directly below the Moon, the tidal force is very strong and points toward the Moon. At the center of Earth, the two forces cancel each other completely. At the point on the opposite side of Earth from the Moon, the tidal force is equal in strength to the tidal force directly under the Moon but directed away from the Moon. The strength of the tidal force at various places is summarized in Figure 15-30b.

Consider the point on Earth closest to the Moon. The strength of the tidal force, F_{tide}, seems to imply that the Moon's gravitational force lifts the water there to create the high tide. However, the force of gravitation from the Moon is not strong enough to raise that water more than a few centimeters. Rather, the high tide closest to the Moon occurs because ocean waters from nearly halfway around to the opposite side of Earth are pulled Moonward by the gravitational force acting on them. This water slides toward the Moon and fills the ocean directly under it, creating a tidal bulge or high tide. Likewise, the net outward force acting on the opposite side of Earth from the Moon pushes the waters that are just over halfway around from the Moon to the side directly away from it, creating a simultaneous, second high tide of equal magnitude on the opposite side of Earth from the Moon. Where the water has been drained away in this process, low tides occur.

Let's spin Earth back up. As it rotates, the geographic locations of the high and low tides continually change. Figure 15-30b indicates that high tides occur when the Moon is high in the sky or below our feet, while low tides occur when the Moon is near either horizon. This means that there are two cycles of high and low tides at most places on Earth each day. These cycles do not span exactly 24 h, however, because the Moon is moving along the celestial sphere, changing the time at which high or low tide occurs at any location from day to day. As a result, the time between consecutive high tides is 12 h, 25 min.

a

b

FIGURE 15-30 Tidal Forces (a) The Moon induces tidal forces, F_{tide}, on Earth. At each point, this force is the difference between the force, F_{out}, created by the orbital motion of the two bodies around their barycenter, and the Moon's gravitational force, F_{grav}, at that point. The magnitude and direction of each arrow represent the strength and direction of each force. (b) Water slides along Earth to create the tides. Ignoring Earth's rotation and the effects of the continents until Figure 15-31, this figure shows how two high tides are created on Earth by the Moon's gravitational pull. The Sun has a weaker, but otherwise identical effect.

Tides (continued)

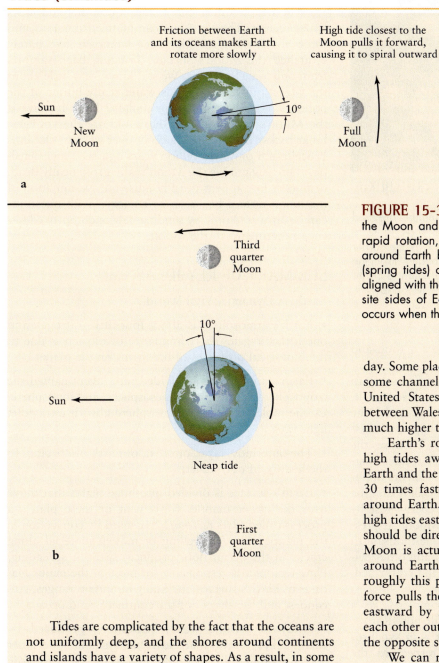

Friction between Earth and its oceans makes Earth rotate more slowly

High tide closest to the Moon pulls it forward, causing it to spiral outward

Sun

New Moon

10°

Full Moon

a

Third quarter Moon

10°

Sun

Neap tide

First quarter Moon

b

FIGURE 15-31 Tides on Earth The gravitational forces of the Moon and the Sun deform the oceans. Due to Earth's rapid rotation, the high tide closest to the Moon leads it around Earth by about 10°. (a) The greatest deformation (spring tides) occurs when the Sun, Earth, and Moon are aligned with the Sun and Moon either on the same or opposite sides of Earth. (b) The least deformation (neap tides) occurs when the Sun, Earth, and Moon form a right angle.

day. Some places have tides that change negligibly, while some channels, such as the Bay of Fundy between the United States and Canada and the Bristol Channel between Wales and England, have two tides that are each much higher than normal.

Earth's rotation also has the effect of moving the high tides away from the line between the centers of Earth and the Moon. Earth's eastward rotation is nearly 30 times faster than the Moon's eastward revolution around Earth. Therefore, as it rotates, Earth drags the high tides eastward with it. As a result, the high tide that should be directly between the centers of Earth and the Moon is actually 10° ahead of the Moon in its orbit around Earth (Figure 15-31). This high tide stays at roughly this position because the Moon's gravitational force pulls the tide westward while it is being dragged eastward by Earth's rotation; the two effects balance each other out. A similar argument applies to the tide on the opposite side of Earth.

We can repeat this analysis for Earth-Sun system and get the same qualitative results.

Tides are complicated by the fact that the oceans are not uniformly deep, and the shores around continents and islands have a variety of shapes. As a result, in some places the oceans have only a single cycle of tides each

direction is reflected back toward its source. Pulses of laser light were fired at the Moon from Earth (Figure 15-32), and the time it took the light to reach the Moon, bounce off a reflector array, and return to Earth was carefully recorded. Knowing the speed of light and the time it takes for the light pulse to make the round trip, astronomers can calculate the distance to the Moon with an error of only a few millimeters. From

such measurements over time, astronomers have established that the Moon is currently spiraling away from Earth at a rate of 3.8 cm per year.

This means, of course, that the Moon used to be closer to Earth. Although the initial distance of the Moon when it coalesced is not known, its present rate of recession tells us that it was at least half its present distance; more accurate calculations suggest one-tenth.

RIVUXG

FIGURE 15-32 Lunar Ranging Beams of laser light are fired through three telescopes at the Observatoire de la Côte d'Azur, France. The light is then reflected back by the corner reflectors placed on the Moon by Apollo astronauts. From the time it takes the light to reach the Moon and return to Earth, astronomers can determine the distance to the Moon to within a few millimeters. (Observatoire de la Côte d'Azur)

At that distance, the tides on Earth were a thousand times higher than they are today.

Where does the Moon get the energy necessary to spiral away from Earth? The answer is from Earth itself. As the tides move westward to try to stay under the Moon, they encounter the eastward-moving landmasses, which disrupt the water's motion. Because the oceans move westward, while the planet rotates eastward, the oceans actually push on the continents opposite to Earth's direction of rotation. The result is that the tides are slowing Earth's rotation. The energy lost by Earth as it spins down is gained by the Moon. At present, the day is slowing down by about one-thousandth of a second each century. Based on fossil and other evidence, geologists calculate that when Earth first formed, the day was only between 5 and 6 h long.

Will the Moon ever leave Earth completely as a result of these dynamics? In fact, it will not. Rather, Earth's rotation will slow down until it is in synchronous rotation with respect to the Moon. Thereafter, the Moon will remain at a fixed distance over one side of Earth. However, billions of years before that happens, the Sun is going to self-destruct!

What effects would the original tides on Earth have had on the geology of our young planet?

15-10 Frontiers yet to be discovered

Earth and the Moon still hold many mysteries. We have yet to understand the different rotation rate of Earth's core from its surface, as well as the details of the dynamo effect that generates Earth's magnetic field. We do not know precisely how fast Earth was rotating initially. We also need to better understand the impact of humans on Earth and its atmosphere.

The Moon still inspires many questions and hides many secrets. The six American-manned missions and three Soviet soft, robotic lunar landings have barely scratched the Moon's surface, bringing back samples from only nine locations. We still know very little about the Moon's far side. Also, is water really buried at the poles? Is the Moon's interior molten? How massive is the Moon's iron-rich core? How old are the youngest lunar rocks? Did lava flows occur over western Oceanus Procellarum only two billion years ago, as crater densities there suggest? Is the Moon really geologically dead, or does it just appear dead because our examination of the lunar surface has been so cursory? Such questions can be answered only by sending new and improved science experiments to the Moon.

SUMMARY OF KEY IDEAS

Earth: A Dynamic, Vital World

• Earth's atmosphere is about four-fifths nitrogen and one-fifth oxygen. This abundance of oxygen is due to the biological processes of life-forms on the planet.

• Earth's atmosphere is divided into layers named the troposphere, stratosphere, mesosphere, and ionosphere. Ozone molecules in the stratosphere absorb ultraviolet light rays.

• The outermost layer, or crust, of Earth offers clues to the history of our planet.

• Earth's surface is divided into huge plates that move over the upper mantle. Movement of these plates, a process called *plate tectonics*, is caused by convection in the mantle. Also, upwelling of molten material along cracks in the ocean floor produces seafloor spreading. Plate tectonics is responsible for most of the major features of Earth's surface, including mountain ranges, volcanoes, and the shapes of the continents and oceans.

• Study of seismic waves (vibrations produced by earthquakes) shows that Earth has a small, solid inner core surrounded by a liquid outer core. The outer core is surrounded by the dense mantle, which in turn is surrounded by the thin, low-density crust on which we live. Earth's inner and outer cores are composed primarily of iron. The mantle is composed of iron-rich minerals.

• Earth's magnetic field produces a magnetosphere that surrounds the planet and blocks the solar wind.

• Some charged particles from the solar wind are trapped in two huge, doughnut-shaped rings called the *Van Allen radiation belts*. A deluge of particles from a coronal mass ejection by the Sun can initiate an auroral display.

The Moon and Tides

• The Moon has light-colored, heavily cratered highlands and dark-colored, smooth-surfaced maria.

• Many lunar rock samples are solidified lava formed largely of minerals also found in Earth rocks.

• Anorthositic rock in the lunar highlands was formed between 4.0 and 4.3 billion years ago, whereas the mare basalts solidified between 3.1 and 3.8 billion years ago. The Moon's surface has undergone very little geologic change over the past 3 billion years.

• Impacts have been the only significant "weathering" agent on the Moon; the Moon's regolith (pulverized rock layer) was formed by meteoritic action. Lunar rocks brought back to Earth contain no water and are depleted of volatile elements.

• Frozen water may have been discovered at the Moon's poles.

• The collision-ejection theory of the Moon's origin, accepted by most astronomers, holds that the young Earth was struck by a huge asteroid, and debris from this collision coalesced to form the Moon.

• The Moon was molten in its early stages, and the anorthositic crust solidified from low-density magma that floated to the lunar surface. The mare basins were created later by the impact of planetesimals and were then filled with lava from the lunar interior.

• Gravitational interactions between Earth and the Moon produce tides in the oceans of Earth and set the Moon into synchronous rotation. The Moon is moving away from Earth, and, consequently, Earth's rotation rate is decreasing.

WHAT DID YOU THINK?

1   *Can Earth's ozone layer, which is now being depleted, be naturally replenished?* Yes. Ozone is created continuously from normal oxygen molecules by their interaction with the Sun's ultraviolet radiation.

2   *Who was the first person to walk on the Moon, and when did this event occur?* Neil Armstrong was the first person to set foot on the Moon. He and Buzz Aldrin flew on the *Apollo 11* spacecraft piloted by Michael Collins. Armstrong and Aldrin set down the Eagle Lander on the Moon on July 20, 1969.

3   *Do we see all parts of the Moon's surface at some time throughout the lunar cycle of phases?* No. Because the Moon's rotation around Earth is synchronous, we always see the same side. The far side of the Moon has been seen only from spacecraft that pass or orbit it.

4   *Does the Moon rotate and, if so, how fast?* The Moon rotates at the same rate that it revolves around Earth. If the Moon did not rotate, then, as it revolved, we would see its entire surface from Earth, which we do not.

5   *What causes the ocean tides?* The tides are created by orbital and gravitational forces, primarily from the Moon and, to a lesser extent, from the Sun.

6   *When does the spring tide occur?* Spring tides occur twice monthly, during each full and new Moon.

Key Terms for Review

anorthosite, 401
capture theory, 405
co-creation theory, 405
collision-ejection theory, 405
continental drift, 390
convection, 394
core, 393
coronal mass ejection, 396
crust, 389
dynamo theory, 395
ejecta blanket, 397
fission theory, 403
highlands, 400
impact breccias, 401
ionosphere (thermosphere), 389
mantle, 393
mare (plural *maria*), 399
mare basalt, 401
mascons, 402
mesosphere, 388

neap tide, 407
northern lights (aurora borealis), 396
ozone layer, 388
planetary differentiation, 393
plate tectonics, 391
regolith, 401
rille, 399
seafloor spreading, 391
seismic waves, 394
seismograph, 394
solar wind, 396
southern lights (aurora australis), 396
spring tide, 407
stratosphere, 388
synchronous rotation, 406
troposphere, 388
Van Allen radiation belts, 396

Review Questions

1. The Moon's surface is best described as: **a.** fine-grained powder, **b.** solid rock, **c.** rocky rubble, **d.** liquid water oceans and dry land, **e.** molten rock.

2. What type of chemical or molecule is most common in Earth's atmosphere? **a.** carbon dioxide, **b.** oxygen, **c.** water, **d.** nitrogen, **e.** hydrogen

3. Why does Earth's albedo change daily? Seasonally?

4. Why is Earth's surface not riddled with craters as is that of the Moon?

5. List the layers of Earth's atmosphere.

6. Describe the process of plate tectonics. Give specific examples of geographic features created by plate tectonics.

7. To review Earth's tectonic plates, do Interactive Exercise 15-1 and print out or sketch the completed diagram.

8. To review the mechanism of plate tectonics, do Interactive Exercise 15-2 and print out or sketch the completed diagram.

9. How do we know about Earth's interior, given that the deepest wells and mines extend only a few kilometers into its crust?

10. Describe the interior structure of Earth.

11. Why is the center of Earth not molten?

12. Describe Earth's magnetosphere.

13. To review Earth's magnetosphere, do Interactive Exercise 15-3 and print out or sketch the completed diagram.

14. What are the Van Allen belts?

15. What kind of features can you see on the Moon with a small telescope?

16. On the basis of lunar rocks brought back by the astronauts, explain why the maria are dark-colored, but the lunar highlands are light-colored.

17. Why are there so few craters on the maria?

18. To review the crater formation history in the solar system, do Interactive Exercise 15-4 and print out or sketch the completed diagram.

19. Briefly describe the main differences and similarities between Moon rocks and Earth rocks.

20. How do we know that the maria were formed after the lunar highlands?

21. What is a tidal force? How do tidal forces produce tides in Earth's oceans?

22. What is the difference between spring tides and neap tides? During which phase(s) do each occur?

23. Why do most scientists support the collision-ejection theory for the Moon's formation?

24. Why are virtually all of the craters on the Moon circular, even though many impacts there were not head-on?

Advanced Questions

25. Explain how the outward flow of energy from Earth's interior drives the process of plate tectonics.

26. Why do some geologists believe that Pangaea was the most recent in a succession of supercontinents?

27. Why are active volcanoes, such as Mount St. Helens, usually located in mountain ranges along the boundaries of tectonic plates?

28. Why is more lunar detail visible through a telescope when the Moon is near quarter phase than when it is at full phase?

29. In *The Tragedy of Pudd'nhead Wilson*, Mark Twain wrote, "Everyone is a Moon, and has a dark side which he never shows to anybody." What, if anything, is wrong with Twain's astronomy?

30. Why are the Moon rocks retrieved by astronauts so much older than typical Earth rocks, even though both worlds formed at nearly the same time?

31. Some people who supported the fission theory proposed that the Pacific Ocean basin is the scar left when the Moon pulled away from Earth. Explain why this idea is wrong.

32. Apollo astronauts left seismometers on the Moon that radioed seismic data back to Earth. The data showed that moonquakes occur more frequently when the Moon is at perigee (closest to Earth) than at other locations along its orbit. Give an explanation for this finding.

33. Why do you think that no Apollo missions landed on the far side of the Moon?

34. How might studying albedo help astronomers locate habitable worlds orbiting other stars?

Discussion Questions

35. If Earth did not have a magnetic field, do you think auroras would be more common or less common than they are today? Explain.

36. Comment on the idea that without the Moon's presence, life would have developed far more slowly.

37. Identify and compare the advantages and disadvantages of lunar exploration by astronauts as opposed to mobile, unmanned instrument packages and robots.

38. When was the last earthquake near your hometown? How far is your hometown from a plate boundary? What kinds of topography (for example, mountains, plains, seashore) dominate the geography of your hometown area? Does that topography and the frequency of earthquakes seem to be consistent with your hometown's proximity to a plate boundary?

39. The ice on the Moon is believed to be mixed with rock just under the Moon's surface. How might that water be economically extracted and purified?

What If . . .

40. Earth had two moons? Assume one is at our Moon's distance and the other is at half that distance. Describe the motion of the two moons in the sky and how they might appear to us. What would be different here on Earth?

41. The Moon orbited Earth in the opposite direction from Earth's rotation, rather than in the same direction? What would be different about Earth and life on it? Assume that today such a counter-revolving Moon is at the same distance as our Moon.

42. Our Moon were one-tenth of its actual distance from Earth? What would be different about Earth and life on it? Hint: The heights of tides vary as $1/r^3$, where r is the distance between the centers of Earth and the Moon.

43. The Moon, located where it actually is, were as massive as Earth? What would be different about Earth and life on it and about the Moon?

44. Earth's interior were entirely solid, rather than partly molten? What would be different about Earth and life on it?

45. Earth were now in synchronous rotation with the Moon? That is, suppose that Earth rotated at the same rate that the Moon orbits Earth. What would be different about Earth and life on it?

Web Questions

46. Use the Web to learn more about plate tectonics. What global changes might accompany the formation and breakup of a supercontinent? How might these changes affect the evolution of life? What life-forms dominated Earth when Pangaea existed some 200 million years ago and also when fragments of the preceding supercontinent were as dispersed as today's continents?

47. Search the Web for information about "Pangaea Ultima," a supercontinent that may form in the distant future. When is it expected to form? How will it compare to Pangaea of 200 million years ago (see Figure 15-6)?

48. Use the Web to determine the status of the Antarctic and Arctic ozone holes. How has the situation changed over the past few years? Explain why most scientists who study this issue blame chemicals called CFCs for the existence of the ozone holes.

49. In 1989, representatives of many countries signed a treaty called the *Montreal Protocols* to protect the ozone layer. Use the Web to learn of the current status of the treaty. How many nations have signed it? Has the treaty been amended? If so, how? List when various substances that destroy the ozone layer are being phased out.

50. Search the Web for current information about upcoming lunar science missions. When is each scheduled to be launched? What new investigations are planned for each mission? What existing scientific issues may these missions resolve?

51. Search the Web for current information about tourist trips into space. For when are these scheduled? How long is the trip? What is its cost? What preparation is involved?

52. Refinements to the mass and impact angle of the planetesimal that struck Earth and created the Moon have recently been made. Search the Web for this information and explain the justification for these new parameters.

53. Some astronomers have observed changes in brightness and color on the Moon, called *Lunar Transient Phenomena.* Search the Web and explain what these events are and what their origins are believed to be.

Observing Projects

  Observing Tips and Tools: You can learn a lot by observing the Moon through binoculars. Note that the Moon will appear right side up through binoculars but inverted through a telescope. So, if you use a map of the Moon to aid your observations, you will need to take this into account. Inexpensive Moon maps are available, such as "Moon Map" published by *Sky & Telescope* magazine. To determine the lunar phase (when you cannot examine the Moon directly), you can usually find it on most calendars; by checking the weather page in your newspaper; by consulting the current issue of *Sky & Telescope* or *Astronomy* magazine; by using the *Starry Night Enthusiast*™ program; or on the Web.

54. Use a telescope or binoculars to observe the Moon. Compare the texture of the lunar surface you see on the maria with that of the lunar highlands. How does the visibility of details vary with distance from the terminator (the boundary between day and night on the Moon)? Why?

55. If you live near the ocean, observe the tides to see how the times of high and low tides are correlated with the position of the Moon in the sky.

56. Observe the Moon through a telescope or binoculars every few nights over a period of two weeks between new Moon and full Moon. Make sketches of various surface features, such as craters, mountain ranges, and maria. How does the appearance of these features change with the Moon's phase? Which features are most easily seen at a low angle of illumination (near the terminator)? Which features show up best with the Sun nearly overhead as seen from the Moon (far from the terminator)?

 57. Use the *Starry Night Enthusiast*™ program to view Earth as it appears from the Moon. Select **Favourites > Solar System > Moon > Eagle has Landed** to view Earth from the *Apollo 11* landing site. Stop time flow and press the "k" key on the keyboard to remove the constellation display from the view. Click the **Now** button in the toolbar. Use the Hand Tool to locate Earth in the lunar sky. If Earth is near the "new" phase, it will be hard to see, but the label will help. With the cursor over the image of Earth in the view, click the right mouse button (**Ctrl–Click** on a Macintosh) to open the object contextual menu for Earth and select **Magnify**. Change the **Time Flow Rate** to **10 minutes** and run time forward for a while. **a.** Observe the surface features of Earth that you see. Do they change? If so, explain why, considering that there is no significant change in the face of the Moon as seen from Earth. **b.** Run time for a much longer period and describe the phases of Earth, ignoring the changes in features.

58. Use *Starry Night Enthusiast*™ to observe the changing appearance of the Moon. Click the **Favourites** tab and select the **Moon phases** view from the **Solar System > Moon** folder. Then click the **Now** button in the toolbar. **a.** What phase is the Moon in? Set the **Time Flow Rate** to between 1 and 6 hours, depending on your computer's speed, and **Start** time flow. **b.** Describe how the phase of the Moon changes. From which side does the shadow come and how does the Moon's shape (phase) change? **c.** Choose **Favourites > Guides > Atlas**. Now select the **Find** tab, type **Moon** in the edit box and hit **Enter**. Zoom until the screen is about 1° of the sky. Set the **Time Flow Rate** to **1 day** and, using the single timestep button ▶ look carefully at the features where the shadows first touch the Moon and last leave it. Are the same lunar features always at these locations? They should be if the Moon's synchronous rotation were the *only* motion that it had, relative to Earth. If not, explain. (Hint: A Web search on the word "libration" [not libation!] may help.) **d.** Does the apparent size of the Moon remain constant? If not, explain what this tells us about the shape of the Moon's orbit around Earth.

59. Exploring Earth. In this exercise, we use the *Starry Night Enthusiast*™ program to examine Earth as seen from space. Select **Favourites > Solar System > Earth > Earth** from the menu. **Stop** time flow. You can use the **Zoom** buttons in the toolbar to zoom in and zoom out. (You can remove the astronaut's feet if desired by selecting **View > Feet**). You can also rotate Earth with the **Location Scroller** by holding down the **Shift** key while holding down the mouse button and moving the mouse (Hold down the left mouse button on a PC.) **a.** Use the **Location Scroller** to move Earth around until you see the Sun almost behind it. From what you now see of Earth's phase, explain how you can

easily find the Sun in searches for it at other planets. Go to Mars by selecting **Favourites > Solar System > Inner Planets > Mars** from the menu. **Stop** time flow and, using what you have just learned, put the Sun behind the planet, thereby confirming your explanation above. **b.** Return to Earth by selecting **Favourites > Solar System > Earth > Earth** from the menu. Use the **Location Scroller** to move Earth around until South America is visible. Set the **Time Flow Rate** to 3000x and allow time to flow until the terminator (the dividing line between the day and night sides of Earth) is over Eastern Brazil, at the eastern edge of South America. **Stop** time. As seen from eastern Brazil, is the sun rising or setting? Explain.

60. Exploring the Moon. In this exercise, we will use *Starry Night Enthusiast*™ to examine the Moon. Click the **Favourites** tab and select the **Moon phases** view from the **Solar System > Moon** folder, and **Stop** time flow when the Moon's phase is full. You can use the **Zoom** buttons in the toolbar to zoom in and out. You can use the **Location Scroller** to rotate the Moon by putting the mouse cursor over the image, holding down the mouse button, and moving the mouse. (Hold down the left mouse button on a PC.) **a.** Based only on what you can see in the image, what evidence can you find that the Moon is geologically inactive? Explain. **b.** Spreading outward from some of the largest craters on the Moon are straight lines of slightly lighter-colored material, called rays, which were caused by material ejected outward by the impact that caused the crater. Grabbing and moving the Moon around might help you see them. Print out the screen, if possible. This will give you a grey disk on which to make a map of the lit side of the Moon. (Otherwise, make a drawing from scratch). Indicate craters that have rays. Place the cursor over these craters and open the contextual menu (**right-click** on a PC, **Ctrl-click** on a Mac). The name of the crater will be found in the command "Mark crater name on surface."

WHAT IF... The Moon Didn't Exist?

Throughout history, people have woven myths about the Moon and its effects on everything from childbirth to stock market activity. Countless romances have begun under a full Moon, and entire nations have dedicated themselves to reaching it.

What if the Moon never existed? Would life even have developed on Earth? If so, how would it be different? Would a self-aware species like ourselves have evolved?

Delayed Origins It would have been much more difficult for life to start evolving on a Moonless Earth. In the first place, the minerals from which life developed in our Earth's young oceans were swept down there, in large part, by gigantic tides, a thousand times higher than the tides of today, created by the Moon when it was some 10 times closer to Earth than at present. The newly formed Earth, spinning more than 4 times faster than at present, caused these tides to move miles inland and back out to sea every 1½ h or so.

A Moonless Earth would still have tides, created by the Sun. At most, however, these tides would never be more than one-third as high as Earth's present tides. Minerals would wash into the oceans incredibly slowly from the flow of rivers. As a result, it is likely that it would have taken much longer, hundreds of millions or even billions of years longer, for enough minerals to be dissolved for life to firmly establish itself.

Harsh Conditions Animal life's transition from oceans to land would also be much harder because of continuous winds, between 50 and 150 mph, created by the planet's rapid rotation. The resulting waves (wind causes waves) would be enormous and perpetual. Fish with legs near shores would almost certainly be pounded to a pulp rather than being able to sedately walk onto land and then back into the water, as apparently happened on Earth.

But we see on Earth how life develops in the most incredible forms and places. It seems plausible, then, that sea life would find a way to make the transfer to dry land and continue to evolve there.

Allowing then for diverse terrestrial life on the Moonless Earth, what would be different about that life compared to life on our Earth? For one thing, creatures evolving there would have to withstand the perpetual pummeling from winds and the debris they carry. Turtle-like shells are one solution.

Obviously, there would be no eclipses or moonlight—all clear nights would be equally dark and starfilled. Therefore, nocturnal animals would be less successful at hunting, foraging, and traveling. Instead, these animals might evolve more enhanced senses to compensate for their inability to see visible light well at night.

Rush Hour Clearly, naked apes do not seem a likely bet on a Moonless Earth, or birds battling the ever-present winds. But given enough time, complex, even self-aware life likely would evolve. After all, we evolved because of the challenges faced by our ancestors, and Earth without a Moon would clearly provide challenges of its own.

The physiology of life on Earth evolved based on a 24-h day. This is most evident in our biological clocks, or circadian rhythms, which are the internal mechanisms that regulate sleeping, waking, eating, and other cyclic activities. Faced with a 6-h day, these circadian rhythms would be hopelessly out of sync with the natural world. To function on such a world, all of its creatures would have to evolve biological clocks based on 6 h, which certainly could have occurred. Considering all that we have to do now, imagine what life would be like with only 6 h in each day!

Wobbling Earth Finally, rapidly rotating terrestrial planets without large moons are unstable, and calculations show that they wobble significantly, dramatically changing the direction of their rotation axes. If Earth had no Moon, this process would create enormous stresses on the planet's surface, leading to catastrophic earthquakes, volcanic activity, and tsunamis, as well as severe atmospheric and magnetic field changes. Our Moon stabilizes Earth's rotation, preventing such phenomena. It would be incomparably harder for complex life to persist on a moonless Earth than it is on our world.

CHAPTER 16

The Other Terrestrial Planets

Photographed by NASA's *Mars Reconnaissance Orbiter*, this shaft descends at least 78 m (255 ft), and probably much deeper, into the side of volcano Arsia Mons. Similar features form on Earth when deep underground caves collapse. (NASA/JPL/University of Arizona)

R I V U X G

WHAT DO YOU THINK?

1 Which terrestrial planet—Mercury, Venus, Earth, or Mars—has the coolest surface temperature?

2 Which planet is most similar in size to Earth?

3 Which terrestrial planet—Mercury, Venus, Earth, or Mars—has the hottest surface temperature?

4 What is the composition of the clouds that surround Venus?

5 Does Mars have liquid water on its surface today? Did it have liquid surface water in the past?

6 Is life known to exist on Mars today?

Answers to these questions appear in the text beside the corresponding numbers in the margins and at the end of the chapter.

Every human being is unique: Each person has his or her own genetic makeup and personal history. On the other hand, people have many similarities: we are either male or female; we all breathe air; and, under normal circumstances, we have common characteristics, such as two eyes, two hands, and ten toes, among other traits. To understand humans fully, biologists and physicians study our common features and then our individual peculiarities, just as psychiatrists and psychologists study our common behaviors and then our differences.

The planets in our solar system also have similarities and differences that astronomers are learning to understand. For example, there are two basic groups of planets: terrestrial worlds (Mercury, Venus, and Mars), similar in size and chemistry to Earth, and the much larger giant or Jovian worlds (Jupiter, Saturn, Uranus, and Neptune).

In this first of two chapters on the rest of the planets, we explore the terrestrial worlds both individually and in comparison to each other and to Earth. The remaining four outer planets are presented in the next chapter.

In this chapter you will discover

- Mercury, a Sun-scorched planet with dormant volcanoes, a heavily cratered surface, and a substantial iron core ?

- Venus, perpetually shrouded in thick, poisonous clouds and mostly covered by gently rolling hills

- Mars, a red, dusty planet that once had running water on its surface and may still have liquid water underground

MERCURY

WEB LINK 16.1 / ANIMATION 16.1 The closest planet to the Sun, Mercury, is a truly inhospitable world of temperature extremes. Its incredibly thin atmosphere and weak magnetic field leave it virtually unprotected from countless impacts and a continuous bath of deadly solar radiation. Despite having a surface nearly as dark as coal, Mercury sometimes appears as one of the brightest objects in our sky, due to its proximity to the Sun. We can see it from Earth only for a few hours before sunrise or a few hours after sunset because its angle from the Sun (its elongation) is always less than 28°.

16-1 Photographs from *Mariner 10* and *Messenger* spacecraft reveal Mercury's lunar-like surface

WEB LINK 16.1 Until late last century, Mercury (Figure 16-1) was a complete mystery. Earth-based telescopic studies provided virtually no information about the planet's surface. The three visits by *Mariner 10* photographed only 79 percent of Mercury's surface. It was only in 2008 that the *Messenger* spacecraft imaged the rest of it. *Messenger* is short for MErcury Surface, Space ENvironment, GEochemistry, and Ranging.

Unlike our Moon, Mercury has a variety of volcanic craters out of which poured large amounts of lava that filled smaller impact craters and created smooth regions between larger ones. Like the Moon, but unlike any other planet, the pounding that Mercury received back then created what we see today. Our strongest evidence for believing this scenario and the timescale of this cratering comes from analyzing and dating Moon rocks (see Chapter 15). Like the Moon, Mercury has an exceptionally low albedo of 0.12 (12% scattering of incoming light). It is very

Mercury   Planet symbol: ☿

R I V U X G

WEB LINK 16.2 **FIGURE 16-1** Mercury Heavily cratered Mercury was visited three times by the *Mariner 10* spacecraft in 1974 and 1975 and twice by *Mercury Messenger* in 2008. This image, taken by *Messenger,* shows the side of Mercury never before seen. It was taken from an altitude of 27,000 km (17,000 mi). *Messenger* will settle into orbit around Mercury on March 18, 2011. (Credit NASA/John Hopkins University Applied Physics Laboratory/Carnegie institution of Washington)

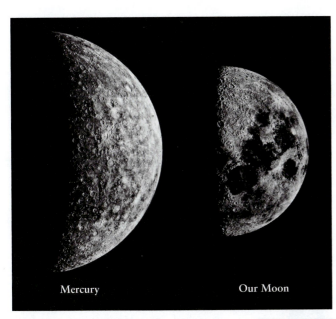

R I V U X G

FIGURE 16-2 Mercury and Our Moon Mercury and our Moon are shown here to the same scale. Mercury's radius is 2439 km and the Moon's is 1738 km. For comparison, the distance from New York to Los Angeles is 3944 km (2451 mi). Mercury's surface is more uniformly cratered than that of the Moon. Daytime temperatures at the equator on Mercury reach 700 K (800°F), hot enough to melt lead or tin. (NASA, UCO/Lick Observatory)

bright as seen from Earth only because the sunlight scattering from Mercury is so intense.

First impressions of Mercury evoke a lunar landscape, and there are several similarities. Craters in the Moon's highlands are densely packed, as are some of the craters on Mercury (Figure 16-2). The impact craters on both worlds have similar features, such as ejecta blankets and central peaks, as discussed for the Moon in Section 15-5.

The most impressive feature discovered by *Mariner 10* is a huge circular region, called *Caloris Basin,* so named because *caloris* is Latin for "heat," and this feature is the closest part of Mercury to the Sun at the planet's perihelion. Caloris basin measures 1300 km (810 mi) in diameter, meaning that it has twice the area of Texas. Lying along the *terminator* (the border between day and night) in Figure 16-3a, you can see that the Caloris Basin is surrounded by a 2-km-high ring of mountains, beyond which are relatively smooth plains. Like the lunar maria, the Caloris Basin was probably gouged out by the impact of a large meteorite that penetrated the planet's crust. *Messenger* revealed small impact craters inside the Caloris Basin that have been filled in by lava from volcanoes that were active long after both the Basin and those craters were formed. A giant volcano 95 km in diameter (bigger than the state of Delaware) has been seen in the Basin.

The Caloris impact was a tumultuous event that shook the entire planet. Indeed, the collision affected the side of Mercury directly opposite of Caloris Basin. That area (Figure 16-3b) has a jumbled, hilly surface

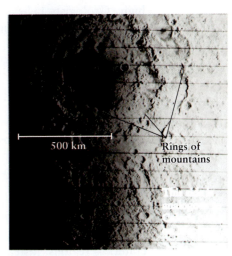

c Mare Orientale on our Moon

R I V U X G

a   b

FIGURE 16-3 Major Impacts on Mercury and on Our Moon (a) Caloris Basin *Mariner 10* sent back this view of a huge impact basin on Mercury's equator. Only about half the Caloris Basin appears because it happened to lie on the terminator when the spacecraft sped past the planet. Although the center of the impact basin is hidden in the shadows (just beyond the left side of the picture), several semicircular rings of mountains reveal its extent. (b) Unusual, Hilly Terrain What look like tiny, fine-grained wrinkles on this picture are actually closely spaced hills, part of a jumbled terrain that covers nearly 500,000 square kilometers on the opposite side of Mercury from the Caloris Basin. The large, smooth-floored crater, Petrarch, has a diameter of 170 km (106 mi). This impact crater was produced more recently than Caloris Basin. (c) Mare Orientale on the Moon This giant impact feature on our Moon is 900 km (560 mi) in diameter and is very similar to Caloris Basin on Mercury (compared to part a). There is also terrain similar to that shown in (b), on the opposite side of the Moon from Mare Orientale. (NASA)

R I V U X G

VIDEO 16.1 **FIGURE 16-4** Mercury's Craters and Plains This view of Mercury's northern hemisphere was taken by *Mariner 10* as it sped past the planet in 1974. Numerous craters on the bottom half of the image and broad intercrater plains on the top half cover an area 480 km (300 mi) wide. (NASA)

covering nearly half a million square kilometers, about twice the size of Wyoming. The hills, which appear as tiny wrinkles that cover most of the photograph, are about 5 to 10 km wide and between 100 and 1800 m high. Geologists believe that energy from the Caloris impact traveled through the planet and became focused, like light through a lens, as it passed through Mercury. As this concentrated energy reached the far surface of the planet, jumbled hills were pushed up. A similar pair of phenomena to Caloris and the jumbled terrain on the other side of Mercury can be seen in our Moon's Mare Orientale (Figure 16-3c) and chaotic hills on the opposite side of that world. Finding such a similar pair of phenomena on a different world supports the connection between the impact and the jumbled surface.

Looking at Mercury in more detail, we start to see a variety of differences from the features on the Moon. Unlike the Moon and its maria, Mercury lacks extensive craterless regions. Instead of maria, Mercury has plains with relatively small craters. Such plains are not found on the Moon. Figure 16-4 shows a typical close-up view of Mercury. As we learned in Section 15-7, the lunar maria were produced by extensive lava flows that occurred between 3.1 and 3.8 billion years ago. Ancient lava flows also probably formed the Mercurian plains. As large meteorites punctured the planet's thin, newly formed crust, lava welled up from the molten inte-

The rings of mountains frozen into Mercury's surface have what *transient* analog on Earth?

1. The floors of these craters were flooded by lava from Mercury's interior.

2. Some time after the lava cooled, Mercury's crust contracted to form this scarp.

3. This crater was distorted when the scarp formed.

50 km

R I V U X G

ANIMATION 16.2 **FIGURE 16-5** Scarps on Mercury A long, meandering cliff, called *Santa Maria Rupes*, runs from north to south across this *Mariner 10* image of a region near Mercury's equator. This cliff, called a *scarp*, by geologists, is more than 1 km high and runs for several hundred kilometers. Note how the crater in the center of the image was distorted vertically when the scarp formed. (NASA)

rior to flood low-lying areas. The existence of craters pitting Mercury's plains suggests that these plains formed just over 3.8 billion years ago, near the end of the era of heavy bombardment. Mercury's plains are therefore older than most of the lunar maria, leaving more time for cratering to eradicate maria-like features.

Mariner 10 and *Messenger* also revealed numerous, long cliffs, called **scarps**, meandering across Mercury's surface (Figure 16-5). The scarps are believed to have developed as the planet cooled. Almost everything that cools, contracts. Therefore, as Mercury's mantle and molten iron core cooled and contracted, its surface moved inward. Because it was solid, Mercury's crust could not collapse uniformly. Instead, it wrinkled as it contracted, forming the scarps. These features and the lack of recent volcanic activity suggest that the planet's interior is solid to a significant depth. Otherwise, lava would have leaked out as the scarps formed.

16-2 Mercury has a higher percentage of iron than does Earth

Mercury's average density of 5430 kg/m^3 is quite similar to Earth's (5520 kg/m^3). As we saw in Section 15-3, typical rocks on Earth's surface have a density of only about 3000 kg/m^3 because they are composed primarily of lightweight elements. The high average densities of both Mercury and our planet are caused by their dense interiors.

Because Mercury is less dense than Earth, you might conclude that Mercury has a lower percentage of iron, a

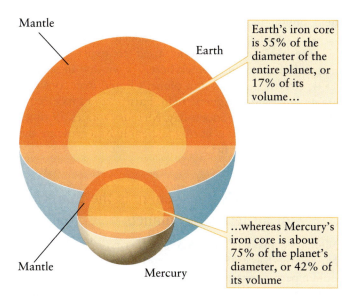

Earth's iron core is 55% of the diameter of the entire planet, or 17% of its volume…

…whereas Mercury's iron core is about 75% of the planet's diameter, or 42% of its volume

FIGURE 16-6 The Interiors of Earth and Mercury Mercury has the highest percentage of iron of any planet in the solar system. Consequently, its iron core occupies an exceptionally large fraction of its interior.

heavy element, than does our planet. Indeed, *Messenger* revealed that its surface has a low iron content. However, in percent of its total mass, Mercury is the most iron-rich planet in the solar system. Only Earth's greater mass, pressing inward and thereby compressing our planet's inner parts, makes it denser than Mercury.

Figure 16-6 shows a scale drawing of Mercury's interior, where an iron core fills 42% of the planet's volume. Surrounding the core is a 600-km-thick rocky mantle. For comparison, Earth's iron core occupies only 17% of the planet's volume. How much of Mercury's core is molten is still unknown.

Events early in Mercury's history must somehow account for its high iron content. We know that the inner regions of the primordial solar nebula were incredibly hot. Perhaps only iron-rich minerals were able to withstand the solar heat there, and these subsequently formed iron-rich Mercury. According to another theory, an especially intense outflow of particles from the young Sun stripped Mercury of its low-density mantle shortly after the Sun formed. A third possibility is that, during the final stages of planet formation, Mercury was struck by a large planetesimal (debris formed as the solar system came into being), just as Earth was struck by a Mars-sized body that led to the formation of our Moon. Computer simulations show that this cataclysmic collision would have ejected much of Mercury's lighter mantle (Figure 16-7).

If Mercury were struck by a large planetesimal, why might that planet not have had a moon form, as Earth did when it was struck?

16-3 Mercury's rotation and revolution are coupled

Mercury has one of the most unusual orbits in the solar system. Recall from Section 15-8 that when Earth was young, its gravitational force created tides on the Moon, thereby forcing the Moon into synchronous rotation. Something similar (but not identical) happened with the Sun playing the role of Earth, and Mercury acting in the place of our Moon. To begin with, the Sun created a significant tidal bulge on young, molten Mercury that locked into place when the planet solidified. Mercury is so close to the Sun (average separation: 0.387 AU) that the Sun's gravitational force on Mercury's tidal bulges changed the planet's rotation rate. However, Mercury's highly eccentric orbit (e = 0.21) prevented the planet from being locked into synchronous orbit like our Moon.

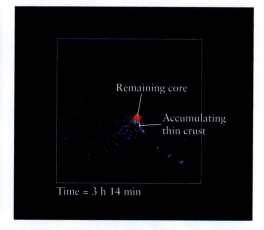

FIGURE 16-7 The Stripping of Mercury's Mantle To account for Mercury's high iron content, one theory proposes that a collision with a massive planetesimal stripped Mercury of most of its rocky mantle. These three images show a computer simulation of a nearly head-on collision between proto-Mercury and a body one-sixth its mass. Both worlds are shattered by the impact, which vaporizes much of their rocky mantles. Mercury eventually re-forms from the remaining iron-rich debris. The rest of the original Mercury and the impactor left this area of the solar system. (a, b, and c: Royal Astronomical Society)

Instead, Mercury developed what is called a **3-to-2 spin-orbit coupling**. This means that Mercury undergoes three sidereal rotations (rotations measured with respect to the distant stars, not the Sun), while undergoing two revolutions around the Sun (Figure 16-8).

The major consequence of spin-orbit coupling is that one or the other of Mercury's regions of high tide is facing the Sun whenever the planet is at perihelion (see Figure 16-8). There is some slight variation in each orbit so that the exact high tide doesn't always point toward the Sun.

A Day on Mercury Is Two Mercury Years Long   The motion of the Sun across Mercury's sky is unique in the solar system. First, a solar day there (noon to noon) is 176 Earth days long, twice the length of a year on Mercury! Furthermore, if you were to set up a camera at the location of high tide on Mercury to take a series of pictures when the planet is passing through perihelion, you would see the Sun start to rise in the east, stop high in the sky, move back toward the east, stop again, and then resume its westward journey. This is analogous to the retrograde motion of the planets that we observe from Earth (see Section 2-2).

Mariner 10 discovered that Mercury has a global weak magnetic field similar in shape to that of the Earth (see Section 15-4). Explaining it presents a challenge to astronomers because the dynamo model used to explain Earth's magnetic field doesn't work here. As we saw in Section 15-4, electric currents in a planet's liquid iron core create a planetwide magnetic field. Earth creates its electric current by rotating once a day. Mercury rotates 59 times more slowly, which is hardly fast enough to generate the observed magnetic field.

> Why isn't Mercury in synchronous rotation with respect to the Sun?

16-4 Mercury's atmosphere is the thinnest of all terrestrial planets

Mercury's mass is only 5.5% that of Earth. Like our Moon, the force of gravity is too weak on Mercury to hold a permanent atmosphere, but trace amounts of five different gases have been detected around Mercury. Scientists think the Sun is the source of the hydrogen and helium gas near it, while sodium and potassium gas escape from rocks inside the planet (a process, called *outgassing*, which also occurs on Earth). Oxygen observed in Mercury's atmosphere may come from polar ice that is slowly evaporating. All of these gases drift into space and are continually being replenished in the atmosphere from their respective sources. The average density of this atmosphere is at least 10^{17} times less dense than the air we breathe.

a

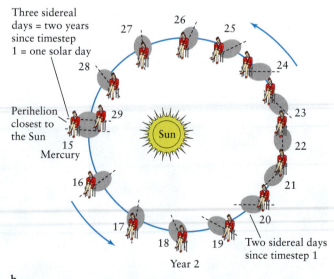

b

FIGURE 16-8 3-to-2 Spin-Orbit Coupling Mercury undergoes three sidereal rotations every 2 years. You can see this by following an observer on Mercury (a) from timestep 1 to timestep 15 and then (b) from timestep 15 to timestep 29. The observer points to the right 4 times during this interval (at times 1, 10, 20, and finally, at 29).

This means that Mercury has rotated three times in exactly two sidereal Mercurian years. Because a sidereal year is 88 Earth days long, a sidereal day on Mercury is 58.7 Earth days long. During the same interval, however, the Sun is at noon as seen from the observer's location only twice: at timestep 1 and timestep 29. Therefore, a solar day is 176 Earth days long.

Mercury's Temperature Range Is the Most Extreme in the Solar System Earth's thick atmosphere stores heat, which helps explain why it feels warm on a cloudy day and at night. Nevertheless, Earth continually loses heat into space. Because of Earth's rapid daily rotation, this lost heat is quickly replaced by the Sun during daylight hours, so the average temperature change between day and night on Earth is only about 11 K (20°F). Because of Mercury's slow rotation and minimal atmosphere, the differences in temperature between day and night there are far more noticeable than on Earth. At noon on Mercury, the surface temperature is 700 K (800°F). At the terminator, where day meets night, the temperature is about 425 K (305°F). On the night side, the temperature decreases to as low as 100 K (−280°F), the coldest temperature on any terrestrial planet! The resulting daily range of temperature is therefore 600 K (1080°F) on Mercury.

Just as astronomers use scattered radio waves to search for water on the Moon (see Section 15-6), they have also sent radio waves to Mercury. In 1992 they made an extraordinary discovery—evidence for ice near Mercury's poles, in craters that are permanently in shadow. Dozens of circular regions, presumably craters, sent back signals with characteristics distinct to ice. Confirmation of the presence of this ice will have to wait until the *Messenger* spacecraft begins orbiting it in 2011. If the ice is there, its origin—whether from cometary impacts, from gases rising from inside the planet and then freezing, or from both sources—also remains to be determined.

INSIGHT INTO SCIENCE

Model-Building In modeling real situations, scientists must consider several different effects simultaneously. Omit any crucial property and you get inaccurate results. For example, consider how long ice can remain at Mercury's poles. Astronomers must take into account the planet's distance from the Sun, the tilt of its axis of rotation, its rotation rate, surface features, whether the ice is exposed or mixed with other material, and the chemical composition of Mercury.

VENUS

One of the few things that Mercury and Venus have in common is that neither has a moon. Venus and Earth, on the other hand, have much in common. They have almost the same mass, the same diameter, and the same average density. Indeed, if Venus were located at the same distance from the Sun as is Earth, then it, too, might well have evolved life. However, Venus is 30% closer to the Sun than Earth, and this one difference

between the two planets leads to a host of others, making Venus inhospitable to life.

16-5 The surface of Venus is completely hidden beneath a permanent cloud cover

At nearly twice the distance from the Sun as Mercury, Venus is often easy to view without interference from the Sun's glare. At its greatest western elongation, Venus is seen high above the western horizon after sunset, where, like Mercury, it is often called the "evening star." Conversely, high in the eastern sky before sunrise, it is called the "morning star."

Venus is easy to identify because it is usually one of the brightest objects in the night sky. Only the Sun and the Moon outshine Venus at its greatest brilliance. Venus is often mistaken for a UFO (Figure 16-9), because when it is low on the horizon, its bright light is strongly refracted by Earth's atmosphere, making it appear to rapidly change color and position. You can see the appearance of the inner planets using *Starry Night Enthusiast*™ (see also *Guided Discovery: The Inner Solar System*).

> What technology would give us the most accurate information about Mercury's interior?

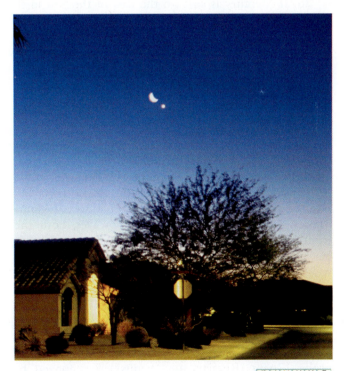

RIVUXG

FIGURE 16-9 UFO? Just Venus Venus is bright and is often seen near the horizon where rising and sinking gases in Earth's atmosphere make it appear to move and change color, like alleged UFOs. (Joe Orman)

The Inner Solar System

 Here we will learn more about the appearances of the inner planets as seen from Earth, using the *Starry Night Enthusiast*™ program.

Mercury Find Mercury (**Find** tab/Mercury). The first step is to see that Mercury is always close to the horizon when it is visible in the night sky.

1. If Mercury is below the horizon: Choose the **Best Time** option when the **not currently visible** warning comes up. If Mercury is to the west (to the right) of the Sun, go to *2a*. If Mercury is to the east (to the left) of the Sun, go to *2b*.

2a. If Mercury is already up in the sky: If Mercury is west (to the right) of the Sun, we want to set up the screen with Mercury as the "morning star." To do this, stop time, reset the time to 5 A.M., relocate Mercury, and run time forward or backward at about 300 × until Mercury is just above the eastern horizon and the Sun is below the horizon. This shows Mercury as the "morning star." Now, run time forward and you will see the white dot of Mercury fade. (If it is too close to the Sun, it won't be visible at all. Change the date and try again.) Now go to section 3.

2b. If Mercury is east (to the left) of the Sun in the sky, then set the time to 6 P.M., relocate Mercury, and run the time forward or backward at about 300 × until the Sun is about to set. Step forward in time by 1-minute intervals and watch Mercury first appear as the "evening star" and then set on the western horizon. (If it is too close to the Sun, it won't be visible at all. Change the date and try again.)

3. Now we explore Mercury's orbit. Go to Atlas mode (**Favourites/Guides/Atlas**) and again find Mercury. Zoom in to about 70°. Make sure you are centered and locked on Mercury (grab Mercury, right-click, and then left-click on **Centre** if it is not already checked). Set the timestep to 1 day (more or less, depending on your computer speed) and press **Run** (the right pointing triangle). Describe and explain the motion of the Sun relative to Mercury. You can learn more about Mercury as seen from Earth by trying *Observing Project 48* at the end of the chapter.

Venus Repeat the same procedures for Venus as we have described for Mercury, starting at step 1 above. What similarities and differences do you see?

Mars Return to Atlas mode, set the date to 9/12/2003, the time to midnight, and lock on Mars. Zoom in on Mars until you see its surface features. What is Mars's phase? Zoom away until you have the Sun on the screen or you have gone the maximum distance (about 100° angle on the screen). Determine how far (in angle) the Sun is away from Mars. If the Sun is not visible, you can make this measurement by "dragging" the celestial sphere around and measuring angles from the grid. Does the Sun ever get this far in angle from Mercury or Venus, as seen from Earth? Explain.

Unlike Mercury, Venus is intrinsically bright because it is completely surrounded by light-colored, highly reflective clouds (Figure 16-10). Because visible light telescopes cannot penetrate this thick, unbroken layer of clouds, we did not even know how fast Venus rotates until 1962. In the 1960s, however, both the United States and the former Soviet Union began sending probes there. The Americans sent fragile, lightweight spacecraft into orbit near the planet. The Soviets, who had more powerful rockets, sent massive vehicles directly into the Venusian atmosphere.

Building spacecraft that could survive the descent into Venus's atmosphere proved to be more frustrating than anyone had expected. Finally, in 1970, the Soviet probe *Venera* (Russian for "Venus") 7 managed to transmit data for 23 min directly from the Venusian surface. Soviet missions continued until 1983, measuring a surface temperature of 750 K (900°F) and a surface air pressure of 90 atm, among other things. This is the same pressure you would feel if you were swimming 0.82 km (2700 ft) underwater on Earth.

In contrast to Earth's present nitrogen- and oxygen-rich atmosphere (see Section 15-1), Venus's thick atmosphere is 96% carbon dioxide, with the remaining 4% mostly nitrogen. This atmosphere is remarkably similar to Earth's early carbon–dioxide-rich atmosphere. These gases were vented from inside Venus through volcanoes and other openings. Unlike Earth, however, Venus has no liquid oceans to dissolve the carbon dioxide or living organisms to convert it into oxygen and carbon compounds, so it remains in the atmosphere today.

Soviet spacecraft also discovered that Venus's clouds are confined to a 20-km-thick layer located 48 to 68 km above the planet's surface. Above and below the clouds are 20-km-thick layers of haze. Beneath the lower level of haze, the Venusian atmosphere is clear all the way down to the surface. Standing on the surface of Venus, you would experience a perpetually cloudy yellow day.

Unlike the clouds on Earth, which appear white from above, the cloud tops of Venus appear yellowish or yellow-orange to the human eye. These colors are typical of sulfur and its compounds. Indeed, spacecraft

Venus   Planet symbol: ♀

RIVUXG

FIGURE 16-10 Venus Venus's thick cloud cover efficiently traps heat from the Sun, resulting in a surface temperature even hotter than that on Mercury. Unlike Earth's clouds, which are made of water droplets, Venus's clouds are very dry and contain droplets of concentrated sulfuric acid. This ultraviolet image was taken by the Pioneer Venus Orbiter in 1979. (NASA)

found substantial amounts of sulfur dust in Venus's upper atmosphere and sulfur dioxide and hydrogen sulfide at lower elevations. The clouds themselves are composed of droplets of concentrated sulfuric acid! Because of the tremendous atmospheric pressure at Venus's surface, the droplets remain suspended as a thick mist, rather than falling as rain.

All of the major chemical compounds spewed into our air by Earth's volcanoes have been detected in Venus's atmosphere. Among these molecules are large quantities of sulfur compounds. Because many of these substances are very short-lived, they must be constantly replenished by new eruptions. In fact, the abundance of sulfur compounds in Venus's atmosphere does vary. These data suggest the possibility that the sulfurous compounds in Venus's atmosphere come from active volcanoes. This idea is supported by the discovery, in 2007, of lightning on Venus. Lightning is a frequent event around active volcanoes here on Earth. The belief that lightning occurs on Venus comes from the detection of variations in the magnetic field around the planet caused by the electrical discharge that is lightning. This discharge creates a short-lived magnetic field that was detected by the Venus Express satellite.

The results of the early probing of Venus's atmosphere are summarized in Figure 16-11. Both pressure and temperature decrease smoothly with increasing altitude.

From the changes in temperature and pressure above a planet's surface, we begin to understand the structure of its atmosphere.

> Why does the day have a yellow hue on Venus?

16-6 The greenhouse effect heats Venus's surface

At first, no one could believe reports that the surface temperature on Venus was higher than the surface temperature on Mercury, which, after all, is closer to the Sun. Setting aside this initial skepticism, astronomers found a straightforward explanation—the **greenhouse effect** (Figure 16-12).

Carbon dioxide is responsible for the warming of Venus's atmosphere. Like your car windows, carbon dioxide is transparent to visible light but absorbs infrared radiation. Although most of the visible sunlight striking the Venusian cloud tops is reflected back into space, enough light reaches the Venusian surface to heat it. The warmed surface, which is also heated from inside the planet, in turn, emits infrared radiation, which is absorbed by Venus's carbon-dioxide-rich atmosphere. Some of this trapped radiation returns to the planet's surface, producing the high temperatures found there. The rest of the heat is radiated into space.

Without the greenhouse effect, the surface of Venus would have a noontime temperature of 465 K.

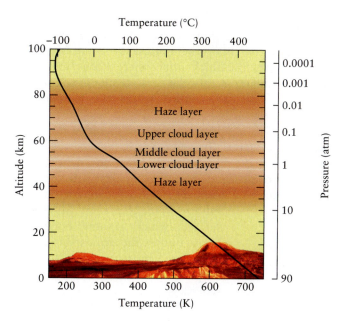

FIGURE 16-11 Temperature and Pressure in the Venusian Atmosphere The pressure at the Venusian surface is a crushing 90 atm (1296 lb/in²). Above the surface, atmospheric pressure decreases smoothly with increasing altitude. Descending into Venus' atmosphere, the temperature increases smoothly from a minimum of about 173 K (−150°F) at an altitude of 100 km to a maximum of nearly 750 K (900°F) on the ground. (NASA/Magellan Images [JPL])

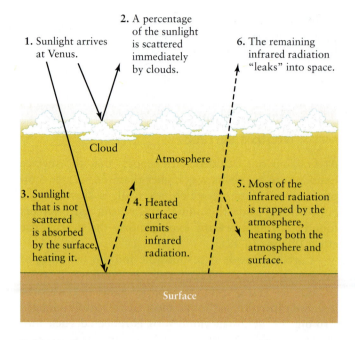

1. Sunlight arrives at Venus.

2. A percentage of the sunlight is scattered immediately by clouds.

6. The remaining infrared radiation "leaks" into space.

Cloud

Atmosphere

3. Sunlight that is not scattered is absorbed by the surface, heating it.

4. Heated surface emits infrared radiation.

5. Most of the infrared radiation is trapped by the atmosphere, heating both the atmosphere and surface.

Surface

FIGURE 16-12 The Greenhouse Effect A portion of the sunlight penetrates through the clouds and atmosphere of Venus, heating its surface. The surface in turn emits infrared radiation, much of which is absorbed by carbon dioxide (and to a much lesser degree, water vapor). The trapped radiation helps increase the average temperatures of the surface and atmosphere. Some infrared radiation does penetrate the atmosphere and leaks into space. In a state of equilibrium, the rate at which the planet loses energy to space in this way is equal to the rate at which it absorbs energy from the Sun.

Because of it, that temperature is actually a sweltering 750 K, hotter than the hottest spot on Mercury! Furthermore, the thick atmosphere keeps the night side of Venus at nearly the same temperature, unlike the night side of Mercury, where the temperature drops precipitously.

This high temperature prevents liquid water from existing on Venus. Just like Earth, Venus was likely to have initially had water both in the debris from which it formed and from the impacts of ice-bearing bodies, such as comets. The high atmospheric temperature has evaporated any water that came to Venus's surface, which is now bone dry. Furthermore, the atmospheric heat has caused most of the water that evaporated into the air to drift into space.

| Where on Earth, besides in a greenhouse, does the greenhouse effect occur? |

16-7 Venus is covered with gently rolling hills, two "continents," and numerous volcanoes

Soviet spacecraft that landed on Venus provided us with intriguing close-up images of the planet's arid surface. Figure 16-13 is a view of Venus's regolith taken by

Venera 13 in 1982. Russian scientists believe that this region was covered with a thin layer of lava that contracted and fractured upon cooling to create the rounded, interlocking shapes seen in the photographs. Indeed, measurements by several spacecraft indicate that Venusian rock is quite similar to lava rocks, called *basalt*, which are common on Earth and the Moon.

 The entire Venusian surface was mapped by the *Magellan* spacecraft, which arrived at Venus in 1990. In orbit about the planet, *Magellan* sent radar signals through the clouds surrounding Venus. Like radar used by police, some of the radar signals bounced off Venus and returned to the spacecraft. By measuring the time delay of the radar echo, scientists determined the heights and depths of Venus's hills and valleys. As a result, astronomers have been able to construct a three-dimensional map of the planet. The detail visible on this map (also called *resolution*) is about 75 m. You could see a football stadium on Venus if there were any. (None were detected.)

Venus is remarkably flat compared to Earth. More than 80% of Venus's surface is covered with volcanic plains and gently rolling hills created by numerous lava flows. Figure 16-14 is an image using the *Magellan* radar data to create a view of a Venusian landscape.

Global radar images revealed just two large highlands, or "continents," rising well above the generally level surface of the planet (Figure 16-15). The continent at high latitudes in the northern hemisphere is called *Ishtar Terra* (after the Babylonian goddess of love) and is approximately the same size as Australia. Ishtar Terra is dominated by a high plateau ringed by towering mountains. The highest mountain is Maxwell Montes, whose summit rises to an altitude of 11 km above the average surface. For comparison, Mount Everest on Earth rises 9 km above sea level.

The larger Venusian continent, *Aphrodite Terra* (named after the Greek goddess Venus by the Romans), is a vast belt of highlands that straddles the equator. Aphrodite is 16,000 km (10,000 mi) in length and 2000 km (1200 mi) wide, giving it an area about one-half that of Africa. The global view of the surface of Venus in Figure 16-16 shows that most of Aphrodite Terra is covered by vast networks of faults and fractures.

Venus has more than 1600 major volcanoes and volcanic features and fewer than a thousand impact craters (Figure 16-17), as compared to the hundreds of thousands seen on the Moon and Mercury. Today, Venus's thick atmosphere heats, and thereby vaporizes, much of the infalling debris that would otherwise create craters. Although pieces larger than around a kilometer in diameter would blast through even this thick atmosphere, most of the rubble entering Venus's atmosphere is turned into dust by friction before it reaches the ground. However, there should have been

a   Image from *Venera 13*

b   Color-corrected image

c

FIGURE 16-13 The Venusian Surface (a) This color photograph, taken by a Soviet spacecraft, shows rocks that appear orange because the light was filtered through the thick, sulfur-rich clouds. (b) By comparing the apparent color of the spacecraft to the color it was known to be, computers can correct for the sulfurous light. The actual color of the rocks is gray. In this view, the rocky plates that cover the ground may be fractured segments of a thin layer of lava. The toothed wheel in each image is part of the landing mechanism that keeps the spherical spacecraft from rolling. (c) By correcting for the curvature of this image and a similar one taken with the camera flipped upside down, the actual surface of Venus near the Venera Lander can be shown. (a: Courtesy of C. M. Pieters and the USSR Academy of Sciences; b and c: Don Mitchell)

a period shortly after Venus formed and before its atmosphere developed (from gases escaping from the planet's interior) when impacts and cratering were common.

Accepting that the young Venus was cratered like every other solid object in the solar system, the small number of craters on it needs to be explained. The problem is that unlike Earth, Venus lacks the large-scale tectonic plates that divide up and refresh Earth's

surface. Put another way, Venus is a one-plate planet. Without plates moving relative to each other, how can the craters be erased? Several mechanisms to explain this are being explored. Most theories posit that the lack of surface motion has forced Venus's crust to become thicker than Earth's crust. In one theory, heat from radioactive elements inside the planet causes mantle convection without tectonic plate motion. The temperatures inside the planet eventually become high

Maat Mons (volcano)

Young lava flows from Maat Mons

Sapas Mons (volcano)

Lava flows from Sapas Mons extend for hundreds of kilometers

500 km

R I V U X G

FIGURE 16-14 A Venusian Landscape A computer combined radio images to yield this perspective view of Venus as you would see it from an altitude of 4 km (2.5 mi). The color results from light being filtered through Venus's thick clouds. The brighter color of the extensive lava flows indicates that the flows reflect radio waves more strongly. The vertical scale has been exaggerated 10 times to show the gentle slopes of Sapas Mons and Maat Mons, volcanoes named for ancient Phoenician and Egyptian goddesses, respectively. (NASA, JPL Multimission Image Processing Laboratory)

enough in the regions of rising mantle to create massive volcanoes out of which pour enormous amounts of magma that cover large areas of the surface. A variation of this theory proposes that the thickening crust is suddenly thinned from below by the horizontal convective motion of the mantle, like stripping layers off of

plywood. When so thinned, heat is able to rise through it and cause many volcanoes to form.

Another theory proposes that, episodically, the one-plate surface of Venus is heated below enough to crack and begin Earthlike tectonic plate motion until it cools down and again becomes a single plate. During the period of plate motion, the surface is erased and replaced.

In yet another theory, the mantle becomes so hot that large sections of the crust melt more or less simultaneously, destroying the old crust and its craters. Indeed, that theory has been expanded to periodic global meltdowns occurring every 700 million years or so. For all of these models, when enough heat is released from the planet's interior, its surface once again solidifies, and, with the possible exception of local volcanic activity, the crust re-thickens.

The sulfur content of the air and the traces of active volcanoes on the surface are compelling indicators that, like Earth, Venus has a molten interior. Because the average density of Venus is similar to that of Earth, its core is predominantly iron. Currents in the molten iron should generate a magnetic field; however, none of the spacecraft sent there has detected one. In general, magnetic fields are created by rotation of currents (see Section 15-4). Therefore, Venus's apparent lack of a magnetic field is plausible only if it rotates exceptionally slowly. In fact, the planet takes 116.8 Earth days to get from one sunrise to the next, if you could see the Sun from Venus's surface.

Unlike Earth and Mercury, Venus exhibits **retrograde rotation**. This means that the direction of Venus's orbit around the Sun (counterclockwise as seen from space far above Earth's North Pole) is *opposite* the direction of its rotation (clockwise as seen from the same vantage point). In other

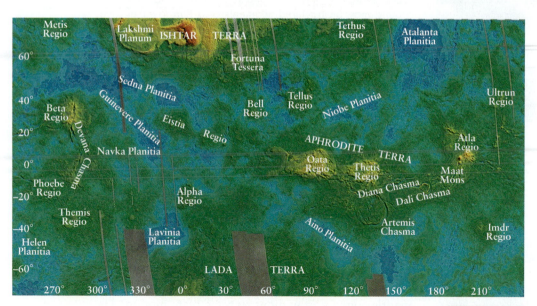

Metis Regio
Lakshmi Planum ISHTAR TERRA
Tethus Regio
Atalanta Planitia
Fortuna Tessera
Sedna Planitia
Beta Regio
Guinevere Planitia
Eistla Regio
Bell Regio
Tellus Regio
Niobe Planitia
Ultrun Regio
Navka Planitia
Devana Chasma
APHRODITE TERRA
Oata Regio
Thetis Regio
Atla Regio
Phoebe Regio
Alpha Regio
Diana Chasma
Maat Mons
Dali Chasma
Themis Regio
Aino Planitia
Artemis Chasma
Imdr Regio
Helen Planitia
Lavinia Planitia
LADA TERRA

270° 300° 330° 0° 30° 60° 90° 120° 150° 180° 210°

FIGURE 16-15 A Map of Venus This false-color radar map of Venus, analogous to a topographic map of Earth, shows the large-scale surface features of the planet. The equator extends horizontally across the middle of the map. Color indicates elevation—red for highest, followed by orange, yellow, green, and blue for lowest. The planet's highest mountain is Maxwell Montes on Ishtar Terra. Scorpion-shaped Aphrodite Terra, a continent-like highland, contains several spectacular volcanoes. Do not confuse the blue and green for oceans and land. (Peter Ford/MIT, NASA/JPL)

R I V U X G

RIVUXG

50 km

RIVUXG

FIGURE 16-16 A "Global" View of Venus A computer using numerous *Magellan* images creates a simulated globe. Color is used to enhance small-scale structures. Extensive lava flows and lava plains cover about 80% of Venus's relatively flat surface. The bright band running almost east-west is the continent-like highland region Aphrodite Terra. (NASA)

FIGURE 16-17 Craters on Venus These three impact craters, with extensive ejecta surrounding each, are located on Venus's southern hemisphere. They were imaged using radar by the *Magellan* spacecraft. The colors are based on the Venera images (see Figure 16-13). (NASA/Magellan Images [JPL])

words, sunrise on Venus occurs in the west. Venus's rotation axis is tilted more than 177°, compared to Earth's 23½° tilt. Because Venus's axis is within 3° of being perpendicular to the plane of its orbit around the Sun, the planet has no seasons. Although we do not know the cause of Venus's retrograde rotation, one likely explanation is that a monumental impact flipped the rotation axis early in the planet's existence.

> Why does little cratering occur on Venus today, even compared to the present low rate of cratering on our Moon?

MARS

Mars's distinctive rust-colored hue (Figure 16-18) makes it stand out in the night sky. For centuries Mars has generated more excitement as a possible home for alien life than any other world in our solar system. The idea began following the first well-documented telescopic observations of Mars in 1659 by the Dutch physicist Christiaan Huygens. Huygens identified a prominent, dark surface feature that reemerged about every 24 h, suggesting a rate of rotation similar to that of Earth.

Belief that advanced life exists on Mars skyrocketed at the end of the nineteenth century after Giovanni Virginio Schiaparelli, an Italian astronomer, reported in

Mars   Planet symbol: ♂

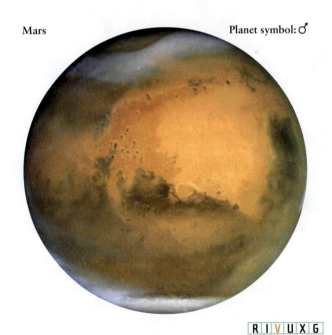

RIVUXG

FIGURE 16-18 Mars This photograph, taken from space, shows the Arabia Terra (in light orange) and carbon dioxide snow at the planet's poles. (NASA, The Hubble Heritage Team)

1877 seeing 40 lines crisscrossing the Martian surface (Figure 16-19). He called these dark features *canali*, an Italian term meaning "channels." It was soon mistranslated into English as *canals*, implying the existence on Mars of intelligent creatures capable of engineering feats. This speculation led Percival Lowell, who came from a wealthy Boston family, to finance a major new observatory near Flagstaff, Arizona. By the end of the nineteenth century, Lowell had allegedly observed 160 Martian "canals."

By the beginning of the twentieth century, it was fashionable to speculate that the Martian canals formed an enormous, planetwide irrigation network to transport water from melting polar ice caps to vegetation near the equator. In view of the planet's reddish, desertlike appearance, Mars was thought to be a dying planet whose inhabitants must go to great lengths to irrigate their farmlands. No doubt the Martians would readily abandon their arid ancestral homeland and invade Earth for its abundant resources.

FIGURE 16-19 The Illusion of Martian Canals Giovanni Schiaparelli examined Mars through a 20-cm (8-in) diameter telescope, the same size used by many amateur astronomers today. His drawings of the red planet showed features perceived by Percival Lowell and others as irrigation canals. Higher resolution images from Earth and spacecraft visiting Mars failed to show the same features. (Michael Hoskin, ed., *The Cambridge Illustrated History of Astronomy*, Cambridge University Press, 1997, p. 286. Illustration by G. V. Schiaparelli. Courtesy of Institute of Astronomy, University of Cambridge, UK)

INSIGHT INTO SCIENCE

Perception and Reality Our tendency to see patterns, even where they do not exist, makes it easy for us to leap to unjustified conclusions. Scientific images and photographs require objective analysis. This is why scientific discoveries are not considered to be authentic until they have been reproduced and verified by independent researchers.

16-8 Mars's global features include plains, canyons, craters, and volcanoes

Spacecraft journeying to Mars have sent back pictures since the 1970s. They show that Mars's surface is completely dry and that it has broad plains, shallow craters, enormous inactive volcanoes, and vast canyons (Figure 16-20). The broad northern plain is called *Vastitas Borealis* (the **northern vastness** or **northern lowlands**) and is shown in blue on Figure 16-20. It is some 5 km (3 mi) below the much more heavily cratered and hilly plains in the south, called the **southern highlands**. The cause of the significant difference in heights of the two hemispheres is not known, but it may stem from either early tectonic activity (see Section 15-2 for a discussion of tectonics) or from a powerful early impact.

Between the two hemispheres, *Mariner 9* discovered a vast canyon now called *Valles Marineris*, which runs roughly parallel to the Martian equator (Figure 16-21). Valles Marineris stretches over 4000 km, about one-fifth the circumference of Mars. If this canyon were located on Earth, it would stretch from New York to Los Angeles. The individual canyons that make up Valles Marineris are up to 6 km (4 mi) deep and 190 km (120 mi) wide. Valles Marineris begins with heavily fractured terrain in the west and ends with ancient-cratered terrain in the east. Geologists believe that Valles Marineris is a large crack that formed as the planet cooled. It was enhanced by nearby rising crust to its west and widened further by erosion. Some of the eastern parts of this system appear to have been formed almost entirely by water flow, similar to Earth's Grand Canyon.

Most impact craters are located on Mars's southern hemisphere (Figure 16-22). This suggests that the northern vastness has been resurfaced by some process that eradicated ancient lowland craters, consistent with either of the earlier explanations of the differences in hemispheres.

Most of the volcanoes on Mars are located in the northern hemisphere. The largest volcano, Olympus Mons (Figure 16-23a), covers an area as big as the state of Missouri and rises 26 km (16 mi) above the surrounding plains—nearly three times the height of

Elevation (km)

RIVUXG

RIVUXG

FIGURE 16-20 The Topography of Mars The color coding on this map of Mars shows elevations above (positive numbers) or below (negative numbers) the planet's average radius. To produce this map, an instrument on board *Mars Global Surveyor* fired pulses of laser light at the planet's surface, then measured how long it took each reflected pulse to return to the spacecraft. The *Viking 1* Lander (VL1), *Viking 2* Lander (VL2), *Mars Pathfinder* (MP), *Opportunity*, and *Spirit* landing sites are each marked with an X. The volcanoes depicted in white are higher than the scale on the right can show. (MOLA Science Team, NASA/GSFC)

FIGURE 16-21 Martian Terrain This high-altitude photograph shows a variety of the features on Mars, including broad, towering volcanoes *(left)* on the highland, called Tharsis Bulge; impact craters *(upper right)*; and vast, windswept plains. The enormous Valles Marineris canyon system crosses horizontally just below the center of the image. *Inset:* Details of the Valles Marineris, which is about 100 km (60 mi) wide. The canyon floor has two major levels. The northern (upper) canyon floor is 8 km (5 mi) beneath the surrounding plateau, whereas the southern canyon floor is only 5 km (3 mi) below the plateau. (USGS/NASA; *inset:* NASA/GSFP/LTP)

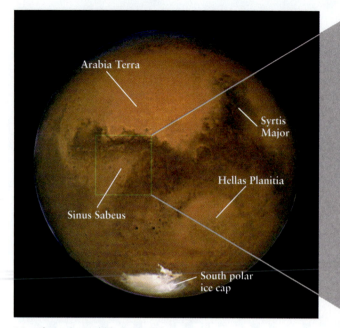

Mars from the Hubble Space Telescope

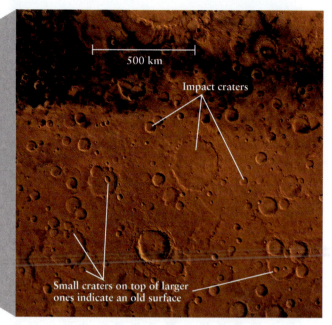

Closeup of Sinus Sabeus region

R I V U X G

FIGURE 16-22 Craters on Mars This image was made during the opposition of 2003. Arabia Terra looks like the Arabian peninsula on Earth. It is an old, highly eroded region dotted with numerous flat-bottom craters. Lava-covered Syrtis Major was first identified by Christiaan Huygens in 1659. A single impact carved out Hellas Planitia, which is five times the size of Texas. *Inset:* This mosaic of images from the *Viking 1* and *2* orbiter spacecraft shows an extensively cratered region located south of the Martian equator. Note how worn down these craters are compared to those on Mercury and our Moon. (NASA; J. Bell, Cornell University; and M. Wolff, SSI; *inset:* USGS)

a

R I V U X G

b

R I V U X G

FIGURE 16-23 The Olympus Caldera (a) This view of the summit of Olympus Mons is based on a mosaic of six pictures taken by one of the Viking orbiters. The caldera consists of overlapping, volcanic craters and measures about 70 km across. The volcano is wreathed in mid-morning clouds brought upslope by cool air currents. The cloud tops are about 8 km below the volcano's peak. (b) These cones on Mars may have been created when lava from Olympus Mons heated underground ice, causing the resulting water and vapor to expand, raise the planet's surface, and burst out. (Peter Lanagan [LPL, U. Arizona] et al., MOC. MGS. NASA)

WEB LINK 16.8

Mount Everest. The highest volcano on Earth, Mauna Loa in the Hawaiian Islands, has a summit only 17 km above the ocean floor.

Around Olympus Mons and other volcanoes, orbiting spacecraft have observed numerous, cratered cone-shaped features that are typically the size of a football field (Figure 16-23b). These cones are believed to have been created by ice below the Martian surface that was liquefied and vaporized by lava from the nearby volcanoes. The expanding water pushed its way up to the surface, deforming the land before exploding out the centers of these "ice" volcanoes. Similar clusters of cones on Earth, formed by the same mechanism, are found in Iceland.

Planetwide, high-resolution photographs of Mars over the past 40 years have failed to show one canal of a size consistent with those allegedly seen from Earth (see Figure 16-19). We now know for certain that Schiaparelli's *canali* were optical illusions, and the science fiction writers who wrote about advanced cultures there were completely off the mark.

INSIGHT INTO SCIENCE

Science Fact and Fiction Science fiction may foreshadow real scientific discoveries. (Can you think of some examples?) But science also reins in the fantasies of the science fiction writer. A century ago, millions of people, including top scientists, believed that technologically advanced life existed on Mars. However, scientific evidence says no.

16-9 Although no canals exist on Mars, it does have some curious natural features

Mars not only lacks canals, but it also has no cities, roads, or other signs of civilization. However, astronomers have photographed several surface features that, at first glance could have been crafted by intelligent life-forms. In 1976, a hundred years after Schiaparelli's alleged discovery of canali, the *Viking 1* orbiter spacecraft photographed a feature that appeared to be a human-like face (Figure 16-24a). However, when the Mars Orbiter photographed the surface in greater detail and from a different angle in 1998 (Figure 16-24b), it found no facial features. If anything, the same spot looked more like a giant heel print.

Other photographs show a collection of pyramids, a skull (Figure 16-24c), and a "happy face" (Figure 16-24d). Were any of these things relics of an advanced civilization? Scientists universally believe that they are all naturally-formed features. The first face is a region of hard rock around which softer debris is being worn away. Similarly, the pyramids and skull are consistent with winds eroding softer rock around harder rock pushed upward from inside Mars billions of years ago. The same thing happens on Earth. The happy face is made from a large impact crater and several small natural features inside it.

What two major activities created Valles Marineris?

Have you ever seen something that you interpreted as one thing, but it turned out to be something else?

INSIGHT INTO SCIENCE

Keep It Simple II Was the Martian "face" made by intelligent life or by wind-sculpted hills? When two or more alternative interpretations present themselves, scientists apply the principle of Occam's razor (page 42) and always choose the explanation that requires the fewest unproven assumptions.

16-10 Mars's interior is less molten than the inside of Earth

Part of the Valles Marineris was apparently formed when large regions of Mars's surface moved apart due to tectonic plate motion. This activity is generated by the flow of rock under the planet's surface (see Section 15-2 for more on plate tectonics). However, there is no evidence that plate tectonic activity occurs on Mars today. This indicates that more of Mars's interior is cooled, solid rock than exists inside Earth. This conclusion is supported by the observation that Mars lacks a global magnetic field (see Section 15-4 for discussion of the field-generating dynamo effect). However, observations made in 2003 revealed that the Sun creates tiny tides (less than a centimeter) on the solid body of Mars. As the land rises and sinks ever so slightly through the day on Mars, friction created by this motion in the planet's interior provides enough heat to keep some of the interior molten. (Rub your hands vigorously together to see how friction creates heat.)

Although a global magnetic field on Mars is absent today, there are remnants of such a field. In 1997 the *Mars Global Surveyor* (which operated until 2006) discovered patterns of local surface magnetic fields in nine places on Mars, similar to places on Earth where tectonic plates separate and lock Earth's magnetic field in the rock (see Section 15-4). Geologists propose that when Mars was young, its internal "dynamo" created a strong magnetic field and that tectonic plates helped craft its surface, while locking traces of its changing magnetic field in the then-molten rock. However, the planet quickly cooled, its global magnetic field vanished, and tectonic-plate activity apparently ceased nearly 4 billion years ago.

a

b

c

d

FIGURE 16-24 In the Eye of the Beholder Images (a) and (b) are of the same site on Mars, taken 22 years apart. They show how at high resolution the apparent face in (a) changed to a more "natural-looking" feature in (b) and (c). In the same region of Mars, other erosion features also appear to be pyramids and skulls. (d) The Galle crater and its interior features combine to give the impression of a "happy face." (a and b: NSSDC/NASA; NASA/JPL/Malin Space Science Systems; c and d: ©ESA/DLR/FU Berlin [G. Neu Kum])

A more molten interior early in its existence is also consistent with the enormous volcanoes that exist on Mars. Volcanoes are places where molten rock oozes or bursts from a body's interior. The volcanoes on Mars are the type where molten rock oozed out, like the volcanoes that created the Hawaiian Islands. Such volcanoes are very different from Mount Saint Helens in Washington State, or Mount Etna in Sicily, where the material from inside Earth is ejected violently. On Earth, the Hawaiian Islands are only the most recent additions to a long chain of volcanoes. They resulted from **hot-spot volcanism**, a process by which molten rock rises to the surface from a fixed hot region far below. The Pacific tectonic plate is slowly moving northwest at a rate of several centimeters per year. As a result, new volcanoes are created above the hot spot, while older ones move off and become extinct, eventually disappearing beneath the ocean.

Because Mars's surface is apparently frozen in place due to the lack of tectonic plate motion, one hot spot can keep pumping lava upward through the same vent for millions of years. One result is Olympus Mons—a single giant volcano rather than a long chain of smaller ones (see Figure 16-23a). This volcano's summit has collapsed to form a volcanic crater, called a **caldera**, large enough to contain the state of Rhode Island. Because Mars's solid crust and mantle (the region directly below the surface) are now so thick, they insulate the surface, preventing any more molten rock from escaping. Therefore, we do not expect to see any more large-scale volcanic activity there.

By observing the orbit of *Mars Global Surveyor*, astronomers have concluded that the Martian crust is about 40 km thick under the northern lowlands and about 70 km thick under the southern highlands. However, they find that the boundary between the thin and thick crusts does not line up with the boundary between high and low terrain. Current models of Mars suggest that it has a core about 3400 km in diameter and that at least some of that core is molten. A detailed understanding of Mars's interior waits for the placing of seismic detectors on the planet's surface.

16-11 Martian air is thin and often filled with dust

To understand more about the activity on Mars's surface, we must examine its atmosphere, which has only 0.6% the pressure of Earth's atmosphere. As on Venus, some 95% of Mars's thin atmosphere is composed of carbon dioxide. The remaining 5% consists of nitrogen, argon, and some traces of oxygen. Unlike the gravitational attraction of tiny Mercury, which is too weak to hold any gases as a permanent atmosphere, Mars's gravitational force is just strong enough to prevent carbon dioxide, nitrogen, argon, and oxygen from escaping into space, but not strong enough to hold down water vapor.

A growing body of evidence indicates that there was once a vast water-filled ocean in the northern hemisphere of Mars. The observations include layers of rock deposited by water and channels in which water from the south flowed northward. However, much of that water has evaporated and drifted into space, never to return. Today, the concentration of water vapor in Mars's atmosphere is 30 times lower than it is in Earth's air. If all of the water vapor could somehow be squeezed out of the Martian atmosphere, it would not fill even one of the five Great Lakes of North America. The remain-

der of the water that was on Mars's surface has turned to ice, much of which remains there today. Since 2002, satellites orbiting Mars have discovered enough water ice under Mars's south pole to cover the entire planet to a depth of 11 m (36 ft). In 2008, the *Phoenix Mars Lander* dug into the Martian regolith and observed water ice just under the surface. Chemical tests made by it confirmed this discovery. Water ice is being discovered by satellites at many latitudes from the polar regions to the equator, and it is certain that much more water remains to be found inside Mars.

 Despite the low density of Mars's air, the red planet experiences Earthlike seasons because of a striking coincidence, first noted in the late 1700s by the German-born English astronomer William Herschel. Just as Earth's equatorial plane is tilted 23½° from the plane of its orbit, Mars's equator makes an angle of about 25° from the plane of its orbit. If, as is now suspected, the northern hemisphere of Mars is lower than the southern hemisphere because the north was struck by a large body, that impact could have also tilted the planet and initiated the seasons there. The Martian seasons last nearly twice as long as Earth's, because Mars takes nearly 2 Earth years to orbit the Sun. When Mars is near opposition, even current telescopes for home use reveal its daily changes. Dark seasonal markings on the Martian surface can be seen to vary, and prominent polar caps shrink noticeably during the spring and summer months (Figure 16-25).

Although sometimes blue (Figure 16-26a), Mars's atmosphere is often pastel red, sometimes turning shades of pink and russet (Figure 16-26b). All of these latter colors are due to the fine dust blown from the planet's desert-like surface during windstorms. The dust is iron oxide, familiar here on Earth as rust. Mars's sky also changes color because the amount of

> If the Hawaiian Islands were fixed over their hot spot, rather than moving, what would happen to them?

FIGURE 16-25 Changing Seasons on Mars During the Martian winter, the temperature decreases so much that carbon dioxide freezes out of the Martian atmosphere. A thin coating of carbon dioxide frost covers a broad region around Mars's north pole. During the summer in the northern hemisphere, the range of this north polar carbon dioxide cap decreases dramatically. During the summer, a ring of dark sand dunes is exposed around Mars's north pole. (S. Lee/J. Bell/M. Wolff/Space Science Institute/NASA)

October 1996 (winter)   March 1997 (summer)

R I V U X G

Hill, 1km (0.6 mi) away

Mars Sojourner rover

R I V U X G

R I V U X G

FIGURE 16-26 The Atmosphere of Mars (a) When Mars's sky is relatively free of dust, it appears similar in color to our sky, as shown in this sunset photo taken by the rover *Opportunity*. The darker, brown color in which the Sun is immersed is due to lingering dust in the sky. When less dust is present, the Sun looks almost white during Martian sunsets. Most of the images we have of Mars's sky show colors like that seen in *(b)*. (b) Taken by the *Mars Pathfinder*, this photograph shows the *Sojourner* rover snuggled against a rock named *Moe* on the Ares Vallis to run tests on the rock. At the top of the image, the pink color of the Martian sky is evident. (NASA/Lunar and Planetary Institute)

dust in the air varies with the season. During the winter, carbon dioxide ice adheres to the dust particles and drags them to the ground. This helps clear and lighten the air. In the summer months, the carbon dioxide is not frozen, and the dust blown by surface winds remains aloft longer.

The sky color also changes over periods of many years. In 1995, for example, the amount of dust was observed to have dropped dramatically compared to that observed in the 1970s. The reason for such long-term changes is still under investigation.

As on Earth, the temperature changes with the seasons on Mars. For example, one day at the landing site of the rover *Spirit*, near the equator, the temperature ranged from 188 K to 243 K, whereas 100 days later it ranged from 200 K to 263 K (see Figure 16-20 for the site location). During the afternoons, heat from the planet's surface warms the air and sometimes creates whirlwinds, called **dust devils** (Figure 16-27). A similar phenomenon occurs in dry or desert terrain on Earth. Martian dust devils reach altitudes of 6 km (20,000 ft). Spacecraft on Mars detect drops in air pressure as dust devils sweep past, just as on Earth. Some dust devils are large enough to be seen by orbiting spacecraft.

The winds on Mars help explain why its surface is eroding. Over the past two centuries, astronomers have seen faint surface markings disappear under a reddish-orange haze as thin Martian winds have stirred up finely powdered dust

Why do you think it is easier to stand up in a dust devil on Mars than in the same equivalent-speed event here on Earth?

from the Martian regolith. Some storms obscure the entire planet, as happened in 2001. Over the ages, winds have worn down crater walls, and deposits of dust have filled in the crater bottoms (Figure 16-28). But the Martian atmosphere is so thin that 3 billion years of sporadic storms have not carried enough of Mars's extremely fine-grained powder to eradicate the craters completely.

16-12 Surface features indicate that water once flowed on Mars

Despite disproving the theory that Mars has broad canals or any other liquid surface water, Mars-orbiting spacecraft did reveal many dried-up riverbeds (Figure 16-29a); lakes (Figure 16-30a); river deltas (Figures 16-30a, b), where water and debris emptied into lakes and oceans; sedimentation laid down by water flow (Figure 16-31); and other water-related features. Some of the riverbeds include intricate branched patterns and delicate channels meandering among flat-bottomed craters. Rivers on Earth invariably follow similarly winding courses (Figure 16-29b).

Surface rovers *Spirit* and *Opportunity* found strong evidence that water did indeed create many of these features (Figure 16-32), including the presence of water-formed hematite—iron-rich rocks that look like "blueberries"—and rocks with extremely high levels of salts that had previously been dissolved in liquid water. Such salts take millions of years to seep into rocks in the quantities they have been found, implying that liquid water

a

328 yards
300 m

b   R I V U X G

FIGURE 16-27 Martian Dust Devil (a) This *Spirit* image is one in a movie sequence of a dust devil moving left to right across the surface of Mars. The rovers have filmed several such events. (b) These dark streaks are the paths of dust devils on the Argyre Planitia of Mars. The tracks cross hills, sand dunes, and boulder fields, among other features on the planet's surface. (a: NASA/JPL; b: NASA/JPL/Malin Space Science Systems)

was on the red planet's surface for at least that length of time. Although water existed on Mars's surface for long periods, there are indications in several places that large bodies of water have been gone for billions of years. The evidence for this is the discovery by the rovers of lava rocks that contain olivine and pyroxene, rocks that dissolve quickly in water. The origin of at least some of the water that was on Mars is believed to have been ice-rich bodies that struck the surface, releasing their water into the atmosphere, and later raining down on the surface.

In 2001, astronomers reported evidence that small-scale volcanic activity may still be causing episodic flows of both water and lava on Mars's surface. *Mars Global Surveyor* images show channels apparently carved by water flowing down the walls of pits or craters, including one event that occurred sometime between 1999 and 2005 (Figure 16-33a, b). What makes these observations especially intriguing is that they appear to be geologically young, indicating that liquid water may still exist under the Martian surface. In 2008, the *Mars Reconnaisance Orbiter* photographed several avalanches in progress. What caused them is still under investigation.

Although the amount of water that created the features shown in Figure 16-33b is relatively modest, large quantities of water—lakes or even oceans—are thought to be still trapped under the Martian surface. If so, volcanic heating or the heat generated by impacts from space debris could open fissures enabling water to melt, escape, and even carve new riverbeds today.

R I V U X G

FIGURE 16-28 Crater Endurance Photographed by the rover *Opportunity*, this crater on Mars is about 130 m (430 ft) across. Rocks are visible in the crater and in vertical cliffs along its walls. By studying such rocks, astronomers hope to understand more of the history of water on Mars. (JPL/NASA)

a

R I V U X G

b

R I V U X G

FIGURE 16-29 Rivers on Mars and Earth (a) Winding canyons on Mars, such as the one in this *Viking 1* orbiter image, appear to be due to sustained water flow. This belief is supported by the terraces seen on the canyon walls in high-resolution Mars Orbiter images. Long periods of water flow require that the planet's atmosphere was once thicker and its climate more Earthlike. (b) Yangtze River near Chongqing, China. Typical of rivers on Earth, it shows the same snakelike curve as the river channels on Mars. (a: NASA; b: TMSC/NASA)

Outlet   Shoreline

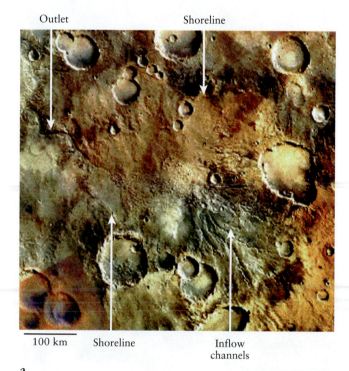

100 km   Shoreline   Inflow
                               channels

a

R I V U X G

Ancient
dried
riverbeds

50 km

b

R I V U X G

FIGURE 16-30 Evidence of Water on Mars (a) This dry Martian lake, photographed by the *Mars Global Surveyor* with a resolution of 1.5 m, is an excellent example of how geology and astronomy overlap. The features of this dry lake are consistent with those found on lakebeds on Earth. (b) This network of dry riverbeds is located on Mars's cratered southern hemisphere. (a: NASA/JPL; b: NASA)

a   R I V U X G

b   R I V U X G

FIGURE 16-31 Ancient Oceans and Lakes on Mars (a) This *Mars Global Surveyor* image of a portion of Valles Marineris reveals terrain with "stair-step layers." Such terrain is likely to have been created by sedimentation at the bottom of an ancient body of water. (b) This image of Burns Cliff, photographed by rover *Opportunity*, shows a close-up of layers of rock laid down on a body of water on Mars that went through wet and dry periods. (a: Malin Space Science Systems/JPL/NASA; b: [NASA/JPL/Cornell])

a   R I V U X G

b   R I V U X G

FIGURE 16-32 Layers of Rock Laid Down by Water (a) This close-up image taken by the rover *Opportunity* shows a small section of rock layers in a location, called *the Dells*. The angled and curved layering seen here is created only on Earth by water flow, strongly suggesting that this sediment was also deposited by water. The nearly spherical rocks, called blueberries, because they are dark, have been chemically identified as hematite, an iron-rich mineral that is usually formed in water. The rovers have found blueberries strewn in a wide variety of locations. (b) This iron meteorite, dubbed the "Heat Shield Rock" because it was discovered behind the Mars rover *Sojourner*'s heat shield, is surrounded by hematite blueberries. (a: NASA/JPL/USGS; b: NASA/JPL/Cornell)

August 1999

September 2005

North

New deposit

300 m

a

b

c

R I V U X G

FIGURE 16-33 *Martian Gullies* (a, b) Two images of the same southern hemisphere crater on Mars taken 6 years apart. Although not incontrovertible, they provide strong evidence for recent water flow from inside the red planet. The reason the deposit occurred now is still under investigation. (NASA/JPL/Malin Space Science Systems)

(c) Polygonal cracks are visible in this *Opportunity* image of Escher Rock. They were believed to have formed when this area was flooded. Water seeped into the rock, which cracked as the water froze and expanded, then the water evaporated away. (NASA/JPL/Cornell)

In 2004 the rover *Opportunity* discovered Martian rock (Figure 16-33c) with polygon-shaped cracks similar to dried mud flats found on Earth. The best explanation for these features on Mars is that long after the lakes and oceans dried there, an impact released water that again briefly flooded the surface rock. The water seeped into the rock and upon drying, it expanded, froze, and cracked the rock. This scenario would require vast amounts of water inside Mars for several reasons. First, heated by the impact, most of the liberated water would have left as vapor. Second, liquid water on the surface wouldn't stay there very long because of the low air pressure. When air pressure is very low, molecules easily escape from a liquid's surface, causing the water to vaporize. Because the pressure of Mars's atmosphere is only 0.6% that of Earth's atmosphere, any liquid water on Mars today would quickly transform into ice or furiously boil and evaporate into the thin Martian air, and then be lost into space. Thus, if liquid water did cause the rock formations seen by *Opportunity*, there must have been a lot of it to start with for enough to cover large areas of the Martian surface that have such cracks.

The total amount of water that existed on Mars's surface is not known. However, by studying flood channels that exist there, geologists estimate that there was once enough water to cover the planet to a depth of 500 m (1500 ft). For comparison, Earth has enough water to cover our planet to a depth of 2700 m (8900 ft), assuming that Earth's surface is everywhere a uniform height.

Further evidence that water once flowed on Mars comes from so-called *SNC meteorites* found on Earth (Figure 16-34a). These space rocks are believed to have once been pieces of Mars because their chemistries are consistent with those of rocks studied on Mars's surface and because they contain trace gases in relative amounts found only in the current Martian atmosphere. The meteorites were ejected into space during especially powerful impacts on that planet's surface. They also contain water-soaked clay, which is not expected to be found on any objects in the solar system besides Mars, Earth, and Jupiter's moons Europa, Ganymede, and Callisto. At least 57 meteorites from Mars have been identified.

Some of the water that was on Mars is still near the surface today, frozen in several places on the red planet from its poles to its equator. Recent satellite measurements indicate that as much as 90% of the ice at the poles is water ice, the remainder being dry ice (carbon dioxide ice). If thawed, this polar ice would create a global ocean 11 m (33 ft) deep. The temperature at Mars's poles, typically 160 K ($-170°F$), keeps the water permanently frozen, like the permafrost found in northern Asia and Antarctica. Seasonal variations at Mars's poles (see Figure 16-25) are created by the freezing and

a

R I V U X G

b

R I V U X G

FIGURE 16-34 A Piece of Mars on Earth Mars's thin atmosphere does little to protect it from impacts. Some of the debris ejected from impact craters there apparently traveled to Earth. (a) SNC meteorite recovered in Antarctica. The meteorite shows strong evidence of having been exposed to liquid water on Mars, perhaps for hundreds of years. (b) Possible fossil remains of primitive bacterial life on Mars, although theories of nonbiological origins have also been presented. (a: NASA; b: NASA/JPL)

evaporating of dry ice, which reaches thickness of 2 m. Indeed, during the winter, about one-third of the carbon dioxide in Mars's atmosphere comes down as dry ice.

The 1997 visit by *Pathfinder* and its little rover (see Figure 16-26b) to the Ares Vallis revealed evidence of several floods at the mouth of a dried flood plain. The evidence includes layers of sediment, clumps of rock and sand stuck together like similar groupings created by water on Earth, rounded rocks worn down as they were dragged by the water, and rocks aligned by water flow. The rover performed some 20 chemical analyses of rocks in the landing area. The soil is a rusty color, laden with sulfur. The source of this sulfur is a mystery, as is the large amount of silicon-rich silica discovered on rocks, such as Barnacle Bill.

There is observational evidence of an ancient shoreline encircling the northern lowlands of Mars. The belief that the northern lowlands were an ocean is supported by the analysis of a 1.2 billion-year-old meteorite from Mars. It has remnants of salt believed to have been deposited there when the Martian ocean dried up.

16-13 Search for microscopic life on Mars continues

The discovery in the 1960s that water once existed on the surface of Mars rekindled speculation about Martian life. Although it was clear that Mars has neither civilizations nor fields of plants, microbial lifeforms still seemed possible. Searching for signs of organic matter was one of the main objectives of the ambitious and highly successful Viking missions.

WEB LINK 16.12

The two Viking spacecraft were launched during the summer of 1975. Each consisted of two modules—an orbiter and a lander. Almost a year later, both Viking landers set down on rocky plains north of the Martian equator.

The landers confirmed the long-held suspicion that the red color of the planet is due to large quantities of iron in its soil. Despite the high iron content of its crust, Mars has a smaller average density (3950 kg/m³) than that of other terrestrial planets (more than 5000 kg/m³ for Mercury, Venus, and Earth). Mars must therefore contain overall a smaller percentage of iron than these other planets.

> Which ice on Mars changes more actively today, carbon dioxide ice or water ice?

Each Viking lander was able to dig into the Martian regolith and retrieve rock samples for analysis, which showed that the rocks at both sites are rich in iron, silicon, and sulfur. The Martian regolith (Figure 16-35) ranges in consistency from clay to sand, although its chemistry is different from clay and sand on Earth. The Viking landers (see Figure 19-1) each carried a compact biological laboratory designed to test for microorganisms in the Martian soil. Three biological experiments were conducted, each based on the idea that living things alter their environment: They eat, they breathe, and they release waste products. During each experiment, a sample of the Martian regolith was placed in a closed container, with or without a nutrient substance. The container was then examined for any changes in its contents.

The first data returned by the Viking biological experiments caused great excitement. In almost every case, rapid and extensive changes were detected inside

Sulfur salts

R I V U X G

FIGURE 16-35 Regolith of Mars The rover *Spirit*'s wheels churned up the regolith of Mars, revealing sulfur-based salts just below the surface. (NASA/ JPL/Cornell)

the sealed containers. However, further analysis showed that these changes were due solely to nonbiological chemical processes. Apparently, the Martian regolith (Figure 16-35) is rich in chemicals that effervesce (fizz) when moistened. A large amount of oxygen is tied up in the regolith in the form of unstable chemicals called *peroxides* and *superoxides*, which break down in the presence of water to release oxygen gas.

The chemical reactivity of the Martian regolith comes from the Sun's ultraviolet radiation and from electric activity inside dust devils. Ultraviolet photons easily break apart molecules of carbon dioxide (CO_2) and water vapor (H_2O) by knocking off oxygen atoms, which then become loosely attached to chemicals in the regolith. Ultraviolet photons also produce highly reactive ozone (O_3) and hydrogen peroxide (H_2O_2) near the planet's surface, which become incorporated in the regolith. Dust devils suck up lots of dust from the planet's surface and transform it into a variety of compounds.

Here on Earth, hydrogen peroxide is commonly used as an antiseptic. When you pour this liquid on a wound, it fizzes and froths as the loosely attached oxygen atoms chemically combine with organic material and destroy germs. The Viking landers may have failed to detect any organic compounds on Mars because the superoxides and peroxides in the Martian regolith act as antiseptics there, making it sterile.

That so much ultraviolet radiation from the Sun gets to the surface of Mars indicates what about the Martian atmosphere?

In 2003, astronomers detected methane in Mars's atmosphere. This is an intriguing discovery, since sunlight breaks down methane within a few centuries. Therefore, the gas needs to be continuously replenished. This simple compound, CH_4, is often a by-product of biological activity (cows come to mind in this regard). Methane is also trapped inside planets, like Earth and possibly Mars, during their formation and thereafter leaks out. Although the methane discovered on Mars is suggestive that simple life might be active under the planet's surface today, it is by no means positive proof that such life exists.

Possible signs of ancient life on Mars have been discovered here on Earth. When cut open, several of the Martian meteorites have shown microscopic features that could be fossils of Martian bacterial life and their excretions (see Figure 16-34b). These features, no larger than 500 nm (1/100 the diameter of a human hair), are 30 times smaller than bacteria found on Earth. Detailed analysis of the meteorites in 2001 revealed the presence of several organically created features, including tiny spheres found with some of Earth's bacteria, and magnetite crystals, which are also used by some bacteria on Earth as compasses to find food. The debate as to whether the features in these meteorites are fossil evidence of life on Mars has raged unabated for several years, and it is not yet resolved. (It is worth noting that similar claims of fossils from Mars in the 1960s have been disproved.)

The search for microscopic life on Mars is not over. If there is liquid water under the Martian surface, then it is entirely possible that life has evolved in it and may still exist. This belief is based on the incredibly wide range of places that life has developed in water on Earth, from volcanic vents on the bottoms of oceans, to geysers, to a lake under the Antarctic polar ice cap.

16-14 Mars's two moons look more like potatoes than spheres

Two tiny moons orbit close to Mars's surface. *Phobos* (Greek, meaning "fear") and *Deimos* ("panic") are so small that they were not discovered until 1877. In mythology, these names are either Mars's sons by Aphrodite or the two horses that pulled Mars's war chariot—take your pick. Potato-shaped Phobos is the inner and larger of the two moons (Figure 16-36). It is heavily cratered, and its surface has been transformed into dust at least 1 m thick by countless tiny impacts over the eons. Football-shaped Deimos is less cratered than Phobos (Figure 16-36).

Both Phobos and Deimos orbit in the same direction that Mars rotates, just as our Moon orbits Earth. However, Phobos is so close to the planet that this moon goes around three times in one Martian day. This rapid revolution creates the unusual situation wherein

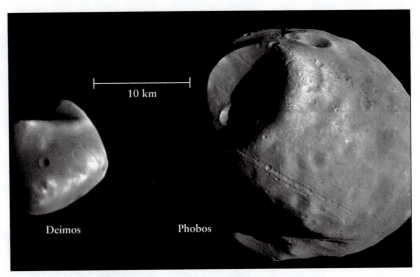

FIGURE 16-36 Phobos and Deimos Phobos, the larger of Mars's two moons, is potato-shaped and measures approximately 28 × 23 × 20 km. It is dominated by crater Stickney, named for discoverer Asaph Hall's wife (Angeline Stickney). Deimos is less cratered than Phobos and measures roughly 16 × 12 × 10 km. (left: European Space Agency); right: Dr. Edwin V. Bell/NSSDC/ Raytheon ITSS)

R I V U X G

Phobos rises in the west, races across the sky in only 5 1/2 h as seen from Mars's equator, and sets in the east. As seen from Mars, Deimos rises in the east and takes about 3 Earth days to creep from one horizon to the other.

Phobos and Deimos were not formed like our Moon, by splashing off Mars. Rather, they are captured planetesimals. Both moons are in synchronous rotation as they orbit the red planet.

INSIGHT INTO SCIENCE

Imagine the Moon The definitions we use for words based on our everyday experience often fail us in astronomy. For example, the word *moon* usually creates an image of a spherical body, like our Moon. In reality, most of the moons in the solar system are unsymmetrical, like Phobos and Deimos.

16-15 Comparisons of planetary features provide new insights

Now that we have examined the terrestrial planets individually, it can be useful to see how their various features compare to each other and to Earth. Table 16-1 (The Inner Planets: A Comparison) summarizes much of this material.

Size and Mass Earth is the largest and most massive of all four terrestrial planets. In this regard, Venus is almost the sister planet to Earth, with nearly 95% of Earth's diameter and 82% of Earth's mass. Although Mars is most similar to Earth in other ways, such as its history of surface water and rotation rate, it is only half (53%) of Earth's diameter and 11% of Earth's mass. Mercury, with 38% of Earth's diameter and a scant 5.5% of Earth's mass, is much more similar to Earth's Moon in dimensions. Mercury is only 1.4 times bigger than the Moon.

Surface Features Although all four terrestrial planets have craters, only Mercury has them in large numbers, similar to what we saw on our Moon. Venus has erased most of its craters by melting them or covering them with magma. Mars has removed many of its craters as a result of erosion and weather. Earth has removed most craters by tectonic plate motion and weathering. Like Earth, Venus has continent-like plateaus and ocean bottomlike lowlands. Mars has a lower northern cap than the rest of the planet, but no real continent-like regions. Mercury has relatively uniform height. While all four terrestrial planets have volcanoes, those of Mars and Mercury are extinct. Water erosion occurred just on Mars and Earth.

> Would our Moon have to be closer or farther away to orbit in the same direction, but rise in the west?

Interior The interior chemistries of the four terrestrial planets are similar, with cores consisting primarily of iron surrounded by rock. Venus, Earth, and Mars are known to have partially molten cores, and it remains to be seen if Mercury's core is also molten. Although Mercury is the smallest terrestrial planet, it has the highest density, meaning that it has the highest percentage of iron of these (and, in fact, of any) planets. This is true probably because Mercury lost more of its outer, rocky layer as a result of impacts than did any other terrestrial planet.

Water Earth contains by far the highest percentage of water of the terrestrial planets. Mars contains water frozen on or near its surface, and we have yet to determine whether it has any liquid water deep inside. Venus contains very little water compared to either Earth or Mars because Venus is so hot that it has evaporated surface water into its atmosphere, and water in its interior has probably been mostly ejected through volcanoes or when the surface periodically melts. Mercury apparently has some water (very little compared to Earth or Mars) frozen at its poles, the result of collisions with water-rich comets. Because its interior is so iron-rich, the water-bearing layers were probably blasted into space by impacts early in Mercury's existence.

Inner Planets: A Comparison

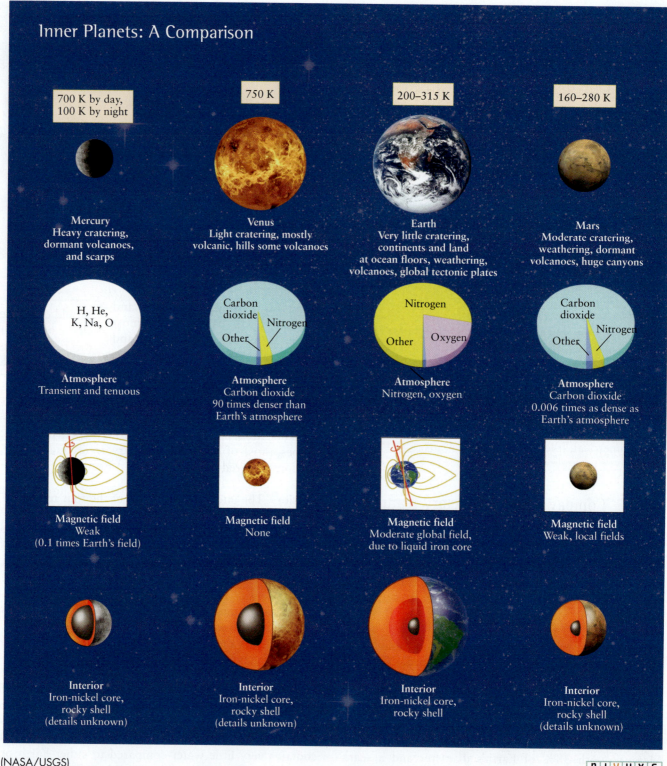

700 K by day, 100 K by night	750 K	200–315 K	160–280 K

Mercury
Heavy cratering, dormant volcanoes, and scarps

Venus
Light cratering, mostly volcanic, hills some volcanoes

Earth
Very little cratering, continents and land at ocean floors, weathering, volcanoes, global tectonic plates

Mars
Moderate cratering, weathering, dormant volcanoes, huge canyons

Atmosphere
Transient and tenuous

Atmosphere
Carbon dioxide 90 times denser than Earth's atmosphere

Atmosphere
Nitrogen, oxygen

Atmosphere
Carbon dioxide 0.006 times as dense as Earth's atmosphere

Magnetic field
Weak (0.1 times Earth's field)

Magnetic field
None

Magnetic field
Moderate global field, due to liquid iron core

Magnetic field
Weak, local fields

Interior
Iron-nickel core, rocky shell (details unknown)

Interior
Iron-nickel core, rocky shell (details unknown)

Interior
Iron-nickel core, rocky shell

Interior
Iron-nickel core, rocky shell (details unknown)

(NASA/USGS)

R I V U X G

Atmosphere Venus has by far the densest atmosphere of the terrestrial planets, with about 90 times as much gas as the air we breathe. Furthermore, Venus's atmosphere is composed primarily of carbon dioxide, with a minor component of nitrogen. The thick atmosphere has protected Venus's surface from all but the most massive infalling space debris. Its thick atmosphere and the possibility that Venus's surface periodically melts, explains why Venus has

few impact craters. Venus's atmosphere is most similar in composition to the air around Mars, although the density of Mars's air is only 0.6% as dense as the air we breathe. Although Mars's thin atmosphere has enabled many pieces of space debris to strike the planet and form craters, these craters are continually being eroded, primarily by wind.

Earth's atmosphere, once very similar to that of Venus, was transformed by water and life into the nitrogen-oxygen atmosphere we have today. As with Venus, Earth's atmosphere protects the surface from most impacts. Many craters have been removed by our planet's plate tectonic motion, which apparently does not occur today on any other planet. Mercury's gravity is too low to hold any gases as a permanent atmosphere. It is surrounded by a very thin atmosphere of transient gases from the Sun and from the planet's interior. As these drift into space, they are replaced by fresh gas. Because its atmosphere is so thin and its surface unchanged by internal activity for billions of years, Mercury is the most heavily cratered of the terrestrial planets.

Temperature Some of the temperatures on the terrestrial worlds are surprising at first glance. Mercury, which is closest to the Sun, has a hot daytime surface of about 700 K (800°F). Its lack of atmosphere allows a lot of this heat to escape at night, bringing its nighttime temperature down to a frigid 100 K (−280°F), much colder than on any other terrestrial planet. Venus's thick atmosphere creates a greenhouse effect that keeps that planet at 750 K (890°F), even hotter than Mercury. Earth's surface temperature ranges from about 330 K (140°F) to 180 K (−130°F), and Mars is only slightly colder, with temperatures ranging from 280 K (45°F) down to 160 K (−170°F).

Rotation and Magnetic Field All of the terrestrial planets rotate, with Earth's solar day being shortest at 24 h. This motion, combined with Earth's liquid iron core, creates a strong magnetic field that surrounds our planet. Mars has virtually the same solar day of 24 h 39 min, but its molten iron core is much smaller than ours and it has no global magnetic field, only local magnetic fields. The solar days of Venus (117 Earth days) and Mercury (176 Earth days) are both extremely long by Earth standards, and, indeed, a solar day on Mercury is 2 Mercurian years long! No magnetic field has been detected around Venus. Mercury has a weak global field that may result from the extremely high amount of iron it contains.

16-16 Frontiers yet to be discovered

All three terrestrial planets other than Earth have much to reveal. We have yet to see all of Mercury's surface. What is the chemical composition of its surface rocks? What is Mercury's cooling history, and how much of its core is molten? Is there really ice at its poles? If so, how did it get there? What will its internal structure reveal?

Does Venus have active volcanoes? If not, what supplies its atmosphere with sulfur compounds? Can we find observational evidence of what caused Venus's rotation axis to flip over? Likewise, can we find further evidence that its surface periodically undergoes significant re-covering?

Some of the most significant questions about our solar system that could be answered are whether life ever existed on Mars, whether there is actually liquid water under the red planet's surface, and whether there is still life there. If so, how far did that life evolve? What are the similarities and differences between such life and life on Earth? Significant similarities might imply a common origin. Furthermore, what is the surface water history of Mars? What causes its local magnetic fields? What does its interior look like? How did its axis get tilted? Where did its moons come from? It is likely that we will have answers to many of these questions in the coming few decades.

> What does the surface of Venus lack, without which its atmosphere cannot transform into one like ours?

SUMMARY OF KEY IDEAS

All four inner planets are composed primarily of rock and metal, and thus they are classified as terrestrial.

Mercury

• Even at its greatest orbital elongations, Mercury can be seen from Earth only briefly after sunset or before sunrise.

• The Mercurian surface is pocked with craters like the Moon's, but extensive, smooth plains lie between these craters. Long cliffs meander across the surface of Mercury. These scarps probably formed as the planet cooled, solidified, and shrank.

• The long-ago impact of a large object formed the huge Caloris Basin on Mercury and shoved up jumbled hills on the opposite side of the planet.

• Mercury has an iron core, which fills more of its interior than Earth's core fills Earth.

Venus

• Venus is similar to Earth in size, mass, and average density, but it is covered by unbroken, highly reflective clouds that conceal its other features from observers using visible-light telescopes.

• Although most of Venus's atmosphere is carbon dioxide, its dense clouds contain droplets of concentrated sulfuric acid mixed with yellowish sulfur dust. Active volcanoes on Venus may be a constant source of this sulfurous veil.

• Venus's exceptionally high temperature is caused by the greenhouse effect, as the dense carbon dioxide atmosphere traps and retains heat emitted by the planet. The surface pressure on Venus is 90 atm, and the surface temperature is 750 K. Both temperature and pressure decrease as altitude increases.

• The surface of Venus is surprisingly flat and mostly covered with gently rolling hills. There are two major "continents" and several large volcanoes. The surface of

Venus shows evidence of local tectonic activity but not the large-scale motions that play a major role in continually reshaping Earth's surface.

Mars

• Earth-based observers found that the Martian solar day is nearly the same as that of Earth, that Mars has polar ice caps that expand and shrink with the seasons, and that the Martian surface undergoes seasonal color changes.

• A century ago, observers reported networks of linear features that many perceived as canals. These observations led to speculation about self-aware life on Mars.

• The Martian surface has many flat-bottomed craters, several huge volcanoes, a vast equatorial canyon, and dried-up riverbeds—but no canals formed by intelligent life. River deltas and dry riverbeds on the Martian surface indicate that large amounts of water once flowed there.

• Liquid water would quickly boil away in Mars's thin present-day atmosphere, but the planet's polar ice caps contain significant quantities of frozen water, and a layer of permafrost exists beneath parts of the regolith.

• The Martian atmosphere is composed mostly of carbon dioxide. The surface pressure is less than 0.01 atm.

• Chemical reactions in the regolith, together with ultraviolet radiation from the Sun, apparently act to sterilize the Martian surface.

• Mars has no global magnetic fields, but local fields pierce its surface in at least nine places.

• Mars has two potato-shaped moons, the captured planetesimals Phobos and Deimos. Both are in synchronous rotation with Mars.

WHAT DID YOU THINK?

1 *Which terrestrial planet—Mercury, Venus, Earth, or Mars—has the coolest surface temperature?* The nighttime side of Mercury, closest planet to the Sun, is the coldest surface of any terrestrial-planet.

2 *Which planet is most similar in size to Earth?* Venus is most similar to Earth in size.

3 *Which terrestrial planet—Mercury, Venus, Earth, or Mars—has the hottest surface temperature?* Venus is hottest, its temperature is above that of Mercury because of the greenhouse effect in Venus' atmosphere.

4 *What is the composition of the clouds that surround Venus?* The clouds are made primarily of sulfuric acid.

5 *Does Mars have liquid water on its surface today? Did it have liquid surface water in the past?* Mars has no liquid surface water today, but there are very strong indications that it had liquid water on its surface in the past.

6 *Is life known to exist on Mars today?* No current life has yet been discovered on Mars, but it may exist in underground water oceans.

Key Terms for Review

3-to-2 spin-orbit coupling, 422
caldera, 434
dust devil, 436
greenhouse effect, 425
hot-spot volcanism, 434

northern vastness (northern lowlands), 430
retrograde rotation, 428
scarp, 420
southern highlands, 430

Review Questions

1. Which pair of planets have atmospheres with the most similar chemical compositions? **a.** Earth and Venus, **b.** Earth and Mars, **c.** Venus and Mars, **d.** Mercury and Mars, **e.** Mercury and Venus

2. Which planet is least likely to have water ice on or just under its surface? **a.** Earth, **b.** Mercury, **c.** Mars, **d.** Venus

3. Which object is most similar to Venus in mass and diameter? **a.** Earth **b.** Mars **c.** Mercury **d.** our Moon

4. Why is Mercury so difficult to observe? When is the best time to see the planet? (Hint: *Guided Discovery: The Inner Solar System*, on page 424, can help.)

5. Compare the surfaces of Mercury and our Moon. How are they similar? How are they different?

6. Compare the interiors of Mercury and Earth. How are they similar? How are they different?

 7. To better understand the interiors of Mercury and Earth, do Interactive Exercise 16-1 on the Web. You can print out the result, if requested.

8. What are the longest features found on Mercury? Why are the examples of this feature probably much older than tectonic features on Earth?

9. Briefly describe a scientific theory explaining why Mercury has such a large iron core.

10. Astronomers often refer to Venus as Earth's twin. What physical properties do the two planets have in common? In what ways are the two planets dissimilar?

11. Why is it hotter on Venus than on Mercury?

12. What evidence exists for active volcanoes on Venus?

13. Describe the Venusian surface. What kinds of geologic features would you see if you could travel around the planet?

14. Why do astronomers believe that Venus's surface was not molded by the kind of continuous tectonic activity that shaped Earth's surface?

15. Why is Mars red?

16. What is the greenhouse effect? What role does it play in the atmospheres of Venus and Earth?

17. During which configuration of planets is it easiest to observe Mars from Earth? (Hint: *Guided Discovery: The Inner Solar System*, on page 424, can help.)

18. Compare the cratered regions of Mercury, the Moon, and Mars. Assuming that the craters on all three worlds originally had equally sharp rims, what can you conclude about the environmental histories of these worlds?

19. How would you tell which craters on Mars were formed by meteoritic impacts and which by volcanic activity?

20. What geologic features indicate that plate tectonic activity once occurred on Mars? What features created by tectonic activity on Earth are not found on Mars?

 21. To better understand the surface features of Mars, do Interactive Exercise 16-2 on the Web. You can print out your results, if requested.

 22. To compare the surfaces of Mercury, Venus, and Mars, do Interactive Exercise 16-3 on the Web. You can print out your results, if requested.

 23. What evidence have astronomers accumulated that liquid water once existed in large quantities on Mars's surface? What evidence is there that water still exists, under its surface?

Advanced Questions

24. What evidence do we have that the surface features on Mercury were not formed during recent geologic history?

25. Venus takes 440 days to move from greatest western elongation to greatest eastern elongation, but it needs only 144 days to go from greatest eastern elongation to greatest western elongation. With the aid of a diagram, explain why.

26. As seen from Earth, the brightness of Venus changes as it moves along its orbit. Describe the main factors that determine Venus's variations in brightness as seen from Earth. (Hint: See the discussion of Venus in Chapter 2.)

27. How might Venus's cloud cover change if all of Venus's volcanic activity suddenly stopped? How might these changes affect the overall Venusian environment?

28. Compare Venus's continents with those on Earth. What do they have in common? How are they different?

29. Explain why Mars has the longest synodic period of all of the planets, although its sidereal period is only 687 days.

30. With carbon dioxide accounting for about 95% of the atmospheres of both Mars and Venus, why is there little greenhouse effect on Mars today?

Discussion Questions

31. If you were planning the *Messenger* mission to Mercury, what features and observations would be of particular interest to you and why?

32. If you were designing a space vehicle to land on Venus, what special features would be necessary? In what ways would this mission and landing craft differ from a spacecraft designed for a similar mission to Mercury?

33. Suppose someone told you that the Viking mission failed to detect life on Mars simply because the tests were designed to detect terrestrial life-forms not Martian life-forms. How would you respond?

34. Compare the scientific opportunities for long-term exploration offered by the Moon and Mars. What difficulties would there be in establishing a permanent base or colony on each of these two worlds?

35. Imagine you are an astronaut living at a base on Mars. Describe your day's activities, what you see, the weather, the spacesuit you are wearing, and so on.

What If . . .

36. Mercury had synchronous rotation? How would the temperatures on such a planet be different than they are today? Where would humans set up camp on such a world?

37. Venus could support self-aware life, but it still had permanent cloud-cover? In what ways would life there be different than it is here? How would their perceptions of the cosmos be different from ours?

38. Mars had the same mass, surface features, and atmosphere as Earth? In what ways would life there be different than it is here?

39. Mars rotated once every 20 days rather than once every 1.026 days? What would be different on the alternate Mars?

40. The carbon dioxide content of Earth's atmosphere increased? What would happen to Earth? (This is not a completely hypothetical question, because carbon dioxide levels are increasing today.)

Web Questions

41. Search the Web for the latest information about the *Messenger* mission to Mercury. Exactly when was it launched and when will it be inserted into orbit around Mercury? What scientific experiments does it carry? What scientific issues are these instruments intended to resolve? How long is it expected to function in orbit around Mercury?

42. Elongations of Mercury Access the animation "Elongations of Mercury" in Chapter 16 of the *Discovering the Universe: From the Stars to the Planets* Web site. **a.** Note the dates of the greatest eastern and western elongations in the animation. Which time interval is greater: from a greatest eastern elongation to a greatest western elongation, or vice versa? **b.** Based on what you observe in the animation, draw a diagram to explain your answer to the question in **a.**

43. Search the Web for the latest information about proposed future missions to Venus. What scientific experiments will they carry? What scientific issues are these instruments intended to resolve?

44. Surface Temperature of Venus Access the Active Integrated Media Module "Wien's Law" in Chapter 4 of the *Discovering the Universe: From the Stars to the Planets* Web site. **a.** Using the Wien's law calculator, determine Venus's approximate temperature if it emits blackbody radiation with a peak wavelength of 3866 nm. **b.** By trial and error, find the wavelength of maximum emission for a surface temperature of 750 K (for present-day Venus) and a surface temperature of 850 K (as it might become if its greenhouse gas density increases). In what part of the electromagnetic spectrum do these wavelengths lie?

45. In 1999, two NASA spacecraft—*Mars Climate Orbiter* and *Mars Polar Orbiter*—failed to reach their destinations. Search the Web for information on these missions. What were their scientific goals? How and why did the missions fail? Which current and future missions, if any, are intended to replace these missions?

46. Search the Web for information about possible manned missions to Mars. How long would such a mission take? How expensive would they be? What are some advantages and disadvantages of a manned mission compared to an unmanned one?

47. Conjunctions of Mars Access and view the animation "The Orbits of Earth and Mars" in Chapter 16 of the *Discovering the Universe: From the Stars to the Planets* Web site. **a.** The animation highlights three dates when Mars is in opposition, so that Earth lies directly between Mars and the Sun. By using the **Stop** and **Play** buttons in the animation, find two times during the animation when Mars is in *conjunction*, so that the Sun lies directly between Mars and Earth (see Figure 2-4). For each conjunction, make a drawing showing the positions of the Sun, Earth, and Mars, and record the month and year when the conjunction occurs. **b.** When Mars is in conjunction, at approximately what time of day does it rise as seen from Earth? At what time of day does it set? Is Mars suitably placed for telescopic observations when it is in conjunction?

Observing Projects

48. Mercury can be seen in the eastern sky at dawn for a few days around greatest western elongation. It can also be seen in the western

sky at dusk for a few days around greatest eastern elongation. The following table shows the dates of the next few greatest elongations of Mercury. Magazines such as *Sky & Telescope* or *Astronomy* will tell you which, if any, of these greatest elongations is going to be especially favorable for viewing the planet. Alternatively you can use your *Starry Night Enthusiast(* planetarium software to examine these elongations. Open **Sky Guide > Guided Tours > Our Solar System, stars and galaxies > The Planets > Mercury > A fast little world**. This view shows Mercury and its orbital path around the Sun. Note how close Mercury stays to the Sun during its orbital journey. Stop the advance of time. Adjust the **Date** to each of the listed dates in the table below for **Greatest western elongation** and click on **Sunrise** to place the Sun on the horizon. You will note that the favorable elongations are those for which the angle of Mercury's orbit with the horizon is largest. Why is Mercury at different altitudes (angles above the horizon) on different dates at these sunrise times? Now Adjust the **Date** to each of the listed dates for **Greatest eastern elongation** and click on **Sunset** and note the favorable dates for observing Mercury in the evening. Set the Time Flow Rate to 1 minute at any of these times and run time forward to observe Mercury's descending motion with respect to the horizon.

If an elongation is favorable, make plans to observe this innermost planet of the solar system, a planet that is rarely seen by anyone but astronomers. Set aside several evenings (or mornings) around the date of the favorable elongation to reduce the chances of being "clouded out." Select an observing site that has a clear, unobstructed view of the horizon where the Sun sets (or rises). Make arrangements to have a telescope at your disposal, if possible. Search for the planet on the dates you have selected and make a drawing of its appearance through your telescope.

Greatest Elongations of Mercury,* 2008–2011

Year	Greatest western elongations	Greatest eastern elongations
2008	March 3, July 1, October 22	January 22, May 14, September 11
2009	February 13, June 13, October 6	January 4, April 26, August 24, December 18
2010	January 27, May 26, September 19	April 8, August 7, December 1
2011	January 9, May 7, September 3, December 23	March 23, July 20, November 14

*Mercury can be seen in the eastern sky at dawn for a few days around greatest western elongation. It can be seen in the western sky at dusk for a few days around greatest eastern elongation.

49. **Venus** can be seen in the eastern sky in the hours before dawn for several months around greatest western elongation. It can also be seen in the western sky in the hours after dusk for several months around greatest eastern elongation. Refer to the following table to see if Venus is currently near a position of greatest elongation. If so, view the planet through a telescope. Make a sketch of the planet's appearance. From your sketch, can you determine whether Venus is closer to us or farther from us than the Sun?

Greatest Elongations of Venus,* 2008–2011

Year	Greatest western elongations	Greatest eastern elongations
2008	(none)	(none)
2009	June 5	January 14
2010	(none)	August 20
2011	January 8	(none)

*Venus can be seen in the eastern sky in the hours before dawn for several months around greatest western elongation. It can be seen in the western sky in the hours after dusk for several months around greatest eastern elongation.

50. Observe Venus through a telescope once a week for a month and make a sketch of the planet's appearance on each occasion. From your sketches, can you determine whether Venus is approaching us or moving away from us?

51. Consult such magazines as *Sky & Telescope* and *Astronomy*, or use your *Starry Night Enthusiast*™ planetarium software to determine Mars's location among the constellations. If Mars is suitably placed for observation, arrange to view the planet through a telescope and sketch this planet, noting particularly its phase. What magnifying power seems to give you the best image? Can you distinguish any surface features? Can you see a polar cap or dark markings? If not, can you offer an explanation for Mars's bland appearance?

52. You can use *Starry Night Enthusiast*™ to follow Mercury's orbit as it nears the Sun. Press the **Home** and **Now** buttons. Open the **Favourites** menu and click on **Guides > Atlas** to place yourself at the center of a transparent Earth. Stop time advance and open the **Find** pane and double-click over the Sun to center upon it. In the **Labels** menus, click on **Planets-Moons** and **Stars** to remove these labels. Set the **Time Flow Rate** to 1 hour (adjust, as necessary). **a.** Locate Mercury by clicking the box to the left of Mercury in the **Find** pane. Watch Mercury's motion against the background stars as you run time forward, and describe its motion. Pay particular attention to the relative speeds when Mercury is passing near to the Sun on east-west and west-east passes respectively. **b.** Stop time advance and set the **Time Flow Rate** to 5 sidereal days. Right-click the mouse button over Mercury and click on **Center** to place Mercury at the center of the view. Single-step time and follow the planet's motion around the Sun, stopping occasionally to zoom in and determine Mercury's phase. Describe the pattern of phase changes and relate these changes to the position of Mercury with respect to the Sun.

53. **Mars Motion as seen from Earth.** You can use *Starry Night Enthusiast*™ to monitor the motion of Mars as seen from the Earth, as the Red planet moves against the background stars. First press the **Home** and **Now** buttons. Open the **Favourites** menu and click on **Guides > Atlas** to position yourself at the center of a transparent Earth. Open the **Labels** menu and click on **Planets-Moons** and **Stars** to remove these labels. Open the **Find** pane and double-click over Mars to center the view on this planet. Set the **Time Flow Rate** to 1 sidereal day and run time forward (press **Play ▶**) to watch Mars move against the background stars. Note that the equatorial coordinate system usually flows smoothly past Mars since you are centered upon that planet. However, you will see that Mars begin to slow down as time progresses. Stop time advance when Mars appears almost to stop in its motion. Right-click on a nearby star and click **Center** to lock your view upon this background star. Run time forward and describe how Mars now moves relative to the coordinate system and the background stars. Run time backward to where Mars is at its stationary point furthest to the east (leftward) position and stop time again. Change the **Time Flow Rate** to 5 sidereal days and single-step the time forward, tracing Mars' motion on tracing paper or transparent film with a soft felt pen. What do we call this odd motion of Mars? What would you expect its phase to be at the mid-point between the two stationary points at which East-West motion ceases for a short time? Move time until Mars is at approximately this mid-position, center upon this planet and check your conclusion by zooming in on Mars.

54. **The Sun as Seen from the Various Terrestrial Planets.** In this exercise, you can use *Starry Night Enthusiast*™ to examine how big the Sun appears in the sky from the surfaces of all of the inner planets, starting with Mercury. In order to record your measurements, draw up a Table with the following columns:—Planet Name, Semi-Major Axis of Orbit, (obtained from Appendix E, Table E-1), Angular Radius of the Sun in arcminutes and arcseconds, Angular Radius in arcseconds, and Ratio of Sun Angular Radius. Open the **Find** pane and ensure that the **Query** box is empty in order to display a list of the planets. Stop time advance. **a.** Click on the **Down** arrow to the left of Mercury to open the drop-down list and click on **Go There** to go to the surface of this planet. In the **Find** tab, double-click on **Sun** to center the view on the Sun. Now **Zoom** in (+) until the Sun fills the view. Use the angle measurement tool of the **Hand Tool** to measure the angular radius of the Sun from this

viewpoint. (Place the **Hand Tool** over the Sun's center, hold down the mouse button and move the cursor to the edge of the Sun.) Record the value of this radius, displayed in arcminutes and arcseconds, in your table. **b.** Noting the apparent size of the Sun in each "sky" compared to your view from each previous planet, repeat the procedure in section **a.** for Venus, Earth and Mars. Qualitatively, how does the Sun change apparent size when going from Mercury to Mars? **c.** We will now find a relationship between the angles you measured and the distances of the planets to the Sun. Convert the angular solar radii you have recorded into arcseconds and enter these values in the appropriate column of your table. (1 arcminute = 60 arcseconds) **d.** divide the solar angular radius as measured from Earth by its radius at Mercury and record this number. Similarly, divide the Sun's radius at Earth by the Sun's radius at Venus, then divide the Sun's radius at Earth by the Sun's radius at Earth (this should be 1, of course), and, finally divide the Sun's radius at Earth by the Sun's radius at Mars. Verify that these ratios are nearly equal to the values of semi-major axes in your table for each of the planets. If the measurements are not exactly the semi-major axes, explain why. (Hint: draw triangles with the Sun's radius as one side and the angle from the planet as the opposite vertex and use simple trigonometry.).

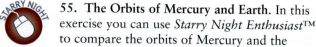

55. The Orbits of Mercury and Earth. In this exercise you can use *Starry Night Enthusiast*™ to compare the orbits of Mercury and the Earth. Open **Favourites > Solar System > Inner Planets > Inner Solar System** and click on the **Stop** (■) button to stop time. From your viewpoint about 5.3 AU from the Sun, you are looking down at the inner planets at an angle to the ecliptic plane. Open the **Find** pane and click on the boxes on either side of the names of Venus and Mars to remove the orbits and labels of these planets from the view. Use the **Location Scroller** (Hold down the **Shift** key while holding down the mouse button and moving the mouse) to move around the Sun on the surface of a sphere of radius 5.3 AU and examine and compare the orbits of Mercury and Earth. **a.** Place yourself directly above the Earth's orbit. (Hint: The orbital path of Earth will appear to be uniformly bright when you are in this position). Which of the orbits of Earth and Mercury is closest to circular? **b.** Use the **Location Scroller** to place yourself in the Earth's orbital plane (i.e., with the ecliptic edge-on). Are the orbits of Earth and Mercury in the same plane? If not, measure the angle between these two orbital planes by zooming in on the Sun to a field of view about 10° wide and tracing the two orbital planes onto tracing paper or transparent film with a soft felt-tipped pen.

CHAPTER 17

The Outer Planets

WEB LINK 17.1 The most complete visualization of Saturn and its ring system, taken by the *Cassini* spacecraft. This image is a combination of ultraviolet, visible light, and infrared observations. (Courtesy NASA/JPL-Caltech)

R I V U X G

WHAT DO YOU THINK?

1 Is Jupiter a "failed star?" Why or why not?

2 What is Jupiter's Great Red Spot?

3 Does Jupiter have continents and oceans?

4 Is Saturn the only planet with rings?

5 Are the rings of Saturn solid ribbons?

Answers to these questions appear in the text beside the corresponding numbers in the margins and at the end of the chapter.

We can pretty easily imagine a trip to Mars, exploring its vast canyons and icy polar regions. Even Venus and Mercury lend themselves to being compared to Earth and the Moon. However, we find few similarities between Earth and the giant outer planets: Jupiter, Saturn, Uranus, and Neptune. Until recently, Pluto was also considered one of the outer planets. That has changed as described in Appendix I. Now Pluto is categorized as a dwarf planet and we explore Pluto in Chapter 18.

The four giant worlds lack solid surfaces, are so much larger, rotate so much faster, and have such different chemical compositions from our planet that upon seeing them one knows, to paraphrase Dorothy, "You're not on Earth anymore." There is nothing on Earth remotely similar to the swirling red and brown clouds of Jupiter, the ever-changing ring system of Saturn, or the blue-green clouds of Uranus and Neptune. The outer four worlds collectively have at least 163 moons, which have an amazing range of shapes, sizes, surfaces, and properties. Some of these moons have water interiors; some have volcanoes; some have surface ice; one, Titan, has a thick atmosphere; and one, Hyperion, is arguably the most bizarre-looking object in the solar system. We begin our exploration of the outer planets with magnificent Jupiter.

In this chapter you will discover

- Jupiter, an active, vibrant, multicolored world more massive than all of the other planets combined

- Jupiter's diverse system of moons

- Saturn, with its spectacular system of thin, flat rings and numerous moons, including bizarre Enceladus and Titan

- what Uranus and Neptune have in common and how they differ from Jupiter and Saturn

JUPITER

Jupiter's multicolored bands, dotted with ovals of white and brown (Figure 17-1), give it the appearance of a world unlike any of the terrestrial planets. Viewed even through a small telescope (see Figure 2-11), you can also see up to four

Planet symbol: ♃

Jupiter   Earth

R I V U X G

FIGURE 17-1 *Jupiter as Seen from a Spacecraft* This view was sent back from *Voyager 1* in 1979. Features as small as 600 km across can be seen in the turbulent cloud tops of this giant planet. The complex cloud motions that surround the Great Red Spot are clearly visible. Also, clouds at different latitudes have different rotation rates. The inset image of Earth shows its size relative to Jupiter. (NASA/JPL; *inset:* NASA)

of its moons. As the high-resolution images in this chapter reveal, Jupiter is a world of breathtaking beauty.

Jupiter is the largest planet in the solar system: more than 1300 Earths could be packed into its volume. Using the orbital periods of its moons in Kepler's laws, astronomers have determined that Jupiter is 318 times more massive than Earth. Indeed, Jupiter has more than 2½ times as much mass as all of the other planets combined. This huge mass has created the urban myth that Jupiter is a failed star, meaning that it has almost enough matter to shine on its own, like the Sun, which we studied in Chapter 5. The fact is that Jupiter would have to be 75 times more massive than it is to generate energy like the Sun, and therefore to be classified as a star. Nevertheless, Jupiter emits approximately twice as much energy as it receives from the Sun. Its extra energy comes from radioactive elements in its core and from an overall contraction amounting to less than 10 cm per century.

17-1 Jupiter's outer layer is a dynamic area of storms and turbulent gases

Jupiter is permanently covered with clouds (see Figure 17-1). Because it rotates about once every 10 h—the fastest of any planet—Jupiter's clouds are in perpetual motion and are confined to narrow bands of latitude.

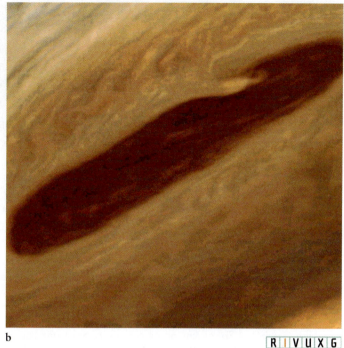

a   R I V U X G   b   R I V U X G

FIGURE 17-2 *Close-ups of Jupiter's Atmosphere* The dynamic winds, rapid rotation, internal heating, and complex chemical composition of Jupiter's atmosphere create its beautiful and complex banded pattern. (a) A *Voyager 2* southern hemisphere image showing a white oval that has existed for over 40 years. (b) A *Voyager 2* northern hemisphere image showing a brown oval. The white feature overlapping the oval is a high cloud. (NASA)

Even through a small telescope, you can see Jupiter's dark, reddish bands called **belts**, alternating with light-colored bands called **zones**. These belts and zones are gases flowing east or west, with very little north-south motion. In contrast, winds on slower-rotating Earth wander over vast ranges of latitude (compare Figure 15-1).

Jupiter's belts and zones provide a framework for turbulent swirling cloud patterns, as well as rotating storms similar in structure to hurricanes or cyclones on Earth. These storms are known as *white ovals* and *brown ovals* (Figure 17-2). The white ovals are observed to be cool clouds higher than the average clouds in Jupiter's atmosphere. The brown ovals are warmer and lower clouds, seen through holes in the normal cloud layer. The various oval features last from hours to centuries. Computers show us how the cloud features on Jupiter would look if the planet's atmosphere were unwrapped like a piece of paper (Figure 17-3).

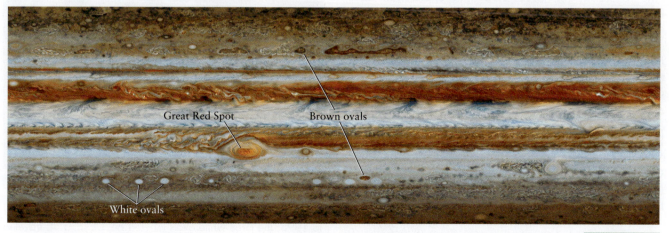

Great Red Spot   Brown ovals

White ovals

R I V U X G

FIGURE 17-3 *Jupiter Unwrapped* *Cassini* images of Jupiter combined and opened to give a maplike representation of the planet. The banded structure is absent near the poles. The Web link will take you to a movie version of this and related images. In them, you will see that the light and dark regions slide by one another, continually moving eastward or westward. (Courtesy NASA/JPL-Caltech)

R I V U X G

FIGURE 17-4 The Great Red Spot
This true color image of the Great Red
Spot, taken by *Galileo* in 1996, shows
what this giant storm would look like if you were traveling over
it in a spacecraft. The counterclockwise circulation of gas in
the Great Red Spot takes about 6 days to make one rotation.
The clouds that encounter it are forced to pass around it, and
when other oval features are near it, the entire system
becomes particularly turbulent, like the batter in a two-bladed
blender. (Courtesy NASA/JPL-Caltech)

2 Jupiter's most striking feature is its **Great Red Spot**
(Figure 17-4), which is so large that it can be seen
through a small telescope. It changes dimensions, and at
present is about 25,000 km long by 12,000 km wide—
large enough so that two Earths could easily fit side by
side inside it. The Great Red Spot was first observed
around 1656, either by the English scientist Robert
Hooke or the Italian astronomer Giovanni Cassini.
Because earlier telescopes were unlikely to have been
able to see it, the Great Red Spot could well have
formed long before that time.

The Great Red Spot is a hurricane or typhoon-like
storm of swirling gases. Heat welling upward from
inside Jupiter has maintained it for more than three centuries. (Consider what life would be like for us if Earth
sustained storms for such long periods!) Between 1998
and 2000, three smaller, white storms on Jupiter merged
(Figure 17-5a–d), creating a larger white storm that
became red in 2006 (Inset: Figure 17-5e). Named *Red
Spot Jr.,* it is similar but somewhat smaller than the
Great Red Spot and located at nearly the same latitude
(Figure 17-5e). As with the Great Red Spot, the cause of
the red color is still being studied.

In 1690, Cassini noticed that the speeds of Jupiter's
clouds vary with latitude, an effect called **differential
rotation**. Near the poles, the rotation period of
Jupiter's atmosphere, 9 h 55 min 30 s, is 5 min longer
than at the equator. Furthermore, clouds at different
latitudes circulate in opposite directions—some east-

ward, some westward. At their boundaries, the clouds
rub against each other, creating beautiful swirling patterns (see Figure 17-2). The interactions of clouds at
different latitudes also help provide stability for storms
like the Great Red Spot.

3 Astronomers first determined Jupiter's overall
chemical composition from its average density—only
1330 kg/m³. Recall from Section 14-5 that average density is mass divided by volume. We determine Jupiter's
mass using Kepler's third law and the orbital periods of
its moons, while the trigonometry of Jupiter's distance
from Earth and angular size in our sky reveal its diameter and hence its volume. This low density implies that
Jupiter is primarily composed of the lightweight
elements hydrogen and helium surrounding a relatively
small core of metal and rock. It has no solid continents,
islands, or water oceans on its surface.

Spectra from Earth-based telescopes and from the
Galileo probe sent into Jupiter's upper atmosphere in
1995 give more detail about the chemistry of this giant

R I V U X G

FIGURE 17-5 Creating Red Spot Jr. (a–d) For 60 years prior
to 1998, the three white ovals labeled *FA, DE,* and *BC* traveled
together at the same latitude on Jupiter. Between 1998 and
2000, they combined into one white oval, labeled *BA,* which (e)
became a red spot, named Red Spot Jr., in 2006. (a–d: NASA/
JPL/WFPC2; e: NASA, ESA, A. Simon-Miller [NASA/GSFC], and I. de
Pater [University of California Berkeley])

world's upper level. About 86% of its atoms are hydrogen and 13% are helium. The remainder consists of molecular compounds, such as methane (CH_4), ammonia (NH_3), and water vapor (H_2O). Keeping in mind that different elements have different masses, we can convert these percentages of atoms into the masses of various substances in Jupiter's atmosphere: 75% hydrogen, 24% helium, and 1% other substances. Because the interior contains heavier elements than the surface, the overall mass distribution in Jupiter has been calculated to be 71% hydrogen, 24% helium, and 5% all heavier elements.

The descent of the probe from the *Galileo* spacecraft into Jupiter's atmosphere revealed wind speeds of up to 600 km/h (375 mph), higher-than-expected air density and temperature, and lower-than-expected concentrations of water, helium, neon, carbon, oxygen, and sulfur. This probably occurred because the probe descended into a particularly arid region of the atmosphere called a *hot spot*, akin to the air over a desert on Earth.

Observations from spacecraft visiting Jupiter and its moons, combined with the scientific model of Jupiter's atmosphere developed to explain the observations of Jupiter's clouds and chemistry, indicate it has three major cloud layers (Figure 17-6a). The uppermost Jovian cloud layer is composed of crystals of frozen ammonia. These crystals and the frozen water in Jupiter's clouds are white, so what chemicals create the subtle tones of brown, red, and orange? The answer is as yet unknown. Some scientists think that sulfur compounds, which can assume many different colors, depending on their temperature, play an important role. Others think that phosphorus is involved, especially in the Great Red Spot. The middle cloud layer is primarily ammonium hydrosulfide, and the bottom cloud layer is mostly composed of water vapor.

Below its cloud layer, Jupiter's mantle is entirely liquid. Here on Earth, the distinction between the gaseous air and the liquid oceans is very clear—jump off a diving board and you know when you hit the water. However, the conditions on Jupiter under which hydrogen liquefies are different from anything we normally experience. As a result, on Jupiter there is no definite boundary between the planet's gaseous atmosphere and its liquid mantle. The hydrogen gradually gets denser until, 1000 km below the cloud tops, the pressure is high enough for the hydrogen to be what we would consider a liquid.

In introducing the solar system, we noted that the young planets heated up as they coalesced. After they formed, radioactive elements continued to heat their interiors. On Earth, this heat leaks out of the surface through volcanoes and other vents. Jupiter loses heat everywhere on its surface, because, unlike Earth, it has no surface landmasses to block the heat loss.

FIGURE 17-6 Jupiter's and Saturn's Upper Layers These graphs display temperature profiles of (a) Jupiter's and (b) Saturn's upper regions, as deduced from measurements at radio and infrared wavelengths. Three major cloud layers are shown in each, along with the colors that predominate at various depths. Data from the *Galileo* spacecraft indicate that Jupiter's cloud layers are not found at all locations around the planet; there are some relatively clear, cloud-free areas.

Heated from deep within Jupiter, blobs of liquid hydrogen and helium move upward inside it. When these blobs reach the cloud tops, they release their heat and descend back into the interior. (The same process, *convection*, drives the motion of Earth's mantle and its tectonic plates, as well as liquid simmering on a stove; see Figure 15-9.) Jupiter's rapid, differential rotation draws the convective gases into bands of winds moving eastward and westward at different speeds around the planet.

Astronomers believe that Jupiter's belts and zones result from the combined actions of the planet's convection and rapid differential rotation. Until recently, they believed that the light-colored zones are regions of hotter, rising gas, while the dark belts are regions of cooler, descending gas (Figure 17-7). However, high-resolution observations by the passing *Cassini* spacecraft on its way to Saturn revealed numerous white clouds that are too small to be seen from Earth rising in the dark belts of Jupiter, while the zones are sinking gas—just the opposite of the original model! The correct explanation for the behavior of gases in the belts and zones may be provided by *Cassini*, when it studies the analogous belts and

What are the two most common elements in Jupiter's upper layers?

FIGURE 17-7 Original Model of Jupiter's Belts and Zones
The light-colored zones and dark-colored belts in Jupiter's atmosphere were believed, until recently, to be regions of rising and descending gases, respectively. In the zones, gases warmed by heat from Jupiter's interior were thought to rise upward and cool, forming high-altitude clouds. In the belts, cooled gases were thought to descend and undergo an increase in temperature; the cloud layers seen there are at lower altitudes than in the zones. Observations by the *Cassini* spacecraft on its way to Saturn suggest that just the opposite may be correct (stay tuned)! In either case, Jupiter's rapid differential rotation shapes the rising and descending gas into bands of winds parallel to the planet's equator. Differential rotation also causes the wind velocities at the boundaries between belts and zones to move predominantly to the east or west.

zones found on Saturn (see Section 17-9) in greater detail in the next few years.

17-2 Jupiter's interior has four distinct regions

WEB LINK 17.5

Descending below Jupiter's clouds, we first encounter liquid molecular hydrogen and helium. As in Earth's oceans, the pressure in this fluid increases with depth. The gravitational force created by Jupiter's enormous mass compresses and heats its interior so much that 20,000 km (12,500 mi) below the cloud tops, the pressure is 3 million atm. Below this depth, the pressure is high enough to transform hydrogen into **liquid metallic hydrogen** (Figure 17-8). In this state, hydrogen acts like a metal (similar to the copper wiring in a house) in its ability to conduct electricity and heat. Electric currents that run through this rotating, metallic region of Jupiter generate a powerful planetary magnetic field (Figure 17-9). At the cloud level of Jupiter, this field is 14 times stronger per square meter than Earth's field is at our planet's surface. Jupiter's magnetic field has a region, like Earth's Van Allen belts, where charged particles are stored (depicted

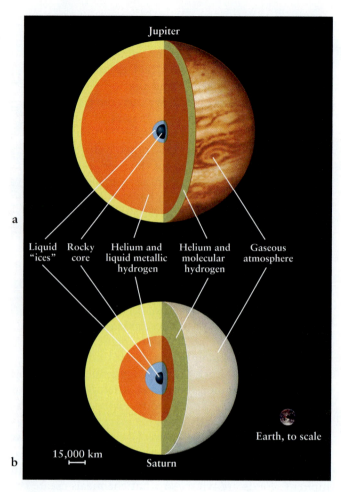

a

b

FIGURE 17-8 Cutaways of Jupiter and Saturn The interiors of both Jupiter and Saturn are believed to have four regions: a terrestrial rocky core, a liquid "ice" shell, a liquid metallic hydrogen shell, and a normal liquid hydrogen mantle. Their atmospheres are thin layers above the normal hydrogen, which boils upward, creating the belts and zones.

in orange in Figure 17-9) and its tail streams outward to Saturn, over 700 million km away.

WEB LINK 17.6

For Earth-based astronomers, the only evidence of Jupiter's magnetosphere is a hiss of radio static, which varies cyclically over a period of 9 h 55 min 30 s, the planet's internal rotation rate. Beginning with the two *Pioneer* and two *Voyager* spacecraft that journeyed past Jupiter in the 1970s, and continuing with the recent *Galileo* and *Cassini* spacecraft, the awesome dimensions of Jupiter's magnetosphere have been revealed: It is nearly 30 million km across. It envelops the orbits of many of its moons. If Jupiter's magnetosphere were visible from Earth, it would cover an area of the sky 16 times larger than the full Moon.

Although the sequence of events in Jupiter's formation is still being worked out, it appears that a terrestrial (rock and metal) protoplanet formed at

Jupiter's orbit and attracted hydrogen and helium to form its outer layers. This terrestrial matter is now Jupiter's core. This core is only 4% of Jupiter's mass, which still amounts to nearly 13 times the mass of the entire Earth. Water, carbon dioxide, methane, and ammonia are likely to have existed as ice in Jupiter's terrestrial protoplanet, and they were also attracted onto the young and growing planet. When astronomers talk in general terms about "ice," they refer to any or all of these four compounds.

The tremendous crushing weight of the bulk of Jupiter above the core—equal to the mass of 305 Earths—compresses the terrestrial core down to a sphere only 10,000 km in radius (see Figure 17-8). By comparison, Earth's diameter is 12,756 km. At the same time, the pressure forced the lighter ices out of the rock and metal, thereby forming a shell of these compounds between the solid core and the liquid metallic hydrogen layer. Calculations reveal that the temperature and pressure inside Jupiter should make these "ices" liquid there! The pressure at Jupiter's very center is calculated to be about 70 million atm, and the temperature there is about 25,000 K, nearly four times hotter than the surface of the Sun. The interior of Jupiter is summarized in Figure 17-8a.

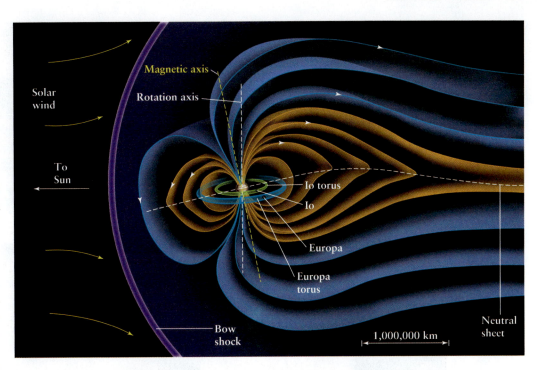

FIGURE 17-9 Jupiter's Magnetosphere Created by the planet's rotation, the ion-trapping regions of Jupiter's magnetosphere (in orange, analogous to the Van Allen belts) extend into the realm of the Galilean moons. Gases from Io and Europa form tori in the magnetosphere. Some of Io's particles are pulled by the field onto the planet. Pushed outward by the Sun, the magnetosphere reaches all the way to Saturn.

17-3 Impacts provide probes into Jupiter's atmosphere

On July 7, 1992, a comet nucleus (a clump of rock and ice a few kilometers across) passed so close to Jupiter that the planet's gravitational tidal force ripped it into at least 21 pieces. The debris from this comet was first observed in March 1993 by comet hunters Carolyn and Gene Shoemaker and David Levy. (Because it was the ninth comet they had found together, it was named Shoemaker-Levy 9 in their honor.) Shoemaker-Levy 9 was an unusual comet that actually orbited Jupiter, rather than just orbiting the Sun. Calculations of the comet's orbit showed that the pieces (Figure 17-10a) would return to strike Jupiter between July 16 and July 22, 1994.

Recall from Section 14-3 that impacts were extremely common in the first 800 million years of the solar system's existence. From more recent times, two chains of impact craters have been discovered on Earth, consistent with pieces of comets having hit our planet within the past 300 million years. However, it is very uncommon for pieces of space debris as large as several kilometers in diameter to collide with planets today. Therefore, the discovery that Shoemaker-Levy 9 would hit Jupiter created great excitement in the astronomical community. Seeing how a planet and a comet respond to such an impact would allow astronomers to deduce information about the planet's atmosphere and interior and also about the striking body's properties.

Why is the terrestrial body inside Jupiter so small compared to what it would be in orbit without the outer layers around it?

The impacts occurred as predicted, with most of Earth's major telescopes—as well as those on several spacecraft—watching closely (Figure 17-10b). At least 20 fragments from Shoemaker-Levy 9 struck Jupiter, and 15 of them had detectable impact sites. The impacts resulted in fireballs some 10 km in diameter with temperatures of 7500 K, which is hotter than the surface of the Sun. Indeed, the largest fragment gave off as much energy as 600 million megatons of TNT, far more than the combined energy that could be released by all nuclear weapons remaining on Earth. Impacts were followed by crescent-shaped ejecta that contained a variety of chemical compounds. Ripples or waves spread out from the impact sites through Jupiter's clouds in splotches that lasted for months.

a   R I V U X G

b   R I V U X G   R I V U X G

Labels on image: Io, Aurora, Great Red Spot, Impact sites, Aurora

FIGURE 17-10 Comet Shoemaker-Levy 9 and its Encounter with Jupiter (a) The comet was torn apart by Jupiter's gravitational force on July 7, 1992, fracturing into at least 21 pieces. This comet originally orbited Jupiter, and its returning debris, shown here in May 1994, struck the planet between July 16 and July 22, 1994. (b) Shown here are visible *(left)* and ultraviolet *(right)* images of Jupiter taken by the Hubble Space Telescope after three pieces of Comet Shoemaker-Levy 9 struck the planet. Astronomers had expected white remnants (the color of condensing ammonia or water vapor); the darkness of the impact sites may have come from carbon compounds in the comet debris. Note the auroras in the ultraviolet image. Auroras and lightning are common on Jupiter, due, in part, to the planet's strong magnetic fields and dynamic cloud motions. (a: H. A. Weaver, T. E. Smith, STScI and NASA; b: NASA)

The observations suggested that the pieces of comet did not penetrate very far into Jupiter's upper cloud layer. This fact, in turn, supports the belief that the pieces were not much larger than a kilometer in diameter. The ejecta from each impact included a dark plume that rose high into Jupiter's atmosphere. The darkness was apparently due to carbon compounds that vaporized from the cometary bodies. Also detected from the comet were water, sulfur compounds, silicon, magnesium, and iron.

> Why did the comet impacts of 1994 not create craters on Jupiter?

JUPITER'S MOONS AND RINGS

Jupiter hosts at least 63 moons. The four largest—Io, Europa, Ganymede, and Callisto—may have formed at the same time as Jupiter from debris orbiting the planet, just as the planets formed from debris orbiting the Sun (see Section 14-3). Jupiter's other moons, as we will see, are probably captured planetesimals and smaller pieces of space debris.

Galileo was the first person to observe the four largest moons, in 1610, seen through his meager telescope as pinpoints of light. He called them the *Medicean* stars to attract the attention of the Medicis, rulers of Florence and wealthy patrons of the arts and sciences. To Galileo, the moons provided evidence supporting the then-controversial Copernican cosmology; at that time, Western theologians asserted that all cosmic bodies orbited Earth. The fact that the Medicean stars orbited Jupiter raised grave concerns in some circles.

To the modern astronomer, these moons are four extraordinary worlds, different both from the rocky terrestrial planets and from hydrogen-rich Jupiter. Now called collectively the **Galilean moons** or **Galilean satellites**, they are named after the mythical lovers and companions of the Greek god Zeus (called Jupiter by the early Romans). From the innermost moon outward, they are Io, Europa, Ganymede, and Callisto.

These four worlds were photographed extensively by the *Voyager 1* and *Voyager 2* flybys and by the *Galileo* spacecraft (Figure 17-11). The two inner Galilean satellites, Io and Europa, are about the same size as our Moon. The two outer satellites, Ganymede and Callisto, are roughly the size of Mercury. Figure 17-11 presents comparative information about these six bodies.

17-4 Io's surface is sculpted by volcanic activity

Sulfurous Io is among the most exotic moons in our solar system (Figure 17-12a). At first glance, its density of 3570 kg/m^3 seems to place its chemical composition between that of the terrestrial and Jovian planets. However, its density has that value in large measure because its mass is so small that it cannot compress its interior nearly as much as can Earth or more massive planets. Allowing for its lower compression, astronomers calculate that Io is mostly rock and iron, like Earth, rather than being composed of lighter elements, as is Jupiter.

Io zooms through its orbit of Jupiter once every 1.8 days. Like our Moon, it is in synchronous rotation with its planet. Images of Io reveal giant plumes emitted by volcanoes (Figure 17-12b) and geysers, similar to Old Faithful on Earth. Most of this ejected material falls back onto Io's surface; the rest is moving fast enough to escape into space. These geysers are usually associated with Io's 300 or so active volcanoes. Observations indicate that these volcanoes emit 10 trillion tons of matter each year in plumes up to 500 km high. That is enough material to resurface Io to a depth of 1 m each century (Figure 17-12c). One eruption in 2002 occurred over an area the size of London, some 1600 sq km (620 sq mi). The volcanoes also emit basaltic lava flows rich in magnesium and iron. Io's volcanoes are named after gods and goddesses associated with fire in Greek, Norse, Hawaiian, and other mythologies. Io also has numerous black "dots" on its surface, which, apparently, are dormant volcanic vents. Old lava flows radiate from many of these locations, which are typically 10 to 50 km in diameter and cover 5% of Io's surface.

INSIGHT INTO SCIENCE

What's in a Name? Following up on the preconceptions that common words create (see *Insight into Science: Imagine the Moon* in Chapter 16), objects with familiar names often have different characteristics than we expect. We tend to envision "moons" as inert, airless, lifeless, dry places similar to our Moon. However, as we will see throughout this chapter, starting with Io, some moons have active volcanoes, atmospheres, and vast, underground, liquid water oceans.

Just before their discovery, the existence of active volcanoes on Io was predicted from analysis of the gravitational forces to which that moon is subjected. As it rapidly orbits Jupiter, Io repeatedly passes between Jupiter and one or another of the remaining Galilean satellites. These moons pull on Io, causing it to slightly change its distance from Jupiter. This change in distance creates a change in the tidal forces (see Section 15-8) acting on it from the planet. As the distance between Io and Jupiter varies, the resulting tidal stresses alternately squeeze and flex the moon. This ongoing motion of Io's interior generates heat through friction, creating as much energy inside Io as the detonation of 2400 tons of TNT every second. Gas and molten rock from inside Io eventually make their way to the moon's surface, where they are ejected through the volcanoes and other vents.

Satellite instruments have identified sulfur and sulfur dioxide in the material erupting from Io's volcanoes. Sulfur is normally bright yellow. If heated and suddenly cooled, however, it forms molecules that assume a range of colors, from orange and red to black, which accounts for Io's tremendous range of colors (see Figure 17-12). Sulfur dioxide (SO_2) is an acrid gas commonly discharged from volcanic vents here on Earth and, apparently, on Venus. When eruptions on Io release this gas into the cold vacuum of space, it crystallizes into white flakes, which fall onto the surface and account for the moon's whitish deposits.

The *Galileo* spacecraft detected an atmosphere around Io. Composed of oxygen, sulfur, and sulfur dioxide, it is only one-billionth as dense as the air we breathe. Io's atmosphere can sometimes be seen to glow blue, red, or green, depending on the gases involved. Gases ejected from Io's volcanoes have also been observed extending out into space and forming a doughnut-shaped region around Jupiter called the *Io torus*, about which we will say more shortly.

What creates most of the heat inside Io that causes it to have volcanoes and geysers?

The Galilean Moons, Mercury, and Earth's Moon: Vital Statistics

	Mean distance from Jupiter (km)	Sidereal period (day)	Diameter (km)	Mass (kg)	Mass (Moon = 1)	Mean density (kg/m³)
Io	421,600	1.77	3630	8.94×10^{22}	1.22	3570
Europa	670,900	3.55	3138	4.80×10^{22}	0.65	2970
Ganymede	1,070,000	7.16	5262	1.48×10^{23}	2.01	1940
Callisto	1,883,000	16.69	4800	1.08×10^{23}	1.47	1860
Mercury	—	—	4878	3.30×10^{23}	4.49	5430
Moon	—	—	3476	7.35×10^{22}	1.00	3340

Io   Europa   Ganymede   Callisto

R I V U X G

FIGURE 17-11 The Galilean Satellites The four Galilean satellites are shown here to the same scale. Io and Europa have diameters and densities comparable to our Moon and are composed primarily of rocky material. Ganymede and Callisto are roughly as big as Mercury, but their low average densities indicate that each contains a thick layer of water and ice. The cross-sectional diagrams of the interiors of the four Galilean moons show the probable internal structures of the moons, based on their average densities and on information from the *Galileo* mission. (NASA and NASA/JPL)

Prometheus Volcano

R I V U X G

a

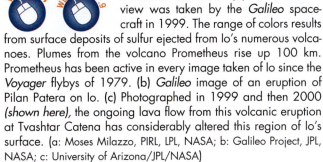

Lava flows

Volcanic calderas

Curtain of lava

50 km

c

R I V U X G

VIDEO 17.4   WEB LINK 17.9

FIGURE 17-12 Io (a) This true-color view was taken by the *Galileo* spacecraft in 1999. The range of colors results from surface deposits of sulfur ejected from Io's numerous volcanoes. Plumes from the volcano Prometheus rise up 100 km. Prometheus has been active in every image taken of Io since the *Voyager* flybys of 1979. (b) *Galileo* image of an eruption of Pilan Patera on Io. (c) Photographed in 1999 and then 2000 *(shown here)*, the ongoing lava flow from this volcanic eruption at Tvashtar Catena has considerably altered this region of Io's surface. (a: Moses Milazzo, PIRL, LPL, NASA; b: Galileo Project, JPL, NASA; c: University of Arizona/JPL/NASA)

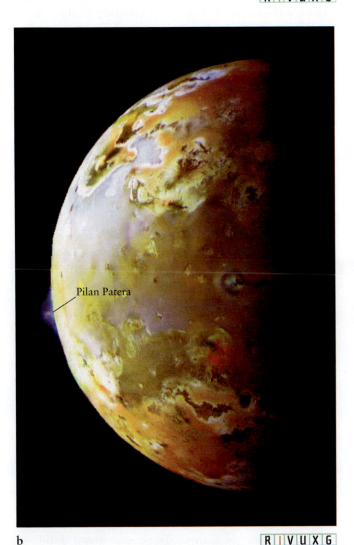

Pilan Patera

b

R I V U X G

17-5 Europa apparently harbors liquid water below its surface

Images of Europa's ice and rock surface from *Voyager 2* and the *Galileo* spacecraft suggest that Jupiter's second-closest Galilean moon contains liquid water (Figure 17-13). Europa orbits Jupiter every 3½ days and, like Io, is in synchronous rotation. The changing gravitational tug of Io creates stress inside Europa similar to the magma-generating distortion that Io undergoes. These stresses apparently cause numerous cracks or fractures seen on Europa's surface (Figure 17-14a). Furthermore, astronomers think this stress creates enough heat inside Europa to keep the water in a liquid state just a few kilometers below the moon's ice and rock surface.

Besides the cracks, numerous other features on Europa support the idea that it has liquid water inside. Just as in the Arctic region on Earth, the movement of surface features on Europa creates ice floes, along with swirls, strips, and ridges, driven by circulating water underneath the moon's surface (Figure 17-14b). Also indicative of liquid water is the moon's reddish color. The coloring may be due to salt deposits left after liquid water rose to the surface and evaporated (Figures 17-14a, c). *Galileo* spacecraft images suggest that some of Europa's features have moved within the past few million years,

R I V U X G

WEB LINK 17.10

FIGURE 17-13 Europa Imaged by the *Galileo* spacecraft, Europa's ice surface is covered by numerous streaks and cracks that give the satellite a fractured appearance. The streaks are typically 20 to 40 km wide. (NASA/JPL)

and perhaps are still in motion. Indeed, the chaotic surface revealed by *Galileo* is interpreted as having formed as a result of water volcanism on Europa, strengthening the belief that this moon still has a liquid water layer. Replenishment of the surface by tectonic plate motion would explain why only a few small impact craters have survived.

Galileo also photographed red and white domes called *lenticulae* on Europa (Figure 17-14c) that geologists think are rising, warmed ice mixed with other material, possibly including organic matter. These lenticulae are typically a 100 m high and 10 km (6 mi) wide. The rising material behaves like the blobs in lava lamps (Figure 17-14d), although the lenticulae are calculated to take 100,000 years to reach the surface from the liquid ocean that is believed to exist inside the moon.

Europa's average density of 2970 kg/m^3 is slightly less than Io's. A quarter of its mass may be water. It also has a metallic core of much higher density and a weak magnetic field. In 1995, astronomers discovered an extremely thin atmosphere containing molecular oxygen surrounding Europa. This gas is about 10^{-11} times less dense than the air we breathe. The oxygen may come from water molecules broken up on the moon's surface by ultraviolet radiation from the Sun.

The underground ocean of Europa is especially interesting to scientists because virtually all liquid water locations on Earth support

What region on Earth does the surface of Europa most resemble?

life, and life may well have evolved in Europa's oceans. If it exists, and if it evolved differently from life on Earth, then life-forms from Earth would likely contaminate it, even wiping it out. This would be analogous to the deadly impact of diseases on Earth, such as introducing smallpox from Europe to the Americas. To prevent the expired spacecraft *Galileo* from contaminating any life that may exist on Europa, NASA sent that spacecraft into Jupiter's thick atmosphere, where the machine was vaporized.

17-6 Ganymede is larger than Mercury

Ganymede is the largest satellite in the solar system (Figure 17-15). Its diameter is greater than Mercury's, but its density of 1940 kg/m^3 is much less than that of Mercury. It also has a permanent magnetic field that is twice as strong as Mercury's field. Ganymede orbits Jupiter in synchronous rotation once every 7.2 days. Like its neighbor Europa, Ganymede has an iron-rich core, a rocky mantle, an underground liquid water ocean, a thin atmosphere, and a covering of dirty ice.

The existence of an underground ocean is implied by the discovery of a second magnetic field around Ganymede, which continually varies. This field has a different origin than the permanent one just mentioned; it is generated by Jupiter's magnetic field. As Ganymede orbits Jupiter, the planet's powerful magnetic field creates an electric current inside the moon, which, in turn, creates Ganymede's varying magnetic field. (The same process is used to create electric currents in power stations here on Earth, where falling water or steam is used to rotate powerful magnets at high speeds. Their magnetic fields cut across wires, thereby pushing electrons in the wire. These moving electrons are the electric current we use.) The best explanation of why current exists inside Ganymede is that liquid salt water exists there, and salt water is a good conductor of electricity. This implies, of course, the presence of a liquid ocean. Furthermore, salts have been observed on Ganymede's surface. They were apparently carried upward and deposited there as water leaked out and froze. As on Europa, liquid water may imply the presence of life.

Like our Moon, Ganymede has two very different kinds of terrain (Figures 17-15, 17-16). Dark, polygon-shaped regions are its oldest surface features, as judged by their numerous craters. Light-colored, heavily grooved terrain is found between the dark, angular islands. These lighter regions are much less cratered and therefore younger. Ganymede's grooved terrain consists of parallel mountain ridges up to 1 km high and spaced 10 to 15 km apart. These features suggest that the process of plate tectonics may have dominated Ganymede's early history. But, unlike Europa, where tectonic activity still occurs today, tectonics on Ganymede bogged down 3 billion years ago as the satellite's crust froze solid.

Smooth ice plains with a network of fractures

Linear ridges (dark colors are caused by minerals in the ice)

500 km

a

R I V U X G

5 km

Ridges produced by folding and faulting of the surface

Smooth area formed by fluid erupting onto the surface

Rugged patch created by a subsurface disturbance

b

R I V U X G

Lenticulae

c

R I V U X G

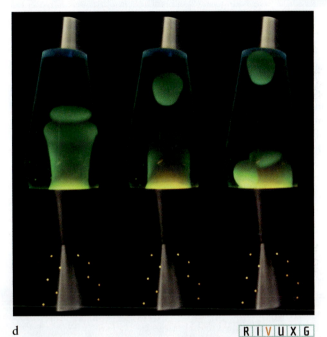

d

R I V U X G

FIGURE 17-14 Surface Features on Europa (a) This false color *Galileo* image of Europa combining visible and infrared observations shows smooth plains of ice, mineral ridges deposited by upwelling water, and numerous fractures believed to be caused by tidal stresses. (b) This region of Europa's surface shows the jumbled, stressed features common to the surface, as well as direct indications of liquid water activity underground. (c) Lenticulae attributed to rising warmed ice and debris travel up from the moon's interior by convection, arriving at and then leaking out at the surface. The white domes are likely to be rising material that has not yet reached the surface. (d) A lava lamp in which warmed material rises through cooler liquid. The rising material in Europa is analogous to the rising motion of the blobs in a lava lamp, except that on Europa, the motion is through ice. (a, b: NASA/JPL; c: NASA/JPL/University of Arizona/University of Colorado; d: Bianca Moscatelli)

Another mechanism that may have created Ganymede's large-scale features is a bizarre property of water. Unlike most liquids, which shrink upon solidifying, water expands when it freezes. Seeping up through cracks in Ganymede's original crust, water may have thus forced apart fragments of that crust. This process could have produced jagged, dark islands of old crust separated by bands of younger, light-colored, heavily grooved ice. Topping off the list of Ganymede's varied features is the discovery that auroras occur there.

Which planet is smaller than Ganymede?

Dark terrain (old, heavily cratered)

Bright terrain (young, fewer craters)

R I V U X G

VIDEO 17.5   WEB LINK 17.11

FIGURE 17-15 Ganymede This side of Ganymede is dominated by the huge, dark, circular region called *Galileo Regio,* which is the largest remnant of Ganymede's ancient crust. Darker areas of the moon are older; lighter areas are younger, tectonically deformed regions. The light white areas in and around some craters indicate the presence of water ice. Large impacts create white craters, filled in by ice from below the surface. (NASA/JPL)

17-7 Callisto bears the scars of a huge asteroid impact

Callisto is Jupiter's outermost Galilean moon. It orbits Jupiter in 16.7 days, and, like the other Galilean moons, its rotation is synchronous. Callisto is 91% as big and 96% as dense as Ganymede, and has a thin atmosphere of hydrogen and carbon dioxide. Like Ganymede, Callisto, apparently, harbors a substantial liquid water ocean. As with Ganymede, this ocean's presence is also inferred by Callisto's changing magnetic field. The heat that keeps the ocean liquid, apparently, comes from energy released by radioactive decay inside the moon.

Although numerous large impact craters are scattered over Callisto's dark, ancient, icy crust, it has very few craters smaller than 100 m across (Figure 17-17). Astronomers speculate that the smaller craters have disintegrated. Unlike Ganymede and Europa, Callisto has no younger, grooved terrain. The absence of grooved terrain suggests that tectonic activity never began there: The satellite simply froze too rapidly. It is bitterly cold on Callisto's surface. *Voyager* instruments measured a noontime temperature of 155 K (−180°F), and the nighttime temperature plunges to 80 K (−315°F).

Like Mercury, our Moon, and other bodies we will encounter shortly, Callisto carries the cold, hard evidence of what happens when one astronomical body strikes another. *Voyager 1* photographed the huge impact basin, named *Valhalla,* on Callisto (see Figure 17-17a). An asteroid-sized object produced Valhalla Basin, which is located on Callisto's Jupiter-facing hemisphere. Like throwing a rock into a calm lake, ripples ran out from the impact

Dark terrain: more craters

Bright terrain: fewer craters

Long grooves

Long, deep furrows

50 km

R I V U X G

FIGURE 17-16 Two Surfaces of Ganymede The older, rougher, more heavily cratered parts of Ganymede are the dark terrain. These regions are surrounded by younger, smoother, less cratered bright terrain. The parallel ridges suggest that the bright terrain has been crafted by tectonic processes. (NASA)

a

5 km

b   R I V U X G

FIGURE 17-17 Callisto The outermost Galilean satellite is almost exactly the same size as Mercury. Numerous craters pockmark Callisto's icy surface. (a) The series of faint, concentric rings that cover much of this image is the result of a huge impact that created the impact basin Valhalla. Valhalla dominates the Jupiter-facing hemisphere of this frozen, geologically inactive world. (b) The two insets in this *Galileo* mission image show spires that contain both ice and some dark material. The spires were probably thrown upward as the result of an impact. The spires erode as dark material in them absorbs heat from the Sun. (a: Courtesy of NASA/JPL; b: NASA/JPL/Arizona State University)

site along Callisto's surface, cracking the surface and freezing into place for eternity. The largest remnant rings surrounding the impact crater have diameters of 3000 km. In 2001, the *Galileo* spacecraft revealed spires 80–100 m high on Callisto (Figure 17-17b). These are also believed to have been created by an impact, perhaps the same one. One thing that is missing from Callisto is the jumbled terrain on the side opposite of Valhalla Basin, as is seen opposite of large impact sites on Mercury and our Moon. The smoothness of Callisto can be explained by a model that shows that a liquid water interior would dampen the impact shock and thereby prevent the opposite side from becoming disturbed. This supports the idea that Callisto has liquid water inside. The probable interiors of the Galilean moons are shown in Figure 17-11.

> What features on our Moon and on Mercury are similar to Callisto's feature Valhalla?

17-8 Other debris orbits Jupiter as smaller moons and ringlets

Besides the four Galilean moons, Jupiter has at least 59 other moons, a set of tenuous ringlets, and two dough-nut-shaped tori of electrically charged gas particles. The non-Galilean moons are all irregular in shape and less than 275 km in diameter. Four of these moons (Figure 17-18) are inside Io's orbit; all of the other known moons are outside Callisto's orbit. The inner moon Amalthea (Figure 17-18) is red-colored, has about the same density as water, and is, apparently, made of pieces of rock and ice barely held together by the moon's own gravity. The Galilean moons, along with the smaller moons closer to Jupiter and six of the outer moons, orbit in the same direction that Jupiter rotates (**prograde orbits**). The remaining outer moons revolve in the opposite direction (**retrograde orbits**). The outer ones appear to be individual, captured planetesimals, while the inner ones are probably smaller pieces broken off a single larger body.

Jupiter is only the first of four planets with rings. As astronomers predicted, the cameras aboard *Voyager 1* discovered three **ringlets** around Jupiter. These are bands of particles, the brightest of which is seen in Figure 17-19a, a *Voyager 2* image. They are the darkest, simplest set of rings in the solar system. They consist of very fine dust particles that are continuously pushed out of orbit by the impact of radiation from Jupiter and the Sun. Therefore, these rings are being replenished by material ejected volcanically from Io and knocked free by impact from the other moons.

At least two doughnut-shaped regions of electrically charged gas particles, called *plasmas*, orbit Jupiter. One is in the same orbit as Io (Figure 17-19b). The Io torus consists of sulfur and oxygen ions (charged atoms) along with free electrons. These particles were ejected by Io's geysers, and they are held in orbit by Jupiter's strong magnetic field. Guided by the field, some of this

R I V U X G

FIGURE 17-18 Irregularly Shaped Inner Moons
The four known inner moons of Jupiter are significantly different from the Galilean satellites. They
are roughly oval-shaped bodies. Although craters have not yet
been resolved on Adrastea and Metis, their irregular shapes
strongly suggest that they are cratered. All four moons are
named for characters in mythology relating to Jupiter (Zeus, in
Greek mythology). (NASA/JPL, Cornell University)

a

R I V U X G

b

FIGURE 17-19 Jupiter's Ring and Torus (a) A
portion of Jupiter's faint ring system, photographed by *Voyager 2*. The ring is probably
composed of tiny rock fragments. The brightest portion of the
ring is about 6000 km wide. The outer edge of the ring is
sharply defined, but the inner edge is somewhat fuzzy. A tenuous sheet of material extends from the ring's inner edge all
the way down to the planet's cloud tops. (b) Quarter images
of Io's and Europa's tori (also called *plasma tori* because the
gas particles in them are charged—plasmas). Io is visible in
its torus (*green*), while Europa is visible in its torus (*blue*). Some
of Jupiter's magnetic field lines are also drawn in. Plasma
from tori flow inward along these field lines toward Jupiter. (a:
NASA/JPL, Cornell University; b: NASA/JPL/Johns Hopkins
University Applied Physics Laboratory)

matter spirals in toward Jupiter, thereby creating the
aurora seen there (see Figure 17-10).

The other torus is in Europa's orbit (Figure
17-19b). It consists of hydrogen and
oxygen ions and electrons created
from water molecules kicked off
Europa's surface by radiation from
Jupiter. The mass of this ring of
gas is calculated to be about 6×10^4 tons.

> What other objects that we
> have already studied in the
> solar system are similar to
> the smaller moons of
> Jupiter?

SATURN

Saturn, with its ethereal rings, presents the most spectacular image of the planets (Figure 17-20). Giant Saturn has
95 times as much mass as Earth, making it second in
mass and size to Jupiter. Like Jupiter, Saturn has a thick,
active atmosphere composed predominantly of hydrogen,
and it radiates more energy than it receives from the
Sun—in this case, three times as much. It also has a
strong magnetic field. Ultraviolet images from the Hubble
Space Telescope reveal auroras around Saturn differing
from those of Jupiter in that Saturn's are spirals (Figure
17-20), while Jupiter's are rings (see Figure 17-10).

17-9 Saturn's atmosphere, surface, and interior are similar to those of Jupiter

Partly obscured by the thick, hazy atmosphere above
them, Saturn's clouds lack the colorful contrast visible
on Jupiter. Nevertheless, photographs do show faint
stripes in Saturn's atmosphere, similar to Jupiter's belts
and zones (Figure 17-21a). Their existence indicates
that Saturn also has internal heat that transports the
cloud gases by convection. Changing features in the
atmosphere show that Saturn, too, has differential rotation—ranging from about 10 h and 14 min at the equator to 10 h and 40 min at high latitudes. As on Jupiter,
some of the belts and zones move eastward, while others move westward. In 2006, the *Cassini* spacecraft

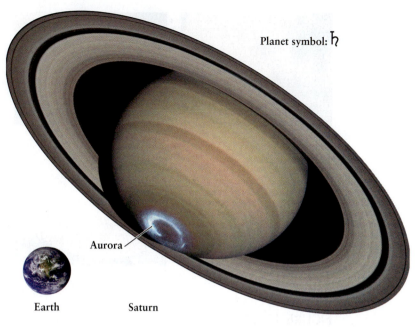

WEB LINK 17.15

FIGURE 17-20 Saturn Combined visible and ultraviolet images from the Hubble Space Telescope reveal spiral arcs of aurora near Saturn's south pole. The image of Earth shows its size relative to Saturn. Note that there is much less contrast between Saturn's clouds than those of Jupiter. (NASA, ESA, J. Clarke [Boston University], and Z. Levay [STScI]; *inset:* NASA)

Planet symbol: ♄

Aurora

Earth   Saturn

R I V U X G

revealed a hexagonal boundary surrounding Saturn's north pole (Figure 17-21b). While belts and zones flow around it, this feature is nearly stationary. The mechanism that creates and maintains it is still under investigation. Although Saturn lacks a long-lived spot like Jupiter's Great Red Spot, it does have storms, including a major one discovered by the Hubble Space Telescope in 1994 and a pair of storms that merged in 2004 (Figure 17-22). In 2004, the *Cassini* spacecraft detected lightning in Saturn's atmosphere.

Saturn's atmosphere is composed of the same basic gases as Jupiter. However, because its mass is smaller than Jupiter's, Saturn's gravitational force on its atmosphere is less. Therefore, Saturn's atmosphere is more spread out than that of its larger neighbor (see Figure 17-6). Saturn has impressive surface winds. At its equator, storms have been clocked moving at speeds of 1600 km/h (1000 mph), which is some ten times faster than Earth's jet streams and three times faster than the fastest winds on Jupiter. The reason for Saturn's extremely high wind speeds is still under investigation.

Astronomers infer that Saturn's interior structure resembles Jupiter's. A layer of molecular hydrogen just below the clouds surrounds a layer of liquid metallic hydrogen, liquid "ices," and a rock and metal core (see Figure 17-8).

Because of its smaller mass, Saturn's interior is also less compressed than Jupiter's, and the pressure inside Saturn is insufficient to convert as much hydrogen into liquid metal. Saturn's rocky core is about 7400 km in radius, while its layer of liquid metallic hydrogen is 12,000 km thick (see Figure 17-8). To alchemists, Saturn was associated with the extremely dense element lead. This is wonderfully ironic in that at 687 kg/m³, Saturn is, on average, the least dense body in the entire solar system; an object with that density on Earth floats in water.

> What two factors cause Saturn's belt and zone system?

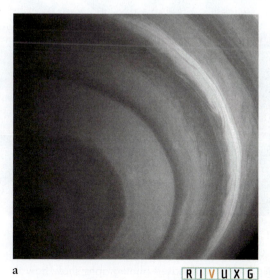

a   R I V U X G

b   R I V U X G

FIGURE 17-21 Belts and Zones on Saturn (a) *Cassini* took this extremely high resolution image of Saturn in 2007. Details as small is 53 km (33 mi) across can be seen. There is less swirling structure between belts and zones on Saturn than on Jupiter. (b) In 2006, *Cassini's* infrared imager took this view of a hexagonal pattern of clouds that rotates much more slowly than the surrounding belts and zones. Its origin is still under investigation. (a: NASA/JPL/Space Science Institute; b: NASA/JPL/University of Arizona)

a   b

c   d

R I V U X G

FIGURE 17-22 *Merging Storms on Saturn* This sequence of *Cassini* images shows two hurricane-like storms merging into one on Saturn in 2004. Each storm is about 1000 km (600 mi) across. (NASA/JPL/Space Science Institute)

17-10 Saturn's spectacular rings are composed of fragments of ice and ice-coated rock

WEB LINK 17.16

Even when viewed through a telescope from the vicinity of Earth, Saturn's magnificent rings are among the most interesting objects in the solar system. They are tilted 27° from the perpendicular to Saturn's plane of orbit, so sometimes they are nearly edge-on to us, making them virtually impossible to see (Figure 17-23). In 1675, Giovanni Cassini discovered an intriguing feature—a dark division in the rings. This 5000-km-wide gap, called the **Cassini division**, separates the dimmer **A ring** from the brighter **B ring**, which lies closer to the planet. By the mid-1800s, astronomers using improved telescopes detected a faint **C ring** just inside the B ring. Figure 17-24 shows a modern perspective of these rings.

The Cassini division occurs because the gravitational force from Saturn's moon Mimas combines with the gravitational force from the planet to keep the region clear of debris. Whenever matter drifts into the Cassini division, Mimas (orbiting at a different rate than the

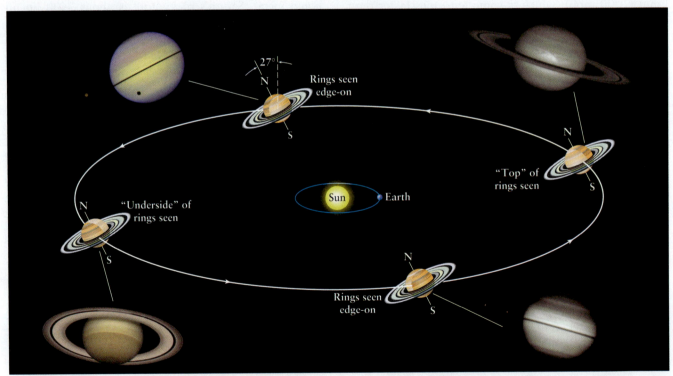

R I V U X G

FIGURE 17-23 *Saturn as Seen from Earth* Saturn's rings are aligned with its equator, which is tilted 27° from the plane of Saturn's orbit around the Sun. Therefore, Earth-based observers see the rings at various angles as Saturn moves around its orbit. The plane of Saturn's rings and equator keeps the same orientation in space as the planet goes around its orbit, just as Earth keeps its 23½° tilt as it orbits the Sun. The accompanying Earth-based photographs show how the rings seem to disappear entirely about every 15 years. (*Top left:* E. Karkoschka/U. of Arizona Lunar and Planetary Lab and NASA; *bottom left:* AURA/STScI/NASA; *bottom right and top right:* A. Bosh/Lowell Obs. and NASA)

FIGURE 17-24 Numerous Thin Ringlets Constitute Saturn's Inner Rings This *Cassini* image shows that Saturn's rings contain numerous ringlets. *Inset:* As moons orbit near or between rings, they cause the ring ices to develop ripples, often like the grooves in an old-fashioned record. (NASA/JPL/Space Science Institute)

the planet faster than the F ring particles. Its gravitational force pulls them forward and nudges the particles back into a higher orbit. The combined effect of these two satellites is to focus the icy particles into a well-defined, narrow band about 100 km wide.

Because of their confining influence, Prometheus and Pandora are called **shepherd satellites** or **shepherd moons**. Among the most curious features of the F ring is that the ringlets are sometimes braided or intertwined and sometimes separate. Although stabilized by the shepherd moons, the F ring does ripple and shows varying brightness. Astronomers theorize that the changes in brightness result from collisions between clumps of matter in the ring.

Beyond the major rings described above lie extensive rings of very fine dust particles, typically smaller than a millionth of a meter (a micron) across. Because such rings are very unstable, meaning that they drift away, there must be small moons that stabilize or renew this debris. The moons Janus, Epimetheus, and Pallene are now associated with these rings. Figure 17-26a shows a *Cassini* image of part of this system, including a ring discovered in the orbit of the moon Pallene. The *Cassini* image in Figure 17-26b gives us our most comprehensive

matter in the Cassini division) periodically exerts a force on this matter, eventually pulling it out of the division. This effect is called a **resonance** and is similar to what happens when you repeatedly push someone on a swing at the right time, enabling them to go higher and higher.

A second gap exists in the outer portion of the A ring, named the **Encke division** (Figure 17-24), after the German astronomer Johann Franz Encke, who allegedly saw it in 1838. (Many astronomers have argued that Encke's report was erroneous, because his telescope was inadequate to resolve such a narrow gap.) The first undisputed observation of the 270-km-wide division was made by the American astronomer James Keeler in the late 1880s, with the newly constructed 36-in refractor at the Lick Observatory in California. A 40-km space near the outer edge of the bright A ring is now named the *Keeler Gap* in honor of his work. Unlike the Cassini division, the Encke division is kept clear because a small moon, Pan, orbits within it. The gravitational tugs of Pan, Daphnis, Prometheus, and other moons also cause the rings to ripple, as shown in the inset of Figure 17-24.

The *Voyager* cameras sent back the first high-quality pictures of the F **ring**, a thin set of ringlets just beyond the outer edge of the A ring (Figure 17-25). Two tiny satellites, Prometheus and Pandora, have orbits on either side of the F ring and serve to keep it intact. The outer of the two satellites orbits Saturn at a slower speed than do the ice particles in the ring, as governed by Kepler's third law. As the ring particles pass near it, they receive a tiny, backward gravitational tug, which slows them down, causing them to fall into orbits a bit closer to Saturn. Meanwhile, the inner satellite orbits

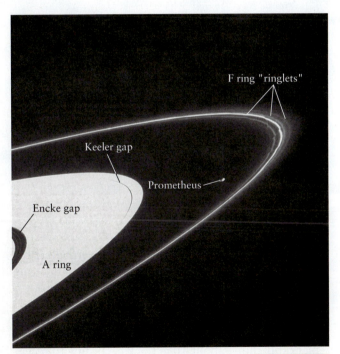

FIGURE 17-25 The F Ring and One of its Shepherds Two tiny satellites, Prometheus and Pandora, each measuring about 50 km across, orbit Saturn on either side of the F ring. Sometimes the ringlets are braided, sometimes parallel to each other. In any case, the passage of the shepherd moons causes ripples in the rings. The gravitational effects of these two shepherd satellites confine the particles in the F ring to a band about 100 km wide. (NASA)

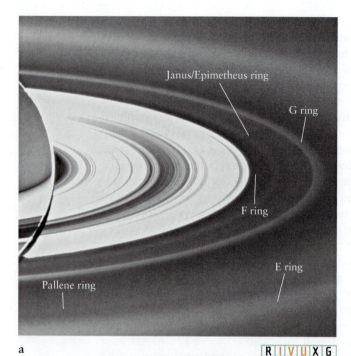

Janus/Epimetheus ring

G ring

F ring

E ring

Pallene ring

a

R I V U X G

FIGURE 17-26 Saturn's Outer Ring System (a) Photographed with the Sun behind Saturn, the inner and intermediate regions of Saturn's ring system are seen to be very different from each other. Beyond the F ring, the particles are dust and pebble-sized. (b) Another view of the inner and intermediate rings, where subtle color differences are indicated. (c) Superimposed on this *Cassini* image are labels that indicate how far the rings extend into the moon system of Saturn. Titan *(off image on right)* is 1.2 million km (750, 000 mi) from the center of Saturn. (a, b: NASA/JPL/Space Science Institute; c: NASA/JPL)

Blue

Reddish

Bluer   Redder

Blue

b

R I V U X G

Janus
Epimetheus

Cassini Saturn Orbit Insertion
Ring Plane Crossing

Cassini
Division

Encke
Division
(Pan)

D
ring   C ring

B ring

A ring
(outer edge: Atlas)

Mimas   Enceladus   Tethys   Dione   Rhea

F ring
(Prometheus,
Pandora)

G ring

E ring   (to Titan)

Titan
Hyperion
Iapetus
Phoebe

c

R I V U X G

view to date of Saturn's ring system. Nevertheless, the rings go out farther still. The scale of the outer ring system (Figure 17-26c) is impressive, extending out a million kilometers, to the moon Titan's orbit.

Pictures from the *Cassini* spacecraft and the Hubble Space Telescope show that Saturn's rings are remarkably thin—about 10 m thick according to recent estimates! This is amazing when you consider that the total ring system has a width (from inner edge to outer edge) of over 380,000 km, more than the distance from Earth to our Moon.

Because Saturn's rings are very bright (their albedo is 0.80), the particles that form them must be highly reflective. Astronomers had long suspected that the rings consist primarily of water ice, along with ice-coated rocks. Spectra taken by Earth-based observatories and by

the spacecraft visiting Saturn have confirmed this suspicion. Because the temperature of the rings ranges from 93 K (−290°F) in the sunshine to less than 73 K (−330°F) in Saturn's shadow, frozen water in them is in no danger of melting or evaporating.

Saturn's rings are slightly salmon-colored. This coloring suggests that they contain traces of organic molecules, which often have such hues. It appears that the rings have gained this material by being bombarded with debris from the outer solar system.

To determine the size of the particles in Saturn's rings, scientists measured the brightness of different rings from many angles as the *Voyager* and *Cassini* spacecraft flew past the planet. Astronomers also measured changes in radio signals received from these spacecraft as they passed behind the rings. Different rings are composed of debris with different sizes, ranging from particles of the sizes found in wood smoke, through those a few centimeters (or inches) across, to particles roughly 10 m in diameter. The outermost main ring, the A ring (see Figure 17-24), is composed of these larger clumps of rubble, which frequently collide with one another and thereby change sizes. This process is analogous to the planet-formation activity described in Chapter 14. Of the particles that would be visible to the naked eye (excluding dust particles), the centimeter-sized particles are most common.

High-resolution images reveal that the ring structure seen from Earth actually consists of thousands of closely spaced ringlets (see Figure 17-24). Even the spaces that we view from Earth as empty—namely Cassini and Encke's divisions—are filled with myriad smoke-sized dust particles. When spacecraft *Cassini* passed through a gap in the ring plane between the F and G rings (see Figure 17-26c), it was peppered with 100,000 impacts from these dust particles in less than 5 min.

Saturn's shell of liquid metallic hydrogen produces a planetwide magnetic field that apparently affects its rings. Saturn's much smaller volume of liquid metallic hydrogen produces a magnetic field at Saturn's surface that is only about two-thirds as strong as the magnetic field here on Earth. Data from spacecraft show that Saturn's magnetosphere contains radiation belts similar to Earth's Van Allen belts. Furthermore, dark **spokes** move around Saturn's rings (Figure 17-27); these are believed to be cre-

ated by electric charges on the ring material interacting with the planet's magnetic field. The electric charge lifts charged particles out of the plane in which the rings orbit. Spreading the particles out decreases the light scattered from them and therefore makes the rings appear darker. The magnetic field causes the regions of spread particles to change, making the spokes appear to revolve around the planet.

Why are there gaps in Saturn's rings?

17-11 Titan has a thick atmosphere, clouds, and lakes filled with liquids

Only 7 of Saturn's 60 known moons are spherical. Several are shown in Figure 17-28a–e. The rest are oblong, suggesting that they are captured asteroids (Figure 17-28f, g). Twelve moons discovered in 2001 move in clumps, indicating that they may be pieces of a larger moon that was broken up by impacts. Saturn's largest moon, Titan, is second in size to Ganymede among the moons of the solar system and is the only moon to have a dense atmosphere. About 10 times more gas lie above each square meter of Titan's surface than lie above each square meter of Earth.

Christiaan Huygens discovered Titan in 1655, the same year he proposed that Saturn has rings. By the early 1900s, several scientists had begun to suspect that Titan might have an atmosphere. This moon is larger than (although less massive than) Mercury and far colder than that planet. Calculations reveal that Titan is cool enough and massive enough to retain heavy gases in its atmosphere. Because of this, Titan was a primary target for the *Voyager* missions (Figure 17-29a). Unexpectedly, its thick haze completely blocked any view of its surface. Observations also reveal that clouds of methane form and dissipate seasonally near both of Titan's poles and in the mid-southern latitudes. Although the causes for the appearance and disappearance of these clouds are not all known, volcanoes, geysers, and evaporation from Titan's lakes are all considered as factors.

Voyager data indicated that roughly 90% of Titan's atmosphere is nitrogen. Most of this gas probably formed from the breakdown of ammonia (NH_3) by the Sun's ultraviolet radiation into hydrogen and nitrogen atoms. Because Titan's gravity is too weak to retain hydrogen, this gas has escaped into space, leaving behind ample nitrogen. The unbreathable atmosphere is about 4 times as dense as that of Earth.

The second most abundant gas on Titan is methane, a major component of natural gas. Sunlight interacting with methane induces chemical reactions that produce a variety of other carbon-hydrogen compounds, or **hydrocarbons**. For example, spacecraft have detected small amounts of ethane (C_2H_6), acetylene (C_2H_2), ethylene (C_2H_4), and propane (C_3H_8) in Titan's atmosphere.

RIVUXG

FIGURE 17-27 Spokes in Saturn's Rings Believed to be caused by Saturn's magnetic field moving electrically charged particles that are lifted out of the ring plane, these dark regions move around the rings like the spokes on a rotating wheel. (NASA)

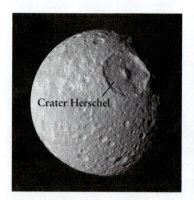

Crater Herschel

Mimas (diameter 392 km)
a

Heavily cratered regions

Regions resurfaced by "lavas"
of water and ammonia

Tethys (diameter 1060 km)
b

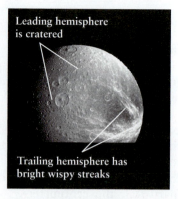

Leading hemisphere
is cratered

Trailing hemisphere has
bright wispy streaks

Dione (diameter 1120 km)
c

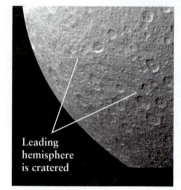

Leading
hemisphere
is cratered

Rhea (diameter 1530 km)
d

Impact basin

Scarps

Ridge – presently
unexplained

Iapetus (diameter 1460 km)
e

Hyperion (diameter 410 km)
f

R I V U X G

Phoebe (diameter 220 km)
g

FIGURE 17-28 Saturn's Diverse Moons (a–e) These *Voyager 1*, *Voyager 2*, and *Cassini* images show the variety of surface features seen on five of Saturn's seven spherical moons. They are not shown to scale (refer to the diameters given below each image). The ridge running along the equator of Iapetus (e) has not yet been explained. (f) Perhaps the most bizarre object photographed in the solar system, Hyperion, shows innumerable impact craters. They are different from craters seen in other objects in that the crater walls here have not filled in the bottom of the craters. The moon's low gravity and the pull of nearby Titan may explain this unusual phenomenon. (g) This *Cassini* image of Phoebe, an irregular moon of Saturn almost as dark as coal, shows how the smaller moons have nonspherical shapes, along with many craters, landslides, grooves, and ridges. Phoebe is barely held in orbit by Saturn. Astronomers believe that it was captured after wandering in from beyond the orbit of Neptune. (a, e, f: NASA/JPL/Space Science Institute)

Methane and ethane condense into droplets in the cold air and fall as rain to Titan's surface. Nitrogen combines with various hydrocarbons to produce other compounds that can exist as liquids on Titan. Although one of these compounds, hydrogen cyanide (HCN), is a poison, some of the others are the building blocks of life's organic molecules. Indeed, biologists who study Titan's atmosphere now think that its chemistry is very similar to that of the young Earth.

Many of the hydrocarbons and carbon-nitrogen compounds in Titan's atmosphere can form complex compounds called **polymers**. These molecules are long repeating chains of atoms; rubber, cellulose, and plastics are the best-known examples on Earth. Scientists hypothesize that droplets of lighter polymers remain suspended in Titan's atmosphere and form a mist, while heavier polymer particles settle down onto Titan's surface.

Images of the surface taken by the *Cassini* spacecraft (Figure 17-29b), which arrived in the Saturn system in 2004, show few craters, indicating that Titan's surface undergoes dynamic change. On the other hand, a mountain range with peaks as high as 1.5 km (1 mi) have been observed on it. They are covered with organic matter and possible methane snow. Some parts of Titan's surface have dunes 100 m high. Their composition is not yet clear, although they may be frozen organic molecules. Using the Keck I telescope in 1999, astronomers observed infrared emissions from Titan's surface consistent with a partially liquid surface. Rain and rivers of methane have been observed (Figure 17-29c). In 2007, radar images from *Cassini* revealed a smooth dark surface of over 10^5 sq km (39,000 sq mi), the largest lake of liquid methane and ethane seen on Titan (Figure 17-29d). Other images reveal similar features elsewhere on that world, including islands in some lakes.

Cassini launched a probe, named Huygens, that successfully parachuted onto Titan. It photographed rocks (Figure 17-29e), highlands, and channels. Upon impact, Huygens vaporized methane that had been just centimeters below the surface. The quantity of methane detected there implies that Titan has reservoirs of methane and a Titan (diameter 5150 km)   R I V U X G

b   R I V U X G

c   R I V U X G   d   R I V U X G   e   R I V U X G

FIGURE 17-29 Surface Features on Titan (diameter 5150 km)
(a) *Voyager* images of Titan's smoggy atmosphere. (b) *Cassini*
image of Titan showing lighter highlands, called *Xanadu*, and dark,
flat, lowlands that may be hydrocarbon seas. Resolution is 4.2 km (2.6 mi).
(c) Riverbeds meandering across the Xanadu highlands of Titan. These
are believed to have been formed by the flow of liquid methane and
ethane. (d) Lake filled with liquid methane and ethane found at Titan's
north pole, with Lake Superior, a Great Lake on Earth, for comparison.
(e) The Huygens spacecraft took this image at Titan's surface on January
14, 2005. What appear like boulders here are actually pebbles strewn
around the landscape. The biggest ones are about 15 cm (6 in.) and 4
cm (1.5 in.) across. (a: NASA/JPL/Space Science Institute; b:
NASA/JPL/University of Arizona; c: NASA/JPL; d: NASA/JPL/GSFC;
e: NASA/JPL/ESA/University of Arizona)

other hydrocarbons, which are continually being broken
down in its atmosphere by the Sun's ultraviolet radiation.

We have little reason to suspect that life exists on
Titan because its surface temperature of 95 K (−288°F)
is prohibitively cold. Nevertheless, the material on
Titan's surface may be similar to what was on the young
Earth, so a more detailed study of the chemistry of Titan
may shed light on the origins of life here.

17-12 Enceladus has an atmosphere and magnetic field

Saturn's sixth largest moon, Enceladus (500 km or
310 mi diameter, Figure 17-30a), moved from the side-
lines toward center stage in 2005, as *Cassini* began
studying it. This bright world has an icy, wrinkled surface similar in appearance to Jupiter's Europa and
Ganymede. The fact that these latter two bodies contain

liquid water suggests that Enceladus
does, too. This theory is supported
by observations that entire regions
of Enceladus's southern hemisphere
are free of craters, replaced, it appears, with ice blocks
the size of large houses, which have apparently erased
many craters due to tectonic plate motion (Figure
17-30b). In addition, the spacecraft revealed that dark
lines on Enceladus, dubbed *tiger stripes* (Figure 17-30a),
are less than 1000 years old. These features are cracks
through which water wells up from inside, freezing into
a smooth, dark surface. The tiger stripes are typically
100 km long and 35 km apart. Current theory proposes that this resurfacing resulted from energy supplied to
this world when its rotation rate was changed. The
cause for this change in spin, probably due to tidal
interactions with Saturn, is still under investigation.

Cassini's detectors also discovered that Enceladus has
a thin atmosphere and a magnetic field (consistent with

> What liquids appear to exist on Titan?

salty liquid water in its interior). Water vapor has been discovered over the moon's south pole (Figure 17-30c). This gas has apparently escaped, along with other vapors and dust, from the interior through ice volcanoes, also called *cryovolcanoes*. Because this world lacks the gravitational attraction to keep the water, it is being replaced, apparently, from warm regions under its surface. How the water is kept liquid is still a mystery. The rest of the atmosphere is composed of hydrogen, carbon dioxide, nitrogen, and carbon monoxide. *Cassini* also detected dust surrounding Enceladus, now believed to be the source of matter for Saturn's E-ring.

> What evidence do astronomers have that Enceladus has liquid water in its interior?

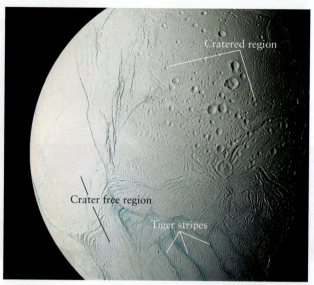

a

R I V U X G

URANUS

Uranus (pronounced *YUR-uh-nus*), so far away that it is virtually invisible to the naked eye, was the first planetary system discovered by a telescopic search of the heavens. It is 19.2 AU from the Sun and takes 84 Earth years to orbit the Sun. Since its discovery in 1781, Uranus has orbited the Sun fewer than three times.

17-13 Uranus sports a hazy atmosphere and clouds

WEB LINK 17.19

Until early 1996, observations of Uranus, the fourth most massive planet in our solar system, revealed few notable features in the visible part of the spectrum. Even the 1986 visit of *Voyager* 2 showed a remarkably bland world. It took the Hubble Space Telescope's infrared camera to find what *Voyager's* visible light camera could not: Uranus has a system of belts and zones. Its hydrogen atmosphere has traces of methane along with a high-altitude haze, under which are clear air and ever-changing methane clouds that dwarf the typical cumulus clouds we see on Earth. Uranus's clouds are towering and huge, each typically as large as Europe (Figure 17-31). Along with the rest of the atmosphere, the clouds go around the planet once every 16½ h, with variation due to differential rotation of the surface clouds.

Uranus contains 14½ times as much mass as Earth and is 4 times bigger in diameter (see Figure 17-31). Its outer layers are composed predominantly of gaseous hydrogen and helium. The temperature in the upper atmosphere of the planet is so low (about 73 K, or −330°F) that the methane and water there condense to form clouds of ice crystals. Because methane freezes at a lower temperature than water, methane forms higher clouds over Uranus. Methane efficiently absorbs red light, giving the planet its blue-green color.

Uranus's rotation axis lies very nearly in the plane of its orbit around the Sun (Figure 17-32). Put another way, recall that Earth's rotation axis is tilted about 23½° from being perpendicular to the ecliptic. Uranus's

b

R I V U X G

c

R I V U X G

FIGURE 17-30 Enceladus (diameter 500 km) (a) *Cassini* view of the two distinct landscapes on Enceladus, one heavily cratered, the other nearly crater-free. The blue "Tiger Stripes" are believed to be due to upwelling of liquid that froze at the surface. (b) The crater-free region near the south pole. The ridges are thought to be created by tectonic flows. *Inset* shows ice boulders. (c) Icy particles leaving Enceladus. (a–c: NASA/JPL/Space Science Institute)

Uranus — planet symbol: ♅

Earth

Neptune — planet symbol: ♆

R I V U X G   R I V U X G   R I V U X G

FIGURE 17-31 *Uranus, Earth, and Neptune* These images of Uranus, Earth, and Neptune are to the same scale. Uranus and Neptune are quite similar in mass, size, and chemical composition. Both planets are surrounded by thin, dark rings, quite unlike Saturn's, which are broad and bright. The clouds on the right of Uranus *(false-color pink)* are each the size of Europe. (NASA)

rotation axis is inclined 98° from a line perpendicular to its plane of orbit (see lower left drawing in Figure 17-32). Therefore, its rotation axis lies within 8° of the plane of orbit.

Because its north pole points below the plane of its orbit around the Sun, Uranus is one of only two planets with retrograde rotation. (Venus is the other.) As Uranus orbits the Sun, its north and south poles alternately point almost directly toward or directly away from the Sun, producing exaggerated seasons (see Figure 17-32). In the summertime, near Uranus's north pole, the Sun is almost directly overhead for many Earth years, at which time southern latitudes are subjected to a continuous, frigid winter night. Forty-two Earth years later, the situation is reversed. For more on this, see the essay "What If . . . Earth's Axis Lay on the Ecliptic?" on page 37.

FIGURE 17-32 *Exaggerated Seasons on Uranus* Uranus's axis of rotation is tilted so steeply that it lies nearly in the plane of its orbit. Seasonal changes on Uranus are thus greatly exaggerated. For example, during midsummer at Uranus's south pole, the Sun appears nearly overhead for many Earth years, during which time the planet's northern regions are subjected to a long, continuous winter night. Half an orbit later, the seasons are reversed.

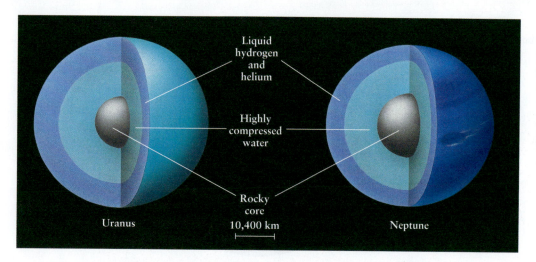

FIGURE 17-33 Cutaways of Uranus and Neptune The interiors of both Uranus and Neptune are believed to have three regions: a terrestrial rocky core surrounded by a liquid water mantle, which is surrounded, in turn, by liquid hydrogen and helium. Their atmospheres are thin layers at the top of their hydrogen and helium layers.

From *Voyager* photographs, planetary scientists have concluded that each of the five largest Uranian moons has probably had at least one shattering impact. A catastrophic collision with an Earth-sized object may also have knocked Uranus on its side, as we see it today.

From its mass and density (1318 kg/m³), astronomers conclude that Uranus's interior has three layers. The outer 30% of the planet is liquid hydrogen and helium; the next 40% inward is highly compressed liquid water (with some methane and ammonia); and the inner 30% is a rocky core (Figure 17-33). Ammonia dissolves easily in water, so an ocean would explain the scarcity of this gas in the planet's atmosphere.

Uranus's surface magnetic field is about three-quarters that of Earth. That strength is reasonable, considering the planet's mass and rotation rate, but everything else about the magnetic field is extraordinary. The field is remarkably tilted—59° from its axis of rotation—and its axis does not even pass through the center of the planet (Figure 17-34).

Because of the large angle between the magnetic field of Uranus and its rotation axis, the magnetos-phere of Uranus wobbles considerably as the planet rotates. Astronomers use this wobble to determine the interior rotation speed of the planet, which rotates once every 17 h, 14 min. This is considerably slower than the rotation rate of the clouds on its surface. A rapidly changing magnetic field like this occurs elsewhere in astronomy. For example, it helps explain the behavior of pulsars, a type of star we studied in Chapter 8.

17-14 A system of rings and satellites revolves around Uranus

Uranus has 13 known rings, each divided into ringlets (Figure 17-35). The first nine of the thin, dark rings were discovered accidentally in 1977, when Uranus passed in front of a star. The star's light was momentarily blocked by each ring, thereby revealing the rings' existence to astronomers (Figure 17-36). Blocking the light of a more distant object, such as the star here, by something between it and us, such as Uranus's rings, is

FIGURE 17-34 The Magnetic Fields of Five Planets This drawing shows how the magnetic fields of Earth, Jupiter, Saturn, Uranus, and Neptune are tilted relative to their rotation axes. Note that the magnetic fields of Uranus and Neptune are offset from the centers of the planets and steeply inclined to their rotation axes. Jupiter, Saturn, and Neptune have north magnetic poles on the hemisphere where Earth has its south magnetic pole.

FIGURE 17-35 The Rings and Moons of Uranus This image of Uranus, its rings, and eight of its moons was taken by the Hubble Space Telescope. *Inset:* Close-up of part of the ring system taken by *Voyager 2* when the spacecraft was in Uranus's shadow looking back toward the Sun. Numerous fine dust particles between the main rings gleam in the sunlight. Uranus's rings are much darker than Saturn's, and this long exposure revealed many very thin rings and dust lanes. The short streaks are star images blurred because of the spacecraft's motion during the exposure. (NASA [inset])

R I V U X G

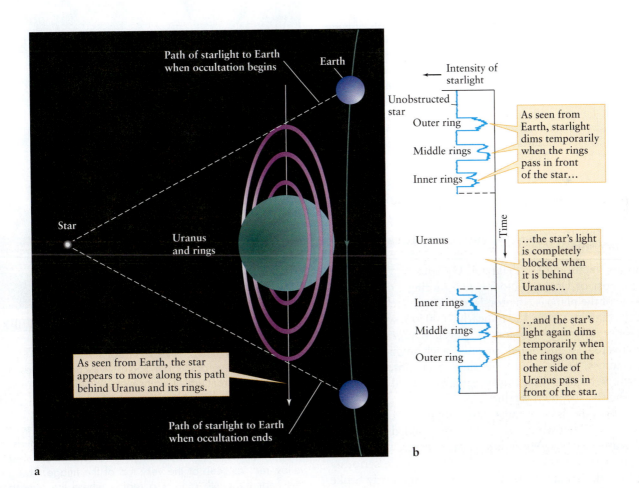

a

b

FIGURE 17-36 Discovery of the Rings of Uranus (a) Light from a star is reduced as the rings move in front of it. (b) With sensitive light detectors, astronomers can detect the variation in light intensity. Such dimming led to the discovery of Uranus's rings. Of course, the star vanishes completely when Uranus occults it.

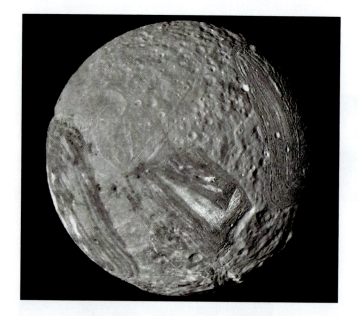

Miranda's core originally consisted of dense rock, while its outer layers were mostly ice. If a powerful impact did occur, blocks of debris broken off from Miranda drifted back together through mutual gravitational attraction. Recolliding with that moon, they formed a chaotic mix of rock and ice. In this scenario, the landscape we see today on Miranda is the result of huge, dense rocks trying to settle toward the satellite's center, forcing blocks of less dense ice upward toward the surface.

NEPTUNE

Neptune is physically similar to Uranus (review Figures 17-31 and 17-33). Neptune has 17.1 times Earth's mass, 3.88 times Earth's diameter, and a densi-

R I V U X G

FIGURE 17-37 Miranda The patchwork appearance of Miranda in this mosaic of *Voyager 2* images suggests that this satellite consists of huge chunks of rock and ice that came back together after an ancient, shattering impact by an asteroid or a neighboring Uranian moon. The curious banded features that cover much of Miranda are parallel valleys and ridges that may have formed as dense, rocky material sank toward the satellite's core. At the very bottom of the image—where a "bite" seems to have been taken out of the satellite—is a range of enormous cliffs that jut upward as high as 20 km, twice the height of Mount Everest. (NASA)

VIDEO 17.8

> Besides a terrestrial world, what dominates the composition of Uranus's interior?

called an **occultation**. The other rings were detected by *Voyager* and the Hubble Space Telescope. The rings are held in orbit by shepherd moons, including Cordelia and Ophelia.

Most of Uranus's moons, like its rings, orbit in the plane of the planet's equator. Five of these satellites, ranging in diameter from 480 to nearly 1600 km, were known before the *Voyager* mission. However, *Voyager*'s cameras discovered 10 additional satellites, each smaller than 50 km across. Still others have since been observed from Earth. Several of these tiny, irregularly shaped moons are among the shepherd satellites.

Miranda is the most fascinating and bizarre of Uranus's 27 known moons. Unusual wrinkled and banded features cover Miranda's surface (Figure 17-37). Its highly varied terrain suggests that it was once seriously disturbed. Perhaps a shattering impact temporarily broke it into several pieces that then recoalesced, or perhaps severe tidal heating, as we saw on Io, moved large pieces of its surface.

Zones (light blue)

Belts (dark blue)

South pole

High-altitude clouds

R I V U X G

FIGURE 17-38 Neptune's Banded Structure Several Hubble Space Telescope images at different wavelengths were combined to create this enhanced-color view of Neptune. The dark blue and light blue areas are the belts and zones, respectively. The dark belt running across the middle of the image lies just south of Neptune's equator. White areas are high-altitude clouds, presumably of methane ice. The very highest clouds are shown in yellow-red, as seen at the very top of the image. The green belt near the south pole is a region where the atmosphere absorbs blue light, probably indicating some differences in chemical composition. (Lawrence Sromovsky, University of Wisconsin-Madison and STScI/NASA)

WEB LINK 17.21

ty of 1640 kg/m³. Its rapid rotation draws its clouds into belts and zones. Unlike Uranus, however, cloud features can readily be discerned on Neptune. Its whitish, cirrus-like clouds consist of methane ice crystals. As on Uranus, the methane absorbs red light, leaving the planet's belts and zones with a banded, bluish appearance (Figure 17-38). Like the other giant planets, the atmosphere of Neptune experiences differential rotation. The winds on Neptune blow at speeds of up to 2000 km/h—among the fastest known in the solar system. Neptune's rotation axis is tilted nearly 30° from the plane of its orbit around the Sun. Recall that the tilt of Earth's axis creates our seasons. Hubble Space Telescope observations of Neptune in 2003 showed seasonal changes near its poles.

17-15 Neptune was discovered because it had to be there

Neptune's discovery is storied because it illustrates a scientific prediction leading to an expected discovery. In 1781, the British astronomer William Herschel discovered Uranus. Its position was carefully plotted, and, by the 1840s, it was clear that even considering the gravitational effects of all of the known bodies in the

solar system, Uranus was not following the path predicted by Newton's and Kepler's laws. Either the theories behind these laws were wrong, or there had to be another, yet-to-be-discovered body in the solar system pulling on Uranus.

> Why do Uranus's rings remain in orbit?

Independent, nearly simultaneous calculations by an English mathematician, John Adams, and a French astronomer, Urbain Leverrier, predicted the same location for the alleged planet. That planet, Neptune, was located in 1846 by German astronomer Johann Galle, within a degree or two of where it had to be to have the observed influence on Uranus.

In August 1989, nearly 150 years after Neptune's discovery, *Voyager 2* arrived at the planet to cap one of NASA's most ambitious space missions. Scientists were overjoyed at the detailed, close-up pictures and wealth of data about Neptune sent back to Earth by the spacecraft.

INSIGHT INTO SCIENCE

Process and Progress Astronomers were able to predict Neptune's position on the basis of theory before it was observed. Likewise, the volcanoes on Io were so predicted. In other cases in science, theories have even correctly predicted results before the technology existed to verify them. Once confirmed, valid theories lead to other predictions and further scientific understanding.

At the time *Voyager 2* passed it, a giant storm raged in Neptune's atmosphere. Called the **Great Dark Spot**, it was about half as large as Jupiter's Great Red Spot. The Great Dark Spot (Figure 17-39) was located at about the same latitude on Neptune and occupied a similar proportion of Neptune's surface as the Great Red Spot does on Jupiter. Although these similarities suggested that similar mechanisms created the spots, the Hubble Space Telescope in 1994 showed that the Great Dark Spot had disappeared. Then, in April 1995, another spot developed in the opposite hemisphere.

Neptune's interior is believed to be very similar in composition and structure to that of Uranus: a rocky core surrounded by ammonia- and methane-laden water (see Figure 17-33). Neptune's surface magnetic field is about 40% that of Earth. As with Uranus, Neptune's magnetic axis (the line connecting its north and south magnetic poles) is tilted sharply from its rotation axis. In this case, the tilt is 47°. Also like Uranus, Neptune's magnetic axis does not pass through the center of the planet (see Figure 17-34).

We saw in Section 17-2 that the magnetic fields of Jupiter and Saturn are believed to be generated by the motions of their liquid metallic hydrogen. However,

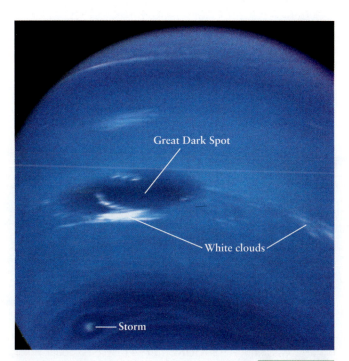

Great Dark Spot

White clouds

Storm

R I V U X G

FIGURE 17-39 Neptune This view from *Voyager 2* looks down on the southern hemisphere of Neptune. The Great Dark Spot's longer dimension at the time was about the same size as Earth's diameter. It has since vanished. Note the white, wispy methane clouds. (NASA/JPL)

VIDEO 17.9

R I V U X G

FIGURE 17-40 Neptune's Rings Two main rings are easily seen in this view alongside overexposed edges of Neptune. In taking this image, the bright planet was hidden so that the dim rings would be visible, hence the black rectangle running down the center of the figure. Careful examination also reveals a faint inner ring. A fainter-still sheet of particles, whose outer edge is located between the two main rings, extends inward toward the planet. (NASA)

Uranus and Neptune lack this material, and their magnetic fields have a different origin. These fields are believed to exist because molecules such as ammonia, which dissolve in their water layers, lose electrons (become ionized). These ions, moving with the planets' rotating, fluid interiors, create the same dynamo effect that produces the magnetic field.

> The "spots" on Neptune are analogous to what on Earth?

17-16 Neptune has rings and has captured most of its moons

Like Uranus, Neptune is surrounded by a system of thin, dark rings (Figure 17-40). It is so cold at these distances from the Sun that both planets' ring particles retain methane ice. Scientists speculate that eons of radiation damage have converted this methane ice into darkish carbon compounds, thus accounting for the low reflectivity of the rings.

Neptune has 13 known moons. Twelve have irregular shapes and highly elliptical orbits, which suggest that Neptune captured them. Triton, discovered in 1846, is spherical and was quickly observed to have a nearly circular, *retrograde* orbit around Neptune. It is difficult to imagine how a planet rotating one way and a satellite revolving the other way could form together. Indeed, only the small outer satellites of Jupiter and Saturn have retrograde orbits, and these bodies are probably captured asteroids. Some scientists have therefore suggested that

Triton may have been captured 3 or 4 billion years ago by Neptune's gravity. Upon being captured, Triton was most likely in a highly elliptical orbit. However, the tides that the moon creates on Neptune's liquid surface would, in turn, have made Triton's orbit more circular.

Triton's average density of about 2100 kg/m³ indicates that it is an equal mix of rock and ice. It has a thin nitrogen atmosphere. When Triton was younger, the tidal force exerted on it by Neptune due to the moon's initially elliptical orbit would have caused Triton to stretch and flex, providing enough energy to melt much of its interior and obliterate its original surface features, including craters. Triton's south polar region is shown in Figure 17-41. Note that very few craters are visible. Calculations based on these observations indicate that Triton's present surface is about 100 million years old.

Triton does exhibit some surface features seen on other icy worlds, such as long cracks resembling those on Europa and Ganymede. Other features unique to Triton are quite puzzling. For example, the top part of Figure 17-41 reveals a wrinkled terrain that resembles the skin of a cantaloupe. Triton also has a few frozen lakes like the one shown in Figure 17-42. Some scientists have speculated that these lakelike features are the calderas of extinct cryovolcanoes. A mixture of methane, ammonia, and water, which can have a melting point far below that of pure water, could have formed a kind of cold lava on Triton.

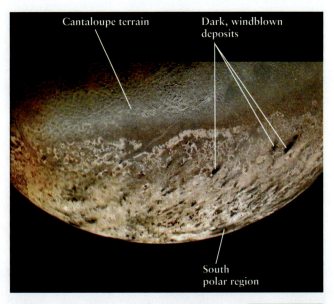

Cantaloupe terrain

Dark, windblown deposits

South polar region

R I V U X G

FIGURE 17-41 Triton's South Polar Cap Approximately a dozen high-resolution *Voyager 2* images were combined to produce this view of Triton's southern hemisphere. The pinkish polar cap is probably made of nitrogen frost. A notable scarcity of craters suggests that Triton's surface was either melted or flooded by icy lava after the era of bombardment that characterized the early history of the solar system. (NASA/JPL)

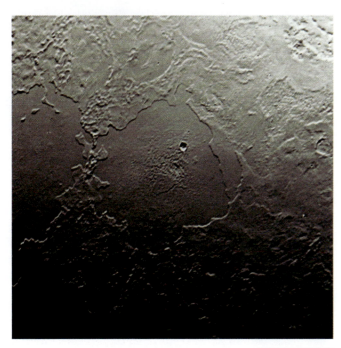

R I V U X G

FIGURE 17-42 *A Frozen Lake on Triton* Scientists think that the feature in the center of this image is a basin filled with water ice. The flooded basin is about 200 km across. (NASA)

Voyager instruments measured a surface temperature of 36 K (−395°F), making Triton the coldest world that our probes have ever visited. Nevertheless, *Voyager* cameras did glimpse two towering plumes of gas extending up to 8 km above the satellite's surface. These are apparently jets of nitrogen gas warmed by interior radioactive decay and escaping through vents or fissures.

Triton continues to create tides on Neptune. Whereas the tides on Earth cause our Moon to spiral outward, the tides on Neptune cause Triton (in its retrograde orbit) to spiral inward. Within the next quarter of a billion years, Triton will reach the **Roche limit**, the distance at which a planet creates tides on its moon's solid surface high enough to pull its moon apart. Pieces of Triton will then literally float into space until the entire moon is demolished! By destroying Triton, Neptune will create a new ring system that will be much more substantial than its present one (Figure 17-43).

17-17 Comparative planetology of the outer planets

Now that we have examined the individual outer planets, it is instructive to compare their various properties. The table, *The Outer Planets: A Comparison* summarizes much of this material.

Size and Mass   The four giant planets, Jupiter, Saturn, Uranus, and Neptune, are all much larger (roughly 4 to 11 times Earth's diameter) and more massive (roughly 14 to 318 times Earth's mass) than Earth.

> Is our Moon inside or outside Earth's Roche limit?

Atmospheres and Rotations   The four outer planets have thick hydrogen and helium-rich atmospheres that are permanently and completely covered with clouds. All of the giant planets rotate rapidly compared to Earth. Their sidereal rotation rates range from about 10 to about 17¼ h. This rapid rotation draws their clouds into parallel bands, called *belts* and *zones*. Earth's slower rotation allows the clouds and winds here to roam over much greater range of latitudes than on the giant planets.

Interiors   All four giant planets have terrestrial bodies as their cores. That is where their internal similarities to Earth end. Jupiter and Saturn have small amounts of "ice" surrounding their cores. These ices are surrounded by thick liquid metallic hydrogen layers, which, in turn, are surrounded by layers of normal hydrogen and helium. In contrast, Uranus's and Neptune's terrestrial cores are surrounded by large amounts of water, which, in turn, are surrounded by liquid hydrogen and helium.

Magnetic Fields and Rings   The four giant planets all have magnetic fields that store particles from the solar wind, similar to Earth's Van Allen belts. Like Earth's magnetic field, those of Jupiter, Uranus, and Neptune are tilted relative to their rotation axes. Saturn's magnetic field is along its rotation axis. The four giants all have rings. Saturn's rings are the most massive and distinctive, followed by those of Uranus, Jupiter, and Neptune.

Moons   Among them, the giant planets have at least 163 moons. Each of the giants has between one and a few spherical moons, like our Moon. Most of the moons in the solar system are much smaller, irregularly shaped bodies. These objects are most likely captured space debris. Unlike our Moon, all of the satellites of the giants are no more than a few hundred thousand times less massive than their planets. Most moons are millions or billions of times less massive than their planets. Recall that our Moon is 81 times less massive than Earth.

> How do the numbers and types of moons found orbiting the giant planets compare to those that orbit the terrestrial planets?

17-18 Frontiers yet to be discovered

The outer planets hold countless new insights into the formation and evolution of the solar system. Considering that most of the known extrasolar planets are Jupiter-like gas giants, our outer planets all have a

The Outer Planets: A Comparison

	Interior	Surface	Rings	Atmosphere	Magnetic Field
Jupiter	Terrestrial core, liquid metallic hydrogen shell, liquid hydrogen mantle	No solid surface, atmosphere gradually thickens to liquid state, belt and zone structure, hurricane-like features	Yes	Primarily H, He	19,000 × Earth's total field; at its cloud layer, 14 × stronger than Earth's surface field
Saturn	Similar to Jupiter, with bigger terrestrial core and less metallic hydrogen	No solid surface, less distinct belt and zone structure than Jupiter	Yes	Primarily H, He	570 × Earth's total field; at its cloud layer, ⅔ × Earth's surface field
Uranus	Terrestrial core, liquid water shell, liquid hydrogen and helium mantle	No solid surface, weak belt and zone system, hurricane-like features, color from methane absorption of red, orange, yellow	Yes	Primarily H, He, some CH_4	50 × Earth's total field; at its cloud layer, 0.73 Earth's surface field
Neptune	Similar to Uranus	Like Uranus	Yes	Primarily H, He, some CH_4	35 × Earth's total field; at its cloud layer, 0.4 × Earth's surface field

For detailed numerical comparisons between planets, see Appendix Tables E-1 and E-2.
*To see the orientations of these magnetic fields relative to the rotation axes of the planets, see Figure 17-34.

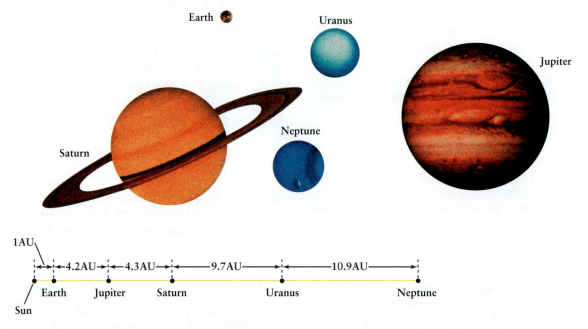

(Stephen P. Meszaros/NASA, M. Buie, K. Horne, and D. Tholen)

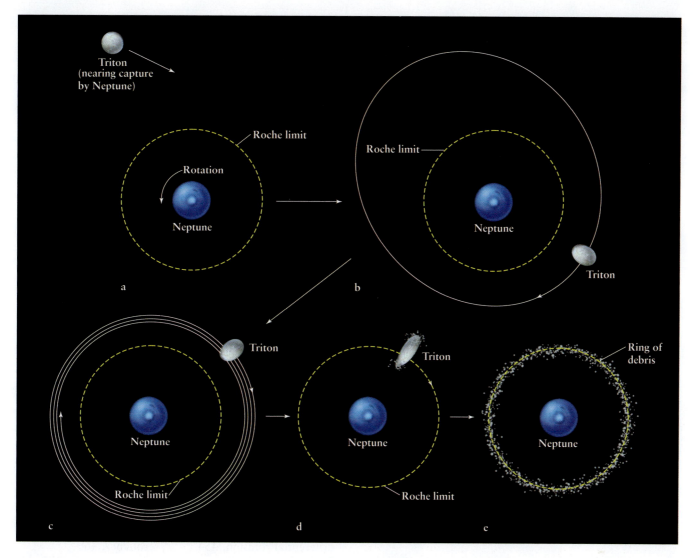

FIGURE 17-43 The Capture and Destruction of Triton This series of drawings depicts how (a) Triton was captured by Neptune in a retrograde orbit. (b) The tides that Triton then created on the planet caused that moon's orbit to become quite circular and (c) to spiral inward. (d) It will eventually reach Neptune's Roche limit and (e) be pulled apart to form a ring.

lot to tell us about planets throughout our Galaxy. Some specific issues include: Why did the *Galileo* probe fail to detect the atmospheric structures believed to exist on Jupiter? How has the Great Red Spot persisted for so long? Is there life in the underground oceans of the outer Galilean moons? What, exactly, do the individual ring particles around each of the planets look like? How long ago did Saturn's ring system form? How long will it last? Are the moving spokes in Saturn's rings really caused by its magnetic field, and, if so, why do the spokes appear as they do? What caused Miranda's surface to become so profoundly disturbed? These are but a few of the questions about the outer planets that are likely to be answered during this century.

SUMMARY OF KEY IDEAS

Jupiter and Saturn

• Jupiter is by far the largest and most massive planet in the solar system.

• Jupiter and Saturn probably have rocky cores surrounded by a thick layer of liquid metallic hydrogen and an outer layer of ordinary liquid hydrogen. Both planets have an overall chemical composition very similar to that of the Sun.

• The visible features of Jupiter exist in the outermost 100 km of its atmosphere. Saturn has similar features, but they are much fainter. Three cloud layers exist in the upper atmospheres of both Jupiter and Saturn. Because Saturn's cloud layers extend through a greater range of altitudes, the colors of the Saturnian atmosphere appear muted.

• The colored ovals visible in the Jovian atmosphere are gigantic storms, some of which (such as the Great Red Spot) are stable and persist for years or even centuries.

• Jupiter and Saturn have strong magnetic fields created by electric currents in their metallic hydrogen layers.

• Four large satellites orbit Jupiter. The two inner Galilean moons, Io and Europa, are roughly the same size as our Moon. The two outer moons, Ganymede and Callisto, are approximately the size of Mercury.

• Io is covered with a colorful layer of sulfur compounds deposited by frequent explosive eruptions from volcanic vents. Europa is covered with a smooth layer of frozen water crisscrossed by an intricate pattern of long cracks.

• The heavily cratered surface of Ganymede is composed of frozen water with large polygons of dark, ancient crust separated by regions of heavily grooved, lighter-colored, younger terrain. Callisto has a heavily cratered ancient crust of frozen water.

• Saturn is circled by a system of thin, broad rings lying in the plane of the planet's equator. Each major ring is composed of a great many narrow ringlets that consist of numerous fragments of ice and ice-coated rock. Jupiter has a much less substantial ring system.

• Titan has a thick atmosphere of nitrogen, methane, and other gases, as well as lakes of methane and ethane.

• Enceladus has areas with very different surface features: an older, heavily cratered region and a newer, nearly crater-free surface created by tectonic activity.

Uranus and Neptune

• Uranus and Neptune are quite similar in appearance, mass, size, and chemical composition. Each has a rocky core surrounded by a thick, watery mantle; the axes of their magnetic fields are steeply inclined to their axes of rotation; and both planets are surrounded by systems of thin, dark rings.

• Uranus is unique in that its axis of rotation lies near the plane of its orbit, producing greatly exaggerated seasons on the planet.

• Uranus has five moderate-sized satellites, the most bizarre of which is Miranda.

• Triton, the largest satellite of Neptune, is an icy world with a tenuous nitrogen atmosphere. Triton moves in a retrograde orbit that suggests it was captured into orbit by Neptune's gravity. It is spiraling down toward Neptune and will eventually break up and form a ring system.

WHAT DID YOU THINK?

1   *Is Jupiter a "failed star"? Why or why not?* No. Jupiter has 75 times too little mass to shine as a star.

2   *What is Jupiter's Great Red Spot?* The Great Red Spot is a long-lived, oval cloud circulation, similar to a hurricane on Earth.

3   *Does Jupiter have continents and oceans?* No. Jupiter is surrounded by a thick atmosphere primarily of hydrogen and helium that gradually becomes liquid as you move inward. The only solid matter in Jupiter is its core.

4   *Is Saturn the only planet with rings?* No. All four outer planets (Jupiter, Saturn, Uranus, and Neptune) have rings.

5   *Are the rings of Saturn solid ribbons?* No. Saturn's rings are all composed of thin, closely spaced ringlets consisting of particles of ice and ice-coated rocks. If they were solid ribbons, Saturn's gravitational tidal force would tear them apart.

Key Terms for Review

A ring, 468	liquid metallic hydrogen, 456
B ring, 468	occultation, 478
belt, 453	polymer, 472
Cassini division, 468	prograde orbit, 465
C ring, 468	resonance, 469
differential rotation, 454	retrograde orbit, 465
Encke division, 469	ringlet, 465
F ring, 469	Roche limit, 481
Galilean moon (satellite), 459	shepherd satellite (moon), 469
Great Dark Spot, 479	spoke, 471
Great Red Spot, 454	zone, 453
hydrocarbon, 471	

Review Questions

1. Which is the most massive planet in the solar system? **a.** Earth, **b.** Neptune, **c.** Saturn, **d.** Jupiter, **e.** Mercury

2. Which of the following planets does *not* have rings? Choose only one. **a.** Mars, **b.** Uranus, **c.** Neptune, **d.** Saturn, **e.** Jupiter

3. Which is the least massive planet in the solar system? **a.** Mercury, **b.** Mars, **c.** Uranus, **d.** Jupiter, **e.** Venus

4. Which planet is presently known to have the most moons? **a.** Mars, **b.** Saturn, **c.** Uranus, **d.** Jupiter, **e.** Neptune

5. Describe the appearance of Jupiter's atmosphere. Which features are long-lived and which are relatively fleeting?

 6. To test your knowledge of Jupiter's belt and zone structure, do Interactive Exercise 17-1 on the Web. You can print out your results, if required.

7. What causes the belts and zones in Jupiter's atmosphere?

 8. To test your knowledge of Jupiter's internal structure, do Interactive Exercise 17-2 on the Web. You can print out your results, if required.

9. What is liquid metallic hydrogen? Which planets contain this substance? What produces this form of hydrogen?

10. Compare and contrast the surface features of the four Galilean satellites, discussing their geologic activity and their evolution.

 11. To test your knowledge of the Galilean moons, do Interactive Exercise 17-3 on the Web. You can print out your results, if required.

12. What energy source powers Io's volcanoes?

13. Why are numerous impact craters found on Ganymede and Callisto but not on Io or Europa?

14. Describe the structure of Saturn's rings. What are they made of?

 15. To test your knowledge of Saturn's rings, do Interactive Exercise 17-4 on the Web. You can print out your results, if required.

16. Why do features in Saturn's atmosphere appear to be much fainter and "washed out" than comparable features in Jupiter's atmosphere?

17. Explain how shepherd satellites affect some planetary rings. Is "shepherd satellite" an appropriate term for these objects? Explain.

18. Describe Titan's atmosphere. What effect does sunlight have on it?

19. Describe Titan's surface.

20. Describe the seasons on Uranus. Why are the Uranian seasons different from those on any other planet?

21. Briefly describe the evidence that supports the idea that Uranus was struck by a large planet-like object several billion years ago.

22. Why are Uranus and Neptune distinctly bluer than Jupiter and Saturn?

23. Compare the ring systems of Saturn and Uranus. Why were Uranus's rings unnoticed until the 1970s?

24. How do the orientations of Uranus's and Neptune's magnetic axes differ from those of the other planets?

 25. To test your knowledge of planetary magnetic fields, do Interactive Exercise 17-5 on the Web. You can print out your results, if required.

26. Explain why Triton will never collide with Neptune, even though Triton is spiraling toward that planet.

Advanced Questions

27. Consult the Internet or such magazines as *Sky & Telescope* or *Astronomy* to determine what space missions are now under way. What data and pictures have they sent back that update information presented in this chapter?

28. Long before the *Voyager* flybys, Earth-based astronomers reported that Io appeared brighter than usual for a few hours after emerging from Jupiter's shadow. Explain this brief brightening of Io.

29. Compare and contrast Valhalla on Callisto with Caloris Basin on Mercury.

30. As seen by Earth-based observers, the intervals between successive edge-on presentations of Saturn's rings alternate between 13 years, 9 months and 15 years, 9 months. Why are these two intervals not equal?

31. Compare and contrast the internal structures of Jupiter and Saturn with the internal structures of Uranus and Neptune. Can you propose an explanation for why the differences between these two pairs of planets occurred?

32. Neptune has more mass than Uranus, but Uranus is the larger of the two planets. Reconcile these two facts.

Discussion Questions

33. Suppose that you were planning a mission to Jupiter, using an airplane-like vehicle that would spend many days, even months, flying through the Jovian clouds. What observations, measurements, and analyses should this aircraft make? What dangers might it encounter, and what design problems would you have to overcome?

34. Discuss why astronomers believe that Europa, Ganymede, and Callisto may harbor some sort of marine life. Why do they not expect any life on the surfaces of these worlds?

35. Suppose you were planning separate missions to each of Jupiter's Galilean moons. What questions would you want these missions to answer, and what kinds of data would you want your spacecraft to send back?

Given the different environments on the four satellites, how would the designs of the four spacecraft differ?

36. NASA and the Jet Propulsion Laboratory have tentative plans to place spacecraft in orbit about Uranus and Neptune in this century. What kinds of data should be collected, and what questions would you like to see answered by these missions?

What If . . .

37. Jupiter, at its present location, were a star? What would Earth be like? (*Hint:* Recall that to be a star, Jupiter would have to have 75 times more mass than it has today.)

38. Jupiter had formed at one-third its present distance of 5.2 AU from the Sun? What would Earth be like?

39. Io were struck by another object of similar size? (*Hint:* You can create a variety of different scenarios by imagining the impacting body striking from different directions, with different speeds, and at different angles.)

40. Jupiter were orbiting in the opposite direction than it actually is? What effects might this have on the other planets? Would this change affect Earth? If so, how?

Web Questions

 41. Moving Weather Systems on Jupiter Access and view the video "The Great Red Spot" in Chapter 17 of the *Discovering the Universe: From the Stars to the Planets* Web site. **a.** Near the bottom of the video window you will see a white oval moving from left to right. By stepping through the video one frame at a time, estimate how long it takes this oval to move a distance equal to its horizontal dimension. (Hint: You can keep track of time by noticing how many frames it takes a feature in the Great Red Spot, at the center of the video window, to move in a complete circle around the center of the spot. The actual time for this feature to complete a circle is about 6 days.) **b.** The horizontal dimension of the white oval is about 4000 km. At what approximate speed (in km/h) does the white oval move? (Speed = distance/time)

42. Search the Web, especially the Web sites at NASA's Jet Propulsion Laboratory and at the European Space Agency, for information about the current status of the *Cassini* mission. When did *Cassini* arrive at Saturn? Where in the Saturn system is it presently? What are the current plans for its tour of Saturn's satellites? What plans have been adopted for the *Cassini* extended mission, which began in 2008?

43. In 2004, astronomers reported the discovery of two new moons of Saturn. Search the Web for information about these. How were they discovered? Have the observations been confirmed? How large are these moons? What sort of orbits do they follow?

 44. The Rotation Rate of Saturn Access and view the video "Saturn from the Hubble Space Telescope" in Chapter 17 of the *Discovering the Universe: From the Stars to the Planets* Web site. The total time that actually elapses in this video is 42.6 h. Using this information, identify and follow an atmospheric feature and determine the rotation period of Saturn.

Observing Projects

 45. Consult such magazines as *Sky & Telescope* and *Astronomy* or your *Starry Night Enthusiast™* software to determine whether Jupiter is currently visible in the night sky. If so, make arrangements to view the planet through a telescope. What magnifying power seems to give you the best view? Draw a picture of what you see. Can you see any belts and zones? How many? Can you see the Great Red Spot?

46. Make arrangements to view Jupiter's Great Red Spot through a telescope. Consult the *Sky & Telescope* Web site, which lists the times when the center of the Great Red Spot passes across Jupiter's central region, as seen from Earth. The Great Red Spot is well placed for viewing for 50 minutes before and after this time. You will need a refractor with an objective lens of at least 15 cm (6 in) diameter or a reflector with an objective of at least 20 cm (8 in) diameter. Using a pale blue or green filter can increase the color contrast and make the spot more visible. For other useful hints, see the article "Tracking Jupiter's Great Red Spot" by Alan MacRobert (*Sky & Telescope*, September 1997). Sketch Jupiter based on your observations.

 47. If Jupiter is visible at night, observe it through a pair of binoculars or a telescope. Can you see all four Galilean moons? Make a drawing of what you observe. Use *Starry Night Enthusiast™* at about the time of your observations to identify the moons. Select **Favourites > Guides > Atlas** from the menu to place yourself at the center of a transparent Earth. Click the **Now** button in the toolbar and click the **Stop** button. Then use the **Find** pane to locate Jupiter. Close the **Find** pane and use the **Zoom** buttons in the toolbar to set a field of view about 25′ wide. You should be able to see all the moons labeled in their present positions. Compare your observations with the locations of the moons on the screen to identify these moons.

 48. Use *Starry Night Enthusiast™* to observe the motion of the Galilean moons of Jupiter. Select **Favourites > Guides > Atlas** from the menu. Click the Now button in the toolbar. Open the **Find** pane and double-click the entry for Jupiter. Use the **Zoom** buttons in the toolbar to change the field of view to about 30′ wide. Change the **Time Flow Rate** to 20 minutes (adjust as needed) and click the **Play** button to start time flow. You will see the four Galilean moons orbiting Jupiter. **a.** Are all four moons ever on the same side of Jupiter? **b.** Observe the moons passing in front of and behind Jupiter (zoom in and slow the **Time Flow Rate** as needed). Explain how your observations tell you that all four satellites orbit Jupiter in the same direction. You can examine these moons more closely by centering and zooming in on each moon in turn.

49. Consult such magazines as *Sky & Telescope* and *Astronomy* or your *Starry Night Enthusiast*™ software to determine whether Saturn is currently visible in the night sky. If so, view Saturn through a small telescope. Make a sketch of what you see. Estimate the angle at which the rings are tilted to your line of sight. Can you see the Cassini division? Can you see any belts or zones in Saturn's clouds? Do you observe a faint, starlike object near Saturn that might be Titan? What observations could you perform to test whether the starlike object is a Saturnian satellite?

50. Use the *Starry Night Enthusiast*™ program to observe the changing appearance of Saturn. Select **Favourites > Guides > Atlas** from the menu and click the **Now** button in the toolbar. Use the **Find** pane to locate and lock the view on Saturn (double-click the entry for Saturn). Zoom in until Saturn and its rings are clearly visible. Change the **Time Flow Rate** to 1 year. Describe qualitatively how the rings change orientation, as seen from Earth. Stop time. Now use the single-step time control buttons to observe the changing aspect or orientation of the rings. During which of the next 30 years will we see the rings edge-on? Why do you see the orientation of the rings change with a period of about 30 years?

51. Determine whether Uranus or Neptune is currently visible in the night sky. If so, make arrangements to view them through a telescope. To help you find these planets, use your *Starry Night Enthusiast*™ software or the star chart published each January in *Sky & Telescope,* showing the paths of Uranus and Neptune against the background stars. For more detailed images than you can get through a telescope, locate these planets with your *Starry Night Enthusiast*™ software and zoom to high resolution. Also, locate Triton with the software.

52. Saturn as Seen from Earth In this exercise we will use *Starry Night Enthusiast*™ to observe Saturn from about 0.001 AU from the planet. Select **Favourites > Solar System > Outer Planets > Saturn** from the menu. Click the Now button in the toolbar and click the stop time button. From the image of Saturn that appears, write down where you expect that the Sun is located (e.g., above, down to the right, etc.). Use the **Location Scroller** (hold down the Shift key while holding down the mouse button and moving the mouse) to change the view so that Saturn's rings are edge on and move around until the Sun appears in the same region of the sky as Saturn from your viewpoint. (You can label the Sun if necessary by clicking the box to the left of the Sun in the Find pane.) Is the Sun in the same plane as the rings of Saturn? What is the phase of Saturn from your viewpoint at this time? Does this phase agree with what you would expect, with the Sun in this position?

WHAT IF... We Lived on a Metal-Poor Earth?

Earth provides a wealth of building blocks necessary for the development and evolution of complex life-forms. More than 80 elements on or near Earth's surface combine in countless ways essential for its diversity of flora and fauna. Especially important for life on Earth are metals, such as iron. The human body typically contains more than 3 g of iron, mostly in the form of hemoglobin that helps transport oxygen through the bloodstream. A slight iron imbalance leads to anemia or toxicity, and without abundant metals, most of our technologies would never have developed.

But what if the solar system formed from an interstellar gas cloud containing fewer metals and high-mass elements—that is, a solar system with less iron, nickel, copper, and other metals crucial for life as we know it?

Metal-Rich Versus Metal-Poor Earth formed with almost 6×10^{21} metric tons of star matter, almost one-third of it iron. If the solar system formed from interstellar gas containing relatively few heavy elements, Earth would contain a much lower fraction of such elements as uranium, lead, iron, and nickel. Earth would be composed of a correspondingly higher fraction of lower-mass elements, such as carbon, nitrogen, oxygen, silicon, and aluminum. That version of Earth—let's call it Lithia—would be profoundly different from our planet.

The Environment of Lithia Without heavy elements like iron, Lithia's density, gravity, and magnetic fields are much lower than Earth's. Let's assume that the density of Lithia is similar to that of the Moon. (We'll assume that the Moon doesn't change in its characteristics or its distance from Lithia.)

If Lithia has the same radius as Earth, the force of gravity on Lithia's surface is only 60% that on Earth. That is, you would weigh 40% less on Lithia. Mobile life-forms that evolve on the metal-poor planet require much less muscle strength to get about and less bone density to resist the planet's gravity.

With fewer radioactive elements to provide heat, the core of Lithia cooled off quickly compared with the core of Earth. As a result, Lithia has much less heat flowing upward through its mantle and crust, so most likely there is no plate tectonic activity. Lithia also has less volcanic activity. But without plate motion to spread the lava flows, volcanoes will grow to higher elevations than on Earth. The lack of crustal motions also means that pockets of high-density material will not rise from below, forming the metal ores mined on Earth.

The Moon's orbit and the tides that it induces are different, too. With Lithia having 60% of Earth's mass, the Moon at its present distance orbits once every 36 days instead of once every 29.5 days. Thus, the cycle of lunar phases is longer, but the cycle of lunar-induced tides is slightly shorter. Surprisingly, the tides have roughly the same height, because the tidal effects depend only on the Moon's gravity and not Earth's, and the distance between Earth and Lithia and their satellite is the same.

Life on Lithia The low abundance of iron and the resulting lack of a planetary magnetic field might preclude the development of complex life-forms on Lithia. And yet, the incredible diversity of life we find on Earth suggests that evolution on Lithia might well occur. If advanced life-forms evolve on Lithia before the Sun uses up its nuclear fuel and becomes a red giant, the chemical differences between Lithia and Earth dictate that they will differ profoundly from the complex life that we find on Earth.

Without the heat of radioactive elements to keep Lithia's core molten for billions of years and with fewer metals, Lithia will have a much weaker magnetic field—if any at all—than Earth. Inhabitants can expect auroras to grace the night sky continuously. On the other hand, high-energy particles and the radiation they create in the atmosphere will bombard any life-forms on the surface continuously and will also adversely affect the ozone layer. A larger amount of the Sun's harmful ultraviolet radiation will reach the surface. Without more protection, life as we know it could not survive.

A Barren Landscape With a thin atmosphere and an abundance of lighter elements, such as silicon (silicon dioxide is sand), Lithia might resemble a barren desert here on Earth. (Photri)

CHAPTER 18

Vagabonds of the Solar System

Comet Hale-Bopp. (WALTER PACHOLKA, ASTROPICS/Photo Researchers)

R I V U X G

WHAT DO YOU THINK?

1. Are the asteroids a former planet that was somehow destroyed? Why or why not?

2. How far apart are the asteroids on average?

3. How are comet tails formed? Of what are they made?

4. In which directions do a comet's tails point?

5. What is a shooting star?

Answers to these questions appear in the text beside the corresponding numbers in the margins and at the end of the chapter.

The formation of the solar system was a messy affair. Countless planetesimals collided to form the planets and their moons. For nearly a billion years, pieces of smaller debris continually struck, scarred, and shaped their surfaces. The impacts of icy cometary bodies may once have provided Earth with water and other material essential to the evolution of life. That era of frequent collisions ended some 3.8 billion years ago, but innumerable pieces of debris still orbit the Sun, and impacts continue.

Earth is not immune to this activity today. Indeed, myriad dust and pebble-sized pieces of rock and metal vaporize daily in Earth's atmosphere. Others slow down in the air but arrive intact on the ground, occasionally striking human-made objects (and humans). Pieces the size of trucks or skyscrapers, which could create huge blast craters, pass disconcertingly close to Earth—even closer than the Moon.

Moderate-sized objects of varying composition orbit in many places throughout the solar system but most notably in the asteroid belt between Mars and Jupiter and in the Kuiper belt beyond Neptune. These leftovers—the asteroids (mostly rock and metal), meteoroids (smaller versions of asteroids), and comets (mostly ice and rock)—are the vagabonds of the solar system. By studying this debris, astronomers better understand how the solar system formed.

In this chapter you will discover

- the relationships between dwarf planets, small solar system bodies, and other classifications of objects in the solar system

- asteroids and meteoroids: pieces of interplanetary rock and metal

- comets: objects containing large amounts of ice and rocky debris

- space debris that falls through Earth's atmosphere

- the asteroid belt and the Kuiper belt, both filled with a variety of debris, including orbiting pairs of objects

- that impacts from space 250 million and 65 million years ago caused mass extinctions of life on Earth

- that a wayward asteroid could again threaten life on Earth

CATEGORIZING SOLAR SYSTEM OBJECTS

Although scientists do not often change the physical properties of objects in space (intentionally slamming a spacecraft into the solid part of comet Tempel 1, described later in this chapter, comes to mind as a counterexample), they do change the classifications of objects when such alterations give them better insights into the types of objects under study. Demoting Pluto to a dwarf planet is not the only example. As we saw in Appendix I *Changing Pluto's Status as a Planet*, the larger asteroids in the asteroid belt were once considered to be planets, and, as you saw in Chapter 6, stars were reclassified at the beginning of the twentieth century.

In the preceding three chapters, we studied what astronomers now call *planets*, and their satellites (moons). In this chapter, we examine the other two major classifications, *dwarf planets* and *small solar-system bodies (SSSBs)*. Complications still arise: First, the objects that were traditionally called *asteroids* do not quite coincide with one or the other of these new groupings. In particular, the largest asteroids are dwarf planets, while the rest are small solar-system bodies. We will try to clarify which asteroids belong in each new category. Second, groups of objects in different orbits beyond Neptune have been assigned different names that don't indicate whether they contain dwarf planets, SSSBs, or both. These groups include *Plutinos* (objects with orbits similar to that of Pluto), *trans-Neptunian Objects* (objects orbiting farther from the Sun than Neptune), *Plutoids* (spherical bodies orbiting the Sun beyond Neptune's orbit, not to be confused with Plutinos), *Kuiper Belt Objects (KBOs;* objects orbiting the Sun in roughly the plane of the ecliptic and located between Neptune and roughly twice its distance from the Sun), and Oort comet cloud objects (objects in a spherical distribution around the Sun beyond Neptune's orbit). Don't be surprised if this seems confusing—many astronomers feel the same way. Figure 18-1 summarizes the relationships among the solar system's vagabonds, as presented here.

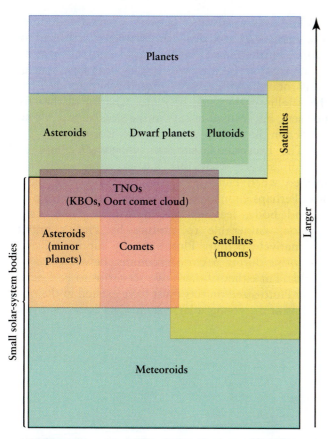

FIGURE 18-1 Different Classifications of Solar System Objects Some of the definitions of the different types of objects in the solar system overlap. For example, the largest asteroids are also being classified as dwarf planets; various trans-Neptunian objects (TNOs) are asteroids or comets; some comets are satellites of Jupiter; some Kuiper belt objects (KBOs) are satellites of other KBOs. Furthermore, TNOs exist in two groups: Kuiper belt objects and Oort comet cloud bodies. Some moons are as large as dwarf planets. Indeed, some moons are as large as small planets!

DWARF PLANETS

By formal definition, a **planet** is a celestial body that is (1) in orbit around the Sun, (2) has sufficient mass for its self-gravity to pull itself into a nearly spherical shape, and (3) has cleared the neighborhood around its orbit. A **dwarf planet** satisfies *(1)* and *(2)*, but it does not have enough gravity to clear its orbital neighborhood of debris. All objects in the solar system that are not planets, dwarf planets, or moons are called **Small Solar-System Bodies** (SSSBs). We consider the dwarf planets first.

So far, just Pluto, Ceres (also an asteroid), Eris, Makemake (pronounced **ma-ki ma-ki**), and Haumea have officially earned the title of dwarf planet. The latter three objects, also designated as KBOs and plutoids, were discovered in 2005. Dozens of other solar-system objects are being evaluated for inclusion in this class. Bear in mind that as the properties of more solar-system objects are pinned down, the definitions of dwarf planets and SSSBs may change.

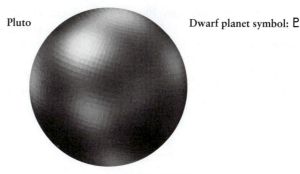

Pluto   Dwarf planet symbol: ♇

RIVUXG

FIGURE 18-2 Pluto This Hubble Space Telescope image of Pluto shows little detail but indicates that the major features of Pluto's surface each cover large amounts of its area. (Alan Stern, Southwest Research Institute; Marc Buie, Lowell Observatory; NASA; and ESA)

18-1 Pluto and its moon, Charon, are about the same size

Pluto (Figure 18-2) is now considered as the archetype of the plutoids. Clyde Tombaugh discovered Pluto because it moved among the background stars from night to night (Figure 18-3), a discovery technique that was and still is standard for finding other dwarf planets and small solar-system bodies.

Pluto
Day 1: Pluto appears between these two stars.

Pluto
Day 2: Pluto has moved relative to the stars.

RIVUXG

FIGURE 18-3 Discovery of Pluto Pluto was discovered in 1930 by searching for a dim, starlike object that slowly moved against the background stars. These two photographs were taken 1 day apart. (UC Regents/Lick Observatory)

Unlike the planets, Pluto's orbital eccentricity of $e = 0.25$ (see Section 2-5) is so great that it is sometimes closer to the Sun than Neptune, as it was from 1979 to 1999. Pluto is now farther away from the Sun than Neptune and will continue to be so for about the next 230 years (Figure 18-4). Pluto's orbit is also more tilted, with respect to the plane of the ecliptic, than that of any planet (Figure 18-4c).

Little was known about Pluto for half a century until astronomers noticed that its image sometimes appears oblong (Figure 18-5). This observation led to the discovery in 1978 of Pluto's largest moon, Charon (pronounced *KER-en,* after the mythical boatman who ferried souls across the River Styx to Hades, the domain ruled by the Roman god Pluto). From 1985 through 1990, the orbit of Charon was oriented so that Earth-based observers could watch it eclipse Pluto. Astronomers used observations of these eclipses to determine that Pluto is only twice as wide as its satellite: Pluto's diameter is 2380 km and Charon's is 1190 km.

The average distance between Charon and Pluto is less than 1/20 the distance between Earth and our Moon. Furthermore, Pluto always keeps the same side facing Charon. In other words, they both have synchronous rotation with respect to each other. As seen from the satellite-facing side of Pluto, Charon neither rises nor sets, but instead hovers in the sky, perpetually suspended in the same place above the horizon.

The exceptional similarities between Pluto and Charon suggest that this binary system may have formed when Pluto collided with a body of similar size. Perhaps chunks of matter were stripped from this second body, leaving behind a mass, now called Charon, vulnerable to capture by Pluto's gravity. Alternatively, perhaps Pluto's gravity captured Charon into orbit during a close encounter between the two worlds. For either of these two scenarios to be feasible, many Pluto-sized objects must have existed in the outer regions of the young solar system. One astronomer esti-

a

b

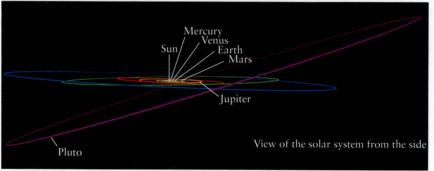

c

FIGURE 18-4 Orbit of Pluto (a) The high-eccentricity orbit of dwarf planet (and KBO) Pluto stands out compared to the orbits of the outer three planets. Notice how many significant events occurred on Earth during Pluto's present orbit of the Sun. (b) Details of Pluto's passage inside the orbit of Neptune. The two bodies will never collide. (c) A nearly edge-on view of the ecliptic and Pluto's orbit compared to it.

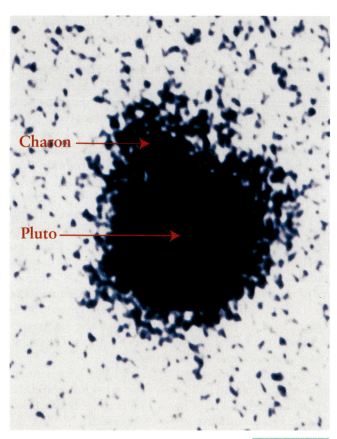

Charon

Pluto

R I V U X G

FIGURE 18-5 Discovery of Charon Long ignored as just a defect in the photographic emulsion, the bump on the upper left of this image of Pluto led astronomer James Christy to discover the moon Charon. (U.S. Naval Observatory)

mates that there had to be at least a thousand Pluto-sized bodies for a collision or close encounter between two of them to have occurred at least once since the solar system formed 4.6 billion years ago.

In 2005, the Hubble Space Telescope detected two other moons orbiting Pluto (Figure 18-6). Named Nix (goddess of the night and mother of Charon) and Hydra (nine-headed serpent living at the gates of Hades), these moons are both much smaller than Charon.

The best pictures of Pluto and Charon have been taken by the Hubble Space Telescope (Figure 18-7). Like Neptune's moon Triton, these worlds are probably both composed of nearly equal amounts of rock and ice. The surface of Pluto has 12 distinct regions—more large-scale features than any object in the solar system other than Earth. Such features on Earth include oceans and continents, while on Pluto they may include regions dominated by rock, regions dominated by ice, major impact regions, or mountains. We just do not know yet. Studies of Pluto's surface temperature show that different regions vary from one another by up to 25 K.

Pluto's spectrum shows that its surface contains frozen nitrogen, methane, and carbon monoxide. The planet is also observed to have a very thin atmosphere of nitrogen and carbon monoxide. In contrast, Charon's surface appears to be covered predominantly with water ice. So intriguing, unusual, and unexplored are Pluto and its companions that NASA launched the *New Horizons* spacecraft in 2006 to visit them (arriving there July 2015) and then continue deeper into the Kuiper belt.

What objects are classified today as planets? Dwarf planets?

18-2 Ceres is a dwarf planet in the asteroid belt and Eris, Makemake, and Haumea are dwarf planets in the Kuiper belt

On New Year's Day, 1801, the Sicilian astronomer Giuseppe Piazzi was carefully mapping faint stars in the constellation Taurus. He noticed a dim, previously uncharted "star" that shifted its position slightly over the next several nights. Uranus had been discovered that way by William Herschel just 20 years before. Later that year, the orbit of this object was determined to lie between Mars and Jupiter. At Piazzi's request, the object was named Ceres (pronounced *See-reez*), after the patron goddess of Sicily. Based on erroneous observations of Ceres's size, Piazzi's object was initially catalogued as a planet. In 1855, further observations revealed that all of the known planets—Mercury, Venus, Earth, Mars, Jupiter, Saturn, and Uranus—were all much larger than Ceres. We now know that Piazzi was the first to discover an asteroid.

R I V U X G

FIGURE 18-6 Pluto's moons Nix and Hydra Observations by the Hubble Space Telescope in 2005 revealed two small moons, each about 5000 times dimmer than Pluto. Named Nix and Hydra, they are between two and three times farther from Pluto than is Charon. The lines radiating from Pluto and Charon are artifacts of the exposure. (NASA)

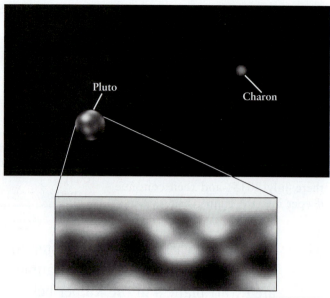

Pluto

Charon

FIGURE 18-7 Pluto and Charon This picture of Pluto and its moon, Charon, is a composite of Hubble Space Telescope images. It shows the greatest detail we have of Pluto's surface. Pluto and Charon are separated here by only 19,700 km. Inset: A map covering 85% of Pluto's surface. The contrast between dark and light strongly suggests regions covered with ice and regions covered with rocky material. (*Pluto:* Alan Stern/SwRI, M. Cuie/Lowell Observatory, NASA, & ESA; *Charon:* R. Albrecht, ESA/ESO Space Telescope European Coordinating Facility, and NASA; *inset:* A. Stern/SwRI, M. Buie/Lowell Observatory, NASA, and ESA)

a. Earth

Moon

Ceres

b. Ceres

FIGURE 18-8 Comparison of Ceres with the Moon and Earth (a) Ceres, the Moon, and Earth are shown here to scale. Dwarf planet Ceres is the largest asteroid. (b) This image suggests that Ceres has regions of ice and rock on its surface. It will be visited by the Dawn spacecraft in 2015. (NASA)

Ceres is spherical like the planets (Figure 18-8), but its diameter is a scant 940 km (585 mi), only one-quarter the diameter of our Moon. In 2006, Ceres was classified as a dwarf planet, along with Pluto and Eris. Makemake became the fourth dwarf planet in 2008. Recent calculations suggest that, like the planets, Ceres is differentiated, containing a rocky core clad in an icy outer layer.

Eris was first observed in 2003, before the present definitions of planets, dwarf planets, and SSSBs were established. It is presently located 97 AU from the Sun, 3 times farther than Pluto. Furthermore, its 557-year orbit is so eccentric that for part of that time it is closer to the Sun than Pluto (Figure 18-9a). A moon orbiting Eris, named Dysnomia (daughter of the Greek goddess Eris), was discovered in 2005 (Figure 18-9b). Moons are invaluable in helping astronomers determine the masses and diameters of the bodies they orbit. Eris has a diameter of about 2400 km (Pluto is 2300 km across) and it is very bright, with an albedo of about 0.86. Observations suggest that one quarter of this world is water ice, greater in volume than all of the fresh water on Earth. The moon Dysnomia is about 300 km across.

Haumea is apparently oval and has a maximum dimension of about 2000 km. Makemake is about 52 AU from the Sun and it has a radius roughly one third that of Pluto. Its hard to be more precise because no moon of Makemake has yet been observed.

While astronomers decide which other bodies will be designated dwarf planets, we turn to the small solar-system bodies, which include the rest of the asteroids, all of the comets, and the smallest rubble of all, the meteoroids.

SMALL SOLAR-SYSTEM BODIES

We know that the solar system formed from a rotating disk of gas and dust. The matter that had too much angular momentum to fall onto the protosun coalesced at varying orbital distances into planetesimals. Many of these chunks of rock and metal eventually collided, forming the planets, dwarf planets, and larger moons of the solar system. Others were captured whole by various planets as small, irregularly shaped moons, such as Phobos and Deimos in orbit around Mars. Many planetesimals still orbit the Sun today in splendid isolation. These are the asteroids, sometimes called *minor planets* (not to be confused with dwarf planets).

18-3 Most asteroids orbit the Sun between Mars and Jupiter

  In 1802, less than 2 years after the asteroid (now also dwarf planet) Ceres was discovered, the German astronomer Heinrich Olbers observed another faint, starlike object that moved against the background stars. He called it *Pallas*, after the Greek goddess of wisdom.

Like Ceres, Pallas orbits the Sun in a nearly circular path between the orbits of Mars and Jupiter. Pallas is even dimmer and smaller than Ceres and has a diameter of only 600 km (375 mi).

Only two more asteroids—Juno and Vesta—were found until the mid-1800s, when telescopes improved. Astronomers then began to stumble across many more asteroids orbiting the Sun at distances from 2 to 3½ AU, between the orbits of Mars and Jupiter. This region of the solar system is now called the **asteroid belt** (Figure 18-10). Asteroids whose orbits lie entirely within this region are called **belt asteroids.**

Spacecraft *Dawn* was sent in 2007 to explore Vesta and Ceres, neither of which can be well resolved from Earth and both of which hold clues to planet formation. The probe will arrive at Vesta in 2011 and Ceres in 2015.

One by one, asteroids were discovered throughout the nineteenth century. The next real breakthrough came in 1891, when the German astronomer Max Wolf applied photographic techniques to the search for them. A total of 300 asteroids had been found up to that time, each painstakingly discovered by scrutinizing the skies for faint, uncharted "stars" whose positions shifted slowly from one night to the next. With the advent of astrophotography, however, the floodgates were opened. Astronomers could simply aim a camera-equipped telescope at the stars and take long exposures. If an asteroid happened to be in the field of view, it left a distinctive trail on the photographic plate (Figure 18-11). Using this technique, Wolf alone discovered 228 asteroids.

a

b

R I V U X G

FIGURE 18-9 Dwarf Planet Eris (a) Three perpendicular views of the orbit of Eris and Dysnomia compared to the planets and Pluto. Their orbit around the Sun ranges from 38 to 98 AU, and has an orbital eccentricity, *e*, of 0.44 and an orbital inclination of 44°. (b) Keck Telescope image of dwarf planet Eris and its moon Dysnomia. (a: Orionist; b: M.E. Brown, W.M. Keck Observatory)

INSIGHT INTO SCIENCE

Confirming Observations New findings in science must be confirmed or replicated by other competent scientists before discoveries or observations are accepted by the scientific community. Therefore, asteroid observations need to be repeated several times to determine exact orbits and eliminate the possibility that a sighting is of a known asteroid or other object.

Ceres, the largest asteroid, alone accounts for about 30% of the mass of all known asteroids combined. Only three asteroids—Ceres, Vesta, and Pallas—have diameters greater than 300 km. Records of all objects smaller than planets in our solar system are kept by the Minor Planet Center in Cambridge, Massachusetts. The center reports that as of September 2008, the existence of 189,407 asteroids had been confirmed and more than 51,272,383 other observations await confirmation as new asteroids. The number of asteroids increases dramatically with *decreasing size*: Only 41 asteroids have diameters between 200 and 300 km; 250 more are bigger than 100 km across; there are thought to be tens of millions that are less than 1 km across.

You may have heard the common belief that the asteroids were formerly a single planet that was somehow destroyed. Indeed, it is

Where is Eris located in the solar system?

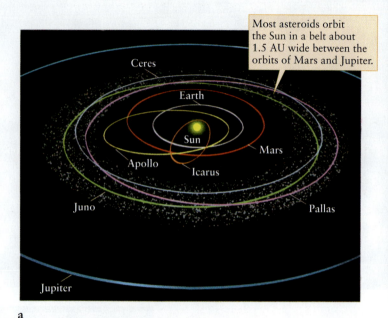

Most asteroids orbit the Sun in a belt about 1.5 AU wide between the orbits of Mars and Jupiter.

a

b

FIGURE 18-10 Asteroid Orbits (a) The orbits of belt asteroids Ceres, Pallas, and Juno are indicated to scale in this diagram. Some asteroids, such as Apollo and Icarus, have highly eccentric paths that cross Earth's orbit. Others, called the Trojan asteroids, follow the same orbit as Jupiter. (b) Actual positions of all known asteroids at Jupiter's orbit or closer. The locations of the belt asteroids are indicated by *green dots*. Objects passing closer than 1.3 AU to the Sun are shown by *red circles*. Objects observed at least twice are indicated by *filled circles,* and objects seen only once are indicated by *outline circles*. Jupiter's Trojan asteroids are *deep blue squares*. Comets are *filled* and *unfilled light-blue squares*. Although the asteroids appear packed together in this drawing, they are typically millions of kilometers apart. The small scale here is deceiving! (Minor Planet Center)

1 plausible that the asteroid belt region once had many more objects in it—possibly several Earth masses worth of debris. However, Jupiter's gravitational force pulled many planetesimals out of this region before large numbers of them could meet, coalesce, and form a single planet-sized object.

If all of the present asteroids had once been part of a single body, it would have had a diameter of only 1500 km, or 12% of Earth's diameter. This is less than two-thirds the diameter of Pluto and half the diameter of our Moon. Considering that Pluto is not a planet, a single body that contains the mass of all of the asteroids would not qualify as a planet, either. Let's now consider Jupiter's gravitational effect on the asteroids in some detail.

18-4 Jupiter's gravity creates gaps in the asteroid belt

In 1867, the American astronomer Daniel Kirkwood called attention to gaps in the asteroid belt. These features, called **Kirkwood gaps,** show the influence of Jupiter's gravitational attraction. Figure 18-12, a graph of asteroid orbital periods, has gaps at simple fractions (⅓, ⅖, ⅗, and ½) of Jupiter's orbital period. The gravitational pull of Jupiter created these empty regions by giving objects in them periodic reinforcing tugs that pulled the orbiting bodies to different orbits. The effect is analogous

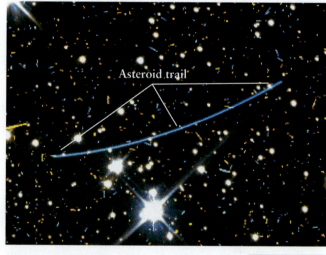

R I V U X G

FIGURE 18-11 Discovering Asteroids In 1998, the Hubble Space Telescope found this asteroid while observing objects in the constellation Centaurus. The exposure, tracking stars, shows the asteroid as a 19-arcsec streak. This asteroid is about 2 km in diameter and was located about 140 million km (87 million mi) from Earth. (R. Evans and K. Stapelfeldt, Jet Propulsion Laboratory and NASA)

2

to the gravitational resonance of Mimas creating the Cassini division in Saturn's rings (see Section 17-10).

It is likely that the asteroid belt contains tens of millions of asteroids, yet their typical separation is an impressive 10 million km. This is quite unlike the image that has been created by innumerable popular movies and television shows of asteroids so close together that you must dodge them as you fly past.

INSIGHT INTO SCIENCE

Get Real Scientists continually apply the "laws of nature" to new situations, problems, observations, and experiments, even to popular culture. For example, applying Newton's law of gravity to the asteroids reveals that they could never swarm, as science fiction movies suggest. At those close quarters, their gravity would cause them either to collide or to pass so close together that they would subsequently fly rapidly and permanently apart.

Despite the large average separation between asteroids, the gravitational influences of Mars and Jupiter have sent some of them smashing into each other at various times over the past 4.6 billion years. A collision between kilometer-sized asteroids must be an awe-inspiring event. Typical collision velocities are estimated to be 3600 to 18,000 km/h (2000 to 11,000 mph), which is more than sufficient to shatter rock. In some collisions, the resulting fragments may not have enough speed to escape from one another's gravitational attraction, and they reassemble. The asteroid Toutatis appears to be composed of two comparably sized pieces connected to each other, as does the asteroid Castalia. These bodies are therefore likely to have been broken apart and re-formed.

Alternatively, several large fragments, such as Ida and Dactyl (Figure 18-13), may end up orbiting each other. Dactyl is a pockmarked asteroid some 1.5 km in diameter that orbits Ida at a distance of 100 km. Petit-Prince is the satellite of Eugenia. Petit-Prince is 13 km across, orbiting 1200 km from the larger body. To date, 151 asteroid/satellite systems have been observed.

In 1918, the Japanese astronomer Kiyotsugu Hirayama drew attention to groups of asteroids that share nearly identical orbits. These fragments are pieces of parent asteroids. Pursuing this line of research, astronomers in 2002 discovered that two large asteroids collided only a few million years ago (very recently in solar system history) and created about 20 separate families or clusters of smaller asteroids. Asteroids in each of these families orbit together. The largest remnant cluster of this impact is named Karin, after its largest member, an asteroid some 20 km (12 mi) across.

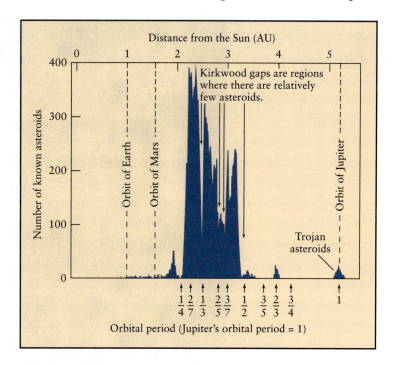

FIGURE 18-12 The Kirkwood Gaps This graph displays the number of asteroids at various distances from the Sun. Note that few asteroids have orbital periods that correspond to such simple fractions as ⅓, ⅖, ⅗, and ½ of Jupiter's orbital period. Repeated alignments with Jupiter have deflected asteroids away from these orbits. The Trojan asteroids accompany Jupiter as it orbits the Sun.

VIDEO 18.3 In the early 1990s, the Jupiter-bound *Galileo* spacecraft passed near two asteroids—Gaspra (see Figure 14-8) and Ida (Figure 18-13)—and sent back close-up views. Because Ida's surface is more heavily cratered than Gaspra's, we deduce that Ida is much older.

> Why do astronomers doubt that the asteroid belt was once made up of a single planet?

18-5 Asteroids occur outside the asteroid belt

While Jupiter's gravitational pull clears out certain areas within the asteroid belt, it actually captures asteroids at two locations in the path of its own orbit. The gravitational forces of the Sun and Jupiter work together to hold asteroids in orbit at these locations, called **stable Lagrange points**, in honor of the French mathematician Joseph Lagrange, whose calculations explained them. One Lagrange point is located 60° ahead of Jupiter, and the other is 60° behind, as shown in Figure 18-14 (see also Figure 18-10b).

The asteroids trapped at Jupiter's Lagrange points are called **Trojan asteroids**, each named after a hero of the Trojan War. As of August 2008, 2543 Trojan asteroids

R I V U X G

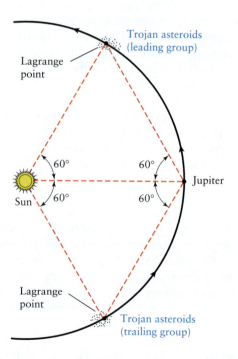

FIGURE 18-13 Ida and Its Satellite The 55-km-long rocky asteroid Ida, shown here with its satellite Dactyl, is about twice the size of the younger asteroid Gaspra (see Figure 14-8). *Inset:* Dactyl is also heavily cratered. (NASA)

orbiting with Jupiter have been catalogued. Neptune has six known Trojan asteroids, Mars has one, and astronomers continue to search for Trojans orbiting with other planets.

Some asteroids have highly elliptical orbits that bring them into the inner regions of the solar system (see red dots in Figure 18-10b). Others have similarly elliptical orbits that extend from the asteroid belt out beyond the farthest reaches of Pluto's orbit. The Amor asteroids cross Mars's orbit, but do not get as close to the Sun as Earth's orbit, whereas the **Apollo asteroids** do cross Earth's orbit. At least *5525 Earth-crossing* asteroids are known, several of which are pairs of asteroids that orbit each other. Not all of these Earth-crossing asteroids pose threats to Earth, but at least 1038 potentially hazardous asteroids (PHAs) may someday strike Earth.

Several close calls took place in recent years. In 1972, space debris was observed to skip off Earth's atmosphere and retreat back into space. On December 9, 1994, asteroid 1994 XM1 (10 m across—the size of a small bus; Figure 18-15) passed within 105,000 km (65,000 mi) of Earth. Two near misses occurred in 2002: one on January 8, when an asteroid passed within 375,000 km (233,000 mi), and the other on June 14, when an asteroid passed only 119,000 km (75,000 mi) away, which is less than a third the distance to the Moon.

FIGURE 18-14 Jupiter's Trojan Asteroids Groups of asteroids orbit at the two stable Lagrange points along Jupiter's orbit, trapped by the combined gravitational forces of Jupiter and the Sun.

FIGURE 18-15 Asteroid 1994 XM1 This image was obtained on December 9, 1994, shortly before the asteroid arrived in Earth's vicinity. When it passed by Earth just 12 h later, asteroid 1994 XM1 was less than half the distance from Earth to the Moon. (Jim Scotti, Spacewatch on Kitt Peak)

During these close encounters, astronomers can examine the details of asteroids. For example, an asteroid's brightness often varies as it rotates because different surface features scatter different amounts of light. Such data show that typical rotation periods for asteroids are between 5 and 20 h, although one asteroid, labeled 1998 KY26, rotates about once every 10.7 min. This is the fastest rotating object known in the solar system.

The proximity of asteroids and the promise of learning more about these ancient and extremely varied members of our solar system prompted NASA to send the *Near Earth Asteroid Rendezvous (NEAR) Shoemaker* spacecraft to visit asteroids Mathilde (Figure 18-16a) and Eros. The Japanese space agency sent the *Hayabusa* spacecraft to the asteroid Itokawa (Figure 18-16b), where the spacecraft attempted to take samples. It will return to Earth in 2010 with its cargo.

NEAR Shoemaker revealed that Mathilde is only 1.3 times denser than water and has an albedo of 0.04, making it darker than charcoal. This heavily cratered, carbon-rich body therefore has only half the density of the other asteroids astronomers have studied, such as Ida. Eros is about 3 times as dense as water and rotates once every 5¼ h. In February 2000, *NEAR Shoemaker* went into orbit around Eros, and for a year the spacecraft's cameras and other sensors sent back a wealth of information about it.

NEAR Shoemaker showed that Eros (Figure 18-17) is a chunk of rock and metal. Whether it is solid or a collection of loosely held pieces is still under investigation. Analyzing Eros's spectra reveals that it is probably much the same as it was when it coalesced 4.6 billion years ago. This means that it was never hot enough to differentiate (separate rock from metal). Infrared observations reveal that like our Moon, Eros has a regolith. It also has several substantial craters and is strewn with boulders

(Figure 18-17c). Astronomers infer from the lack of high crater walls that recent impacts have caused the asteroid to vibrate, collapsing earlier craters. On February 12, 2001, NASA engineers landed *NEAR Shoemaker* on Eros so gently that the spacecraft continued to transmit data after landing. Details of Eros as small as 1.4 m across were imaged by *NEAR Shoemaker*.

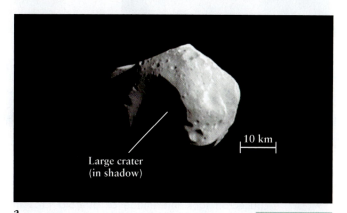

Large crater (in shadow)

10 km

a

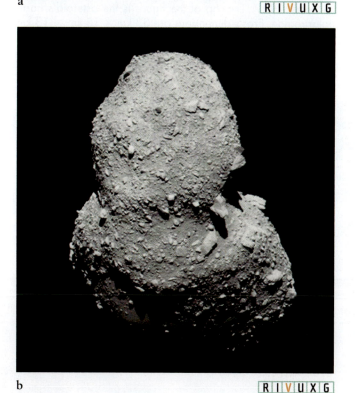

b

FIGURE 18-16 Asteroids (a) *Mathilde* Reflecting only half as much light as a charcoal briquette, Mathilde is half as dense as typical stony asteroids. Slightly larger than Ida, irregularly shaped Mathilde measures 66 km × 48 km × 46 km, rotates once every 17.4 days, and has a mass equivalent to 110 trillion tons. The part of the asteroid shown is about 59 km × 47 km. The large crater in shadow is about 20 km across. (b) *Itokawa* A near Earth asteroid visited by the Japanese space probe *Hayabusa*. Samples were taken, but it is not known how much of the asteroid was captured by the spacecraft, which will return to Earth in 2010. (a: Johns Hopkins University, Applied Physics Laboratory; b: ISAS, JAXA)

WEB LINK 18.6

a   b   c

R I V U X G

FIGURE 18-17 Asteroid Eros (a) *The Near-Earth Asteroid Rendezvous (NEAR) Shoemaker* spacecraft took this image of asteroid Eros in February 1999. The top of the figure is the asteroid's north polar region. Eros's dimensions are 33 km × 13 km × 13 km (21 mi × 8 mi × 8 mi) and it rotates every 5¼ h. Its density is 2700 kg/m³, close to the average density of Earth's crust and twice as dense as asteroid Mathilde. (b) Looking into the large crater near the top of *(a)*, which is 5.3 km (3.3 mi) across. (c) This is the penultimate image taken by *NEAR Shoemaker* before it gently landed on Eros. Taken from an altitude of 250 m (820 ft), the image is only 12 m across. You can see rocks and boulders buried to different depths in the regolith. (Johns Hopkins Applied Physics Laboratory)

Describe the appearance of small asteroids.

The Sun is not the only star with an asteroid belt. In 2001, astronomers discovered that the star Zeta Leporis, 70 ly from Earth in the constellation Lepus (the Hare), has a disk of debris that appears to contain asteroids. This star system is less than 0.5 billion years old; astronomers hope it will provide insights into the early evolution of the asteroids and other objects in the disk of gas and dust that surrounded the early Sun.

COMETS

While asteroids consist primarily of rock and metal, other space debris is composed of frozen water, along with rock, metal, and ices of other compounds. We have seen in earlier chapters that water ice, along with carbon dioxide, methane, and ammonia ices, was locked up in planets and moons. Ices in the young solar system also condensed with roughly equal amounts of small rocky and metallic debris into bodies that still remain in orbit around the Sun. These dirty snowballs in space are the **comets**.

18-6 Comets come from far out in the solar system

In the first few hundred million years of the solar system's existence, comets formed in its outer reaches at roughly the distances of Saturn, Uranus, and Neptune. In that region, water was plentiful and the temperature was low enough for the ices to condense into chunks several kilometers across. After they formed, gravitational tugs from Uranus and Neptune flung the comets in every direction.

In 2001, astronomers were able to measure the temperature at which ammonia ice formed in Comet LINEAR. This temperature determines the ammonia's solid structure and, so by determining the structure of the ammonia in Comet LINEAR, they were able to calculate the temperature when it formed and how far the comet was from the Sun at that time. That comet coalesced between the orbits of Saturn and Uranus.

Today, the solar system is thought to contain two reservoirs of comets: the Kuiper belt and the Oort comet cloud. Most comets that eventually enter the inner solar system are believed to come from a doughnut-shaped region beyond the orbit of Pluto, called the **Kuiper belt**. Comets head inward as a result of collisions or near misses with other cometary bodies. Named after the American astronomer Gerard Kuiper, who first proposed its existence in 1951, the Kuiper belt is centered on the plane of the ecliptic. Its main body of objects extends out some 50 AU from the Sun (Figure 18-18).

a

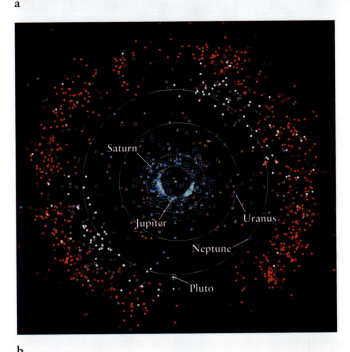

b

FIGURE 18-18 The Kuiper Belt and Oort Comet Cloud   (a) The classical Kuiper belt of comets spreads from Neptune out 50 AU from the Sun. Most of the estimated 200 million belt comets are believed to orbit in or near the plane of the ecliptic. The spherical Oort comet cloud extends outward from the Kuiper belt. More detail of Sedna's orbit is given in Figure 18-20. (b) The current positions of the known dwarf planets and SSSBs in the outer solar system are shown in this diagram. Objects with unusually high eccentricity orbits are shown as *cyan triangles.* Objects roaming among the outer planets, Centaur objects, are *orange triangles.* Plutinos are *white circles.* Miscellaneous objects are *magenta circles,* and classical KBOs are *red circles.* Objects observed only once are denoted by *open* symbols; objects with two separate observations are denoted by *filled* symbols. Comets are *filled* and *unfilled light-blue squares.* (a: NASA and A. Field/Space Telescope Science Institute; b: Courtesy of Gareth Williams, Minor Planet Center)

The Kuiper belt consists of two groups of objects. Those with roughly circular orbits are called *classical KBOs* (Kuiper belt objects) and are found between 30 AU (Neptune's orbit) and 50 AU from the Sun. The second group, called *scattered KBOs,* have more elliptical orbits that range from 35 AU to at least 200 AU from the Sun. At least 1471 KBOs have been observed (Figure 18-18 and Figure 18-19a, b). Although most of these objects are tens of kilometers across, astronomers have also begun to observe some KBOs only 10 to 100 m in diameter. Based on these latter observations, astronomers estimate that 10^{15} such objects exist out there.

At least 1% of the KBOs are orbiting pairs (Figure 18-19c), including Quaoar, which is 1250 km across, some 400 km larger than Ceres or about half the size of Pluto. At least 95 of the KBOs orbit the Sun in the same region as Pluto. Because these latter bodies are smaller, they are called *Plutinos.* The periods of the orbits of both Pluto and the Plutinos are characterized by being one and a half (often written as 3:2) times as long as the period of Neptune's orbit. Indeed, it appears that Pluto and the Plutinos were forced into their present orbit by Neptune's gravity. Combining these last two points, we say that they are locked into a *3:2 resonance* with Neptune.

The vast majority of comets are believed to lie even farther from the Sun. Unlike the Kuiper belt comets and the rest of the solar system, these bodies are believed to have a spherical distribution around the Sun, called the **Oort comet cloud** (sometimes shortened to Oort cloud; Figure 18-18a), named after the Dutch astronomer Jan Oort, who first proposed its existence in the 1950s.

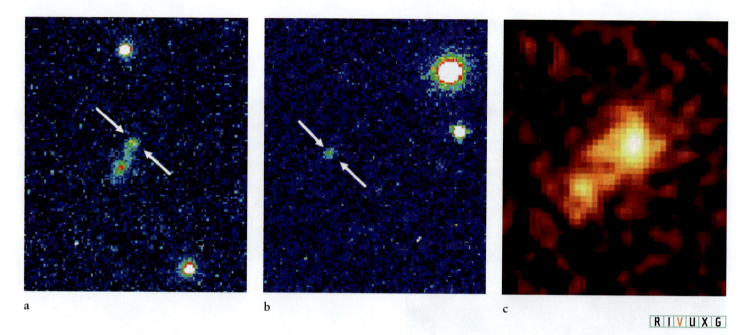

a

b

c

R I V U X G

FIGURE 18-19 Kuiper Belt Objects (a, b) These 1993 images show the discovery (white arrows) of one of at least 1277 known KBOs. These two images of KBO 1993 SC were taken 4.6 hours apart, during which time the object moved against the background stars. (c) The KBO 1998 WW31 and its moon (lower left). (a & b: Alan Fitzsimmons, Queen's University of Belfast; c: C. Veillet/CFHT)

In 2003, astronomers discovered an object larger than Quaoar. Called Sedna (see Section 14-4), this body is roughly 1800 km (1100 mi) across and is presently 86 AU from the Sun. It has a highly elliptical orbit (Figure 18-20) that takes it from the outer realm of the main Kuiper belt and possibly into the Oort comet cloud. Because the distance from the Sun to the inner edge of the Oort cloud is not known, astronomers are still debating whether it is another KBO or the first known Oort comet cloud object. Sedna has a sidereal rotation period of 10 h.

Astronomers calculate that the Oort comet cloud extends out at least 50,000 AU, one-fifth the distance to the nearest stars. Most of these comets have orbits so nearly circular that they never even get as close as Pluto is to the Sun. However, occasionally, a collision between two comet nuclei or the gravitational tug from a passing star or passing giant molecular cloud nudges a distant comet toward the inner solar system. As a result of its inward plunge, comets from the Oort comet cloud, such as Hale-Bopp and Hyakutake, have highly elliptical, parabolic, or even slightly hyperbolic orbits (see Section 2-8). Often comets also have orbits that are highly tilted, even perpendicular, to the plane of the ecliptic.

Because the comets in the Kuiper belt and Oort comet cloud are far from the Sun, they are completely frozen. Solid cometary bodies, called **nuclei** (*singular:* **nucleus**), are typically between 1 and 10 km across, although some, like that of Comet Halley, are 15 km or more in diameter. The first pictures of a comet's nucleus were obtained when a fleet of spacecraft flew past Comet Halley in 1986 (Figure 18-21a). Halley's potato-shaped nucleus is darker than coal, probably because of carbon-rich compounds left behind as ice evaporates. In 2001, the *Deep Space* 1 spacecraft took high-resolution images of Comet Borrelly's nucleus (Figure 18-21b). The ends of the comet are very rugged, while the center has long, rolling plains, from which jets of gas appear to originate.

To better determine the composition of the gas emitted by comets, the *Stardust* probe visited Comet Wild (pronounced *Vilt*) 2 (Figure 18-22a), where it collected samples (Figure 18-22b, c) and returned them to Earth in 2006. Comet Wild 2 showed many unexpected features, including towering spires, many craters and cliffs, and more than 24 jets, rather than the 2 or 3 jets seen on other comets.

As a comet nucleus comes within 20 AU of the Sun, solar radiation begins to vaporize the ices on its surface. The liberated gases form an atmosphere, or **coma**, around the nucleus. Because the coma scatters sunlight, it appears as a fuzzy, luminous ball. The largest coma ever measured was more than a million kilometers across—nearly as large as the Sun. Not visible to the human eye is the **hydrogen envelope**, a sphere of tenuous gas surrounding the comet's nucleus and measuring as much as 20 million km in diameter (Figure 18-23).

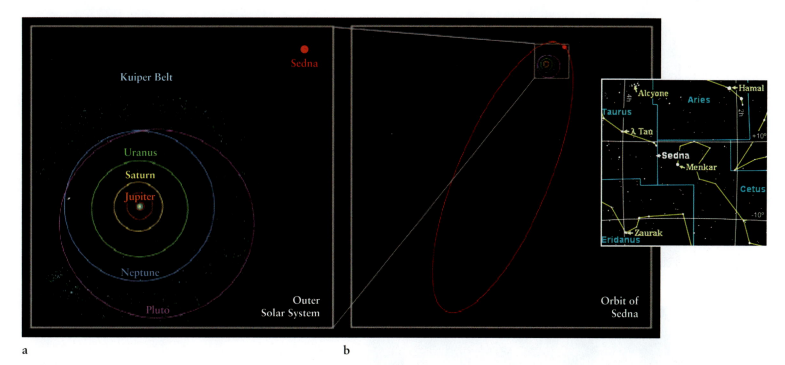

a   b

FIGURE 18-20 Sedna's Orbit (a) The farthest known body in the solar system is in a highly elliptical orbit (b) that ranges from the outer reaches of the Kuiper belt and possibly extends to the inner Oort comet cloud. (NASA/Caltech)

18-7 Comet tails develop from gases and dust pushed outward by the Sun

The most visible and inspiring features of comets are their long, flowing, diaphanous tails (Figure 18-24). Comet tails develop from coma gases and dust pushed outward by radiation and particles from the Sun. This means that comet tails do not trail behind the nucleus, as the exhaust from a jet plane does in Earth's atmosphere. Rather, *at the comet's nucleus, the tails always*

point away from the Sun (Figure 18-25), regardless of the direction of the comet's motion. The implication that something from the Sun "blows" the comet's gases radially outward led Ludwig Biermann to predict the existence of the solar wind (see Sections 11-4 and 5-3). This stream of particles from the Sun was actually discovered in 1962, a full decade after it was predicted, by instruments on the spacecraft *Mariner 2*. In 2001, NASA launched

> What is the solid part of a comet called?

a   R I V U X G

b   R I V U X G

FIGURE 18-21 Comet Nuclei (a) The nucleus of Comet Halley. This image, taken by the *Giotto* spacecraft, shows the potato-shaped nucleus of the comet. Its dark nucleus measures 15 km in its longest dimension and about 8 km in its shortest. The numerous bright areas on the nucleus are icy out-croppings that reflect more sunlight than surrounding areas of the comet. Two jets of gas can be seen emanating from the left side of the nucleus. (b) The nucleus of Comet Borrelly, in an image taken by *Deep Space* 1. The nucleus is 8 km (5 mi) long. (a: Max-Planck-Institut für Aeronomie; b: *Deep Space* 1 Team, JPL, NASA)

a   R I V U X G

Ejecta in aerogel

Tiny crater in aluminum
foil holding aerogel.

b   R I V U X G

c   R I V U X G

FIGURE 18-22 Comet Wild 2 (a) This picture shows two images combined. One is a high-resolution photograph showing the surprisingly heavily cratered comet. The other image is a longer photograph showing gas and dust jetting from the comet. Its tails are millions of kilometers long. (b) A substance called aerogel was used to capture particles from Comet Wild 2's dust tail. A piece of space debris pierced the aluminum foil holding the aerogel and embedded in it, along with pieces of the foil. (c) A 2-μm piece of comet dust, composed of a mineral, called *forsterite*. On Earth this mineral is used to make gems called *peridot*. (b: NASA/JPL; c: NASA/JPL-Caltech/ University of Washington.)

the *Genesis* spacecraft, which collected pristine solar wind particles for two and a half years. Although it crash-landed back on Earth in September 2004, some of the particles that it collected were salvaged and are now being studied.

INSIGHT INTO SCIENCE

Starting with Observations Science often advances by working from observations and experiments that require scientific explanations. For example, seeing comet tails and how they behave led Biermann to wonder what causes them. The simplest physical explanation is that matter from each cometary body is being evaporated and pushed upon by something from the Sun—the solar wind. Remember Occam's razor (Section 2-1).

Outflowing particles and radiation from the Sun produce two comet tails: a **gas** (or **ion** or **plasma**) **tail** and a **dust tail** (Figure 18-26 and the figure opening this chapter). Positively charged ions (atoms missing one or more electrons) from the coma are swept directly away from the Sun by the solar wind to form the gas tail. This tail often appears blue, because, like the gas in our air, the tail's gas strongly scatters blue light. The relatively straight gas tail can change dramatically from night to night (see Figure 18-26). This occurs because the solar wind that pushes the gases changes rapidly and the gases themselves are very light and easily moved about, like the smoke from a fire (Figure 18-26b).

The dust tail is formed when sunlight strikes dust particles that have been freed from the comet's evaporating nucleus. Light exerts pressure on any object that absorbs or scatters it. While this pressure, called **radiation pressure** or **photon pressure**, is quite weak, fine-

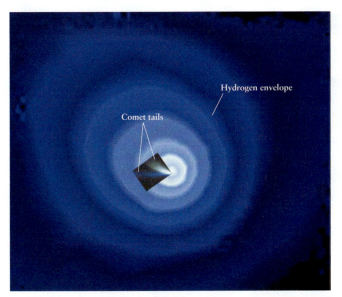

FIGURE 18-23 Comet Hale-Bopp In 1997, Hale-Bopp had a hydrogen envelope 1 AU in diameter (blue ovals). This gas was observed in the ultraviolet. The visible-light inset shows the scale of the visible tails (see also the image at the beginning of the chapter). (a: Johns Hopkins University and Naval Research Laboratory; b. Mike Combi)

grained dust particles in a comet's coma are sufficiently light to be pushed from the vicinity of the comet, thus producing a dust tail. The dust tail is often white or light gray. The dust particles are massive enough not to flow straight away from the Sun; rather, the dust tail arches in a path that lies between the gas tail and the direction from which the comet came (Figure 18-25 and Figure 18-27).

Figure 18-27 outlines the elements of comets, although many have bigger or smaller features. Some comets, like the one shown in Figure 18-28, have a large, bright coma but short, stubby tails. Others have an inconspicuous coma but one or more tails of astonishing length. The gas tail in Figure 18-29 stretches more than 150 million km (1 AU) in length.

> Which comet tail points directly away from the Sun, and why?

18-8 Comets are fragile, yet durable

To better understand the chemistry and structure of comets, astronomers sent the *Deep Impact* spacecraft, roughly the size of a small refrigerator, slamming into Comet Tempel 1 at 36,000 km/h (22,000 mph) in

FIGURE 18-24 Comet West Astronomer Michael M. West first noticed this comet on a photograph taken with a telescope in 1975. After passing near the Sun, Comet West became one of the brightest comets of the 1970s. This photograph shows the comet in the predawn sky in March 1976. (Hans Vehrenberg)

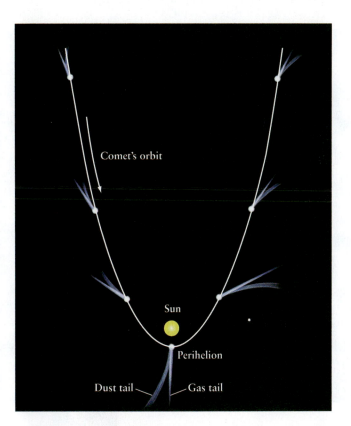

FIGURE 18-25 The Orbit and Tails of a Comet The sunlight and solar wind blow a comet's dust particles and ionized atoms away from the Sun. Consequently, comets' tails always point away from the Sun.

Gas tail

Dust tail

a   August 22   August 24   August 26   b

R I V U X G

FIGURE 18-26 The Two Tails of Comet Mrkos (a) Comet Mrkos dominated the evening sky in August 1957. These three views, taken at 2-day intervals, show dramatic changes in the comet's gas tail. In contrast, the slightly curved dust tail remained fuzzy and featureless. (b) Wind blowing smoke from this forest fire causes the smoke column to change shape and direction, just like the solar wind and sunlight cause the tails of comets to change shape and direction. (a: Palomar Observatory; b: Paul Von Baich/The Image Works)

2005 (Figure 18-30). Observations indicate that the projectile gouged a crater 100 m across and 30 m deep. It sent up a cloud of vapor and melted rock estimated to have had a mass of 10 million kg (22 million lb), still less than a ten-millionth of the comet's total mass. The particles ejected were typically smaller than 100 microns (μm).

> What is the composition of a comet?

The size of the impact crater and the volume of matter ejected provide strong evidence that this comet is composed of very weakly bound fragments of rock and ice, rather than solid pieces of each. Spectra of the debris thrown upward revealed many molecules containing carbon that had been part of organic compounds. This supports the conjecture by some astronomers and geologists that comet impacts brought such compounds to Earth early in our planet's existence.

Even though this comet has an extremely low gravitational force with which to hold itself together, it managed to pull over 90% of the dust that it ejected back onto itself. The craters on this comet and others show that they have been hit by other pieces of space debris throughout their histories. Despite being a pile of rubble that is barely held together, the comet's ability to essentially reform itself after this and other collisions helps explain why such objects have not all been annihilated over time.

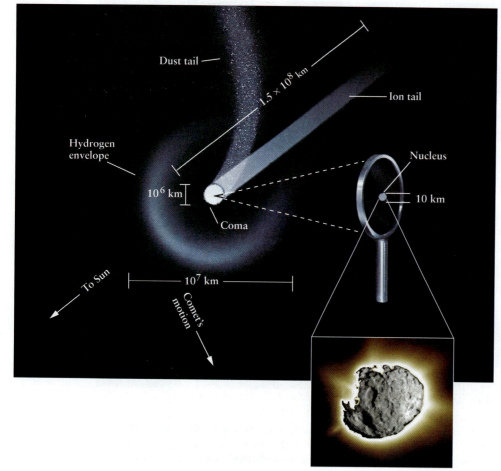

Dust tail

1.5×10^8 km

Ion tail

Hydrogen envelope

Nucleus

10^6 km

10 km

Coma

To Sun

10^7 km

Comet's motion

FIGURE 18-27 The Structure of a Comet The solid part of a typical comet (the nucleus) is roughly 10 km in diameter. The coma can be as large as 10^5 to 10^6 km across, and the hydrogen envelope is typically 10^7 km in diameter. A comet's tail can be enormous—as long as 1 AU. Comet Wild 2 (inset) is examined further in Figure 18-22. (This drawing is not to scale.) (NASA/JPL [inset])

RIVUXG

FIGURE 18-28 The Head of Comet Brooks This comet had an exceptionally large, bright coma. As are all comets, it was named after its discoverer, in this case William R. Brooks. It dominated the night skies in October 1911. (UCO/Lick Observatory)

RIVUXG

FIGURE 18-29 The Tail of Comet Ikeya-Seki Named after its codiscoverers in Japan, this comet dominated the predawn skies in late October 1965. The yellow in the tail comes from emission by sodium atoms in the dust that was released by the comet. Although its coma was tiny, its tail spanned over 1 AU. (Roger Lynds/NOAO/AURA/NSF)

18-9 Comets do not last forever

Astronomers, many of them amateurs, typically discover at least a dozen new comets each year. Falling Sunward from the Kuiper belt or Oort comet cloud, most are **long-period comets**, which have such eccentric orbits that they leave the inner solar system after one pass by the Sun and typically take 1 million to 30 million years to return, if they ever do.

However, sometimes a comet passes so closely to a giant planet that the planet's gravitational force changes the comet's orbit, slowing it down and trapping it in the inner solar system (Figure 18-31). The long-period comet then becomes a **short-period comet**, orbiting the Sun in fewer than 200 years. Like Comet Halley, short-period comets appear again and again at predictable intervals. Your next chance to see Comet Halley is in 2061.

Comets cannot survive an infinite number of passages near the Sun. A typical comet is estimated to lose between 1/60th and 1/100th of its mass at each perihelion (closest approach to the Sun). Therefore, a typical comet survives, at most, 100 close passes to the Sun before its ices evaporate completely. Comet Halley has been seen at least 27 times, meaning that it has fewer than 73 more close passes to the Sun before it dissipates. Because it returns every 76 years or so, it will survive for about $73 \times 76 = 5548$ more years. A comet can also be torn apart when it comes too close to a planet or the Sun, or it can be destroyed completely by striking a planet, a moon, or the Sun. A spectacular example was Comet Shoemaker-Levy 9. As discussed in Section 17-3, that comet fragmented under the tidal force from Jupiter in 1992. Two years later, with the world's astronomers watching carefully, the pieces returned and struck the planet (see Figure 17-10).

Like Comet Tempel 1, Shoemaker-Levy 9 broke up because its nucleus was held together very weakly. As discussed previously, it may have been composed of separate pieces that had stuck together until Jupiter's tidal force pulled them apart. Since then, astronomers have observed at least five other comets break up, in 1999, 2000 (Comet LINEAR), 2001, 2002 (Comet Toit-Neujmin-Delporte), and 2006 (Schwassmann-Wachmann 3). Figure 18-32 shows a piece of comet Schwassmann-Wachmann 3 breaking up shortly after it passed perihelion. Soon thereafter, its remaining dust and rock fragments spread out in a loose collection of debris that continues to circle the Sun along the comet's orbital path.

Comets occasionally lose mass quickly by ejecting it in bursts. Several times, starting in September 1995, astronomers observed Comet Hale-Bopp eject 7 to 10 times more mass than usual (Figure 18-33). Astronomers think that this event resulted from surface ice and dust being heated by the Sun and then being rapidly ejected from a local region on the comet's surface, called a *vent*. Hale-Bopp rotates with a period of about 12 h. The ejected matter therefore formed a

a

b

c

d

R I V U X G

e

FIGURE 18-30 Comet Tempel 1 (a) This composite image of Comet Tempel 1 has higher resolution at the bottom, as the projectile from *Deep Impact* headed in that direction. The smooth regions on the comet have yet to be explained. (b) Thirty seconds before the projectile struck the comet. (c) Seconds after impact, hot debris explodes away from the comet nucleus. The white horizontal half-ellipses are areas where the CCDs were overloaded with light from the event. (d) Moments later, the gases and dust were expanding outward. (e) Sixty-seven seconds after impact. Within minutes, the cloud of debris became much larger than the entire nucleus. (a: NASA/JPL/UMD; b–e: NASA/JPL-Caltech/UMD)

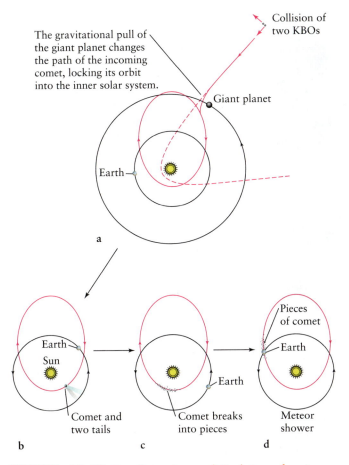

FIGURE 18-31 Transformation and Evolution of a Long-Period Comet (a) The gravitational force of a giant planet can change a comet's orbit. Comets initially on highly elliptical orbits are sometimes deflected into more circular paths that keep them in the inner solar system. (b-d) These figures show the evolution of a comet into gas, dust, and rubble, and why some of this debris may strike Earth. (See also Figure 9-37.)

RIVUXG

FIGURE 18-32 The Fragmentation of Comet Schwassmann-Wachmann-3 This comet, with a 5.4-year orbit, has been coming apart for decades. In 2006, it further fragmented after passing perihelion. One piece, Fragment B, shed at least 30 smaller pieces, as shown here. (NASA; ESA; H. Weaver [APL/JHU]; M. Mutchler and Z. Levay [STLScI])

pinwheel-shaped distribution spiraling from the comet's nucleus at a speed of 109 km/h (68 mph). Although the ejected matter did not represent a significant fraction of the comet's mass, its light-reflecting dust made it look like a huge comet fragment.

For thousands of years, astronomers have observed comets that pass very closely to the Sun. This was difficult, of course, because of the Sun's brilliance, so only a few dozen had been carefully observed prior to the mid-1990s. Called *sungrazing comets*, they come within about 0.01 AU (1.46×10^6 km) of the Sun's surface and are typically vaporized or broken into smaller pieces near perihelion. At the end of the nineteenth century, Heinrich Kreutz observed that sungrazers all have the same orbit, indicating that they are pieces of a single comet that broke up perhaps centuries ago. They are called *Kreutz comets* in his honor. In 1995, the *Solar and Heliospheric Observatory (SOHO)* began a dedicated mission of observing the Sun and, among other things, locating sungrazing comets. To date, it has observed over 1000 Kreutz comets (Figure 18-34).

The Sun is not the only star that comets orbit. A star labeled *CW Leonis*, about 500 ly from Earth, is presently enlarging in size and vaporizing billions of comets around it. Astronomers have observed the spectra of the water vapor from these cometary bodies.

What evidence do we have that comet Shoemaker-Levy 9 was not one solid chunk of rock surrounded by ice?

METEOROIDS, METEORS, AND METEORITES

As noted earlier, asteroids occasionally collide with each other, sending fragments into interplanetary space. These smaller pieces, along with rocky and metallic debris from evaporating comets, and material that never coalesced with larger bodies, are still orbiting the Sun. **Meteoroids** are rocky and metallic debris smaller than asteroids scattered throughout the solar system. Although no official size standard distinguishes the two, meteoroids are no more than about 10 m across, and the vast majority of them are smaller than 1 mm. **Meteors** are meteoroids burning up in the atmosphere. **Meteorites** are remnants of meteors that land intact.

RIVUXG

Meteorite impact, Poughkeepsie, NY, 1992

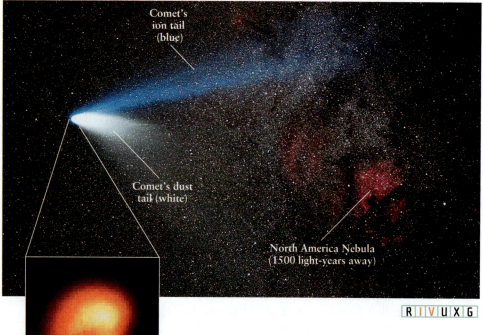

Comet's ion tail (blue)

Comet's dust tail (white)

North America Nebula (1500 light-years away)

R I V U X G

FIGURE 18-33 Comet Hale-Bopp Discovered on July 23, 1995, this comet was at its breathtaking best in mid-1997. Inset: Jets of gas and debris were observed shooting out from Comet Hale-Bopp several times. This image shows the comet nucleus (lower bright region), an ejected piece of the comet's surface (upper bright region), and a spiral tail. The ejected piece eventually disintegrated, following the same spiral pattern as the tail. (Tony and Daphne Hallas, Astrophotos)

WEB LINK 18.12

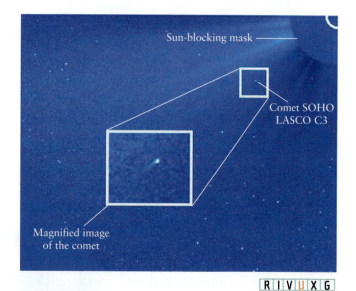

Sun-blocking mask

Comet SOHO LASCO C3

Magnified image of the comet

R I V U X G

FIGURE 18-34 Sungrazing Comet Comet SOHO LASCO C3 is shown in the smaller box and magnified in the larger one. Discovered in March 2004, it was the 750th sungrazing comet discovered from the SOHO data. It vaporized near perihelion. (ESA/NASA-SOHO/LASCO)

By studying the chemistry of various asteroids, astronomers are now beginning to identify which ones were the origins for other ones and for meteoroids. For example, the asteroid Braille (named in honor of Louis Braille, developer of the alphabet for the blind) is a piece of debris 2 km long that blasted off the 500-km-diameter asteroid Vesta.

18-10 Small, rocky debris peppers the solar system

Passing meteoroids are often pulled by gravity toward Earth's atmosphere. As they move through the air, they compress and heat the gases in front of them so much that the air glows. The glowing gas, accompanied by vaporized pieces of the infalling matter, creates the trail we see, and the meteoroid becomes a **meteor** (Figure 18-35). Common names for these dramatic streaks of light flashing across the sky include *shooting stars*, *fireballs*, and *bolides*. Fireballs are meteors at least as bright as Venus; bolides are bright meteors that explode in the air. Therefore, "shooting stars" are not stars of any kind or dying stars.

Most meteors vaporize completely before they can strike Earth. Their dust settles to the ground, often carried by raindrops. (This is not the source of acid rain, however, which comes from natural and human-made gases ejected into the atmosphere.) Any part of a meteor that survives its fiery descent to Earth may leave an **impact crater**. Weather and water erosion are wearing away all of the 200 or so impact craters now known on Earth, and thousands more have long since been drawn into Earth by the motion of Earth's tectonic plates. Indeed, the craters that exist today are all less than 500 million years old because of forces reshaping Earth's surface.

One of Earth's least-weathered impact craters is Meteor (or Barringer) Crater near Winslow, Arizona (Figure 18-36). Measuring 1.2 km across and 200 m deep, it formed approximately 50,000 years ago when an iron-rich meteoroid some 50 m across (about half the length of a football field) struck the ground at 40,000 km/h (25,000 mph). The blast was like the detonation of a 20-megaton hydrogen bomb.

On a typical clear night, you can expect to see a meteor about every 10 min. However, at predictable times throughout each year, Earth is inundated with them. These **meteor showers** occur when Earth passes through the orbit of debris left behind by a comet (Figures 18-37 and 18-38).

Some 30 meteor showers can be seen each year. Because the meteors in each shower appear to come from

R I V U X G

FIGURE 18-35 Meteor This brilliant meteor is seen lighting up the dark desert skies of the California desert area of Joshua Tree National Park. To the right of it are the Pleiades. (Wally Pacholka; http://www.astropics.com)

a fixed region of the sky, they are named after the constellation from which the meteors appear to radiate (Figure 18-38). For example, meteors in the Perseid shower appear to originate in the constellation Perseus. More than one meteor can be seen each minute at the peaks of such prodigious meteor showers as the Perseids, which take place in the summertime and are among the most exciting light shows in astronomy. Except for the Lyrids, meteor showers are best seen after midnight.

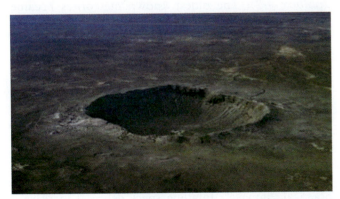

R I V U X G

WEB LINK 18.14

FIGURE 18-36 Meteor Crater An iron meteor measuring 50 m across struck the ground in Arizona 50,000 years ago. The result was this beautifully symmetric impact crater. (D. J. Roddy and K. Zeller/USGS)

The Moon also moves through the debris that creates meteor showers on Earth. Whereas our atmosphere vaporizes most of the meteors before they strike Earth, the Moon's thin atmosphere provides no protection, so the Moon is struck by numerous meteoroids during each shower. The Leonid meteor shower of 1999 provided astronomers with an excellent opportunity to observe contemporary impacts on the Moon. Some of these impacts were even bright enough to see with the naked eye (Figure 18-39). When a 10-kg meteorite strikes the Moon, the impact is as powerful as 10^4 pounds of TNT, and the resulting cloud is momentarily between 5×10^4 and 10×10^4 K, which is hotter than the surface of the Sun. Subsequent Leonids have provided further impact sightings on the Moon.

> Why are the times when meteor showers occur so predictable?

18-11 Meteorites are space debris that land intact

Although most meteors that enter Earth's atmosphere completely vaporize or are destroyed on impact, some land without totally disintegrating. Such pieces of debris are called meteorites, and they come from a variety of sources. Many are thought to have been broken off from asteroids by collisions; some are debris that were never part of larger bodies; still others come from the Moon, Mars, Vesta, or comets. We saw in Section 16-12,

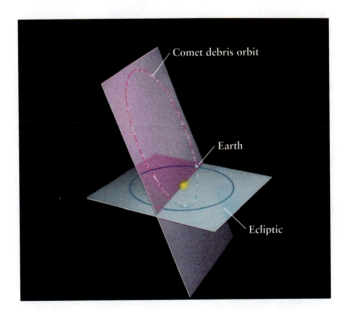

FIGURE 18-37 The Origin of Meteor Showers As comets dissipate, (see Figure 18-31) they leave debris behind that spreads out along their orbits. When Earth plows through such material, many meteors can be seen emanating from the same place within a very short time—a meteor shower. As shown in this diagram, many comets have high orbital inclinations.

PROMINENT YEARLY METEOR SHOWERS

Shower	Date of maximum intensity	Typical hourly rate	Constellation
Quadrantids	January 3	40	Boötes
Lyrids	April 22	15	Lyra
Eta Aquarids	May 4	20	Aquarius
Delta Aquarids	July 30	20	Aquarius
Perseids	August 12	80	Perseus
Orionids	October 21	20	Orion
Taurids	November 4	15	Taurus
Leonids	November 16	15	Leo Major
Geminids	December 13	50	Gemini
Ursids	December 22	15	Ursa Minor

FIGURE 18-38 Meteor Shower The time exposure, taken in 1998, shows meteors streaking away from the constellation Leo Major. They are part of the Leonid meteor shower. This shower occurs because Earth is moving through debris left by comet Temple-Tuttle. The table lists highly active, annual meteor showers, which last for several days. (Tony and Daphne Hallas/Photo Researchers, Inc.)

for example, how SNC meteorites offer clues to the chemistry of Mars. By studying the chemistry and age of one meteorite from the Moon that was found on Earth, geologists have determined that it was ejected into space when the mare Imbrium impact basin formed. Meteorites from the asteroid Vesta show signs of having been molten rock on that world. They were ejected when an impact occurred there, creating a 460

FIGURE 18-39 Recent Impacts on the Moon The locations *A–F* are places on the Moon where impacts were observed from Earth in 1999 during the Leonid meteor shower. The impacting bodies hit the Moon at around 260,000 km/h (160,000 mph) and had masses of between 1 and 10 kg. Each impact created a short-lived cloud that momentarily heated to between 5×10^4 and 10×10^4 K, much hotter than the surface of the Sun. (NASA)

km (280 mi) wide, 13 km (8 mi) deep crater, which has been seen by the Hubble Space Telescope.

Meteorites tell us the age of the solar system. Measuring the relative amounts of various radioactive elements in them allows astronomers to determine how long ago they formed (see *An Astronomer's Toolbox 4-2*). The oldest known meteorites became solid bodies about 4.57 billion years ago, and impacts on Earth as early as 4 billion years ago have been identified, based on how they disturbed preexisting rock here. Therefore, the solar system is at least 4.57 billion years old, and we will use 4.6 billion years as its age throughout the book.

People have been picking up debris from space for thousands of years. The descriptions of meteorites in historical Chinese, Indian, Islamic, Greek, and Roman literature show that early peoples placed special significance on these "rocks from heaven." They also have a practical "impact": Infalling space debris is increasing Earth's mass by nearly 300 tons per day on average (Figure 18-40).

Stony Meteorites Often Look like Ordinary Rocks

Meteorites are classified as stones (or stony), stony-irons, and irons. Most **stony meteorites** look much like ordinary rocks, although some are covered with a dark *fusion*

RIVUXG

FIGURE 18-40 The Mass of Impacts on Earth The Vatican Obelisk is about 300 tons, the amount of mass that strikes Earth daily. As a result, Earth's mass increases by this amount every day. (John and Dallas Heaton/Corbis)

are more likely to be noticed. Consequently, the easily found iron and stony-iron meteorites dominate most museum collections.

Iron Meteorites Are Very Dense and Look Noticeably Different from Rocks Iron meteorites (Figure 18-42a) may also contain from 10 to 20% nickel by weight. Iron is moderately abundant in the universe as well as being one of the most common rock-forming elements, so it is not surprising that iron is an important constituent of asteroids and meteoroids. Another element, iridium, is common in the iron-rich minerals of meteorites but rare in ordinary rocks on Earth's surface because most of the iridium that collected here when Earth formed settled deep into Earth eons ago. Measurements of iridium in Earth's crust can thus tell us the rate at which meteoritic material has been deposited over the ages.

In 1808, Count Alois von Widmanstätten, director of the Imperial Porcelain Works, in Vienna, discovered a conclusive test for the most common type of iron meteorite. Most iron meteorites have a unique structure of long nickel-iron crystals, called **Widmanstätten patterns**, which become visible when the meteorites are cut, polished, and briefly dipped into a dilute solution of acid (Figure 18-42b). Because nickel-iron crystals can grow to lengths of several centimeters only if the molten metal cools slowly over many millions of years, Widmanstätten patterns are never found in counterfeit meteorites, or "meteorwrongs."

Stony-Iron Meteorites Are the Most Exotic of All Space Debris on Earth The final category of meteorites are the **stony-iron meteorites**, which consist of roughly equal

crust (Figure 18-41a), created when the meteorite's outer layer melts during its fiery descent through the atmosphere (see Figure 18-38). When a stony meteorite is cut in two and polished, tiny flecks of iron are sometimes found in the rock (Figure 18-41b).

Meteorites with a high iron content can be located with a metal detector. They also look unusual and hence

a   RIVUXG

b   RIVUXG

FIGURE 18-41 Stony Meteorites (a) Most meteorites that fall to Earth are stones. Many freshly discovered specimens, like the one shown here, are coated with thin, dark crusts. This stony meteorite fell in Morocco. (b) Some stony meteorites con-

tain tiny specks of iron, which can be seen when the stones are cut and polished. This specimen was discovered in Ohio. (The R. N. Hartman Collection)

FIGURE 18-42 Iron Meteorites (a) Irons are composed almost entirely of iron-nickel minerals. The surface of a typical iron is covered with thumbprint-like depressions created as the meteorite's outer layers vaporized during its high-speed descent through the atmosphere. This specimen was found in Argentina. (b) When cut, polished, and etched with a weak acid solution, most iron meteorites exhibit interlocking crystals in designs, called *Widmanstätten patterns*. This meteorite was found in Australia. (a: The R. N. Hartman Collection; b: R. A. Oriti)

amounts of rock and iron. Figure 18-43, for example, shows the greenish mineral olivine suspended in a matrix of iron.

To understand why different types of meteorites exist, we consider their formation. Many meteorites were once pieces of asteroids. Heat impacts and from the rapid decay of radioactive isotopes melted newly formed asteroid interiors. Over the next few million years, differentiation occurred, just as in young Earth. Iron sank toward the asteroid's center, while lighter rock floated up to its surface. Iron meteorites are fragments of asteroid cores, and stony-irons come from the boundary regions between the iron cores and stony crusts.

Stony meteorites have a variety of origins. Some are from the outer layers of asteroids. Other stony meteorites, ordinary **chondrites**, show no evidence of ever having been melted as parts of asteroids. They may therefore be primordial material from which our solar system was created. **Carbonaceous chondrites** are rare chondrites that contain small glass-rich beads called *chondrules*. These meteorites also contain complex carbon compounds, including simple sugars and glycerin. They also have as much as 20% water bound into their minerals. The organic compounds would have been broken down and the water driven out if these meteorites had been significantly heated. Undifferentiated asteroid Mathilde, shown in Figure 18-16a, has a very dark gray color and virtually the same spectrum as a carbonaceous chondrite meteorite, so it is likely composed of primordial material.

Amino acids, the building blocks of proteins upon which terrestrial life is based, are among the organic compounds occasionally found inside carbonaceous chondrites. These amino acids may be contaminants acquired after the meteoroids entered Earth's atmosphere. However, space scientists have actually created them in the laboratory under the conditions found in deep space, showing that amino acids may exist out there. Indeed, some scientists suspect that amino-acid–rich carbonaceous chondrites may have played a role in the origin of life on Earth.

Not All Meteorite Impacts ("Falls") Lead to Meteorite "Finds" Stony meteorites account for about 95% of all meteoritic material that fall on Earth. Most stony

FIGURE 18-43 Stony-Iron Meteorite Stony-irons account for about 1% of all meteorites that fall to Earth. This specimen, a variety called a pallasite, was found in Antarctica. It has been thinly cut and appears to glow because of a light located behind it. (James C. Hartman)

meterorites are not identified as meteorites because, well, they look like stones. Indeed, the percentages of the stony, iron, and stony-iron meteorites that are discovered, called *finds*, are quite different from the percentages that actually land, called *falls*. Nevertheless, astronomers and geologists have a good idea about how many of each type strike land. They get the correct percentages of impacts by carefully surveying areas in which only meteorites land, namely snow and ice-covered regions, such as Antarctica, or on deserts. By counting all of the debris under the surface using metal detectors and other technologies, the actual number of impacts of each type of meteorite is determined.

a

R I V U X G

b

R I V U X G

FIGURE 18-44 Pieces of the Allende Meteorite (a) This carbonaceous chondrite fell near Chihuahua, Mexico, in February 1969. Note the meteorite's dark color, caused by a high abundance of carbon. Geologists believe that this meteorite is a specimen of primitive planetary material. The ruler is 15 cm long. (b) Sliced open, the Allende meteorite shows round, rocky *inclusions* called *chondrules* in a matrix of dark rock. a: J. A. Wood; b: The R. N. Hartman Collection)

18-12 The Allende meteorite and Tunguska mystery provide evidence of catastrophic explosions

A chance to study debris immediately after impact came shortly after midnight on February 8, 1969, when a brilliant, blue-white light shot across the night sky over Chihuahua, Mexico. Hundreds of people witnessed the dazzling display. The light disappeared in a spectacular and deafening explosion that dropped thousands of rocks and pebbles over the terrified onlookers. Within hours, teams of scientists were on their way to collect specimens of carbonaceous chondrites, collectively named the *Allende meteorite*, after the locality in which they fell (Figure 18-44).

One of the most significant discoveries to come from the Allende meteorite was evidence of the detonation of a nearby supernova 4.6 billion years ago. Among nature's most violent and spectacular phenomena, a *supernova explosion* occurs when a massive star blows apart in a cataclysm that hurls matter outward at tremendous speeds, as we explored in Chapter 8. During this detonation, violent collisions between atomic nuclei produce a host of radioactive elements, including a short-lived radioactive isotope of aluminum. Based on its decay products, scientists found unmistakable evidence that this isotope once lay within the Allende meteorite. Some astronomers interpret this as evidence for a supernova in our vicinity at about the time the Sun was born. By compressing interstellar gas and dust, the supernova's shock wave may have helped stimulate the birth of our solar system.

Sixty years before the meteoritic event in Chihuahua, Mexico, at 7:14 A.M. local time on June 30, 1908, a much more spectacular explosion occurred, this one over the Tunguska region of Siberia. The blast, comparable to a nuclear detonation of several megatons, knocked a man off his porch some 60 km away and was audible more than 1000 km away. Millions of tons of dust were injected into the atmosphere, darkening the air as far away as California.

Preoccupied with wars, along with political and economic upheaval, neither Russia nor its successor, the former Soviet Union, sent a scientific expedition to the site until 1927. At that time, Soviet researchers found that trees had been seared and

Finding a meteorite in Antarctica (NASA Johnson Space Center)

Why are stony-iron meteorites so rare compared to stony or iron meteorites?

Examining growth rings in a tree exposed to the Tunguska event (Tunguska Page of Bologna University: http://www-th.bo.infn.it/tunguska/)

What evidence do astronomers have from the Allende meteorite that a supernova explosion occurred at about the time the solar system formed?

felled radially outward in an area about 30 km in diameter (Figure 18-45). There was no clear evidence of a crater. In fact, the trees at "ground zero" were left standing upright, although they were completely stripped of branches and leaves. Because no significant meteorite samples were found, for many years scientists assumed that a small comet had exploded in the atmosphere.

Recently, however, several teams of astronomers have maintained that a small comet, composed of primarily light elements and ice, breaks up too high above the ground to cause significant damage. They argue that the Tunguska explosion was actually caused by a small asteroid or large meteoroid traveling at supersonic speed. The Tunguska event is consistent with an explosion of an asteroid about 80 m (260 ft) in diameter entering Earth's atmosphere at 79,000 km/h (50,000 mph) and exploding in the air as a result of becoming exceedingly hot. The resulting shock wave, slamming into the ground, would cause the damage seen without creating a crater. The debate on the origin of the Tunguska event continues.

18-13 Asteroid impacts with Earth have caused mass extinctions

Luis *(left)* and Walter Alvarez examining the iridium layer. (LBNL, Image Library)

In the late 1970s, the geologist Walter Alvarez and his father, physicist Luis Alvarez, discovered the remnants of a far more powerful impact. Working at a site of exposed marine limestone in the Apennine Mountains in Italy that had been on Earth's surface 65 million years ago, the Alvarez team discovered an exceptionally high abundance of iridium in a dark-colored layer of clay between limestone strata (Figure 18-46).

Since 1979, when this discovery was announced, a comparable layer of iridium-rich material has been uncovered at numerous sites around the world. In every case, geologic dating reveals that this apparently worldwide iridium-rich layer was deposited about 65 million years ago. Paleontologists were quick to realize the significance of this date, because it was back then that all of the dinosaurs rather suddenly became extinct. In fact, two-thirds of the species on Earth disappeared within a brief span of time 65 million years ago.

The Alvarez discovery supported an astronomical explanation for the dramatic extinction of so much of the life that once inhabited our planet—an asteroid impact. There is no universal agreement on the disasters that befell Earth

during this episode. One scenario has an asteroid 10 km in diameter slamming into Earth at high speed and throwing enough dust into the atmosphere to block out sunlight for several years. As the temperature dropped drastically and plants died for lack of sun-

R I V U X G

FIGURE 18-45 *Aftermath of the Tunguska Event* In 1908, a stony asteroid traveling at supersonic speed struck Earth's atmosphere and exploded over the Tunguska region of Siberia. Trees were blown down for many kilometers in all directions from the impact site. (SOVFOTO)

R I V U X G

FIGURE 18-46 *Iridium-Rich Layer of Clay* This photograph of strata in the Apennine Mountains of Italy shows a dark-colored layer of iridium-rich clay sandwiched between white limestone *(bottom)* from the late Mesozoic era and grayish limestone *(top)* from the early Cenozoic era. The coin is the size of a U.S. quarter. (W. Alvarez) shine, the dinosaurs would have perished, along with many other creatures in the food chain that were highly dependent on vegetation. The dust eventually settled, depositing an iridium-rich layer over Earth. Another theory posits that the impact created enough heat to cause planetwide fires, followed by changes in the oceans that killed many species of life in them. Whatever scenario turns out to be correct, small creatures capable of ferreting out seeds and nuts were among the animals that managed to survive this holocaust, setting the stage for the rise of mammals and, consequently, the evolution of humans.

In 1992, a team of geologists discovered that the hypothesized asteroid crashed into a site on the eastern edge of Mexico. They based this conclusion on glassy debris and violently shocked grains of rock ejected from the multiringed, 195-km-diameter Chicxulub (pronounced Chih-Chuh-Lube) Crater buried under the Yucatán Peninsula and western Caribbean Sea in Mexico (Figure 18-47). From the known rate at which radioactive potassium decays, the scientists have pinpointed the date when the asteroid struck—64.98 million years ago. In 1998, geologists digging on the Pacific Ocean floor discovered a piece of meteoritic debris with precisely the same age, apparently a piece of the offending asteroid. While some geologists and paleontologists are not yet convinced that an asteroid impact led to the extinction of the dinosaurs, most agree that this hypothesis better fits the available evidence than any other explanation that has been offered so far.

Another Impact Led to an Earlier, More Devastating Mass Extinction   The end of the dinosaurs' reign was not the only mass extinction caused by an impact. In 2001, evidence came to light concerning the Permian-Triassic boundary of 250 million years ago that suggests another devastating blow from space. That time was called the "Great Dying," a mass extinction during which some 80% of the species of life living on land and 90% of those living in the oceans perished. Rocks from that time discovered in places from Japan to Hungary show evidence, in the form of *fullerenes*—soccer ball-shaped molecules containing at least 60 carbon atoms—of an impact from space. Trapped inside these fullerenes were gases that could only have been forced into them by exploding stars. (This signature of gas-filled fullerenes from space has now also been discovered in the layer of rock that existed on Earth's surface 65 million years ago.)

A 250-million-year-old crater that may have been the relevant impact site was found in 2004 off the northwestern coast of modern-day Australia. Keep in mind that both this impact and the one that ended the

reign of the dinosaurs occurred on a world whose surface is in continual motion. Therefore, their present locations are not the same as their initial impact sites. The Chicxulub impact actually occurred in the Pacific Ocean, whereas the Australian impact occurred before

R I V U X G

FIGURE 18-47 *Confirming an Extinction-Level Impact Site* By measuring slight variations in the gravitational attraction of different materials under Earth's surface, geologists create images of underground features. Concentric rings of the underground Chicxulub Crater *(right inset)* lie under a portion of the Yucatán Peninsula. This crater has been dated to 65 million years ago and is believed to be the site of the impact that led to the extinction of the dinosaurs. A piece of 65-million-year-old meteorite discovered in the middle of the Pacific Ocean in 1998 is believed to be a fragment of that meteorite. The fragment, about 0.3 cm (0.1 in) in. long, was cut into two pieces for study *(left inset)*. (Virgil L. Sharpton, Lunar and Planetary Institute; digital image by Peter W. Sloss, NOAA-NESDIS-NGDCD [right inset]; Frank T. Kyte, UCLA [left inset]) the continents were separated from Pangaea (see Section 15-2). The impact explanation of the Great Dying is not universally accepted. Another possible explanation is a massive amount of volcanic eruption that darkened the skies and changed the climate.

Could a catastrophic impact happen again? So many asteroids cross Earth's path that scientists agree that it is a matter of "when" rather than "if." The good news is that studies of craters show that larger asteroids strike Earth significantly less often than do smaller ones. Although asteroids large enough to create Meteor Crater strike Earth about once every 10,000 years, killer asteroids, like the one that killed off the dinosaurs, collide with Earth only once every 100 million years. The threat of a catastrophic impact by an asteroid or comet in our lifetimes, thankfully, is remote.

> How did the Chicxulub crater move from the Pacific to the Caribbean?

18-14 Frontiers yet to be discovered

The years of the *Pioneer* and *Voyager* spacecraft were the first golden era of solar system research. We are now in another such period, with spacecraft either in development or already en route to planets, asteroids, comets, and even, as we will see shortly, out observing the weather in space. From studies of solar system debris, astronomers hope to learn whether life on Earth was brought here from elsewhere by asteroid or meteoroid impacts. They also hope to answer such questions as the evolutionary history of the space debris; how much of Earth's water came here during the planet's formation and how much landed afterward from comet impacts; whether the asteroids have sufficiently valuable compositions to justify mining them; whether comets can be harvested to supply water and other materials for people colonizing the solar system; and whether the Oort comet cloud really exists.

SUMMARY OF KEY IDEAS

- Astronomical objects smaller than the eight planets are classified as dwarf planets or small solar-system bodies (SSSBs).

- A variety of other names, including asteroids, comets, meteoroids, trans-Neptunian objects, Kuiper belt objects (KBOs), and Oort comet cloud objects, overlap with "dwarf planet" and "SSSB."

- KBOs and Oort comet cloud objects are trans-Neptunian objects—they orbit farther from the Sun than the outermost planet.

- To date, five objects—Pluto, Ceres, Eris, Makemake, and Haumea—have been classified as dwarf planets.

- Pluto, a KBO and dwarf planet, is an icy world that may well resemble the moon Triton.

- Other objects orbit the Sun beyond Neptune. For example, at least 1471 KBO's have been observed. A few potential Oort comet cloud objects have also been identified.

Asteroids

- Tens of thousands of belt asteroids with diameters larger than a kilometer are known to orbit the Sun between the orbits of Mars and Jupiter. The gravitational attraction of Jupiter depletes certain orbits within the asteroid belt. The resulting Kirkwood gaps occur at simple fractions of Jupiter's orbital period.

- Jupiter's and the Sun's gravity combine to capture Trojan asteroids in two locations, called stable Lagrange points, along Jupiter's orbit.

- The Apollo asteroids move in highly elliptical orbits that cross the orbit of Earth. Many of these asteroids will eventually strike the inner planets.

Comets

- Many comet nuclei orbit the Sun in the Kuiper belt, a doughnut-shaped region beyond Pluto. Billions of cometary nuclei are also believed to exist in the spherical Oort comet cloud located far beyond Pluto.

- Comet nuclei are fragments of ice and rock often orbiting at a great inclination to the plane of the ecliptic. In the Kuiper belt and Oort comet cloud they have fairly circular orbits. When close to the Sun, they generally move in highly elliptical orbits.

- As an icy comet nucleus approaches the Sun, it develops a luminous coma surrounded by a vast hydrogen envelope. A gas (or ion) tail and a dust tail extend from the comet, pushed away from the Sun by the solar wind and radiation pressure.

Meteoroids, Meteors, and Meteorites

- Boulder-sized and smaller pieces of rock and metal in space are called meteoroids. When a meteoroid enters Earth's atmosphere, it produces a fiery trail, and it is then called a meteor. If part of the object survives the fall, the fragment that reaches Earth's surface is called a meteorite.

- Meteorites are grouped in three major classes according to their composition: iron, stony-iron, and stony meteorites. Rare stony meteorites, called carbonaceous chondrites, may be relatively unmodified material from the primitive solar nebula. These meteorites often contain organic hydrocarbon compounds, including amino acids.

- Fragments of rock from "burned-out" comets produce meteor showers.

• An analysis of the Allende meteorite suggests that a nearby supernova explosion may have been involved in the formation of the solar system some 4.6 billion years ago.

• An asteroid that struck Earth 65 million years ago probably contributed to the extinction of the dinosaurs and many other species. Another impact caused the "Great Dying" of life 250 million years ago. Such devastating impacts occur on average every 100 million years.

WHAT DID YOU THINK?

1 *Are the asteroids a former planet that was somehow destroyed? Why or why not?* No. The gravitational pull from Jupiter prevented a planet from ever forming in the asteroid belt. Also, the total mass of the asteroids is much less than even the mass of tiny Pluto, a dwarf planet.

2 *How far apart are the asteroids on average?* The distance between asteroids averages 10 million km.

3 *How are comet tails formed? Of what are they made?* Ices in comet nuclei are turned into gas by absorbing energy from the Sun. Debris is released in this process. Sunlight and the solar wind push on the gas and dust, creating the tails.

4 *In which directions do a comet's tails point?* Comets' gas tails point directly away from the Sun; their dust tails make arcs pointing away from the Sun.

5 *What is a shooting star?* A shooting star is a piece of space debris plunging through Earth's atmosphere—a meteor. It is not a star.

Key Terms for Review

amino acid, 514
Apollo asteroid, 498
asteroid belt, 495
belt asteroid, 495
carbonaceous chondrite, 514
chondrites, 514
coma (of a comet), 502
comet, 500
dust tail (of a comet), 504
dwarf planet, 491
gas (ion) tail, 504
hydrogen envelope, 502

impact crater, 510
iron meteorite, 513
Kirkwood gaps, 496
Kuiper belt, 500
long-period comet, 507
meteor, 509
meteor shower, 510
meteorite, 509
meteoroid, 509
nucleus (of a comet), 502
Oort comet cloud, 501
planet, 491
radiation (photon) pressure, 504

short-period comet, 507
Small Solar-System Bodies (SSSBs), 491
stable Lagrange points, 497

stony meteorite, 512
stony-iron meteorite, 513
Trojan asteroid, 497
Widmanstätten patterns, 513

Review Questions

1. A piece of space debris that you pick up from the ground is called a(n): **a.** asteroid, **b.** meteoroid, **c.** meteor, **d.** meteorite, **e.** comet.

2. Space debris that is a roughly equal mix of rock and ice is called a(n): **a.** asteroid, **b.** comet, **c.** meteoroid, **d.** meteorite, **e.** meteor.

3. Which is the rarest type of meteorite in space? **a.** iron, **b.** stony-iron, **c.** stony meteorite

4. Which part of a comet is solid? **a.** nucleus, **b.** halo, **c.** gas tail, **d.** dust tail, **e.** coma

5. Suppose you were standing on Pluto. Describe the motions of Charon relative to the horizon. Under what circumstances would you never see Charon?

6. Describe the circumstantial evidence that supports the idea that Pluto is one of a thousand similar icy worlds that once occupied the outer regions of the solar system.

7. What role did Charon play in enabling astronomers to determine Pluto's mass?

8. Why are asteroids, meteoroids, and comets of special interest to astronomers who want to understand the early history of the solar system?

9. Describe the objects in the asteroid belt, including their sizes, orbits, and separation.

 10. To test your understanding of the asteroid belt, do Interactive Exercise 18-1 on the Web. You can print out your results, if required.

11. Why are there many small asteroids but only a few very large ones?

12. Describe the different chemistries of the two tails of a comet.

13. In what directions do comet tails point, and why?

14. What are the Kirkwood gaps, and what causes them?

15. What are the Trojan asteroids, and where are they located?

16. Describe the three main classifications of meteorites. How do astronomers believe that these different types of meteorites originated?

17. Why do astronomers believe that the debris that creates many isolated meteors comes from asteroids, whereas the debris that creates meteor showers is related to comets?

18. To test your understanding of comets, do Interactive Exercise 18-2 on the Web. You can print out your results, if required.

19. Why is the phrase "dirty snowball" an appropriate characterization of a comet's nucleus?

20. What and where is the Kuiper belt, and how is it related to debris left over from the formation of the solar system?

21. Why do scientists think the Tunguska event was caused by a large meteoroid and not a comet?

22. What evidence in Figure 18-33 supports the labeling of the gas and dust tails?

Advanced Questions

The answers to computational problems, which are preceded by an asterisk (*), appear at the end of the book.

23. Would you expect the surfaces of Pluto and Charon to be heavily cratered? Explain.

24. How did the regolith on asteroid Eros form?

25. Why are comets generally brighter after passing perihelion (closest approach to the Sun) than before reaching perihelion?

26. Can you think of another place in the solar system where a phenomenon similar to the Kirkwood gaps in the asteroid belt is likely to exist? Explain.

27. Where on Earth might you find large numbers of stony meteorites that have not been significantly changed by weathering?

*28. Assuming a constant rate of meteor infall, how much mass has Earth gained in the past 4.6 billion years?

Discussion Questions

29. Suppose it was discovered that the asteroid Hermes had been perturbed in such a way as to put it on a collision course with Earth. Describe what you would do to counter such a catastrophe using present technology.

30. From the abundance of craters on the Moon and Mercury, we know that numerous asteroids and meteoroids struck the inner planets during the very early history of the solar system. Is it reasonable to suppose that numerous comets also pelted the planets 4.0 to 4.5 billion years ago? What effects would such a cometary bombardment have had, especially with regard to the evolution of the primordial atmospheres of the terrestrial planets and oceans on Earth?

What If . . .

31. An ocean on Earth were struck by a comet nucleus a kilometer across? What physical effects would occur to Earth?

32. We passed through the tail of a comet? What would happen to Earth and life on it?

33 As some astronomers have recently argued, passage of the solar system through an interstellar cloud of gas could perturb the Oort cloud, causing many comets to deviate slightly from their original orbits? What might be the consequences for Earth?

34. All of the space debris (asteroids, meteoroids, and comets) had been cleared out of the solar system 3.8 billion years ago? What would have been different in the history of Earth and in the history of life on Earth?

Web Questions

35. The discovery of Charon (see Figure 18-5) was made by an astronomer at the U.S. Naval Observatory. Search the Web to find out why the U.S. Navy carries out work in astronomy.

36. Search the Web to find out why some scientists disagree with the idea that a tremendous impact led to the demise of the dinosaurs. (They do not dispute that the impact occurred, only what its consequences were.) What are their arguments? From what you have learned, what is your opinion?

37. Several scientific research programs are dedicated to the search for near-Earth objects (NEOs), especially those that might someday strike our planet. Search the Web for information about at least one of these programs. How does the program search for NEOs? How many NEOs are now known, and how many has this program found? Will any of these NEOs pose a threat in your lifetime?

38 . Search the Web to learn whether there are any comets visible at present. List them. What constellations are they presently in? Are any visible to the naked eye? (Recall that the unaided human eye sees objects brighter than about sixth magnitude.)

Observing Projects

39. Use the *Starry Night Enthusiast*™ program to observe Pluto and Charon. First go to Atlas mode (**Favourites/Guides/Atlas**), **Stop Time,** and find Charon (**Find** tab/Charon). Zoom in so that Charon and Pluto are separated by a few cm or in Set the timestep to 3 hours. Use the single-step time button to step through enough time to determine the period of Charon's orbit. How does your answer compare with Pluto's sidereal rotation period given in Appendix E-2? Explain.

40. Pluto, the Sun, and Asterisms In this exercise, we will use *Starry Night Enthusiast*™ to locate the Sun as seen from Pluto. First, turn off labels for Sun and planets by selecting **Options > Solar System** and unclick **labels** to the right of **planets-**

moons. Now select **Favourites > Guides > Atlas**, opening the **Options** pane and clicking on **Constellations > Boundaries** and **Constellations > Labels**. Next, open the **Find** pane and type "Pluto" in the search box. Click the menu button at the extreme left of the listing for Pluto and select **Go There**. This will place you about 5000 km above the dwarf planet's surface. Click once on the Up ▲ Elevation button to display this altitude in the Viewing Location. You can use the **Location Scroller** (hold down the **Shift** key as well as the mouse button while moving the mouse) to move over the surface of a sphere surrounding Pluto at this elevation. Do this until an object that you believe is the Sun lies just to one side of Pluto. (*Hint*: Think about what Pluto's phase will be when you are in this position, and move the Location Scroller accordingly.) Place the cursor over that object and use the Heads-up display to see if you are correct. If not, move around with the **Location Scroller** until you find the Sun. With the Sun next to Pluto, make a simple drawing of Pluto, the Sun, and a few constellations showing the Sun's location in Pluto's sky. Is the Sun in the same constellation as seen from Earth? Explain. Next, locate the constellation of Orion. Does it look the same from Pluto as it does from Earth? Explain why or why not.

41. Make arrangements to view an asteroid. At opposition, some of the largest asteroids are bright enough to be seen through a modest telescope. You can use *Starry Night Enthusiast™* to see which bright asteroids are visible Update asteroids on-line by opening the **LiveSky** menu at the top of the program and clicking on **Update Comets/Asteroids/Satellites**. Close and restart the program. Click the **Home** button in the toolbar followed by the **Stop** button to stop time flow. Set the time in the toolbar to your expected observing time. Open the **Find** pane and clear any data in the search box to bring up a list of solar system objects. Click the "+" button to the left of the **Asteroids** entry to expand the list of asteroids. Asteroids that are visible from your location at the time of your planned observation will appear in black typeface while those that are not visible will appear in a light gray font. Double-click on one of the asteroids that will be visible at the time of your planned observation. Zoom in on the asteroid progressively and print a chart of its position in the sky at various zoom factors. To print a chart, move mouse to asteroid (right-click on a PC or Ctrl-click on a Mac), and select **Print Chart. . .** In the **Print Settings** dialog, select the **3 Pane** layout, choose 3 different fields of view, such as 90°, 50°, and 5°, and select **Use current settings** options. As an alternative to using *Starry Night Enthusiast™*, you can check the "Minor Planets" section of the current issue of the *Astronomical Almanac* to see if any bright asteroids are near to opposition, or check the current issue, as well as the most recent January issue, of *Sky & Telescope* for star charts showing the paths of bright asteroids among the constellations. You will need an appropriate sky chart to distinguish the asteroid from background stars.

Observe the asteroid on at least two occasions separated by a few days. On each night, draw a star chart of the objects in your telescope's field of view. Has the position of one star-like object shifted between observing sessions? Does the position of the moving object agree with the path plotted on your star charts? Do you feel confident that you have in fact observed an asteroid? Explain.

42. Make arrangements to view a comet through a telescope. Because astronomers discover roughly a dozen comets each year, a comet is usually visible somewhere in the sky. Unfortunately, because most comets are quite dim, you will need access to a moderately large telescope. You can use *Starry Night Enthusiast™* to see which comets are visible. First, update comets on-line by opening the **LiveSky** menu at the top of the program and clicking on **Update Comets/Asteroids/Satellites**. Close and restart the program. Set the time in the toolbar to your expected observing time, **Stop Time** flow, and then find a comet that is visible by opening the **Find** pane, clearing the search box and expanding the **Comets** list. The names of those comets visible at your selected observing time will be in black font. Find its apparent magnitude by clicking the menu button at the extreme left of the listing for the comet and select **Show Info**. Find the comet's apparent magnitude under the **Other Data** layer of the **Info** pane. A comet fainter than about 12th magnitude will not be easily visible. If the brightness of the comet you have chosen is too low, try other comets. Remember that the larger the apparent magnitude value, the lower the brightness. Double-click on one and zoom in on it to see the direction of its tail. You can print finder charts as described in the previous Observing Project on asteroids. Alternatively, consult the Web or recent issues of the *IAU Circular*, published by the International Astronomical Union's Central Bureau for Astronomical Telegrams, which contains predicted positions and the anticipated brightness of comets in the sky. Also, if there is an especially bright comet in the sky, the latest issue of *Sky & Telescope* might contain useful information. Observe the comet through a telescope. Can you distinguish it from background stars? Can you see its coma? How many tails do you see? (Note that a telescope with a low magnification or a pair of binoculars is more suitable for viewing a bright, nearby comet with an extended tail, because this comet is a diffuse source of light.)

43. Make arrangements to view a meteor shower. Figure 18-38 lists the dates of maximum intensity of the major showers. Often, a good display can be seen a day or two before or after the maximum date. Ideally, you need a clear, moonless sky. The Moon's glow in the sky will hide many of the faint meteor trails.

44. Use *Starry Night Enthusiast™* software to observe several comets. **a.** Change the date in the toolbar to March 22, 1997 and click the **Stop** button to stop time flow. Change the viewing location to either the North or South Pole by selecting **Options > Viewing Location. . .** from the menu. This is the date of the equinox and the Sun is somewhere on the horizon. Open the **Find** pane and expand the **Comets** list. Locate Comet Hale-Bopp in the expanded list and double-click on it. Zoom in (if necessary) until you can see the comet and its tail. Predict in what direction the Sun is located relative to the comet and explain how you made your prediction. To verify your hypothesis, use **Find > Sun** to locate it. Were you correct? **b.** Using the Comet List again, double-click on the name of another visible comet to center on it, but do not zoom in. Then move the sky with the hand cursor and locate the Sun. From your observations, predict the direction of the comet's tail on the sky, and explain how you made your prediction. Center again on the comet and zoom in until you can see its tail. (If this comet does not have a visible tail, repeat this part with yet another comet.) Was your prediction correct?

 45. Halley's Comet In this exercise we will use *Starry Night Enthusiast*™ to locate Halley's comet Select **Favourites > Solar System > Comets and Asteroids > Comet Halley's Orbit** from the menu. Click the **Stop** button to stop time flow. Explain the reason for the direction in which the comet's tail is pointed. From the direction of the tail, can you predict the direction of Halley's motion when this image was taken? If so, give its direction and explain how you deduced it. If not, explain why not.

Click the **Play** button to start time flow and watch the orbital motion of Halley's comet. Does the comet orbit the Sun in the same or in the opposite sense to the planets visible in the view? Right-click over the Sun, click on **Centre** and then use the **Location Scroller** (hold down the **Shift** key as well as the mouse button while moving the mouse) move around the Sun until Earth's orbital plane is flat on the screen. Run time back and forward to watch Halley's comet as it crosses this plane in its close approach to the Sun, noting particularly the direction of its tail.

 46. Dwarf Planet Ceres In this exercise we will use the *Starry Night Enthusiast*™ program to locate locate the largest asteroid, Ceres, now designated a dwarf planet. Select **Favourites > Solar System > Comets and Asteroids > Ceres and the Main Belt** from the menu and click the **Stop** button to stop time flow. Ceres will not yet be labeled. Write down the general region of the solar system in which you expect to find it. Now turn on the label for Ceres by opening the **Find** pane, clearing the search box and expanding the **Dwarf Planets** listing. Click the box to the left of the entry for Ceres so it is checked. This will label Ceres in the view. In the **Find** pane, click the check box to the right of the label for Ceres in the list to display this object's orbit. Use the Location Scroller to move the screen around. Does Ceres move in the plane of the ecliptic?

CHAPTER 19

Astrobiology

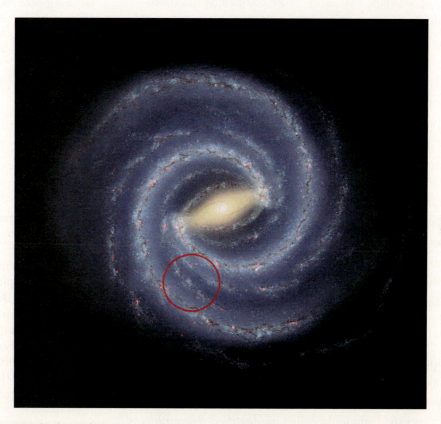

Centered on Earth, the red oval shows the region in which astronomers are searching for extraterrestrial intelligence. (NASA/Space Telescope Science Institute)

WHAT DO YOU THINK?

1 Why is water so important to the formation of life?

2 What element is uniquely suited to be the foundation of life as we know it, and why?

3 How do astronomers search for extraterrestrial intelligence?

4 Have astronomers located any extraterrestrial civilizations?

5 If advanced alien civilizations exist, is there any way they might know of our existence?

Answers to these questions appear in the text beside the corresponding numbers in the margins and at the end of the chapter.

If, as we said in the first chapter of this book, the starry night sky inspires us to look beyond ourselves, the study of astronomy often motivates us to contemplate the formation of Earth, the nature of the stars, and even the creation of the universe. Perhaps two of the most compelling questions in this realm of nature are how life began here on Earth and whether there is other life in the universe.

Scientists are searching for the origins of life on Earth by doing experiments in hopes of recreating steps in that process. A key part of these experiments involves trying to determine the physical environment of Earth and the materials available at the time when life began.

Understanding the origins of life may be helped forward by discovering primitive life elsewhere in our solar system, such as inside Mars, Europa, Ganymede, and, possibly, other moons of the solar system. Astronomers have begun looking for it, developing life-seeking space missions to other worlds, including two that went to Mars in the 1970s (Figure 19-1; also see Section 16-13).

The search for advanced extraterrestrial life has gone in several directions, starting right here on Earth. Conceivably, advanced civilizations from elsewhere have sent spacecraft to visit. In fact, some people believe that they have seen extraterrestrials or, at least, their spacecraft, commonly called *unidentified flying objects* (UFOs). This leads to the question of whether life on Earth was "seeded" intentionally or accidentally by alien visitors. If Earth was visited, that event is possible, but Occam's razor requires us to explore simpler avenues for the beginning of life here.

Are there other advanced civilizations elsewhere in the cosmos? Some people find comfort in the belief that Earth is the only home for life. Others look at the vastness of the universe and the size and diversity of the objects in the cosmos and thereby find reason to expect there may be other worlds suitable for sustaining life. Some scientists are now undertaking high-tech searches for distant extraterrestrial life, trying to answer this question of whether or not we are alone in the universe or whether others are out there, perhaps searching just as we are, looking for company in the cosmos.

In this chapter you will discover

- what qualities scientists believe a world must have in order to support life
- why many scientists are open to the possibility that primitive life exists elsewhere in the solar system
- how scientists estimate the number of planets orbiting other stars that could support complex life
- how scientists search for life beyond our own solar system—and the results of those searches
- how we are trying to communicate with extraterrestrial life

19-1 Astrobiology connects the cosmos and the origins of life

The amazing advances in all realms of science that began in the seventeenth century and accelerated to tremendous heights by the mid-twentieth century have finally enabled scientists to meaningfully explore issues surrounding the beginnings of life on Earth. **Astrobiology** combines all of the natural sciences in an effort to understand the formation, evolution, and future of life here and throughout space. As we will explore shortly, efforts to replicate the conditions under which life formed on Earth have been carried out since the 1950s. The discovery of strong evidence that liquid water, essential for life as we know

R I V U X G

FIGURE 19-1 *Viking Mars Lander* Astronomer and renowned popularizer Carl Sagan poses by a model of the Viking lander. This image was taken in Death Valley, California, where the background creates the feel of a Martian landscape. Sagan was instrumental in choosing some of the experiments flown on the Viking spacecraft. (NASA Jet Propulsion Laboratory) it, exists in several worlds in the solar system has added fuel to the astrobiology effort.

Astrobiology as a unified discipline received a big boost in 1996, when a meteorite from Mars, labeled ALH84001, was opened to reveal organic-looking structures that *could* have been formed by life on Mars. This possibility sparked the interest of many people, including those who fund NASA and other organizations worldwide. Thousands of scientists now work in this field.

Finding other potential homes for life in the solar system is one facet of astrobiology. Finding homes beyond the solar system is another. The number of known extrasolar planets is greater than 310. Although most are Jupiter-like, that is, hydrogen-rich bodies, Earthlike planets containing water on their surfaces, such as the one orbiting the star Gliese 581, are now being discovered.

The search for advanced civilizations is also under way. It began on Earth, in the form of people who believe they have seen aliens or, more commonly, their UFO spacecraft. Beliefs are one thing, scientific evidence is something else. Despite all of the alleged UFO sightings and personal encounters that have been reported, no one has produced a single piece of physical evidence to convince scientists that an intelligent, extraterrestrial life-form has *ever* visited Earth. Furthermore, despite all of the wonderful science-fiction books, movies, and TV shows, there is no known way for spacecraft to travel faster than the speed of light. Therefore, unless they come from planets around the few stars within a few light-years of Earth, aliens would have to travel for centuries, or even millennia, to span the void of space between their worlds and ours. After all that travel time, it would seem illogical for space visitors not to land here.

Searches for *extraterrestrial intelligence* (SETI) in space began in 1960, when American astronomer Frank Drake carried out Project Ozma (named after the land in L. Frank Baum's Oz series of books). Drake used a radio telescope at the National Radio Astronomy Observatory in West Virginia to "listen" for radio signals from civilizations that might have existed around two Sunlike stars: τ (tau) Ceti and ϵ (epsilon) Eridani. He found nothing.

19-2 The existence of life depends on chemical and physical properties of matter

To build a physical foundation for astrobiology, let us begin by considering some of the fundamental requirements for the existence of life as we know it. First is the presence of liquid water. Water, neither acidic nor basic, enables many kinds of atoms and molecules immersed in it to bond to one another or separate from one another in large numbers. While water molecules combine with some atoms and molecules, water also allows many interactions to occur in it without chemical interference. Such

$$Z - X - X - X - X - Y$$

a   Linear molecule

b   Glucose

FIGURE 19-2 Creating Complex Molecules (a) Atoms (denoted by capital letters) that can bond strongly to only two other atoms can make linear chains, as depicted here. However, when such atoms bond to atoms that can only make one bond, denoted Y and Z, the chain stops. In no case can atoms X, Y, and Z combine to create nonlinear chains other than loops. (b) When an atom, like carbon (C), can share electrons with more than two other atoms (in carbon's case, with four atoms), then the complex, nonlinear chains essential for life can form. Chains of carbon atoms form the backbone of organic molecules. For example, glucose, with carbon, oxygen (O), and hydrogen (H), is a nonlinear molecule that serves as a nutrient for many life-forms; it is a sugar. The lines indicate bonds between atoms.

diverse activity is essential for the creation and evolution of the complex molecules and systems of molecules that function in living organisms. Given all of these factors, biologists believe that life on Earth began to develop in primordial bodies of water.

> How would most scientists respond to the statement: "Aliens might be here, but we just haven't seen them yet"?

The second requirement is an element that can bond strongly, but not too strongly, with at least three other atoms to allow for complex molecules necessary for life. Atoms able to make only two bonds can form chains (Figure 19-2a) or loops of atoms but cannot create complex structures (Figure 19-2b) that contain atoms of other elements, as is necessary for making the variety of molecules essential for life.

Five elements can bond by sharing electrons (that is, bond covalently) with three or more elements: boron (B), carbon (C), nitrogen (N), silicon (Si), and phosphorus (P). Bonding by donating or receiving an electron (ionic bonding) creates bonds that are unsuitable for the formation of the complex, flexible, rapidly-changing molecules that life requires. Carbon is unique among the five elements listed here in that its bonds are flexible yet strong. As counterexamples, silicon-silicon bonds come apart under the slightest disturbance; silicon-oxygen bonds create gels and liquids that are very hard to alter (Figure 19-3a); and silicon-oxygen-oxygen bonds are so strong that they create rocks (Figure 19-3b). The other three elements that might serve as the backbone of life—boron,

a   R I V U X G   b   R I V U X G

FIGURE 19-3 Non-Carbon Organic Molecules? When any element other than carbon that can make three or more strong bonds combines, it makes compounds that are too soft, too hard, too reactive, or too inert to be useful in supporting life. Consider silicon. (a) The silicon-oxygen pair that creates the backbone of silicone is too inert to allow such molecules to react rapidly and thereby serve as organic molecules. Furthermore, these bonds produce gel or liquid compounds, as shown. (b) When the backbone is silicon-oxygen-oxygen, the bonds are rigid, as in this quartz rock. (a: Richard Megna/Fundamental Photographs, New York; b: Mark A. Schneider/Visuals Unlimited)

nitrogen, and phosphorus—have similar bonding problems. Despite the enormous range of conditions under which terrestrial life exists, all of it is based on the unique properties of carbon.

Carbon atoms form chemical bonds that can combine to create especially long, complex molecules. These molecules can be further linked in elaborate chains, lattices, and fibers. It is for these reasons that carbon-based compounds, called **organic molecules**, are the stuff of life. Indeed, biologists have already identified more than six million carbon compounds.

The variety and stability of living organisms depend on the complex, self-regulating, chemical reactions of organic molecules. Happily, carbon is also among the most abundant elements in the universe. The unique versatility and abundance of carbon imply that extraterrestrial biology would also be based on organic chemistry.

Third, a life-supporting environment must also have a variety of other elements essential for developing life. Other constituents of organic molecules, such as hydrogen, oxygen, nitrogen, and sulfur, are plentiful in our Galaxy. All of the organic molecules necessary to form

life may develop naturally in the oceans of Earth and similar planets. Nevertheless, some astronomers have proposed that after planets came into existence, organic material from space rained down on these worlds, thereby expediting the evolution of life. This belief is supported by the fact that spectra taken of comets in our solar system reveal that they contain an assortment of organic compounds.

Further evidence for the existence of organic molecules in interplanetary space was collected by, among others, the spacecraft *Giotto* as it passed close to the nucleus of Comet Halley. In 1997, NASA launched the *Stardust* mission, which rendezvoused with Comet Wild 2 in January 2004, collecting dust and other debris from the comet. The spacecraft returned safely to Earth in January 2006. It revealed the presence of organic molecules that had been released from the comet, supporting the belief that comets may have helped seed compounds useful in the formation of life on Earth.

Evidence of extraterrestrial organic molecules also comes from meteorites, called *carbonaceous chondrites* (Figure 19-4). Many of the carbonaceous chondrites that have fallen to Earth contain a variety of organic substances. As noted in Section 18-11, carbonaceous chondrites are ancient meteorites dating from the formation of the solar system. If they contained organic molecules while in space, it seems reasonable to conclude that, even from their earliest days, the planets have been continually bombarded with organic compounds.

In interstellar clouds, carbon atoms have combined with other elements to produce an impressive variety of organic compounds. Since the 1960s, radio astronomers have detected tell-tale microwave emission lines from interstellar clouds that help identify over 80 of these carbon-based compounds. Examples include polycyclic aromatic hydrocarbons (PAHs), a type of molecule composed of just carbon and hydrogen atoms, as well as ethyl alcohol (CH_3CH_2OH), formaldehyde (H_2CO), acetic acid ($HCOOCH_3$), and acetaldehyde (CH_3CHO). In 2003, scientists detected the amino acid glycine in space. In 2005, PAHs were observed over 9 billion light-years away. These were the first observations of complex organic molecules from that far in the past. Because stars and planets condense from interstellar gas and dust enriched in these elements, these compounds can help provide the building blocks of life elsewhere in the Galaxy. Indeed, in 2004, astronomers who use the Spitzer Space Telescope detected some of the ingredients for life, including methanol, water, and carbon dioxide in orbit around newly forming stars.

Interstellar space is not the only source of organic material. In a classic experiment performed in 1952, American chemists Stanley Miller and Harold Urey

a

Organic molecules

b

RIVUXG

FIGURE 19-4 Carbonaceous Chondrite (a) Carbonaceous chondrites are meteorites that date back to the formation of the solar system. This sample is a piece of the Allende meteorite, a large carbonaceous chondrite that fell in Mexico in 1969. (b) Chemical analysis of newly fallen carbonaceous chondrites discloses that they are rich in organic compounds. (a: From the collection of Ronald A. Oriti; b: Harvard Smithsonian Center for Astrophysics)

demonstrated that simple chemicals can combine on liquid-water-bearing planets to form organic compounds. Hoping to capture conditions on primitive Earth and in its atmosphere, as then understood, Miller and Urey mixed hydrogen, ammonia, methane, and water vapor in a closed container. They then subjected their mixture to an electric arc to simulate lightning bolts. At the end of a week, the inside of the container had become coated with a reddish-brown substance rich in compounds essential to life (Figure 19-5).

Since that time, geochemists have come to realize that Earth's early atmosphere was more likely to have been a mixture of carbon dioxide, nitrogen, and hydrogen, along with water vapor "outgassed" from volcanoes and deposited by comet impacts. Modern versions

of the Miller-Urey experiment have therefore used these common gases, and, once again, organic compounds were produced. Could these compounds be forming elsewhere in the universe today?

Keep in mind that scientists have not yet created life in a test tube. Biologists have yet to figure out, among other things, how simple organic molecules gathered into cells and developed systems for self-replication.

The fourth issue surrounding the existence of life elsewhere relates to physical properties of planets and life—the environments supporting life-forms must have suitable temperature, radiation, and other factors. The fundamental factor here is that water must be liquid either on or in a world for life to evolve there. The region around a star or planet where this happens is called the **habitable zone**. Earth is in the Sun's habitable zone, while Europa is in Jupiter's.

If a planet is too close to a star, the temperature and deadly ultraviolet radiation levels will be too high for life to evolve. If the planet is too far away, it will be too cold to have liquid water. If a planet's environment is otherwise hostile—such as having too much radiation, being too windy, or being too seismically active—life may either not be able to spread or may quickly become extinct. If the star is too massive, it will explode before advanced life on any of its planets has a chance to evolve. If the star is too small, the planet would have to be very close to it to be warm enough for life, but then tidal forces from the star would lock the planet in synchronous rotation, making most of its surface either too hot (daytime side) or too cold (nighttime side) for life to flourish. These issues are summarized in Figure 19-6.

> Synchronous rotation occurred in what other situation that we explored in this book?

19-3 Evidence is mounting that life might exist elsewhere in our solar system

Despite all of these potential limitations, scientists have strong evidence that everything does not have to be optimal for life to form. Geologists and biologists have discovered life on Earth in some incredibly challenging environments, such as on the ocean bottom, in a lake far under the Antarctic ice pack, deep inside our planet's crust, and even in hot geothermal vents (Figure 19-7). It therefore seems reasonable to believe that life could have originated off Earth under similarly challenging conditions. Scientists consider at least four places in our solar system as possible habitats for life past or present. They are Mars and Jupiter's moons Europa, Ganymede, and Callisto.

Astronomers have discovered (see Chapter 16) that Mars once had surface liquid water and that

> How is water inside worlds like Europa kept liquid?

R I V U X G

FIGURE 19-5 Miller-Urey Experiment Updated
(a) Modern versions of this classic experiment prove that numerous organic compounds important to life can be synthesized from gases that were present in Earth's primordial atmosphere. This experiment supports the hypothesis that life on Earth arose as a result of ordinary chemical reactions. (b) This photograph shows a slime of organic material (in brown vials) created in the original Miller-Urey experiment. (b: NASA)

it may still have underground bodies of liquid water. We saw in Sections 17-5 through 17-7 that Europa, Ganymede, and Callisto are also likely to have liquid water under their frozen surfaces. These four bodies, then, are candidates for having evolved simple life-forms. Indeed, we saw in Chapter 16 what may be fossilized bacterial life from Mars (see Figure 16-34).

Confirmation that life evolved on any of these worlds would demonstrate that the formation of complex molecules and structures is possible off Earth. However, given the limited chemical resources available on other worlds in our solar system, these locations are unsuitable for the evolution of sufficiently complex life, like ourselves, that is self-aware, and able to create advanced civilizations.

19-4 Searches for advanced civilizations try to detect their radio signals

We have not yet discovered Earthlike planets orbiting Sunlike stars. The closest case was the discovery in 2007 of a 5 Earth-mass planet orbiting the M3 star Gliese 581, which has about a third the Sun's mass. Furthermore, we saw in Section 14-6 that many stars have Jupiter-mass planets that orbit very close to them. The presence of massive planets near stars would make it very hard for those stars to support habitable planets because the gravitational tugs from the massive planets would pull Earthlike planets into highly elliptical orbits. During parts of such orbits, terrestrial worlds would be overheated and overradiated by their stars, whereas during other parts of their orbits, the planets would freeze. However, not all Sunlike stars have close-orbiting massive planets.

Billions of Sunlike stars in our Galaxy remain to be studied, and most astronomers and many other people expect that we will eventually discover habitable planets around some of them. This belief helps motivate SETI. Unlike the fossil evidence for life on Mars, the effort to locate life outside of the solar system is based on searching for high-tech evidence of advanced civilizations.

How might we ascertain whether extraterrestrial civilizations exist, given the tremendous distances that separate us from other stars and the huge number of stars with the potential to support life-bearing planets? Looking for individual, habitable, extrasolar planets is technically demanding and time-consuming. As we saw in Section 14-6, it requires careful observation of data available only in the past 25 years. Furthermore, the discovery of a planet does not imply that it harbors life. SETI astronomers therefore need a way to bypass the daunting search for habitable planets. Starting in the 1950s and 1960s, they identified a promising approach to searching directly for extraterrestrial intelligence—they listen for radio transmissions from distant civilizations.

As we have seen, radio waves can travel immense distances without being significantly altered by the interstellar medium. Because they penetrate gas and dust, radio waves are a logical choice for interstellar communication. Even if alien civilizations are not trying to communicate with us, they probably use radio waves in their own technology. But at what frequencies

Habitable Zones for Life

Intelligent civilizations in our Milky Way Galaxy can evolve only in a certain region called the galactic habitable zone. In that zone, a suitable planet must lie within the planetary habitable zone of its parent star. (From C. H. Lineweaver, Y. Fenner, and B. K. Gibson)

Galactic habitable zone

Too close to the center of the Milky Way Galaxy:
• The distances between stars are small, so there can be close encounters between stars that would disrupt a planetary system.
• There are also frequent outbursts of potentially lethal radiation from supernovae and from the supermassive black hole at the very center of the Galaxy.

Too far from the center of the Milky Way Galaxy:
• Stars are deficient in elements heavier than hydrogen and helium, so they lack both the materials needed to form Earthlike planets and the chemical substances required for life as we know it.

Planetary habitable zone

The Star:
• Must have a mass that is neither too large, or too small.
• If the star's mass is *too large*, it will use up its hydrogen fuel so rapidly that it will burn out before life can evolve on any of its planets.
• If the star's mass is *too small*, the habitable planet would be so close that it would be in synchronous rotation. Water would vaporize on one side and freeze on the other.

The Neighborhood:
• There needs to be one or more large Jovian planets whose gravitational forces will clear away comets and meteors.

The Planet:
• Must be a terrestrial planet with a solid surface.
• Must have enough mass to provide the gravity needed to retain an atmosphere and oceans.
• Must be at a comfortable distance from the star so that water can be liquid on its surface.
• Must be in a stable, nearly circular orbit. (A highly elliptical orbit would cause excessively large temperature swings as the planet moved toward and away from the star.)

FIGURE 19-6 Zone For Habitable Planets This figure summarizes the locations in the Galaxy and in orbit around stars where habitable planets might be found. Earth, of course, is in such a location.

Living organisms called
hyperthermophiles

a Grand prismatic spring, Yellowstone National Park   R I V U X G

b   R I V U X G

FIGURE 19-7 Hyperthermophiles (a) These microscopic thermophiles (heat-loving organisms) live in water that is between 80°C and 100°C (85°F–140°F). (b) Tube worms (light-green tubes) with hemoglobin-rich red plumes. They reside around black smokers—vents in the ocean bottom that are in the same temperature range as the hot springs shown in (a). These vents are over 3 km (2 mi) under water. (inset: NASA; a: Jim Peaco; July 2001 Yellowstone National Park Image by NPS Photo; b: Fisheries and Oceans of Canada)

should we look to maximize the odds of detecting alien signals?

WEB LINK 19.5   We could try random frequencies in the hopes of hearing something—anything. But the number of directions in which to look and frequencies to check are both daunting. We need to find some way to improve the odds. It turns out that some radio wavelengths travel farther without being absorbed by interstellar gas and dust, or without competing with noise from other sources, than do other radio wavelengths. The SETI pioneer Bernard Oliver pointed out that a range of relatively clear frequencies exists in the neighborhood of the microwave emission lines of hydrogen and the hydroxyl radical (OH) (Figure 19-8). This region of the microwave part of the radio spectrum is called the **water hole,** a humorous reference to the H and OH lines being so close together. If extraterrestrial beings are purposefully sending messages into space (to communicate with spacecraft or other worlds), it seems reasonable that they would choose to transmit on these frequencies.

Even if only a few alien civilizations are scattered across the Galaxy, we have the technology to detect radio transmissions from them. Since Project Ozma was carried out, the number of organized searches for signals from extraterrestrials approaches 100. SETI searches use radio telescopes (Figure 19-9) along with sophisticated electronic equipment and powerful computers to conduct both a targeted search and an all-sky survey.

FIGURE 19-8 Water Hole The so-called water hole is a range of radio wavelengths from about 3 to 30 centimeters (cm) that happens to have relatively little cosmic noise. Some scientists suggest that this noise-free region would be well-suited for interstellar communication.

FIGURE 19-9 Radio Telescope Used for SETI The Arecibo Observatory's radio telescope, with a diameter of 305 m (1000 ft) is the largest in the world. It is located in Arecibo, Puerto Rico. In the past decade, it was used in an all-sky survey, along with an antenna located in the Mojave Desert in California to search for extraterrestrial intelligence. In 1996, an antenna in Canberra, Australia, joined the network. (Courtesy of the NAIC-Arecibo Observatory, a facility of the NSF)

Although NASA has created astrobiology centers to study issues regarding life in the solar system and Galaxy, most of the SETI programs in operation today are privately funded. Several are actively listening and analyzing enormous volumes of data.

 Indeed, so much data are available to analyze that one SETI project, called SETI@home (setiathome.berkeley.edu), has enlisted the help of home-computer users. These SETI researchers provide actual data from radio telescopes to be analyzed, along with a data-analysis program. While the computer's screensaver is on, the program runs, the data are analyzed, and the results are sent back to the researchers to be combined with data from other home-computer users. The processed data are replaced with fresh data to be analyzed. Another SETI program uses an array of small, inexpensive TV satellite antennas to look for signals from space.

Figure 19-10 shows the region of the Galaxy from which radio signals similar to the kinds of transmissions we humans make could be detected. Occasionally, a SETI project does detect powerful or unusual signals from space. However, because none of these signals are ever repeated, SETI researchers do not yet believe they have discovered any extraterrestrial civilization.

The detection of a message from an alien civilization would be one of the greatest events in human history. Communicating with aliens could dramatically change the course of civilization on Earth through the sharing of scientific information or an awakening of social or humanistic enlightenment. In only a few years, our industry and social structure could advance centuries into the future. The effects of such change would touch every person on Earth.

> If we detect incidental, rather than directed, signals from alien civilizations, what information might they contain?

19-5 The Drake equation: How many civilizations are likely to exist in the Milky Way?

Just how many planets throughout our Galaxy are likely to harbor complex life? The first person to tackle this question quantitatively was Frank Drake. He proposed that the number of technologically advanced civilizations in the Galaxy (designated by the letter N) could be estimated by what is now called the **Drake equation:**

$$N = R^* f_p n_e f_l f_i f_c L,$$

in which

R^* = rate at which solar-type stars form in the Galaxy

f_p = fraction of solar-type stars that have planets

n_e = number of planets per solar-type star system suitable for life

f_l = fraction of those habitable planets on which life actually arises

f_i = fraction of those life-forms that evolve into intelligent species

f_c = fraction of those species that develop adequate technology and then choose to send messages out into space

L = lifetime of that technologically advanced civilization

The Drake equation expresses quantitatively the number of extraterrestrial civilizations as a product of terms, some of which can be estimated from what we know about stars and stellar evolution. For instance, thanks to new evidence for extrasolar planets, astronomers can now hope to determine the first two terms, R^* and f_p, by observation. We should probably exclude stars larger than about 1.5 $M_\odot$, because they have main-sequence lifetimes shorter than the time it took for intelligent life to develop on Earth—some 3.8 to 4.0 billion years. If that is typical of the time needed to evolve higher life-forms, then a massive star becomes a giant or even explodes as a supernova before self-aware creatures can evolve on any of its planets.

Although low-mass stars have much longer lifetimes, they are less suited for supporting life on their

FIGURE 19-10 Region of the Search for Extraterrestrial Intelligence The red circle, centered on the solar system, shows the region of the Milky Way in which SETI searches can reasonably expect to detect radio emissions from alien civilizations. This assumes that the signals are similar in nature to the kinds of radio emissions we generate on Earth. (NASA/Space Telescope Science Institute)

planets because they are so cool. As noted earlier, only planets very near a low-mass star would be sufficiently warm for water to be a liquid. But, as also mentioned earlier, a planet that close would become tidally coupled to the star, developing synchronous rotation. One side would have continuous daylight, leading to the evaporation of oceans, while the other side would be in perpetual, frigid darkness. The only place that life could survive on such planets is in the narrow ring at the boundary between day and night. Such a small fraction of the land available greatly reduces the likelihood that life would be able to evolve to the complexity necessary for technological civilizations to develop.

As "ideal" life-supporting stars, this leaves main-sequence stars with masses near those of the Sun. These are stars in spectral types between F5 and M0. Based on statistical studies of star formation in the Milky Way, astronomers calculate that roughly one of these Sunlike stars forms in the Galaxy each year, yielding a value of $R^* = 1$ per year.

We learned in Sections 14-2 and 14-3 that the planets in our solar system formed in conjunction with the birth of the Sun. We have also seen evidence that similar processes of planetary formation may be commonplace around isolated stars. Many astronomers, therefore, give f_p a value of 1, meaning they believe it is likely that most Sunlike stars have planets.

Unfortunately, the rest of the factors in the Drake equation are very hypothetical. The chances that a planetary system has an Earthlike world are not known. Were we to consider our own solar system as represen-

tative, we could put n_e at 1. Let us be more conservative, however, and suppose that 1 in 10 solar-type stars is orbited by a habitable planet, making $n_e = 0.1$.

From what we know about the evolution of life on Earth, we might assume that, given appropriate conditions, the development of life is a certainty, which would make $f_l = 1$. This is, of course, an area of intense interest to biologists. For the sake of argument, we might also assume that evolution naturally leads to the development of intelligence (a conjecture that is hotly debated) and also make $f_i = 1$. It is anyone's guess as to whether these intelligent extraterrestrial beings would attempt communication with other civilizations in the Galaxy, but, if we assume they all would, f_c would also be put at 1.

The last variable, L, the longevity of technological civilization, is the most uncertain of all—it cannot be tested! Looking at our own example, we see a planet whose atmosphere and oceans are increasingly polluted, potentially destroying the food chain. When we add in how close we have come to destroying ourselves with weapons of mass destruction, it may be that we humans are among the lucky few technological civilizations to squeak through its first years. In other words, L may be as short as 100 years. Putting all of these numbers together, we arrive at

$$N = 1 \times 1 \times 0.1 \times 1 \times 1 \times 1 \times 100 = 10$$

Therefore, out of the hundreds of billions of stars in the Galaxy, there may be only 10 civilizations technologically advanced enough to communicate with us. Of course, the numbers used here are just estimates. A wide

"AS I READ IT, WE'RE RECEIVING A MESSAGE FROM OUTER SPACE TELLING US TO STOP BOMBARDING THEM WITH UNINTELLIGIBLE MESSAGES."

© Sidney Harris

Hyperfine transition of
neutral hydrogen

Silhouette of
spacecraft

Binary equivalent
of decimal 8

a

b

c

Position of sun
relative to 14
pulsars and the
center of the galaxy

Planets of solar
system and binary
relative distances

R I V U X G

FIGURE 19-11 Human Memorabilia in Space (a) Humans have beamed radio signals into space, hoping that the message will someday be intercepted by an alien civilization. This is a visual version of the signal sent in 1974 from the Arecibo radio telescope toward the globular cluster M13. The *Pioneer* and *Voyager* spacecraft, now in interstellar space, also carry messages from Earth. (b) The plaques on *Pioneer 10* and *Pioneer 11* provide information about where we are, what we look like, and some of the science we know. (c) Images and sounds sent on *Voyager 1* and *Voyager 2* were stored on phonographic records, long before DVDs were even a twinkle in an engineer's eye. There are also instructions for playing the record, which contains information about our biology, our technology, and our knowledge base. Each record also contains the sounds of children's voices. It is remotely possible that another race might someday discover the spacecraft. (a: Frank Drake/UCSC, et al., Arecibo Observatory/Cornell/NAIC: b and c: NASA)

range of values has been proposed for the terms in the Drake equation, and these various numbers produce vastly different estimates of N. Increasing the average lifetime of advanced civilizations significantly increases the total number, N. Some scientists argue that there is exactly one advanced civilization in the Galaxy and that we are it. Others speculate that there may be tens of millions of planets inhabited by intelligent creatures. Although we do not know yet, science enables us to home in on the number.

19-6 Humans have been sending signals into space for more than a century

5 We have been doing more than just listening passively for voices from the cosmos. Astronomers have inten-

tionally broadcast well-focused signals through radio telescopes toward star systems likely to harbor advanced life, including a 1974 "hello" message aimed at the globular cluster M13 (Figure 19-11a) and a 2008 broadcast of the Beatle's "Across the Universe." Unintentional broadcasts into space have occurred for longer than that. Since 1895, with the first radio transmission by the Italian engineer Guglielmo Marconi, we have been sending messages about ourselves into space. The spherical region within some 115 light-years of the solar system is now filled with signals from radio and television shows. If other advanced civilizations fall

With what two numbers used in the Drake equation calculation presented previously do you disagree most? Using your numbers, how many civilizations do you estimate exist?

within that sphere, they could very well be listening to radio programs or watching television shows from decades past, trying to make sense of our species.

Four spacecraft also carry information about the human race. Launched in 1972 and 1973, respectively, the *Pioneer* 10 and *Pioneer* 11 spacecraft carry plaques about us (Figure 19-11b). They are now making their way into interstellar space. In 1977, the two *Voyager* spacecraft were sent toward the outer solar system. They carry old-fashioned phonograph records encoded with human images and voices. Because space is so vast and the spacecraft are so small, the likelihood of their being discovered is truly remote. And yet millions, perhaps billions, of years from now, one of the spacecraft just might reach another race of intelligent creatures. From its path and the information on board, those creatures might be able to determine where the *Pioneer* and *Voyager* spacecraft came from. If their travel budgets are sufficiently large, they might even decide to come and visit us. At the least, they could send us radio messages. Will our descendants be here to receive them?

19-7 Frontiers yet to be discovered

We still have much to learn about astrobiology. Myriad steps in the formation of life on Earth remain to be understood. Although the evidence for liquid water once having existed on Mars is overwhelming, we still need to confirm whether life ever existed there. Does life exist on Mars, Europa, Ganymede, or Callisto today? Astronomers are still working to refine their ideas of how organic compounds formed in space. A paramount frontier is detecting life elsewhere in the universe.

> What negative consequence could arise from an alien civilization discovering the *Pioneer* 10 or 11 spacecraft?

SUMMARY OF KEY IDEAS

• The chemical building blocks of life exist throughout the Milky Way Galaxy.

• Organic molecules and water have been discovered in interstellar clouds, in some meteorites, in comets, and in newly forming star and planet systems.

• The Drake equation is used to estimate the number of technologically advanced civilizations in the Galaxy whose radio transmissions we might discover. Estimates of this number vary from 1 to millions.

• Astronomers are using radio telescopes to search for signals from other self-aware life in the Galaxy. This effort is called the search for extraterrestrial intelligence,

or SETI. SETI is primarily done at frequencies where radio waves pass most easily through the interstellar medium. So far, these searches have not detected any life outside Earth.

• Everyday radio and television transmissions from Earth, along with intentional broadcasts into space, may be detected by other life-forms.

WHAT DID YOU THINK?

1 *Why is water so important to the formation of life?* Water allows many interactions between atoms and molecules dissolved in it.

2 *What element is uniquely suited to be the foundation of life as we know it, and why?* Carbon is the only atom that can bond flexibly (but not too flexibly) with three or more other atoms.

3 *How do astronomers search for extraterrestrial intelligence?* They search for radio signals from other advanced civilizations.

4 *Have astronomers located any extraterrestrial civilizations?* No, extraterrestrial civilizations have yet been discovered.

5 *If advanced alien civilizations exist, is there any way they might know of our existence?* Yes, they might detect our intentional radio broadcasts into space, our everyday radio and television broadcasts, or, possibly, one of our spacecraft traveling in interstellar space.

Key Terms for Review

astrobiology, 524
Drake equation, 531
habitable zone, 527
organic molecule, 526
SETI, 525
water hole, 530

Discussion Questions

1. Which stellar spectral type is most likely to have a planet on which advanced life exists? **a.** O **b.** B **c.** A **d.** G **e.** M

2. Which element serves as the foundation or "backbone" of life? **a.** oxygen **b.** carbon **c.** silicon **d.** hydrogen **e.** nitrogen

3. What arguments could you make against sending messages via radio or spacecraft into interstellar space?

4. Try using the Drake equation with values that you find reasonable. How many civilizations do you estimate there are in our Galaxy?

5. Discuss the possible biological effects of our probes that visit other life-sustaining worlds.

6. What social effects might probes from other worlds have on us?

7. List some of the pros and cons in the argument that alien spacecraft have visited Earth.

What If . . .

8. Our planet were the moon of a much larger, Jupiter-like planet that orbits the Sun at 1 AU? What might Earth and life inhabiting it be like?

9 Our Moon were Earth-sized and also had life on it?

10. We discover living organisms in Europa's subsurface oceans? What precautions should we take in the search for such life, and why?

11. We develop life in a Miller-Urey-like experiment? Explore some of the implications that this event would have.

Web Questions

12. What information would you have put on the record sent out on the Voyager spacecraft? Compare your answer to the results listed in the Web link associated with that record.

 13. To test your understanding of the Miller-Urey experiment, do Interactive Exercise 19-1 on the Web. You can print out your answers, if required.

 14. To test your understanding of the Drake equation, do Interactive Exercise 19-2 on the Web. You can print out your answers, if required.

APPENDIX A: Powers-of-Ten Notation

Astronomy is a science of extremes. As we examine various cosmic environments, we find an astonishing range of conditions—from the incredibly hot, dense centers of stars to the frigid, near-perfect vacuum of interstellar space. To describe such divergent conditions accurately, we need a wide range of both large and small numbers. Astronomers avoid such confusing terms as "a million billion billion" (1,000,000,000,000,000,000,000,000) by using a standard shorthand system. All the cumbersome zeros that accompany such a large number are consolidated into one term consisting of 10 followed by an *exponent,* which is written as a superscript and called the **power of ten.** The exponent merely indicates how many zeros you would need to write out the long form of the number. Thus,

$$10^0 = 1$$
$$10^1 = 10$$
$$10^2 = 100$$
$$10^3 = 1000$$
$$10^4 = 10,000$$

and so forth. Equivalently, the exponent tells you how many tens must be multiplied together to yield the desired number. For example, ten thousand can be written as 10^4 ("ten to the fourth") because $10^4 = 10 \times 10 \times 10 \times 10 = 10,000$. Similarly, 273,000,000 can be written as 2.73×10^8.

In scientific notation, numbers are written as a figure between 1 and 10 multiplied by the appropriate power of 10. The distance between Earth and the Sun, for example, can be written as 1.5×10^8 km. Once you get used to it, you will find this notation more convenient than writing "150,000,000 km" or "one hundred and fifty million kilometers."

This powers-of-ten system can also be applied to numbers that are less than 1 by using a minus sign in front of the exponent. A negative exponent tells you that the location of the decimal point is as follows:

$$10^0 = 1.0$$
$$10^{-1} = 0.1$$
$$10^{-2} = 0.01$$
$$10^{-3} = 0.001$$
$$10^{-4} = 0.0001$$

and so forth. For example, the diameter of a hydrogen atom approximately is 1.1×10^{-8} cm. That notation is more convenient than stating "0.000000011 cm" or "11 billionths of a centimeter." Similarly, 0.000728 equals 7.28×10^{-4}.

Using the powers-of-ten shorthand, one can write large or small numbers like these compactly:

$$3,416,000 = 3.416 \times 10^6$$
$$0.000000807 = 8.07 \times 10^{-7}$$

Because powers-of-ten notation bypasses all the cumbersome zeros, a wide range of circumstances can be numerically described conveniently:

$$\text{one thousand} = 10^3$$
$$\text{one million} = 10^6$$
$$\text{one billion} = 10^9$$
$$\text{one trillion} = 10^{12}$$

and also

$$\text{one thousandth} = 10^{-3} = 0.001$$
$$\text{one millionth} = 10^{-6} = 0.000001$$
$$\text{one billionth} = 10^{-9} = 0.000000001$$
$$\text{one trillionth} = 10^{-12} = 0.000000000001$$

Try these questions: Write 3,141,000,000 and 0.0000000031831 in scientific notation. Write 2.718282×10^{10} and 3.67879×10^{-11} in standard notation.

(Answers appear at the end of the book.)

APPENDIX B: Guidelines for Solving Math Problems and Reading Graphs

Astronomy relies on mathematics. Although we have distilled the concepts into words, you can gain further insights into the distances, intensities, and other astronomical quantities from the equations presented in this book. In addition, much scientific information is presented in the form of graphs, a technique that presents data very compactly and often provides a good way to see the trends that the data reveal. This appendix thus provides guidance on three things: setting up problems to be solved analytically (using equations), solving algebra equations, and reading graphs.

Setting Up and Solving Analytical Problems

At the end of most chapters, there are questions, indicated by an asterisk (*), that require mathematical solutions. You will need to translate the question into an equation or two in order to solve for some value. This is best done systematically. In what follows, we will work a question, appropriate to Chapter 2: How fast does a meteoroid (small piece of rocky space debris) of mass 5 kg (kilograms) move if it has a kinetic energy of 10 J (1 J = 1 kg × m²/s²)? This approach should work for most of the questions in the book.

Step 1. *Write down all the information you are given in terms of the variable names used in the book.* For example, denoting mass by m and kinetic energy by KE, write: mass $m = 5$ kg, kinetic energy $KE = 10$ J.

Step 2. *Identify and write down the thing you are trying to find in terms of its variable name.* In this case: speed, $v = ?$.

Step 3. *Find (usually) one or (sometimes) two equations that are needed to solve for the unknown.* All the equations presented in the book are listed in Appendix C. For our problem, use the kinetic energy equation $KE = \frac{1}{2}mv^2$, because this equation has the unknown, v, along with only variables and constants that are given, namely KE and m. (As in this case, you will often need only one equation, but sometimes you will have to solve first one equation and then another to get the answer. As an example, you might be given the information above but asked for the value of a momentum, p, of the meteoroid. The equation for momentum, also in Chapter 2, is $p = mv$. First you would solve for the speed, as we are now doing, and then use that v in this last equation.)

Step 4. *Manipulate the equation(s) until you have the unknown variable on one side and all the other variables and constants on the other side.* Variables can be added, subtracted, multiplied, and divided in the same way that numbers are combined. For our example, we want to find v, so we start with $KE = \frac{1}{2}mv^2$, multiply by 2, and divide by m. This gives $v^2 = 2\ KE/m$. Because we want v, we now take the square root of both sides or $v = \sqrt{2\ KE/m}$.

Step 5. *Make sure that all the units match up.* Units on both sides of the equation must be identical. Period. If you have one value in kilometers (km) and another value in meters (m), then the result you will get by combining numbers is not meaningful. If you end up with inconsistent units on opposite sides of an equation, convert one unit to the other. For example, change kilometers into meters or vice versa. In this case, use the fact that 1 km = 10^3 m.

Step 6: *Plug in all the numbers and solve for the unknown.* In our case, $v = \sqrt{2 \times 10\ \text{J}/5\ \text{kg}}$ or $v = 2$ m/s.

Reading Graphs

The graphs you will encounter in this book are compact ways of displaying patterns of information relating two variables, like the temperature and luminosity of stars or the temperature at different altitudes in an atmosphere. The relevant values of one of the variables are presented along the horizontal or x axis and the relevant values of the other variable are presented along the vertical or y axis. The word "relevant" here indicates that often graphs do not start at 0. Consider three examples. First is data presented in Chapter 16 concerning the temperature of Venus's atmosphere at different altitudes (**Figure 1**).

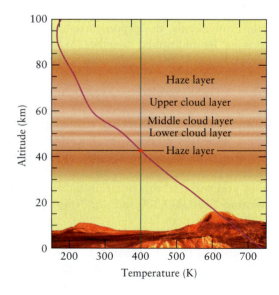

FIGURE 1 Linear Graph

The horizontal axis of the graph indicates the temperature in degrees Kelvin (K). The vertical axis denotes the altitude above Venus's surface in kilometers (km). Note that both axes are always labeled with a name (temperature or altitude, here) and units (K or km, respectively). Bear in mind that some variables such as temperature and altitude in Figure 1 increase to the right or upward (these are more common), but some variables will increase to the left or downward.

To read a graph, note that a value on the horizontal axis is then transferred directly upward through the graph. For example, all points on the blue line in Figure 1 (which extends upward from 400 K) have a temperature of 400 K. Equivalently, the value given on the vertical axis is transferred to all points horizontally across from this value. All the points on the black line in Figure 1 are at an altitude of about 43 km above Venus's surface. We have interpolated between 40 and 45 to get this value (**Figure 2**).

A curve or a set of points on the graph presents the *relationship* between the variable represented on the horizontal axis and the variable on the vertical axis. Each point on a curve or each separate point relates the two variables. Choose a point, say the dot on Figure 1. It represents the temperature at a certain altitude above Venus's surface. To find the temperature at that point, you slide directly down (along the blue line in this example) from the point and read the value of the horizontal variable under it. To find the altitude for that point, you slide directly over to the side (along the black line) and read the value of the vertical variable there. In our example, sliding along the blue line leads to 400 K on the temperature line. Therefore, this point corresponds to a temperature of 400 K. Moving horizontally from the point, you encounter, by interpolation, the 43-km indicator. Combining the data, you conclude that *the temperature of Venus's atmosphere 43 km above its surface is 400 K.*

The purple curve in Figure 1 provides the relationship between altitude and temperature for a wide range of locations above Venus in a representation that is much more informative than a table of heights and tempera-

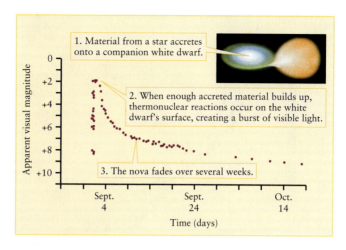

FIGURE 3 The Brightness of a Nova

1. Material from a star accretes onto a companion white dwarf.

2. When enough accreted material builds up, thermonuclear reactions occur on the white dwarf's surface, creating a burst of visible light.

3. The nova fades over several weeks.

tures. Specifically, this curve shows you the trend of temperature with altitude. Note that variables are occasionally placed on the right side and on the top of graphs.

Try these questions: What is the temperature at 20 km? What is the altitude at which the temperature is 300 K? For most of this graph, what is the general *trend* of the temperature with height? (Answers appear at the end of the book.)

Sometimes, the known information is not a curve, but rather a set of points, as in our second example (**Figure 3**), from Chapter 8. In this case, the graph connects the apparent magnitude of a nova (how bright it appears to be as seen from Earth regardless of its distance or other factors) and time. Each dot indicates how bright the nova (an explosion on the surface of certain stars) was at different times. For example, the peak brightness of the nova was an apparent magnitude of about +2 and it occurred on September 2. Noting that time increases to the right, you can immediately see that the trend of the nova's brightness is to increase rapidly and decrease more slowly.

Try these questions: What is the apparent magnitude on September 24? October 9? On what two days was the apparent magnitude +6? (Answers appear at the end of the book.)

The graphs so far have shown variables that change uniformly along the axes. For example, the distance on Figure 1 from 300 K to 400 K is the same as the distance from 400 K to 500 K, and so on. Many graphs you will encounter have variables that do not change uniformly (that is, linearly) along the axes. That means that the change in value going along each axis varies—there is not the same amount of change per centimeter along the axis. In **Figure 4**, from Chapter 7 and typical of graphs in the second half of the book, the temperature decreases going from left to right nonuniformly and the luminosity (total energy emitted per second) increases upward nonuniformly. These are called logarithmic scales.

The purpose of logarithmic and other nonlinear axes is to present in compact form data that varies

This line is about $\frac{3}{5}$ the distance from 40 to 45. Because there are 5 integers between 40 and 45, this line is at about $40 + \frac{3}{5} \times 5 = 43$.

FIGURE 2 Interpolation

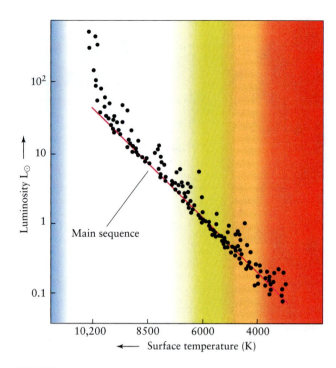

FIGURE 4 Logarithmic Graph

Area of a Circle $A = \pi d^2/4$, where π is approximately 3.14, and d is the diameter of the circle (Chapter 3).

Average Density $\rho = m/V$, where ρ is the average density, m is the total mass, and V is the volume of the object (Chapter 14).

Distance from Parallax $d = 1/p$ where d is the distance in parsecs and p is the parallax angle in arcseconds, or $d_{ly} = 3.26/p$, where d_{ly} is the distance in light-years (Chapter 6).

Distance-Magnitude Relationship $M = m - 5 \log(d/10)$, where M is the star's absolute magnitude, m is its apparent magnitude, and d is its distance in parsecs (Chapter 6).

Doppler Shift $\Delta\lambda/\lambda_o = v/c$, where $\Delta\lambda$ is the change in wavelength $(\lambda-\lambda_0)$, λ is the observed wavelength, λ_0 is the wavelength the object emits as seen by someone not moving toward or away from it, v is the speed of the object toward or away from the observer, and c is the speed of light (Chapter 4).

Drake Equation $N = R^* f_p\, n_e\, f_l\, f_i\, f_c\, L$, where N is the number of advanced civilizations in our Galaxy estimated by the equation, R^* is the rate at which solar-type stars form in the Galaxy, f_p is the fraction of stars that have planets, n_e is the number of planets per star system suitable for life, f_l is the fraction of habitable planets on which life arises, f_i is the fraction of life-forms that develop advanced intelligence, f_c is the frac-

very widely in values. For example, the dimmest star represented in Figure 4 is just less than 0.1 times as luminous as the Sun, while the brightest star is nearly 1000 times as luminous. The process of getting information from logarithmic graphs is the same as linear graphs. You must just be careful not to think of values as doubling or tripling when you go over or up two or three intervals. **Figure 5** shows how a logarithmic scale varies over one decade of values. The same numbering intervals apply for any decade of values, for example, 1 to 10 or 10^5 to 10^6. As you can see, the numbers bunch up near the highest value, so you need to interpolate these graphs more carefully than linear graphs.

Referring to Figure 4, the luminosity of a star with surface temperature of 4000 K is about 0.1 $L_\odot$. Note also that some graphs are linear on one axis and logarithmic on the other axis.

Try these questions: Approximately what is the luminosity of a star with surface temperature 8500 K? What is the surface temperature of a star with the same luminosity as the Sun (1 $L_\odot$)? (Answers appear at the end of the book.)

APPENDIX C: Key Formulas

Angular Momentum $L = I\omega$, where L is the angular momentum, I is the object's moment of inertia, and ω is the object's angular velocity (Chapter 2).

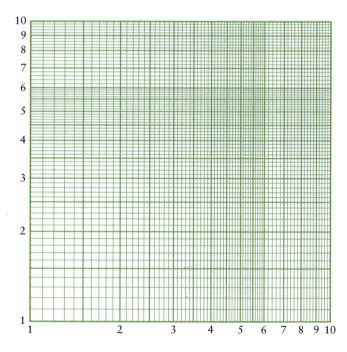

FIGURE 5 Logarithmic Scale

tion of species that develop technology and send signals into space, and L is the lifetime of technological civilizations (Chapter 19).

Energy Flux $F = \sigma T^4$, where σ is Stefan's constant and T is the blackbody's temperature in K (Chapter 4).

Energy-Mass Equation $E = mc^2$, where c is the speed of light and E is the energy released when mass m is converted to energy (Chapter 5).

Gravitational Force $F = G(m_1 m_2 / r^2)$, where G is the gravitational constant, m_1 and m_2 are the masses of the two interacting objects, and r is the distance between them (Chapter 2).

Gravitational Potential Energy (always true) $PE = GmM/r$, where PE is the potential energy, G is the gravitational constant, m is the mass of the object whose gravitational potential energy you are measuring, M is the mass of the object creating the gravitational potential, and r is the distance between the centers of the two (Chapter 2).

Gravitational Potential Energy (near Earth) $PE = mgh$, where m is the mass of the object whose gravitational potential energy you are measuring, $g = 9.8$ m/s^2, and h is the height of the object above the Earth's surface (Chapter 2).

Kepler's Third Law $P^2 = a^3$, where P is the sidereal period in Earth years and a is its semimajor axis in AU (Chapter 2).

Kinetic Energy $KE = \frac{1}{2} mv^2 = p^2/2m = L^2/2I$, where KE is its kinetic energy, m is its mass, v is its speed, p is the momentum, L is the angular momentum, and I is the object's moment of inertia. The first two equalities apply if the object is moving in a straight line. The last equality applies if the object is revolving or rotating (Chapter 2).

Luminosity $L = F \cdot 4\pi r^2$, where L is the luminosity (total energy per second) emitted, F is the energy flux, and r is the radius of the object (Chapter 4).

Magnification $M = f_o / f_e$, where f_o is the focal length of the primary mirror or objective lens and f_e is the focal length of the eyepiece (Chapter 3).

Momentum $p = mv$, where p is the momentum, m is the mass, and v is the velocity (Chapter 2).

Newton's Force Law $F = ma$, where F is the force acting on an object, m is its mass, and a is its acceleration (F and a are in boldface to denote that they act in some direction or other) (Chapter 2).

Newton's Version of Kepler's Third Law for Binary Star Systems $M_1 + M_2 = a^3/P^2$, where M_1 and M_2 are the masses of the stars, a is the average distance between them, and P is the period of their orbit in years (Chapter 6).

Photon Energy $E = hc/\lambda$, where E is the photon's energy, h is Planck's constant, c is the speed of light, and λ is the wavelength of the light (Chapter 3).

Pressure $P = F/A$, where P is the pressure, F is the force, and A is the area over which the force acts (Chapter 15).

Recessional Velocity of Galaxies $v = H_0 \times d$, where v is the recessional velocity in km/s, H_0 is the Hubble constant, and d is the distance to the galaxy in Mpc (Chapter 11).

Schwarzschild Radius of a Black Hole $R_{Sch} = 2 GM/c^2$, where R_{Sch} is the Schwarzschild radius in meters, G is the gravitational constant, M is the mass of the black hole in kg, and c is the speed of light (Chapter 9).

Size of a Distant Object $D = R \times \tan \theta$, where D is the physical diameter of an object, R is the distance to it, and $\tan \theta$ is the tangent of the angle it makes in the sky (Chapter 1).

Wien's law $\lambda_{max} = 2.9 \times 10^3/T$, where λ_{max} is the peak wavelength of the blackbody and T is the blackbody's temperature in K (Chapter 4).

Work $W = Fd$, where W is the work, F is the force acting, and d is the distance moved in the direction that the force acts (Chapter 2).

APPENDIX D: Temperature Scales

Three temperature scales are in common use. Throughout most of the world, temperatures are expressed in degrees Celsius (°C), named in honor of the Swedish astronomer Anders Celsius, who proposed it in 1742. The **Celsius temperature scale** (also known as the "centigrade scale") is based on the behavior of water, which freezes at 0°C and boils at 100°C at sea level on Earth.

Scientists usually prefer the **Kelvin scale**, named after the British physicist Lord Kelvin (William Thomson), who made many important contributions to our knowledge about heat and temperature. On the Kelvin temperature scale, water freezes at 273 K and boils at 373 K. Note that we do not use the degree symbol with the Kelvin temperature scale.

Because water must be heated by 100 K or 100°C to go from its freezing point to its boiling point, you can see that the size of a Kelvin is the same as the size of a degree Celsius. When considering temperature *changes*, measurements in Kelvins and in degrees Celsius lead to the same number.

A temperature expressed in Kelvins is always equal to the temperature in degrees Celsius plus 273. Scientists prefer the Kelvin scale because it is closely related to the physical meaning of temperature. All substances are made of atoms, which are very tiny (a typical atom has a diameter of about 10^{-10} m) and constantly in motion. The temperature of a substance is directly related to the average speed of its atoms. If something is hot, its atoms are moving at high speeds. If a substance is cold, its atoms are moving much more slowly.

The coldest possible temperature is the temperature at which atoms move as slowly as possible (they can never quite stop completely). This minimum possible temperature, called *absolute zero*, is the starting point for the Kelvin scale. Absolute zero is 0 K, or $-273°C$. Because it is impossible for anything to be colder than 0 K, there are no negative temperatures on the Kelvin scale.

In the United States, many people still use the now archaic Fahrenheit scale, which expresses temperature in degrees Fahrenheit (°F). When the German physicist Gabriel Fahrenheit introduced this scale in the early 1700s, he intended 0°F to represent the coldest temperature then achievable (with a mixture of ice and saltwater) and 100°F to represent the temperature of a healthy human body. On the Fahrenheit scale, water freezes at 32°F and boils at 212°F. Because there are 180 degrees Fahrenheit between the freezing and boiling points of water, a degree Fahrenheit is only 100/180 (= 5/9) the size of the other scales.

The following equation converts from degrees Fahrenheit to degrees Celsius:

$$T_C = 5/9 \, (T_F - 32)$$

To convert from Celsius to Fahrenheit, a simple rearrangement of terms gives the relationship

$$T_F = 9/5 \, T_C + 32$$

where T_F is the temperature in degrees Fahrenheit and T_C is the temperature in degrees Celsius.

For example, consider a typical room temperature of 68°F. Using the first equation, we can convert this measurement to the Celsius scale as follows:

$$T_C = 5/9 \, (68 - 32) = 20°C$$

To arrive at the Kelvin scale, we simply add 273 degrees to the value in degrees Celsius. Thus, 68°F = 20°C = 293 K. The accompanying figure displays the relationships among these three temperature scales.

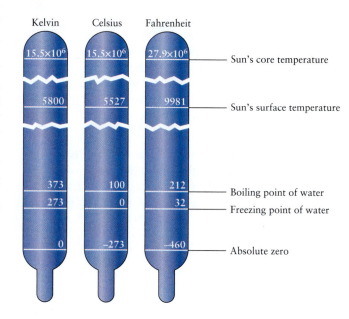

Try these questions: The Sun's surface temperature is about 5800 K. What is its temperature in Celsius and Fahrenheit? The temperature of empty space is about 3 K. What is its temperature in Celsius and Fahrenheit? (Answers appear at the end of the book.)

APPENDIX E: Data Tables

TABLE E-1 The Planets: Orbital Data

Planet	Semimajor axis (AU)	Semimajor axis (10⁶ km)	Sidereal period (year)	Sidereal period (day)	Synodic period (day)	Mean orbital speed (km/s)	Orbital eccentricity	Inclination of orbit to ecliptic (°)
Mercury	0.3871	57.9	0.2408	87.97	115.88	47.9	0.206	7.00
Venus	0.7233	108.2	0.6152	224.70	583.92	35.0	0.007	3.39
Earth	1.0000	149.6	1.0000	365.26	—	29.8	0.017	0.00
Mars	1.5237	227.9	1.8808	686.98	779.94	24.1	0.093	1.85
Jupiter	5.2034	778.6	11.862	4,332.6	398.9	13.1	0.048	1.31
Saturn	9.5371	1433.5	29.457	10,759	378.1	9.7	0.054	2.48
Uranus	19.1913	2872.5	84.01	30,685	369.7	6.8	0.047	0.77
Neptune	30.0690	4495.1	164.79	60,189	367.5	5.4	0.009	1.77

TABLE E-2 The Planets: Physical Data

Planet	Equatorial diameter (km)	Equatorial diameter (Earth = 1)	Mass (kg)	Mass (Earth = 1)	Mean density (kg/m³)	Rotation period* (days)	Inclination of equator to orbit (°)	Surface gravity (Earth = 1)	Albedo	Escape speed (km/s)
Mercury	4,879	0.383	3.302×10^{23}	0.055	5427	58.646	0.0	0.38	0.12	4.3
Venus	12,104	0.949	4.869×10^{24}	0.815	5243	243.02^R	177.4	0.91	0.59	10.4
Earth	12,756	1.000	5.974×10^{24}	1.000	5515	0.996	23.45	1.000	0.37	11.2
Mars	6,794	0.533	6.419×10^{23}	0.107	3933	1.025	25.19	0.38	0.16	5.0
Jupiter	142,984	11.209	1.899×10^{27}	317.83	1326	0.413	3.13	2.5	0.52	59.5
Saturn	120,536	9.449	5.685×10^{26}	95.16	687	0.445	26.73	1.1	0.47	35.5
Uranus	51,118	4.007	8.683×10^{25}	14.54	1270	0.717^R	97.77	0.91	0.56	21.3
Neptune	49,528	3.883	1.024×10^{26}	17.147	1638	0.671	28.32	1.1	0.41	23.5

* For Jupiter, Saturn, Uranus, and Neptune, the internal rotation period is given. A superscript R means that the rotation is retrograde (opposite the planet's orbital motion).

TABLE E-3 Satellites of the Planets

Planet	Satellite	Discoverers of satellite	Average distance from center (km)	Orbital (sidereal) period* (days)	Orbital eccentricity	Diameter of satellite (km)	Mass (kg)
EARTH	Moon	—	384,400	27.322	0.0549	3476	7.349×10^{22}
MARS	Phobos	Hall (1877)	9,378	0.319	0.02	$28 \times 23 \times 2$	1.1×10^{16}
	Deimos	Hall (1877)	23,459	1.262	0.00	$16 \times 12 \times 10$	2.4×10^{15}
JUPITER	Metis	Synott (1980)	127,960	0.2948	0.00	44	1×10^{17}
	Adrastea	Jewitt et al. (1979)	128,980	0.2983	0.00	$24 \times 16 \times 20$	1.9×10^{16}
	Amalthea	Barnard (1892)	181,300	0.4982	0.00	$270 \times 170 \times 150$	7.5×10^{18}
	Thebe	Synott (1979)	221,900	0.6745	0.02	98	8×10^{17}
	Io	Galileo (1610)	421,600	1.769	0.00	3643	8.93×10^{22}
	Europa	Galileo (1610)	670,900	3.551	0.01	3138	4.80×10^{22}
	Ganymede	Galileo (1610)	1,070,000	7.155	0.00	5268	1.48×10^{23}
	Callisto	Galileo (1610)	1,883,000	16.689	0.01	4821	1.08×10^{23}
	Themisto	Kowal (1975)	7,435,000	130.02	0.24	8	(?)
	Leda	Kowal (1974)	11,094,000	238.72	0.15	18	6×10^{15}
	Himalia	Perrine (1904)	11,480,000	250.57	0.16	170	9.5×10^{18}
	Lysithea	Nicholson (1938)	11,720,000	259.22	0.11	38	8×10^{16}
	Elara	Perrine (1905)	11,737,000	259.65	0.21	80	8×10^{17}
	S/2000/J11	Sheppard et al. (2000)	12,654,000	287	0.25	4	(?)
	Euporie	Sheppard et al. (2001)	19,017,000	553.1^R	0.16	2	(?)
	Chaldene	Sheppard et al. (2000)	20,375,000	723.8^R	0.24	4	(?)
	Iocaste	Sheppard et al. (2000)	20,733,000	632^R	0.22	5	(?)
	Kale	Sheppard et al. (2001)	20,804,000	721^R	0.48	2	(?)
	Orthosie	Sheppard et al. (2001)	20,876,000	623^R	0.27	2	(?)
	Thyone	Sheppard et al. (2001)	20,876,000	632^R	0.30	4	(?)
	Euanthe	Sheppard et al. (2001)	20,947,000	620^R	0.18	3	(?)
	Harpalyke	Sheppard et al. (2000)	21,019,000	623^R	0.23	4	(?)
	Praxidike	Sheppard et al. (2000)	21,162,000	625^R	0.22	7	(?)
	Ananke	Nicholson (1951)	21,200,000	631^R	0.17	28	4×10^{16}
	Hermippe	Sheppard et al. (2001)	21,376,000	632^R	0.25	4	(?)
	Taygete	Sheppard et al. (2000)	21,734,000	732^R	0.25	5	(?)
	Erinome	Sheppard et al. (2000)	21,948,000	728^R	0.27	3	(?)
	Carme	Nicholson (1938)	22,600,000	692^R	0.21	46	9×10^{16}
	Isonoe	Sheppard et al. (2000)	22,806,000	726	0.26	4	(?)
	Pasithee	Sheppard et al. (2001)	22,949,000	715^R	0.29	2	(?)
	Eurydome	Sheppard et al. (2001)	23,378,000	721^R	0.35	3	(?)
	Aitne	Sheppard et al. (2001)	23,449,000	741^R	0.29	3	(?)
	Pasiphae	Melotte (1908)	23,500,000	735^R	0.38	36	2×10^{17}
	Megaclite	Sheppard et al. (2000)	23,521,000	753^R	0.43	5	(?)
	Sponde	Sheppard et al. (2001)	23,592,000	749^R	0.45	2	(?)
	Sinope	Nicholson (1914)	23,700,000	758^R	0.28	28	8×10^{16}
	Autonoe	Sheppard et al. (2001)	23,979,000	765^R	0.42	4	(?)
	Kalyke	Sheppard et al. (2000)	24,164,000	743^R	0.24	5	(?)
	Callirrhoe	Scotti et al. (1999)	24,200,000	759^R	0.28	8	(?)

TABLE E-3 Satellites of the Planets (continued)

Planet	Satellite	Discoverers of satellite	Average distance from center (km)	Orbital (sidereal) period* (days)	Orbital eccentricity	Diameter of satellite (km)	Mass (kg)
SATURN	Pan	Showalter (1990)	133,570	0.573	0.00	20	(?)
	Atlas	Terrile (1980)	137,670	0.602	0	$40 \times 30 \times 30$	8×10^{17}
	Prometheus	Collins et al. (1980)	139,353	0.613	0.00	$140 \times 80 \times 100$	8×10^{17}
	Pandora	Collins et al. (1980)	141,700	0.629	0.00	$110 \times 70 \times 100$	2×10^{17}
	Epimetheus	Walker (1966)	151,422	0.694	0.01	$140 \times 100 \times 100$	5×10^{17}
	Janus	Dolfuss (1966)	151,472	0.695	0.01	$220 \times 160 \times 200$	2×10^{18}
	Mimas	Herschel (1789)	185,520	0.942	0.02	392	3.8×10^{19}
	Enceladus	Herschel (1789)	238,020	1.370	0.00	444	7.3×10^{19}
	Tethys	Cassini (1684)	294,660	1.888	0.00	1050	6.3×10^{20}
	Calypso	Pascu et al. (1980)	294,660	1.888	0 (?)	$30 \times 20 \times 25$	8×10^{17}
	Telesto	Smith et al. (1980)	294,660	1.888	0 (?)	24	8×10^{17}
	Dione	Cassini (1684)	377,400	2.737	0.00	1120	1.1×10^{21}
	Helene	Laques et al. (1980)	377,400	2.737	0.01	$40 \times 30 \times 30$	8×10^{17}
	Rhea	Cassini (1672)	527,040	4.518	0.00	1528	2.3×10^{21}
	Titan	Huygens (1655)	1,221,830	15.945	0.03	5150	1.3×10^{23}
	Hyperion	Bond et al. (1848)	1,481,100	21.277	0.10	$410 \times 260 \times 220$	8×10^{17}
	Iapetus	Cassini (1671)	3,561,300	79.330	0.03	1436	1.6×10^{21}
	Phoebe	Pickering (1898)	12,952,000	550.56^R	0.16	220	4×10^{17}
URANUS	Cordelia	*Voyager 2* (1986)	49,752	0.335	0.00	26	8×10^{17}
	Ophelia	*Voyager 2* (1986)	53,763	0.376	0.01	30	8×10^{17}
	Bianca	*Voyager 2* (1986)	59,166	0.435	0.00	42	8×10^{17}
	Cressida	*Voyager 2* (1986)	61,767	0.464	0.00	66	8×10^{17}
	Desdemona	*Voyager 2* (1986)	62,658	0.474	0.00	54	8×10^{17}
	Juliet	*Voyager 2* (1986)	64,358	0.493	0.00	84	8×10^{17}
	Portia	*Voyager 2* (1986)	66,097	0.513	0.00	108	8×10^{17}
	Rosalind	*Voyager 2* (1986)	69,927	0.558	0.00	54	8×10^{17}
	Mab	Showalter et al. (2003)	97,734	(?)	0 (?)	26	(?)
	Belinda	*Voyager 2* (1986)	75,236	0.624	0.00	66	8×10^{17}
	Perdita	*Voyager 2* (1986)	76,420	0.638	0 (?)	20	(?)
	Puck	*Voyager 2* (1986)	86,004	0.762	0.00	144	8×10^{17}
	Cupid	Showalter et al. (2003)	74,800	(?)	0 (?)	10	(?)
	Miranda	Kuiper (1948)	129,872	1.413	0.00	~470	6.6×10^{19}
	Ariel	Lassell (1851)	190,945	2.520	0.00	~1160	1.4×10^{21}
	Umbriel	Lassell (1851)	265,998	4.144	0.01	1169	1.2×10^{21}
	Titania	Herschel (1787)	436,298	8.706	0.00	1578	3.5×10^{21}
	Oberon	Herschel (1787)	583,519	13.463	0.00	1523	3.0×10^{21}
	Caliban	Gladman et al. (1997)	7,169,000	579^R	0.16	80	8×10^{17}
	Sycorax	Gladman et al. (1997)	12,213,000	1284^R	0.52	160	8×10^{17}
	Stephano	Kavelaars et al. (1999)	>7,979,000	677^R	0.23	32	8×10^{17}
	Prospero	Kavelaars et al. (1999)	>16,665,000	1993^R	0.43	50	8×10^{17}
	Setebos	Kavelaars et al. (1999)	>17,879,000	2194^R	0.59	47	8×10^{17}

TABLE E-3 Satellites of the Planets (continued)

Planet	Satellite	Discoverers of satellite	Average distance from center (km)	Orbital (sidereal) period* (days)	Orbital eccentricity	Diameter of satellite (km)	Mass (kg)
NEPTUNE	Naiad	*Voyager 2* (1989)	48,230	0.294	0.00	58	8×10^{17}
	Thalassa	*Voyager 2* (1989)	50,070	0.311	0.00	80	8×10^{17}
	Despina	*Voyager 2* (1989)	52,530	0.335	0.00	148	8×10^{17}
	Galatea	*Voyager 2* (1989)	61,950	0.429	0.05	178	8×10^{17}
	Larissa	*Voyager 2* (1989)	73,550	0.553	0.00	~190	8×10^{17}
	Proteus	*Voyager 2* (1989)	117,650	1.122	0.00	~415	8×10^{17}
	Triton	Lassell (1846)	354,760	5.877^{R}	0.00	2704	2.14×10^{22}
	Nereid	Kuiper (1949)	5,513,400	360.1	0.75	340	2×10^{19}

* A superscript R means that the satellite orbits in a retrograde direction (opposite to the planet's rotation).

TABLE E-4 Dwarf Planets

	Pluto	Ceres	Eris	Makemake	Haumea
Year of discovery	1930	1801	2003	2005	2004
Semimajor axis (AU)	39.4817	2.766	67.668	45.79	43.34
Semimajor axis (10^6 km)	5906.4	413.7	10210	6848.6	6483.6
Sidereal period (year)	247.7	4.599	557	310	285
Sidereal period (day)	90470	1680	2.03×10^5	113,183	104,234
Mean orbital speed (km/s)	4.67	17.88	3.44	4.48	4.48
Orbital eccentricity	0.249	0.080	0.442	0.159	0.189
Inclination of orbit to ecliptic (°)	17.14	10.59	44.19	28.96	28.19
Equatorial diameter (km)	2390	941	2400	1900	1960^1
Equatorial diameter (Earth = 1)	0.19	0.074	0.19	0.151	0.156
Mass (kg)	1.3×10^{22}	9.5×10^{20}	1.7×10^{22}	4.0×10^{21}	4.2×10^{21}
Mass (Earth = 1)	2.2×10^{-3}	1.6×10^{-4}	2.8×10^{-3}	6.77×10^{-4}	7.11×10^{-4}
Mean density (kg/m³)	2030	2080	2100	~2000	~3000
Rotation period (days)	6.388^{R}	0.3781	?	?	0.163
Inclination of equator to orbit (°)	122.5	4	?	?	?
Surface gravity (Earth = 1)	0.06	0.028	0.07	0.048	0.045
Escape velocity (km/s)	1.2	0.51	1.3	0.84	0.84
Location	Kuiper belt (TNO)	Asteroid belt	TNO	Kuiper belt (TNO)	Kuiper belt (TNO)
Number of known satellites	3	0	1	0	2

[1] Haumea is very oblong, with dimensions 1960 km × 1518 km × 996 km

TABLE E-5 The Nearest Stars

Name*	Parallax (arcsec)	Distance (ly)	Spectral type	Radial velocity** (km/s)	Proper motion (arcsec/year)	Apparent visual magnitude	Absolute visual magnitude	Luminosity (Sun = 1)
Sun			G2 V			−26.7	+4.85	1.00
Proxima Centauri	0.769	4.22	M5.5 V	−22	3.853	+11.09	+15.53	8.2×10^{-4}
Alpha Centauri A	0.747	4.40	G2 V	−25	3.710	−0.01	+4.38	1.77
Alpha Centauri B	0.747	4.40	K0 V	−21	3.724	+1.34	+5.71	0.55
Barnard's Star	0.547	5.94	M4 V	−111	10.358	+9.53	+13.22	3.6×10^{-3}
Wolf 359	0.419	7.80	M6 V	+13	4.696	+13.44	+16.6	3.5×10^{-4}
Lalande 21185	0.393	8.32	M2 V	−84	4.802	+7.47	+10.44	0.023
L 726-8 A	0.374	8.56	M5.5 V	+29	3.368	+12.54	+15.4	9.4×10^{-4}
L 726-8 B	0.374	8.56	M6 V	+32	3.368	+12.99	+15.9	5.6×10^{-4}
Sirius A	0.380	8.61	A1 V	−9	1.339	−1.43	+1.47	26.1
Sirius B	0.380	8.61	white dwarf	−9	1.339	+8.44	+11.34	2.4×10^{-3}
Ross 154	0.337	9.71	M3.5 V	−12	0.666	+10.43	+13.07	4.1×10^{-3}
Ross 248	0.316	10.32	M5.5 V	−78	1.617	+12.29	+14.8	1.5×10^{-3}
Epsilon Eridani	0.310	10.49	K2 V	+17	0.977	+3.73	+6.19	0.40
Lacaille 9352	0.304	10.73	M1.5 V	+10	6.896	+7.34	+9.75	0.051
Ross 128	0.299	10.87	M4 V	−31	1.361	+11.13	+13.51	2.9×10^{-3}
L 789-6	0.294	11.09	M5 V	−60	3.259	+12.33	+14.7	1.3×10^{-3}
61 Cygni A	0.286	11.36	K5 V	−65	5.281	+5.21	+7.49	0.16
61 Cygni B	0.286	11.44	K7 V	−64	5.172	+6.03	+8.31	0.095
Procyon A	0.286	11.40	F5 IV−V	−4	1.259	+0.38	+2.66	7.73
Procyon B	0.286	11.40	white dwarf	−4	1.259	+10.7	+12.98	5.5×10^{-4}
BD +59° 1915 A	0.281	11.61	M3 V	−1	2.238	+8.94	+11.18	0.020
BD +59° 1915 B	0.281	11.61	M3.5 V	+1	2.313	+9.70	+11.97	0.010
Groombridge 34 A	0.281	11.65	M1.5 V	+12	2.918	+8.08	+10.32	0.030
Groombridge 34 B	0.281	11.65	M3.5 V	+11	2.918	+11.06	+13.3	3.1×10^{-3}
Epsilon Indi	0.276	11.82	K5 V	−40	4.704	+4.69	+6.89	0.27
GJ 1111	0.276	11.82	M6.5 V	−5	1.290	+14.78	+16.98	2.7×10^{-4}
Tau Ceti	0.274	11.90	G8 V	−17	1.922	+3.49	+5.68	0.62
GJ 1061	0.272	12.08	M5.5 V	−20	0.826	+13.09	+15.26	1.0×10^{-3}
L 725-32	0.269	12.12	M4.5 V	+28	1.372	+12.10	+14.25	1.7×10^{-3}
BD +05° 1668	0.263	12.40	M3.5 V	+18	3.738	+9.84	+11.94	0.011
Kapteyn's star	0.255	12.79	M1.5 V	+246	8.670	+8.84	+10.87	0.013
Lacaille 8760	0.253	12.89	M0 V	+28	3.455	+6.67	+8.69	0.094
Krüger 60 A	0.248	13.05	M3 V	−33	0.990	+9.79	+11.76	0.010
Krüger 60 B	0.248	13.05	M4 V	−32	0.990	+11.41	+13.38	3.4×10^{-3}

* Stars that are components of binary systems are labeled A and B.
** A positive radial velocity means the star is receding; a negative radial velocity means the star is approaching.
Compiled from the Hipparcos General Catalogue and from data reported by the Research Consortium on Nearby Stars. The table lists all known stars within 4.00 parsecs (13.05 light-years).

TABLE E-6 The Visually Brightest Stars

Name	Designation	Distance (ly)	Spectral type	Radial velocity* (km/s)	Proper motion (arcsec/year)	Apparent visual magnitude	Apparent visual brightness** (Sirius = 1)	Absolute visual magnitude	Luminosity (Sun = 1)
Sirius A	α CMa A	8.61	A1 V	−9	1.339	−1.43	1.000	+1.47	26.1
Canopus	α Car	313	F0 I	+21	0.031	−0.62	0.470	−5.53	1.4×10^4
Arcturus	α Boo	36.7	K2 III	−5	2.279	−0.05	0.278	−0.31	190
Rigil Kentaurus	α Cen A	4.4	G2 V	−25	3.71	−0.01	0.268	+4.38	1.77
Vega	α Lyr	25.3	A0 V	−14	0.035	+0.03	0.258	+0.58	61.9
Capella	α Aur	42.2	G8 III	+30	0.434	+0.08	0.247	−0.48	180
Rigel	β Ori A	773	B8 Ia	+21	0.002	+0.18	0.225	−6.69	7.0×10^5
Procyon	α CMi A	11.4	F5 IV−V	−4	1.259	+0.38	0.184	+2.66	7.73
Achernar	α Eri	144	B3 IV	+19	0.097	+0.45	0.175	−2.77	5250
Betelgeuse	α Ori	427	M2 Iab	+21	0.029	+0.45	0.175	−5.14	4.1×10^4
Hadar	β Cen	525	B1 II	−12	0.042	+0.61	0.151	−5.42	8.6×10^4
Altair	α Aql	16.8	A7 IV−V	−26	0.661	+0.77	0.132	+2.2	11.8
Aldebaran	α Tau A	65.1	K5 III	+54	0.199	+0.87	0.119	−0.63	370
Spica	α Vir	262	B1 V	+1	0.053	+0.98	0.108	−3.55	2.5×10^4
Antares	α Sco A	604	M1 Ib	−3	0.025	+1.06	0.100	−5.28	3.7×10^4
Pollux	β Gem	33.7	K0 III	+3	0.627	+1.16	0.091	+1.09	46.6
Fomalhaut	α PsA	25.1	A3 V	+7	0.368	+1.17	0.090	+1.74	18.9
Deneb	α Cyg	3230	A2 Ia	−5	0.002	+1.25	0.084	−8.73	3.2×10^5
Mimosa	β Cru	353	B0.5 III	+20	0.05	+1.25	0.084	−3.92	3.4×10^4
Regulus	α Leo A	77.5	B7 V	+4	0.249	+1.36	0.076	−0.52	331

Data in this table was compiled from the Hipparcos General Catalogue.
* A positive radial velocity means the star is receding; a negative radial velocity means the star is approaching.
** This is the ratio of the star's apparent brightness to that of Sirius, the brightest star in the night sky
Note: Acrux, or α Cru (the brightest star in Crux, the Southern Cross), appears to the naked eye as a star of apparent magnitude +0.87, the same as Aldebaran, but it does not appear in this table because Acrux is actually a binary star system. The blue-white component stars of this binary system have apparent magnitudes of +1.4 and +1.9, and thus they are dimmer than any of the stars listed here.

TABLE E-7 The Constellations

Name	Meaning	R.A.	Dec.	Genitive*	Abbreviation
Andromeda	proper name; princess	1	+40	Andromedae	And
Antlia	air pump	10	−35	Antliae	Ant
Apus	bee	16	−75	Apodis	Aps
Aquarius[1,2]	waterman	22	−10	Aquarii	Aqr
Aquila	eagle	20	+15	Aquilae	Aql
Ara	altar	17	−55	Arae	Ara
Aries[2]	ram	3	+20	Arietis	Ari
Auriga	charioteer	6	+40	Aurigae	Aur
Boötes	proper name; herdsman, wagoner	15	+30	Boötis	Boo
Caelum	engraving tool	5	−40	Caeli	Cae
Camelopardalis	giraffe	6	+70	Camelopardalis	Cam
Cancer[2]	crab	8.5	+15	Cancri	Cnc
Canes Venatici	hunting dogs	13	+40	Canum Venaticorum	CVn
Canis Major	larger dog	7	−20	Canis Majoris	CMa
Canis Minor	smaller dog	8	+5	Canis Minoris	CMi
Capricornus[1,2]	water-goat	21	−20	Capricornii	Cap
Carina	keel	9	−60	Carinae	Car
Cassiopeia	proper name; queen	1	+60	Cassiopeiae	Cas
Centaurus	centaur	13	−45	Centauri	Cen
Cepheus	proper name; king	22	+65	Cephei	Cep
Cetus	whale	2	−10	Ceti	Cet
Chamaeleon	chameleon	10	−80	Chamaeleontis	Cha
Circinus	compasses	15	−65	Circini	Cir
Columba	dove	6	−35	Columbae	Col
Coma Berenices	Berenice's hair	13	+20	Comae Berenices	Com
Corona Australis[3]	southern crown	19	+40	Coronae Australis	CrA
Corona Borealis[4]	northern crown	16	+30	Coronae Borealis	CrB
Corvus[5]	crow, raven	12	−20	Corvi	Crv
Crater	cup	11	−15	Crateris	Crt
Crux[6]	southern cross	12	−60	Crucis	Cru
Cygnus	swan	21	+40	Cygni	Cyg
Delphinus[1]	dolphin	21	+10	Delphini	Del
Dorado[7]	swordfish	6	−55	Doradus	Dor
Draco[8]	dragon	15	+60	Draconis	Dra
Equuleus	little horse	21	+10	Equulei	Equ
Eridanus	proper name; river	4	−30	Eridani	Eri
Fornax	furnace	3	−30	Fornacis	For
Gemini[2]	twins	7	+20	Geminorum	Gem
Grus	crane	22	−45	Gruis	Gru
Hercules[9]	proper name; hero	17	+30	Herculis	Her
Horologium	clock	3	−55	Horologii	Hor

TABLE E-7 The Constellations (continued)

Name	Meaning	R.A.	Dec.	Genitive*	Abbreviation
Hydra	water serpant	12	−25	Hydrae	Hya
Hydrus	water snake	2	−70	Hydri	Hyi
Indus	Indian	22	−70	Indi	Ind
Lacerta	lizard	22	+45	Lacertae	Lac
Leo[2]	lion	11	+15	Leonis	Leo
Leo Minor	smaller lion	10	+35	Leonis Minoris	LMi
Lepus	hare	6	−20	Leporis	Lep
Libra[2,10]	scales	15	−15	Librae	Lib
Lupus	wolf	15	−45	Lupi	Lup
Lynx	lynx	8	+45	Lyncis	Lyn
Lyra[5]	lyre	19	+35	Lyrae	Lyr
Microscopium	microscope	21	−40	Microscopii	Mic
Monoceros	unicorn	7	0	Monocerotis	Mon
Mensa	table	6	−75	Mensae	Men
Musca (Australis)[11]	(southern) fly	12	−70	Muscae	Mus
Norma	square	16	−50	Normae	Nor
Octans[12]	octant	—	−90	Octantis	Oct
Ophiuchus[13]	serpent-bearer	17	0	Ophiuchi	Oph
Orion	proper name; hunter, giant	6	0	Orionis	Ori
Pavo	peacock	20	−70	Pavonis	Pav
Pegasus	proper name; winged horse	23	+20	Pegasi	Peg
Perseus	proper name; hero	3	+45	Persei	Per
Phoenix	phoenix	1	−50	Phoenicis	Phe
Pictor	easel	6	−55	Pictoris	Pic
Pisces[1,2]	fishes	1	+10	Piscium	Psc
Piscis Austrinus[1]	southern fish	22	−30	Piscis Austrini	PsA
Puppis	stern	8	−30	Puppis	Pup
Pyxis	compass	9	−30	Pyxidis	Pyx
Reticulum	net	4	−60	Reticuli	Ret
Sagitta	arrow	20	+20	Sagittae	Sge
Sagittarius[2,14]	archer	19	−25	Sagittarii	Sgr
Scorpius[2]	scorpion	17	−30	Scorpii	Sco
Sculptor[15]	sculptor's workshop	1	−30	Sculptoris	Scl
Scutum[16]	shield	19	−10	Scuti	Sct
Serpens[13]	serpent	17	0	Serpentis	Ser
Sextans	sextant	10	0	Sextantis	Sex
Taurus[2]	bull	5	+20	Tauri	Tau
Telescopium	telescope	19	−50	Telescopii	Tel
Triangulum	triangle	2	+30	Trianguli	Tri
Triangulum Australe	southern triangle	16	−65	Trianguli Australis	TrA
Tucana[17]	toucan	0	−65	Tucanae	Tuc
Ursa Major	larger bear	11	+60	Ursae Majoris	UMa
Ursa Minor[18]	smaller bear	16	+80	Ursae Minoris	UMi

TABLE E-7 The Constellations (continued)

Name	Meaning	R.A.	Dec.	Genitive*	Abbreviation
Vela	sails	10	−45	Velorum	Vel
Virgo[2]	Virgin	13	0	Virginis	Vir
Volans	flying fish	8	−70	Volantis	Vol
Vulpecula	fox	20	+25	Vulpeculae	Vul

* Genitive is the grammatical case denoting possession. For example, astronomers denote the brightest or α (alpha) star in Orion (Betelgeuse) as α Orionis.
[1] Constellations of the area of the sky known as the wet quarter for its many watery images.
[2] A zodiac constellation.
[3] Sometimes considered as Sagittarius's crown.
[4] Ariadne's crown.
[5] Corvus was Orpheus's companion, Lyra his harp.
[6] Originally a part of Centaurus.
[7] Contains the Large Magellanic Cloud and the south ecliptic pole.
[8] Contains the north ecliptic pole.
[9] One of the oldest constellations known.
[10] Originally the claws of Scorpius.
[11] Originally named Musca Australis to distinguish it from Musca Borealis, the northern fly, which is now defunct; "Australis" is now dropped.
[12] Contains the south celestial pole.
[13] Ophiuchus is identified with the physician Aesculapius, and Serpens with the caduceus.
[14] Contains the galactic center.
[15] Originally named l'Atelier du Sculpteur (in Latin, Apparatus Sculptoris) by Lacaille; now known simply as Sculptor. Contains the south galactic pole.
[16] Shield of the Polish hero John Sobieski.
[17] Contains the Small Magellanic Cloud.
[18] Contains the north celestial pole.

TABLE E-8 Some Useful Astronomical Quantities

Astronomical unit:	1 AU = 1.496×10^{11} m
Light-year:	1 ly = 9.461×10^{15} m = 63,240 AU
Parsec:	1 pc = 3.086×10^{16} m = 3.262 ly
Solar mass:	1 $M_\odot$ = 1.989×10^{30} kg
Solar radius:	1 $R_\odot$ = 6.960×10^{8} m
Solar luminosity:	1 $L_\odot$ = 3.827×10^{26} W
Earth's mass:	1 $M_\oplus$ = 5.974×10^{24} kg
Earth's equatorial radius:	1 $R_\oplus$ = 6.378×10^{6} m
Moon's mass:	1 M_{Moon} = 7.349×10^{22} kg
Moon's equatorial radius:	1 R_{Moon} = 1.738×10^{6} m

TABLE E-9 Some Useful Physical Constants

Speed of light:	$c = 2.998 \times 10^{8}$ m/s
Gravitational constant:	$G = 6.668 \times 10^{-11}$ N m^2 kg^{-2}
Planck constant:	$h = 6.626 \times 10^{-34}$ J s = 4.136×10^{-15} eV s
Boltzmann constant:	$k = 1.380 \times 10^{-23}$ J K^{-1} = 8.617×10^{-5} eV K^{-1}
Stefan–Boltzmann constant:	$\sigma = 5.670 \times 10^{-8}$ W m^{-2} K^{-4}
Mass of electron:	$m_e = 9.109 \times 10^{-31}$ kg
Mass of 1H atom:	$m_H = 1.673 \times 10^{-27}$ kg

TABLE E-10 Common Conversions between British and Metric Units

1 inch = 2.54 centimeters (cm)
1 cm = 0.394 inch (in)
1 yard = 0.914 meter (m)
1 meter = 1.09 yards = 39.37 inches
1 mile = 1.61 kilometers (km)
1 km = 0.621 mile (mi)

TABLE E-11 Spiral Galaxies and Interacting Galaxies

Spiral galaxy	R.A.	Decl.	Hubble type	Interacting galaxies	R.A.	Decl.
M31 (NGC 224)	0^h 42.7^m	+41° 16′	Sb	M51 (NGC 5194)	13^h 29.9^m	+47°12′
M58 (NGC 4579)	12 37.7	+11 49	Sb	M51b (NGC 5195)	13 30.0	+47 16
M61 (NGC 4303)	12 21.9	+ 4 28	Sc	M65 (NGC 3623)	11 18.9	+13 05
M63 (NGC 5055)	13 15.8	+42 02	Sb	M66 (NGC 3627)	11 20.2	+12 59
M64 (NGC 4826)	12 56.7	+21 41	Sb	M81 (NGC 3031)	9 55.6	+69 04
M74 (NGC 628)	1 36.7	+15 47	Sc	M82 (NGC 3034)	9 55.8	+69 41
M83 (NGC 5236)	13 37.0	−29 52	Sc	M95 (NGC 3351)	10 44.0	+11 42
M88 (NGC 4501)	12 32.0	+14 25	Sb	M96 (NGC 3368)	10 46.8	+11 49
M90 (NGC 4569)	12 36.8	+13 10	Sb	M105 (NGC 3379)	10 47.8	+12 35
M94 (NGC 4736)	12 50.9	+41 07	Sb			
M98 (NGC 4192)	12 13.8	+14 54	Sb			
M99 (NGC 4254)	12 18.8	+14 25	Sc			
M100 (NGC 4321)	12 22.9	+15 49	Sc			
M101 (NGC 5457)	14 03.2	+54 21	Sc			
M104 (NGC 4594)	12 40.0	−11 37	Sa			
M108 (NGC 3556)	11 11.5	+55 40	Sc			

TABLE E-12 Mass and Energy Inventory for the Universe

	Individual Contribution	Section Total
Dark matter and dark energy contributions		0.954 ± 0.003*
Dark energy	0.72 ± 0.03	
Dark matter	0.23 ± 0.03	
Primeval Graviational Radiation	$\leq 10^{-10}$	
Contributions from Big Bang era		0.0010 ± 0.0005
Electromagnetic Radiation	$10^{-4.3 \pm 0.000001}$	
Neutrinos	$10^{-2.9 \pm 0.1}$	
Normal particle (baryon) rest mass		0.045 ± 0.003
Charged particles (plasma) between stars and galaxies	0.0418 ± 0.003	
Main sequence stars in elliptical galaxies and nuclear bulges	0.0015 ± 0.0004	
Neutral hydrogen & helium	0.00062 ± 0.00010	
Main sequence stars in galactic disks and in irregular galaxies	0.00055 ± 0.00014	
White dwarfs	0.00036 ± 0.00008	
Molecular gas	0.00016 ± 0.00006	
Substellar objects	0.00014 ± 0.00007	
Black holes	0.00007 ± 0.00002	
Neutron stars	0.00005 ± 0.00002	
Planets	$10^{-6 \pm 0.1}$	

*The numbers after the plus or minus (±) symbol indicate the possible errors in the given numbers.
This table lists the major contributions to the mass and energy of the universe.

APPENDIX F: More to Know

For expanded coverage of these key concepts in the text, visit our More to Know resource at www.whfreeman.com/dtustars.

Understanding Astronomy

Chapter 1

- Earth Satellites
- Stars' Technical Names
- Height of the Sun
- Maximum Number of Eclipses Annually

Chapter 2

- The Conservation of Angular Momentum

Chapter 3

- Seeing Nonvisible Radiation

Chapter 4

- Radioactive Age-Dating

Understanding the Stars

Chapter 5

- Zeeman Effect
- Fusion

Chapter 6

- Exact Spectra
- Stellar Sizes

Chapter 8

- Origins of Cosmic Rays
- Brightest Planetary Nebulae

Chapter 9

- Gravitational Waves

Understanding the Universe

Chapter 11

- Computers in Astronomical Research

Chapter 12

- Black Hole Masses or How Plausible Are Extremely Massive Black Holes?

Understanding the Solar System

Chapter 14

- Computer Simulations of Planet Formation
- Spectra

Chapter 15

- Waves in and on Earth
- The Origin of Earth's Magnetic Field
- Radioactive Age-Dating

Chapter 16

- *Mariner 10* and the Exploration of Mercury
- Radar Doppler Measurements

Chapter 17

- Roche Limit

Chapter 18

- Orbital Resonances
- Discovering Comets
- Meteors

APPENDIX G: Using the *Starry Night Enthusiast*™ CD-ROM

The *Starry Night Manual* is available on the *Starry Night Enthusiast*™ CD-ROM. It contains detailed instructions on how to install the *Starry Night Enthusiast*™ software and how to use its various features. Once the software has been installed, this manual is available from the Help menu.

It might be useful, and it is certainly good scientific practice, to keep a notebook specifically for these exercises in which you can write down the date and time of each observation as well as notes on set-up procedures as you become familiar with them. This appendix introduces some of the major features of the software to jumpstart your effective use of it for making astronomical observations.

The first time that you launch *Starry Night Enthusiast*™ after installing it, you will be prompted to select a home location. Once selected, the program will use this home location as the default viewing location when the program opens or a new file is generated. You can use the **File > Set Home Location. . .** menu command to change the default home location whenever necessary. The Home Location dialog window allows you to select your default viewing location by selecting from a list of cities, clicking points on a map, or entering a specific latitude and longitude.

Once the Home Location has been set, the default behavior of *Starry Night Enthusiast*™ when it opens is to present a dialog with a list of interesting astronomical events that may be visible from your home location. You can explore these events with the **View Event** button. To dismiss the dialog window click the **Close** button.

The *Starry Night Enthusiast*™ screen contains four sections:

1. The menu bar across the top of the screen, which contains commands for handling and opening files and altering various options in the display.

2. The toolbar, which is below the menu bar. The toolbar displays the various parameters that are in effect for the current view such as the time, date, viewing location, gaze direction and field of view as well as controls for changing these parameters and for animating the view at various speeds.

3. The main view window comprises the majority of the screen and shows a virtual view of the sky consistent with the parameters indicated in the toolbar.

4. The tabs on the left edge of the main view window open side panes from which you can quickly find objects in the sky, adjust various viewing options,

expedite file handling and obtain more information about the parameters of the observation and details on astronomical objects. You can also obtain further assistance in using the *Starry Night Enthusiast*™ interface through guided tutorials from the appropriate side pane.

Starry Night Enthusiast™ opens with an image of the ground and sky centered on your home location at the date and time specified by the system clock of your computer. You control the time of day or night with the **Time and Date** information on the toolbar and the rate of time flow with the buttons in the Time Flow Rate section of it. The buttons allow you to stop time flow, run time forward (the Play button) or backward, step time forward or backward in specific intervals, and change the rate of time flow. A list of time options will appear when you click the downward pointing arrow to the right of the Time Flow Rate display. You should experiment with these controls.

Clicking a field in the **Time and Date** section of the toolbar allows you to change its value directly from the keyboard or increment and decrement the value with the + and − keys. The **Viewing Location** section of the toolbar contains a panel indicating the current location from which you are observing the sky. The up and down wedge buttons below the viewing location display decrease and increase the current elevation of your viewing location, respectively. The **Home** button resets the view to the current date and time from your home location. Clicking the − and + buttons in the **Zoom** section at the right of the toolbar allow you to zoom in and out.

To find a specific astronomical object in the sky, select **Edit > Find. . .** from the menu or click the **Find** tab on the left of the main view window to open the **Find** pane. The Find pane contains a list of solar system objects when the edit box at the top of the pane is empty. A button containing a + sign to the left of the name of an object in the list indicates that the list for that object can be expanded by clicking the button. To the extreme left of an entry in the object list is a blue button with a downward pointing arrowhead. This menu button contains several important menu commands that allow you to locate and center the object in the view and **Magnify** the object in the view as well as several other interesting commands. To find an object in the sky that is not in the list, type its name in the search box at the top of the **Find** pane.

There are several ways to adjust the view directly from the main view window by using the mouse. The default mouse icon is the Hand Tool. When this mouse icon is visible, you can grab the screen by clicking the mouse (left button on a PC) and dragging the view. You can change the behavior of the mouse by holding down the Shift key on the keyboard. The mouse cursor changes to a small square of four arrowheads to become the Location Scroller. To use the location scroller, click the mouse button (left button on a PC) and drag the mouse while holding down the **Shift** key. Unlike the hand tool, which changes the view by altering the gaze direction, rather like turning your head, the Location Scroller changes the view by altering the viewing location. As you move the four arrowheads with the mouse, you can see your viewing location change on the toolbar.

When the mouse cursor is over a specific object in the sky, it changes its appearance to an arrow with an attached list. Clicking the mouse button (right mouse button on a PC) when the mouse icon is an arrow allows you, among other things, to label and select the specific object at which the mouse is pointing. Pressing **ctrl** when the mouse icon is an arrow reveals details about the object you are viewing. If you click and hold the mouse button (left mouse button on a PC) when it is an arrow and drag the mouse, the Selection Tool can be used to measure angular and physical distances between objects in the sky.

Try the practice exercises below to help to familiarize yourself with the *Starry Night Enthusiast*™ interface.

1. Click the **Home** button to reset the view. **Stop time** and then change the **Time** in the toolbar to **11:00:00** PM on today's date. Select **Labels > Planets-Moons** from the menu. Use the Hand Tool to look around the sky. Will any planets be visible from your home location tonight at 11 PM? If you are using a notebook, record the date and time of your observation and the planets that are visible.

2. Click the **Home** button to reset the view and **Stop time.** Click and hold the **Increase current elevation** button (the second button from the left in the **Viewing Location** section of the toolbar) until the Viewing Location panel indicates that you are about **0.001 au** above your home location. Earth should be visible near the bottom of the view. Position the mouse cursor over the image of Earth so that the cursor changes from the Hand Tool to the Selection Tool. Open the object contextual menu by right-clicking the Selection Tool (Ctrl-click on a Mac) and select Centre from the list of commands. This will center Earth in the view. Select **Labels > Planets-Moons** from the menu. Now use the Location Scroller (hold down the Shift key while clicking and dragging the mouse) to roll Earth around like a ball

in the view until the Moon becomes visible in the view. Position the mouse cursor over the Moon so that it turns into a Selection Tool. Now click and hold the mouse button (the left button on a PC) and drag the mouse from the Moon toward Earth. The Selection Tool will draw a line from the Moon to Earth and display the angular distance between Earth and the Moon from your current viewing location as well as the actual physical distance separating the two bodies. Use the Location Scroller to change your viewing location and repeat the measurement of the angular and physical distances between Earth and the Moon. Which of these values has changed and why?

3. Click the **Home** button to reset the view. **Stop time.** Find the Andromeda Galaxy by opening the **Find** pane and typing "Andromeda Galaxy" into the edit box at the top of the pane. The list will narrow down as you type. Double-click the entry for the Andromeda Galaxy in the Found items list. The view will label and center this object. Use the **Zoom in** button to change the field of view to about 2° wide. Position the mouse cursor over any other interesting objects that you see in this magnified view and use the information displayed on the screen when you position the mouse cursor over these objects to identify them. What are the names of these objects? Use the object contextual menu over these objects to select the **Show Info** command. This will open the Info side pane in which a description and other data on the object are displayed. What types of objects are those that you identified?

4. Click the **Home** button to reset the view. **Stop time.** Click and hold the **Increase current elevation** button (the second button from the left in the **Viewing Location** section of the toolbar) until the Viewing Location panel indicates that you are about **0.001 au** above your home location, as you did in section 2 above. Now continue to increase your altitude above Earth in factors of about 10 as you move away from our location in the universe, noting first the local stars in our neighborhood, then the Milky Way galaxy in which the solar system is located, then the nearby galaxies and finally move outward into the realm of galaxies and note the soap-bubble structure formed by these very distant galaxies. You can stop at any distance and center upon an object and use the location scroller to move around the selected object to see its overall shape.

APPENDIX H: Periodic Table of the Elements

1	2	3	4	5	6	7	8	9	10	11	12	13	14	15	16	17	18
1 **H** Hydrogen																	2 **He** Helium
3 **Li** Lithium	4 **Be** Beryllium											5 **B** Boron	6 **C** Carbon	7 **N** Nitrogen	8 **O** Oxygen	9 **F** Fluorine	10 **Ne** Neon
11 **Na** Sodium	12 **Mg** Magnesium											13 **Al** Aluminum	14 **Si** Silicon	15 **P** Phosphorus	16 **S** Sulfur	17 **Cl** Chlorine	18 **Ar** Argon
19 **K** Potassium	20 **Ca** Calcium	21 **Sc** Scandium	22 **Ti** Titanium	23 **V** Vanadium	24 **Cr** Chromium	25 **Mn** Manganese	26 **Fe** Iron	27 **Co** Cobalt	28 **Ni** Nickel	29 **Cu** Copper	30 **Zn** Zinc	31 **Ga** Gallium	32 **Ge** Germanium	33 **As** Arsenic	34 **Se** Selenium	35 **Br** Bromine	36 **Kr** Krypton
37 **Rb** Rubidium	38 **Sr** Strontium	39 **Y** Yttrium	40 **Zr** Zirconium	41 **Nb** Niobium	42 **Mo** Molybdenum	43 **Tc** Technetium	44 **Ru** Ruthenium	45 **Rh** Rhodium	46 **Pd** Palladium	47 **Ag** Silver	48 **Cd** Cadmium	49 **In** Indium	50 **Sn** Tin	51 **Sb** Antimony	52 **Te** Tellurium	53 **I** Iodine	54 **Xe** Xenon
55 **Cs** Cesium	56 **Ba** Barium	57 **La** Lanthanum	72 **Hf** Hafnium	73 **Ta** Tantalum	74 **W** Tungsten	75 **Re** Rhenium	76 **Os** Osmium	77 **Ir** Iridium	78 **Pt** Platinum	79 **Au** Gold	80 **Hg** Mercury	81 **Tl** Thallium	82 **Pb** Lead	83 **Bi** Bismuth	84 **Po** Polonium	85 **At** Astatine	86 **Rn** Radon
87 **Fr** Francium	88 **Ra** Radium	89 **Ac** Actinium	104 **Rf** Rutherfordium	105 **Db** Dubnium	106 **Sg** Seaborgium	107 **Bh** Bohrium	108 **Hs** Hassium	109 **Mt** Meitnerium	110 **Ds** Darmstadtium	111 **Rg** Roentgenium							

58 **Ce** Cerium	59 **Pr** Praseodymium	60 **Nd** Neodymium	61 **Pm** Promethium	62 **Sm** Samarium	63 **Eu** Europium	64 **Gd** Gadolinium	65 **Tb** Terbium	66 **Dy** Dysprosium	67 **Ho** Holmium	68 **Er** Erbium	69 **Tm** Thulium	70 **Yb** Ytterbium	71 **Lu** Lutetium
90 **Th** Thorium	91 **Pa** Protactinium	92 **U** Uranium	93 **Np** Neptunium	94 **Pu** Plutonium	95 **Am** Americium	96 **Cm** Curium	97 **Bk** Berkelium	98 **Cf** Californium	99 **Es** Einsteinium	100 **Fm** Fermium	101 **Md** Mendelevium	102 **No** Nobelium	103 **Lr** Lawrencium

APPENDIX 1: Changing Pluto's Status as a Planet

Pluto has been removed from the pantheon of planets and relegated to the status of a "dwarf planet," leaving eight surviving planets. Although this change requires everyone to re-learn the number of planets, it places Pluto into a new category that will eventually help astronomers better understand all the types of bodies that orbit the Sun.

Changing categories in astronomy is not new. Imagine that the year is 1780. Six planets are known: Mercury, Venus, Earth, Mars, Jupiter, and Saturn. Now move ahead a year: Uranus has just been discovered, requiring people to learn that there are seven planets, not six. Fast forward nineteen years. . . . It is New Year's Day, 1801 and the Swiss/Italian amateur astronomer Giuseppe Piazzi discovers the eighth planet, which he names Ceres (**Figure 1**). It is located between Mars and Jupiter. During the following year, German astronomer H. Wilhelm Olbers discovers the ninth planet, Pallas, in the same region of the solar system. The discovery of *Pallas* is followed within five years by the discovery of the tenth and eleventh planets—Juno by German astronomer Karl Harding and Vesta by Olbers (**Figure 2**). These bodies (Ceres, Pallas, Juno, and Vesta) also orbit the Sun between Mars and Jupiter.

The fifth such body, Astraea, located between Mars and Jupiter was discovered in 1845 by amateur German astronomer Karl Hencke. It is significantly smaller than any previously-discovered planet, just like all but one[1] of the next eighteen planets discovered by 1852.

FIGURE 2 Asteroid Vesta, between Mars and Jupiter (Ben Zellner [Georgia Southern Univ.], Peter Thomas [Cornell Univ.], and NASA)

Furthermore, all these smaller bodies are located in the same general region of the solar system—between Mars and Jupiter. Seeing a physically meaningful pattern emerging, some astronomers began listing in journals the bodies smaller than Vesta (namely, Astraea and all the subsequently-discovered small bodies) as *asteroids*, or equivalently, *minor planets* or *planetoids*. The number of planets plunged from over two dozen back to twelve. The terms *Asteroids* and *minor planets* are still in common use today.

Prior to 1855, the sizes of Ceres, Pallas, Juno, and Vesta had been grossly over-estimated. During that year, another measurement scheme was used that actually underestimated their sizes. Based on these smaller sizes, these four bodies were added to the list of asteroids, which resulted in classifying eight bodies as planets: Mercury, Venus, Earth, Mars, Jupiter, Saturn, Uranus, and Neptune. This categorization supports the idea that the smaller objects—the asteroids—are fundamentally different than the eight planets.[2]

Modern observational techniques have finally established the sizes of the four largest asteroids as being smaller than the sizes originally measured and larger than the sizes measured in 1855. Nevertheless, these asteroids are all very small compared to even the smallest of the eight planets, Mercury. Indeed, the largest of the asteroids, Ceres, is only a quarter the diameter of our Moon.

Zoom ahead to 1930. During that year, Pluto is discovered by American astronomer Clyde Tombaugh. Pluto was considered to be at the edge of the known solar system and has an orbit that sometimes takes it closer to the Sun than Neptune. Indeed, Pluto's orbit is much more elliptical than the orbits of the eight planets, but similar to the very elliptical orbits of many asteroids.

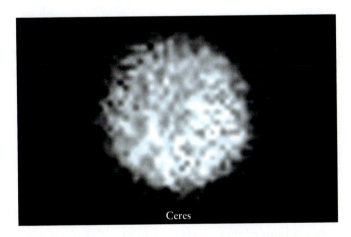

FIGURE 1 Ceres, the Largest Asteroid in Orbit between Mars and Jupiter (NASA, ESA, J. Parker [Southwest Research Institute], P. Thomas [Cornell University], L. McFadden [University of Maryland, College Park], and M. Mutchler and Z. Levay [STScI])

---

[1] The one exception is giant Neptune, discovered in 1846 out beyond Uranus.
[2] We honestly don't yet know all the differences between planets and asteroids or between the various asteroids.

FIGURE 3 Pluto and Its Moon Charon (Dr. R. Albrecht, ESA/ESO Space Telescope European Coordinating Facility; NASA)

However, unlike the asteroids, Pluto is not located between Mars and Jupiter. Pluto's size is decidedly unusual, too. With a diameter eighteen percent that of Earth, just less than half that of Mercury, just two-thirds that of our Moon, and just four times that of Ceres, Pluto doesn't fit well as a planet or as an asteroid. It was grouped with the eight planets, despite debate about whether it was more like them or like the ever-growing number of known asteroids.

Although asteroids continued to be discovered throughout the twentieth century, the status of Pluto as a planet was unthreatened until 1978, when Pluto was discovered to have a moon, Charon, which is almost as large as Pluto (**Figure 3**). (Pluto has a diameter of 2300 km while Charon has a diameter of 1190 km.) For comparison, Earth and our Moon, which are the pair of planet and moon that is most similar to Pluto and Charon, have diameters of 12,800 km and 3500 km, respectively. Pluto and Charon are a pair of nearly equal mass bodies orbiting each other, which is unique in the solar system. The status of Pluto as a planet was further questioned as similar-sized bodies, such as Sedna and Eris, were discovered in the same remote area of the solar system.

Over a hundred thousand asteroids and bodies orbiting beyond Neptune have now been identified. Their sizes and locations suggest that they are all quite different from the eight planets and similar to Pluto. Pluto's size is more similar to the larger asteroids than to any planet and its orbit is more similar to those of other bodies orbiting the Sun in space beyond Neptune than to the orbit of any planet. Pluto has finally been reclassified to better reflect these connections. These classifications were done during the summer of 2006 at a meeting of the International Astronomical Union, the group charged with classifying and naming objects in space. That meeting was filled with tension, as numer-

ous competing definitions of planets and other objects in the solar system were proposed and rejected. In the end, the following definitions were adopted.

A **planet** is a celestial body that (a) is in orbit around the Sun, (b) has sufficient mass for its own gravity to overcome rigid body forces so that it assumes a *hydrostatic equilibrium* (i.e., the object is nearly spherical), and (c) it has cleared the neighborhood around its orbit (of smaller bodies).

Although (a) is self-explanatory and Pluto indeed orbits the Sun, the conditions, (b) and (c), deserve some explanation. If an object has its shape because of the bonds between its atoms, then it can have essentially any shape. Rocks look like rocks, potatoes look like potatoes, and you and I look as we do because of such bonding. However, if an object has enough mass, then the gravitational attraction between its particles are strong enough to reshape the object—pulling down high places and pushing up low ones—until the object becomes nearly spherical. The resulting balance between the force of gravity pulling inward and the pressure created by the matter pressing on itself resulting in the matter pushing outward is called hydrostatic equilibrium. There are about three dozen objects in the solar system that meet this criterion, including Mercury, Venus, Earth, Mars, Jupiter, Saturn, Uranus, Neptune, Pluto, the asteroid Ceres, our Moon, several moons of each of the large planets, Pluto's moon Charon, and several of the Pluto-like objects orbiting beyond Neptune that have been discovered, including Sedna Makemake, Haumea and Eris. Conversely, there are well over a hundred thousand objects in the solar system that look more like potatoes than anything else. These bodies include most of the moons and asteroids and all of the comets.

The third condition to be considered a planet, (c) above, means that the object must also have enough gravitational attraction to pull onto itself, or to fling far away, the myriad smaller pieces of debris that orbit in its neighborhood. The inner eight planets satisfy this condition, but observations reveal numerous objects still in Pluto's vicinity. Pluto does not have enough gravitational attraction to clear its surrounding area; so, by using the new criteria for determining planets, Pluto is not a planet.

Nevertheless, Pluto and numerous other objects are still a part of the solar system, so they need to be classified. Because we know very little about the chemical compositions of these bodies (including Pluto), we don't yet know if they are a single class of objects or several distinct classes of objects. For example, some of them may be composed mostly of rock and metal, while others may be mostly rock and ice; some may have sheaths of ice, while others have rocky surfaces. Until such details are known, all the other spherical objects

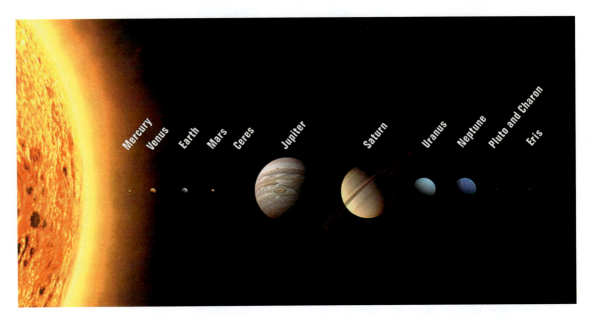

FIGURE 4 Scales of the Solar System All these bodies are presented to scale. (International Astronomical Union/NASA)

orbiting the Sun that have *not* cleared their neighborhoods in space of debris and that are not satellites[3] of other bodies are classified as **dwarf planets.** Dwarf planets that orbit farther from the Sun, on average, than Neptune are called **plutoids.** Pluto is both a dwarf planet and a plutoid, but its moon Charon is *neither* because Charon orbits the slightly larger body, Pluto. The asteroid Ceres is now classified as a dwarf planet. Plutoids Eris Makemake, and Haumea are also dwarf planets.

Finally, the panoply of nonsatellite, nonspherical objects in the solar system are called **small solar-system bodies** (SSSBs). Neither this name nor the name dwarf planets is very satisfying. The four inner planets, Mercury, Venus, Earth, and Mars, are much smaller than the four outer planets, so it is easy to confuse the inner "terrestrial planets" with "dwarf planets." Just as the name planetoid was dropped in favor of asteroid, the two names—dwarf planets and small solar-system bodies—may eventually be changed to better reflect the objects they represent.

As has occurred several times since the eighteenth century, people now have to go through the uncomfortable process of learning new names for old astronomical objects. Clearly, changing Pluto's designation to "dwarf planet (134340) Pluto" does not change it physical properties. Indeed, astronomers hope that the new designations will *better* represent Pluto's properties, along with the properties of the other small objects orbiting the Sun. Although the renaming will be confusing for a while, it will eventually help clarify our perception of the solar system (**Figure 4**).

---

[3] i.e., moons

A ring   The outermost ring of Saturn visible from Earth; it is located just beyond Cassini's division.

absolute magnitude   The apparent magnitude that a star would have if it were 10 parsecs from Earth.

absorption line   A dark line in a continuous spectrum created when photons of a certain energy are absorbed by atoms or molecules.

absorption line spectrum   Dark lines superimposed on a continuous spectrum.

acceleration   A change in the direction or magnitude of a velocity.

accretion   The gradual accumulation of matter by an astronomical body, usually caused by gravity.

accretion disk   An orbiting disk of matter spiraling in toward a star or black hole.

achromatic lens   A compound lens designed to minimize the effect of chromatic aberration.

active galactic nuclei (AGN)   Supermassive black holes in the cores of some galaxies that emit particles and radiation which, when viewed from different angles, create Seyfert galaxies, radio galaxies, double radio sources, BL Lacertae objects, and quasars.

active galaxy   A very luminous galaxy, often containing an active galactic nucleus.

active optics   A system that adjusts a reflecting telescope in response to changes in temperature and shape of the mount; it helps optimize an image.

adaptive optics   Primary telescope mirrors that are continuously and automatically adjusted to compensate for the distortion of starlight due to the motion of Earth's atmosphere.

AGB star   *See* asymptotic giant branch (AGB) star.

albedo   The fraction of sunlight that a planet, asteroid, or satellite scatters directly back into space.

amino acids   A class of chemical compounds that are the building blocks of proteins.

angle   The opening between two straight lines that meet at a point.

angular diameter (angular size)   The arc angle across an object.

angular momentum   A measure of how much energy an object has stored in its rotation and/or revolution.

angular resolution   The angular size of the smallest detail of an astronomical object that can be distinguished with a telescope.

annular eclipse   An eclipse of the Sun in which the Moon is too distant to cover the Sun completely so that a ring of sunlight is seen around the Moon at mideclipse.

anorthosite   A light-colored rock found throughout the lunar highlands and in some very old mountains on Earth.

aphelion   The point in its orbit where a planet or other solar system body is farthest from the Sun.

Apollo asteroid   An asteroid that is sometimes closer to the Sun than Earth is.

apparent magnitude   A measure of the brightness of light from a star or other object as seen from Earth.

arc angle   The measurement of the angle between two objects or two parts of the same object.

arcminute (arcmin)   One-sixtieth of a degree of arc.

arcsecond (arcsec)   One-sixtieth of a minute of arc.

asteroid (minor planet)   Any of the rocky objects larger than a few hundred meters in diameter (and not classified as a planet or moon) that orbits the Sun.

asteroid belt   A 1½-astronomical-unit-wide region between the orbits of Mars and Jupiter in which most of the asteroids are found.

astrobiology   The study of life in the universe.

astronomical unit (AU)   The average distance between Earth and the Sun: 1.5×10^8 km = 93 million mi.

astronomy   The branch of science dealing with objects and phenomena that lie beyond Earth's atmosphere.

astrophysics   That part of astronomy dealing with the physics of astronomical objects and phenomena.

asymptotic giant branch (AGB) star   A red giant star that has completed core helium fusion and has re-expanded for a second time.

atom   The smallest particle of an element that has the properties characterizing that element.

atomic number   The number of protons in the nucleus of an atom.

aurora (*plural* **aurorae**)   Light radiated by atoms and ions formed by the solar wind in Earth's upper atmosphere; seen most commonly in the polar regions.

autumnal equinox   The intersection of the ecliptic and the celestial equator where the Sun crosses the equator moving from north to south. The beginning of autumn (around September 23).

average density   The mass of an object divided by its volume.

B ring   The brightest of the three rings of Saturn visible from Earth; it lies just inside the Cassini division.

barred spiral galaxy   A spiral galaxy in which the spiral arms begin from the ends of a bar running through the central bulge.

belt asteroid   An asteroid whose orbit lies in the asteroid belt.

belts (of Jupiter)   Dark, reddish bands in Jupiter's cloud cover.

Big Bang   An explosion that took place roughly 15 billion years ago, creating all space, time, matter, and energy in which the universe emerged.

binary star   Two stars revolving about each other; a double star.

birth line   A line on the Hertzsprung-Russell diagram corresponding to where stars with different masses transform from protostars to pre–main-sequence stars.

BL Lacertae (BL Lac) object   A type of active galaxy; a blazar.

black hole   An object whose gravity is so strong that the escape velocity from it exceeds the speed of light.

blackbody   A hypothetical perfect radiator that absorbs and reemits all radiation falling upon it.

blackbody curve   The curve obtained when the intensity of radiation from a blackbody at a particular temperature is plotted against wavelength.

blackbody radiation   Electromagnetic radiation emitted by a blackbody.

blazar   A BL Lacertae object.

blueshift   A shift of all spectral features toward shorter wavelengths; the Doppler shift of light from an approaching source.

Bohr atom   A model of the atom, described by Niels Bohr, in which electrons revolve about the nucleus in various circular orbits.

Bok globule   A small, roundish, dark nebula in which stars are forming.

brown dwarf   Any of the planetlike bodies with less than $0.08 \, M_\odot$ and more than about $13 \, M_{Jupiter}$; such bodies do not have enough mass to sustain fusion in their cores.

C ring   The faint, inner portion of Saturn's main ring system.

caldera   The crater at the summit of a volcano.

capture theory   The idea that the Moon was created at a different location in the solar system and subsequently captured by Earth's gravity.

carbon fusion   The thermonuclear fusion of carbon nuclei to produce oxygen and neon.

carbonaceous chondrites   A class of extremely ancient, carbon-rich meteorites.

Cassegrain focus   An optical arrangement in a reflecting telescope in which light rays are reflected by a secondary mirror through a hole in the primary mirror.

Cassini division   A prominent gap between Saturn's A and B rings discovered in 1675 by J. D. Cassini.

celestial equator   A great circle on the celestial sphere 90° from the celestial poles.

celestial poles   Points about which the celestial sphere appears to rotate.

celestial sphere   A hypothetical sphere of very large radius centered on the observer; the apparent sphere of the night sky.

Celsius temperature scale   *See* **temperature (Celsius)**.

center of mass   The point around which a rigid system is perfectly balanced in a gravitational field; also, the point in space around which mutually orbiting bodies have elliptical orbits.

central bulge   A distribution of stars in the shape of a flattened sphere that surrounds the nucleus of a spiral galaxy like the Milky Way.

Cepheid variable star   One of two types of yellow, supergiant, pulsating stars.

Cerenkov radiation   Radiation produced by particles traveling through a substance faster than light can.

Ceres   The largest known asteroid and the first to be discovered.

Chandrasekhar limit   The maximum mass of a white dwarf, about $1.4 \, M_\odot$.

charge-coupled device (CCD)   A type of solid-state silicon wafer designed to detect photons.

chromatic aberration   An optical property whereby different colors of light passing through a lens are focused at different distances from it.

chromosphere   The layer in the solar atmosphere between the photosphere and the corona.

circumpolar stars   All the stars that never set at a given latitude; all the stars between Polaris and the northern horizon.

close binary   A binary star whose members are separated by a few stellar diameters.

closed universe   A universe that contains enough matter to cause it to recollapse. It is finite in extent and has no "outside."

cluster (of galaxies)   A collection of a few hundred to a few thousand galaxies bound by gravity.

co-creation theory   The theory that the Moon formed simultaneously with the Earth and in orbit around it.

collision-ejection theory   The theory that the Moon was created by the impact of a planet-sized object with Earth; presently considered the most plausible theory of the Moon's formation.

color-magnitude diagram   A plot of the surface temperatures (colors) versus the absolute magnitudes of stars.

coma (of a comet)   The nearly spherical, diffuse gas surrounding the nucleus of a comet near the Sun.

comet   A small body of ice and dust in orbit about the Sun. While passing near the Sun, a comet's vaporized ices give rise to a coma, tails, and a hydrogen envelope.

conduction (thermal)   The transfer of heat by passing energy directly from atom to atom.

configuration (of a planet)   A particular geometric arrangement of Earth, a planet, and the Sun.

confinement   The moment shortly after the Big Bang when quarks bound together to form particles like protons and neutrons.

conic section   The curve of intersection between a circular cone and a plane. This curve can be a circle, ellipse, parabola, or hyperbola.

conjunction   The alignment of two bodies in the solar system so that they appear in the same part of the sky as seen from Earth.

conservation of angular momentum   The law of physics stating that the total amount of angular momentum in an isolated system remains constant.

conservation of linear momentum   If the sum of the external forces on a system remains zero, the total linear momentum of the system remains constant.

constellation   Any of the 88 contiguous regions that cover the entire celestial sphere, including all the objects in each region; also, a configuration of stars often named after an object, a person, or an animal.

contact binary   A close binary system in which both stars fill or overflow their Roche lobes.

continental drift   The gradual movement of the continents over the surface of Earth due to plate tectonics.

continuous spectrum (continuum)   A spectrum of light over a range of wavelengths without any spectral lines.

convection   The transfer of energy by moving currents of fluid or gas containing that energy.

convective zone   A layer in a star where energy is transported outward by means of convection; also known as the *convective envelope* or *convection zone*.

core   The central portion of any astronomical object.

core-accretion model   The traditional theory of giant planet formation that begins with terrestrial planets slowly pulling abundant amounts of water, hydrogen, and helium onto themselves.

core helium fusion   The fusion of helium to form carbon and oxygen at the center of a star.

core hydrogen fusion   The fusion of hydrogen to form helium at the center of a star.

corona   The Sun's outer atmosphere.

coronagraph   A specially designed telescope with a baffle that blocks out the solar disk so that the corona can be photographed.

coronal hole   A dark region of the Sun's inner corona as seen at X-ray wavelengths.

coronal mass ejection   Large volumes of high energy gas released from the Sun's corona.

cosmic censorship   The belief that the only connection between a black hole and the universe is the black hole's event horizon.

cosmic light horizon   A sphere, centered on Earth, whose radius equals the distance traveled by light since the Big Bang.

cosmic microwave background   Photons from every part of the sky with a blackbody spectrum at 2.73 K; the cooled-off radiation from the primordial fireball that originally filled all space.

cosmic ray (primary cosmic ray)   High-speed particles traveling through space.

cosmic ray shower (secondary cosmic rays)   Groups of particles from Earth's atmosphere propelled Earthward by the impact of a cosmic ray.

cosmological constant   A number sometimes inserted in the equations of general relativity that represents a pressure that opposes gravity throughout the universe.

cosmological redshift   An increase in wavelength from distant galaxies and quasars caused by the expansion of the universe.

cosmology   The study of the formation, organization, and evolution of the universe.

coudé focus   A reflecting telescope in which a series of mirrors direct light to a remote focus away from the moving parts of the telescope.

crater   A circular depression on a celestial body caused by the impact of a meteoroid, asteroid, or comet, or by a volcano.

crescent Moon   A lunar phase during which the Moon appears less than half full.

crust   The solid surface layer of some astronomical bodies, including the terrestrial planets, the moons, the asteroids, and some stellar remnants.

dark ages   The age of the universe between the time of decoupling and the first burst of star formation.

dark energy   A repulsive gravitational effect that is causing the universe to accelerate outward.

dark matter (missing mass)   The as-yet-undetected matter in the universe that is underluminous and probably quite different from ordinary matter.

dark nebula   A cloud of interstellar gas and dust that obscures the light of more distant stars.

declination (dec)   The coordinate on the celestial sphere exactly analogous to latitude on Earth; measured north and south of the celestial equator.

decoupling   The epoch in the early universe when electrons and ions first combined to create stable atoms; the time when electromagnetic radiation ceased to dominate over matter.

deferent   A fixed circle in the Earth-centered universe along which a smaller circle (an epicycle) moves carrying a planet, the Sun, or the Moon.

degree (°)   A unit of angular measure or a temperature measure.

dense core   Any of the regions of interstellar gas clouds that are slightly denser than normal and destined to collapse to form one or a few stars.

density   The ratio of the mass of an object to its volume.

density wave   A spiral-shaped compression of the gas and dust in a spiral galaxy.

density-wave theory   An explanation of spiral arms in galaxies elaborated by C. C. Lin and F. Shu. (See **spiral density wave**.)

detached binary   A binary system in which the surfaces of both stars are inside their Roche lobes.

differential rotation   The rotation of a nonrigid object in which parts at different latitudes or different radial distances move at different speeds.

differentiation   See **planetary differentiation**.

diffraction grating   An optical device consisting of closely spaced lines ruled on a piece of glass that is used like a prism to disperse light into a spectrum.

direct motion   The gradual, eastward apparent motion of a planet against the background stars as seen from Earth.

disk (of a galaxy)   A flattened assemblage of stars, gas, and dust in a spiral galaxy like the Milky Way.

distance modulus   The difference between the apparent and absolute magnitudes of an object.

diurnal motion   Cyclic motion with a 1-day period.

Doppler effect (or **Doppler shift**)   The change in wavelength of radiation due to relative motion between the source and the observer along the line of sight.

double-line spectroscopic binary   A spectroscopic binary whose spectrum exhibits spectral lines of both stars.

double radio source   An extragalactic radio source characterized by two large lobes of radio emission, often located on either side of an active galaxy.

Drake equation   A mathematical equation used to estimate the number of extraterrestrial civilizations that may exist in our Galaxy.

dust devil   Whirlwind found in dry or desert areas on both Earth and Mars.

dust tail   A comet tail caused by dust particles escaping from the comet's nucleus.

dwarf elliptical galaxy   A small elliptical galaxy with far fewer stars than a typical galaxy.

dwarf planet   A celestial body that is in orbit around the Sun and has sufficient mass for its self-gravity to pull the body into a nearly spherical shape, but does not have enough gravity to clear its orbital neighborhood of all the small debris orbiting there.

dwarf star   Any star smaller than a giant, such as a main-sequence star or a white dwarf.

dynamo theory   The generation of a magnetic field by circulating electric charges.

eccentricity   See **orbital eccentricity**.

eclipse   The blocking of part or all of the light from the Moon by Earth (lunar eclipse) or from the Sun by the Moon (solar eclipse).

eclipse path   The track of the tip of the Moon's shadow along Earth's surface during a total or annular solar eclipse.

eclipsing binary   A double star system in which stars periodically pass in front of each other as seen from Earth.

ecliptic   The annual path of the Sun on the celestial sphere; the plane of Earth's orbit around the Sun.

Einstein cross   The appearance of four images of the same galaxy or quasar due to gravitational lensing by an intervening galaxy.

Einstein ring   The circular or arc-shaped image of a distant galaxy or quasar created by gravitational lensing by an intervening galaxy.

ejecta blanket   The ring of material surrounding a crater that was ejected during the crater-forming impact.

electromagnetic force   The interaction between charged particles, the second of four fundamental forces in nature.

electromagnetic radiation   Radiation consisting of oscillating electric and magnetic fields, namely gamma rays, X rays, visible light, ultraviolet and infrared radiation, and radio waves.

electromagnetic spectrum   The entire array of electromagnetic radiation.

electron   A negatively charged subatomic particle usually found in orbit about the nucleus of an atom.

electron degeneracy pressure   A powerful pressure produced by repulsion of closely packed (degenerate) electrons.

element   A substance that cannot be decomposed by chemical means into simpler substances. Every atom of the same element contains the same number of protons.

ellipse   A closed curve obtained by cutting completely through a circular cone with a plane; the shape of planetary orbits.

elliptical galaxy   A galaxy with an elliptical shape, little interstellar matter, and no spiral arms.

elongation   The angle between a planet and the Sun as seen from Earth.

emission line   A bright line of electromagnetic radiation.

emission line spectrum   A spectrum that contains only bright emission lines.

emission nebula   A glowing gaseous nebula whose light comes from fluorescence caused by a nearby star.

Encke division   A thin gap in Saturn's A ring, possibly first seen by J. F. Encke in 1838.

energy   The ability to do work.

energy flux   The amount of energy emitted from each square meter of an object's surface per second.

energy level (in an atom)   A particular amount of energy possessed by an electron in orbit around a nucleus.

epicycle   In the Earth-centered universe, a moving circle about which planets revolve.

equations of stellar structure   A set of relationships that describe the interactions of matter, energy, and gravity inside a star.

equinox   Either of the two days of the year when the Sun crosses the celestial equator and is therefore directly over Earth's equator; *see also* **autumnal equinox** *and* **vernal equinox**.

era of recombination   The time, roughly 500,000 years after the Big Bang, when the universe became transparent.

ergoregion   The region of space immediately outside the event horizon of a rotating black hole where it is impossible to remain at rest.

event horizon   The location around a black hole where the escape velocity equals the speed of light; the boundary of a black hole.

evolutionary track   On the Hertzsprung-Russell diagram, the path followed by a point representing an evolving star.

excited state   Orbit of an electron with energy greater than the lowest energy orbit (or state) available to that election.

expanding universe   The motion of the superclusters of galaxies away from each other.

eyepiece lens   A magnifying lens used to view the image produced at the focus of a telescope.

F ring   A thin ring just beyond the outer edge of Saturn's main ring system.

Fahrenheit scale   *See* **temperature (Fahrenheit)**.

filament   A dark curve seen above the Sun's photosphere that is the top view of a solar prominence.

first quarter Moon   A phase of the waxing Moon when Earth-based observers see half of the Moon's illuminated hemisphere.

fission theory   The theory that the Moon formed from matter flung off Earth because the planet was rotating extremely fast.

flare   *See* **solar flare**.

flocculent spiral galaxy   A spiral galaxy whose spiral arms are broad, fuzzy, and poorly demarcated.

focal length   The distance from a lens or concave mirror to where converging light rays meet.

focal plane   The plane at the focal length of a lens or concave mirror on which an extended object is focused.

focal point   *See* **focus**.

focus (of a lens or concave mirror)   The place at the focal length where light rays from a point object (that is, one that is too distant or tiny to resolve) are converged by a lens or concave mirror.

focus (*plural* **foci**) (of an ellipse)   The two points inside an ellipse, the sum of whose distances from any point on the ellipse is constant.

force   That which can change the momentum of an object.

frequency   The number of peaks or troughs of a wave that pass a fixed point each second. Also, the number of complete vibrations or oscillations per second.

full Moon   A phase of the Moon during which its full daylight hemisphere can be seen from Earth.

galactic cannibalism   A collision between two galaxies of unequal mass and size in which the smaller galaxy is absorbed by the larger galaxy.

galactic merger   A collision and subsequent merger of two roughly equal-sized galaxies.

galactic nucleus   The center of a galaxy; the center of the Milky Way Galaxy.

galaxy   A large assemblage of stars, gas, and dust bound together by their mutual gravitational attraction.

Galilean satellite (Galilean moon)   Any one of the four large moons of Jupiter (Callisto, Europa, Ganymede, Io) that is visible from Earth through a small telescope.

gamma ray   The most energetic form of electromagnetic radiation.

gamma-ray burst   A short burst of gamma rays; the sources of the bursts are outside our Galaxy.

gas (ion) tail   The relatively straight tail of a comet produced by the solar wind acting on ions in a comet's coma.

geocentric cosmology   The belief that Earth is at the center of the universe.

giant elliptical galaxy   A very large, extremely massive elliptical galaxy, usually located near the center of a rich cluster of galaxies.

giant molecular cloud   A large interstellar cloud of cool gas and dust in a galaxy.

giant star   A star whose diameter is roughly 10 to 100 times that of the Sun.

gibbous Moon   A phase of the Moon in which more than half, but not all, of the Moon's daylight hemisphere is visible from Earth.

glitch   A sudden speedup in the period of a pulsar.

globular cluster   A large spherical cluster of gravitationally bound stars usually found in the outlying regions of a galaxy.

grand-design spiral galaxy   A spiral galaxy whose spiral arms are thin, graceful, and well defined.

grand unified theory (GUT)   A theory that describes and explains the four physical forces.

granulation   The rice-grain–like structure of the solar photosphere due to convection of solar gases.

granules   Lightly colored convection features about 1000 km in diameter seen constantly in the solar photosphere.

gravitation (gravity)   The tendency of all matter to attract all other matter.

gravitational instability model   The theory of giant planet formation in which gases collapse together quickly, without the need for a seed terrestrial planet to form first.

gravitational lensing   The distortion of the appearance of an object by a source of gravity between it and the observer.

gravitational redshift   The redshift of photons leaving the gravitational field of any massive object, such as a star or black hole.

gravitational waves (gravitational radiation)   Ripples in the overall geometry of space produced by nonspherical moving objects.

gravity   *See* **gravitation.**

Great Dark Spot   A large, dark, oval-shaped storm that used to be in Neptune's southern hemisphere.

Great Red Spot   A large, red-orange, oval-shaped storm in Jupiter's southern hemisphere.

Great Wall   A huge arc of galaxies between two voids in the cosmos.

greatest elongation   The largest possible angle between the Sun and an inferior planet.

greenhouse effect   The trapping of infrared radiation near a planet's surface by the planet's atmosphere.

ground state   The lowest energy level of an atom.

H II region   A region of ionized hydrogen in interstellar space.

habitable zone   The region around any star wherein water can exist in liquid form and, hence, life as we know it can conceivably exist.

halo (of a galaxy)   A spherical distribution of globular clusters, isolated stars, and possibly dark matter that surrounds a galaxy.

Hawking process   The formation of real particles from virtual ones just outside a black hole's event horizon; the means by which black holes evaporate.

head-tail source   A radio galaxy whose radio emission is deflected from the galaxy.

heliocentric cosmology   A theory of the formation and evolution of the solar system with the Sun at the center.

helioseismology   The study of vibrations of the solar surface.

helium flash   The explosive ignition of helium fusion in the core of a low-mass, giant star.

helium fusion   The thermonuclear fusion of helium to produce carbon and oxygen.

helium shell flash   The explosive ignition of helium fusion in a thin shell surrounding the core of a low-mass star.

helium shell fusion   Helium fusion that occurs in a thin shell surrounding the core of a star.

Hertzsprung-Russell (H-R) diagram   A plot of the absolute magnitude or luminosity of stars versus their surface temperatures or spectral classes.

highlands   Heavily cratered, mountainous regions of the lunar surface.

homogeneity   The property of the universe being smooth or uniform as measured over suitably large distance intervals.

horizon problem (isotropy problem)   The difficulty in explaining why seemingly disconnected regions of the universe have the same temperature.

horizontal-branch stars   A group of post–helium-flash stars near the main sequence on the Hertzsprung-Russell diagram of a typical globular cluster.

hot-spot volcanism   The creation of volcanoes on a planet's surface caused by a reservoir of hot magma in the planet's mantle under a thin part of the crust.

Hubble classification   A system of classifying galaxies according to their appearance into one of four broad categories: spirals, barred spirals, ellipticals, and irregulars.

Hubble constant (H_0)   The constant of proportionality in the relation between the recessional velocities of remote galaxies and their distances; the correct value will determine the age of the universe.

Hubble flow   The recession of the galaxies caused by the expansion of the universe.

Hubble law   The relationship that states that the redshifts of remote galaxies are directly proportional to their distances from Earth.

hydrocarbon   A molecule based on hydrogen and carbon.

hydrogen envelope   An extremely large, tenuous sphere of hydrogen gas surrounding the head of a comet.

hydrogen fusion (hydrogen burning)   The thermonuclear fusion of hydrogen to produce helium.

hydrogen shell fusion   Hydrogen fusion that occurs in a thin shell surrounding the core of a star.

hydrostatic equilibrium   A balance between the weight of a layer in a star and the pressure that supports it.

hyperbola   An open curve obtained by cutting a cone with a plane.

impact breccia   A rock consisting of various fragments cemented together by the impact of a meteoroid.

impact crater   A crater on the surface of a planet or moon produced by the impact of an asteroid, meteoroid, or comet.

inferior conjunction   The configuration when Mercury or Venus is directly between the Sun and Earth.

inflation   A sudden expansion of space.

inflationary epoch   A brief period shortly after the Big Bang during which the scale of the universe increased very rapidly.

infrared radiation   Electromagnetic radiation of a wavelength longer than visible light but shorter than radio waves.

initial mass function   The numbers of stars on the main sequence at all different masses.

instability strip   A region on the Hertzsprung-Russell diagram occupied by pulsating stars.

interferometry   A method of increasing resolving power by combining electromagnetic radiation obtained by two or more telescopes.

intergalactic gas   Gas located between the galaxies within a cluster of galaxies.

intermediate-mass black hole   A black hole with a mass between a few hundred and few thousand solar masses.

interstellar dust   Microscopic solid grains of various compounds in interstellar space usually encased in ice.

interstellar extinction   The dimming of starlight as it passes through the interstellar medium.

interstellar gas   Sparse gas in interstellar space.

interstellar medium   Interstellar gas and dust.

interstellar reddening   The reddening of starlight passing through the interstellar medium resulting from the scattering of short-wavelength light more than long-wavelength light.

inverse-square law   The gravitational attraction between two objects and the apparent brightness of a light source are both inversely proportional to the square of its distance.

ion   An atom that has become electrically charged due to the loss or addition of one or more electrons.

ionization   The process by which an atom loses or gains electrons.

ionosphere (thermosphere)   Region of Earth's atmosphere, above the mesosphere, in which sunlight ionizes many atoms.

iron meteorite   A meteorite composed primarily of iron with an admixture of nickel; also called an *iron*.

irregular cluster (of galaxies)   An unevenly distributed group of galaxies bound together by their mutual gravitational attraction.

irregular galaxy   An asymmetrical galaxy having neither spiral arms nor an elliptical shape.

isotope   Atoms that all have the same number of protons (atomic number) but different numbers of neutrons. Their nuclear properties often differ greatly.

isotropy   The fact that the average number of galaxies at different distances from Earth is the same in all directions; also, the fact that the temperature of the cosmic microwave background is essentially the same in all directions.

isotropy problem (horizon problem)   The difficulty in explaining why seemingly disconnected regions of the universe have the same temperature.

Jeans instability   The condition under which gravitational forces overcome thermal forces to cause part of an interstellar cloud to collapse and form stars and planets.

Kelvin   *See* **temperature (Kelvin)**.

Kepler's laws   Three statements, formulated by Johannes Kepler, that describe the motions of the planets.

Kerr black hole   Any rotating, uncharged black hole.

kiloparsec (kpc)   One thousand parsecs; about 3260 light-years.

kinetic energy   The energy an object has as a result of its motion.

Kirchhoff's laws   Three statements formulated by Gustav Kirchhoff describing what physical conditions produce each type of spectra.

Kirkwood gaps   Gaps in the spacing of asteroid orbits discovered by Daniel Kirkwood caused by gravitational attractions of planets.

Kuiper belt   A doughnut-shaped ring of space around the Sun beyond Pluto that contains many frozen comet bodies, some of which are occasionally deflected toward the inner solar system.

Large Magellanic Cloud (LMC)   An irregular galaxy, companion to the Milky Way about 179,000 ly away.

last quarter Moon   A phase of the waning Moon when Earth-based observers see half of the Moon's illuminated hemisphere.

law of equal areas   Kepler's second law.

law of inertia (Newton's first law of motion)   The physical law that an object will stay at rest or move at a constant speed in a fixed direction unless acted upon by an outside force.

law of universal gravitation   Newton's law of gravitation, which describes how the gravitational force between two bodies depends on their masses and separation.

lenticular galaxy   A disk-shaped galaxy without spiral arms.

light   Electromagnetic radiation, which travels in packets called photons.

light curve   A graph that displays variations in the brightness of a star or other astronomical object over time.

light-gathering power   A measure of how much light a telescope intercepts and brings to a focus.

light-year (ly)   The distance that light travels in a vacuum in 1 year.

lighthouse model   The explanation that a pulsar pulses by rotating and funneling energy outward via magnetic fields that are not aligned with the rotation axis.

limb (of the Sun)   The apparent edge of the Sun as seen in the sky.

limb darkening   The phenomenon whereby the Sun is darker near its limb than near the center of its disk.

line of nodes   The line along which the plane of the Moon's orbit intersects the plane of the ecliptic.

liquid metallic hydrogen   A metal-like form of hydrogen that is produced under extreme pressure.

Local Group   The cluster of about 40 galaxies, of which our own Galaxy is a member.

long-period comet   A comet that takes tens of thousands of years or more to orbit the Sun once.

luminosity   The rate at which electromagnetic radiation is emitted from a star or other object.

luminosity class   The classification of a star of a given spectral type according to its luminosity and density; the classes are supergiant, bright giant, giant, subgiant, and main sequence.

lunar   Referring to the Moon.

lunar eclipse   An eclipse during which Earth blocks light that would have struck the Moon.

lunar month   *See* **synodic month**.

lunar phases   The names given to the apparent shapes of the Moon as seen from Earth.

Lyman series   A series of spectral lines of hydrogen produced by electron transitions to and from the lowest energy state of the hydrogen atom.

magnetar   Especially hot, rapidly-rotating neutron stars whose motion helps generate extra-strong magnetic fields.

magnetic dynamo   A theory that explains phenomena of the solar cycle as a result of periodic winding and unwinding of the Sun's magnetic field in the solar atmosphere.

magnetic field   A region of space near a magnetized body within which magnetic forces can be detected.

magnetosphere   The region around a planet occupied by its magnetic field.

magnification (magnifying power)   The number of times larger in angular diameter an object appears through a telescope than when it is seen by the naked eye.

magnitude   A measure of brightness. *See* **absolute magnitude**; **apparent magnitude**.

magnitude scale   The system of denoting magnitudes.

main sequence   A grouping of stars on the Hertzsprung-Russell diagram extending diagonally across the graph from the hottest, brightest stars to the dimmest, coolest stars.

main-sequence star   A star, fusing hydrogen to helium in its core, whose surface temperature and luminosity place it on the main sequence on the Hertzsprung-Russell diagram.

mantle (of a planet)   That portion of a terrestrial planet located between its crust and core.

mare (*plural* **maria**)   Latin for "sea," a large, relatively crater-free plain on the Moon.

mare basalt   Dark, solidified lava that covers the lunar maria.

mascons   Regions of high-density matter near the surface of the Moon.

mass   A measure of the total amount of material in an object.

mass-luminosity relation   The direct relationship between the masses and luminosities of main-sequence stars.

matter-dominated universe   A universe in which the radiation field that fills all space is unable to prevent the existence of neutral atoms.

megaparsec (Mpc)   One million parsecs.

mesosphere   The layer in Earth's atmosphere above the stratosphere.

metal   All elements except hydrogen and helium.

metal-poor star   *See* **Population II star.**

metal-rich star   *See* **Population I star.**

meteor   The streak of light seen when any space debris vaporizes in Earth's atmosphere; a "shooting star."

meteor shower   Frequent meteors that seem to originate from a common point in the sky.

meteorite   A fragment of space debris that has survived passage through Earth's atmosphere.

meteoroid   A small rock in interplanetary space.

microlensing   The gravitational focusing of light from a distant star by a closer object to give a brighter image of the star.

Milky Way Galaxy   The galaxy in which our solar system resides.

minor planet   *See* **asteroid.**

missing mass   *See* **dark matter.**

model   A hypothesis that has withstood observational or experimental tests.

molecule   Two or more atoms bonded together.

molecular clouds   Nebulae that are often embedded in much larger bodies of gas and dust.

momentum   A measure of the inertia of an object; an object's mass multiplied by its velocity.

moment of inertia (about an axis)   A measure of the inertial resistance of an object to changes in the object's rotational motion about the axis.

moons (natural satellites)   Bodies that orbit larger objects, which in turn orbit stars.

neap tide   The least change from high to low tide during a day; it occurs during the first-and third-quarter phases of the Moon.

nebula (*plural* **nebulae**)   A cloud of interstellar gas and dust.

neon fusion   The thermonuclear fusion of neon nuclei.

neutrino   A subatomic particle, with no electric charge and little mass, that is important in many nuclear reactions and in supernovae.

neutron   A nuclear particle with no electric charge and with a mass nearly equal to that of the proton.

neutron degeneracy pressure   A powerful pressure produced by degenerate neutrons.

neutron star   A very compact, dense stellar remnant composed almost entirely of neutrons.

New General Catalogue **(NGC)**   A catalog of star clusters, nebulae, and galaxies, first published in 1888.

new Moon   The phase of the Moon when it is nearest the Sun in the sky.

Newton's laws of motion   Newton's equations that describe the motion of matter as a result of forces acting on it.

Newtonian reflector   An optical arrangement in a reflecting telescope in which a small, flat mirror reflects converging light rays to a focus on one side of the telescope tube.

nonthermal radiation   *See* **synchrotron radiation.**

north celestial pole   The location on the celestial sphere directly above Earth's northern rotation pole.

northern lights (aurora borealis)   Light radiated by atoms and ions in Earth's upper atmosphere due to high-energy particles from the Sun and seen mostly in the northern polar regions.

northern vastness (northern lowlands)   Relatively young, crater-free terrain in the northern hemisphere of Mars.

nova (*plural* **novae**)   A star in a binary system that experiences a sudden outburst of radiant energy, temporarily increasing its luminosity by a factor of between 10^4 and 10^6.

nuclear   Referring to the nucleus of an atom.

nuclear density   The density of matter in the nucleus of an atom; about 10^{17} kilograms per cubic meter (kg/m^3).

nucleosynthesis   The formation, by fusion, of higher-mass elements from lower-mass ones.

nucleus (of an atom)   The massive part of an atom, composed of protons and neutrons; electrons surround a nucleus.

nucleus (of a comet)   A collection of ices and dust that constitute the solid part of a comet.

nucleus (of a galaxy)   *See* **galactic nucleus.**

OB association   An unbound group of very young, massive stars predominantly of spectral types O and B.

OBAFGKM sequence   The sequence of stellar spectral classifications from hottest to coolest stars.

objective lens   The principal lens of a refracting telescope.

observable universe   All space that is nearer to us than the distance traveled by light since the time of the Big Bang.

Occam's razor   The principle of choosing the simplest scientific theory that correctly explains any phenomenon.

occultation   The eclipsing of an astronomical object other than the Moon or Sun by another astronomical body.

Oort comet cloud   A hollow spherical region of the solar system beyond the Kuiper belt where most comets are believed to spend most of their time.

open cluster   A loosely bound group of young stars in the disk of the galaxy; a galactic cluster.

open universe   A universe with a hyperbolic shape; lacks the mass necessary to someday stop expanding and recollapse. It will expand forever.

opposition   The configuration of a planet when it is at an elongation of 180° and thus appears opposite the Sun in the sky.

optical double   A pair of stars that appear to be near each other but are unbound and at very different distances from Earth.

optics   The branch of physics dealing with the behavior and properties of light.

orbit   The path of an object that is moving about a second object.

orbital eccentricity   A measure between 0 and 1 indicating how close to circular a planet's orbit is (the eccentricity of a circular orbit is 0).

orbital inclination   The tilt or angle of an object's orbital plane around the Sun compared to the ecliptic.

organic molecule   A carbon-based compound.

overcontact binary   A close binary system in which the two stars share a common atmosphere.

oxygen fusion   The thermonuclear fusion of oxygen nuclei.

ozone layer   The lower stratosphere, where most of the ozone in the air exists.

pair production   The creation of a particle and an antiparticle from energetic photons.

parabola   An open curve formed by cutting a circular cone at an angle parallel to the sides of the cone.

parallax   The apparent displacement of an object relative to more distant objects caused by viewing it from different locations.

parsec (pc)   A unit of distance equal to 3.26 light-years.

partial eclipse   A lunar or solar eclipse in which the eclipsed object does not appear completely covered.

Pauli exclusion principle   A principle of quantum mechanics that says that two identical particles cannot simultaneously have the same position and momentum.

peculiar galaxy   Any Hubble class of galaxy that appears to be blowing apart.

penumbra   The portion of a shadow in which only part of the light source is covered by the shadow-making body.

penumbral eclipse   A lunar eclipse in which the Moon passes only through Earth's penumbra.

perihelion   The point in its orbit where a planet is nearest the Sun.

period   The interval of time between successive repetitions of a periodic phenomenon.

period-luminosity relation   A relationship between the period and average luminosity of a pulsating star.

periodic table   A listing of the chemical elements according to their properties; created by D. Mendeleev.

phase (of the Moon)   The appearance of the Moon at different points in its orbit of Earth.

photodisintegration   The breakup of nuclei in the core of a massive star due to the effects of energetic gamma rays.

photometry   The measurement of light intensities.

photon   A discrete unit of electromagnetic energy.

photon pressure   The force per unit area exerted by photons on stellar or interstellar gas.

photosphere   The region in the solar atmosphere from which most of the visible light escapes into space.

physics   Basic principles that govern the behavior of physical reality.

pixel   A contraction of the term "picture element"; usually refers to one square of a grid into which the light-sensitive component of a charge-coupled device is divided.

plage   A bright spot on the Sun believed to be associated with an emerging magnetic field.

Planck era   Time from the Big Bang until the Planck time (10^{-43} s).

Planck time   The brief interval of time, about 10^{-43} s, immediately after the Big Bang, when all four forces (gravity, electromagnetism, weak, strong) had the same strength.

Planck's law   A relationship between the energy carried by a photon and its wavelength.

planet   An object orbiting a star that is held together by its own gravitational force in a nearly spherical shape, that is able to clear its neighborhood of debris, and is not the moon (or satellite) of a larger orbiting body.

planetary differentiation   The process early in the life of each planet whereby denser elements sank inward and lighter ones rose.

planetary nebula   A luminous shell of gas ejected from an old, low-mass star.

planetesimal   Primordial asteroidlike object from which the planets accreted.

plasma   A hot, ionized gas.

plate tectonics   The motions of large segments (plates) of Earth's surface caused by convective motions in the underlying mantle.

polymer   A long molecule composed of many smaller molecules.

poor cluster (of galaxies)   A cluster of galaxies with only a few members.

Population I star   A star, such as the Sun, whose spectrum exhibits spectral lines of many elements heavier than helium; a metal-rich star.

Population II star   A star whose spectrum exhibits comparatively few spectral lines of elements heavier than helium; a metal-poor star.

positron   An electron with a positive rather than negative electric charge; an antielectron.

potential energy   The energy stored in an object as a result of its location in space.

powers of ten   A convenient method of writing large and small numbers that uses a number between 1 and 10 multiplied by a power of 10.

pre–main-sequence star   The stage of star formation just before the main sequence; it involves slow contraction of the young star.

precession (of Earth)   A slow, conical motion of Earth's axis of rotation caused by the gravitational pull of the Moon and Sun on Earth's equatorial bulge.

precession of the equinoxes   The slow westward motion of the equinoxes along the ecliptic because of Earth's precession.

primary mirror   The large, concave, light-gathering mirror in a reflecting telescope, analogous to the objective lens on a refracting telescope.

prime focus   The point in a reflecting telescope where the primary mirror focuses light.

primordial black hole   A relatively low-mass black hole hypothetically formed at the beginning of the universe.

primordial fireball   The extremely hot gas that filled the universe immediately following the Big Bang.

primordial nucleosynthesis   The transformation by fusion of protons and electrons into hydrogen isotopes, helium, and some lithium in the first few minutes of the existence of the universe.

prism   A wedge-shaped piece of glass used to disperse white light into a spectrum.

prograde orbit   An orbit of a moon or satellite around a planet that is in the same direction as the planet's rotation.

prominence   Flamelike protrusion seen near the limb of the Sun and extending into the solar corona. The side view of a filament.

proper motion   The change in the location of a star on the celestial sphere.

proton   A heavy, positively charged nuclear particle.

protoplanet   The embryonic stage of a planet when it is growing because of collisions with planetesimals.

protoplanetary disk (proplyd)   A disk of material encircling a protostar or a newborn star.

protostar   The earliest stage of a star's life before fusion commences and when gas is rapidly falling onto it.

protosun   The Sun prior to the time when hydrogen fusion began in its core.

pulsar   A pulsating source associated with a rapidly rotating neutron star with an off-axis magnetic field.

quantum mechanics   The branch of physics dealing with the structure and behavior of atoms and their interactions with each other and with light.

quark   A particle that is a building block of the heavy nuclear particles such as protons and neutrons.

quarter Moon   A phase of the Moon when it is located 90° from the Sun in the sky; halfway between the new and full phases.

quasar (quasi-stellar radio source)   A starlike object with a very large redshift.

quasi-stellar object (QSO)   A quasar.

quintessence   One of the explanations of the dark energy causing the universe to accelerate outward.

radial velocity   That portion of an object's velocity parallel to the line of sight.

radial-velocity curve   A plot showing the variation of radial velocity with time for a binary star or variable star.

radiation   Electromagnetic energy; photons.

radiation-dominated universe   The time at the beginning of the universe when the electromagnetic radiation prevented ions and electrons from combining to make neutral atoms.

radiation (photon) pressure   The transfer of momentum carried by radiation to an object on which the radiation falls.

radiative zone   A region inside a star where energy is transported outward by the movement of photons through a gas from a hot location to a cooler one.

radio astronomy   The branch of astronomy dealing with observations at radio wavelengths.

radio galaxy   A galaxy that emits an unusually large amount of radio waves.

radio lobes   Vast regions of radio emission on opposite sides of a radio galaxy.

radio telescope   A telescope designed to detect radio waves.

radio wave   Long-wavelength electromagnetic radiation.

radioactive   The property whereby certain atomic nuclei naturally decompose by spontaneously emitting particles.

red dwarf   A low-mass main-sequence star.

red giant   A large, cool star of high luminosity.

red supergiant   An extremely large, cool star of high luminosity; a star in the upper right corner of the Hertzsprung-Russell diagram.

redshift   The shifting to longer wavelengths of the light from remote galaxies and quasars; the Doppler shift of light from any receding source.

reflecting telescope (reflector)   A telescope in which the principal light-gathering component is a concave mirror.

reflection   The rebounding of light rays off a smooth surface.

reflection nebula   A comparatively dense cloud of gas and dust in interstellar space that is illuminated by a star between it and Earth.

refracting telescope (refractor)   A telescope in which the principal light-gathering component is a lens.

refraction   The bending of light rays when they pass from one transparent medium to another.

regolith   The powdery, lifeless material on the surface of a moon or planet.

regular cluster (of galaxies)   An evenly distributed group of galaxies bound together by mutual gravitational attraction.

resolution   The degree to which fine details in an optical image can be distinguished.

resolving power   A measure of the ability of an optical system to distinguish fine details in the image it produces.

resonance   The large response of an object to a small periodic gravitational tug from another object.

retrograde motion   The occasional backward (that is, westward) apparent motion of a planet against the background stars as seen from Earth. Retrograde motion is an optical illusion.

retrograde orbit   The orbit of a moon or satellite around a planet that is in the direction opposite to the planet's rotation.

retrograde rotation   The rotation of a planet opposite to its direction of revolution around the Sun. Only Pluto, Uranus, and Venus have retrograde rotation.

revolution   The orbit of one body about another.

rich cluster (of galaxies)   A cluster of galaxies with many members.

right ascension (r.a.)   The celestial coordinate analogous to longitude on Earth and measured around the celestial equator from the vernal equinox.

rille   A winding crack or depression in the lunar surface caused by the collapse of a solidified lava tube.

ringlet   Any one of numerous, closely spaced, thin bands of particles in planetary ring systems.

Roche limit   The shortest distance from a planet or other object at which a second object can be held together by its own gravitational forces.

Roche lobe   The teardrop-shaped regions around each star in a binary star system inside of which gas is gravitationally bound to that star.

rotation   The spinning of a body about an axis passing through it.

rotation curve (of a galaxy)   A graph showing how the orbital speed of material in a galaxy depends on the distance from the galaxy's center.

RR Lyrae variable star   A type of pulsating star with a period of less than 1 day.

Sa, Sb, Sc   Categories of spiral galaxies determined by the sizes of their central bulges or how tightly wound their spiral arms are; an Sa galaxy is the most tightly wound. Spiral arms are directly attached to the central bulge.

Sagittarius A   The strong radio source associated with the nucleus of the Milky Way Galaxy.

satellite   A body that revolves about a larger one.

SBa, SBb, SBc   Categories of barred spiral galaxies determined by how tightly wound their spiral arms are; SBa galaxies are the most tightly wound.

scarp   A cliff on Mercury believed to have formed when the planet cooled and shrank.

Schmidt corrector plate   A specially shaped lens used with spherical mirrors that corrects for spherical aberration and provides an especially wide field of view.

Chapter 18

MARGIN QUESTION: What objects are today classified as planets? Dwarf planets? ANSWERS: Planets are Mercury, Venus, Earth, Mars, Jupiter, Saturn, Uranus, and Neptune. Dwarf planets are Pluto, Ceres, Eris, Makemake, and Haumea.

MARGIN QUESTION: Why do astronomers doubt that the asteroid belt was once made up of a single planet? ANSWER: The total mass of all the objects in the asteroid belt combined is much less than that of any planet.

MARGIN QUESTION: What evidence do we have that comet Shoemaker-Levy 9 was not one solid chunk of rock surrounded by ice? ANSWER: Shoemaker-Levy 9 broke into numerous pieces under the relatively weak tidal influence of Jupiter. A solid comet would not have been so easily pulled apart.

MARGIN QUESTION: Why are stony-iron meteorites so rare compared to stony or iron meteorites? ANSWER: Stony-iron meteorites come from the boundaries between the rocky and metal parts of asteroids. These are very thin layers; hence, the meteorites from them are rare.

29. 5.0×10^{14} tons

Chapter 19

MARGIN QUESTION: Synchronous rotation occurred in what other situation that we explored in this book? ANSWER: Examples of synchronous rotation that we have considered are the orbits of most of the moons in the solar system around their respective planets, including our Moon orbiting Earth, and the synchronous rotation of Pluto with respect to its moon Charon.

MARGIN QUESTION: What negative consequence could arise from an alien civilization discovering the *Pioneer 10* or *11* spacecraft? ANSWER: If an alien civilization discovered *Pioneer 10* or *11*, they could trace them back to Earth. Visits from them might introduce diseases for which we have no defense, or they might lead to interplanetary war, if we can't get along with our new neighbors.

Appendix A

3.141×10^9, 3.1831×10^{-9}; 27182820000, 0.0000000000367879

Appendix B

580 K, 55 km, temperature decreases with height; +8, +10, September 1, September 6; $8L_\odot$, 6000 K

Appendix D

5527°C, 9980°F; –270°C, –454°F

Page numbers in italics indicate figures.

THE NIGHT SKY IN SEPTEMBER

Chart time (Daylight Savings Time):

11 pm...First of September
10 pm...Middle of September
9 pm...Last of September

THE NIGHT SKY IN NOVEMBER

Chart time (Local Standard Time):

10 pm...First of November
9 pm...Middle of November
8 pm...Last of November

NORTHERN HORIZON

EASTERN HORIZON

WESTERN HORIZON

SOUTHERN HORIZON

THE NIGHT SKY IN JANUARY

Chart time (Local Standard Time):

10 pm...First of January
9 pm...Middle of January
8 pm...Last of January

THE NIGHT SKY IN MARCH

Chart time (Local Standard Time):

10 pm...First of March
9 pm...Middle of March
8 pm...Last of March

THE NIGHT SKY IN MAY

Chart time (Daylight Savings Time):

11 pm ... First of May
10 pm ... Middle of May
9 pm ... Last of May

NORTHERN HORIZON

EASTERN HORIZON

WESTERN HORIZON

SOUTHERN HORIZON

THE NIGHT SKY IN JULY

Chart time (Daylight Savings Time):

11 pm...First of July

10 pm...Middle of July

9 pm...Last of July